Petroleum Reservoir Simulation

The Engineering Approach

Second Edition

Petroleum Reservoir Simulation

The Engineering Approach

Second Edition

Jamal H. Abou-Kassem

M. Rafiqul Islam

S.M. Farouq Ali

ELSEVIER

Gulf Professional Publishing
An imprint of Elsevier

Gulf Professional Publishing is an imprint of Elsevier
50 Hampshire Street, 5th Floor, Cambridge, MA 02139, United States
The Boulevard, Langford Lane, Kidlington, Oxford, OX5 1GB, United Kingdom

Library of Congress Cataloging-in-Publication Data
A catalog record for this book is available from the Library of Congress

British Library Cataloguing-in-Publication Data
A catalogue record for this book is available from the British Library

ISBN: 978-0-12-819150-7

For information on all Gulf Professional publications
visit our website at https://www.elsevier.com/books-and-journals

Publisher: Brian Romer
Senior Acquisition Editor: Katie Hammon
Editorial Project Manager: Naomi Robertson
Production Project Manager: Surya Narayanan Jayachandran
Cover Designer: Miles Hitchen

Typeset by SPi Global, India

Dedication

We dedicate this book to our parents.

Contents

6. Well representation in simulators

7. Single-phase flow equation for various fluids

8. Linearization of flow equations

Preface

The "Information Age" promises infinite transparency, unlimited productivity, and true access to knowledge. Knowledge, quite distinct and apart from "know-how," requires a process of thinking, or imagination—the attribute that sets human beings apart. Imagination is necessary for anyone wishing to make decisions based on science. Imagination always begins with visualization—actually, another term for simulation. Of course, subjective imagination has no meaning unless backed with objective facts. In fact, subjective knowledge of the truth has nothing to do with objective facts, but everything to do with the theory used by the subject to cognize. No other discipline has contributed to collecting objective facts (data) than the petroleum industry, so the onus is on modelers who must bring their perception or imagination as close to objective reality as possible. This is where this book makes a big contribution. By eliminating steps that are redundant, convoluted, and potentially misleading, the book makes it easier to keep the big picture transparent.

Under normal conditions, we simulate a situation prior to making any decision; that is, we abstract absence and start to fill in the gaps. Reservoir simulation is no exception. The two most important points that must not be overlooked in simulation are science and the multiplicity of solutions. Science is the essence of knowledge, and acceptance of the multiplicity of solutions is the essence of science. It is so because today's mathematics is not capable of producing a single solution to a nonlinear equation. Science, on the other hand, is limited to governing "laws" that are often a collection of simplistic assumptions. Science, not restricted by the notion of a single solution to every problem, must follow knowledge-based perception. Multiplicity of solutions has been promoted as an expression of uncertainty. This leads not to science or to new authentic knowledge, but rather to creating numerous models that generate "unique" solutions that fit a predetermined agenda of the decision-makers. This book re-establishes the essential features' real phenomena in their original form and applies them to reservoir engineering problems. This approach, which reconnects with the old—or in other words, time-tested—concept of knowledge, is refreshing and novel in the Information Age.

The petroleum industry is known as the biggest user of computer models. Even though space research and weather prediction models are robust and are often tagged as "the mother of all simulation," the fact that a space probe device or a weather balloon can be launched—while a vehicle capable of moving around in a petroleum reservoir cannot—makes modeling more vital for tackling problems in the petroleum reservoir than in any other discipline. Indeed, from the advent of computer technology, the petroleum industry pioneered the use of computer simulations in virtually all aspects of decision-making. This revolutionary approach required significant investment in long-term research and advancement of science. That time, when the petroleum industry was the energy provider of the world, was synonymous with its reputation as the most aggressive investor in engineering and science. More recently however, as the petroleum industry transited into its "middle age" in a business sense, the industry could not keep up its reputation as the biggest sponsor of engineering and long-term research. A recent survey by the US Department of Energy showed that none of the top ten breakthrough petroleum technologies in the last decade could be attributed to operating companies. If this trend continues, major breakthroughs in the petroleum industry over the next two decades are expected to be in the areas of information technology and materials science. When it comes to reservoir simulators, this latest trend in the petroleum industry has produced an excessive emphasis on the tangible aspects of modeling, namely, the number of blocks used in a simulator, graphics, computer speed, etc. For instance, the number of blocks used in a reservoir model has gone from thousands to millions in just a few years. Other examples can be cited, including graphics in which flow visualization has leapt from 2-D, to 3-D, to 4-D and computer processing speeds that make it practically possible to simulate reservoir activities in real time. While these developments outwardly appear very impressive, the lack of science and, in essence, true engineering render the computer revolution irrelevant and quite possibly dangerous. In the last decade, most investments have been made in software dedicated to visualization and computer graphics with little being invested in physics or mathematics. Engineers today have little appreciation of what physics and mathematics provide for the very framework of all the fascinating graphics that are generated by commercial reservoir simulators. As companies struggle to deal with scandals triggered by Enron's collapse, few have paid attention to the lack of any discussion in engineering education regarding what could be characterized as scientific fundamentals. Because of this lack, little has been done to promote innovation in reservoir simulation, particularly in the areas of physics and mathematics, the central topical content of reservoir engineering.

This book provides a means of understanding the underlying principles of petroleum reservoir simulation. The focus is on basic principles because understanding these principles is a prerequisite to developing more accurate advanced models. Once the fundamentals are understood, further development of more useful simulators is only a matter of time. The book takes a truly engineering approach and elucidates the principles behind formulating the governing equations. In contrast to cookbook-type recipes of step-by-step procedures for manipulating a black box, this approach is full of insights. To paraphrase the caveat about computing proposed by R.W. Hamming, the inventor of the Hamming code, the purpose of simulation must be insight, not just numbers. The conventional approach is more focused on packaging than on insight, making the simulation process more opaque than transparent. The formulation of governing equations is followed by elaborate treatment of boundary conditions. This is one aspect that is usually left to the engineers to "figure out" by themselves, unfortunately creating an expanding niche for the select few who own existing commercial simulators. As anyone who has ever engaged in developing a reservoir simulator well knows, this process of figuring out by oneself is utterly confusing. In keeping up with the same rigor of treatment, this book presents the discretization scheme for both block-centered and point-distributed grids. The difference between a well and a boundary condition is elucidated. In the same breadth, we present an elaborate treatment of radial grid for single-well simulation. This particular application has become very important due to the increased usage of reservoir simulators to analyze well test results and the use of well pseudofunctions. This aspect is extremely important for any reservoir engineering study. The book continues to give insight into other areas of reservoir simulation. For instance, we discuss the effect of boundary conditions on material-balance-check equations and other topics with unparalleled lucidity.

This is a basic book and is time honored. As such, it can hardly be altered or updated. So, why come up with a second edition? It turns out that none of the existing books on the topic covers several crucial aspects of modeling. Ever since the publication of the first edition in 2006, a number of research articles have been published praising the engineering approach that we introduced. After 13 years of the first publication, it was high time for us to introduce a comprehensive comparison between the conventional mathematical approach and the engineering approach that we introduced. This will enable the readership to appreciate the fact that the engineering approach is much easier to implement, bolstered with a number of advantages over the mathematical approach, without the scarifying accuracy of the solutions. Finally, a glossary was added to help the readership with a quick lookup of terms, which might not be familiar or which might have been misunderstood.

Even though the book is written principally for reservoir simulation developers, it takes an engineering approach that has not been taken before. Topics are discussed in terms of science and mathematics, rather than with graphical representation in the backdrop. This makes the book suitable and in fact essential for every engineer and scientist engaged in modeling and simulation. Even those engineers and scientists who wish to limit their activities to field applications will benefit greatly from this book, which is bound to prepare them better for the Information Age. The additions made in the second editions are both timely and comprehensive.

J.H. Abou-Kassem
M.R. Islam
S.M. Farouq Ali

Introduction

In this book the basics of reservoir simulation are presented through the modeling of single-phase fluid flow and multiphase flow in petroleum reservoirs using the engineering approach. This text is written for senior-level BSc students and first-year MSc students studying petroleum engineering. The aim of this book is to restore engineering and physics sense to the subject. In this way the misleading impact of excess mathematical glitter, which has dominated reservoir simulation books in the past, is challenged. The engineering approach, used in this book, uses mathematics extensively, but it injects engineering meaning to differential equations and to boundary conditions used in reservoir simulation. It does not need to deal with differential equations as a means for modeling, and it interprets boundary conditions as fictitious wells that transfer fluids across reservoir boundaries. The contents of the book can be taught in two consecutive courses. The first undergraduate senior-level course includes the use of block-centered grid in rectangular coordinates in single-phase flow simulation. The material is mainly included in Chapters 2, 3, 4, 6, 7, and 9. The second graduate-level course deals with block-centered grid in radial-cylindrical coordinates, point-distributed grid in both rectangular and radial-cylindrical coordinates, and the simulation of multiphase flow in petroleum reservoirs. The material is covered in Chapters 5, 8, 10, and 11 in addition to specific sections in Chapters 2, 4, 5, 6, and 7 (Sections 2.7, 4.5, 5.5, 6.2.2, 7.3.2, and 7.3.3).

Chapter 1 provides an overview of reservoir simulation and the relationship between the mathematical approach presented in simulation books and the engineering approach presented in this book. In Chapter 2, we present the derivation of single-phase, multidimensional flow equations in rectangular and radial-cylindrical coordinate systems. In Chapter 3, we introduce the control volume finite difference (CVFD) terminology as a means to writing the flow equations in multidimensions in compact form. Then, we write the general flow equation that incorporates both (real) wells and boundary conditions, using the block-centered grid (in Chapter 4) and the point-distributed grid (in Chapter 5), and present the corresponding treatments of boundary conditions as fictitious wells and the exploitation of symmetry in practical reservoir simulation

Chapter 6 deals with wells completed in a single layer and in multilayers and presents fluid flow rate equations for different well operating conditions. Chapter 7 presents the explicit, implicit, and Crank-Nicolson formulations of single-phase, slightly compressible, and compressible flow equations and introduces the incremental and cumulative material balance equations as internal checks to monitor the accuracy of generated solutions. In Chapter 8, we introduce the space and time treatments of nonlinear terms encountered in single-phase flow problems. Chapter 9 presents the basic direct and iterative solution methods of linear algebraic equations used in reservoir simulation. Chapter 10 presents differences between the engineering approach and the mathematical approach in derivation, treatment of wells and boundary conditions, and linearization. Chapter 11 is entirely devoted to multiphase flow in petroleum reservoirs and its simulation. The book concludes with Appendix A that presents a user's manual for a single-phase simulator. The folder available at www.emertec.ca includes a single-phase simulator written in FORTRAN 95, a compiled version, and data and output files for four solved problems. The single-phase simulator provides users with intermediate results and solutions to single-phase flow problems so that a user's solution can be checked and errors are identified and corrected. Educators may use the simulator to make up new problems and obtain their solutions.

Nomenclature

a_n	coefficient of unknown $x_{n+n_xn_y}$, defined by Eq. (9.46f)
A	parameter, defined by Eq. (9.28) in Tang's algorithm
$[\mathbf{A}]$	square coefficient matrix
A_x	cross-sectional area normal to x-direction, $ft^2[m^2]$
$A_x\|_x$	cross-sectional area normal to x-direction at x, $ft^2[m^2]$
$A_x\|_{x+\Delta x}$	cross-sectional area normal to x-direction at $x+\Delta x$, $ft^2[m^2]$
$A_x\|_{x_i\mp 1/2}$	cross-sectional area normal to x-direction at block boundary $x_{i\mp 1/2}$, $ft^2[m^2]$
b	reservoir boundary
b_E	reservoir east boundary
b_L	reservoir lower boundary
b_N	reservoir north boundary
b_s	reservoir south boundary
b_U	reservoir upper boundary
b_W	reservoir west boundary
b_n	coefficient of unknown $x_{n-n_xn_y}$, defined by Eq. (9.46a)
B	parameter, defined by Eq. (9.29) in Tang's algorithm
B	fluid formation volume factor, RB/STB $[m^3/std\,m^3]$
$\overline{B}$	average fluid formation volume factor in wellbore, RB/STB $[m^3/std\,m^3]$
B_g	gas formation volume factor, RB/scf $[m^3/std\,m^3]$
B_i	fluid formation volume factor for block i, RB/STB $[m^3/std\,m^3]$
B_o	oil formation volume factor, RB/STB $[m^3/std\,m^3]$
B_{ob}	oil formation volume factor at bubble-point pressure, RB/STB $[m^3/std\,m^3]$
B_{pi}	formation volume factor of phase p in block i
B_w	water formation volume factor, RB/B $[m^3/std\,m^3]$
B°	fluid formation volume factor at reference pressure p° and reservoir temperature, RB/STB $[m^3/std\,m^3]$
c	fluid compressibility, psi^{-1} $[kPa^{-1}]$
c_i	coefficient of unknown of block i in Thomas' algorithm
c_n	coefficient of unknown x_n, defined by Eq. (9.46g)
c_N	coefficient of unknown x_N in Thomas' or Tang's algorithm
c_o	oil-phase compressibility, psi^{-1} $[kPa^{-1}]$
c_ϕ	porosity compressibility, psi^{-1} $[kPa^{-1}]$

c_μ	rate of fractional viscosity change with pressure change, psi^{-1} [kPa^{-1}]
C	parameter, defined by Eq. (9.30) in Tang's algorithm
C_{MB}	cumulative material balance check, dimensionless
C_{op}	coefficient of pressure change over time step in expansion of oil accumulation term, STB/D-psi [std m^3/(d.kPa)]
C_{ow}	coefficient of water saturation change over time step in expansion of oil accumulation term, STB/D [std m^3/d]
C_{wp}	coefficient of pressure change over time step in expansion of water accumulation term, B/D-psi [std m^3/(d kPa)]
C_{ww}	coefficient of water saturation change over time step in expansion of water accumulation term, B/D [std m^3/d]
$\vec{d}$	vector of known values
D	parameter, defined by Eq. (9.31) in Tang's algorithm
d_i	known RHS of equation for block i in Thomas' algorithm
$d_{\max}$	maximum absolute difference between two successive iterations
d_n	RHS of equation for gridblock n, defined by Eq. (9.46h)
e_i	coefficient of unknown of block $i+1$ in Thomas' algorithm
e_n	coefficient of unknown x_{n+1}, defined by Eq. (9.46d)
e_N	coefficient of unknown x_1 in Tang's algorithm
f()	function of
f_p	the pressure-dependent term in transmissibility
$f^{n+1}_{p_{i\mp 1/2}}$	nonlinearity, defined by Eq. (8.17)
$F(t)$	argument of an integral at time t
F_i	ratio of wellblock i area to theoretical area from which well withdraws its fluid (in Chapter 6), fraction
F^m	argument of an integral evaluated at time t^m
$F(t^m)$	argument of an integral evaluated at time t^m
F^n	argument of an integral evaluated at time t^n
$F(t^n)$	argument of an integral evaluated at time t^n
F^{n+1}	argument of an integral evaluated at time t^{n+1}
$F(t^{n+1})$	argument of an integral evaluated at time t^{n+1}
$F^{n+1/2}$	argument of an integral evaluated at time $t^{n+1/2}$
$F(t^{n+1/2})$	argument of an integral evaluated at time $t^{n+1/2}$
g	gravitational acceleration, ft/s^2 [m/s^2]
g_i	element i of a temporary vector $\left(\vec{g}\right)$ generated in Thomas' algorithm
G	geometric factor
G_w	well geometric factor, RB-cp/D-psi [m^3 mPa s/(d kPa)]
G_{w_i}	well geometric factor for wellblock i, defined by Eq. (6.32), RB-cp/D-psi [m^3 mPa s/(d kPa)]
$G^*_{w_i}$	well geometric factor of the theoretical well for wellblock i, RB-cp/D-psi [m^3 mPa s/(d kPa)]

$G_{x_{i\mp1/2}}$	interblock geometric factor between block i and block $i\mp1$ along the x-direction, defined by Eq. (8.4)
$G_{x_{1,2}}$	interblock geometric factor between blocks 1 and 2 along the x-direction
$G_{y_{2,6}}$	interblock geometric factor between blocks 2 and 6 along the y-direction
$G_{r_{i\mp1/2,j,k}}$	interblock geometric factor between block (i,j,k) and block $(i\mp1,j,k)$ along the r-direction in radial-cylindrical coordinates, defined in Table 4.2, 4.3, 5.2, and 5.3
$G_{x_{i\mp1/2,j,k}}$	interblock geometric factor between block (i,j,k) and block $(i\mp1,j,k)$ along the x-direction in rectangular coordinates, defined in Tables 4.1 and 5.1
$G_{y_{i,\mp1/2,k}}$	interblock geometric factor between block (i,j,k) and block $(i,j\mp1,k)$ along the y-direction in rectangular coordinates, defined in Tables 4.1 and 5.1
$G_{z_{i,j,k\mp1/2}}$	interblock geometric factor between block (i,j,k) and block $(i,j,k\mp1)$ along the z-direction in rectangular coordinates, defined in Tables 4.1 and 5.1
$G_{z_{i,j,k\mp1/2}}$	interblock geometric factor between block (i,j,k) and block $(i,j,k\mp1)$ along the z-direction in radial–cylindrical coordinates, defined in Tables 4.2, 4.3, 5.2, and 5.3
$G_{\theta_{i,j\mp1/2,k}}$	interblock geometric factor between block (i,j,k) and block $(i,j\mp1,k)$ along the θ-direction in radial-cylindrical coordinates, defined in Tables 4.2, 4.3, 5.2, and 5.3
h	thickness, ft [m]
h_i	thickness of wellblock i, ft [m]
h_l	thickness of wellblock l, ft [m]
I_{MB}	incremental material balance check, dimensionless
k_H	horizontal permeability, md [μm^2]
k_{H_i}	horizontal permeability of wellblock i, md [μm^2]
k_r	permeability along the r-direction in radial flow, md [μm^2]
k_{rg}	relative permeability to gas phase, dimensionless
k_{ro}	relative permeability to oil phase, dimensionless
k_{rocw}	relative permeability to oil phase at irreducible water saturation, dimensionless
k_{rog}	relative permeability to oil phase in gas/oil/irreducible water system, dimensionless
k_{row}	relative permeability to oil phase in oil/water system, dimensionless
k_{rp}	relative permeability to phase p, dimensionless
$k_{rp}\|_{x_{i\mp1/2}}$	relative permeability phase p between point i and point $i\mp1$ along the x-axis, dimensionless
k_{rw}	relative permeability to water phase, dimensionless
k_V	vertical permeability, md [μm^2]

k_x	permeability along the x-axis, md [μm^2]
$k_x\|_{i\mp1/2}$	permeability between point i and point $i\mp1$ along the x-axis, md [μm^2]
k_y	permeability along the y-axis, md [μm^2]
k_z	permeability along the z-axis, md [μm^2]
k_θ	permeability along the θ-direction, md [μm^2]
$\log_e$	natural logarithm
L	reservoir length along the x-axis, ft [m]
[L]	lower triangular matrix
L_x	reservoir length along the x-axis, ft [m]
m_a	mass accumulation, lbm [kg]
m_{a_i}	mass accumulation in block i, lbm [kg]
m_{ca_i}	mass accumulation of component c in block i, lbm [kg]
m_{ci}	mass of component c entering reservoir from other parts of reservoir, lbm [kg]
$m_{ci}\|_{x_{i-1/2}}$	mass of component c entering block i across block boundary $x_{i-1/2}$, lbm [kg]
$m_{co}\|_{x_{i+1/2}}$	mass of component c leaving block i across block boundary $x_{i+1/2}$, lbm [kg]
m_{cs_i}	mass of component c entering (or leaving) block i through a well, lbm [kg]
$m_{cv_i}^n$	mass of component c per unit volume of block i at time t^n, 1bm/ft^3 [kg/m^3]
$m_{cv_i}^{n+1}$	mass of component c per unit volume of block i at time t^{n+1}, 1bm/ft^3 [kg/m^3]
$\dot{m}_{cx}$	x-component of mass flux of component c, 1bm/D-ft^2 [kg/(d m^2)]
m_{fgv}	mass of free-gas component per unit volume of reservoir rock, 1bm/ft^3[kg/m^3]
$\dot{m}_{fgx}$	x-component of mass flux of free-gas component, 1bm/D-ft^2 [kg/(d m^2)]
m_i	mass of fluid entering reservoir from other parts of reservoir, lbm [kg]
$m_i\|_x$	mass of fluid entering control volume boundary at x, lbm [kg]
$m_i\|_r$	mass of fluid entering control volume boundary at r, lbm [kg]
$m_i\|_{x_{i-1/2}}$	mass of fluid entering block i across block boundary $x_{i-1/2}$, lbm [kg]
$m_i\|_\theta$	mass of fluid entering control volume boundary at θ, lbm [kg]
m_o	mass of fluid leaving reservoir to other parts of reservoir, lbm [kg]
$m_o\|_{r+\Delta r}$	mass of fluid leaving control volume boundary at $r+\Delta r$, lbm [kg]
m_{ov}	mass of oil component per unit volume of reservoir rock, 1bm/ft^3 [kg/m^3]
$\dot{m}_{ox}$	x-component of mass flux of oil component, lbm/D-ft [kg/(d m^2)]
$m_o\|_{x+\Delta x}$	mass of fluid leaving control volume boundary at $x+\Delta x$, lbm [kg]

$m_o\|_{x_{i+1/2}}$	mass of fluid leaving block i across block boundary $x_{i+1/2}$, lbm [kg]
$m_o\|_{\theta+\Delta\theta}$	mass of fluid leaving control volume boundary at $\theta+\Delta\theta$, lbm [kg]
m_s	mass of fluid entering (or leaving) reservoir through a well, lbm [kg]
m_{sgv}	mass of solution-gas component per unit volume of reservoir rock, 1bm/ft^3 [kg/m^3]
$\dot{m}_{sgx}$	x-component of mass flux of solution-gas component, lbm/D-ft^2 [kg/(d m^2)]
m_{s_i}	mass of fluid entering (or leaving) block i through a well, lbm [kg]
m_v	mass of fluid per unit volume of reservoir rock, lbm/ft [kg/m^3]
$m_{v_i}^n$	mass of fluid per unit volume of block i at time t^n, lbm/ft^3 [kg/m^3]
$m_{v_i}^{n+1}$	mass of fluid per unit volume of block i at time t^{n+1}, lbm/ft^3 [kg/m^3]
m_{wv}	mass of water component per unit volume of reservoir rock, lbm/ft^3 [kg/m^3]
$\dot{m}_{wx}$	x-component of mass flux of water component, lbm/D-ft^2 [kg/(d m^2)]
$\dot{m}_x$	x-component of mass flux, lbm/D-ft^2 [kg/(d m^2)]
$\dot{m}_x\|_x$	x-component of mass flux across control volume boundary at x, lbm/D-ft^2 [kg/(d m^2)]
$\dot{m}_x\|_{x+\Delta x}$	x-component of mass flux across control volume boundary at $x+\Delta x$, lbm/D-ft^2 [kg/(d m^2)]
$\dot{m}_x\|_{x_{i\mp1/2}}$	x-component of mass flux across block boundary $x_{i\mp1/2}$, lbm/D-ft^2 [kg/(d m^2)]
M	gas molecular weight, lbm/lb mole [kg/kmole]
M_{p_i}	mobility of phase p in wellblock i, defined in Table 11.4
n_n	coefficient of unknown x_{n+n_x}, defined by Eq. (9.46e)
n_r	number of reservoir gridblocks (or gridpoints) along the r-direction
n_{vps}	number of vertical planes of symmetry
n_x	number of reservoir gridblocks (or gridpoints) along the x-axis
n_y	number of reservoir gridblocks (or gridpoints) along the y-axis
n_z	number of reservoir gridblocks (or gridpoints) along the z-axis
n_θ	number of reservoir gridblocks (or gridpoints) in the θ-direction
N	number of blocks in reservoir
p	pressure, psia [kPa]
p°	reference pressure, psia [kPa]
$\overline{p}$	average value pressure, defined by Eq. (8.21), psia [kPa]
p_b	oil bubble-point pressure, psia [kPa]
p_g	gas-phase pressure, psia [kPa]
p_i	pressure of gridblock (gridpoint) or wellblock i, psia [kPa]
p_i^m	pressure of gridblock (gridpoint) i at time t^m, psia [kPa]
$p_{i\mp1}^m$	pressure of gridblock (gridpoint) $i\mp1$ at time t^m, psia [kPa]
$p_{i,j,k}^m$	pressure of gridblock (gridpoint) (i,j,k) at time t^m, psia [kPa]

$p^m_{i\mp1,j,k}$	pressure of gridblock (gridpoint) $(i\mp1,j,k)$ at time t^m, psia [kPa]
$p^m_{i,j\mp1,k}$	pressure of gridblock (gridpoint) $(i,j\mp1,k)$ at time t^m, psia [kPa]
$p^m_{i,j,k\mp1}$	pressure of gridblock (gridpoint) $(i,j,k\mp1)$ at time t^m, psia [kPa]
p^n_i	pressure of gridblock (gridpoint) i at time t^n, psia [kPa]
p^{n+1}_i	pressure of gridblock (gridpoint) i at time t^{n+1}, psia [kPa]
$p_i^{\overset{(\nu+1)}{n+1}}$	pressure of gridblock (gridpoint) i at time level t^{n+1} and iteration $\nu+1$, psia [kPa]
$\delta p_i^{\overset{(\nu+1)}{n+1}}$	change in pressure of gridblock (gridpoint) i over an iteration at time level $n+1$ and iteration $\nu+1$, psi [kPa]
p_{i-1}	pressure of gridblock (gridpoint) $i-1$, psia [kPa]
p_{i+1}	pressure of gridblock (gridpoint) $i+1$, psia [kPa]
p^n_{i+1}	pressure of gridblock (gridpoint) $i+1$ at time t^n, psia [kPa]
p^{n+1}_{i+1}	pressure of gridblock (gridpoint) $i+1$ at time t^{n+1}, psia [kPa]
$p^{n+1}_{i\mp1}$	pressure of gridblock (gridpoint) $i\mp1$ at time t^{n+1}, psia [kPa]
$p_{i,j,k}$	pressure of gridblock (gridpoint) or wellblock (i,j,k), psia [kPa]
p_l	pressure of neighboring gridblock (gridpoint) l, psia [kPa]
p_n	pressure of gridblock (gridpoint) or wellblock n, psia [kPa]
p^0_n	initial pressure of gridblock (gridpoint) n, psia [kPa]
p^n_n	pressure of gridblock (gridpoint) or wellblock n at time level n, psia [kPa]
$p_i^{\overset{(\nu)}{n+1}}$	pressure of gridblock (gridpoint) i at time level t^{n+1} and iteration ν, psia [kPa]
p^{n+1}_n	pressure of gridblock (gridpoint) or wellblock n at time level $n+1$, psia [kPa]
$p^{(v)}_n$	pressure of gridblock (gridpoint) n at old iteration v, psia [kPa]
$p^{(v+1)}_n$	pressure of gridblock (gridpoint) n at new iteration $\nu+1$, psia [kPa]
p_{p_i}	pressure of phase p in gridblock (gridpoint) i, psia [kPa]
$p_{p_{i\mp1}}$	pressure of phase p in gridblock (gridpoint) $i\mp1$, psia [kPa]
p_o	oil pressure, psia [kPa]
p_{ref}	pressure at reference datum, psia [kPa]
p_{sc}	standard pressure, psia [kPa]
p_w	water-phase pressure, psia [kPa]
p_{wf}	well flowing bottom-hole pressure, psia [kPa]
$p_{wf_{est}}$	estimated well flowing bottom-hole pressure at reference depth, psia [kPa]
p_{wf_i}	well flowing bottom-hole pressure opposite wellblock i, psia [kPa]
$p_{wf_{ref}}$	well flowing bottom-hole pressure at reference depth, psia [kPa]
$p_{wf_{sp}}$	specified well flowing bottom-hole pressure at reference depth, psia [kPa]
P_{cgo}	gas/oil capillary pressure, psi [kPa]
P_{cgw}	gas/water capillary pressure, psi [kPa]

P_{cow}	oil/water capillary pressure, psi [kPa]
q	well production rate at reservoir conditions, RB/D [m^3/d]
q_{cm_i}	mass rate of component c entering block i through a well, lbm/D [kg/d]
q_{fg}	production rate of free-gas component at reservoir conditions, RB/D [std m^3/d]
q_{fgm}	mass production rate of free-gas component, lbm/D [kg/d]
q_{fgsc}	production rate of free-gas component at standard conditions, scf/D [std m^3/d]
q_m	mass rate entering control volume through a well, lbm/D [kg/d]
q_{m_i}	mass rate entering block i through a well, lbm/D [kg/d]
q_o	production rate of oil phase at reservoir conditions, RB/D [std m^3/d]
q_{om}	mass production rate of oil component, lbm/D [kg/d]
q_{osc}	production rate of oil phase at standard conditions, STB/D [std m^3/d]
q_{sc}	well production rate at standard conditions, STB/D or scf/D [std m^3/d]
q_{sc_i}	production rate at standard conditions from wellblock i, STB/D or scf/D [std m^3/d]
$q^m_{sc_i}$	production rate at standard conditions from wellblock i at time t^m, STB/D or scf/D [std m^3/d]
$q^m_{sc_n}$	production rate at standard conditions from wellblock n at time t^m, STB/D or scf/D [std m^3/d]
$q^m_{sc_{i,j,k}}$	production rate at standard conditions from wellblock (i,j,k) at time t^m, STB/D or scf/D [std m^3/d]
$q^{n+1}_{sc_i}$	production rate at standard conditions from wellblock i at time level $n+1$, STB/D or scf/D [std m^3/d]
$q^{(\nu)}_{sc_i}{}^{n+1}$	production rate at standard conditions from wellblock i at time t^{n+1} and iteration ν, STB/D or scf/D [std m^3/d]
$q^m_{sc_{l,(i,j,k)}}$	volumetric rate of fluid at standard conditions crossing reservoir boundary l to block (i,j,k) at time t^m, STB/D or scf/D [std m^3/d]
$q_{sc_{l,n}}$	volumetric rate of fluid at standard conditions crossing reservoir boundary l to block n, STB/D or scf/D [std m^3/d]
$q^m_{sc_{l,n}}$	volumetric rate of fluid at standard conditions crossing reservoir boundary l to block n at time t^m, STB/D or scf/D [std m^3/d]
q_{sc_n}	production rate at standard conditions from wellblock n, STB/D or scf/D [std m^3/d]
$q_{sc_{i\mp1/2}}$	interblock volumetric flow rate at standard conditions between gridblock (gridpoint) i and gridblock (gridpoint) $i\mp1$, STB/D or scf/D [std m^3/d]
$q_{sc_{b,bB}}$	volumetric flow rate at standard conditions across reservoir boundary to boundary gridblock bB, STB/D or scf/D [std m^3/d]
$q_{sc_{b,bP}}$	volumetric flow rate at standard conditions across reservoir boundary to boundary gridpoint bP, STB/D or scf/D [std m^3/d]

$q_{sc_{b_W,1}}$	volumetric flow rate at standard conditions across reservoir west boundary to boundary gridblock (gridpoint) 1, STB/D or scf/D [std m^3/d]
$q_{sc_{b_E,n_x}}$	volumetric flow rate at standard conditions across reservoir east boundary to boundary gridblock (gridpoint) n_x, STB/D or scf/D [std m^3/d]
q_{sgm}	mass production rate of solution-gas component, lbm/D[kg/d]
q_{spsc}	specified well rate at standard conditions, STB/D or scf/D [std m^3/d]
q_{wm}	mass production rate of water component, lbm/D[kg/d]
q_{wsc}	production rate of water phase at standard conditions, B/D [std m^3/d]
q_x	volumetric rate at reservoir conditions along the x-axis, RB/D [m^3/d]
r	distance in the r-direction in the radial-cylindrical coordinate system, ft [m]
r_e	extemal radius in Darcy's law for radial flow, ft [m]
r_{eq}	equivalent wellblock radius, ft [m]
r_{eq_n}	equivalent radius of the area from which the theoretical well for block n withdraws its fluid, ft [m]
$r_{i\mp1}$	r-direction coordinate of point $i\mp1$, ft [m]
$r^L_{i\mp1/2}$	radii for transmissibility calculations, defined by Eqs. (4.82b) and (4.83b) (or Eqs. 5.75b and 5.76b), ft [m]
$r^2_{i\mp1/2}$	radii squared for bulk volume calculations, defined by Eqs. (4.84b) and (4.85b) (or Eqs. 5.77b and 5.78b), ft^2 [m^2]
r_n	residual for block n, defined by Eq. (9.61)
r_w	well radius, ft [m]
Δr_i	size of block (i, j, k) along the r-direction, ft [m]
R_s	solution GOR, scf/STB [std m^3/std m^3]
s	skin factor, dimensionless
S	fluid saturation, fraction
S_g	gas-phase saturation, fraction
S_{iw}	irreducible water saturation, fraction
s_n	coefficient of unknown x_{n-n_x}, defined by Eq. (9.46b)
S_o	oil-phase saturation, fraction
S_w	water-phase saturation, fraction
t	time, day
T	reservoir temperature, °R[K]
Δt	time step, day
t^m	time at which the argument F of integral is evaluated at, Eq. (2.30), day
Δt_m	m^{th} time step, day
t^n	old time level, day

Δt_n	old time step, day
t^{n+1}	new or current time level, day
Δt_{n+1}	current (or new) time step, day
$T^m_{b,bB}$	transmissibility between reservoir boundary and boundary gridblock at time t^m
$T^m_{b,bP}$	transmissibility between reservoir boundary and boundary gridpoint at time t^m
T^m_{b,bP^*}	transmissibility between reservoir boundary and gridpoint immediately inside reservoir boundary at time t^m
T_{gx}	gas-phase transmissibility along the x-direction, scf/D-psi [std m^3/(d kPa)]
$T^m_{l,(i,j,k)}$	transmissibility between gridblocks (gridpoints) l and (i,j,k) at time t^m
$T^m_{l,n}$	transmissibility between gridblocks (gridpoints) l and n at time t^m
T_{ox}	oil-phase transmissibility along the x-direction, STB/D-psi [std m^3/(d kPa)]
$T_{r_{i\mp1/2,j,k}}$	transmissibility between point (i,j,k) and point $(i\mp1,j,k)$ along the r-direction, STB/D-psi or scf/D-psi [std m^3/(d kPa)]
$T^m_{r_{i\mp1/2,j,k}}$	transmissibility between point (i,j,k) and point $(i\mp1,j,k)$ along the r-direction at time t^m, STB/D-psi or scf/D-psi [std m^3/(d kPa)]
T_{sc}	standard temperature, °R[K]
T_{wx}	water-phase transmissibility along the x-direction, B/D-psi [std m^3/(d kPa)]
$T_{x_{i\mp1/2}}$	transmissibility between point i and point $i\mp1$ along the x-axis, STB/D-psi or scf/D-psi [std m^3/(d kPa)]
$T^{n+1}_{x_{i\mp1/2}}$	transmissibility between point i and point $i\mp1$ along the x-axis at time t^{n+1}, STB/D-psi or scf/D-psi [std m^3/(d kPa)]
$\overset{(\nu)}{T}{}^{n+1}_{x_{i\mp1/2}}$	transmissibility between point i and point $i\mp1$ along the x-axis at time t^{n+1} and iteration ν, STB/D-psi or scf/D-psi [std m^3/(d kPa)]
$T_{x_{i\mp1/2,j,k}}$	transmissibility between point (i,j,k) and point $(i\mp1,j,k)$ along the x-axis, STB/D-psi or scf/D-psi [std m^3/(d kPa)]
$T^m_{x_{i\mp1/2,j,k}}$	transmissibility between point (i,j,k) and point $(i\mp1,j,k)$ along the x-axis at time t^m, STB/D-psi or scf/D-psi [std m^3/(d kPa)]
$T_{y_{i,j\mp1/2,k}}$	transmissibility between point (i,j,k) and point $(i,j\mp1,k)$ along the y-axis, STB/D-psi or scf/D-psi [std m^3/(d kPa)]
$T^m_{y_{i,j\mp1/2,k}}$	transmissibility between point (i,j,k) and point $(i,j\mp1,k)$ along the y-axis at time t^m, STB/D-psi or scf/D-psi [std m^3/(d kPa)]
$T_{z_{i,j,k\mp1/2}}$	transmissibility between point (i,j,k) and point $(i,j,k\mp1)$ along the z-axis, STB/D-psi or scf/D-psi [std m^3/(d kPa)]
$T^m_{z_{i,j,k\mp1/2}}$	transmissibility between point (i,j,k) and point $(i,j,k\mp1)$ along the z-axis at time t^m, STB/D-psi or scf/D-psi [std m^3/(d kPa)]
$T_{\theta_{i,j\mp1/2,k}}$	transmissibility between point (i,j,k) and point $(i,j\mp1,k)$ along the θ-direction, STB/D-psi or scf/D-psi [std m^3/(d kPa)]

$T^{m}_{\theta_{i,j\mp 1/2,k}}$	transmissibility between point (i,j,k) and point $(i,j\mp 1,k)$ along the θ-direction at time t^m STB/D-psi or scf/D-psi [std m^3/(d kPa)]
[U]	upper triangular matrix
u_{gx}	x-component of volumetric velocity of gas phase at reservoir conditions, RB/D-ft^2 [m^3/(d m^2)]
u_i	element i of a temporary vector $\left(\vec{u}\right)$ generated in Thomas' algorithm
u_{ox}	x-component of volumetric velocity of oil phase at reservoir conditions, RB/D-ft^2 [m^3/(d m^2)]
$u_{px}\vert_{x_{i\mp 1/2}}$	x-component of volumetric velocity of phase p at reservoir conditions between point i and point $i\mp 1$, RB/D-ft^2 [m^3/(d m^2)]
u_{wx}	x-component of volumetric velocity of water phase at reservoir conditions, RB/D-ft^2 [m^3/(d m^2)]
u_x	x-component of volumetric velocity at reservoir conditions, RB/D-ft^2 [m^3/(d m^2)]
V_b	bulk volume, ft^3 [m^3]
V_{b_i}	bulk volume of block i, ft^3[m^3]
$V_{b_{i,j,k}}$	bulk volume of block (i,j,k), ft^3 [m^3]
V_{b_n}	bulk volume of block n, ft^3[m^3]
$w_{ci}\vert_{x_{i-1/2}}$	mass rate of component c entering block i across block boundary $x_{i-1/2}$, lbm/D [kg/d]
$w_{ci}\vert_{x_{i+1/2}}$	mass rate of component c leaving block i across block boundary $x_{i+1/2}$, lbm/D [kg/d]
w_{cx}	x-component of mass rate of component c, lbm/D [kg/d]
w_i	coefficient of unknown of block $i-1$ in Thomas' algorithm
w_n	coefficient of unknown x_{n-1}, defined by Eq. (9.46c)
w_N	coefficient of unknown x_{N-1} in Thomas' or Tang's algorithm
w_x	x-component of mass rate, lbm/D [kg/d]
$w_x\vert_x$	x-component of mass rate entering control volume boundary at x, lbm/D [kg/d]
$w_x\vert_{x+\Delta x}$	x-component of mass rate leaving control volume boundary at $x+\Delta x$, lbm/D [kg/d]
$w_x\vert_{x_{i\mp 1/2}}$	x-component of mass rate entering (or leaving) block i across block boundary $x_{i\mp 1/2}$, lbm/D [kg/d]
x	distance in the x-direction in the Cartesian coordinate system, ft [m]
Δx	size of block or control volume along the x-axis, ft [m]
$\vec{x}$	vector of unknowns (in Chapter 9)
x_i	x-direction coordinate of point i, ft [m]
x_i	unknown for block i in Thomas' algorithm
Δx_i	size of block i along the x-axis, ft [m]
δx_{i-}	distance between gridblock (gridpoint) i and block boundary in the direction of decreasing i along the x-axis, ft [m]

δx_{i^+}	distance between gridblock (gridpoint) i and block boundary in the direction of increasing i along the x-axis, ft [m]
$x_{i\mp1}$	x-direction coordinate of point $i\mp1$, ft [m]
$x_{i\mp1}$	unknown for block $i\mp1$ in Thomas' algorithm (in Chapter 9)
$\Delta x_{i\mp1}$	size of block $i\mp1$ along the x-axis, ft [m]
$x_{i\mp1/2}$	x-direction coordinate of block boundary $x_{i\mp1/2}$, ft [m]
$\Delta x_{i\mp1/2}$	distance between point i and point $i\mp1$ along the x-axis, ft [m]
x_n	unknown for block n (in Chapter 9)
$x_n^{(v)}$	unknown for block n at old iteration v (in Chapter 9)
$x_n^{(v+1)}$	unknown for block n at new iteration $v+1$ (in Chapter 9)
x_{n_x}	x-direction coordinate of gridblock (gridpoint) n_x, ft [m]
y	distance in the y-direction in the Cartesian coordinate system, ft [m]
Δy	size of block or control volume along the y-axis, ft [m]
Δy_j	size of block j along the y-axis, ft [m]
z	gas compressibility factor, dimensionless
z	distance in the z-direction in the Cartesian coordinate system, ft [m]
Δz	size of block or control volume along the z-axis, ft [m]
Δz_k	size of block k along the z-axis, ft [m]
$\Delta z_{i,j,k}$	size of block (i,j,k) along the z-axis, ft [m]
Z	elevation below datum, ft [m]
Z_b	elevation of center of reservoir boundary below datum, ft [m]
Z_{bB}	elevation of center of boundary gridblock bB below datum, ft [m]
Z_{bP}	elevation of boundary gridpoint bP below datum, ft [m]
Z_i	elevation of gridblock (gridpoint) i or wellblock i, ft [m]
$Z_{i\mp1}$	elevation of gridblock (gridpoint) $i\mp1$, ft [m]
$Z_{i,j,k}$	elevation of gridblock (gridpoint) (i,j,k), ft [m]
Z_l	elevation of gridblock (gridpoint) l, ft [m]
Z_n	elevation of gridblock (gridpoint) n, ft [m]
Z_{ref}	elevation of reference depth in a well, ft [m]
$\frac{\partial p}{\partial x}$	pressure gradient in the x-direction, psi/ft [kPa/m]
$\left.\frac{\partial p}{\partial x}\right\vert_b$	pressure gradient in the x-direction evaluated at reservoir boundary, psi/ft [kPa/m]
$\left(\frac{\partial p}{\partial x}\right)_{i\mp1/2}$	pressure gradient in the x-direction evaluated at block boundary $x_{i\mp1/2}$, psi/ft [kPa/m]
$\left.\frac{\partial p}{\partial r}\right\vert_{r_w}$	pressure gradient in the r_w r-direction evaluated at well radius, psi/ft [kPa/m]
$\frac{\partial \Phi}{\partial x}$	potential gradient in the x-direction, psi/ft [kPa/m]
$\frac{\partial Z}{\partial x}$	elevation gradient in the x-direction, dimensionless

$\left.\frac{\partial Z}{\partial x}\right\vert_b$	elevation gradient in the x-direction evaluated at reservoir boundary, dimensionless
α_c	volume conversion factor whose numerical value is given in Table 2.1
α_{lg}	logarithmic spacing constant, defined by Eq. (4.86) (or Eq. 5.79), dimensionless
β_c	transmissibility conversion factor whose numerical value is given in Table 2.1
β_i	element i of a temporary vector $\left(\vec{\beta}\right)$ generated in Tang's algorithm (in Chapter 9)
γ	fluid gravity, psi/ft[kPa/m]
γ_i	element i of a temporary vector $\left(\vec{\gamma}\right)$ generated in Tang's algorithm (in Chapter 9)
γ_c	gravity conversion factor whose numerical value is given in Table 2.1
γ_g	gravity of gas phase at reservoir conditions, psi/ft [kPa/m]
$\gamma_{i\mp1/2}$	fluid gravity between point i and point $i\mp1$ along the x-axis, psi/ft [kPa/m]
$\gamma^m_{i\mp1/2,j,k}$	fluid gravity between point (i,j,k) and neighboring point $(i\mp1,j,k)$ along the x-axis at time t^m, psi/ft [kPa/m]
$\gamma^m_{i,j\mp1/2,k}$	fluid gravity between point (i,j,k) and neighboring point $(i,j\mp1,k)$ along the y-axis at time t^m, psi/ft [kPa/m]
$\gamma^m_{i,j,k\mp1/2}$	fluid gravity between point (i,j,k) and neighboring point $(i,j,k\mp1)$ along the z-axis at time t^m, psi/ft [kPa/m]
$\gamma^m_{l,(i,j,k)}$	fluid gravity between point (i,j,k) and neighboring point l at time t^m, psi/ft [kPa/m]
$\gamma^m_{l,n}$	fluid gravity between point n and neighboring point l at time t^m, psi/ft [kPa/m]
$\gamma_{l,(i,j,k)}$	fluid gravity between point (i,j,k) and neighboring point l, psi/ft [kPa/m]
$\gamma_{l,n}$	fluid gravity between point n and neighboring point l, psi/ft [kPa/m]
γ_o	gravity of oil phase at reservoir conditions, psi/ft [kPa/m]
$\gamma_{p_{i\mp1/2}}$	gravity of phase p between point i and point $i\mp1$ along the x-axis, psi/ft [kPa/m]
$\gamma_{p_{l,n}}$	gravity of phase p between point l and point n, psi/ft [kPa/m]
γ_w	gravity of water phase at reservoir conditions, psi/ft [kPa/m]
$\overline{\gamma}_{wb}$	average fluid gravity in wellbore, psi/ft [kPa/m]
ε	convergence tolerance
η_{inj}	set of phases in determining mobility of injected fluid $=\{o,w,g\}$
η_{prd}	set of phases in determining mobility of produced fluids, defined in Table 10.4
θ	angle in the θ-direction, rad
$\Delta\theta_j$	size of block (i,j,k) along the θ-direction, rad
$\Delta\theta_{j\mp1/2}$	angle between point (i,j,k) and point $(i,\ j\mp1,k)$ along the θ-direction, rad

ϕ	porosity, fraction
$\phi_{i,j,k}$	porosity of gridblock (gridpoint) (i,j,k), fraction
ϕ_n	porosity of gridblock (gridpoint) n, fraction
ϕ°	porosity at reference pressure p°, fraction
Φ	potential, psia [kPa]
Φ_g	potential of gas phase, psia [kPa]
Φ_i	potential of gridblock (gridpoint) i, psia [kPa]
Φ_i^m	potential of gridblock (gridpoint) i at time t^m, psia [kPa]
Φ_i^n	potential of gridblock (gridpoint) i at time t^n, psia [kPa]
Φ_i^{n+1}	potential of gridblock (gridpoint) i at time t^{n+1}, psia [kPa]
$\Phi_{i\mp1}$	potential of gridblock (gridpoint) $i\mp1$, psia [kPa]
$\Phi_{i\mp1}^m$	potential of gridblock (gridpoint) $i\mp1$ at time t^m, psia [kPa]
$\Phi_{i\mp1}^n$	potential of gridblock (gridpoint) $i\mp1$ at time t^n, psia [kPa]
$\Phi_{i\mp1}^{n+1}$	potential of gridblock (gridpoint) $i\mp1$ at time t^{n+1}, psia [kPa]
$\Phi_{i,j,k}^m$	potential of gridblock (gridpoint) (i,j,k) at time t^m, psia [kPa]
Φ_l^m	potential of gridblock (gridpoint) l at time t^m, psia [kPa]
Φ_o	potential of oil phase, psia [kPa]
Φ_{p_i}	potential of phase p in gridblock (gridpoint) i, psia [kPa]
Φ_{ref}	potential at reference depth, psia [kPa]
Φ_w	potential of water phase, psia [kPa]
μ	fluid viscosity, cP [mPa s]
μ_i	viscosity of fluid in gridblock (gridpoint) i, cP [mPa s]
μ°	fluid viscosity at reference pressure p°, cP [mPa s]
μ_g	gas-phase viscosity, cP [mPa s]
$\mu_p\vert_{x_{i\mp1/2}}$	viscosity of phase p between point i and point $i\mp1$ along the x-axis, cP [mPa s]
μ_o	oil-phase viscosity, cP [mPa s]
μ_{ob}	oil-phase viscosity at bubble-point pressure, cP [mPa s]
μ_w	water-phase viscosity, cP [mPa s]
$\mu\vert_{x_{i\mp1/2}}$	fluid viscosity between point i and point $i\mp1$ along the x-axis, cP [mPa s]
ψ	a set containing gridblock (or gridpoint) numbers
ψ_b	the set of gridblocks (or gridpoints) sharing the same reservoir boundary b
$\psi_{i,j,k}$	the set of existing gridblocks (or gridpoints) that are neighbors to gridblock (gridpoint) (i,j,k)
ψ_n	the set of existing gridblocks (or gridpoints) that are neighbors to gridblock (gridpoint) n
ψ_{r_n}	the set of existing gridblocks (or gridpoints) that are neighbors to gridblock (gridpoint) n along the r-direction
ψ_{x_n}	the set of existing gridblocks (or gridpoints) that are neighbors to gridblock (gridpoint) n along the x-axis
ψ_{y_n}	the set of existing gridblocks (or gridpoints) that are neighbors to gridblock (gridpoint) n along the y-axis
ψ_{z_n}	the set of existing gridblocks (or gridpoints) that are neighbors to gridblock (gridpoint) n along the z-axis

ψ_{θ_n}	the set of existing gridblocks (or gridpoints) that are neighbors to gridblock (gridpoint) n along the θ-direction
ψ_w	the set that contains all wellblocks penetrated by a well
ρ	fluid density at reservoir conditions, lbm/ft^3[kg/m^3]
ρ°	fluid density at reference pressure p° and reservoir temperature, lbm/ft^3[kg/m^3]
ρ_g	gas-phase density at reservoir conditions, lbm/ft^3[kg/m^3]
ρ_{GS}	Gauss-Seidel spectral radius
ρ_{gsc}	gas-phase density at standard conditions, lbm/ft^3[kg/m^3]
ρ_o	oil-phase density at reservoir conditions, lbm/ft^3[kg/m^3]
ρ_{osc}	oil-phase density at standard conditions, lbm/ft^3[kg/m^3]
ρ_{sc}	fluid density at standard conditions, lbm/ft^3[kg/m^3]
ρ_w	water-phase density at reservoir conditions, lbm/ft^3[kg/m^3]
ρ_{wsc}	water-phase density at standard conditions, lbm/ft^3[kg/m^3]
$\overline{\rho}_{wb}$	average fluid density in wellbore, lbm/ft^3[kg/m^3]
$\sum_{l\in\psi}$	summation over all members of set ψ
$\sum_{l\in\psi_{i,j,k}}$	summation over all members of set $\psi_{i,j,k}$
$\sum_{l\in\psi_n}$	summation over all members of set ψ_n
$\sum_{i\in\psi_w}$	summation over all members of set ψ_w
$\sum_{l\in\psi_w}$	summation over all members of set ψ_w
$\sum_{l\in\xi_n}$	summation over all members of set ξ_n
ζ_j	element i of a temporary vector $\left(\vec{\zeta}\right)$ generated in Tang's algorithm
$\xi_{i,j,k}$	set of all reservoir boundaries that are shared with gridblock (gridpoint) (i,j,k)
ξ_n	set of all reservoir boundaries that are shared with gridblock (gridpoint) n
ω	overrelaxation parameter
ω_{opt}	optimum overrelaxation parameter
{ }	empty set or a set that contains no elements
$\cup$	union operator

Subscripts

1,2	between gridpoints 1 and 2
b	bulk, boundary, or bubble point
bB	boundary gridblock
bB^{**}	gridblock next to reservoir boundary but falls outside the reservoir
bP	boundary gridpoint

bP^*	gridpoint next to reservoir boundary but falls inside the reservoir
bP^{**}	gridpoint next to reservoir boundary but falls outside the reservoir
c	component c, $c=o, w, fg, sg$; conversion; or capillary
ca	accumulation for component c
ci	entering (in) for component c
cm	mass for component c
co	leaving (out) for component c
cv	per unit bulk volume for component c
E	east
est	estimated
fg	free-gas component
g	gas phase
i	index for gridblock, gridpoint, or point along the x- or r-direction
$i\mp 1$	index for neighboring gridblock, gridpoint, or point along the x- or r-direction
$i\mp 1/2$	between i and $i\mp 1$
$i,i\mp 1/2$	between block (or point) i and block boundary $i\mp 1/2$ along the x-direction
(i,j,k)	index for gridblock, gridpoint, or point in x-y-z (or r-θ-z) space
iw	irreducible water
j	index for gridblock, gridpoint, or point along the y- or θ-direction
$j\mp 1$	index for neighboring gridblock, gridpoint, or point along the y- or θ-direction
$j\mp 1/2$	between j and $j\mp 1$
$j,j\mp 1/2$	between block (or point) j and block boundary $j\mp 1/2$ along the y-direction
k	index for gridblock, gridpoint, or point along the z-direction
$k\mp 1$	index for neighboring gridblock, gridpoint, or point along the z-direction
$k\mp 1/2$	between k and $k\mp 1$
$k,k\mp 1/2$	between block (or point) k and block boundary $k\mp 1/2$ along the z-direction
l	index for neighboring gridblock, gridpoint, or point
L	lower
lg	logarithmic
l,n	between gridblocks (or gridpoints) l and n
m	mass
n	index for gridblock (or gridpoint) for which a flow equation is written
N	north
n_x	last gridblock (or gridpoint) in the x-direction for a parallelepiped reservoir
n_y	last gridblock (or gridpoint) in the y-direction for a parallelepiped reservoir

n_z	last gridblock (or gridpoint) in the z-direction for a parallelepiped reservoir
o	oil phase or oil component
opt	optimum
p	phase p, $p = o,w,g$
r	r-direction
ref	reference
$r_{i \mp 1/2}$	between i and $i \mp 1$ along the r-direction
s	solution
S	south
sc	standard conditions
sg	solution-gas
sp	specified
U	upper
v	per unit volume of reservoir rock
w	water phase or water component
W	west
wb	wellbore
wf	flowing well
x	x-direction
$x_{i \mp 1/2}$	between i and $i \mp 1$ along the x-direction
y	y-direction
$y_{j \mp 1/2}$	between j and $j \mp 1$ along the y-direction
z	z-direction
$z_{k \mp 1/2}$	between k and $k \mp 1$ along the z-direction
θ	θ-direction
$\theta_{j \mp 1/2}$	between j and $j \mp 1$ along the θ-direction

Superscripts

m	time level m
n	time level n (old time level)
$n+1$	time level $n+1$ (new time level, current time level)
$\overset{(v)}{n+1}$	time level $n+1$ and old iteration v
$\overset{(v+1)}{n+1}$	time level $n+1$ and current iteration $v+1$
(ν)	old iteration v
$(\nu+1)$	current iteration $v+1$
*	intermediate value before SOR acceleration
$\circ$	reference
-	average
$'$	derivative with respect to pressure
$\rightarrow$	vector

Chapter 1

Introduction

Chapter outline

1.1 Background

Reservoir simulation in the oil industry has become the standard for solving reservoir engineering problems. Simulators for various recovery processes have been developed and continue to be developed for new oil recovery processes. *Reservoir simulation* is the art of combining physics, mathematics, reservoir engineering, and computer programming to develop a tool for predicting hydrocarbon reservoir performance under various operating strategies. Fig. 1.1 depicts the major steps involved in the development of a reservoir simulator: formulation, discretization, well representation, linearization, solution, and validation (Odeh, 1982). In this figure, *formulation* outlines the basic assumptions inherent to the simulator, states these assumptions in precise mathematical terms, and applies them to a control volume in the reservoir. The result of this step is a set of coupled, nonlinear partial differential equations (PDEs) that describes fluid flow through porous media.

The PDEs derived during the formulation step, if solved analytically, would give reservoir pressure, fluid saturations, and well flow rates as continuous functions of space and time. Because of the highly nonlinear nature of the PDEs, however, analytical techniques cannot be used, and solutions must be obtained with numerical methods. In contrast to analytical solutions, numerical solutions give the values of pressure and fluid saturations only at discrete points in the reservoir and at discrete times. *Discretization* is the process of converting PDEs into algebraic equations. Several numerical methods can be used to discretize the PDEs; however, the most common approach in the oil industry today is the finite-difference method. The most commonly used finite-difference approach essentially builds on Taylor series expansion and neglects terms that are considered to be small when small difference in space parameters is considered. This expanded form is a set of algebraic equations. Finite element method, on the other hand,

Petroleum Reservoir Simulation. https://doi.org/10.1016/B978-0-12-819150-7.00001-3

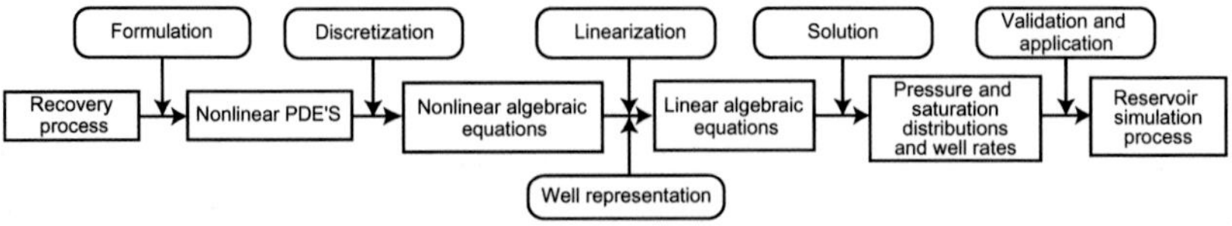

FIG. 1.1 Major steps used to develop reservoir simulators. *(Modified from Odeh, A.S., 1982. An overview of mathematical modeling of the behavior of hydrocarbon reservoirs. SIAM Rev. 24(3), 263.)*

uses various functions to express variables in the governing equation. These functions lead to the development of an error function that is minimized in order to generate solutions to the governing equation. To carry out discretization, a PDE is written for a given point in space at a given time level. The choice of time level (old time level, current time level, or intermediate time level) leads to the explicit, implicit, or Crank-Nicolson formulation method. The discretization process results in a system of nonlinear algebraic equations. These equations generally cannot be solved with linear equation solvers, and the linearization of such equations becomes a necessary step before solutions can be obtained. *Well representation* is used to incorporate fluid production and injection into the nonlinear algebraic equations. *Linearization* involves approximating nonlinear terms (transmissibilities, production and injection, and coefficients of unknowns in the accumulation terms) in both space and time. Linearization results in a set of linear algebraic equations. Any one of several linear equation solvers can then be used to obtain the *solution*, which comprises pressure and fluid saturation distributions in the reservoir and well flow rates. *Validation* of a reservoir simulator is the last step in developing a simulator, after which the simulator can be used for practical field applications. The validation step is necessary to make sure that no errors were introduced in the various steps of development or in computer programming. This validation is distinct from the concept of conducting experiments in support of a mathematical model. Validation of a reservoir simulator merely involves testing the numerical code.

There are three methods available for the discretization of any PDE: the Taylor series method, the integral method, and the variational method (Aziz and Settari, 1979). The first two methods result in the finite-difference method, whereas the third results in the variational method. The "mathematical approach" refers to the methods that obtain the nonlinear algebraic equations through deriving and discretizing the PDEs. Developers of simulators relied heavily on mathematics in the mathematical approach to obtain the nonlinear algebraic equations or the finite-difference equations. However, Abou-Kassem (2006) recently has presented a new approach that derives the finite-difference equations without going through the rigor of PDEs and discretization and that uses fictitious wells to represent boundary conditions. This new tactic is termed the "engineering approach" because it is closer to the engineer's thinking and to the physical meaning of the terms in the flow equations. The engineering approach is simple and yet general and rigorous, and both the engineering and mathematical approaches treat boundary conditions with the same accuracy if the mathematical approach uses second-order approximations. In addition, the engineering approach results in the same finite-difference equations for any hydrocarbon recovery process. Because the engineering approach is independent of the mathematical approach, it reconfirms the use of central differencing in space discretization and highlights the assumptions involved in choosing a time level in the mathematical approach.

1.2 Milestones for the engineering approach

The foundations for the engineering approach have been overlooked all these years. Traditionally, reservoir simulators were developed by first using a control volume (or elementary volume), such as that shown in Fig. 1.2 for 1-D flow or in Fig. 1.3 for 3-D flow that was visualized by mathematicians to develop fluid flow equations. Note that point x in 1-D and point (x, y, z) in 3-D fall on the edge of control volumes. The resulting flow equations are in the form of PDEs. Once the PDEs are derived, early pioneers of simulation looked to mathematicians to provide solution methods. These methods started with the description of the reservoir as a collection of gridblocks, represented by points that fall within them (or gridpoints representing blocks that surround them), followed by the replacement of the PDEs and boundary conditions by algebraic equations, and finally the solution of the resulting algebraic equations. Developers of simulators were all the time occupied by finding the solution and, perhaps, forgot that they were solving an engineering problem. The engineering approach can be realized should one try to relate the terms in the discretized flow equations for any block to the block itself and to all its neighboring blocks.

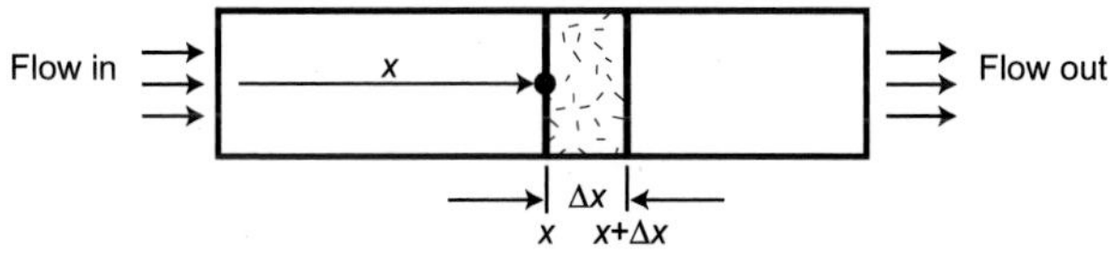

FIG. 1.2 Control volume used by mathematicians for 1-D flow.

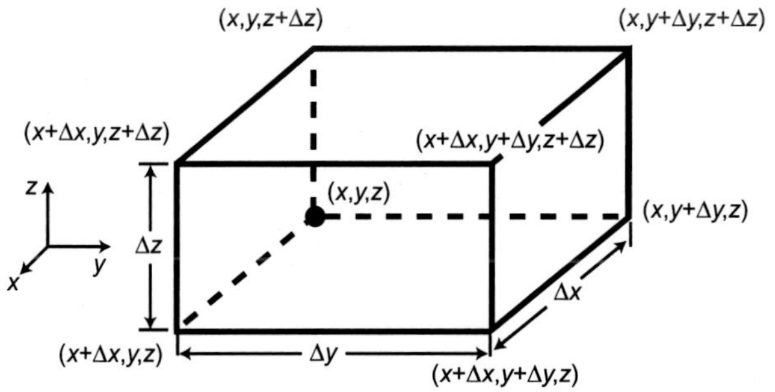

FIG. 1.3 Control volume used by mathematicians for 3-D flow. *(Modified from Bear, J., 1988. Dynamics of Fluids in Porous Media. Dover Publications, New York.)*

A close inspection of the flow terms in a discretized flow equation of a given fluid (oil, water, or gas) in a black-oil model for a given block reveals that these terms are nothing but Darcy's law describing volumetric flow rates of the fluid at standard conditions between the block and its neighboring blocks. The accumulation term is the change in the volume at standard conditions of the fluid contained in the block itself at two different times.

Farouq Ali (1986) observed that the flow terms in the discretized form of governing equations were nothing but Darcy's law describing volumetric flow rate between any two neighboring blocks. Making use of this observation coupled with an assumption related to the time level at which flow terms are evaluated, he developed the forward-central-difference equation and the backward-central-difference equation without going through the rigor of the mathematical approach in teaching reservoir simulation to undergraduate students. Ertekin et al. (2001) were the first to use a control volume represented by a point at its center in the mathematical approach as shown in Fig. 1.4 for 1-D flow and Fig. 1.5 for 3-D flow. This control volume is closer to engineer's thinking of representing blocks in reservoirs. The observation by Farouq Ali in the early 1970s and the introduction of the new control volume by Ertekin et al. have been the two milestones that contributed significantly to the recent development of the engineering approach.

Overlooking the engineering approach has kept reservoir simulation closely tied with PDEs. From a mathematician's point of view, this is a blessing because researchers in reservoir simulation have devised advanced methods for solving

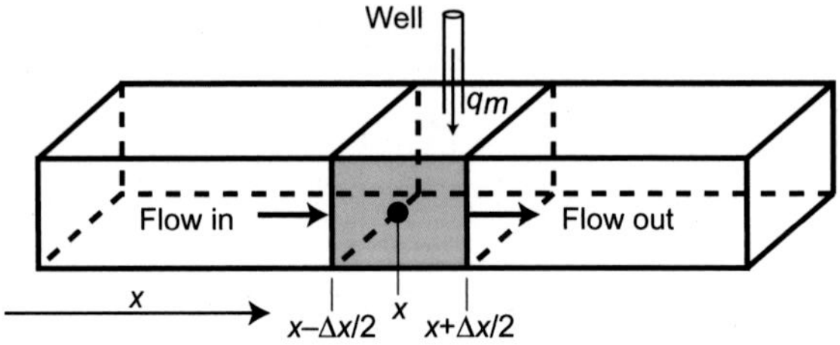

FIG. 1.4 Control volume for 1-D flow.

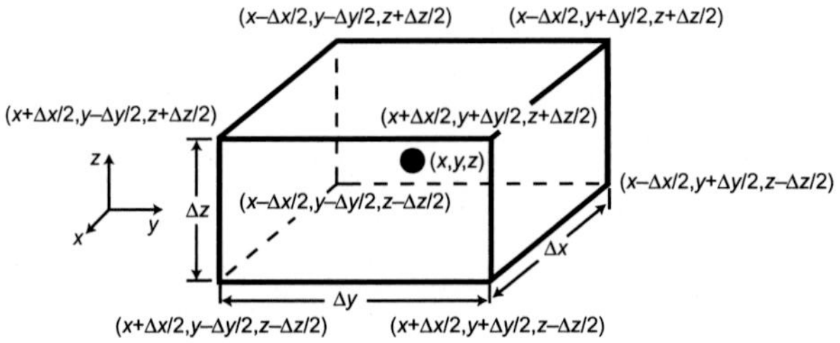

FIG. 1.5 Control volume for 3-D flow.

highly nonlinear PDEs, and this enriched the literature in mathematics in this important area. Contributions of reservoir simulation to solving PDEs in multiphase flow include the following:

- Treating nonlinear terms in space and time (Settari and Aziz, 1975; Coats et al., 1977; Saad, 1989; Gupta, 1990)
- Devising methods of solving systems of nonlinear PDEs, such as the IMPES (Breitenbach et al., 1969), SEQ (Spillette et al., 1973; Coats, 1978), fully implicit SS (Sheffield, 1969), and adaptive implicit (Thomas and Thurnau, 1983) methods
- Devising advanced iterative methods for solving systems of linear algebraic equations, such as the Block Iterative (Behie and Vinsome, 1982), Nested Factorization (Appleyard and Cheshire, 1983), and Orthomin (Vinsome, 1976) methods

1.3 Importance of the engineering and mathematical approaches

The importance of the engineering approach lies in being close to the engineer's mindset and in its capacity to derive the algebraic flow equations easily and without going through the rigor of PDEs and discretization. In reality, the development of a reservoir simulator can do away with the mathematical approach because the objective of this approach is to obtain the algebraic flow equations for the process being simulated. In addition, the engineering approach reconfirms the use of central-difference approximation of the second-order space derivative and provides interpretation of the approximations involved in the forward-, backward-, and central-difference of the first-order time derivative that are used in the mathematical approach.

The majority, if not all, of available commercial reservoir simulators were developed without even looking at an analysis of truncation errors, consistency, convergence, or stability. The importance of the mathematical approach, however, lies within its capacity to provide analysis of such items. Only in this case do the two approaches complement each other and both become equally important in reservoir simulation.

1.4 Summary

The traditional steps involved in the development of a reservoir simulator include formulation, discretization, well representation, linearization, solution, and validation. The mathematical approach involves formulation to obtain a differential equation, followed by reservoir discretization to describe the reservoir, and finally the discretization of the differential equation to obtain the flow equation in algebraic form. In contrast, the engineering approach involves reservoir discretization to describe the reservoir, followed by formulation to obtain the flow equations in integral form, which, when approximated, produce the flow equations in algebraic form. The mathematical approach and engineering approach produce the same flow equations in algebraic form but use two unrelated routes. The seeds for the engineering approach existed long time ago but were overlooked by pioneers in reservoir simulation because modeling petroleum reservoirs has been considered a mathematical problem rather than an engineering problem. The engineering approach is both easy and robust. It does not involve differential equations, discretization of differential equations, or discretization of boundary conditions.

1.5 Exercises

1.1 Name the major steps used in the development of a reservoir simulator using the mathematical approach.

1.2 Indicate the input and the expected output for each major step in Exercise 1.1.

1.3 How does the engineering approach differ from the mathematical approach in developing a reservoir simulator?

1.4 Name the major steps used in the development of a reservoir simulator using the engineering approach.

1.5 Indicate the input and the expected output for each major step in Exercise 1.4.

1.6 Draw a sketch, similar to Fig. 1.1, for the development of a reservoir simulator using the engineering approach.

1.7 Using your own words, state the importance of the engineering approach in reservoir simulation.

Chapter 2

Single-phase fluid flow equations in multidimensional domain

Chapter outline

2.1 Introduction

The development of flow equations requires an understanding of the physics of the flow of fluids in porous media; the knowledge of fluid properties, rock properties, fluid-rock properties, and reservoir discretization into blocks; and the use of basic engineering concepts. We have seen in the previous chapter that the description of the process within the engineering approach is simplified because casting of equations into partial differential equations is avoided. In practical term, it means savings of many man months of company time. In this chapter, single-phase flow is used to show the effectiveness of the engineering approach. Discussions of fluid-rock properties are postponed until Chapter 11, which deals with the simulation of multiphase flow. The engineering approach is used to derive a fluid flow equation. This approach involves three consecutive steps: (1) discretization of the reservoir into blocks; (2) derivation of the algebraic

Petroleum Reservoir Simulation. https://doi.org/10.1016/B978-0-12-819150-7.00002-5

flow equation for a general block in the reservoir using basic engineering concepts such as material balance, formation volume factor (FVF), and Darcy's law; and (3) approximation of time integrals in the algebraic flow equation derived in the second step. Even though petroleum reservoirs are geometrically three dimensional, fluids may flow in one direction (1-D flow), two directions (2-D flow), or three directions (3-D flow). This chapter presents the flow equation for single phase in 1-D reservoir. Then, it extends the formulation to 2-D and 3-D in Cartesian coordinates. In addition, this chapter presents the derivation of the single-phase flow equation in 3-D radial-cylindrical coordinates for single-well simulation.

2.2 Properties of single-phase fluid

Fluid properties that are needed to model single-phase fluid flow include those that appear in the flow equations, namely, density (ρ), formation volume factor (B), and viscosity (μ). Fluid density is needed for the estimation of fluid gravity (γ) using:

$$\gamma = \gamma_c \rho g \tag{2.1}$$

where γ_c = the gravity conversion factor and g = acceleration due to gravity. In general, fluid properties are a function of pressure. Mathematically, the pressure dependence of fluid properties is expressed as:

$$\rho = \mathrm{f}(p) \tag{2.2}$$

$$B = \mathrm{f}(p) \tag{2.3}$$

and

$$\mu = \mathrm{f}(p) \tag{2.4}$$

The derivation of the general flow equation in this chapter does not require more than the general dependence of fluid properties on pressure as expressed by Eqs. (2.2) through (2.4). In Chapter 7, the specific pressure dependence of fluid properties is required for the derivation of the flow equation for each type of fluid.

2.3 Properties of porous media

Modeling single-phase fluid flow in reservoirs requires the knowledge of basic rock properties such as porosity and permeability or, more precisely, effective porosity and absolute permeability. Other rock properties include reservoir thickness and elevation below sea level. *Effective porosity* is the ratio of interconnected pore spaces to bulk volume of a rock sample. Petroleum reservoirs usually have heterogeneous porosity distribution; that is, porosity changes with location. A reservoir is described as homogeneous if porosity is constant independent of location. Porosity depends on reservoir pressure because of solid and pore compressibilities. It increases as reservoir pressure (pressure of the fluid contained in the pores) increases and vice versa. This relationship can be expressed as

$$\phi = \phi^{\circ}\left[1 + c_{\phi}(p - p^{\circ})\right] \tag{2.5}$$

where ϕ°= porosity at reference pressure (p°) and c_{ϕ}= porosity compressibility. *Permeability* is the capacity of the rock to transmit fluid through its connected pores when the same fluid fills all the interconnected pores. Permeability is a directional rock property. If the reservoir coordinates coincide with the principal directions of permeability, then permeability can be represented by k_x, k_y, and k_z. The reservoir is described as having isotropic permeability distribution if $k_x = k_y = k_z$; otherwise, the reservoir is anisotropic if permeability shows directional bias. Usually, $k_x = k_y = k_H$, and $k_z = k_V$ with $k_V < k_H$ because of depositional environments.

2.4 Reservoir discretization

Reservoir discretization means that the reservoir is described by a set of gridblocks (or gridpoints) whose properties, dimensions, boundaries, and locations in the reservoir are well defined. Chapter 4 deals with reservoirs discretized using a block-centered grid, and Chapter 5 discusses reservoirs discretized using a point-distributed grid. Fig. 2.1 shows reservoir discretization in the x-direction as one focuses on block i.

The figure shows how the blocks are related to each other—block i and its neighboring blocks (blocks $i-1$ and $i+1$)—block dimensions (Δx_i, Δx_{i-1}, Δx_{i+1}),

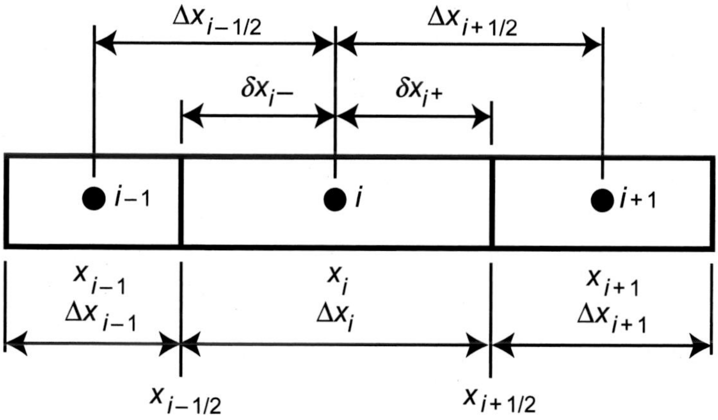

FIG. 2.1 Relationships between block i and its neighboring blocks in 1-D flow.

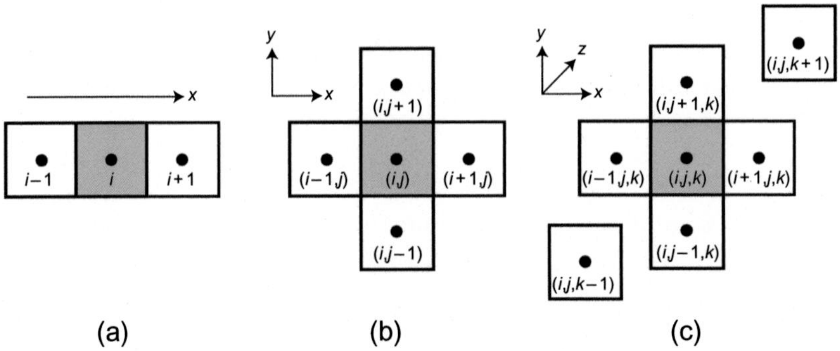

FIG. 2.2 A block and its neighboring blocks in (a) 1-D, (b) 2-D, and (c) 3-D flow using engineering notation.

block boundaries ($x_{i-1/2}$, $x_{i+1/2}$), distances between the point that represents the block and block boundaries (δx_{i^-},δx_{i^+}), and distances between the points representing the blocks ($\Delta x_{i-1/2}$, $\Delta x_{i+1/2}$). The terminology presented in Fig. 2.1 is applicable to both block-centered and point-distributed grid systems in 1-D flow in the direction of the x-axis. Reservoir discretization in the y- and z-directions uses similar terminology. In addition, each gridblock (or gridpoint) is assigned elevation and rock properties such as porosity and permeabilities in the x-, y-, and z-directions. The transfer of fluids from one block to the rest of reservoir takes place through the immediate neighboring blocks. When the whole reservoir is discretized, each block is surrounded by a set (group) of neighboring blocks. Fig. 2.2a shows that there are two neighboring blocks in 1-D flow along the x-axis, Fig. 2.2b shows that there are four neighboring blocks in 2-D flow in the x-y plane, and Fig. 2.2c shows that there are six neighboring blocks in 3-D flow in x-y-z space.

It must be made clear that once the reservoir is discretized and rock properties are assigned to gridblocks (or gridpoints), space is no longer a variable and functions that depend on space, such as interblock properties, become well defined. In other words, reservoir discretization removes space from being a variable in the formulation of the problem. More elaboration follows in Section 2.6.2.

2.5 Basic engineering concepts

The basic engineering concepts include mass conservation, equation of state, and constitutive equation. The principle of *mass conservation* states that the total mass of fluid entering minus the fluid leaving a volume element of the reservoir, shown in Fig. 2.3 as block i, must equal the net increase in the mass of the fluid in the reservoir volume element, that is,

$$m_i - m_o + m_s = m_a \tag{2.6}$$

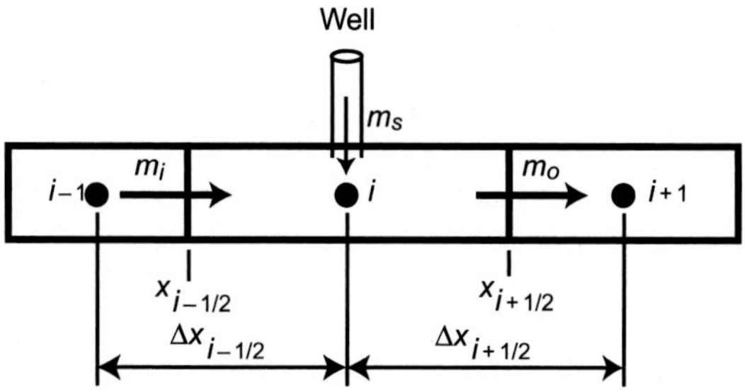

FIG. 2.3 Block *i* as a reservoir volume element in 1-D flow.

where m_i= the mass of fluid entering the reservoir volume element from other parts of the reservoir, m_o= the mass of fluid leaving the reservoir volume element to other parts of the reservoir, m_s= the mass of fluid entering or leaving the reservoir volume element externally through wells, and m_a= the mass of excess fluid stored in or depleted from the reservoir volume element over a time interval.

An *equation of state* describes the density of fluid as a function of pressure and temperature. For single-phase fluid,

$$B = \rho_{sc}/\rho \tag{2.7a}$$

for oil or water,

$$B_g = \frac{\rho_{gsc}}{\alpha_c \rho_g} \tag{2.7b}$$

for gas, where ρ and ρ_g= fluid densities at reservoir conditions, ρ_{sc} and ρ_{gsc}= fluid densities at standard conditions, and α_c= the volume conversion factor.

A *constitutive equation* describes the rate of fluid movement into (or out of) the reservoir volume element. In reservoir simulation, Darcy's law is used to relate fluid flow rate to potential gradient. The differential form of Darcy's law in a 1-D inclined reservoir is

$$u_x = q_x/A_x = -\beta_c \frac{k_x}{\mu} \frac{\partial \Phi}{\partial x} \tag{2.8}$$

where β_c= the transmissibility conversion factor, k_x= absolute permeability of rock in the direction of the *x*-axis, μ= fluid viscosity, Φ= potential, and u_x= volumetric (or superficial) velocity of fluid defined as fluid flow rate (q_x) per unit cross-sectional area (A_x) normal to flow direction *x*. The potential is related to pressure through the following relationship:

$$\Phi - \Phi_{ref} = (p - p_{ref}) - \gamma (Z - Z_{ref}) \tag{2.9}$$

where Z=elevation from datum, with positive values downward.

Therefore,

$$\frac{\partial \Phi}{\partial x} = \left(\frac{\partial p}{\partial x} - \gamma \frac{\partial Z}{\partial x} \right) \tag{2.10}$$

and the potential differences between block i and its neighbors, block $i-1$ and block $i+1$, are

$$\Phi_{i-1} - \Phi_i = (p_{i-1} - p_i) - \gamma_{i-1/2}(Z_{i-1} - Z_i) \tag{2.11a}$$

and

$$\Phi_{i+1} - \Phi_i = (p_{i+1} - p_i) - \gamma_{i+1/2}(Z_{i+1} - Z_i) \tag{2.11b}$$

2.6 Multidimensional flow in Cartesian coordinates

2.6.1 Block identification and block ordering

Before writing the flow equation for a 1-D, 2-D, or 3-D reservoir, the blocks in the discretized reservoir must be identified and ordered. Any block in the reservoir can be identified either by engineering notation or by the number the block holds in a given ordering scheme. Engineering notation uses the order of the block in the x-, y-, and z-directions, that is, it identifies a block as (i,j,k), where i, j, and k are the orders of the block in the three directions x, y, and z, respectively. The engineering notation for block identification is the most convenient for entering reservoir description (input) and for printing simulation results (output). Fig. 2.4 shows the engineering notation for block identification in a 2-D reservoir consisting of 4×5 blocks. Block ordering not only serves to identify blocks in the reservoir but also minimizes matrix computations in obtaining the solution of linear equations.

There are many block-ordering schemes, including natural ordering, zebra ordering, diagonal (D2) ordering, alternating diagonal (D4) ordering, cyclic ordering, and cyclic-2 ordering. If the reservoir has inactive blocks within its external boundaries, such blocks will be skipped, and ordering of active blocks will continue (Abou-Kassem and Ertekin, 1992). For multidimensional

(1,5)	(2,5)	(3,5)	(4,5)
(1,4)	(2,4)	(3,4)	(4,4)
(1,3)	(2,3)	(3,3)	(4,3)
(1,2)	(2,2)	(3,2)	(4,2)
(1,1)	(2,1)	(3,1)	(4,1)

FIG. 2.4 Engineering notation for block identification.

reservoirs, natural ordering is the simplest to program but is the least efficient in solving linear equations, whereas D4 ordering requires complicated programming but is the most efficient in obtaining the solution when the number of blocks is large. If the number of blocks is very large, however, the zebra ordering scheme becomes twice as efficient as D4 ordering in obtaining the solution (McDonald and Trimble, 1977). Fig. 2.5 shows the various block-ordering schemes for the 2-D reservoir shown in Fig. 2.4. Given the engineering notation for block identification, block ordering is generated internally in a simulator. Any ordering scheme becomes even more efficient computationally if the ordering is performed along the shortest direction, followed by the intermediate direction, and finally the longest direction (Abou-Kassem and Ertekin, 1992).

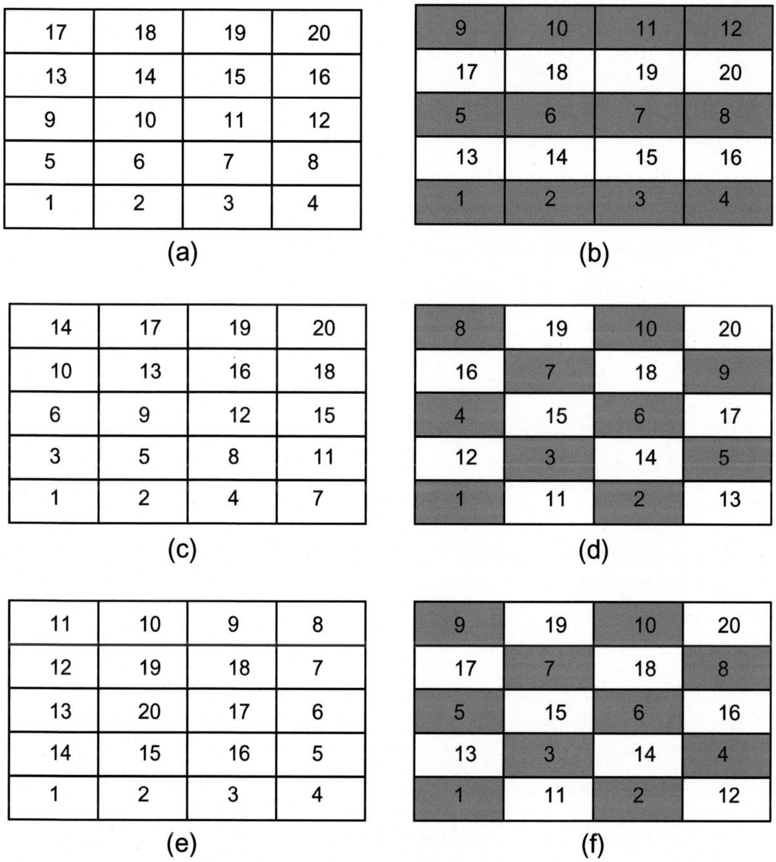

FIG. 2.5 Block-ordering schemes used in reservoir simulation. (a) Natural ordering, (b) zebra ordering, (c) diagonal (D2) ordering, (d) alternating diagonal (D4) ordering, (e) cyclic ordering, and (f) cyclic-2 ordering.

Details related to various ordering schemes and computational efficiency in solving linear equations are not discussed further in this book but can be found elsewhere (Woo et al., 1973; Price and Coats, 1974; McDonald and Trimble, 1977). The natural ordering scheme is used throughout this book because it produces equations that are readily solvable with handheld calculators and easily programmable for computer usage. The following three examples demonstrate the use of engineering notation and natural ordering to identify blocks in multidimensions.

Example 2.1 Consider the 1-D reservoir shown in Fig. 2.6a. This reservoir is discretized using four blocks in the x-direction as shown in the figure. Order the blocks in this reservoir using natural ordering.

Solution

We first choose one of the corner blocks (say the left corner block), identify it as block 1, and then move along a given direction to the other blocks, one block at a time. The order of the next block is obtained by incrementing the order of the previous block by one. The process of block ordering (or numbering) continues until the last block in that direction is numbered. The final ordering of blocks in this reservoir is shown in Fig. 2.6b.

Example 2.2 Consider the 2-D reservoir shown in Fig. 2.7a. This reservoir is discretized using 4×3 blocks as shown in the figure. Identify the blocks in this reservoir using the following:

1. Engineering notation
2. Natural ordering

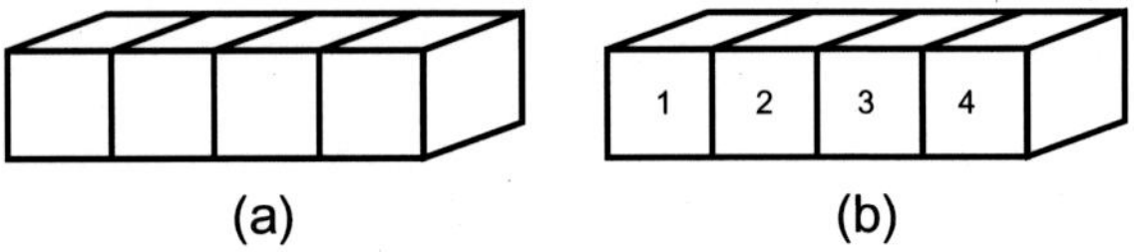

FIG. 2.6 1-D reservoir representation in Example 2.1. (a) Reservoir representation and (b) natural ordering of blocks.

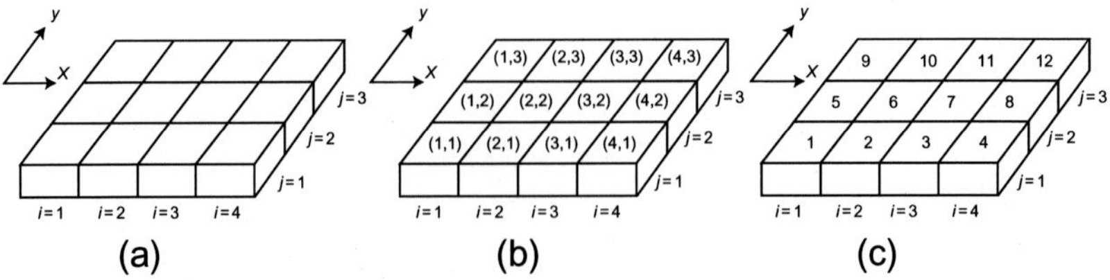

FIG. 2.7 2-D reservoir representation in Example 2.2. (a) Reservoir representation, (b) engineering notation, and (c) natural ordering of blocks.

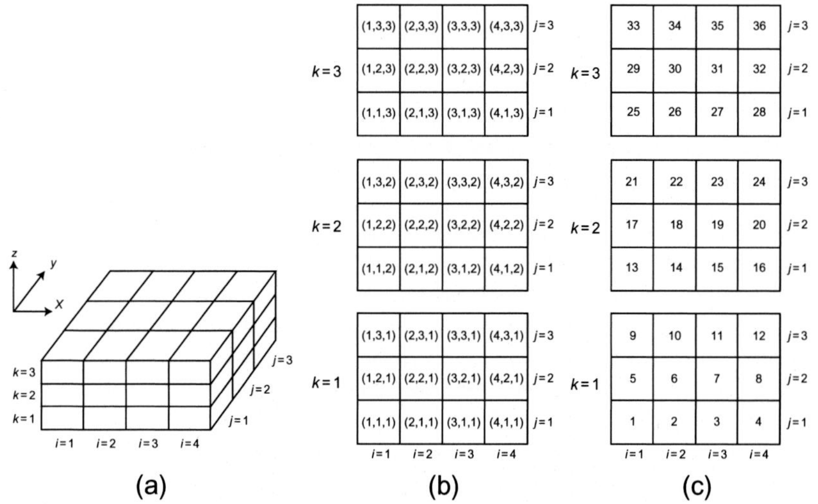

FIG. 2.8 3-D reservoir representation in Example 2.3. (a) Reservoir representation, (b) engineering notation, and (c) natural ordering of blocks.

Solution

1. The engineering notation for block identification is shown in Fig. 2.7b.
2. We start by choosing one of the corner blocks in the reservoir. In this example, we arbitrarily choose the lower-left corner block, block (1,1), and identify it as block 1. In addition, we choose to order blocks along rows. The rest of the blocks in the first row ($j=1$) are numbered as explained in Example 2.1. Block (1,2) in the first column ($i=1$) and second row ($j=2$) is numbered next as block 5, and block numbering along this row continues as in Example 2.1. Block numbering continues row by row until all the blocks are numbered. The final ordering of blocks in this reservoir is shown in Fig. 2.7c.

Example 2.3 Consider the 3-D reservoir shown in Fig. 2.8a. This reservoir is discretized into $4 \times 3 \times 3$ blocks as shown in the figure. Identify the blocks in this reservoir using the following:

1. Engineering notation
2. Natural ordering.

Solution

1. The engineering notation for block identification in this reservoir is shown in Fig. 2.8b.
2. We arbitrarily choose the bottom-lower-left corner block, block (1,1,1), and identify it as block 1. In addition, we choose to order blocks layer by layer and along rows. The blocks in the first (bottom) layer ($k=1$) are ordered as shown in Example 2.2. Next, block (1,1,2) is numbered as block 13, and the ordering of blocks in this second layer is carried out as in the first layer. Finally, block (1,1,3) is numbered as block 25, and the ordering of blocks

in this third layer ($k=3$) is carried out as before. Fig. 2.8c shows the resulting natural ordering of blocks in this reservoir.

2.6.2 Derivation of the one-dimensional flow equation in Cartesian coordinates

Fig. 2.3 shows block i and its neighboring blocks (block $i-1$ and block $i+1$) in the x-direction. At any instant in time, fluid enters block i, coming from block $i-1$ across its $x_{i-1/2}$ face at a mass rate of $w_x|_{x_{i-1/2}}$, and leaves to block $i+1$ across its $x_{i+1/2}$ face at a mass rate of $w_x|_{x_{i+1/2}}$. The fluid also enters block i through a well at a mass rate of q_{m_i}. The mass of fluid contained in a unit volume of rock in block i is m_{v_i}. Therefore, the material balance equation for block i written over a time step $\Delta t = t^{n+1} - t^n$ can be rewritten as

$$m_i|_{x_{i-1/2}} - m_o|_{x_{i+1/2}} + m_{s_i} = m_{a_i} \tag{2.12}$$

Terms like $w_x|_{x_{i-1/2}}$, $w_x|_{x_{i+1/2}}$ and q_{m_i} are functions of time only because space is not a variable for an already discretized reservoir as discussed in Section 2.4. Further justification is presented later in this section. Therefore,

$$m_i|_{x_{i-1/2}} = \int_{t^n}^{t^{n+1}} w_x|_{x_{i-1/2}} dt \tag{2.13}$$

$$m_o|_{x_{i+1/2}} = \int_{t^n}^{t^{n+1}} w_x|_{x_{i+1/2}} dt \tag{2.14}$$

and

$$m_{s_i} = \int_{t^n}^{t^{n+1}} q_{m_i} dt \tag{2.15}$$

Using Eqs. (2.13) through (2.15), Eq. (2.12) can be rewritten as

$$\int_{t^n}^{t^{n+1}} w_x|_{x_{i-1/2}} dt - \int_{t^n}^{t^{n+1}} w_x|_{x_{i+1/2}} dt + \int_{t^n}^{t^{n+1}} q_{m_i} dt = m_{a_i} \tag{2.16}$$

The mass accumulation is defined as

$$m_{a_i} = \Delta_t (V_b m_v)_i = V_{b_i}\left(m_{v_i}^{n+1} - m_{v_i}^n\right) \tag{2.17}$$

Note that mass rate and mass flux are related through

$$w_x = \dot{m}_x A_x \tag{2.18}$$

Mass flux ($\dot{m}_x$) can be expressed in terms of fluid density and volumetric velocity:

$$\dot{m}_x = \alpha_c \rho u_x \tag{2.19}$$

mass of fluid per unit volume of rock (m_v) can be expressed in terms of fluid density and porosity:

$$m_v = \phi \rho \tag{2.20}$$

and mass of injected or produced fluid (q_m) can be expressed in terms of well volumetric rate (q) and fluid density:

$$q_m = \alpha_c \rho q \tag{2.21}$$

Substitution of Eqs. (2.17) and (2.18) into Eq. (2.16) yields:

$$\int_{t^n}^{t^{n+1}} (\dot{m}_x A_x)|_{x_{i-1/2}} \mathrm{d}t - \int_{t^n}^{t^{n+1}} (\dot{m}_x A_x)|_{x_{i+1/2}} \mathrm{d}t + \int_{t^n}^{t^{n+1}} q_{m_i} \mathrm{d}t = V_{b_i}\left(m_{v_i}^{n+1} - m_{v_i}^{n}\right) \tag{2.22}$$

Substitution of Eqs. (2.19) through (2.21) into Eq. (2.22) yields:

$$\int_{t^n}^{t^{n+1}} (\alpha_c \rho u_x A_x)|_{x_{i-1/2}} \mathrm{d}t - \int_{t^n}^{t^{n+1}} (\alpha_c \rho u_x A_x)|_{x_{i+1/2}} \mathrm{d}t + \int_{t^n}^{t^{n+1}} (\alpha_c \rho q)_i \mathrm{d}t = V_{b_i}\left[(\phi\rho)_i^{n+1} - (\phi\rho)_i^{n}\right] \tag{2.23}$$

Substitution of Eq. (2.7a) into Eq. (2.23), dividing by $\alpha_c \rho_{sc}$ and noting that $q/B = q_{sc}$, yields

$$\int_{t^n}^{t^{n+1}} \left(\frac{u_x A_x}{B}\right)\Bigg|_{x_{i-1/2}} \mathrm{d}t - \int_{t^n}^{t^{n+1}} \left(\frac{u_x A_x}{B}\right)\Bigg|_{x_{i+1/2}} \mathrm{d}t + \int_{t^n}^{t^{n+1}} q_{sc_i} \mathrm{d}t = \frac{V_{b_i}}{\alpha_c}\left[\left(\frac{\phi}{B}\right)_i^{n+1} - \left(\frac{\phi}{B}\right)_i^{n}\right] \tag{2.24}$$

Fluid volumetric velocity (flow rate per unit cross-sectional area) from block $i-1$ to block i ($u_x|_{x_{i-1/2}}$) at any time instant t is given by the algebraic analog of Eq. (2.8):

$$u_x|_{x_{i-1/2}} = \beta_c \frac{k_x|_{x_{i-1/2}}}{\mu|_{x_{i-1/2}}}\left[\frac{(\Phi_{i-1} - \Phi_i)}{\Delta x_{i-1/2}}\right] \tag{2.25a}$$

where $k_x|_{x_{i-1/2}}$ is rock permeability between blocks $i-1$ and i that are separated by a distance $\Delta x_{i-1/2}$, Φ_{i-1}, and Φ_i are the potentials of blocks $i-1$ and i, and $\mu|_{x_{i-1/2}}$ is viscosity of the fluid between blocks $i-1$ and i.

Likewise, fluid flow rate per unit cross-sectional area from block i to block $i+1$ is:

$$u_x|_{x_{i+1/2}} = \beta_c \frac{k_x|_{x_{i+1/2}}}{\mu|_{x_{i+1/2}}}\left[\frac{(\Phi_i - \Phi_{i+1})}{\Delta x_{i+1/2}}\right] \tag{2.25b}$$

Substitution of Eq. (2.25) into Eq. (2.24) and grouping terms results in

$$\int_{t^n}^{t^{n+1}} \left[\left(\beta_c \frac{k_x A_x}{\mu B \Delta x} \right) \bigg|_{x_{i-1/2}} (\Phi_{i-1} - \Phi_i) \right] dt - \int_{t^n}^{t^{n+1}} \left[\left(\beta_c \frac{k_x A_x}{\mu B \Delta x} \right) \bigg|_{x_{i+1/2}} (\Phi_i - \Phi_{i+1}) \right] dt + \int_{t^n}^{t^{n+1}} q_{sc_i} dt = \frac{V_{b_i}}{\alpha_c} \left[\left(\frac{\phi}{B} \right)_i^{n+1} - \left(\frac{\phi}{B} \right)_i^n \right] \tag{2.26}$$

or

$$\int_{t^n}^{t^{n+1}} \left[T_{x_{i-1/2}} (\Phi_{i-1} - \Phi_i) \right] dt + \int_{t^n}^{t^{n+1}} \left[T_{x_{i+1/2}} (\Phi_{i+1} - \Phi_i) \right] dt + \int_{t^n}^{t^{n+1}} q_{sc_i} dt = \frac{V_{b_i}}{\alpha_c} \left[\left(\frac{\phi}{B} \right)_i^{n+1} - \left(\frac{\phi}{B} \right)_i^n \right] \tag{2.27}$$

where

$$T_{x_{i\mp1/2}} = \left(\beta_c \frac{k_x A_x}{\mu B \Delta x} \right) \bigg|_{x_{i\mp1/2}} \tag{2.28}$$

is transmissibility in the x-direction between block i and the neighboring block $i \mp 1$. The derivation of Eq. (2.27) is rigorous and involves no assumptions other than the validity of Darcy's law (Eq. 2.25) to estimate fluid volumetric velocity between block i and its neighboring block $i \mp 1$. The validity of Darcy's law is well accepted. Note that similar derivation can be made even if Darcy's law is replaced by another flow equation, such as Brinkman's equation, etc. (Islam, 1992; Mustafiz et al., 2005a, b). For heterogeneous block permeability distribution and irregular grid blocks (neither constant nor equal Δx, Δy, and Δz), the part $\left(\beta_c \frac{k_x A_x}{\Delta x} \right) \Big|_{x_{i\mp1/2}}$ of transmissibility $T_{x_{i\mp1/2}}$ is derived in Chapter 4 for a block-centered grid and in Chapter 5 for a point-distributed grid. Note that for a discretized reservoir, blocks have defined dimensions and permeabilities; therefore, interblock geometric factor $\left[\left(\beta_c \frac{k_x A_x}{\Delta x} \right) \Big|_{x_{i\mp1/2}} \right]$ is constant, independent of space and time. In addition, the pressure-dependent term $(\mu B)|_{x_{i\mp1/2}}$ of transmissibility uses some average viscosity and formation volume factor (FVF) of the fluid contained in block i and the neighboring block $i \mp 1$ or some weight (upstream weighting and average weighting) at any instant of time t. In other words, the term $(\mu B)|_{x_{i\mp1/2}}$ is not a function of space but a function of time as block pressures change with time. Hence, transmissibility $T_{x_{i\mp1/2}}$ between block i and its neighboring block $i \mp 1$ is a function of time only; it does not depend on space at any instant of time.

Again, the accumulation term in Eq. (2.27) can be expressed in terms of the change in the pressure of block i as shown in Eq. (2.29a):

$$\int_{t^n}^{t^{n+1}} \left[T_{x_{i-1/2}}(\Phi_{i-1}-\Phi_i)\right]dt + \int_{t^n}^{t^{n+1}} \left[T_{x_{i+1/2}}(\Phi_{i+1}-\Phi_i)\right]dt + \int_{t^n}^{t^{n+1}} q_{sc_i}dt$$
$$= \frac{V_{b_i}}{\alpha_c}\frac{d}{dp}\left(\frac{\phi}{B}\right)_i\left[p_i^{n+1}-p_i^n\right] \quad (2.29a)$$

or after substituting Eq. (2.11) for potential,

$$\int_{t^n}^{t^{n+1}} \left\{T_{x_{i-1/2}}\left[(p_{i-1}-p_i)-\gamma_{i-1/2}(Z_{i-1}-Z_i)\right]\right\}dt$$
$$+\int_{t^n}^{t^{n+1}} \left\{T_{x_{i+1/2}}\left[(p_{i+1}-p_i)-\gamma_{i+1/2}(Z_{i+1}-Z_i)\right]\right\}dt \quad (2.29b)$$
$$+\int_{t^n}^{t^{n+1}} q_{sc_i}dt = \frac{V_{b_i}}{\alpha_c}\frac{d}{dp}\left(\frac{\phi}{B}\right)_i\left[p_i^{n+1}-p_i^n\right]$$

where $\frac{d}{dp}\left(\frac{\phi}{B}\right)_i$ = the chord slope of $\left(\frac{\phi}{B}\right)_i$ between p_i^{n+1} and p_i^n.

2.6.3 Approximation of time integrals

If the argument of an integral is an explicit function of time, the integral can be evaluated analytically. This is not the case for the integrals appearing on the left-hand side (LHS) of either Eq. (2.27) or Eq. (2.29). If Eq. (2.29b) is written for every block $i = 1, 2, 3\ldots n_x$, then the solution can be obtained by one of the ODE methods (Euler's method, the modified Euler method, the explicit Runge-Kutta method, or the implicit Runge-Kutta method) reviewed by Aziz and Settari (1979). ODE methods, however, are not efficient for solving reservoir simulation problems. Therefore, performing these integrations necessitates making certain assumptions.

Consider the integral $\int_{t^n}^{t^{n+1}} F(t)dt$ shown in Fig. 2.9. This integral is equal to the area under the curve $F(t)$ in the interval $t^n \le t \le t^{n+1}$. This area is also equal to the

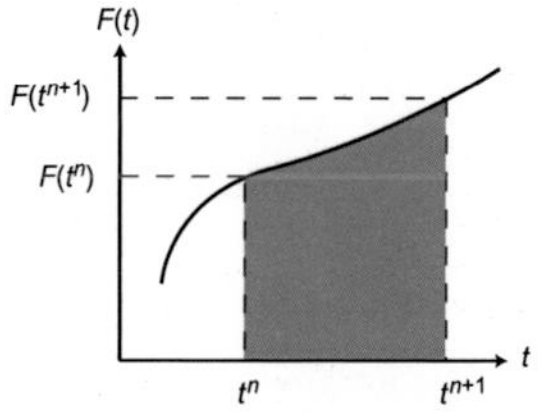

FIG. 2.9 Representation of the integral function as the area under the curve.

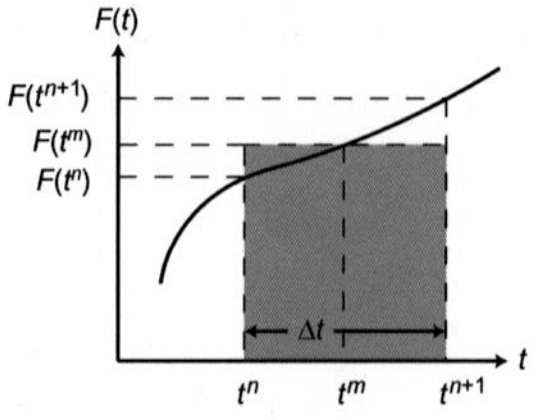

FIG. 2.10 Representation of the integral of a function as $F(t^m)\times\Delta t$.

area of a rectangle with the dimensions of $F(t^m)$, where F is evaluated at time t^m, where $t^n \le t^m \le t^{n+1}$ and Δt, where $\Delta t=(t^{n+1}-t^n)$, as shown in Fig. 2.10. Therefore,

$$\int_{t^n}^{t^{n+1}} F(t)\mathrm{d}t = \int_{t^n}^{t^{n+1}} F(t^m)\mathrm{d}t = \int_{t^n}^{t^{n+1}} F^m \mathrm{d}t = F^m \int_{t^n}^{t^{n+1}} \mathrm{d}t = F^m \times t\Big|_{t^n}^{t^{n+1}} = F^m \times (t^{n+1}-t^n) = F^m \times \Delta t \tag{2.30}$$

The value of this integral can be calculated using the previous equation provided that the value of F^m or $F(t^m)$ is known. In reality, however, F^m is not known, and therefore, it needs to be approximated. The area under the curve in Fig. 2.9 can be approximated by one of the following four methods: (1) $F(t^n)\times\Delta t$ as shown in Fig. 2.11a, (2) $F(t^{n+1})\times\Delta t$ as shown in Fig. 2.11b, (3) $^1/_2[F(t^n)+F(t^{n+1})]\times\Delta t$ as shown in Fig. 2.11c, or (4) numerical integration. The argument F in Eq. (2.30) stands for $[T_{x_{i-1/2}}(\Phi_{i-1}-\Phi_i)]$, $[T_{x_{i+1/2}}(\Phi_{i+1}-\Phi_i)]$, or q_{sc_i} that appears on the LHS of Eq. (2.27), and $F^m=$ value of F at time t^m.

Therefore, Eq. (2.27) after this approximation becomes:

$$\left[T^m_{x_{i-1/2}}\left(\Phi^m_{i-1}-\Phi^m_i\right)\right]\Delta t + \left[T^m_{x_{i+1/2}}\left(\Phi^m_{i+1}-\Phi^m_i\right)\right]\Delta t + q^m_{sc_i}\Delta t = \frac{V_{b_i}}{\alpha_c}\left[\left(\frac{\phi}{B}\right)^{n+1}_i - \left(\frac{\phi}{B}\right)^n_i\right] \tag{2.31}$$

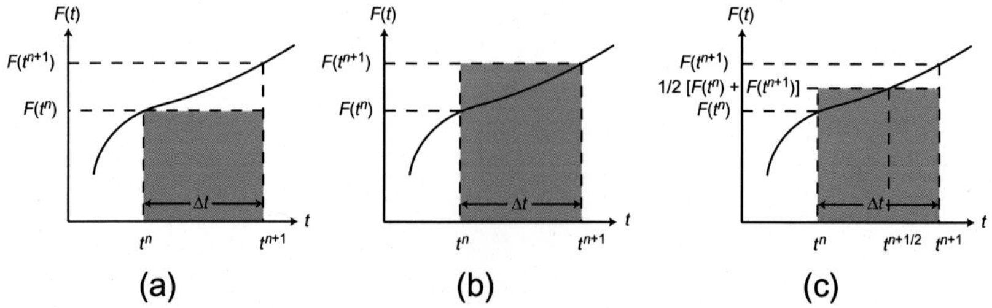

FIG. 2.11 Approximations of the time integral of function.

Dividing the previous equation by Δt gives:

$$T_{x_{i-1/2}}^m\left(\Phi_{i-1}^m-\Phi_i^m\right)+T_{x_{i+1/2}}^m\left(\Phi_{i+1}^m-\Phi_i^m\right)+q_{sc_i}^m=\frac{V_{b_i}}{\alpha_c\Delta t}\left[\left(\frac{\phi}{B}\right)_i^{n+1}-\left(\frac{\phi}{B}\right)_i^n\right] \tag{2.32}$$

Substituting Eq. (2.11) into Eq. (2.32), we obtain the flow equation for block i:

$$T_{x_{i-1/2}}^m\left[\left(p_{i-1}^m-p_i^m\right)-\gamma_{i-1/2}^m(Z_{i-1}-Z_i)\right]+T_{x_{i+1/2}}^m\left[\left(p_{i+1}^m-p_i^m\right)-\gamma_{i+1/2}^m(Z_{i+1}-Z_i)\right]$$
$$+q_{sc_i}^m=\frac{V_{b_i}}{\alpha_c\Delta t}\left[\left(\frac{\phi}{B}\right)_i^{n+1}-\left(\frac{\phi}{B}\right)_i^n\right] \tag{2.33}$$

The right-hand side (RHS) of the flow equation expressed as Eq. (2.33), known as the fluid accumulation term, vanishes in problems involving the flow of incompressible fluid ($c=0$) in an incompressible porous medium ($c_\phi=0$). This is the case where both B and ϕ are constant independent of pressure. Reservoir pressure in this type of flow problems is independent of time. Example 2.4 demonstrates the application of Eq. (2.33) for an interior block in a 1-D reservoir using a regular grid. In Chapter 7, the explicit, implicit, and Crank-Nicolson formulations are derived from Eq. (2.33) by specifying the approximation of time t^m as t^n, t^{n+1}, or $t^{n+1/2}$, which are equivalent to using the first, *sec*ond, and third integral approximation methods mentioned previously. The fourth integration method mentioned previously leads to the Runge-Kutta solution methods of ordinary differential equations. Table 2.1 presents the units of all the quantities that appear in flow equations.

Example 2.4 Consider single-phase fluid flow in a 1-D horizontal reservoir. The reservoir is discretized using four blocks in the x-direction, as shown in Fig. 2.12. A well located in block 3 produces at a rate of 400 STB/D. All grid blocks have $\Delta x=250$ ft, $w=900$ ft, $h=100$ ft, and $k_x=270$ md. The FVF and the viscosity of the flowing fluid are 1.0 RB/STB and 2 cP, respectively. Identify the interior and boundary blocks in this reservoir. Write the flow equation for block 3 and give the physical meaning of each term in the equation.

Solution

Blocks 2 and 3 are interior blocks, whereas blocks 1 and 4 are boundary blocks. The flow equation for block 3 can be obtained from Eq. (2.33) for $i=3$, that is,

$$T_{x_{3-1/2}}^m\left[\left(p_2^m-p_3^m\right)-\gamma_{3-1/2}^m(Z_2-Z_3)\right]+T_{x_{3+1/2}}^m\left[\left(p_4^m-p_3^m\right)-\gamma_{3+1/2}^m(Z_4-Z_3)\right]$$
$$+q_{sc_3}^m=\frac{V_{b_3}}{\alpha_c\Delta t}\left[\left(\frac{\phi}{B}\right)_3^{n+1}-\left(\frac{\phi}{B}\right)_3^n\right] \tag{2.34}$$

TABLE 2.1 Quantities used in flow equations in different systems of units.

Quantity	Symbol	System of units		
		Customary units	*SPE metric units*	*Lab units*
Length	x, y, z, r, Z	ft	m	cm
Area	$A, A_x, A_y, A_z, A_r, A_\theta$	ft^2	m^2	cm^2
Permeability	$k, k_x, k_y, k_z, k_r, k_\theta$	md	μm^2	darcy
Phase viscosity	μ, μ_o, μ_w, μ_g	cP	mPa.s	cP
Gas FVF	B, B_g	RB/scf	m^3/std m^3	cm^3/std cm^3
Liquid FVF	B, B_o, B_w	RB/STB	m^3/std m^3	cm^3/std cm^3
Solution GOR	R_s	scf/STB	std m^3/std m^3	std cm^3/std cm^3
Phase pressure	p, p_o, p_w, p_g	psia	kPa	atm
Phase potential	$\Phi, \Phi_o, \Phi_w, \Phi_g$	psia	kPa	atm
Phase gravity	$\gamma, \gamma_o, \gamma_w, \gamma_g$	psi/ft	kPa/m	atm/cm
Gas flow rate	q_{sc}, q_{gsc}	scf/D	std m^3/d	std cm^3/s
Oil flow rate	q_{sc}, q_{osc}	STB/D	std m^3/d	std cm^3/s
Water flow rate	q_{sc}, q_{wsc}	B/D	std m^3/d	std cm^3/s
Volumetric velocity	u	RB/D-ft^2	m/d	cm/s
Phase density	$\rho, \rho_o, \rho_w, \rho_g$	lbm/ft^3	kg/m^3	g/cm^3
Block bulk volume	V_b	ft^3	m^3	cm^3

Compressibility	c, c_o, c_ϕ	psi^{-1}	kPa^{-1}	atm^{-1}
Compressibility factor	z	Dimensionless	Dimensionless	Dimensionless
Temperature	T	°R	K	K
Porosity	ϕ	Fraction	Fraction	Fraction
Phase saturation	S, S_o, S_w, S_g	Fraction	Fraction	Fraction
Relative permeability	k_{ro}, k_{rw}, k_{rg}	Fraction	Fraction	Fraction
Gravitational acceleration	g	32.174 ft/s^2	9.806635 m/s^2	980.6635 cm/s^2
Time	t, Δt	day	day	sec
Angle	θ	rad	rad	rad
Transmissibility conversion factor	β_c	0.001127	0.0864	1
Gravity conversion factor	γ_c	0.21584×10^{-3}	0.001	0.986923×10^{-6}
Volume conversion factor	α_c	5.614583	1	1

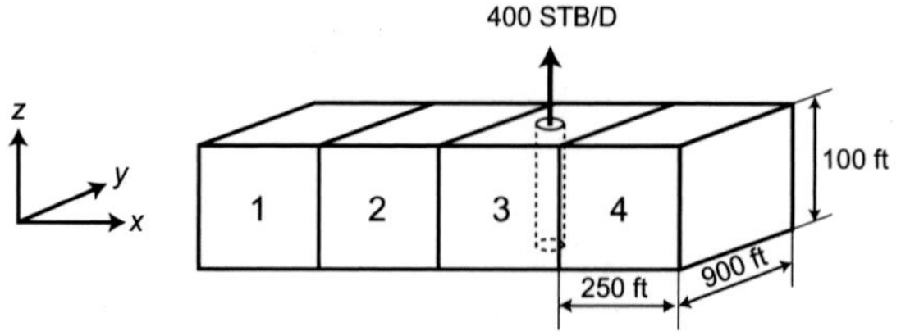

FIG. 2.12 1-D reservoir representation in Example 2.4.

For block 3, $Z_2=Z_3=Z_4$ for horizontal reservoir and $q^m_{sc_3}=-400\,\text{STB/D}$. Because $\Delta x_{3\mp1/2}=\Delta x$ and because μ and B are constant,

$$T^m_{x_{3-1/2}}=T^m_{x_{3+1/2}}=\beta_c\frac{k_xA_x}{\mu B\Delta x}=0.001127\times\frac{270\times(900\times100)}{2\times1\times250}$$
$$=54.7722\,\text{STB/D-psi} \tag{2.35}$$

Substitution of Eq. (2.35) into Eq. (2.34) gives

$$(54.7722)\left(p^m_2-p^m_3\right)+(54.7722)\left(p^m_4-p^m_3\right)-400=\frac{V_{b_3}}{\alpha_c\Delta t}\left[\left(\frac{\phi}{B}\right)^{n+1}_3-\left(\frac{\phi}{B}\right)^n_3\right] \tag{2.36}$$

The LHS of Eq. (2.36) comprises three terms. The first term represents the rate of fluid flow from block 2 to block 3, the second term represents the rate of fluid flow from block 4 to block 3, and the third term represents the rate of fluid production from the well in block 3. The RHS of Eq. (2.36) represents the rate of fluid accumulation in block 3. All terms have the units of STB/D.

2.6.4 Flow equations in multidimensions using engineering notation

A close inspection of the flow equation expressed as Eq. (2.33) reveals that this equation involves three different groups: the interblock flow terms between block i and its two neighboring blocks in the x-direction $\{T^m_{x_{i-1/2}}[(p^m_{i-1}-p^m_i)-\gamma^m_{i-1/2}(Z_{i-1}-Z_i)]$ and $T^m_{x_{i+1/2}}[(p^m_{i+1}-p^m_i)-\gamma^m_{i+1/2}(Z_{i+1}-Z_i)]\}$, the source term due to injection or production ($q^m_{sc_i}$), and the accumulation term $\left\{\frac{V_{b_i}}{\alpha_c\Delta t}\left[\left(\frac{\phi}{B}\right)^{n+1}_i-\left(\frac{\phi}{B}\right)^n_i\right]\right\}$. Any block in the reservoir has one source term and one accumulation term, but the number of interblock flow terms equals the number of its neighboring blocks. Specifically, any block has a maximum of two neighboring blocks in 1-D flow (Fig. 2.2a), four neighboring blocks in 2-D flow (Fig. 2.2b), and six neighboring blocks in 3-D flow (Fig. 2.2c). Therefore, for 2-D flow, the flow equation for block (i,j) in the x-y plane is:

$$
\begin{aligned}
&T^m_{y_{i,j-1/2}}\left[\left(p^m_{i,j-1}-p^m_{i,j}\right)-\gamma^m_{i,j-1/2}\left(Z_{i,j-1}-Z_{i,j}\right)\right] \\
&+T^m_{x_{i-1/2,j}}\left[\left(p^m_{i-1,j}-p^m_{i,j}\right)-\gamma^m_{i-1/2,j}\left(Z_{i-1,j}-Z_{i,j}\right)\right] \\
&+T^m_{x_{i+1/2,j}}\left[\left(p^m_{i+1,j}-p^m_{i,j}\right)-\gamma^m_{i+1/2,j}\left(Z_{i+1,j}-Z_{i,j}\right)\right] \\
&+T^m_{y_{i,j+1/2}}\left[\left(p^m_{i,j+1}-p^m_{i,j}\right)-\gamma^m_{i,j+1/2}\left(Z_{i,j+1}-Z_{i,j}\right)\right] \\
&+q^m_{sc_{i,j}}=\frac{V_{b_{i,j}}}{\alpha_c\Delta t}\left[\left(\frac{\phi}{B}\right)^{n+1}_{i,j}-\left(\frac{\phi}{B}\right)^{n}_{i,j}\right]
\end{aligned}
\tag{2.37}
$$

For 3-D flow, the flow equation for block (i,j,k) in the x-y-z space is:

$$
\begin{aligned}
&T^m_{z_{i,j,k-1/2}}\left[\left(p^m_{i,j,k-1}-p^m_{i,j,k}\right)-\gamma^m_{i,j,k-1/2}\left(Z_{i,j,k-1}-Z_{i,j,k}\right)\right] \\
&+T^m_{y_{i,j-1/2,k}}\left[\left(p^m_{i,j-1,k}-p^m_{i,j,k}\right)-\gamma^m_{i,j-1/2,k}\left(Z_{i,j-1,k}-Z_{i,j,k}\right)\right] \\
&+T^m_{x_{i-1/2,j,k}}\left[\left(p^m_{i-1,j,k}-p^m_{i,j,k}\right)-\gamma^m_{i-1/2,j,k}\left(Z_{i-1,j,k}-Z_{i,j,k}\right)\right] \\
&+T^m_{x_{i+1/2,j,k}}\left[\left(p^m_{i+1,j,k}-p^m_{i,j,k}\right)-\gamma^m_{i+1/2,j,k}\left(Z_{i+1,j,k}-Z_{i,j,k}\right)\right] \\
&+T^m_{y_{i,j+1/2,k}}\left[\left(p^m_{i,j+1,k}-p^m_{i,j,k}\right)-\gamma^m_{i,j+1/2,k}\left(Z_{i,j+1,k}-Z_{i,j,k}\right)\right] \\
&+T^m_{z_{i,j,k+1/2}}\left[\left(p^m_{i,j,k+1}-p^m_{i,j,k}\right)-\gamma^m_{i,j,k+1/2}\left(Z_{i,j,k+1}-Z_{i,j,k}\right)\right] \\
&+q^m_{sc_{i,j,k}}=\frac{V_{b_{i,j,k}}}{\alpha_c\Delta t}\left[\left(\frac{\phi}{B}\right)^{n+1}_{i,j,k}-\left(\frac{\phi}{B}\right)^{n}_{i,j,k}\right]
\end{aligned}
\tag{2.38}
$$

where,

$$
T_{x_{i\mp1/2,j,k}}=\left(\beta_c\frac{k_xA_x}{\mu B\Delta x}\right)\bigg|_{x_{i\mp1/2,j,k}}=\left(\beta_c\frac{k_xA_x}{\Delta x}\right)_{x_{i\mp1/2,j,k}}\left(\frac{1}{\mu B}\right)_{x_{i\mp1/2,j,k}}=G_{x_{i\mp1/2,j,k}}\left(\frac{1}{\mu B}\right)_{x_{i\mp1/2,j,k}}
\tag{2.39a}
$$

$$
T_{y_{i,j\mp1/2,k}}=\left(\beta_c\frac{k_yA_y}{\mu B\Delta y}\right)\bigg|_{y_{i,j\mp1/2,k}}=\left(\beta_c\frac{k_yA_y}{\Delta y}\right)_{y_{i,j\mp1/2,k}}\left(\frac{1}{\mu B}\right)_{y_{i,j\mp1/2,k}}=G_{y_{i,j\mp1/2,k}}\left(\frac{1}{\mu B}\right)_{y_{i,j\mp1/2,k}}
\tag{2.39b}
$$

and

$$
T_{z_{i,j,k\mp1/2}}=\left(\beta_c\frac{k_zA_z}{\mu B\Delta z}\right)\bigg|_{z_{i,j,k\mp1/2}}=\left(\beta_c\frac{k_zA_z}{\Delta z}\right)_{z_{i,j,k\mp1/2}}\left(\frac{1}{\mu B}\right)_{z_{i,j,k\mp1/2}}=G_{z_{i,j,k\mp1/2}}\left(\frac{1}{\mu B}\right)_{z_{i,j,k\mp1/2}}
\tag{2.39c}
$$

Expressions for the geometric factors G for irregular grids in heterogeneous reservoirs are presented in Chapters 4 and 5. It should be mentioned that the interblock flow terms in the flow equations for 1-D (Eq. 2.33), 2-D (Eq. 2.37), or 3-D (Eq. 2.38) problems appear in the sequence shown in Fig. 2.13 for neighboring blocks. As will be shown in Chapter 9, the sequencing of neighboring blocks as

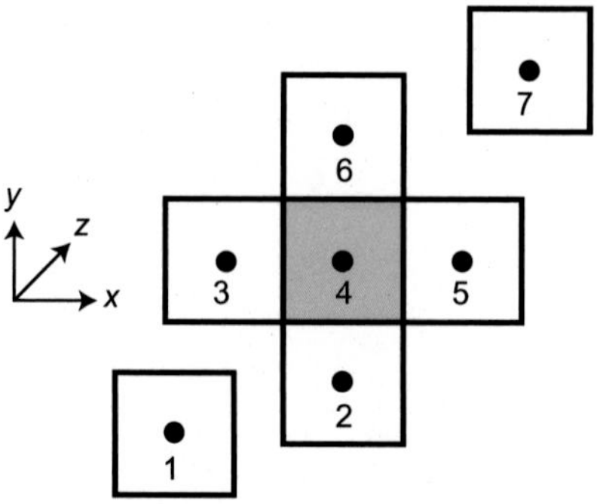

FIG. 2.13 The sequence of neighboring blocks in the set $\psi_{i,j,k}$ or ψ_n.

in Fig. 2.13 produces flow equations with unknowns already ordered as they appear in the vector of unknowns for the whole reservoir.

The following two examples demonstrate the application of Eqs. (2.37) and (2.38) for interior blocks in multidimensional anisotropic reservoirs using regular grids.

Example 2.5 Consider single-phase fluid flow in a 2-D horizontal reservoir. The reservoir is discretized using 4×3 blocks as shown in Fig. 2.14. A well that is located in block (3,2) produces at a rate of 400 STB/D. All gridblocks have $\Delta x = 250$ ft, $\Delta y = 300$ ft, $h = 100$ ft, $k_x = 270$ md, and $k_y = 220$ md. The FVF and the viscosity of the flowing fluid are 1.0 RB/STB and 2 cP, respectively. Identify the interior and boundary blocks in this reservoir. Write the flow equation for block (3,2) and give the physical meaning of each term in the flow equation. Write the flow equation for block (2,2).

Solution

Interior blocks in this reservoir include reservoir blocks that are located in the second and third columns in the second row. Other reservoir blocks are boundary blocks. In explicit terms, blocks (2,2) and (3,2) are interior blocks, whereas blocks (1,1), (2,1), (3,1), (4,1), (1,2), (4,2), (1,3), (2,3), (3,3), and (4,3) are boundary blocks.

The flow equation for block (3,2) can be obtained from Eq. (2.37) for $i = 3$ and $j = 2$, that is,

$$\begin{aligned} &T^m_{y_{3,2-1/2}}\left[\left(p^m_{3,1} - p^m_{3,2}\right) - \gamma^m_{3,2-1/2}(Z_{3,1} - Z_{3,2})\right] \\ &+ T^m_{x_{3-1/2,2}}\left[\left(p^m_{2,2} - p^m_{3,2}\right) - \gamma^m_{3-1/2,2}(Z_{2,2} - Z_{3,2})\right] \\ &+ T^m_{x_{3+1/2,2}}\left[\left(p^m_{4,2} - p^m_{3,2}\right) - \gamma^m_{3+1/2,2}(Z_{4,2} - Z_{3,2})\right] \\ &+ T^m_{y_{3,2+1/2}}\left[\left(p^m_{3,3} - p^m_{3,2}\right) - \gamma^m_{3,2+1/2}(Z_{3,3} - Z_{3,2})\right] \\ &+ q^m_{sc_{3,2}} = \frac{V_{b_{3,2}}}{\alpha_c \Delta t}\left[\left(\frac{\phi}{B}\right)^{n+1}_{3,2} - \left(\frac{\phi}{B}\right)^{n}_{3,2}\right] \end{aligned} \tag{2.40}$$

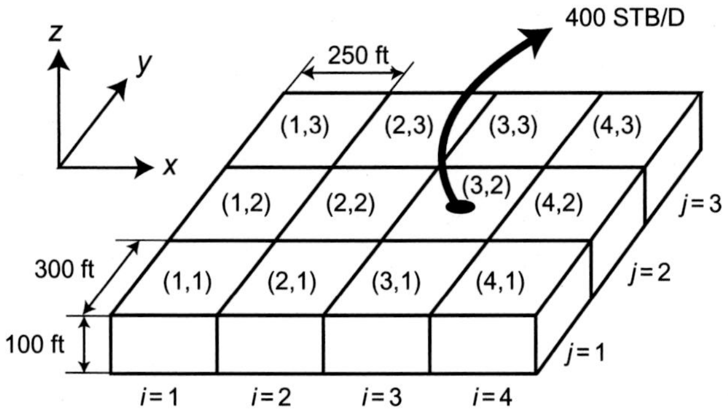

FIG. 2.14 2-D reservoir representation in Example 2.5.

For block (3,2), $Z_{3,1}=Z_{2,2}=Z_{3,2}=Z_{4,2}=Z_{3,3}$ for a horizontal reservoir and $q^m_{sc_{3,2}}=-400\,\text{STB/D}$. Because $\Delta x_{3\mp1/2,2}=\Delta x=250\,\text{ft}$, $\Delta y_{3,2\mp1/2}=\Delta y=300\,\text{ft}$, and μ and B are constant,

$$T^m_{x_{3-1/2,2}}=T^m_{x_{3+1/2,2}}=\beta_c\frac{k_xA_x}{\mu B\Delta x}=0.001127\times\frac{270\times(300\times100)}{2\times1\times250}$$
$$=18.2574\,\text{STB/D-psi} \tag{2.41a}$$

and

$$T^m_{y_{3,2-1/2}}=T^m_{y_{3,2+1/2}}=\beta_c\frac{k_yA_y}{\mu B\Delta y}=0.001127\times\frac{220\times(250\times100)}{2\times1\times300}$$
$$=10.3308\,\text{STB/D-psi} \tag{2.41b}$$

Substitution into Eq. (2.40) gives

$$(10.3308)\left(p^m_{3,1}-p^m_{3,2}\right)+(18.2574)\left(p^m_{2,2}-p^m_{3,2}\right)+(18.2574)\left(p^m_{4,2}-p^m_{3,2}\right)$$
$$+(10.3308)\left(p^m_{3,3}-p^m_{3,2}\right)-400=\frac{V_{b_{3,2}}}{\alpha_c\Delta t}\left[\left(\frac{\phi}{B}\right)^{n+1}_{3,2}-\left(\frac{\phi}{B}\right)^{n}_{3,2}\right] \tag{2.42}$$

The LHS of Eq. (2.42) comprises five terms. The first term represents the rate of fluid flow from block (3,1) to block (3,2), the second term from block (2,2) to block (3,2), the third from block (4,2) to block (3,2), and the fourth from block (3,3) to block (3,2). Finally, the fifth term represents the rate of fluid production from the well in block (3,2). The RHS of Eq. (2.42) represents the rate of fluid accumulation in block (3,2). All terms have the units STB/D.

The flow equation for block (2,2) can be obtained from Eq. (2.37) for $i=2$ and $j=2$; that is,

$$\begin{aligned}&T^m_{y_{2,2-1/2}}\left[\left(p^m_{2,1}-p^m_{2,2}\right)-\gamma^m_{2,2-1/2}(Z_{2,1}-Z_{2,2})\right]\\&+T^m_{x_{2-1/2,2}}\left[\left(p^m_{1,2}-p^m_{2,2}\right)-\gamma^m_{2-1/2,2}(Z_{1,2}-Z_{2,2})\right]\\&+T^m_{x_{2+1/2,2}}\left[\left(p^m_{3,2}-p^m_{2,2}\right)-\gamma^m_{2+1/2,2}(Z_{3,2}-Z_{2,2})\right]\\&+T^m_{y_{2,2+1/2}}\left[\left(p^m_{2,3}-p^m_{2,2}\right)-\gamma^m_{2,2+1/2}(Z_{2,3}-Z_{2,2})\right]\\&+q^m_{sc_{2,2}}=\frac{V_{b_{2,2}}}{\alpha_c\Delta t}\left[\left(\frac{\phi}{B}\right)^{n+1}_{2,2}-\left(\frac{\phi}{B}\right)^{n}_{2,2}\right]\end{aligned}\tag{2.43}$$

For block (2,2), $Z_{2,2}=Z_{2,1}=Z_{1,2}=Z_{2,2}=Z_{3,2}=Z_{2,3}$ for a horizontal reservoir, $q^m_{sc_{2,2}}=0$ STB/D because block (2,2) does not host a well, $T^m_{x_{2-1/2,2}}=T^m_{x_{2+1/2,2}}=18.2574$ STB/D-psi, and $T^m_{y_{2,2-1/2}}=T^m_{y_{2,2+1/2}}=10.3308$ STB/D-psi.

Substitution into Eq. (2.43) gives:

$$\begin{aligned}&(10.3308)\left(p^m_{2,1}-p^m_{2,2}\right)+(18.2574)\left(p^m_{1,2}-p^m_{2,2}\right)+(18.2574)\left(p^m_{3,2}-p^m_{2,2}\right)\\&+(10.3308)\left(p^m_{2,3}-p^m_{2,2}\right)=\frac{V_{b_{2,2}}}{\alpha_c\Delta t}\left[\left(\frac{\phi}{B}\right)^{n+1}_{2,2}-\left(\frac{\phi}{B}\right)^{n}_{2,2}\right]\end{aligned}\tag{2.44}$$

Example 2.6 Consider single-phase fluid flow in a 3-D horizontal reservoir. The reservoir is discretized using $4\times3\times3$ blocks as shown in Fig. 2.15a. A well

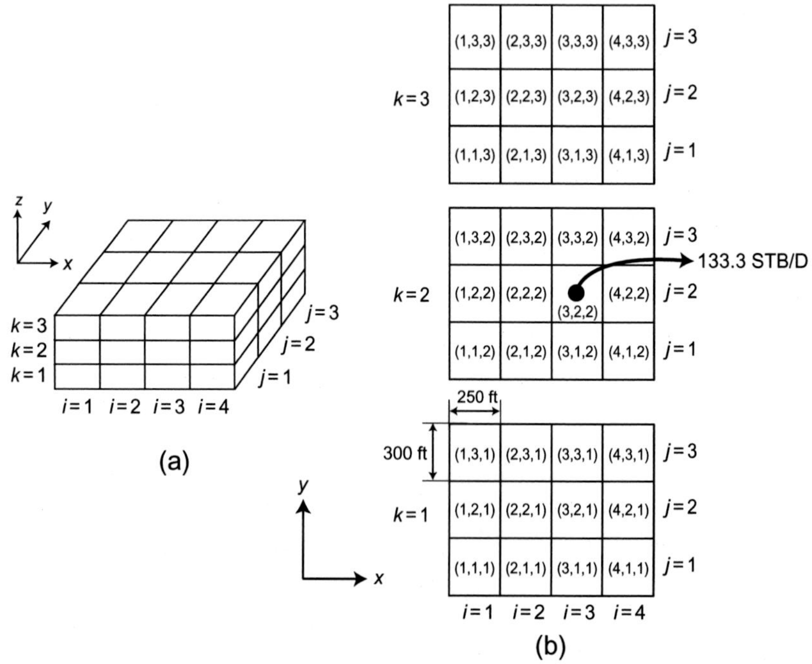

FIG. 2.15 3-D reservoir representation in Example 2.6. (a) Reservoir representation and (b) engineering notation.

that is located in block (3,2,2) produces at a rate of 133.3 STB/D. All grid blocks have $\Delta x = 250\,\text{ft}$, $\Delta y = 300\,\text{ft}$, $\Delta z = 33.333\,\text{ft}$, $k_x = 270\,\text{md}$, $k_y = 220\,\text{md}$, and $k_z = 50\,\text{md}$. The FVF, density, and viscosity of the flowing fluid are 1.0 RB/STB, 55 lbm/ft^3, and 2 cP, respectively. Identify the interior and boundary blocks in this reservoir. Write the flow equation for block (3,2,2).

Solution

As can be seen in Fig. 2.15b, interior blocks include reservoir blocks that are located in the second and third columns in the second row in the second layer, that is, blocks (2,2,2) and (3,2,2). All other reservoir blocks are boundary blocks.

The flow equation for block (3,2,2) can be obtained from Eq. (2.38) for $i=3$, $j=2$, and $k=2$, that is,

$$\begin{aligned}
&T^m_{z_{3,2,2-1/2}}\left[\left(p^m_{3,2,1}-p^m_{3,2,2}\right)-\gamma^m_{3,2,2-1/2}(Z_{3,2,1}-Z_{3,2,2})\right]\\
&+T^m_{y_{3,2-1/2,2}}\left[\left(p^m_{3,1,2}-p^m_{3,2,2}\right)-\gamma^m_{3,2-1/2,2}(Z_{3,1,2}-Z_{3,2,2})\right]\\
&+T^m_{x_{3-1/2,2,2}}\left[\left(p^m_{2,2,2}-p^m_{3,2,2}\right)-\gamma^m_{3-1/2,2,2}(Z_{2,2,2}-Z_{3,2,2})\right]\\
&+T^m_{x_{3+1/2,2,2}}\left[\left(p^m_{4,2,2}-p^m_{3,2,2}\right)-\gamma^m_{3+1/2,2,2}(Z_{4,2,2}-Z_{3,2,2})\right]\\
&+T^m_{y_{3,2+1/2,2}}\left[\left(p^m_{3,3,2}-p^m_{3,2,2}\right)-\gamma^m_{3,2+1/2,2}(Z_{3,3,2}-Z_{3,2,2})\right]\\
&+T^m_{z_{3,2,2+1/2}}\left[\left(p^m_{3,2,3}-p^m_{3,2,2}\right)-\gamma^m_{3,2,2+1/2}(Z_{3,2,3}-Z_{3,2,2})\right]\\
&+q^m_{sc_{3,2,2}}=\frac{V_{b_{3,2,2}}}{\alpha_c\Delta t}\left[\left(\frac{\phi}{B}\right)^{n+1}_{3,2,2}-\left(\frac{\phi}{B}\right)^{n}_{3,2,2}\right]
\end{aligned} \tag{2.45}$$

For block (3,2,2), $Z_{3,1,2}=Z_{2,2,2}=Z_{3,2,2}=Z_{4,2,2}=Z_{3,3,2}$, $Z_{3,2,1}-Z_{3,2,2}=33.333\,\text{ft}$, $Z_{3,2,3}-Z_{3,2,2}=-33.333\,\text{ft}$, and $q^m_{sc_{3,2,2}}=-133.3\,\text{STB/D}$. Because $\Delta x_{3\mp1/2,2,2}=\Delta x=250\,\text{ft}$, $\Delta y_{3,2\mp1/2,2}=\Delta y=300\,\text{ft}$, $\Delta z_{3,2,2\mp1/2}=\Delta z=33.333\,\text{ft}$ and because μ, ρ, and B are constant, $\gamma^m_{3,2,2-1/2}=\gamma^m_{3,2,2+1/2}=\gamma_c\rho g=0.21584\times10^{-3}\times55\times32.174=0.3819\,\text{psi/ft}$,

$$T^m_{x_{3\mp1/2,2,2}}=\beta_c\frac{k_xA_x}{\mu B\Delta x}=0.001127\times\frac{270\times(300\times33.333)}{2\times1\times250}=6.0857\,\text{STB/D-psi} \tag{2.46a}$$

$$T^m_{y_{3,2\mp1/2,2}}=\beta_c\frac{k_yA_y}{\mu B\Delta y}=0.001127\times\frac{220\times(250\times33.333)}{2\times1\times300}=3.4436\,\text{STB/D-psi} \tag{2.46b}$$

and

$$T^m_{z_{3,2,2\mp1/2}}=\beta_c\frac{k_zA_z}{\mu B\Delta z}=0.001127\times\frac{50\times(250\times300)}{2\times1\times33.333}=63.3944\,\text{STB/D-psi} \tag{2.46c}$$

Substitution into Eq. (2.45) gives:

$$(63.3944)\left[\left(p_{3,2,1}^m - p_{3,2,2}^m\right) - 12.7287\right] + (3.4436)\left(p_{3,1,2}^m - p_{3,2,2}^m\right)$$
$$+ (6.0857)\left(p_{2,2,2}^m - p_{3,2,2}^m\right) + (6.0857)\left(p_{4,2,2}^m - p_{3,2,2}^m\right) + (3.4436)\left(p_{3,3,2}^m - p_{3,2,2}^m\right)$$
$$+ (63.3944)\left[\left(p_{3,2,3}^m - p_{3,2,2}^m\right) + 12.7287\right] - 133.3 = \frac{V_{b_{3,2,2}}}{\alpha_c \Delta t}\left[\left(\frac{\phi}{B}\right)_{3,2,2}^{n+1} - \left(\frac{\phi}{B}\right)_{3,2,2}^{n}\right] \tag{2.47}$$

2.7 Multidimensional flow in radial-cylindrical coordinates

2.7.1 Reservoir discretization for single-well simulation

Single-well simulation uses radial-cylindrical coordinates. A point in space in radial-cylindrical coordinates is identified as point (r, θ, z) as shown in Fig. 2.16. A cylinder with the well coinciding with its longitudinal axis represents the reservoir in single-well simulation. Reservoir discretization involves dividing the cylinder into n_r concentric radial segments with the well passing through the center. Rays from the center divide the radial segments into n_θ cake-like slices. Planes normal to the longitudinal axis divide the cake-like slices into n_z segments.

A reservoir block in a discretized reservoir is identified as block (i,j,k), where i, j, and k are, respectively, the orders of the block in r-, θ-, and z-directions with $1 \le i \le n_r$, $1 \le j \le n_\theta$, $1 \le k \le n_z$. This block has the shape shown in Fig. 2.17.

Fig. 2.18a shows that block (i,j,k) is surrounded by blocks $(i-1,j,k)$ and $(i+1,j,k)$ in the r-direction and by blocks $(i,j-1,k)$ and $(i,j+1,k)$ in the θ direction. In addition, the figure shows the boundaries between block (i,j,k) and its neighboring blocks: block boundaries $(i-{}^1/_2,j,k)$, $(i+{}^1/_2,j,k)$, $(i,j-{}^1/_2,k)$, and $(i,j+{}^1/_2,k)$. Fig. 2.18b shows that block (i,j,k) is surrounded by blocks

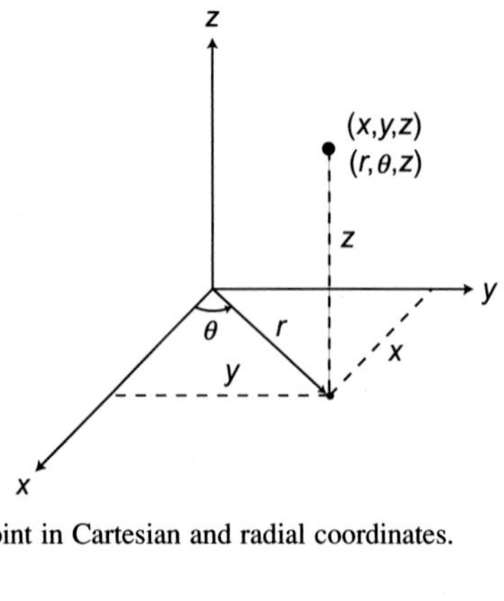

FIG. 2.16 Graphing a point in Cartesian and radial coordinates.

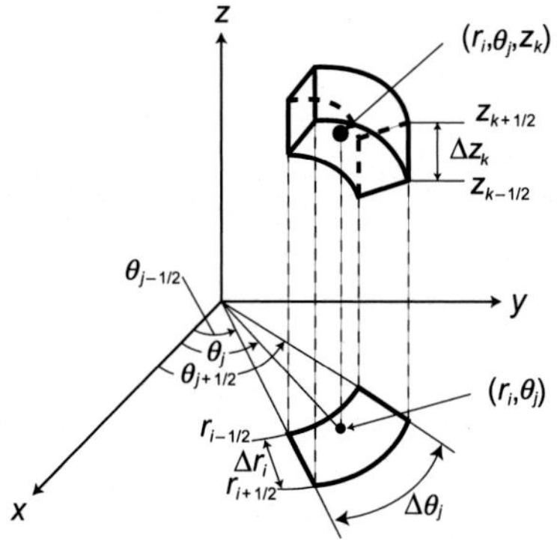

FIG. 2.17 Block (i,j,k) in single-well simulation.

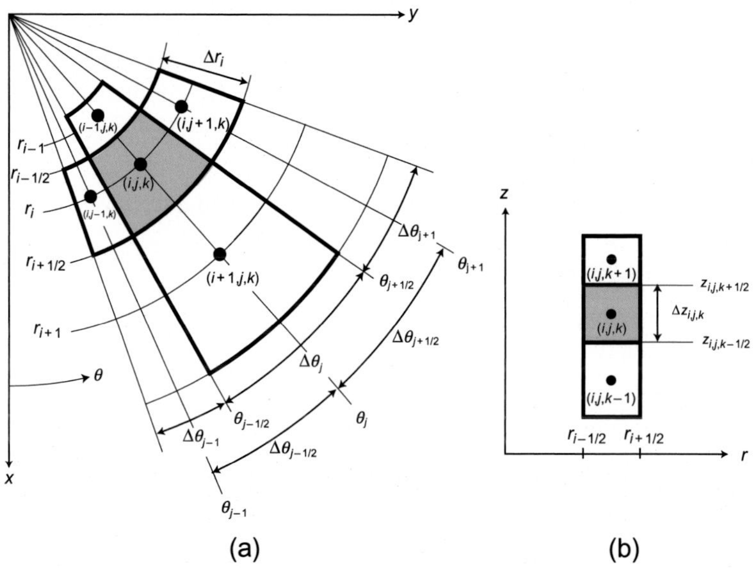

FIG. 2.18 Block (i,j,k) and its neighboring blocks in single-well simulation. (a) Block (i,j,k) and its neighboring blocks in horizontal plane and (b) block (i,j,k) and its neighboring blocks in the z-direction.

$(i, j, k-1)$ and $(i,j,k+1)$ in the z-direction. The figure also shows block boundaries $(i,j,k-{}^1/_2)$ and $(i,j,k+{}^1/_2)$. We will demonstrate block identification and ordering in single-well simulation in the following two examples. In the absence of fluid flow in the θ-direction, block ordering and identification in radial and rectangular coordinates are identical.

Example 2.7 In single-well simulation, a reservoir is discretized in the r-direction into four concentric cylindrical blocks as shown in Fig. 2.19a. Order blocks in this reservoir using natural ordering.

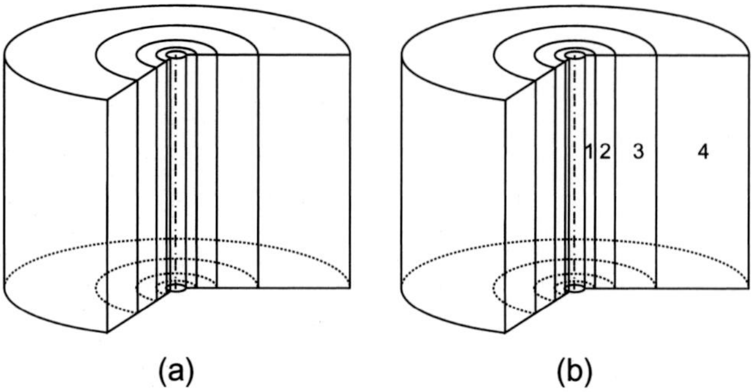

FIG. 2.19 1-D radial-cylindrical reservoir representation in Example 2.7. (a) Reservoir representation and (b) natural ordering of blocks.

Solution

We identify the innermost block enclosing the well as block 1. Then we move to other blocks, one block at a time, in the direction of increasing radius. The order of the next block is obtained by incrementing the order of the previous block by one. We continue the process of block ordering (or numbering) until the outermost block is numbered. The final ordering of blocks in this reservoir is shown in Fig. 2.19b.

Example 2.8 Let the reservoir in Example 2.7 consists of three layers as shown in Fig. 2.20a.

Identify the blocks in this reservoir using the following:

1. The engineering notation
2. Natural ordering

Solution

1. The engineering notation for block identification in this reservoir is shown in Fig. 2.20b.
2. We arbitrarily choose to order blocks in each layer along rows. Blocks in the first layer ($k=1$) are numbered as explained in Example 2.7. Block (1,2) in first column ($i=1$) and second plane ($k=2$) is numbered next as block 5, and block numbering continues as in Example 2.7. Block numbering continues (layer by layer) until all blocks are numbered. The final ordering of blocks in this reservoir is shown in Fig. 2.20c.

2.7.2 Derivation of the multidimensional flow equation in radial-cylindrical coordinates

To write the material balance for block (i,j,k) in Fig. 2.18 over a time step $\Delta t = t^{n+1} - t^n$, we assume that the fluid coming from neighboring blocks enters block (i,j,k) through block boundaries $(i-{}^1/_2,j,k)$, $(i,j-{}^1/_2,k)$, and

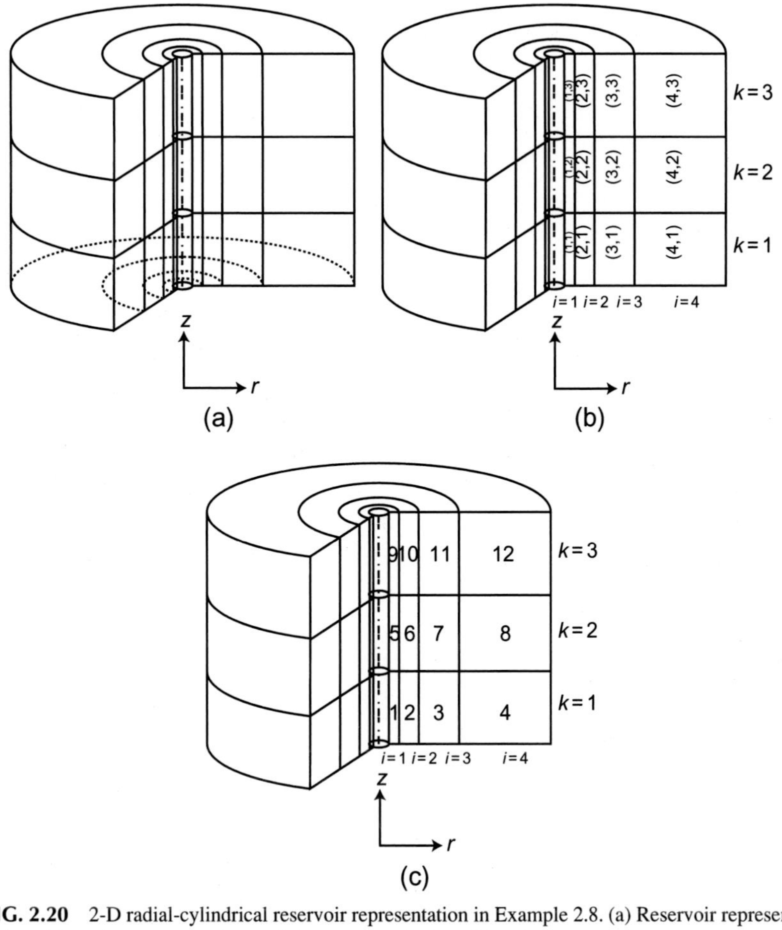

FIG. 2.20 2-D radial-cylindrical reservoir representation in Example 2.8. (a) Reservoir representation, (b) engineering notation, and (c) natural ordering of blocks.

$(i, j, k-{}^1/_2)$ leaves through block boundaries $(i+{}^1/_2, j, k)$, $(i, j+{}^1/_2, k)$, and $(i, j, k+{}^1/_2)$. The application of Eq. (2.6) results in

$$\left(m_i\big|_{r_{i-1/2,j,k}} - m_o\big|_{r_{i+1/2,j,k}}\right) + \left(m_i\big|_{\theta_{i,j-1/2,k}} - m_o\big|_{\theta_{i,j+1/2,k}}\right) + \left(m_i\big|_{z_{i,j,k-1/2}} - m_o\big|_{z_{i,j,k+1/2}}\right) + m_{s_{i,j,k}} = m_{a_{i,j,k}} \tag{2.48}$$

Terms like mass rates, $w_r\big|_{r_{i-1/2,\,j,\,k}}$, $w_\theta\big|_{\theta_{i,\,j-1/2,\,k}}$, $w_z\big|_{z_{i,\,j,\,k-1/2}}$, $w_r\big|_{r_{i+1/2,\,j,\,k}}$, $w_\theta\big|_{\theta_{i,\,j+1/2,\,k}}$, $w_z\big|_{z_{i,\,j,\,k+1/2}}$, and well mass rate, $q_{m_{i,\,j,\,k}}$, are functions of time only (see justification in Section 2.6.2); therefore,

$$m_i\big|_{r_{i-1/2,j,k}} = \int_{t^n}^{t^{n+1}} w_r\big|_{r_{i-1/2,j,k}} \mathrm{d}t \tag{2.49a}$$

$$m_i|_{\theta_{i,j-1/2,k}} = \int_{t^n}^{t^{n+1}} w_\theta|_{\theta_{i,j-1/2,k}} \mathrm{d}t \tag{2.49b}$$

$$m_i|_{z_{i,j,k-1/2}} = \int_{t^n}^{t^{n+1}} w_z|_{z_{i,j,k-1/2}} \mathrm{d}t \tag{2.49c}$$

$$m_o|_{r_{i+1/2,j,k}} = \int_{t^n}^{t^{n+1}} w_r|_{r_{i+1/2,j,k}} \mathrm{d}t \tag{2.50a}$$

$$m_o|_{\theta_{i,j+1/2,k}} = \int_{t^n}^{t^{n+1}} w_\theta|_{\theta_{i,j+1/2,k}} \mathrm{d}t \tag{2.50b}$$

$$m_o|_{z_{i,j,k+1/2}} = \int_{t^n}^{t^{n+1}} w_z|_{z_{i,j,k+1/2}} \mathrm{d}t \tag{2.50c}$$

and

$$m_{s_{i,j,k}} = \int_{t^n}^{t^{n+1}} q_{m_{i,j,k}} \mathrm{d}t \tag{2.51}$$

In addition, mass accumulation is defined as:

$$m_{a_{i,j,k}} = \Delta_t (V_b m_v)_{i,j,k} = V_{b_{i,j,k}} \left(m_{v_{i,j,k}}^{n+1} - m_{v_{i,j,k}}^{n} \right) \tag{2.52}$$

Mass rates and mass fluxes are related through

$$w_r|_r = \dot{m}_r A_r \tag{2.53a}$$

$$w_\theta|_\theta = \dot{m}_\theta A_\theta \tag{2.53b}$$

and

$$w_z|_z = \dot{m}_z A_z \tag{2.53c}$$

mass fluxes can be expressed in terms of fluid density and volumetric velocities:

$$\dot{m}_r = \alpha_c \rho u_r \tag{2.54a}$$

$$\dot{m}_\theta = \alpha_c \rho u_\theta \tag{2.54b}$$

and

$$\dot{m}_z = \alpha_c \rho u_z \tag{2.54c}$$

and m_v can be expressed in terms of fluid density and porosity:

$$m_{v_{i,j,k}} = (\phi\rho)_{i,j,k} \tag{2.55}$$

Also, the well mass rate can be expressed in terms of well volumetric rate and fluid density:

$$q_{m_{i,j,k}} = (\alpha_c \rho q)_{i,j,k} \tag{2.56}$$

Substitution of Eq. (2.54) into Eq. (2.53) yields:

$$w_r|_r = \alpha_c \rho u_r A_r \tag{2.57a}$$

$$w_\theta|_\theta = \alpha_c \rho u_\theta A_\theta \tag{2.57b}$$

and

$$w_z|_z = \alpha_c \rho u_z A_z \tag{2.57c}$$

Substitution of Eq. (2.57) into Eqs. (2.49) and (2.50) yields:

$$m_i|_{r_{i-1/2,j,k}} = \int_{t^n}^{t^{n+1}} \alpha_c (\rho u_r A_r)|_{r_{i-1/2,j,k}} \mathrm{d}t \tag{2.58a}$$

$$m_i|_{\theta_{i,j-1/2,k}} = \int_{t^n}^{t^{n+1}} \alpha_c (\rho u_\theta A_\theta)|_{\theta_{i,j-1/2,k}} \mathrm{d}t \tag{2.58b}$$

$$m_i|_{z_{i,j,k-1/2}} = \int_{t^n}^{t^{n+1}} \alpha_c (\rho u_z A_z)|_{z_{i,j,k-1/2}} \mathrm{d}t \tag{2.58c}$$

$$m_o|_{r_{i+1/2,j,k}} = \int_{t^n}^{t^{n+1}} \alpha_c (\rho u_r A_r)|_{r_{i+1/2,j,k}} \mathrm{d}t \tag{2.59a}$$

$$m_o|_{\theta_{i,j+1/2,k}} = \int_{t^n}^{t^{n+1}} \alpha_c (\rho u_\theta A_\theta)|_{\theta_{i,j+1/2,k}} \mathrm{d}t \tag{2.59b}$$

and

$$m_o|_{z_{i,j,k+1/2}} = \int_{t^n}^{t^{n+1}} \alpha_c (\rho u_z A_z)|_{z_{i,j,k+1/2}} \mathrm{d}t \tag{2.59c}$$

Substitution of Eq. (2.56) into Eq. (2.51) yields:

$$m_{s_{i,j,k}} = \int_{t^n}^{t^{n+1}} (\alpha_c \rho q)_{i,j,k} \mathrm{d}t \tag{2.60}$$

Substitution of Eq. (2.55) into Eq. (2.52) yields:

$$m_{a_{i,j,k}} = V_{b_{i,j,k}} \left[(\phi\rho)^{n+1}_{i,j,k} - (\phi\rho)^{n}_{i,j,k} \right] \tag{2.61}$$

Substitution of Eqs. (2.58) through (2.61) into Eq. (2.48) results in:

$$\begin{aligned}
&\int_{t^n}^{t^{n+1}} \alpha_c (\rho u_r A_r)|_{r_{i-1/2,j,k}} \mathrm{d}t - \int_{t^n}^{t^{n+1}} \alpha_c (\rho u_r A_r)|_{r_{i+1/2,j,k}} \mathrm{d}t + \int_{t^n}^{t^{n+1}} \alpha_c (\rho u_\theta A_\theta)|_{\theta_{i,j-1/2,k}} \mathrm{d}t \\
&- \int_{t^n}^{t^{n+1}} \alpha_c (\rho u_\theta A_\theta)|_{\theta_{i,j+1/2,k}} \mathrm{d}t + \int_{t^n}^{t^{n+1}} \alpha_c (\rho u_z A_z)|_{z_{i,j,k-1/2}} \mathrm{d}t - \int_{t^n}^{t^{n+1}} \alpha_c (\rho u_z A_z)|_{z_{i,j,k+1/2}} \mathrm{d}t \\
&+ \int_{t^n}^{t^{n+1}} (\alpha_c \rho q)_{i,j,k} \mathrm{d}t = V_{b_{i,j,k}} \left[(\phi\rho)^{n+1}_{i,j,k} - (\phi\rho)^{n}_{i,j,k} \right]
\end{aligned} \tag{2.62}$$

Substitution of Eq. (2.7a) into Eq. (2.62), dividing by $\alpha_c \rho_{sc}$ and noting that $q_{sc} = q/B$, yields:

$$\begin{aligned}
&\int_{t^n}^{t^{n+1}} \left(\frac{u_r A_r}{B} \right)\Bigg|_{r_{i-1/2,j,k}} \mathrm{d}t - \int_{t^n}^{t^{n+1}} \left(\frac{u_r A_r}{B} \right)\Bigg|_{r_{i+1/2,j,k}} \mathrm{d}t + \int_{t^n}^{t^{n+1}} \left(\frac{u_\theta A_\theta}{B} \right)\Bigg|_{\theta_{i,j-1/2,k}} \mathrm{d}t \\
&- \int_{t^n}^{t^{n+1}} \left(\frac{u_\theta A_\theta}{B} \right)\Bigg|_{\theta_{i,j+1/2,k}} \mathrm{d}t + \int_{t^n}^{t^{n+1}} \left(\frac{u_z A_z}{B} \right)\Bigg|_{z_{i,j,k-1/2}} \mathrm{d}t - \int_{t^n}^{t^{n+1}} \left(\frac{u_z A_z}{B} \right)\Bigg|_{z_{i,j,k+1/2}} \mathrm{d}t \\
&+ \int_{t^n}^{t^{n+1}} q_{sc_{i,j,k}} \mathrm{d}t = \frac{V_{b_{i,j,k}}}{\alpha_c} \left[\left(\frac{\varphi}{B} \right)^{n+1}_{i,j,k} - \left(\frac{\varphi}{B} \right)^{n}_{i,j,k} \right]
\end{aligned} \tag{2.63}$$

Fluid volumetric velocities in the r, θ, and z-directions are given by the algebraic analogs of Eq. (2.8); i.e.,

$$u_r|_{r_{i-1/2,j,k}} = \beta_c \frac{k_r|_{r_{i-1/2,j,k}}}{\mu|_{r_{i-1/2,j,k}}} \left[\frac{(\Phi_{i-1,j,k} - \Phi_{i,j,k})}{\Delta r_{i-1/2,j,k}} \right] \tag{2.64a}$$

and

$$u_r|_{r_{i+1/2,j,k}} = \beta_c \frac{k_r|_{r_{i+1/2,j,k}}}{\mu|_{r_{i+1/2,j,k}}} \left[\frac{(\Phi_{i,j,k} - \Phi_{i+1,j,k})}{\Delta r_{i+1/2,j,k}} \right] \tag{2.64b}$$

Likewise,

$$u_z\big|_{z_{i,j,k-1/2}} = \beta_c \frac{k_z\big|_{z_{i,j,k-1/2}}}{\mu\big|_{z_{i,j,k-1/2}}}\left[\frac{\left(\Phi_{i,j,k-1} - \Phi_{i,j,k}\right)}{\Delta z_{i,j,k-1/2}}\right] \tag{2.65a}$$

and

$$u_z\big|_{z_{i,j,k+1/2}} = \beta_c \frac{k_z\big|_{z_{i,j,k+1/2}}}{\mu\big|_{z_{i,j,k+1/2}}}\left[\frac{\left(\Phi_{i,j,k} - \Phi_{i,j,k+1}\right)}{\Delta z_{i,j,k+1/2}}\right] \tag{2.65b}$$

Similarly,

$$u_\theta\big|_{\theta_{i,j-1/2,k}} = \beta_c \frac{k_\theta\big|_{\theta_{i,j-1/2,k}}}{\mu\big|_{\theta_{i,j-1/2,k}}}\left[\frac{\left(\Phi_{i,j-1,k} - \Phi_{i,j,k}\right)}{r_{i,j,k}\Delta\theta_{i,j-1/2,k}}\right] \tag{2.66a}$$

and

$$u_\theta\big|_{\theta_{i,j+1/2,k}} = \beta_c \frac{k_\theta\big|_{\theta_{i,j+1/2,k}}}{\mu\big|_{\theta_{i,j+1/2,k}}}\left[\frac{\left(\Phi_{i,j,k} - \Phi_{i,j+1,k}\right)}{r_{i,j,k}\Delta\theta_{i,j+1/2,k}}\right] \tag{2.66b}$$

Substitution of Eqs. (2.64) through (2.66) into Eq. (2.63) and grouping terms results in:

$$\begin{aligned}
&\int_{t^n}^{t^{n+1}} \left[\left(\beta_c \frac{k_r A_r}{\mu B \Delta r}\right)\bigg|_{r_{i-1/2,j,k}} \left(\Phi_{i-1,j,k} - \Phi_{i,j,k}\right)\right] dt \\
&+ \int_{t^n}^{t^{n+1}} \left[\left(\beta_c \frac{k_r A_r}{\mu B \Delta r}\right)\bigg|_{r_{i+1/2,j,k}} \left(\Phi_{i+1,j,k} - \Phi_{i,j,k}\right)\right] dt \\
&+ \int_{t^n}^{t^{n+1}} \left[\frac{1}{r_{i,j,k}}\left(\beta_c \frac{k_\theta A_\theta}{\mu B \Delta\theta}\right)\bigg|_{\theta_{i,j-1/2,k}} \left(\Phi_{i,j-1,k} - \Phi_{i,j,k}\right)\right] dt \\
&+ \int_{t^n}^{t^{n+1}} \left[\frac{1}{r_{i,j,k}}\left(\beta_c \frac{k_\theta A_\theta}{\mu B \Delta\theta}\right)\bigg|_{\theta_{i,j+1/2,k}} \left(\Phi_{i,j+1,k} - \Phi_{i,j,k}\right)\right] dt \\
&+ \int_{t^n}^{t^{n+1}} \left[\left(\beta_c \frac{k_z A_z}{\mu B \Delta z}\right)\bigg|_{z_{i,j,k-1/2}} \left(\Phi_{i,j,k-1} - \Phi_{i,j,k}\right)\right] dt \\
&+ \int_{t^n}^{t^{n+1}} \left[\left(\beta_c \frac{k_z A_z}{\mu B \Delta z}\right)\bigg|_{z_{i,j,k+1/2}} \left(\Phi_{i,j,k+1} - \Phi_{i,j,k}\right)\right] dt \\
&+ \int_{t^n}^{t^{n+1}} q_{sc_{i,j,k}} dt = \frac{V_{b_{i,j,k}}}{\alpha_c}\left[\left(\frac{\phi}{B}\right)_{i,j,k}^{n+1} - \left(\frac{\phi}{B}\right)_{i,j,k}^{n}\right]
\end{aligned} \tag{2.67}$$

Eq. (2.67) can be rewritten as:

$$\int_{t^n}^{t^{n+1}} \left[T_{z_{i,j,k-1/2}}\left(\Phi_{i,j,k-1}-\Phi_{i,j,k}\right)\right]dt + \int_{t^n}^{t^{n+1}} \left[T_{\theta_{i,j-1/2,k}}\left(\Phi_{i,j-1,k}-\Phi_{i,j,k}\right)\right]dt$$
$$+ \int_{t^n}^{t^{n+1}} \left[T_{r_{i-1/2,j,k}}\left(\Phi_{i-1,j,k}-\Phi_{i,j,k}\right)\right]dt + \int_{t^n}^{t^{n+1}} \left[T_{r_{i+1/2,j,k}}\left(\Phi_{i+1,j,k}-\Phi_{i,j,k}\right)\right]dt$$
$$+ \int_{t^n}^{t^{n+1}} \left[T_{\theta_{i,j+1/2,k}}\left(\Phi_{i,j+1,k}-\Phi_{i,j,k}\right)\right]dt + \int_{t^n}^{t^{n+1}} \left[T_{z_{i,j,k+1/2}}\left(\Phi_{i,j,k+1}-\Phi_{i,j,k}\right)\right]dt$$
$$+ \int_{t^n}^{t^{n+1}} q_{sc_{i,j,k}} dt = \frac{V_{b_{i,j,k}}}{\alpha_c}\left[\left(\frac{\phi}{B}\right)^{n+1}_{i,j,k} - \left(\frac{\phi}{B}\right)^{n}_{i,j,k}\right] \tag{2.68}$$

where

$$T_{r_{i\mp1/2,j,k}} = \left(\beta_c \frac{k_r A_r}{\mu B \Delta r}\right)\Bigg|_{r_{i\mp1/2,j,k}} = \left(\beta_c \frac{k_r A_r}{\Delta r}\right)_{r_{i\mp1/2,j,k}} \left(\frac{1}{\mu B}\right)_{r_{i\mp1/2,j,k}}$$
$$= G_{r_{i\mp1/2,j,k}} \left(\frac{1}{\mu B}\right)_{r_{i\mp1/2,j,k}} \tag{2.69a}$$

$$T_{\theta_{i,j\mp1/2,k}} = \frac{1}{r_{i,j,k}} \left(\beta_c \frac{k_\theta A_\theta}{\mu B \Delta \theta}\right)\Bigg|_{\theta_{i,j\mp1/2,k}} = \left(\beta_c \frac{k_\theta A_\theta}{r_{i,j,k}\Delta \theta}\right)_{\theta_{i,j\mp1/2,k}} \left(\frac{1}{\mu B}\right)_{\theta_{i,j\mp1/2,k}}$$
$$= G_{\theta_{i,j\mp1/2,k}} \left(\frac{1}{\mu B}\right)_{\theta_{i,j\mp1/2,k}} \tag{2.69b}$$

and

$$T_{z_{i,j,k\mp1/2}} = \left(\beta_c \frac{k_z A_z}{\mu B \Delta z}\right)\Bigg|_{z_{i,j,k\mp1/2}} = \left(\beta_c \frac{k_z A_z}{\Delta z}\right)_{z_{i,j,k\mp1/2}} \left(\frac{1}{\mu B}\right)_{z_{i,j,k\mp1/2}} = G_{z_{i,j,k\mp1/2}} \left(\frac{1}{\mu B}\right)_{z_{i,j,k\mp1/2}} \tag{2.69c}$$

Expressions for geometric factors G for irregular grids in heterogeneous reservoirs are presented in Chapters 4 and 5.

2.7.3 Approximation of time integrals

Using Eq. (2.30) to approximate integrals in Eq. (2.68) and dividing by Δt, the flow equation in radial–cylindrical coordinates becomes:

$$
\begin{aligned}
&T^m_{z_{i,j,k-1/2}}\left[\left(\Phi^m_{i,j,k-1}-\Phi^m_{i,j,k}\right)\right]+T^m_{\theta_{i,j-1/2,k}}\left[\left(\Phi^m_{i,j-1,k}-\Phi^m_{i,j,k}\right)\right]\\
&\quad+T^m_{r_{i-1/2,j,k}}\left[\left(\Phi^m_{i-1,j,k}-\Phi^m_{i,j,k}\right)\right]+T^m_{r_{i+1/2,j,k}}\left[\left(\Phi^m_{i+1,j,k}-\Phi^m_{i,j,k}\right)\right]\\
&\quad+T^m_{\theta_{i,j+1/2,k}}\left[\left(\Phi^m_{i,j+1,k}-\Phi^m_{i,j,k}\right)\right]+T^m_{z_{i,j,k+1/2}}\left[\left(\Phi^m_{i,j,k+1}-\Phi^m_{i,j,k}\right)\right]\\
&\quad+q^m_{sc_{i,j,k}}=\frac{V_{b_{i,j,k}}}{\alpha_c\Delta t}\left[\left(\frac{\phi}{B}\right)^{n+1}_{i,j,k}-\left(\frac{\phi}{B}\right)^{n}_{i,j,k}\right]
\end{aligned}
\tag{2.70}
$$

Using the definition of potential difference, Eq. (2.70) becomes:

$$
\begin{aligned}
&T^m_{z_{i,j,k-1/2}}\left[\left(p^m_{i,j,k-1}-p^m_{i,j,k}\right)-\gamma^m_{i,j,k-1/2}\left(Z_{i,j,k-1}-Z_{i,j,k}\right)\right]\\
&\quad+T^m_{\theta_{i,j-1/2,k}}\left[\left(p^m_{i,j-1,k}-p^m_{i,j,k}\right)-\gamma^m_{i,j-1/2,k}\left(Z_{i,j-1,k}-Z_{i,j,k}\right)\right]\\
&\quad+T^m_{r_{i-1/2,j,k}}\left[\left(p^m_{i-1,j,k}-p^m_{i,j,k}\right)-\gamma^m_{i-1/2,j,k}\left(Z_{i-1,j,k}-Z_{i,j,k}\right)\right]\\
&\quad+T^m_{r_{i+1/2,j,k}}\left[\left(p^m_{i+1,j,k}-p^m_{i,j,k}\right)-\gamma^m_{i+1/2,j,k}\left(Z_{i+1,j,k}-Z_{i,j,k}\right)\right]\\
&\quad+T^m_{\theta_{i,j+1/2,k}}\left[\left(p^m_{i,j+1,k}-p^m_{i,j,k}\right)-\gamma^m_{i,j+1/2,k}\left(Z_{i,j+1,k}-Z_{i,j,k}\right)\right]\\
&\quad+T^m_{z_{i,j,k+1/2}}\left[\left(p^m_{i,j,k+1}-p^m_{i,j,k}\right)-\gamma^m_{i,j,k+1/2}\left(Z_{i,j,k+1}-Z_{i,j,k}\right)\right]\\
&\quad+q^m_{sc_{i,j,k}}=\frac{V_{b_{i,j,k}}}{\alpha_c\Delta t}\left[\left(\frac{\phi}{B}\right)^{n+1}_{i,j,k}-\left(\frac{\phi}{B}\right)^{n}_{i,j,k}\right]
\end{aligned}
\tag{2.71}
$$

Eq. (2.38), the flow equation in Cartesian coordinates (x-y-z), is used for field simulation, whereas Eq. (2.71), the flow equation in radial-cylindrical coordinates (r-θ-z), is used for single-well simulation. These two equations are similar in form. The RHS of both equations represents fluid accumulation in block (i,j,k). On the LHS, both equations have a source term represented by well production or injection and six flow terms representing interblock flow between block (i,j,k) and its six neighboring blocks: blocks $(i-1,j,k)$ and $(i+1,j,k)$ in the x-direction (or r-direction), blocks $(i,j-1,k)$ and $(i,j+1,k)$ in the y-direction (or θ-direction), and blocks $(i,j,k-1)$ and $(i,j,k+1)$ in the z-direction. The coefficients of potential differences are transmissibilities T_x, T_y, and T_z in the x-y-z space and T_r, T_θ, and T_z in the r-θ-z space. Eqs. (2.39) and (2.69) define these transmissibilities. The geometric factors in these equations are presented in Chapters 4 and 5.

2.8 Summary

In this chapter, we reviewed various engineering steps involved in rendering governing equations into algebraic equations. Governing equations, involving both the rock and fluid properties are discretized without conventional finite-difference or finite element approximation of PDEs. Fluid properties such as density, FVF, and viscosity are, in general, functions of pressure. Reservoir porosity depends on pressure and has heterogeneous distribution, and reservoir permeability is usually anisotropic. The basic knowledge of material balance, FVF, potential difference, and Darcy's law are necessary for deriving flow equations in petroleum reservoirs. Rectangular coordinates and radial coordinates are two ways of describing reservoirs in space. Although it is common to study reservoirs using rectangular coordinates, there are a few applications that require using radial-cylindrical coordinates. Using the engineering approach, the single-phase flow equation can be derived in any coordinate system. In this approach, the reservoir first is discretized into blocks, which are identified using the engineering notation or any block-ordering scheme. The second step involves writing the fluid material balance for a general reservoir block in a multidimensional reservoir over the time interval $t^n \le t \le t^{n+1}$ and combining it with Darcy's law and the formation volume factor. The third step provides for an evaluation method of the time integrals in the flow equation that was obtained in the second step. The result is a flow equation in algebraic form with all functions evaluated at time t^m, where $t^n \le t^m \le t^{n+1}$. In Chapter 7, we demonstrate how the choice of time t^m as old time level t^n, new time level t^{n+1}, or intermediate time level $t^{n+1/2}$ gives rise to the explicit formulation, implicit formulation, or the Crank-Nicolson formulation of the flow equation.

2.9 Exercises

2.1 List the physical properties of rock and fluid necessary for the derivation of single-phase flow equation.

2.2 Enumerate the three basic engineering concepts or equations used in the derivation of a flow equation.

2.3 Eq. (2.33) has four major terms, three on the LHS and one on the RHS. What is the physical meaning of each major term? What are units of each major term in the three systems of units? Using customary units, state the units of each variable or function that appears in Eq. (2.33).

2.4 Compare Eq. (2.33) with Eq. (2.37), that is, identify the similar major terms and the extra major terms in Eq. (2.37). What is the physical meaning of each of these extra terms and to which direction do they belong?

2.5 Compare Eq. (2.33) with Eq. (2.38), that is, identify the similar major terms and the extra major terms in Eq. (2.38). What is the physical meaning of each of these extra terms? Group the extra terms according to the direction they belong.

2.6 Compare the 3-D flow equation in rectangular coordinates (x-y-z) in Eq. (2.38) with the 3-D flow equation in radial-cylindrical coordinates

(r-θ-z) in Eq. (2.71). Elaborate on the similarities and differences in these two equations. Note the differences in the definition of geometric factors.

2.7 Consider the 2-D reservoir shown in Fig. 2.21. This reservoir is discretized using 5×5 blocks but it has irregular boundaries, as shown in the figure.

Use the following schemes to identify and order the blocks in this reservoir:

a. Engineering notation
b. Natural ordering by rows
c. Natural ordering by columns
d. Diagonal (D2) ordering
e. Alternating diagonal (D4) ordering
f. Zebra ordering
g. Cyclic ordering
h. Cyclic-2 ordering

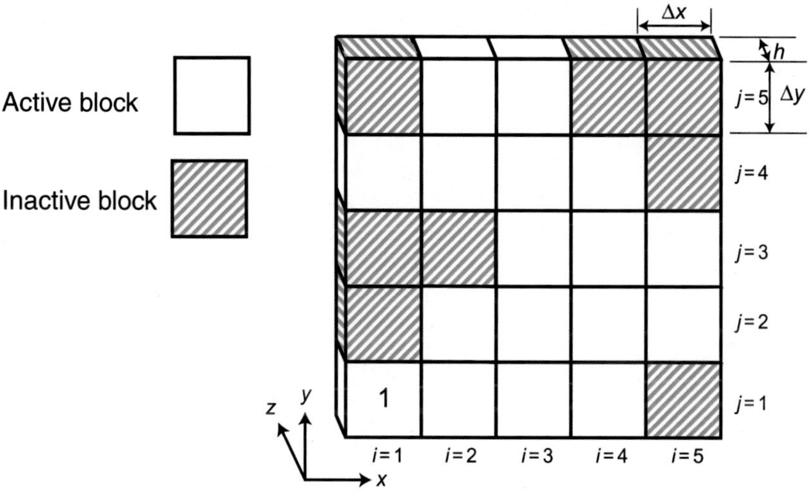

FIG. 2.21 2-D reservoir representation in Exercise 2.7.

2.8 Consider single-phase flow in a 1-D inclined reservoir. The flow equation for block i in this reservoir is expressed as Eq. (2.33).

a. Write Eq. (2.33) for block i assuming $t^m = t^n$. The resulting equation is the explicit formulation of the flow equation for block i.
b. Write Eq. (2.33) for block i assuming $t^m = t^{n+1}$. The resulting equation is the implicit formulation of the flow equation for block i.
c. Write Eq. (2.33) for block i assuming $t^m = t^{n+1/2}$. The resulting equation is the Crank-Nicolson formulation of the flow equation for block i.

2.9 Consider single-phase flow of oil in a 1-D horizontal reservoir. The reservoir is discretized using six blocks as shown in Fig. 2.22. A well that is located in block 4 produces at a rate of 600 STB/D. All blocks have $\Delta x = 220$ ft, $\Delta y = 1000$ ft, $h = 90$ ft, and $k_x = 120$ md. The oil FVF, viscosity, and compressibility are 1.0 RB/STB, 3.5 cP, and 1.5×10^{-5} psi^{-1}, respectively.

a. Identify the interior and boundary blocks in this reservoir.
b. Write the flow equation for every interior block. Leave the RHS of flow equation without substitution of values.
c. Write the flow equation for every interior block assuming incompressible fluid and porous medium.

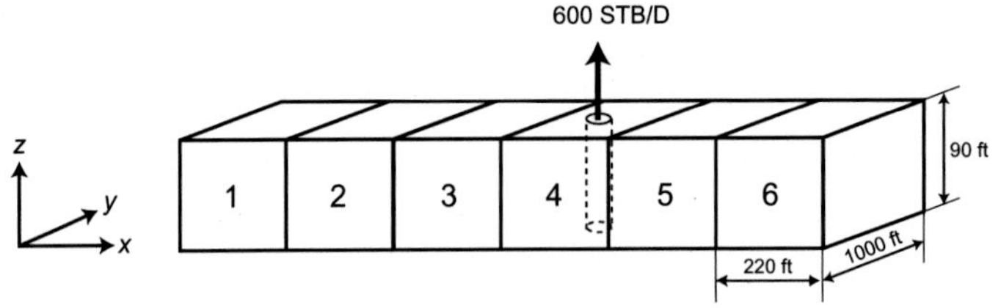

FIG. 2.22 1-D reservoir representation in Exercise 2.9.

2.10 Consider single-phase flow of water in a 2-D horizontal reservoir. The reservoir is discretized using 4×4 blocks as shown in Fig. 2.23. Two wells are located in blocks (2,2) and (3,3), and each produces at a rate of 200 STB/D. All blocks have $\Delta x = 200$ ft, $\Delta y = 200$ ft, $h = 50$ ft, and $k_x = k_y = 180$ md. The oil FVF, viscosity, and compressibility are 1.0 RB/STB, 0.5 cP, and 1×10^{-6} psi^{-1}, respectively.

a. Identify the interior and boundary blocks in this reservoir.
b. Write the flow equation for every interior block. Leave the RHS of flow equation without substitution of values.
c. Write the flow equation for every interior block assuming incompressible fluid and porous medium.

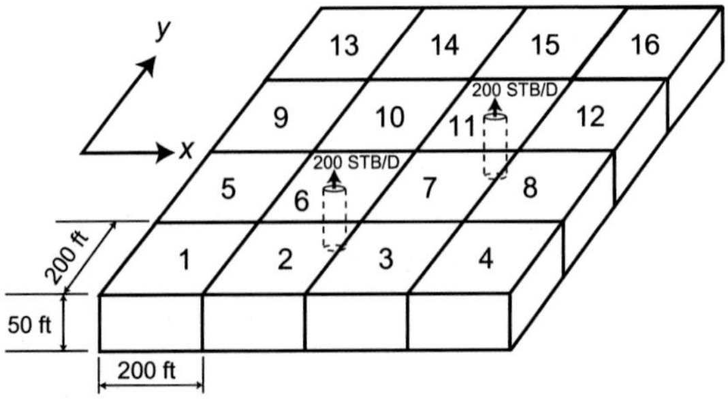

FIG. 2.23 2-D reservoir representation in Exercise 2.10.

2.11 Consider the 2-D horizontal reservoir presented in Fig. 2.21. All blocks have same dimensions ($\Delta x = 300$ ft, $\Delta y = 300$ ft, and $h = 20$ ft) and rock properties ($k_x = 140$ md, $k_y = 140$ md, and $\phi = 0.13$). The oil FVF and viscosity are 1.0 RB/STB and 3 cP, respectively. Write the flow equations for the interior blocks in this reservoir assuming incompressible fluid flow in incompressible porous medium. Order the blocks using natural ordering along the rows.

2.12 Consider the 1-D radial reservoir presented in Fig. 2.19. Write the flow equations for the interior blocks in this reservoir. Do not estimate interblock radial transmissibility. Leave the RHS of flow equations without substitution.

2.13 Consider the 2-D radial reservoir presented in Fig. 2.20b. Write the flow equations for the interior blocks in this reservoir. Do not estimate interblock radial or vertical transmissibilities. Leave the RHS of flow equations without substitution.

2.14 A single-phase oil reservoir is described by five equal blocks as shown in Fig. 2.24. The reservoir is horizontal and has homogeneous and isotropic rock properties, $k = 210$ md and $\phi = 0.21$. Block dimensions are $\Delta x = 375$ ft, $\Delta y = 450$ ft, and $h = 55$ ft. Oil properties are $B = 1$ RB/STB and $\mu = 1.5$ cP. The pressure of blocks 1 and 5 is 3725 and 1200 psia, respectively. Block 4 hosts a well that produces oil at a rate of 600 STB/D. Find the pressure distribution in the reservoir assuming that the reservoir rock and oil are incompressible. Estimate the rates of oil loss or gain across the right boundary of block 5 and that across the left boundary of block 1.

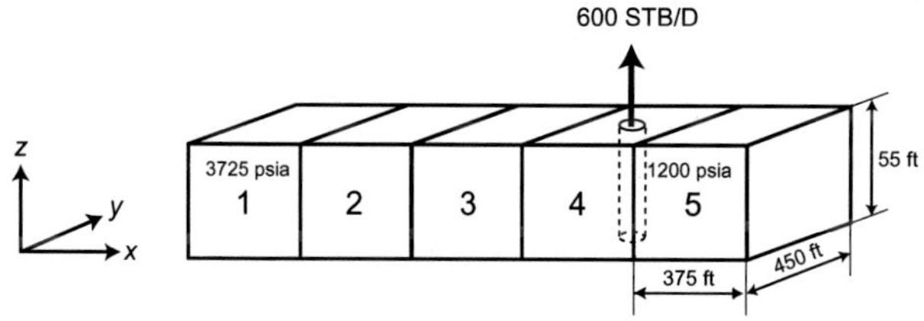

FIG. 2.24 1-D reservoir representation in Exercise 2.14.

2.15 A single-phase water reservoir is described by five equal blocks as shown in Fig. 2.25. The reservoir is horizontal and has $k = 178$ md and $\phi = 0.17$. Block dimensions are $\Delta x = 275$ ft, $\Delta y = 650$ ft, and $h = 30$ ft. Water properties are $B = 1$ RB/B and $\mu = 0.7$ cP. The pressure of blocks 1 and 5 is maintained at 3000 and 1000 psia, respectively. Block 3 hosts a well that

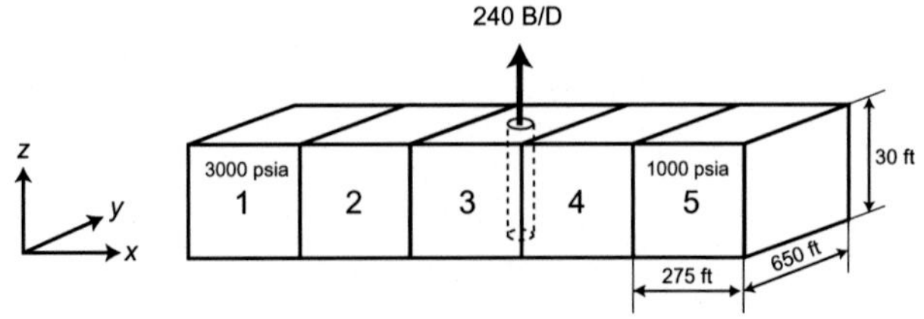

FIG. 2.25 1-D reservoir representation in Exercise 2.15.

produces water at a rate of 240 B/D. Find the pressure distribution in the reservoir assuming that the reservoir water and rock are incompressible.

2.16 Consider the reservoir presented in Fig. 2.14 and the flow problem described in Example 2.5. Assuming that both the reservoir fluid and rock are incompressible and given that a strong aquifer keeps the pressure of all boundary blocks at 3200 psia, estimate the pressure of blocks (2,2) and (3,2).

2.17 Consider single-phase flow of water in a 2-D horizontal reservoir. The reservoir is discretized using 4×4 equal blocks as shown in Fig. 2.26. Block 7 hosts a well that produces 500 B/D of water. All blocks have $\Delta x = \Delta y = 230$ ft, $h = 80$ ft, and $k_x = k_y = 65$ md. The water FVF and viscosity are 1.0 RB/B and 0.5 cP, respectively. The pressure of reservoir boundary blocks is specified as $p_2 = p_3 = p_4 = p_8 = p_{12} = 2500$, $p_1 = p_5 = p_9 = p_{13} = 4000$, and $p_{14} = p_{15} = p_{16} = 3500$ psia. Assuming that the reservoir water and rock are incompressible, calculate the pressure of blocks 6, 7, 10, and 11.

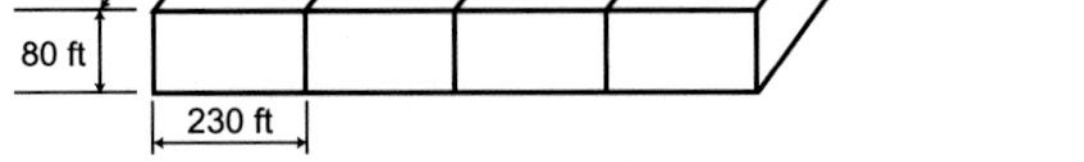

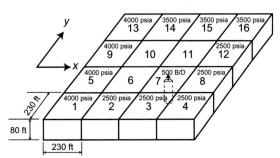

FIG. 2.26 2-D reservoir representation in Exercise 2.17.

Chapter 3

Flow equations using CVFD terminology

Chapter outline

3.1 Introduction

The importance of the control volume finite difference (CVFD) method lies in its capacity to use the same form of flow equation for 1-D, 2-D, and 3-D flow problems regardless of the ordering scheme of blocks. The same theme applies to energy balance equations for solutions to nonisothermal problems (Liu et al., 2013). The only difference among 1-D, 2-D, and 3-D flow equations is the definition of the elements for the set of neighboring blocks. The CVFD method is mainly used to write flow equations in a compact form, which is independent of the dimensionality of flow, the coordinate system used, or the block ordering scheme. This chapter introduces the terminology used in the CVFD method and the relationship between this method and the traditional way of writing finite-difference equations presented in Chapter 2.

3.2 Flow equations using CVFD terminology

In petroleum engineering, Aziz (1993) was the first author to refer to the CVFD method. However, the method had been developed and used by others without giving it a name (Abou-Kassem, 1981; Lutchmansingh, 1987; Abou-Kassem and Farouq Ali, 1987). The terminology presented in this section is based on a 2001 work published by Ertekin, Abou-Kassem, and King. With this

Petroleum Reservoir Simulation. https://doi.org/10.1016/B978-0-12-819150-7.00003-7

terminology, we can write the equations for 1-D, 2-D, and 3-D flow in compact form, using Cartesian or radial–cylindrical coordinates. For the flow equation in Cartesian space, we define ψ_{x_n}, ψ_{y_n}, and ψ_{z_n} as the sets whose members are the neighboring blocks of block n in the directions of the x-axis, y-axis, and z-axis, respectively. Then, we define ψ_n as the set that contains the neighboring blocks in all flow directions as its members; that is,

$$\psi_n = \psi_{x_n} \cup \psi_{y_n} \cup \psi_{z_n} \tag{3.1a}$$

If there is no flow in a given direction, then the set for that direction is the empty set, { }. For the flow equation in radial-cylindrical space, the equation that corresponds to Eq. (3.1a) is

$$\psi_n = \psi_{r_n} \cup \psi_{\theta_n} \cup \psi_{z_n} \tag{3.1b}$$

where ψ_{r_n}, ψ_{θ_n}, and ψ_{z_n} are the sets whose members are the neighboring blocks of block n in the r-direction, θ-direction, and z-axis, respectively.

The following sections present the flow equations for blocks identified by engineering notation or by block ordering using the natural ordering scheme.

3.2.1 Flow equations using CVFD terminology and engineering notation

For 1-D flow in the direction of the x-axis, block n is termed in engineering notation as block i (i.e., $n \equiv i$) as shown in Fig. 3.1a. In this case,

$$\psi_{x_n} = \{(i-1), (i+1)\} \tag{3.2a}$$

$$\psi_{y_n} = \{\} \tag{3.2b}$$

and

$$\psi_{z_n} = \{\} \tag{3.2c}$$

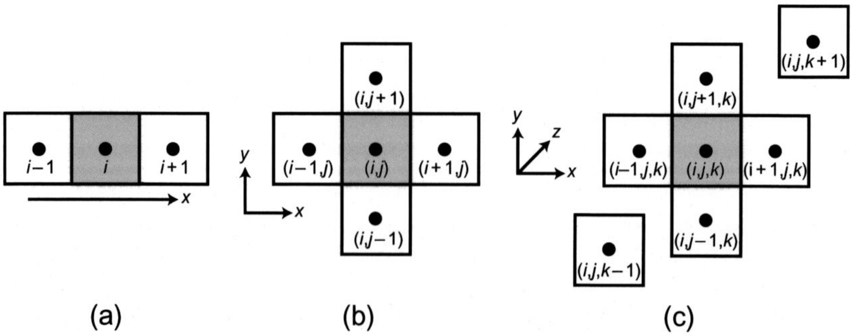

FIG. 3.1 A block and its neighboring blocks in 1-D, 2-D, and 3-D flow using engineering notation.
(a) $\psi_i = \{(i-1), (i+1)\}$
(b) $\psi_{i,j} = \{(i,j-1), (i-1,j), (i+1,j), (i,j+1)\}$
(c) $\psi_{i,j,k} = \{(i,j,k-1), (i,j-1,i), (i-1,j,k), (i+1,j,k), (i,j+1,k), (i,j,k+1)\}$

Substitution of Eq. (3.2) into Eq. (3.1a) results in

$$\psi_n = \psi_i = \{(i-1),(i+1)\} \cup \{\} \cup \{\} = \{(i-1),(i+1)\} \tag{3.3}$$

The flow equation for block i in 1-D flow reservoir is expressed as Eq. (2.33):

$$T^m_{x_{i-1/2}}\left[\left(p^m_{i-1}-p^m_i\right)-\gamma^m_{i-1/2}(Z_{i-1}-Z_i)\right]+T^m_{x_{i+1/2}}\left[\left(p^m_{i+1}-p^m_i\right)-\gamma^m_{i+1/2}(Z_{i+1}-Z_i)\right]$$
$$+q^m_{sc_i}=\frac{V_{b_i}}{\alpha_c\Delta t}\left[\left(\frac{\phi}{B}\right)^{n+1}_i-\left(\frac{\phi}{B}\right)^n_i\right] \tag{3.4a}$$

which can be written in CVFD form as

$$\sum_{l\in\psi_i} T^m_{l,i}\left[\left(p^m_l-p^m_i\right)-\gamma^m_{l,i}(Z_l-Z_i)\right]+q^m_{sc_i}=\frac{V_{b_i}}{\alpha_c\Delta t}\left[\left(\frac{\phi}{B}\right)^{n+1}_i-\left(\frac{\phi}{B}\right)^n_i\right] \tag{3.4b}$$

where

$$T^m_{i\mp1,i}=T^m_{i,i\mp1}\equiv T^m_{x_{i\mp1/2}} \tag{3.5}$$

and transmissibilities $T^m_{x_{i\mp1/2}}$ are defined by Eq. (2.39a). In addition,

$$\gamma^m_{i\mp1,i}=\gamma^m_{i,i\mp1}\equiv\gamma^m_{i\mp1/2} \tag{3.6}$$

For 2-D flow in the x-y plane, block n is termed in engineering notation as block (i,j), that is, $n\equiv(i,j)$, as shown in Fig. 3.1b. In this case,

$$\psi_{x_n}=\{(i-1,j),(i+1,j)\} \tag{3.7a}$$

$$\psi_{y_n}=\{(i,j-1),(i,j+1)\} \tag{3.7b}$$

and

$$\psi_{z_n}=\{\} \tag{3.7c}$$

Substitution of Eq. (3.7) into Eq. (3.1a) results in

$$\begin{aligned}\psi_n=\psi_{i,j}&=\{(i-1,j),(i+1,j)\}\cup\{(i,j-1),(i,j+1)\}\cup\{\}\\&=\{(i,j-1),(i-1,j),(i+1,j),(i,j+1)\}\end{aligned} \tag{3.8}$$

Eq. (2.37) expresses the flow equation for block (i,j) as

$$T^m_{y_{i,j-1/2}}\left[\left(p^m_{i,j-1}-p^m_{i,j}\right)-\gamma^m_{i,j-1/2}(Z_{i,j-1}-Z_{i,j})\right]$$
$$+T^m_{x_{i-1/2,j}}\left[\left(p^m_{i-1,j}-p^m_{i,j}\right)-\gamma^m_{i-1/2,j}(Z_{i-1,j}-Z_{i,j})\right]$$
$$+T^m_{x_{i+1/2,j}}\left[\left(p^m_{i+1,j}-p^m_{i,j}\right)-\gamma^m_{i+1/2,j}(Z_{i+1,j}-Z_{i,j})\right]$$
$$+T^m_{y_{i,j+1/2}}\left[\left(p^m_{i,j+1}-p^m_{i,j}\right)-\gamma^m_{i,j+1/2}(Z_{i,j+1}-Z_{i,j})\right]+q^m_{sc_{i,j}}=\frac{V_{b_{i,j}}}{\alpha_c\Delta t}\left[\left(\frac{\phi}{B}\right)^{n+1}_{i,j}-\left(\frac{\phi}{B}\right)^n_{i,j}\right] \tag{3.9a}$$

which can be written in CVFD form as

$$\sum_{l\in\psi_{i,j}} T^m_{l,(i,j)}\left[\left(p^m_l - p^m_{i,j}\right) - \gamma^m_{l,(i,j)}\left(Z_l - Z_{i,j}\right)\right] + q^m_{sc_{i,j}} = \frac{V_{b_{i,j}}}{\alpha_c \Delta t}\left[\left(\frac{\phi}{B}\right)^{n+1}_{i,j} - \left(\frac{\phi}{B}\right)^{n}_{i,j}\right] \tag{3.9b}$$

where

$$T^m_{(i\mp1,j),(i,j)} = T^m_{(i,j),(i\mp1,j)} \equiv T^m_{x_{i\mp1/2,j}} \tag{3.10a}$$

and

$$T^m_{(i,j\mp1),(i,j)} = T^m_{(i,j),(i,j\mp1)} \equiv T^m_{y_{i,j\mp1/2}} \tag{3.10b}$$

Transmissibilities $T^m_{x_{i\mp1/2,j}}$ and $T^m_{y_{i,j\mp1/2}}$ have been defined by Eqs. (2.39a) and (2.39b), respectively. In addition,

$$\gamma^m_{(i\mp1,j),(i,j)} = \gamma^m_{(i,j),(i\mp1,j)} \equiv \gamma^m_{i\mp1/2,j} \tag{3.11a}$$

and

$$\gamma^m_{(i,j\mp1),(i,j)} = \gamma^m_{(i,j),(i,j\mp1)} \equiv \gamma^m_{i,j\mp1/2} \tag{3.11b}$$

For 3-D flow in the *x-y-z* space, block n is termed in engineering notation as block (i,j,k); that is, $n\equiv(i,j,k)$, as shown in Fig. 3.1c. In this case,

$$\psi_{x_n} = \{(i-1,j,k),(i+1,j,k)\} \tag{3.12a}$$

$$\psi_{y_n} = \{(i,j-1,k),(i,j+1,k)\} \tag{3.12b}$$

and

$$\psi_{z_n} = \{(i,j,k-1),(i,j,k+1)\} \tag{3.12c}$$

Substitution of Eq. (3.12) into Eq. (3.1a) results in

$$\begin{aligned}\psi_n &= \psi_{i,j,k}\\ &= \{(i-1,j,k),(i+1,j,k)\}\cup\{(i,j-1,k),(i,j+1,k)\}\cup\{(i,j,k-1),(i,j,k+1)\}\\ &= \{(i,j,k-1),(i,j-1,k),(i-1,j,k),(i+1,j,k),(i,j+1,k),(i,j,k+1)\}\end{aligned} \tag{3.13}$$

The flow equation for block (i,j,k) in 3-D flow reservoir is expressed as Eq. (2.38):

$$\begin{aligned}&T^m_{z_{i,j,k-1/2}}\left[\left(p^m_{i,j,k-1} - p^m_{i,j,k}\right) - \gamma^m_{i,j,k-1/2}\left(Z_{i,j,k-1} - Z_{i,j,k}\right)\right]\\ &+T^m_{y_{i,j-1/2,k}}\left[\left(p^m_{i,j-1,k} - p^m_{i,j,k}\right) - \gamma^m_{i,j-1/2,k}\left(Z_{i,j-1,k} - Z_{i,j,k}\right)\right]\\ &+T^m_{x_{i-1/2,j,k}}\left[\left(p^m_{i-1,j,k} - p^m_{i,j,k}\right) - \gamma^m_{i-1/2,j,k}\left(Z_{i-1,j,k} - Z_{i,j,k}\right)\right]\\ &+T^m_{x_{i+1/2,j,k}}\left[\left(p^m_{i+1,j,k} - p^m_{i,j,k}\right) - \gamma^m_{i+1/2,j,k}\left(Z_{i+1,j,k} - Z_{i,j,k}\right)\right]\end{aligned}$$

$$+T^m_{y_{i,j+1/2,k}}\left[\left(p^m_{i,j+1,k}-p^m_{i,j,k}\right)-\gamma^m_{i,j+1/2,k}\left(Z_{i,j+1,k}-Z_{i,j,k}\right)\right]$$
$$+T^m_{z_{i,j,k+1/2}}\left[\left(p^m_{i,j,k+1}-p^m_{i,j,k}\right)-\gamma^m_{i,j,k+1/2}\left(Z_{i,j,k+1}-Z_{i,j,k}\right)\right]$$
$$+q^m_{sc_{i,j,k}}=\frac{V_{b_{i,j,k}}}{\alpha_c\Delta t}\left[\left(\frac{\phi}{B}\right)^{n+1}_{i,j,k}-\left(\frac{\phi}{B}\right)^{n}_{i,j,k}\right]\tag{3.14a}$$

which can be written in CVFD form as

$$\sum_{l\in\psi_{i,j,k}}T^m_{l,(i,j,k)}\left[\left(p^m_l-p^m_{i,j,k}\right)-\gamma^m_{l,(i,j,k)}\left(Z_l-Z_{i,j,k}\right)\right]+q^m_{sc_{i,j,k}}$$
$$=\frac{V_{b_{i,j,k}}}{\alpha_c\Delta t}\left[\left(\frac{\phi}{B}\right)^{n+1}_{i,j,k}-\left(\frac{\phi}{B}\right)^{n}_{i,j,k}\right]\tag{3.14b}$$

where

$$T^m_{(i\mp1,j,k),(i,j,k)}=T^m_{(i,j,k),(i\mp1,j,k)}\equiv T^m_{x_{i\mp1/2,j,k}}\tag{3.15a}$$
$$T^m_{(i,j\mp1,k),(i,j,k)}=T^m_{(i,j,k),(i,j\mp1,k)}\equiv T^m_{y_{i,j\mp1/2,k}}\tag{3.15b}$$

and

$$T^m_{(i,j,k\mp1),(i,j,k)}=T^m_{(i,j,k),(i,j,k\mp1)}\equiv T^m_{z_{i,j,k\mp1/2}}\tag{3.15c}$$

As mentioned earlier, transmissibilities $T^m_{x_{i\mp1/2,j,k}}$, $T^m_{y_{i,j\mp1/2,k}}$, and $T^m_{z_{i,j,k\mp1/2}}$ have been defined in Eq. (2.39). Also,

$$\gamma^m_{(i\mp1,j,k),(i,j,k)}=\gamma^m_{(i,j,k),(i\mp1,j,k)}\equiv\gamma^m_{i\mp1/2,j,k}\tag{3.16a}$$
$$\gamma^m_{(i,j\mp1,k),(i,j,k)}=\gamma^m_{(i,j,k),(i,j\mp1,k)}\equiv\gamma^m_{i,j\mp1/2,k}\tag{3.16b}$$

and

$$\gamma^m_{(i,j,k\mp1),(i,j,k)}=\gamma^m_{(i,j,k),(i,j,k\mp1)}\equiv\gamma^m_{i,j,k\mp1/2}\tag{3.16c}$$

Eq. (3.4b) for 1-D flow, Eq. (3.9b) for 2-D flow, and Eq. (3.14b) for 3-D flow reduce to

$$\sum_{l\in\psi_n}T^m_{l,n}\left[\left(p^m_l-p^m_n\right)-\gamma^m_{l,n}(Z_l-Z_n)\right]+q^m_{sc_n}=\frac{V_{b_n}}{\alpha_c\Delta t}\left[\left(\frac{\phi}{B}\right)^{n+1}_n-\left(\frac{\phi}{B}\right)^n_n\right]\tag{3.17}$$

where, as mentioned before, $n\equiv i$ for 1-D flow, $n\equiv(i,j)$ for 2-D flow, and $n\equiv(i,j,k)$ for 3-D flow, and the elements of set ψ_n are defined accordingly (Eq. 3.3, 3.8, or 3.13).

Note that the elements of the sets that contain the neighboring blocks given by Eqs. (3.3), (3.8), and (3.13) for 1-D, 2-D, and 3-D, respectively, are ordered as shown in Fig. 3.2. The following examples demonstrate the use of CVFD

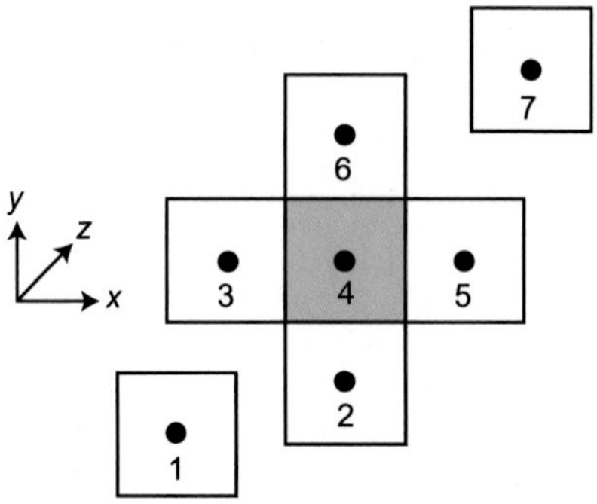

FIG. 3.2 The sequence of neighboring blocks in the set $\psi_{i,j,k}$ or ψ_n.

terminology to write the flow equations for an interior block identified by engineering notation in 1-D and 2-D reservoirs.

Example 3.1 Consider the reservoir described in Example 2.4. Write the flow equation for interior block 3 using CVFD terminology.

Solution

We make use of Fig. 2.12, which gives block representation of this reservoir. For block 3, $\psi_{x_3}=\{2,4\}$, $\psi_{y_3}=\{\}$, and $\psi_{z_3}=\{\}$. Substitution into Eq. (3.1a) gives $\psi_3=\{2,4\}\cup\{\}\cup\{\}=\{2,4\}$. The application of Eq. (3.17) for $n\equiv3$ produces

$$\sum_{l\in\psi_3} T_{l,3}^m\left[\left(p_l^m-p_3^m\right)-\gamma_{l,3}^m(Z_l-Z_3)\right]+q_{sc_3}^m=\frac{V_{b_3}}{\alpha_c\Delta t}\left[\left(\frac{\phi}{B}\right)_3^{n+1}-\left(\frac{\phi}{B}\right)_3^n\right] \tag{3.18}$$

which can be expanded as

$$T_{2,3}^m\left[\left(p_2^m-p_3^m\right)-\gamma_{2,3}^m(Z_2-Z_3)\right]+T_{4,3}^m\left[\left(p_4^m-p_3^m\right)-\gamma_{4,3}^m(Z_4-Z_3)\right]$$
$$+q_{sc_3}^m=\frac{V_{b_3}}{\alpha_c\Delta t}\left[\left(\frac{\phi}{B}\right)_3^{n+1}-\left(\frac{\phi}{B}\right)_3^n\right] \tag{3.19}$$

For this flow problem,

$$T_{2,3}^m=T_{4,3}^m=\beta_c\frac{k_xA_x}{\mu B\Delta x}=0.001127\times\frac{270\times(900\times100)}{2\times1\times250}$$
$$=54.7722\ \text{STB/D-psi} \tag{3.20}$$

$Z_2=Z_3=Z_4$ for a horizontal reservoir, and $q_{sc_3}^m=-400$ STB/D.

Substitution into Eq. (3.19) yields

$$(54.7722)\left(p_2^m-p_3^m\right)+(54.7722)\left(p_4^m-p_3^m\right)-400=\frac{V_{b_3}}{\alpha_c\Delta t}\left[\left(\frac{\phi}{B}\right)_3^{n+1}-\left(\frac{\phi}{B}\right)_3^n\right] \tag{3.21}$$

Eq. (3.21) is identical to Eq. (2.36), obtained in Example 2.4.

Example 3.2 Consider the reservoir described in Example 2.5. Write the flow equation for interior block (3,2) using CVFD terminology.

Solution

We make use of Fig. 2.14, which gives block representation of this reservoir. For block (3,2), $\psi_{x_{3,2}}=\{(2,2),(4,2)\}$, $\psi_{y_{3,2}}=\{(3,1),(3,3)\}$, and $\psi_{z_{3,2}}=\{\}$. Substitution into Eq. (3.1a) gives $\psi_{3,2}=\{(2,2),(4,2)\}\cup\{(3,1),(3,3)\}\cup\{\}=\{(3,1),(2,2),(4,2),(3,3)\}$. The application of Eq. (3.17) for $n\equiv(3,2)$ produces

$$\sum_{l\in\psi_{3,2}} T^m_{l,(3,2)}\left[\left(p^m_l - p^m_{3,2}\right) - \gamma^m_{l,(3,2)}(Z_l - Z_{3,2})\right] + q^m_{sc_{3,2}} = \frac{V_{b_{3,2}}}{\alpha_c \Delta t}\left[\left(\frac{\phi}{B}\right)^{n+1}_{3,2} - \left(\frac{\phi}{B}\right)^{n}_{3,2}\right] \quad (3.22)$$

which can be expanded as

$$\begin{aligned} &T^m_{(3,1),(3,2)}\left[\left(p^m_{3,1} - p^m_{3,2}\right) - \gamma^m_{(3,1),(3,2)}(Z_{3,1} - Z_{3,2})\right] \\ &+T^m_{(2,2),(3,2)}\left[\left(p^m_{2,2} - p^m_{3,2}\right) - \gamma^m_{(2,2),(3,2)}(Z_{2,2} - Z_{3,2})\right] \\ &+T^m_{(4,2),(3,2)}\left[\left(p^m_{4,2} - p^m_{3,2}\right) - \gamma^m_{(4,2),(3,2)}(Z_{4,2} - Z_{3,2})\right] \\ &+T^m_{(3,3),(3,2)}\left[\left(p^m_{3,3} - p^m_{3,2}\right) - \gamma^m_{(3,3),(3,2)}(Z_{3,3} - Z_{3,2})\right] \\ &+q^m_{sc_{3,2}} = \frac{V_{b_{3,2}}}{\alpha_c \Delta t}\left[\left(\frac{\phi}{B}\right)^{n+1}_{3,2} - \left(\frac{\phi}{B}\right)^{n}_{3,2}\right] \end{aligned} \quad (3.23)$$

For this flow problem,

$$T^m_{(2,2),(3,2)} = T^m_{(4,2),(3,2)} = \beta_c \frac{k_x A_x}{\mu B \Delta x} = 0.001127 \times \frac{270 \times (300 \times 100)}{2 \times 1 \times 250} = 18.2574\,\text{STB/D-psi} \quad (3.24)$$

$$T^m_{(3,1),(3,2)} = T^m_{(3,3),(3,2)} = \beta_c \frac{k_y A_y}{\mu B \Delta y} = 0.001127 \times \frac{220 \times (250 \times 100)}{2 \times 1 \times 300} = 10.3308\,\text{STB/D-psi} \quad (3.25)$$

$Z_{3,1}=Z_{2,2}=Z_{3,2}=Z_{4,2}=Z_{3,3}$ for a horizontal reservoir, and $q^m_{sc_{3,2}}=-400$ STB/D. Substitution into Eq. (3.23) yields

$$\begin{aligned} &(10.3308)\left(p^m_{3,1} - p^m_{3,2}\right) + (18.2574)\left(p^m_{2,2} - p^m_{3,2}\right) + (18.2574)\left(p^m_{4,2} - p^m_{3,2}\right) \\ &+(10.3308)\left(p^m_{3,3} - p^m_{3,2}\right) - 400 = \frac{V_{b_{3,2}}}{\alpha_c \Delta t}\left[\left(\frac{\phi}{B}\right)^{n+1}_{3,2} - \left(\frac{\phi}{B}\right)^{n}_{3,2}\right] \end{aligned} \quad (3.26)$$

Eq. (3.26) is identical to Eq. (2.42), obtained in Example 2.5.

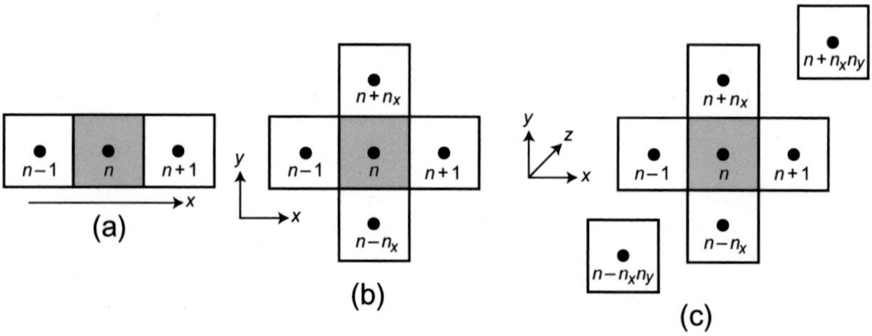

FIG. 3.3 A block and its neighboring blocks in 1-D, 2-D, and 3-D flow using natural ordering.
(a) $\psi_n = \{(n-1), (n+1)\}$
(b) $\psi_n = \{(n-n_x), (n-1), (n+1), (n+n_x)\}$
(c) $\psi_n = \{(n-n_xn_y), (n-n_x), (n-1), (n+1), (n+n_x), (n+n_xn_y)\}$

3.2.2 Flow equations using CVFD terminology and the natural ordering scheme

The flow equation in this case has one generalized form that is given by Eq. (3.17) with the corresponding definition of ψ_n for 1-D, 2-D, or 3-D flow. Blocks in natural ordering can be ordered along rows or along columns. In this book, we adopt natural ordering along rows (with rows being parallel to the x-axis) and refer to it, for short, as natural ordering. From this point on, all related discussions will use only natural ordering.

Fig. 3.3a shows block n in 1-D flow in the direction of the x-axis. In this case,

$$\psi_{x_n} = \{(n-1), (n+1)\} \tag{3.27a}$$

$$\psi_{y_n} = \{\} \tag{3.27b}$$

and

$$\psi_{z_n} = \{\} \tag{3.27c}$$

Substitution of Eq. (3.27) into Eq. (3.1a) results in

$$\begin{aligned} \psi_n &= \{(n-1), (n+1)\} \cup \{\} \cup \{\} \\ &= \{(n-1), (n+1)\} \end{aligned} \tag{3.28}$$

Fig. 3.3b shows block n in 2-D flow in the x-y plane. In this case,

$$\psi_{x_n} = \{(n-1), (n+1)\} \tag{3.29a}$$

$$\psi_{y_n} = \{(n-n_x), (n+n_x)\} \tag{3.29b}$$

and

$$\psi_{z_n} = \{\} \tag{3.29c}$$

Substitution of Eq. (3.29) into Eq. (3.1a) results in

$$\begin{aligned}\psi_n &= \{(n-1),(n+1)\} \cup \{(n-n_x),(n+n_x)\} \cup \{\} \\ &= \{(n-n_x),(n-1),(n+1),(n+n_x)\}\end{aligned} \tag{3.30}$$

Fig. 3.3c shows block n in 3-D flow in the x-y-z space. In this case,

$$\psi_{x_n} = \{(n-1),(n+1)\} \tag{3.31a}$$

$$\psi_{y_n} = \{(n-n_x),(n+n_x)\} \tag{3.31b}$$

and

$$\psi_{z_n} = \left\{\left(n-n_xn_y\right),\left(n+n_xn_y\right)\right\} \tag{3.31c}$$

Substitution of Eq. (3.31) into Eq. (3.1a) results in

$$\begin{aligned}\psi_n &= \{(n-1),(n+1)\} \cup \{(n-n_x),(n+n_x)\} \cup \left\{\left(n-n_xn_y\right),\left(n+n_xn_y\right)\right\} \\ &= \left\{\left(n-n_xn_y\right),(n-n_x),(n-1),(n+1),(n+n_x),\left(n+n_xn_y\right)\right\}\end{aligned} \tag{3.32}$$

Note that the elements of the sets containing the neighboring blocks given by Eqs. (3.28), (3.30), and (3.32) for 1-D, 2-D, and 3-D are ordered as shown in Fig. 3.2. Now, the flow equation for block n in 1-D, 2-D, or 3-D can be written in CVFD form again as Eq. (3.17),

$$\sum_{l\in\psi_n} T_{l,n}^m\left[\left(p_l^m - p_n^m\right) - \gamma_{l,n}^m(Z_l - Z_n)\right] + q_{sc_n}^m = \frac{V_{b_n}}{\alpha_c \Delta t}\left[\left(\frac{\phi}{B}\right)_n^{n+1} - \left(\frac{\phi}{B}\right)_n^n\right] \tag{3.17}$$

where transmissibility $T_{l,n}^m$ is defined as

$$T_{n\mp1,n}^m = T_{n,n\mp1}^m \equiv T_{x_{i\mp1/2,j,k}}^m \tag{3.33a}$$

$$T_{n\mp n_x,n}^m = T_{n,n\mp n_x}^m \equiv T_{y_{i,j\mp1/2,k}}^m \tag{3.33b}$$

and

$$T_{n\mp n_xn_y,n}^m = T_{n,n\mp n_xn_y}^m \equiv T_{z_{i,j,k\mp1/2}}^m \tag{3.33c}$$

In addition, fluid gravity $\gamma_{l,n}^m$ is defined as

$$\gamma_{n,n\mp1}^m = \gamma_{n\mp1,n}^m \equiv \gamma_{i\mp1/2,j,k}^m \tag{3.34a}$$

$$\gamma_{n,n\mp n_x}^m = \gamma_{n\mp n_x,n}^m \equiv \gamma_{i,j\mp1/2,k}^m \tag{3.34b}$$

and

$$\gamma_{n,n\mp n_xn_y}^m = \gamma_{n\mp n_xn_y,n}^m \equiv \gamma_{i,j,k\mp1/2}^m \tag{3.34c}$$

We should mention here that, throughout this book, we use subscript n to refer to block order while superscripts n and $n+1$ refer to old and new time levels, respectively. The following examples demonstrate the use of CVFD terminology to write the flow equations for an interior block identified by natural ordering in 2-D and 3-D reservoirs.

Example 3.3 As we did in Example 2.5, write the flow equations for interior block (3,2) using CVFD terminology, but this time, use natural ordering of blocks as shown in Fig. 3.4.

Solution

Block (3,2) in Fig. 2.14 corresponds to block 7 in Fig. 3.4. Therefore, $n=7$. For $n=7$, $\psi_{x_7}=\{6,8\}$, $\psi_{y_7}=\{3,11\}$, and $\psi_{z_7}=\{\}$. Substitution into Eq. (3.1a) results in $\psi_7=\{6,8\}\cup\{3,11\}\cup\{\}=\{3,6,8,11\}$.

The application of Eq. (3.17) produces

$$\sum_{l\in\psi_7} T_{l,7}^m\left[\left(p_l^m-p_7^m\right)-\gamma_{l,7}^m(Z_l-Z_7)\right]+q_{sc_7}^m=\frac{V_{b_7}}{\alpha_c\Delta t}\left[\left(\frac{\phi}{B}\right)_7^{n+1}-\left(\frac{\phi}{B}\right)_7^n\right] \tag{3.35}$$

which can be expanded as

$$\begin{aligned}&T_{3,7}^m\left[\left(p_3^m-p_7^m\right)-\gamma_{3,7}^m(Z_3-Z_7)\right]+T_{6,7}^m\left[\left(p_6^m-p_7^m\right)-\gamma_{6,7}^m(Z_6-Z_7)\right]\\&+T_{8,7}^m\left[\left(p_8^m-p_7^m\right)-\gamma_{8,7}^m(Z_8-Z_7)\right]+T_{11,7}^m\left[\left(p_{11}^m-p_7^m\right)-\gamma_{11,7}^m(Z_{11}-Z_7)\right]\\&+q_{sc_7}^m=\frac{V_{b_7}}{\alpha_c\Delta t}\left[\left(\frac{\phi}{B}\right)_7^{n+1}-\left(\frac{\phi}{B}\right)_7^n\right]\end{aligned} \tag{3.36}$$

Here again,

$$\begin{aligned}T_{6,7}^m=T_{8,7}^m=\beta_c\frac{k_xA_x}{\mu B\Delta x}&=0.001127\times\frac{270\times(300\times100)}{2\times1\times250}\\&=18.2574\,\text{STB/D-psi}\end{aligned} \tag{3.37}$$

$$\begin{aligned}T_{3,7}^m=T_{11,7}^m=\beta_c\frac{k_yA_y}{\mu B\Delta y}&=0.001127\times\frac{220\times(250\times100)}{2\times1\times300}\\&=10.3308\,\text{STB/D-psi}\end{aligned} \tag{3.38}$$

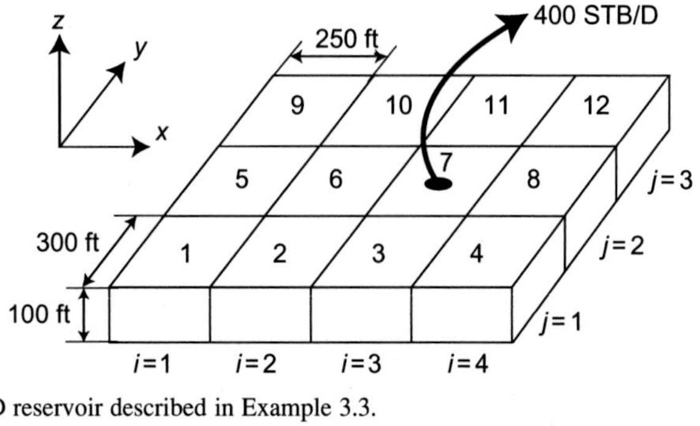

FIG. 3.4 2-D reservoir described in Example 3.3.

$Z_3 = Z_6 = Z_7 = Z_8 = Z_{11}$ for a horizontal reservoir, and $q_{sc_7}^m = -400$ STB/D. Substitution into Eq. (3.36) gives

$$\begin{aligned}&(10.3308)\left(p_3^m - p_7^m\right) + (18.2574)\left(p_6^m - p_7^m\right) + (18.2574)\left(p_8^m - p_7^m\right)\\&+(10.3308)\left(p_{11}^m - p_7^m\right) - 400 = \frac{V_{b_7}}{\alpha_c \Delta t}\left[\left(\frac{\phi}{B}\right)_7^{n+1} - \left(\frac{\phi}{B}\right)_7^n\right]\end{aligned} \tag{3.39}$$

Eq. (3.39) corresponds to Eq. (2.42) in Example 2.5, which uses engineering notation.

Example 3.4 Consider single-phase fluid flow in the 3-D horizontal reservoir in Example 2.6. Write the flow equation for interior block (3,2,2) using CVFD terminology, but this time, use natural ordering of blocks as shown in Fig. 3.5.

Solution

Block (3,2,2) in Fig. 2.15 is block 19 in Fig. 3.5. Therefore, $n = 19$. For $n = 19$, $\psi_{x_{19}} = \{18, 20\}$, $\psi_{y_{19}} = \{15, 23\}$, and $\psi_{z_{19}} = \{7, 31\}$. Substitution into Eq. (3.1a) gives $\psi_{19} = \{18, 20\} \cup \{15, 23\} \cup \{7, 31\} = \{7, 15, 18, 20, 23, 31\}$. The application of Eq. (3.17) produces

$$\sum_{l \in \psi_{19}} T_{l,19}^m \left[\left(p_l^m - p_{19}^m\right) - \gamma_{l,19}^m (Z_l - Z_{19})\right] + q_{sc_{19}}^m = \frac{V_{b_{19}}}{\alpha_c \Delta t}\left[\left(\frac{\phi}{B}\right)_{19}^{n+1} - \left(\frac{\phi}{B}\right)_{19}^n\right] \tag{3.40}$$

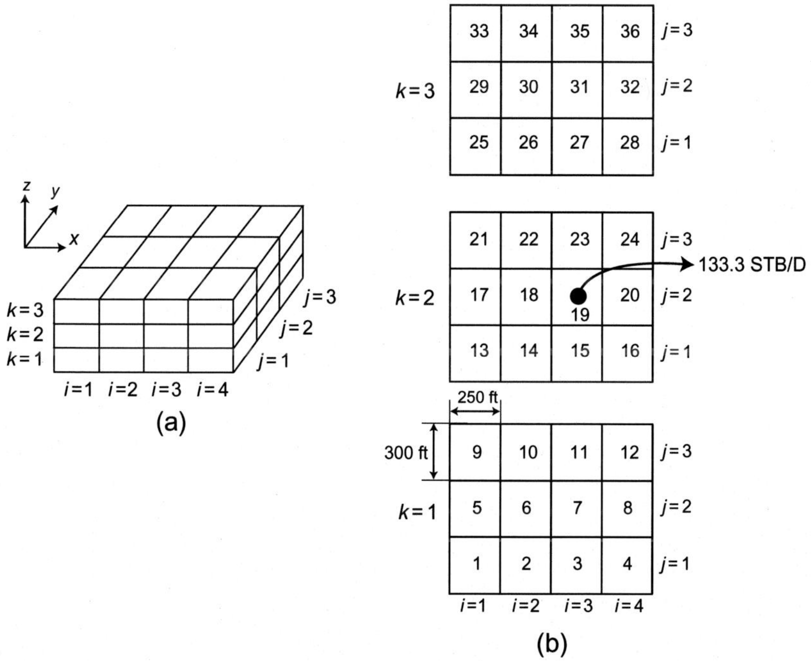

FIG. 3.5 3-D reservoir described in Example 3.4. (a) Reservoir representation and (b) natural ordering of blocks.

This equation can be expanded as

$$T_{7,19}^m\left[\left(p_7^m-p_{19}^m\right)-\gamma_{7,19}^m(Z_7-Z_{19})\right]+T_{15,19}^m\left[\left(p_{15}^m-p_{19}^m\right)-\gamma_{15,19}^m(Z_{15}-Z_{19})\right]$$
$$+T_{18,19}^m\left[\left(p_{18}^m-p_{19}^m\right)-\gamma_{18,19}^m(Z_{18}-Z_{19})\right]+T_{20,19}^m\left[\left(p_{20}^m-p_{19}^m\right)-\gamma_{20,19}^m(Z_{20}-Z_{19})\right]$$
$$+T_{23,19}^m\left[\left(p_{23}^m-p_{19}^m\right)-\gamma_{23,19}^m(Z_{23}-Z_{19})\right]+T_{31,19}^m\left[\left(p_{31}^m-p_{19}^m\right)-\gamma_{31,19}^m(Z_{31}-Z_{19})\right]$$
$$+q_{sc_{19}}^m=\frac{V_{b_{19}}}{\alpha_c\Delta t}\left[\left(\frac{\phi}{B}\right)_{19}^{n+1}-\left(\frac{\phi}{B}\right)_{19}^{n}\right] \tag{3.41}$$

For block 19, $Z_1=Z_{15}=Z_{18}=Z_{19}=Z_{20}=Z_{23}$, $Z_7-Z_{19}=33.33$ ft, $Z_{31}-Z_{19}=-33.33$ ft, and $q_{sc_{19}}^m=-133.3$ STB/D. Since $\Delta x_{18,19}=\Delta x_{20,19}=\Delta x=250$ ft, $\Delta y_{15,19}=\Delta y_{23,19}=\Delta y=300$ ft, $\Delta z_{7,19}=\Delta z_{31,19}=\Delta z=33.33$ ft, and μ, ρ, and B are constants, then $\gamma_{7,19}^m=\gamma_{31,19}^m=\gamma_c\rho g=0.21584\times10^{-3}\times55\times32.174=0.3819$ psi/ft,

$$T_{18,19}^m=T_{20,19}^m=\beta_c\frac{k_xA_x}{\mu B\Delta x}=0.001127\times\frac{270\times(300\times33.33)}{2\times1\times250}$$
$$=6.0857\ \text{STB/D-psi} \tag{3.42}$$

$$T_{15,19}^m=T_{23,19}^m=\beta_c\frac{k_yA_y}{\mu B\Delta y}=0.001127\times\frac{220\times(250\times33.33)}{2\times1\times300}$$
$$=3.4436\ \text{STB/D-psi} \tag{3.43}$$

and

$$T_{7,19}^m=T_{31,19}^m=\beta_c\frac{k_zA_z}{\mu B\Delta z}=0.001127\times\frac{50\times(250\times300)}{2\times1\times33.33}$$
$$=63.3944\ \text{STB/D-psi} \tag{3.44}$$

Substitution into Eq. (3.41) gives

$$(63.3944)\left[\left(p_7^m-p_{19}^m\right)-12.7287\right]+(3.4436)\left(p_{15}^m-p_{19}^m\right)+(6.0857)\left(p_{18}^m-p_{19}^m\right)$$
$$+(6.0857)\left(p_{20}^m-p_{19}^m\right)+(3.4436)\left(p_{23}^m-p_{19}^m\right)+(63.3944)\left[\left(p_{31}^m-p_{19}^m\right)+12.7287\right]$$
$$-133.3=\frac{V_{b_{19}}}{\alpha_c\Delta t}\left[\left(\frac{\phi}{B}\right)_{19}^{n+1}-\left(\frac{\phi}{B}\right)_{19}^{n}\right] \tag{3.45}$$

Eq. (3.45) corresponds to Eq. (2.47) in Example 2.6, which uses engineering notation.

3.3 Flow equations in radial-cylindrical coordinates using CVFD terminology

The equations presented in Sections 3.2.1 and 3.2.2 use Cartesian coordinates. The same equations can be made specific to radial-cylindrical coordinates by

TABLE 3.1 Functions in Cartesian and radial-cylindrical coordinates.

	Function in Cartesian coordinates	Function in radial-cylindrical coordinates
Coordinate	x	r
	y	θ
	z	z
Transmissibility	T_x	T_r
	T_y	T_θ
	T_z	T_z
Set of neighboring blocks along a direction	ψ_x	ψ_r
	ψ_y	ψ_θ
	ψ_z	ψ_z
Number of blocks along a direction	n_x	n_r
	n_y	n_θ
	n_z	n_z

replacing the directions (and subscripts) x and y with the directions (and subscripts) r and θ, respectively. Table 3.1 lists the corresponding functions for the two coordinate systems. As such, we can obtain the generalized 3-D flow equation in the r-θ-z space for block n—termed block (i,j,k) in engineering notation, meaning $n \equiv (i,j,k)$—from those in the x-y-z space, Eqs. (3.12) through (3.16). Keep in mind that i, j, and k are counting indices in the r-direction, θ-direction, and z-axis, respectively. Therefore, Eq. (3.12) becomes

$$\psi_{r_n} = \{(i-1,j,k),(i+1,j,k)\} \tag{3.46a}$$

$$\psi_{\theta_n} = \{(i,j-1,k),(i,j+1,k)\} \tag{3.46b}$$

and

$$\psi_{z_n} = \{(i,j,k-1),(i,j,k+1)\} \tag{3.46c}$$

Substitution of Eq. (3.46) into Eq. (3.1b) produces

$$\begin{aligned}\psi_n &= \psi_{i,j,k}\\ &= \{(i-1,j,k),(i+1,j,k)\} \cup \{(i,j-1,k),(i,j+1,k)\} \cup \{(i,j,k-1),(i,j,k+1)\}\\ &= \{(i,j,k-1),(i,j-1,k),(i-1,j,k),(i+1,j,k),(i,j+1,k),(i,j,k+1)\}\end{aligned} \tag{3.47}$$

which is identical to Eq. (3.13).

The flow equation for block (i,j,k), represented by Eq. (3.14a), becomes

$$\begin{aligned}
&T^m_{z_{i,j,k-1/2}}\left[\left(p^m_{i,j,k-1}-p^m_{i,j,k}\right)-\gamma^m_{i,j,k-1/2}\left(Z_{i,j,k-1}-Z_{i,j,k}\right)\right]\\
&+T^m_{\theta_{i,j-1/2,k}}\left[\left(p^m_{i,j-1,k}-p^m_{i,j,k}\right)-\gamma^m_{i,j-1/2,k}\left(Z_{i,j-1,k}-Z_{i,j,k}\right)\right]\\
&+T^m_{r_{i-1/2,j,k}}\left[\left(p^m_{i-1,j,k}-p^m_{i,j,k}\right)-\gamma^m_{i-1/2,j,k}\left(Z_{i-1,j,k}-Z_{i,j,k}\right)\right]\\
&+T^m_{r_{i+1/2,j,k}}\left[\left(p^m_{i+1,j,k}-p^m_{i,j,k}\right)-\gamma^m_{i+1/2,j,k}\left(Z_{i+1,j,k}-Z_{i,j,k}\right)\right]\\
&+T^m_{\theta_{i,j+1/2,k}}\left[\left(p^m_{i,j+1,k}-p^m_{i,j,k}\right)-\gamma^m_{i,j+1/2,k}\left(Z_{i,j+1,k}-Z_{i,j,k}\right)\right]\\
&+T^m_{z_{i,j,k+1/2}}\left[\left(p^m_{i,j,k+1}-p^m_{i,j,k}\right)-\gamma^m_{i,j,k+1/2}\left(Z_{i,j,k+1}-Z_{i,j,k}\right)\right]\\
&+q^m_{sc_{i,j,k}}=\frac{V_{b_{i,j,k}}}{\alpha_c\Delta t}\left[\left(\frac{\phi}{B}\right)^{n+1}_{i,j,k}-\left(\frac{\phi}{B}\right)^{n}_{i,j,k}\right]
\end{aligned}\tag{3.48a}$$

Eq. (3.14b), the flow equation in CVFD terminology, retains its form:

$$\begin{aligned}
&\sum_{l\in\psi_{i,j,k}} T^m_{l,(i,j,k)}\left[\left(p^m_l-p^m_{i,j,k}\right)-\gamma^m_{l,(i,j,k)}\left(Z_l-Z_{i,j,k}\right)\right]+q^m_{sc_{i,j,k}}\\
&=\frac{V_{b_{i,j,k}}}{\alpha_c\Delta t}\left[\left(\frac{\phi}{B}\right)^{n+1}_{i,j,k}-\left(\frac{\phi}{B}\right)^{n}_{i,j,k}\right]
\end{aligned}\tag{3.48b}$$

Eq. (3.15), which defines transmissibilities, becomes

$$T^m_{(i\mp1,j,k),(i,j,k)}=T^m_{(i,j,k),(i\mp1,j,k)}\equiv T^m_{r_{i\mp1/2,j,k}}\tag{3.49a}$$

$$T^m_{(i,j\mp1,k),(i,j,k)}=T^m_{(i,j,k),(i,j\mp1,k)}\equiv T^m_{\theta_{i,j\mp1/2,k}}\tag{3.49b}$$

and

$$T^m_{(i,j,k\mp1),(i,j,k)}=T^m_{(i,j,k),(i,j,k\mp1)}\equiv T^m_{z_{i,j,k\mp1/2}}\tag{3.49c}$$

Transmissibilities in radial-cylindrical coordinates, $T^m_{r_{i\mp1/2,j,k}}$, $T^m_{\theta_{i,j\mp1/2,k}}$, and $T^m_{z_{i,j,k\mp1/2}}$, are defined by Eq. (2.69). Note that gravity terms, as described by Eq. (3.16), remain intact for both coordinate systems:

$$\gamma^m_{(i\mp1,j,k),(i,j,k)}=\gamma^m_{(i,j,k),(i\mp1,j,k)}\equiv\gamma^m_{i\mp1/2,j,k}\tag{3.50a}$$

$$\gamma^m_{(i,j\mp1,k),(i,j,k)}=\gamma^m_{(i,j,k),(i,j\mp1,k)}\equiv\gamma^m_{i,j\mp1/2,k}\tag{3.50b}$$

and

$$\gamma^m_{(i,j,k\mp1),(i,j,k)}=\gamma^m_{(i,j,k),(i,j,k\mp1)}\equiv\gamma^m_{i,j,k\mp1/2}\tag{3.50c}$$

For 3-D flow in the r-θ-z space, if we desire to obtain the equations in CVFD terminology for block n with the blocks being ordered using natural ordering, we must write the equations that correspond to Eqs. (3.31) through (3.34) with the aid of Table 3.1 and then use Eq. (3.17). The resulting equations are listed as follows:

$$\psi_{r_n} = \{(n-1), (n+1)\} \tag{3.51a}$$

$$\psi_{\theta_n} = \{(n-n_r), (n+n_r)\} \tag{3.51b}$$

and

$$\psi_{z_n} = \{(n-n_r n_\theta), (n+n_r n_\theta)\} \tag{3.51c}$$

Substitution of Eq. (3.51) into Eq. (3.1b) results in

$$\begin{aligned}\psi_n &= \{(n-1), (n+1)\} \cup \{(n-n_r), (n+n_r)\} \cup \{(n-n_r n_\theta), (n+n_r n_\theta)\} \\ &= \{(n-n_r n_\theta), (n-n_r), (n-1), (n+1), (n+n_r), (n+n_r n_\theta)\}\end{aligned} \tag{3.52}$$

Now, the flow equation for block n in 3-D flow can be written again as Eq. (3.17):

$$\sum_{l\in\psi_n} T^m_{l,n}\left[\left(p^m_l - p^m_n\right) - \gamma^m_{l,n}(Z_l - Z_n)\right] + q^m_{sc_n} = \frac{V_{b_n}}{\alpha_c \Delta t}\left[\left(\frac{\phi}{B}\right)^{n+1}_n - \left(\frac{\phi}{B}\right)^n_n\right] \tag{3.17}$$

where transmissibility $T^m_{l,n}$ is defined as

$$T^m_{n\mp1,n} = T^m_{n,n\mp1} \equiv T^m_{r_{i\mp1/2,j,k}} \tag{3.53a}$$

$$T^m_{n\mp n_r,n} = T^m_{n,n\mp n_r} \equiv T^m_{\theta_{i,j\mp1/2,k}} \tag{3.53b}$$

and

$$T^m_{n\mp n_r n_\theta,n} = T^m_{n,n\mp n_r n_\theta} \equiv T^m_{z_{i,j,k\mp1/2}} \tag{3.53c}$$

In addition, fluid gravity $\gamma^m_{l,n}$ is defined as

$$\gamma^m_{n,n\mp1} = \gamma^m_{n\mp1,n} \equiv \gamma^m_{i\mp1/2,j,k} \tag{3.54a}$$

$$\gamma^m_{n,n\mp n_r} = \gamma^m_{n\mp n_r,n} \equiv \gamma^m_{i,j\mp1/2,k} \tag{3.54b}$$

and

$$\gamma^m_{n,n\mp n_r n_\theta} = \gamma^m_{n\mp n_r n_\theta,n} \equiv \gamma^m_{i,j,k\mp1/2} \tag{3.54c}$$

There are two distinct differences, however, between the flow equations in Cartesian (x-y-z) coordinates and radial-cylindrical (r-θ-z) coordinates. First, while reservoir external boundaries exist along the y-axis at $j=1$ and $j=n_y$, there are no external boundaries in the θ-direction because the blocks in this direction form a ring of blocks; that is, block $(i,1,k)$ is preceded by block (i,n_θ,k), and block (i,n_θ,k) is followed by block $(i,1,k)$. Second, any block in Cartesian coordinates is a candidate to host (or contribute to) a well, whereas in radial-cylindrical coordinates, only one well penetrates the inner circle of blocks parallel to the z-direction, and only blocks $(1, j, k)$ are candidates to contribute to this well.

3.4 Flow equations using CVFD terminology in any block ordering scheme

The flow equation using CFVD terminology for block n in any block ordering scheme is given by Eq. (3.17), where ψ_n is expressed by Eq. (3.1). The elements contained in sets ψ_{x_n}, ψ_{y_n}, and ψ_{z_n} are, respectively, the neighboring blocks of block n along the x-axis, y-axis, and z-axis for Cartesian coordinates, and the elements contained in sets ψ_{r_n}, ψ_{θ_n}, and ψ_{z_n} are, respectively, the neighboring blocks of block n in the r-direction, θ-direction, and z-axis for radial-cylindrical coordinates. The only difference between one ordering scheme and another is that the blocks in each scheme have different orders. Once reservoir blocks are ordered, the neighboring blocks are defined for each block in the reservoir, and finally, the flow equation for any reservoir block can be written. This is in relation to writing the flow equations in a given reservoir; the method of solving the resulting set of equations is however another matter (see Chapter 9).

3.5 Summary

A flow equation in CVFD terminology has the same form regardless of the dimensionality of the flow problem or the coordinate system; hence, the objective of CVFD terminology is to write flow equations in compact form only. In CVFD terminology, the flow equation for block n can be made to describe flow in 1-D, 2-D, or 3-D reservoirs by defining the appropriate set of neighboring blocks (ψ_n). In Cartesian coordinates, Eqs. (3.3), (3.8), and (3.13) define the elements of ψ_n for 1-D, 2-D, and 3-D reservoirs, respectively. Eq. (3.17) gives the flow equation, and transmissibilities and gravities are defined by Eqs. (3.15) and (3.16). Equivalent equations can be written for radial-cylindrical coordinates if subscript x is replaced with subscript r and subscript y is replaced with subscript θ.

3.6 Exercises

3.1 Is 0 the same as $\{\}$? If not, how does it differ?

3.2 Write the answers for $2+3$ and $\{2\} \cup \{3\}$.

3.3 Using your own words, give the physical meanings conveyed by Eqs. (3.2a) and (3.2b).

3.4 Consider the 1-D reservoir representation in Fig. 2.6b. Find ψ_1, ψ_2, ψ_3, and ψ_4.

3.5 Consider the 2-D reservoir representation in Fig. 3.4. Find ψ_n for $n=1, 2, 3, \ldots 12$.

3.6 Consider the 3-D reservoir representation in Fig. 2.8c. Find ψ_n for $n=1, 2, 3, \ldots 36$.

3.7 Consider the 3-D reservoir representation in Fig. 2.8b. Find $\psi_{(1,1,1)}$, $\psi_{(2,2,1)}$, $\psi_{(3,2,2)}$, $\psi_{(4,3,2)}$, $\psi_{(4,1,3)}$, $\psi_{(3,2,3)}$, and $\psi_{(1,3,3)}$.

3.8 Using the definitions of ψ_n, ψ_{x_n}, ψ_{y_n}, and ψ_{z_n} along with the aid of Fig. 3.3c, prove that $\psi_n=\psi_{x_n}\cup\psi_{y_n}\cup\psi_{z_n}$.

3.9 Consider fluid flow in a 1-D horizontal reservoir along the x-axis. The reservoir left and right boundaries are closed to fluid flow. The reservoir consists of three blocks as shown in Fig. 3.6.

a. Write the appropriate flow equation for a general block n in this reservoir.

b. Write the flow equation for block 1 by finding ψ_1 and then using it to expand the equation in (a).

c. Write the flow equation for block 2 by finding ψ_2 and then using it to expand the equation in (a).

d. Write the flow equation for block 3 by finding ψ_3 and then using it to expand the equation in (a).

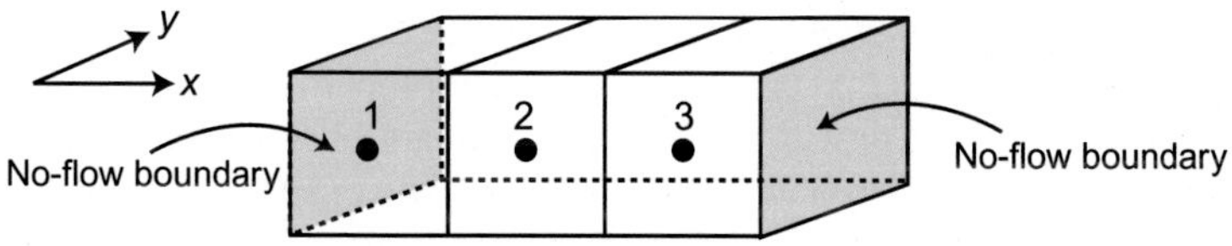

FIG. 3.6 1-D reservoir for Exercise 3.9.

3.10 Consider fluid flow in a 2-D, horizontal, closed reservoir. The reservoir consists of nine blocks as shown in Fig. 3.7.

a. Write the appropriate flow equation for a general block n in this reservoir.

b. Write the flow equation for block 1 by finding ψ_1 and then using it to expand the equation in (a).

c. Write the flow equation for block 2 by finding ψ_2 and then using it to expand the equation in (a).

d. Write the flow equation for block 4 by finding ψ_4 and then using it to expand the equation in (a).

e. Write the flow equation for block 5 by finding ψ_5 and then using it to expand the equation in (a).

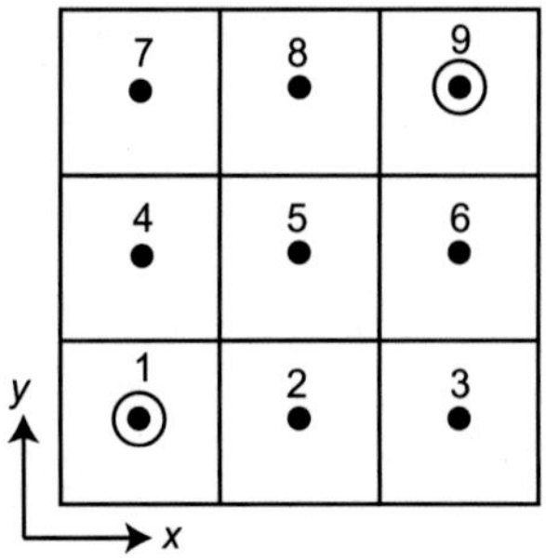

FIG. 3.7 2-D reservoir for Exercise 3.10.

3.11 A 2-D oil reservoir is discretized into 4 × 4 blocks.

a. Order the blocks in this reservoir using the natural ordering scheme, letting block 1 be the lower left corner block.

b. Write the flow equation for each interior block in this reservoir.

3.12 A 2-D oil reservoir is discretized into 4 × 4 blocks.

a. Order the blocks in this reservoir using the D4 ordering scheme, letting block 1 be the lower left corner block.

b. Write the flow equation for each interior block in this reservoir.

3.13 A single-phase oil reservoir is described by four equal blocks as shown in Fig. 3.8. The reservoir is horizontal and has homogeneous and isotropic rock properties, $k=150$ md and $\phi=0.21$. Block dimensions are $\Delta x=400$ ft, $\Delta y=600$ ft, and $h=25$ ft. Oil properties are $B=1$ RB/STB and $\mu=5$ cP. The pressures of blocks 1 and 4 are 2200 and 900 psia, respectively. Block 3 hosts a well that produces oil at a rate of 100 STB/D. Find the pressure distribution in the reservoir assuming that the reservoir rock and oil are incompressible.

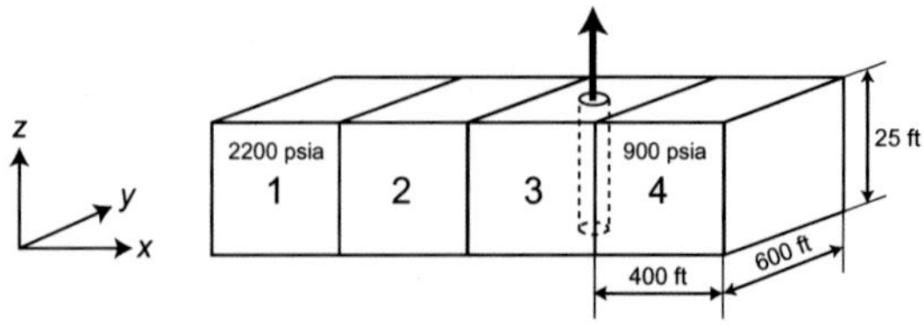

FIG. 3.8 1-D reservoir representation in Exercise 3.13.

3.14 A single-phase oil reservoir is described by five equal blocks as shown in Fig. 3.9. The reservoir is horizontal and has $k=90$ md and $\phi=0.17$. Block dimensions are $\Delta x=500$ ft, $\Delta y=900$ ft, and $h=45$ ft. Oil properties are $B=1$ RB/STB and $\mu=3$ cP. The pressures of blocks 1 and 5 are maintained at 2700 and 1200 psia, respectively. Gridblock 4 hosts a well that produces oil at a rate of 325 STB/D. Find the pressure distribution in the reservoir assuming that the reservoir oil and rock are incompressible.

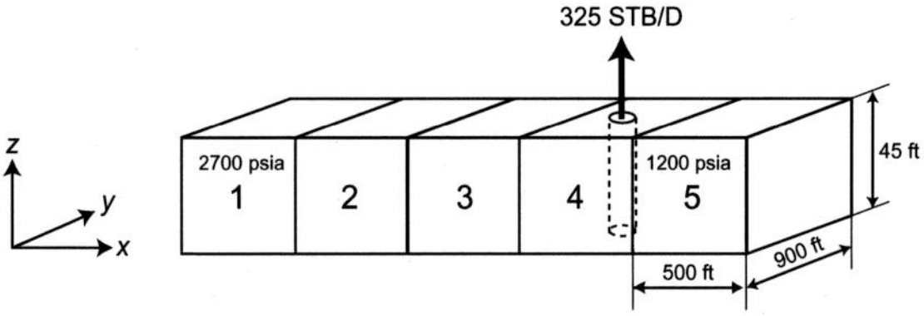

FIG. 3.9 1-D reservoir representation in Exercise 3.14.

3.15 Consider single-phase flow of oil in a 2-D horizontal reservoir. The reservoir is discretized using 4×4 equal blocks as shown in Fig. 3.10. Block (2,3) hosts a well that produces 500 STB/D of oil. All blocks have $\Delta x=\Delta y=330$ ft, $h=50$ ft, and $k_x=k_y=210$ md. The oil FVF and viscosity are 1.0 RB/B and 2 cP, respectively. The pressures of reservoir boundary blocks are specified in Fig. 3.10. Assuming that the reservoir oil and rock are incompressible, calculate the pressures of blocks (2,2), (3,2), (2,3), and (3,3).

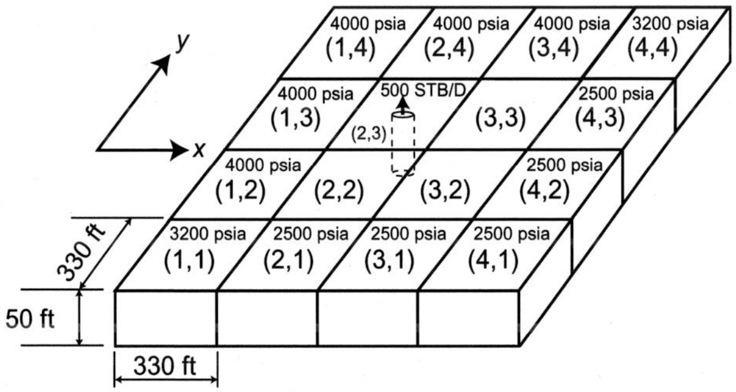

FIG. 3.10 2-D reservoir representation in Exercise 3.15.

3.16 Consider single-phase flow of oil in a 2-D horizontal reservoir. The reservoir is discretized using 4×4 equal blocks as shown in Fig. 3.11. Each of blocks 6 and 11 hosts a well that produces oil at the rate shown in the figure. All blocks have $\Delta x=200$ ft, $\Delta y=250$ ft, $h=60$ ft, $k_x=80$ md, and $k_y = 65$ md. The oil FVF and viscosity are 1.0 RB/STB and 2 cP, respectively. The pressures of reservoir boundary blocks are specified in Fig. 3.11. Assuming that the reservoir oil and rock are incompressible, calculate the pressures of blocks 6, 7, 10, and 11.

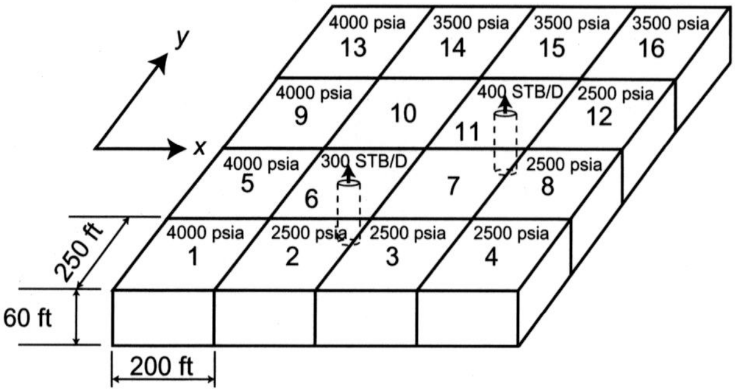

FIG. 3.11 2-D reservoir representation in Exercise 3.16.

Chapter 4

Simulation with a block-centered grid

Chapter outline

4.1 Introduction

This chapter presents discretization of 1-D, 2-D, and 3-D reservoirs using block-centered grids in Cartesian and radial-cylindrical coordinate systems. As the name implies, the gridblock dimensions are selected first, followed by the placement of points in central locations of the blocks. In this, the distance between block boundaries is the defining variable in space. In contrast, the grid-points (or nodes) are selected first in the point-distributed grid, which is discussed in Chapter 5. Chapter 2 introduced the terminology for reservoir discretization into blocks. This chapter describes the construction of a block-centered grid for a reservoir and the relationships between block sizes, block boundaries, and distances between points representing blocks. The resulting gridblocks can be classified into interior and boundary gridblocks. Chapter 2 also derived the flow equations for interior gridblocks. However, the boundary gridblocks are subject to boundary conditions and thus require special treatment. This chapter presents the treatment of various boundary conditions and introduces a general flow equation that is applicable for interior blocks and boundary blocks. This chapter also presents the equations for directional transmissibilities in both Cartesian and radial-cylindrical coordinate systems and discusses the use of symmetry in reservoir simulation.

Petroleum Reservoir Simulation. https://doi.org/10.1016/B978-0-12-819150-7.00004-9

4.2 Reservoir discretization

Reservoir discretization means that the reservoir is described by a set of gridblocks whose properties, dimensions, boundaries, and locations in the reservoir are well defined. Fig. 4.1 shows a block-centered grid for a 1-D reservoir in the direction of the x-axis. The grid is constructed by choosing n_x gridblocks that span the entire reservoir length in the x-direction. The gridblocks are assigned predetermined dimensions (Δx_i, $i=1, 2, 3\ldots n_x$) that are not necessarily equal. Then, the point that represents each gridblock is subsequently located at the center of that gridblock. Fig. 4.2 focuses on gridblock i and its neighboring gridblocks in the x-direction. It shows how the gridblocks are related to each other, gridblock dimensions (Δx_{i-1}, Δx_i, Δx_{i+1}), gridblock boundaries ($x_{i-1/2}$, $x_{i+1/2}$), distances between the point that represents gridblock i and gridblock boundaries (δx_{i^-},δx_{i^+}), and distances between the points representing these gridblocks ($\Delta x_{i-1/2}$, $\Delta x_{i+1/2}$).

Gridblock dimensions, boundaries, and locations satisfy the following relationships:

$$
\begin{aligned}
&\sum_{i=1}^{n_x} \Delta x_i = L_x,\\
&\delta x_{i^-} = \delta x_{i^+} = {}^1/_2 \Delta x_i,\ i=1,2,3\ldots n_x,\\
&\Delta x_{i-1/2} = \delta x_{i^-} + \delta x_{i-1^+} = {}^1/_2(\Delta x_i + \Delta x_{i-1}),\ i=2,3\ldots n_x,\\
&\Delta x_{i+1/2} = \delta x_{i^+} + \delta x_{i+1^-} = {}^1/_2(\Delta x_i + \Delta x_{i+1}),\ i=1,2,3\ldots n_x-1,\\
&x_{i+1} = x_i + \Delta x_{i+1/2},\ i=1,2,3\ldots n_x-1,\ x_1 = {}^1/_2 \Delta x_1,\\
&x_{i-1/2} = x_i - \delta x_{i^-} = x_i - {}^1/_2 \Delta x_i,\ i=1,2,3\ldots n_x,\\
&x_{i+1/2} = x_i + \delta x_{i^+} = x_i + {}^1/_2 \Delta x_i,\ i=1,2,3\ldots n_x
\end{aligned}
\tag{4.1}
$$

Fig. 4.3 shows the discretization of a 2-D reservoir into a 5×4 irregular grid. An irregular grid implies that block sizes in the direction of the x-axis (Δx_i) and the y-axis (Δy_j) are neither equal nor constant. Discretization using a regular grid means that block sizes in the x- and y-directions are constants but not necessarily equal. The discretization in the x-direction uses the procedure just mentioned and the relationships presented in Eq. (4.1). The discretization in the y-direction uses a procedure and relationships similar to those for the x-direction, and the same can be said for the z-direction for a 3-D reservoir. Inspection of Figs. 4.1 and 4.3 shows that the point that represents a gridblock falls in the

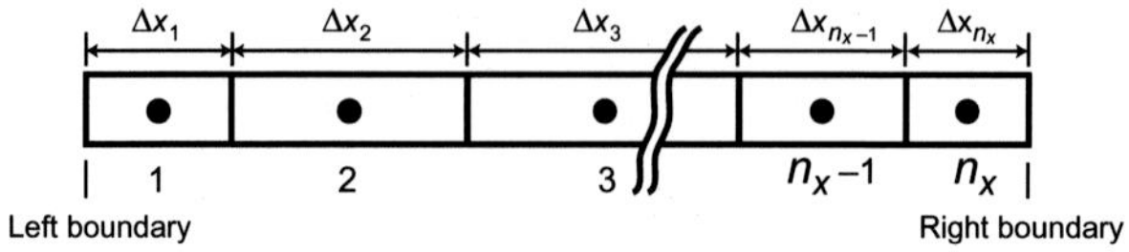

FIG. 4.1 Discretization of a 1-D reservoir using a block-centered grid.

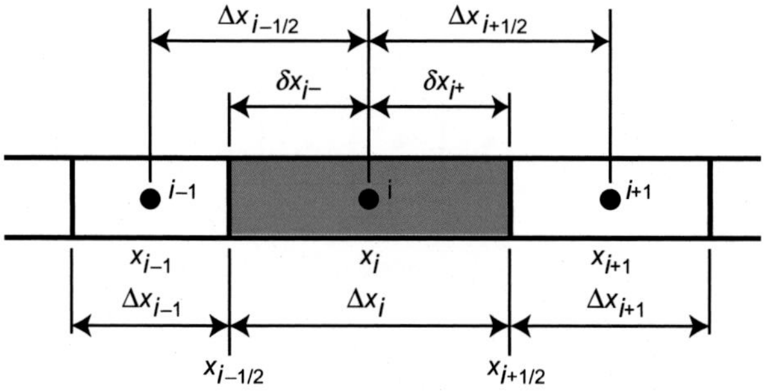

FIG. 4.2 Gridblock i and its neighboring gridblocks in the x-direction.

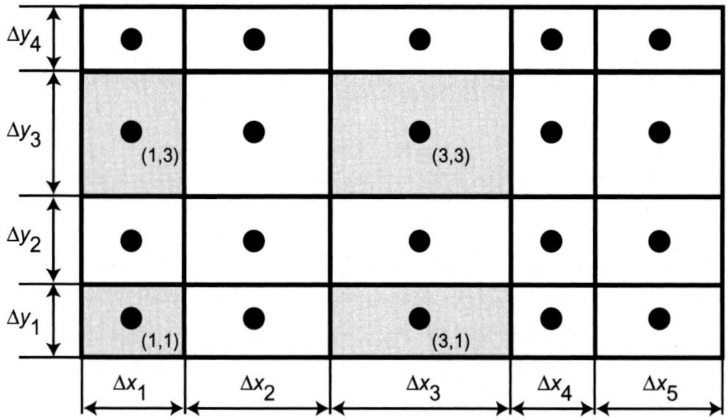

FIG. 4.3 Discretization of a 2-D reservoir using a block-centered grid.

center of that block and that all points representing gridblocks fall inside reservoir boundaries.

Example 4.1 A 5000 × 1200 × 75 ft horizontal reservoir contains oil that flows along its length. The reservoir rock porosity and permeability are 0.18 and 15 md, respectively. The oil FVF and viscosity are 1 RB/STB and 10 cP, respectively. The reservoir has a well located at 3500 ft. from the reservoir left boundary and produces oil at a rate of 150 STB/D. Discretize the reservoir into five equal blocks using a block-centered grid and assign properties to the gridblocks comprising this reservoir.

Solution

Using a block-centered grid, the reservoir is divided along its length into five equal blocks. Each block is represented by a point at its center. Therefore, $n_x = 5$, and $\Delta x = L_x/n_x = 5000/5 = 1000$ ft. Gridblocks are numbered from 1 to 5 as shown in Fig. 4.4. Now, the reservoir is described through assigning properties to its five gridblocks ($i = 1, 2, 3, 4, 5$). All the gridblocks (or the points that represent them) have the same elevation because the reservoir is horizontal. Each gridblock has the dimensions of $\Delta x = 1000$, $\Delta y = 1200$, and $\Delta z = 75$ and properties of $k_x = 15$ md and $\phi = 0.18$. The points representing gridblocks are equally spaced; that is,

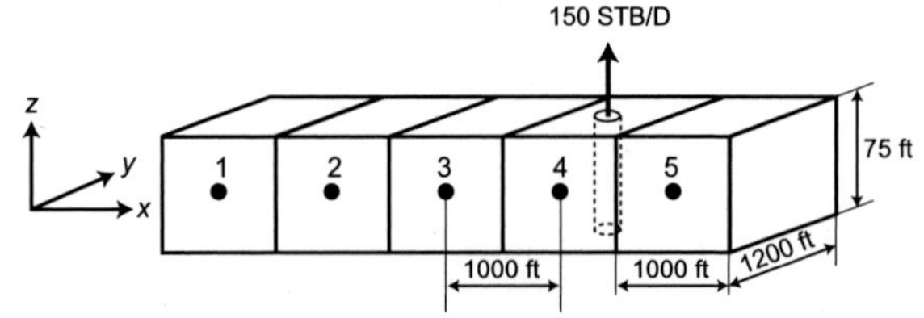

FIG. 4.4 Discretized 1-D reservoir in Example 4.1.

$\Delta x_{i\mp1/2} = \Delta x = 1000$ ft and $A_{x_{i\mp1/2}} = A_x = \Delta y \times \Delta z = 1200 \times 75 = 90{,}000$ ft^2. Gridblock 1 falls on the reservoir left boundary, and gridblock 5 falls on the reservoir right boundary. Gridblocks 2, 3, and 4 are interior gridblocks. In addition, gridblock 4 hosts a well with $q_{sc_4} = -150$ STB/D. Fluid properties are $B = 1$ RB/STB and $\mu = 10$ cP.

4.3 Flow equation for boundary gridblocks

In this section, we present a form of the flow equation that applies to interior blocks and boundary blocks. This means that the proposed flow equation reduces to the flow equations presented in Chapters 2 and 3 for interior blocks, but it also includes the effects of boundary conditions for boundary blocks. Fig. 4.1 shows a discretized 1-D reservoir in the direction of the x-axis. Gridblocks 2, 3, … $n_x - 1$ are interior blocks, whereas gridblocks 1 and n_x are boundary blocks that each falls on one reservoir boundary. Fig. 4.3 shows a discretized 2-D reservoir. This figure highlights an interior gridblock, gridblock (3,3); two boundary gridblocks that each falls on one reservoir boundary, gridblocks (1,3) and (3,1); and a gridblock that falls on two reservoir boundaries, gridblock (1,1). In 3-D reservoirs, there are interior gridblocks and boundary gridblocks. Boundary gridblocks may fall on one, two, or three reservoir boundaries. Fig. 4.5 demonstrates the terminology used in this book for the reservoir boundaries in the negative and positive directions of the x-, y-, and z-axes. Reservoir boundaries along the x-axis are termed reservoir west boundary (b_W) and reservoir east boundary (b_E), and those along the y-axis are termed reservoir south boundary (b_S) and reservoir north boundary (b_N). Reservoir boundaries along the z-axis are termed reservoir lower boundary (b_L) and reservoir upper boundary (b_U).

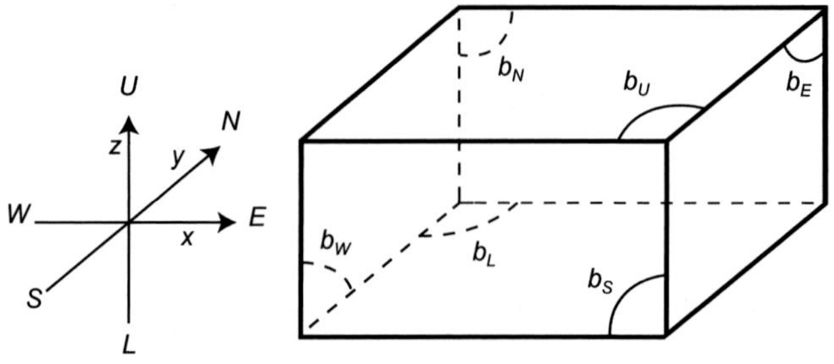

FIG. 4.5 Definition of left and right boundaries in 3-D reservoirs.

The characteristic forms of the difference equations for interior and boundary gridblocks differ in the terms of dealing with space variables; that is, the flow terms. The production (injection) term and the accumulation term are the same for both interior and boundary gridblocks. The engineering approach involves replacing the boundary condition with a no-flow boundary plus a fictitious well having a flow rate $q^m_{sc_{b,bB}}$ that reflects fluid transfer between the reservoir boundary itself (b) and the boundary block (bB). In other words, a fictitious well having flow rate of $q^m_{sc_{b,bB}}$ replaces the flow term that represents fluid transfer across a reservoir boundary between a boundary block and a block exterior to the reservoir. The number of flow terms in the flow equation for an interior gridblock equals the number of neighboring gridblocks (two, four, or six terms for 1D-, 2-D, or 3-D reservoir, respectively). For the flow equation for a boundary gridblock, the number of flow terms equals the number of existing neighboring gridblocks in the reservoir and the number of fictitious wells equals the number of reservoir boundaries adjacent to the boundary gridblock.

A general form of the flow equation that applies to boundary gridblocks and interior gridblocks in 1-D, 2-D, or 3-D flow in both Cartesian and radial-cylindrical coordinates can be expressed best using CVFD terminology. The use of summation operators in CVFD terminology makes it flexible and suitable for describing flow terms in the equation of any gridblock sharing none or any number of boundaries with the reservoir. The general form for gridblock n can be written as:

$$\sum_{l\in\psi_n} T^m_{l,n}\left[\left(p^m_l - p^m_n\right) - \gamma^m_{l,n}(Z_l - Z_n)\right] + \sum_{l\in\xi_n} q^m_{sc_{l,n}} + q^m_{sc_n} = \frac{V_{b_n}}{\alpha_c \Delta t}\left[\left(\frac{\phi}{B}\right)^{n+1}_n - \left(\frac{\phi}{B}\right)^n_n\right] \tag{4.2a}$$

or, in terms of potentials, as

$$\sum_{l\in\psi_n} T^m_{l,n}\left(\Phi^m_l - \Phi^m_n\right) + \sum_{l\in\xi_n} q^m_{sc_{l,n}} + q^m_{sc_n} = \frac{V_{b_n}}{\alpha_c \Delta t}\left[\left(\frac{\phi}{B}\right)^{n+1}_n - \left(\frac{\phi}{B}\right)^n_n\right] \tag{4.2b}$$

where ψ_n= the set whose elements are the existing neighboring gridblocks in the reservoir, ξ_n= the set whose elements are the reservoir boundaries (b_L, b_S, b_W, b_E, b_N, b_U) that are shared by gridblock n, and $q^m_{sc_{l,n}}$ = flow rate of the fictitious well representing fluid transfer between reservoir boundary l and gridblock n as a result of a boundary condition. For a 3-D reservoir, ξ_n is either an empty set for interior gridblocks or a set that contains one element for gridblocks that fall on one reservoir boundary, two elements for gridblocks that fall on two reservoir boundaries, or three elements for gridblocks that fall on three reservoir boundaries. An empty set implies that the gridblock does not fall on any reservoir boundary; that is, gridblock n is an interior gridblock and hence $\sum_{l\in\xi_n} q^m_{sc_{l,n}} = 0$. In engineering notation, $n \equiv (i,j,k)$ and Eq. (4.2a) becomes:

$$\sum_{l\in\psi_{i,j,k}} T^m_{l,(i,j,k)}\left[\left(p^m_l - p^m_{i,j,k}\right) - \gamma^m_{l,(i,j,k)}\left(Z_l - Z_{i,j,k}\right)\right] + \sum_{l\in\xi_{i,j,k}} q^m_{sc_{l,(i,j,k)}} + q^m_{sc_{i,j,k}}$$

$$= \frac{V_{b_{i,j,k}}}{\alpha_c \Delta t}\left[\left(\frac{\phi}{B}\right)^{n+1}_{i,j,k} - \left(\frac{\phi}{B}\right)^{n}_{i,j,k}\right] \tag{4.2c}$$

It must be mentioned that reservoir blocks have a three-dimensional shape whether fluid flow is 1-D, 2-D, or 3-D. The number of existing neighboring gridblocks and the number of reservoir boundaries shared by a reservoir gridblock add up to six as is the case in 3-D flow. Existing neighboring gridblocks contribute to flow to or from the gridblock, whereas reservoir boundaries may or may not contribute to flow depending on the dimensionality of flow and the prevailing boundary conditions. The dimensionality of flow implicitly defines those reservoir boundaries that do not contribute to flow at all. In 1-D flow problems, all reservoir gridblocks have four reservoir boundaries that do not contribute to flow. In 1-D flow in the x-direction, the reservoir south, north, lower, and upper boundaries do not contribute to flow to any reservoir gridblock, including boundary gridblocks. These four reservoir boundaries (b_L, b_S, b_N, b_U) are discarded as if they did not exist. As a result, an interior reservoir gridblock has two neighboring gridblocks and no reservoir boundaries, whereas a boundary reservoir gridblock has one neighboring gridblock and one reservoir boundary. In 2-D flow problems, all reservoir gridblocks have two reservoir boundaries that do not contribute to flow at all. For example, in 2-D flow in the x-y plane, the reservoir lower and upper boundaries do not contribute to flow to any reservoir gridblock, including boundary gridblocks. These two reservoir boundaries (b_L, b_U) are discarded as if they did not exist. As a result, an interior reservoir gridblock has four neighboring gridblocks and no reservoir boundaries, a reservoir gridblock that falls on one reservoir boundary has three neighboring gridblocks and one reservoir boundary, and a reservoir gridblock that falls on two reservoir boundaries has two neighboring gridblocks and two reservoir boundaries. In 3-D flow problems, any of the six reservoir boundaries may contribute to flow depending on the specified boundary condition. An interior gridblock has six neighboring gridblocks. It does not share any of its boundaries with any of the reservoir boundaries. A boundary gridblock may fall on one, two, or three of the reservoir boundaries. Therefore, a boundary gridblock that falls on one, two, or three reservoir boundaries has five, four, or three neighboring gridblocks, respectively. The earlier discussion leads to a few conclusions related to the number of elements contained in sets ψ and ξ.

(1) For an interior reservoir gridblock, set ψ contains two, four, or six elements for a 1-D, 2-D, or 3-D flow problem, respectively, and set ξ contains no elements or, in other words, is empty.

(2) For a boundary reservoir gridblock, set ψ contains less than two, four, or six elements for a 1-D, 2-D, or 3-D flow problem, respectively, and set ξ is not empty.

(3) The sum of the number of elements in sets ψ and ξ for any reservoir gridblock is a constant that depends on the dimensionality of flow. This sum is two, four, or six for a 1-D, 2-D, or 3-D flow problem, respectively.

For 1-D reservoirs, the flow equation for interior gridblock i is given by Eq. (2.32) or (2.33):

$$T^m_{x_{i-1/2}}\left(\Phi^m_{i-1}-\Phi^m_i\right)+T^m_{x_{i+1/2}}\left(\Phi^m_{i+1}-\Phi^m_i\right)+q^m_{sc_i}=\frac{V_{b_i}}{\alpha_c \Delta t}\left[\left(\frac{\phi}{B}\right)^{n+1}_i-\left(\frac{\phi}{B}\right)^n_i\right] \tag{4.3}$$

The above flow equation can be obtained from Eq. (4.2b) for $n=i$, $\psi_i=\{i-1, i+1\}$, and $\xi_i=\{\}$, and by observing that $\sum_{l\in\xi_i} q^m_{sc_{l,i}}=0$ for an interior gridblock and $T^m_{i\mp1,i}=T^m_{x_{i\mp1/2}}$.

The flow equation for boundary gridblock 1, which falls on the reservoir west boundary in Fig. 4.6, can be written as

$$T^m_{x_{1-1/2}}\left(\Phi^m_0-\Phi^m_1\right)+T^m_{x_{1+1/2}}\left(\Phi^m_2-\Phi^m_1\right)+q^m_{sc_1}=\frac{V_{b_1}}{\alpha_c \Delta t}\left[\left(\frac{\phi}{B}\right)^{n+1}_1-\left(\frac{\phi}{B}\right)^n_1\right] \tag{4.4a}$$

The first term on the LHS of Eq. (4.4a) represents the rate of fluid flow across the reservoir west boundary (b_W). This term can be replaced with the flow rate of a fictitious well ($q^m_{sc_{b_W,1}}$) that transfers fluid across the reservoir west boundary to gridblock 1; that is,

$$q^m_{sc_{b_W,1}}=T^m_{x_{1-1/2}}\left(\Phi^m_0-\Phi^m_1\right) \tag{4.5a}$$

Substitution of Eq. (4.5a) into Eq. (4.4a) yields

$$q^m_{sc_{b_W,1}}+T^m_{x_{1+1/2}}\left(\Phi^m_2-\Phi^m_1\right)+q^m_{sc_1}=\frac{V_{b_1}}{\alpha_c \Delta t}\left[\left(\frac{\phi}{B}\right)^{n+1}_1-\left(\frac{\phi}{B}\right)^n_1\right] \tag{4.4b}$$

The above flow equation can be obtained from Eq. (4.2b) for $n=1, \psi_1=\{2\}$, and $\xi_1=\{b_W\}$, and by observing that $\sum_{l\in\xi_1} q^m_{sc_{l,1}}=q^m_{sc_{b_W,1}}$ and $T^m_{2,1}=T^m_{x_{1+1/2}}$.

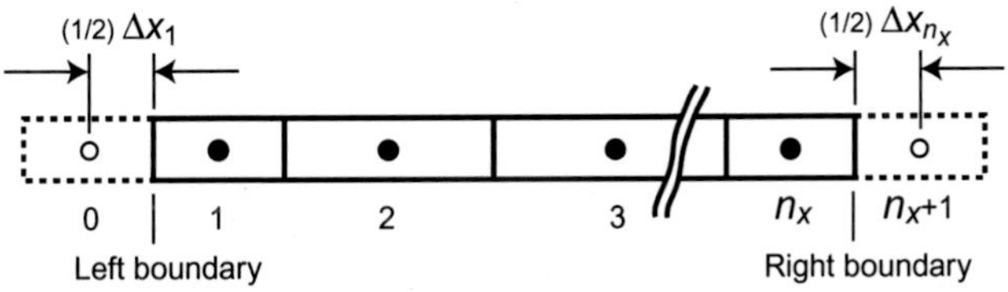

FIG. 4.6 Boundary gridblocks at the left and right boundaries of a 1-D reservoir (*dashed lines* represent fictitious reflective blocks).

The flow equation for boundary gridblock n_x, which falls on the reservoir east boundary in Fig. 4.6, can be written as

$$T^m_{x_{n_x-1/2}}\left(\Phi^m_{n_x-1}-\Phi^m_{n_x}\right)+T^m_{x_{n_x+1/2}}\left(\Phi^m_{n_x+1}-\Phi^m_{n_x}\right)+q^m_{sc_{n_x}} = \frac{V_{b_{n_x}}}{\alpha_c\Delta t}\left[\left(\frac{\phi}{B}\right)^{n+1}_{n_x}-\left(\frac{\phi}{B}\right)^{n}_{n_x}\right] \tag{4.6a}$$

The second term on the LHS of Eq. (4.6a) represents the rate of fluid flow across the reservoir east boundary (b_E). This term can be replaced with the flow rate of a fictitious well ($q^m_{sc_{b_E,n_x}}$) that transfers fluid across the reservoir east boundary to gridblock n_x; that is,

$$q^m_{sc_{b_E,n_x}}=T^m_{x_{n_x+1/2}}\left(\Phi^m_{n_x+1}-\Phi^m_{n_x}\right) \tag{4.7a}$$

Substitution of Eq. (4.7a) into Eq. (4.6a) yields

$$T^m_{x_{n_x-1/2}}\left(\Phi^m_{n_x-1}-\Phi^m_{n_x}\right)+q^m_{sc_{b_E,n_x}}+q^m_{sc_{n_x}}=\frac{V_{b_{n_x}}}{\alpha_c\Delta t}\left[\left(\frac{\phi}{B}\right)^{n+1}_{n_x}-\left(\frac{\phi}{B}\right)^{n}_{n_x}\right] \tag{4.6b}$$

The above flow equation can also be obtained from Eq. (4.2b) for $n=n_x$, $\psi_{n_x}=\{n_x-1\}$, and $\xi_{n_x}=\{b_E\}$, and by observing that $\sum\limits_{l\in\xi_{n_x}} q^m_{sc_{l,n_x}}=q^m_{sc_{b_E,n_x}}$ and $T^m_{n_x-1,n_x}=T^m_{x_{n_x+1/2}}$.

For 2-D reservoirs, the flow equation for interior gridblock (i,j) is given by Eq. (2.37):

$$T^m_{y_{i,j-1/2}}\left(\Phi^m_{i,j-1}-\Phi^m_{i,j}\right)+T^m_{x_{i-1/2,j}}\left(\Phi^m_{i-1,j}-\Phi^m_{i,j}\right)+T^m_{x_{i+1/2,j}}\left(\Phi^m_{i+1,j}-\Phi^m_{i,j}\right)$$
$$+T^m_{y_{i,j+1/2}}\left(\Phi^m_{i,j+1}-\Phi^m_{i,j}\right)+q^m_{sc_{i,j}}=\frac{V_{b_{i,j}}}{\alpha_c\Delta t}\left[\left(\frac{\phi}{B}\right)^{n+1}_{i,j}-\left(\frac{\phi}{B}\right)^{n}_{i,j}\right] \tag{4.8}$$

The above flow equation can be obtained from Eq. (4.2b) for $n\equiv(i,j)$, $\psi_{i,j}=\{(i,j-1),(i-1,j),(i+1,j),(i,j+1)\}$, and $\xi_{i,j}=\{\}$, and by observing that $\sum\limits_{l\in\xi_{i,j}} q^m_{sc_{l,(i,j)}}=0$ for an interior gridblock, $T^m_{(i,j\mp1),(i,j)}=T^m_{y_{i,j\mp1/2}}$, and $T^m_{(i\mp1,j),(i,j)}=T^m_{x_{i\mp1/2,j}}$.

For a gridblock that falls on one reservoir boundary, like gridblock (3,1), which falls on the reservoir south boundary in Fig. 4.3, the flow equation can be written as

$$T^m_{y_{3,1-1/2}}\left(\Phi^m_{3,0}-\Phi^m_{3,1}\right)+T^m_{x_{3-1/2,1}}\left(\Phi^m_{2,1}-\Phi^m_{3,1}\right)+T^m_{x_{3+1/2,1}}\left(\Phi^m_{4,1}-\Phi^m_{3,1}\right)$$
$$+T^m_{y_{3,1+1/2}}\left(\Phi^m_{3,2}-\Phi^m_{3,1}\right)+q^m_{sc_{3,1}}=\frac{V_{b_{3,1}}}{\alpha_c\Delta t}\left[\left(\frac{\phi}{B}\right)^{n+1}_{3,1}-\left(\frac{\phi}{B}\right)^{n}_{3,1}\right] \tag{4.9a}$$

The first term on the LHS of Eq. (4.9a) represents the rate of fluid flow across the reservoir south boundary (b_S). This term can be replaced with the flow rate of a fictitious well $\left(q^m_{sc_{b_S},(3,1)}\right)$ that transfers fluid across the reservoir south boundary to gridblock (3,1); that is,

$$q^m_{sc_{b_S},(3,1)} = T^m_{y_{3,1-1/2}}\left(\Phi^m_{3,0} - \Phi^m_{3,1}\right) \tag{4.10}$$

Substitution of Eq. (4.10) into Eq. (4.9a) yields

$$q^m_{sc_{b_S},(3,1)} + T^m_{x_{3-1/2,1}}\left(\Phi^m_{2,1} - \Phi^m_{3,1}\right) + T^m_{x_{3+1/2,1}}\left(\Phi^m_{4,1} - \Phi^m_{3,1}\right)$$
$$+T^m_{y_{3,1+1/2}}\left(\Phi^m_{3,2} - \Phi^m_{3,1}\right) + q^m_{sc_{3,1}} = \frac{V_{b_{3,1}}}{\alpha_c \Delta t}\left[\left(\frac{\phi}{B}\right)^{n+1}_{3,1} - \left(\frac{\phi}{B}\right)^{n}_{3,1}\right] \tag{4.9b}$$

The above flow equation can be obtained from Eq. (4.2b) for $n \equiv (3,1)$, $\psi_{3,1} = \{(2,1),(4,1),(3,2)\}$, and $\xi_{3,1} = \{b_S\}$, and by observing that $\sum_{l \in \xi_{3,1}} q^m_{sc_{l,(3,1)}} = q^m_{sc_{b_S},(3,1)}$, $T^m_{(2,1),(3,1)} = T^m_{x_{3-1/2,1}}$, $T^m_{(4,1),(3,1)} = T^m_{x_{3+1/2,1}}$, and $T^m_{(3,2),(3,1)} = T^m_{y_{3,1+1/2}}$.

For a gridblock that falls on two reservoir boundaries, like boundary gridblock (1,1), which falls on the reservoir south and west boundaries in Fig. 4.3, the flow equation can be written as

$$T^m_{y_{1,1-1/2}}\left(\Phi^m_{1,0} - \Phi^m_{1,1}\right) + T^m_{x_{1-1/2,1}}\left(\Phi^m_{0,1} - \Phi^m_{1,1}\right) + T^m_{x_{1+1/2,1}}\left(\Phi^m_{2,1} - \Phi^m_{1,1}\right)$$
$$+T^m_{y_{1,1+1/2}}\left(\Phi^m_{1,2} - \Phi^m_{1,1}\right) + q^m_{sc_{1,1}} = \frac{V_{b_{1,1}}}{\alpha_c \Delta t}\left[\left(\frac{\phi}{B}\right)^{n+1}_{1,1} - \left(\frac{\phi}{B}\right)^{n}_{1,1}\right] \tag{4.11a}$$

The first term on the LHS of Eq. (4.11a) represents fluid flow rate across the reservoir south boundary (b_S). This term can be replaced with the flow rate of a fictitious well $\left(q^m_{sc_{b_S},(1,1)}\right)$ that transfers fluid across the reservoir south boundary to gridblock (1,1); that is,

$$q^m_{sc_{b_S},(1,1)} = T^m_{y_{1,1-1/2}}\left(\Phi^m_{1,0} - \Phi^m_{1,1}\right) \tag{4.12}$$

The second term on the LHS of Eq. (4.11a) represents fluid flow rate across the reservoir west boundary (b_W). This term can also be replaced with the flow rate of another fictitious well $\left(q^m_{sc_{b_W},(1,1)}\right)$ that transfers fluid across the reservoir west boundary to gridblock (1,1); that is,

$$q^m_{sc_{b_W},(1,1)} = T^m_{x_{1-1/2,1}}\left(\Phi^m_{0,1} - \Phi^m_{1,1}\right) \tag{4.13}$$

Substitution of Eqs. (4.12) and (4.13) into Eq. (4.11a) yields

$$q^m_{sc_{b_S},(1,1)} + q^m_{sc_{b_W},(1,1)} + T^m_{x_{1+1/2,1}}\left(\Phi^m_{2,1} - \Phi^m_{1,1}\right)$$
$$+T^m_{y_{1,1+1/2}}\left(\Phi^m_{1,2} - \Phi^m_{1,1}\right) + q^m_{sc_{1,1}} = \frac{V_{b_{1,1}}}{\alpha_c \Delta t}\left[\left(\frac{\phi}{B}\right)^{n+1}_{1,1} - \left(\frac{\phi}{B}\right)^{n}_{1,1}\right] \tag{4.11b}$$

The earlier flow equation can also be obtained from Eq. (4.2b) for $n \equiv (1,1)$, $\psi_{1,1} = \{(2,1),(1,2)\}$, and $\xi_{1,1} = \{b_S, b_W\}$, and by observing that $\sum\limits_{l \in \xi_{1,1}} q^m_{sc_{l,(1,1)}} = q^m_{sc_{b_S,(1,1)}} + q^m_{sc_{b_W,(1,1)}}$, $T^m_{(2,1),(1,1)} = T^m_{x_{1+1/2,1}}$, and $T^m_{(1,2),(1,1)} = T^m_{y_{1,1+1/2}}$.

The following example demonstrates the use of the general equation, Eq. (4.2a), to write the flow equations for interior gridblocks in a 1-D reservoir.

Example 4.2 For the 1-D reservoir described in Example 4.1, write the flow equations for interior gridblocks 2, 3, and 4.

Solution

The flow equation for gridblock n, in a 1-D horizontal reservoir, is obtained by neglecting the gravity term in Eq. (4.2a), yielding

$$\sum_{l \in \psi_n} T^m_{l,n}\left(p^m_l - p^m_n\right) + \sum_{l \in \xi_n} q^m_{sc_{l,n}} + q^m_{sc_n} = \frac{V_{b_n}}{\alpha_c \Delta t}\left[\left(\frac{\phi}{B}\right)^{n+1}_n - \left(\frac{\phi}{B}\right)^n_n\right] \tag{4.14}$$

For interior gridblocks n, $\psi_n = \{n-1, n+1\}$ and $\xi_n = \{\}$. Therefore, $\sum\limits_{l \in \xi_n} q^m_{sc_{l,n}} = 0$. The gridblocks in this problem are equally spaced; therefore, $T^m_{l,n} = T^m_{x_{n \mp 1/2}} = T^m_x$, where

$$T^m_x = \beta_c \frac{k_x A_x}{\mu B \Delta x} = 0.001127 \times \frac{15 \times (1200 \times 75)}{10 \times 1 \times 1000} = 0.1521\,\text{STB/D-psi} \tag{4.15}$$

For gridblock 2, $n=2$, $\psi_2 = \{1,3\}$, $\xi_2 = \{\}$, $\sum\limits_{l \in \xi_2} q^m_{sc_{l,2}} = 0$, and $q^m_{sc_2} = 0$. Therefore, Eq. (4.14) becomes

$$(0.1521)\left(p^m_1 - p^m_2\right) + (0.1521)\left(p^m_3 - p^m_2\right) = \frac{V_{b_2}}{\alpha_c \Delta t}\left[\left(\frac{\phi}{B}\right)^{n+1}_2 - \left(\frac{\phi}{B}\right)^n_2\right] \tag{4.16}$$

For gridblock 3, $n=3$, $\psi_3 = \{2,4\}$, $\xi_3 = \{\}$, $\sum\limits_{l \in \xi_3} q^m_{sc_{l,3}} = 0$, and $q^m_{sc_3} = 0$. Therefore, Eq. (4.14) becomes

$$(0.1521)\left(p^m_2 - p^m_3\right) + (0.1521)\left(p^m_4 - p^m_3\right) = \frac{V_{b_3}}{\alpha_c \Delta t}\left[\left(\frac{\phi}{B}\right)^{n+1}_3 - \left(\frac{\phi}{B}\right)^n_3\right] \tag{4.17}$$

For gridblock 4, $n=4$, $\psi_4 = \{3,5\}$, $\xi_4 = \{\}$, $\sum\limits_{l \in \xi_4} q^m_{sc_{l,4}} = 0$, and $q^m_{sc_4} = -150$ STB/D. Therefore, Eq. (4.14) becomes

$$(0.1521)\left(p^m_3 - p^m_4\right) + (0.1521)\left(p^m_5 - p^m_4\right) - 150 = \frac{V_{b_4}}{\alpha_c \Delta t}\left[\left(\frac{\phi}{B}\right)^{n+1}_4 - \left(\frac{\phi}{B}\right)^n_4\right] \tag{4.18}$$

4.4 Treatment of boundary conditions

A reservoir boundary can be subject to one of four conditions: (1) no-flow boundary, (2) constant-flow boundary, (3) constant pressure gradient boundary, and (4) constant pressure boundary. In fact, the first three boundary conditions reduce to the specified pressure gradient condition (the Neumann boundary condition), and the fourth boundary condition is the Dirichlet boundary condition (constant pressure value). This section presents in detail the treatment of boundary conditions for 1-D flow in the x-direction, followed by generalizations for the treatment of boundary conditions in multidimensional reservoirs. In this section, we refer to reservoir boundaries as left and right boundaries because the lower, south, and west boundaries can be considered left boundaries, while the east, north, and upper boundaries can be considered right boundaries in 3-D reservoirs. The flow rate of the fictitious well ($q^m_{sc_{b,bB}}$) reflects fluid transfer between the boundary block (bB) (e.g., gridblock 1 for the reservoir left boundary and gridblock n_x for the reservoir right boundary in Fig. 4.1) and the reservoir boundary itself (b), or between the boundary block and the block next to the reservoir boundary that falls outside the reservoir (bB^{**}) (e.g., gridblock 0 for the reservoir left boundary and gridblock n_x+1 for the reservoir right boundary in Fig. 4.6). Eq. (4.4b) expresses the flow equation for boundary gridblock 1, which falls on the reservoir left boundary, and Eq. (4.6b) expresses the equation for boundary gridblock n_x, which falls on the reservoir right boundary.

For boundary gridblock 1, which falls on the reservoir left boundary, the rate of fictitious well is expressed by Eq. (4.5a), which states

$$q^m_{sc_{b_W,1}} = T^m_{x_{1-1/2}}\left(\Phi^m_0 - \Phi^m_1\right) \tag{4.5a}$$

Since there is no geologic control for areas outside the reservoir, including aquifers, it is not uncommon to assign reservoir rock properties to those areas in the neighborhood of the reservoir under consideration. Therefore, we use the reflection technique at left boundary of the reservoir, shown in Fig. 4.6, with regard to transmissibility only (i.e., $T^m_{0,b_W}=T^m_{b_W,1}$) and evaluate $T^m_{x_{1-1/2}}$ in terms of the transmissibilities between gridblock 0 and reservoir west boundary b_W and between gridblock 1 and reservoir west boundary b_W. The result is:

$$\begin{aligned} T^m_{x_{1/2}} &= \left[\beta_c \frac{k_x A_x}{\mu B \Delta x}\right]^m_{1/2} = \left[\beta_c \frac{k_x A_x}{\mu B \Delta x_1}\right]^m_1 = {}^1\!/_2\left[\beta_c \frac{k_x A_x}{\mu B (\Delta x_1/2)}\right]^m_1 = {}^1\!/_2 T^m_{b_W,1} \\ &= {}^1\!/_2 T^m_{0,b_W} \end{aligned} \tag{4.19a}$$

or

$$T^m_{0,b_W} = T^m_{b_W,1} = 2T^m_{x_{1/2}} \tag{4.19b}$$

Substitution of Eq. (4.19b) into Eq. (4.5a) gives

$$q^m_{sc_{b_W,1}} = {}^1\!/_2 T^m_{b_W,1}\left(\Phi^m_0 - \Phi^m_1\right) \tag{4.5b}$$

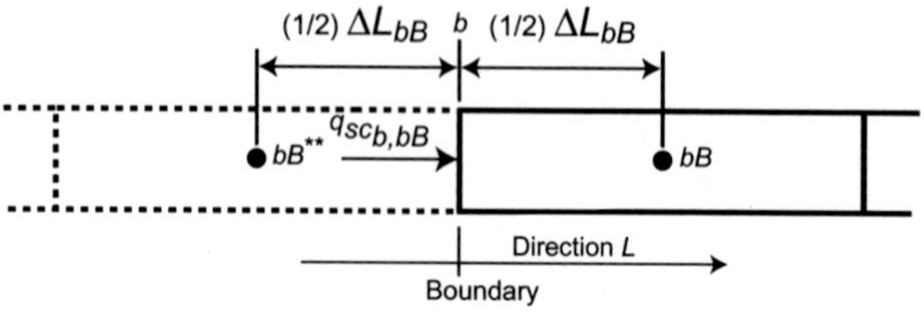

FIG. 4.7 Definition of terminology used in Eq. (4.20).

Similarly, for boundary gridblock n_x, which falls on the reservoir right boundary,

$$q^m_{sc_{b_E,n_x}} = T^m_{x_{n_x+1/2}}\left(\Phi^m_{n_x+1} - \Phi^m_{n_x}\right) \tag{4.7a}$$

and

$$q^m_{sc_{b_E,n_x}} = {}^1/_2 T^m_{b_E,n_x}\left(\Phi^m_{n_x+1} - \Phi^m_{n_x}\right) \tag{4.7b}$$

In other words, the flow term between a boundary gridblock and the gridblock located immediately on the other side of the reservoir boundary can be replaced by a fictitious well having a flow rate $q^m_{sc_{b,bB}}$. The general form for $q^m_{sc_{b,bB}}$ is

$$q^m_{sc_{b,bB}} = T^m_{bB,bB^{**}}\left(\Phi^m_{bB^{**}} - \Phi^m_{bB}\right) \tag{4.20a}$$

or

$$q^m_{sc_{b,bB}} = {}^1/_2 T^m_{b,bB}\left(\Phi^m_{bB^{**}} - \Phi^m_{bB}\right) \tag{4.20b}$$

where, as shown in Fig. 4.7, $q^m_{sc_{b,bB}}$ = flow rate of a fictitious well representing flow across reservoir boundary (b) between boundary block (bB) and the block that is exterior to the reservoir and located immediately next to reservoir boundary (bB^{**}), $T^m_{bB,\,bB^{**}}$ = transmissibility between boundary gridblock bB and gridblock bB^{**}, and $T^m_{b,bB}$ = transmissibility between reservoir boundary (b) and boundary gridblock bB.

In the following sections, we derive expressions for $q^m_{sc_{b,bB}}$ under various boundary conditions for a block-centered grid in Cartesian coordinates. We stress that this rate must produce the same effects as the specified boundary condition. In Cartesian coordinates, real wells have radial flow, and fictitious wells have linear flow, whereas in radial-cylindrical coordinates in single-well simulation both real wells and fictitious wells have radial flow. Therefore, in single-well simulation, (1) the equations for the flow rate of real wells presented in Sections 6.2.2 and 6.3.2 can be used to estimate the flow rate of fictitious wells representing boundary conditions in the radial direction only, (2) the flow rate equations of fictitious wells in the z-direction are similar to those presented next in this section because flow in the vertical direction is linear, and (3) there are no reservoir boundaries and hence no fictitious wells in the θ-direction. The flow

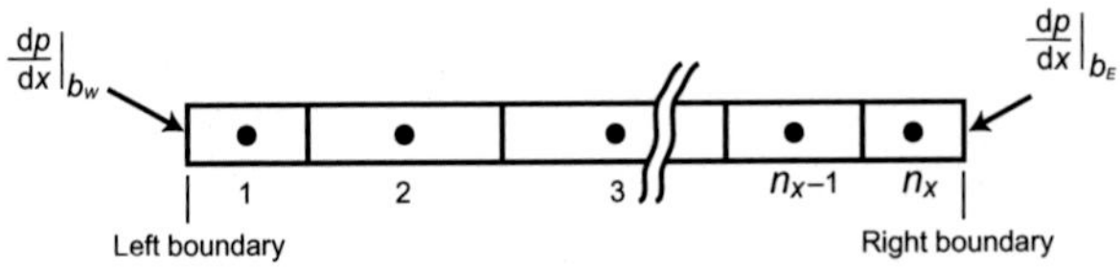

FIG. 4.8 Specified pressure gradient condition at reservoir boundaries in a block-centered grid.

rate of a fictitious well is positive for fluid gain (injection) or negative for fluid loss (production) across a reservoir boundary.

4.4.1 Specified pressure gradient boundary condition

For boundary gridblock 1 shown in Fig. 4.8, which falls on the left boundary of the reservoir, Eq. (4.20a) reduces to Eq. (4.5a) that can be rewritten as:

$$q^m_{sc_{bW},1} = T^m_{x_{1/2}}\left(\Phi^m_0 - \Phi^m_1\right) = \left[\beta_c \frac{k_x A_x}{\mu B \Delta x}\right]^m_{1/2}\left(\Phi^m_0 - \Phi^m_1\right) = \left[\beta_c \frac{k_x A_x}{\mu B}\right]^m_{1/2}\frac{\left(\Phi^m_0 - \Phi^m_1\right)}{\Delta x_{1/2}}$$
$$\cong \left[\beta_c \frac{k_x A_x}{\mu B}\right]^m_{1/2}\left[-\frac{\partial \Phi}{\partial x}\bigg|^m_{bW}\right] = -\left[\beta_c \frac{k_x A_x}{\mu B}\right]^m_{1/2}\frac{\partial \Phi}{\partial x}\bigg|^m_{bW} = -\left[\beta_c \frac{k_x A_x}{\mu B}\right]^m_{1}\frac{\partial \Phi}{\partial x}\bigg|^m_{bW} \tag{4.21}$$

Note that in arriving at the above equation, we used the reflection technique shown in Fig. 4.6 with respect to transmissibility and used the central-difference approximation of first-order derivative of potential.

Similarly for gridblock n_x, which falls on the reservoir right boundary, Eq. (4.20a) reduces to Eq. (4.7a) that can be rewritten as

$$q^m_{sc_{bE},n_x} = T^m_{x_{n_x+1/2}}\left(\Phi^m_{n_x+1} - \Phi^m_{n_x}\right) = \left[\beta_c \frac{k_x A_x}{\mu B \Delta x}\right]^m_{n_x+1/2}\left(\Phi^m_{n_x+1} - \Phi^m_{n_x}\right)$$
$$= \left[\beta_c \frac{k_x A_x}{\mu B}\right]^m_{n_x+1/2}\frac{\left(\Phi^m_{n_x+1} - \Phi^m_{n_x}\right)}{\Delta x_{n_x+1/2}} \cong \left[\beta_c \frac{k_x A_x}{\mu B}\right]^m_{n_x+1/2}\left[\frac{\partial \Phi}{\partial x}\bigg|^m_{bE}\right]$$
$$= \left[\beta_c \frac{k_x A_x}{\mu B}\right]^m_{n_x+1/2}\frac{\partial \Phi}{\partial x}\bigg|^m_{bE} = \left[\beta_c \frac{k_x A_x}{\mu B}\right]^m_{n_x}\frac{\partial \Phi}{\partial x}\bigg|^m_{bE} \tag{4.22}$$

Here again, we used the reflection technique shown in Fig. 4.6 with respect to transmissibility and used the central-difference approximation of first-order derivative of potential.

In general, for specified pressure gradient at the reservoir left (lower, south, or west) boundary,

$$q^m_{sc_{b,bB}} \cong -\left[\beta_c \frac{k_l A_l}{\mu B}\right]^m_{bB}\frac{\partial \Phi}{\partial l}\bigg|^m_{b} \tag{4.23a}$$

or after combining with Eq. (2.10),

$$q^m_{sc_{b,bB}} \cong -\left[\beta_c \frac{k_l A_l}{\mu B}\right]^m_{bB} \left[\frac{\partial p}{\partial l}\bigg|^m_b - \gamma^m_{bB} \frac{\partial Z}{\partial l}\bigg|_b\right] \tag{4.23b}$$

and at the reservoir right (east, north, or upper) boundary,

$$q^m_{sc_{b,bB}} \cong \left[\beta_c \frac{k_l A_l}{\mu B}\right]^m_{bB} \frac{\partial \Phi}{\partial l}\bigg|^m_b \tag{4.24a}$$

or after combining with Eq. (2.10),

$$q^m_{sc_{b,bB}} \cong \left[\beta_c \frac{k_l A_l}{\mu B}\right]^m_{bB} \left[\frac{\partial p}{\partial l}\bigg|^m_b - \gamma^m_{bB} \frac{\partial Z}{\partial l}\bigg|_b\right] \tag{4.24b}$$

where l is the direction normal to the boundary.

4.4.2 Specified flow rate boundary condition

The specified flow rate boundary condition arises when the reservoir near the boundary has higher or lower potential than that of a neighboring reservoir or aquifer. In this case, fluids move across the reservoir boundary. Methods such as water influx calculations and classical material balance in reservoir engineering can be used to estimate fluid flow rate, which we term here as specified (q_{spsc}). Therefore, Eq. (4.5a) for boundary gridblock 1 becomes

$$q^m_{sc_{b_W,1}} = T^m_{x_{1/2}}\left(\Phi^m_0 - \Phi^m_1\right) = q_{spsc} \tag{4.25}$$

and Eq. (4.7a) for boundary gridblock n_x becomes

$$q^m_{sc_{b_E,n_x}} = T^m_{x_{n_x+1/2}}\left(\Phi^m_{n_x+1} - \Phi^m_{n_x}\right) = q_{spsc} \tag{4.26}$$

In general, for a specified flow rate boundary condition, Eq. (4.20a) becomes

$$q^m_{sc_{b,bB}} = q_{spsc} \tag{4.27}$$

In multidimensional flow with q_{spsc} specified for the whole reservoir boundary, $q^m_{sc_{b,bB}}$ for each boundary gridblock is obtained by prorating q_{spsc} among all boundary gridblocks that share that boundary; that is,

$$q^m_{sc_{b,bB}} = \frac{T^m_{b,bB}}{\sum\limits_{l \in \psi_b} T^m_{b,l}} q_{spsc} \tag{4.28}$$

where ψ_b is the set that contains all boundary gridblocks that share the reservoir boundary in question; $T_{b,l}$ = transmissibility between the reservoir boundary and boundary gridblock l, which is a member of the set ψ_b, and $T^m_{b,bB}$ is defined as

$$T^m_{b,bB} = \left[\beta_c \frac{k_l A_l}{\mu B(\Delta l/2)}\right]^m_{bB} \tag{4.29}$$

The length l and subscript l in Eq. (4.29) are replaced with x, y, or z depending on the boundary face of boundary block. It should be mentioned that Eq. (4.28) incorporates the assumption that the potential drops across the reservoir boundary for all gridblocks sharing that boundary are equal.

4.4.3 No-flow boundary condition

The no-flow boundary condition results from vanishing permeability at a reservoir boundary (e.g., $T^m_{x_{1/2}}=0$ for the left boundary of gridblock 1 and $T^m_{x_{n_x+1/2}}=0$ for the right boundary of gridblock n_x) or because of symmetry about the reservoir boundary (e.g., $\Phi^m_0=\Phi^m_1$ for gridblock 1 and $\Phi^m_{n_x}=\Phi^m_{n_x+1}$ for gridblock n_x). In either case, Eq. (4.5a) for boundary gridblock 1 reduces to

$$q^m_{sc_{b_W,1}}=T^m_{x_{1/2}}\left(\Phi^m_0-\Phi^m_1\right)=0\left(\Phi^m_0-\Phi^m_1\right)=T^m_{x_{1/2}}(0)=0 \tag{4.30}$$

and Eq. (4.7a) for boundary gridblock n_x reduces to

$$q^m_{sc_{b_E,n_x}}=T^m_{x_{n_x+1/2}}\left(\Phi^m_{n_x+1}-\Phi^m_{n_x}\right)=0\left(\Phi^m_{n_x+1}-\Phi^m_{n_x}\right)=T^m_{x_{n_x+1/2}}(0)=0 \tag{4.31}$$

In general, for a reservoir no-flow boundary, Eq. (4.20a) becomes

$$q^m_{sc_{b,bB}}=0 \tag{4.32}$$

For multidimensional flow, $q^m_{sc_{b,bB}}$ is set to zero, as Eq. (4.32) implies, for each boundary gridblock that falls on a no-flow boundary in the x-, y-, or z-direction.

4.4.4 Specified boundary pressure condition

This condition arises when the reservoir is in communication with a strong water aquifer or when wells on the other side of the reservoir boundary operate to maintain voidage replacement and as a result keep boundary pressure (p_b) constant. Fig. 4.9 shows this boundary condition at the reservoir left and right boundaries.

Eq. (4.5a) for boundary gridblock 1 can be rewritten as

$$\begin{aligned}q^m_{sc_{b_W,1}}&=T^m_{x_{1/2}}\left(\Phi^m_0-\Phi^m_1\right)=T^m_{x_{1/2}}\left[\Phi^m_0-\Phi_{b_W}+\Phi_{b_W}-\Phi^m_1\right]\\&=T^m_{x_{1/2}}\left[\left(\Phi^m_0-\Phi_{b_W}\right)+\left(\Phi_{b_W}-\Phi^m_1\right)\right]=T^m_{x_{1/2}}\left(\Phi^m_0-\Phi_{b_W}\right)+T^m_{x_{1/2}}\left(\Phi_{b_W}-\Phi^m_1\right)\end{aligned} \tag{4.33}$$

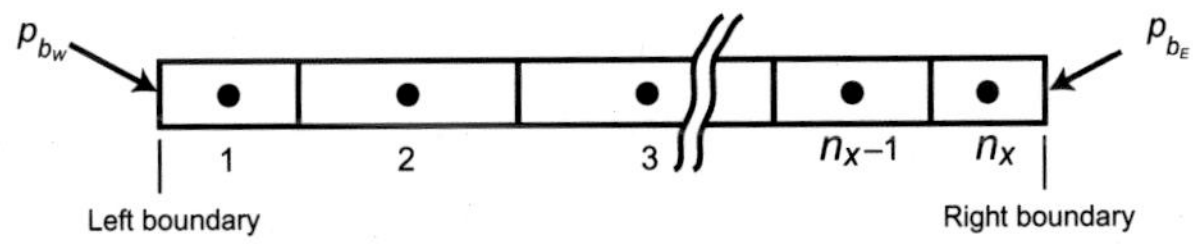

FIG. 4.9 Specified pressure condition at reservoir boundaries in a block-centered grid.

Combining the above equation and Eq. (4.19b) yields:

$$q^m_{sc_{bW},1} = {}^1/_2 T^m_{0,bW}\left(\Phi^m_0 - \Phi_{bW}\right) + {}^1/_2 T^m_{bW,1}\left(\Phi_{bW} - \Phi^m_1\right) \quad (4.34)$$

To keep the potential at the left boundary of gridblock 1 constant, the fluid leaving the reservoir boundary to one side (point 1) has to be equal to the fluid entering the reservoir boundary from the other side (point 0); see Fig. 4.6. That is,

$$T^m_{0,bW}\left(\Phi^m_0 - \Phi_{bW}\right) = T^m_{bW,1}\left(\Phi_{bW} - \Phi^m_1\right) \quad (4.35)$$

Substitution of Eq. (4.35) into Eq. (4.34) and making use of Eq. (4.19b) yield:

$$q^m_{sc_{bW},1} = T^m_{bW,1}\left(\Phi_{bW} - \Phi^m_1\right) \quad (4.36)$$

Keeping the potential at any point constant implies the pressure is kept constant because potential minus pressure is constant as required by Eq. (2.11).

In general, for a specified pressure boundary, Eq. (4.20a) becomes

$$q^m_{sc_{b,bB}} = T^m_{b,bB}\left(\Phi_b - \Phi^m_{bB}\right) \quad (4.37a)$$

Eq. (4.37a) can be rewritten in terms of pressure as

$$q^m_{sc_{b,bB}} = T^m_{b,bB}\left[\left(p_b - p^m_{bB}\right) - \gamma^m_{b,bB}(Z_b - Z_{bB})\right] \quad (4.37b)$$

where $\gamma^m_{b,bB}$ is nothing but fluid gravity in boundary block bB and $T^m_{b,bB}$ = transmissibility between the reservoir boundary and the point representing the boundary gridblock and is given by Eq. (4.29):

$$T^m_{b,bB} = \left[\beta_c \frac{k_l A_l}{\mu B(\Delta l/2)}\right]^m_{bB} \quad (4.29)$$

Combining Eqs. (4.29) and (4.37b) gives

$$q^m_{sc_{b,bB}} = \left[\beta_c \frac{k_l A_l}{\mu B(\Delta l/2)}\right]^m_{bB} \left[\left(p_b - p^m_{bB}\right) - \gamma^m_{b,bB}(Z_b - Z_{bB})\right] \quad (4.37c)$$

Substitution of Eq. (4.37c) in the flow equation for boundary gridblock bB maintains a second-order correct finite-difference flow equation in the mathematical approach (see Exercise 4.7). Abou-Kassem et al. (2007) proved that such a treatment of this boundary condition is second-order correct. In multidimensional flow, $q^m_{sc_{b,bB}}$ for a boundary gridblock falling on a specified pressure boundary in the x-, y-, or z-direction is estimated using Eq. (4.37c) with the corresponding x, y, or z replacing l.

4.4.5 Specified boundary block pressure

This condition arises if one makes the mathematical assumption that the boundary pressure is displaced half a block to coincide with the center of the boundary gridblock; that is, $p_1 \cong p_{b_W}$ or $p_{n_x} \cong p_{b_E}$. This approximation is first-order correct and produces results that are less accurate than the treatment that uses Eq. (4.37c). Currently available books on reservoir simulation use this treatment to deal with the specified boundary pressure condition. Following this treatment, the problem reduces to finding the pressure of other gridblocks in the reservoir as demonstrated in Example 7.2 in Chapter 7.

The following examples demonstrate the use of the general equation, Eq. (4.2a), and the appropriate expressions for $q_{sc_{b,bB}}^m$ to write the flow equations for boundary gridblocks in 1-D and 2-D reservoirs that are subject to various boundary conditions.

Example 4.3 For the 1D reservoir described in Example 4.1, the reservoir left boundary is kept at a constant pressure of 5000 psia, and the reservoir right boundary is a no-flow (sealed) boundary as shown in Fig. 4.10. Write the flow equations for boundary gridblocks 1 and 5.

Solution

The flow equation for gridblock n in a 1-D horizontal reservoir is obtained from Eq. (4.2a) by neglecting the gravity term, resulting in

$$\sum_{l \in \psi_n} T_{l,n}^m \left(p_l^m - p_n^m\right) + \sum_{l \in \xi_n} q_{sc_{l,n}}^m + q_{sc_n}^m = \frac{V_{b_n}}{\alpha_c \Delta t}\left[\left(\frac{\phi}{B}\right)_n^{n+1} - \left(\frac{\phi}{B}\right)_n^n\right] \tag{4.14}$$

From Example 4.2, $T_{l,n}^m = T_x^m = 0.1521$ STB/D-psi.

For boundary gridblock 1, $n=1$, $\psi_1=\{2\}$, $\xi_1=\{b_W\}$, $\sum_{l \in \xi_1} q_{sc_{l,1}}^m = q_{sc_{b_W,1}}^m$, and $q_{sc_1}^m = 0$.

Therefore, Eq. (4.14) becomes

$$0.1521\left(p_2^m - p_1^m\right) + q_{sc_{b_W,1}}^m = \frac{V_{b_1}}{\alpha_c \Delta t}\left[\left(\frac{\phi}{B}\right)_1^{n+1} - \left(\frac{\phi}{B}\right)_1^n\right] \tag{4.38}$$

where the rate of flow across the reservoir left boundary is given by Eq. (4.37c):

$$\begin{aligned} q_{sc_{b_W,1}}^m &= \left[\beta_c \frac{k_x A_x}{\mu B(\Delta x/2)}\right]_1^m \left[\left(p_{b_W} - p_1^m\right) - \gamma_{b_W,1}\left(Z_{b_W} - Z_1\right)\right] \\ &= 0.001127 \times \frac{15 \times (1200 \times 75)}{10 \times 1 \times (1000/2)}\left[\left(5000 - p_1^m\right) - \gamma_{b_W,1} \times 0\right] \end{aligned} \tag{4.39}$$

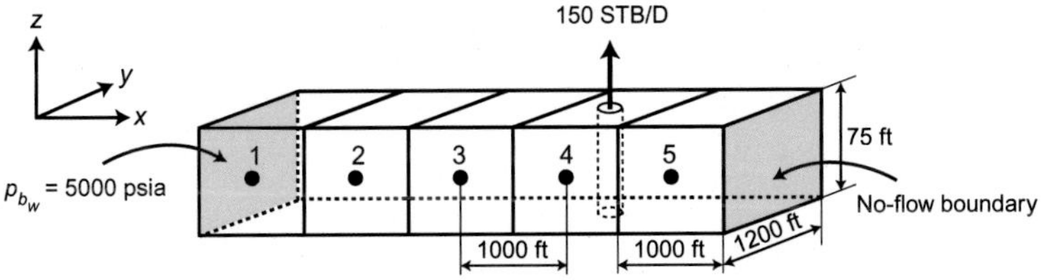

FIG. 4.10 Discretized 1-D reservoir in Example 4.3.

or

$$q^m_{sc_{b_W,1}} = (0.3043)(5000 - p^m_1) \tag{4.40}$$

Substitution of Eq. (4.40) into Eq. (4.38) results in the flow equation for boundary gridblock 1:

$$(0.1521)(p^m_2 - p^m_1) + (0.3043)(5000 - p^m_1) = \frac{V_{b_1}}{\alpha_c \Delta t}\left[\left(\frac{\phi}{B}\right)^{n+1}_1 - \left(\frac{\phi}{B}\right)^n_1\right] \tag{4.41}$$

For boundary gridblock 5, $n=5$, $\psi_5=\{4\}$, $\xi_5=\{b_E\}$, $\sum_{l\in\xi_5} q^m_{sc_{l,5}} = q^m_{sc_{b_E,5}}$, and $q^m_{sc_5}=0$. Therefore, Eq. (4.14) becomes

$$(0.1521)(p^m_4 - p^m_5) + q^m_{sc_{b_E,5}} = \frac{V_{b_5}}{\alpha_c \Delta t}\left[\left(\frac{\phi}{B}\right)^{n+1}_5 - \left(\frac{\phi}{B}\right)^n_5\right] \tag{4.42}$$

where the flow rate across the reservoir right boundary (no-flow boundary) is given by Eq. (4.32). For the reservoir right boundary, $b \equiv b_E$, $bB \equiv 5$, and

$$q^m_{sc_{b_E,5}} = 0 \tag{4.43}$$

Substitution into Eq. (4.42) results in the flow equation for boundary gridblock 5:

$$(0.1521)(p^m_4 - p^m_5) = \frac{V_{b_5}}{\alpha_c \Delta t}\left[\left(\frac{\phi}{B}\right)^{n+1}_5 - \left(\frac{\phi}{B}\right)^n_5\right] \tag{4.44}$$

Example 4.4 For the 1-D reservoir described in Example 4.1, the reservoir left boundary is kept at a constant pressure gradient of −0.1 psi/ft and the reservoir right boundary is supplied with fluid at a rate of 50 STB/D as shown in Fig. 4.11. Write the flow equations for boundary gridblocks 1 and 5.

Solution

The flow equation for gridblock n in a 1-D horizontal reservoir is obtained from Eq. (4.2a) by neglecting the gravity term, resulting in

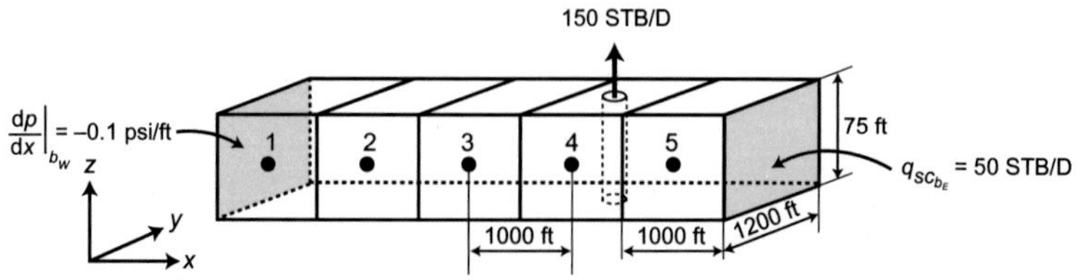

FIG. 4.11 Discretized 1-D reservoir in Example 4.4.

$$\sum_{l\in\psi_n} T_{l,n}^m \left(p_l^m - p_n^m\right) + \sum_{l\in\xi_n} q_{sc_{l,n}}^m + q_{sc_n}^m = \frac{V_{b_n}}{\alpha_c \Delta t}\left[\left(\frac{\phi}{B}\right)_n^{n+1} - \left(\frac{\phi}{B}\right)_n^n\right] \tag{4.14}$$

From Example 4.2, $T_{l,n}^m = T_x^m = 0.1521$ STB/D-psi.

For boundary gridblock 1, $n=1$, $\psi_1=\{2\}$, $\xi_1=\{b_W\}$, $\sum_{l\in\xi_1} q_{sc_{l,1}}^m = q_{sc_{b_W,1}}^m$, and $q_{sc_1}^m=0$. Therefore, Eq. (4.14) becomes

$$(0.1521)\left(p_2^m - p_1^m\right) + q_{sc_{b_W,1}}^m = \frac{V_{b_1}}{\alpha_c \Delta t}\left[\left(\frac{\phi}{B}\right)_1^{n+1} - \left(\frac{\phi}{B}\right)_1^n\right] \tag{4.38}$$

where the flow rate of a fictitious well for the specified pressure gradient at the reservoir left boundary is estimated using Eq. (4.23b):

$$q_{sc_{b_W,1}}^m = -\left[\beta_c \frac{k_x A_x}{\mu B}\right]_1^m \left[\left.\frac{\partial p}{\partial x}\right|_{b_W} - \gamma_1^m \left.\frac{\partial Z}{\partial x}\right|_{b_W}\right]$$

$$= -\left[0.001127 \times \frac{15 \times (1200 \times 75)}{10 \times 1}\right][-0.1 - 0] = -152.145 \times (-0.1) \tag{4.45}$$

or

$$q_{sc_{b_W,1}}^m = 15.2145 \tag{4.46}$$

Substitution of Eq. (4.46) into Eq. (4.38) results in the flow equation for boundary gridblock 1:

$$(0.1521)\left(p_2^m - p_1^m\right) + 15.2145 = \frac{V_{b_1}}{\alpha_c \Delta t}\left[\left(\frac{\phi}{B}\right)_1^{n+1} - \left(\frac{\phi}{B}\right)_1^n\right] \tag{4.47}$$

For boundary gridblock 5, $n=5$, $\psi_5=\{4\}$, $\xi_5=\{b_E\}$, $\sum_{l\in\xi_5} q_{sc_{l,5}}^m = q_{sc_{b_E,5}}^m$, and $q_{sc_5}^m=0$. Therefore, Eq. (4.14) becomes

$$(0.1521)\left(p_4^m - p_5^m\right) + q_{sc_{b_E,5}}^m = \frac{V_{b_5}}{\alpha_c \Delta t}\left[\left(\frac{\phi}{B}\right)_5^{n+1} - \left(\frac{\phi}{B}\right)_5^n\right] \tag{4.42}$$

where the flow rate of a fictitious well for a specified rate boundary is estimated using Eq. (4.27); that is,

$$q_{sc_{b_E,5}}^m = 50\,\text{STB/D} \tag{4.48}$$

Substitution of Eq. (4.48) into Eq. (4.42) results in the flow equation for boundary gridblock 5:

$$(0.1521)\left(p_4^m - p_5^m\right) + 50 = \frac{V_{b_5}}{\alpha_c \Delta t}\left[\left(\frac{\phi}{B}\right)_5^{n+1} - \left(\frac{\phi}{B}\right)_5^n\right] \tag{4.49}$$

Example 4.5 Consider single-phase fluid flow in the 2-D horizontal reservoir shown in Fig. 4.12. A well located in gridblock 7 produces at a rate of 4000 STB/D. All gridblocks have $\Delta x = 250$ ft, $\Delta y = 300$ ft, $h = 100$ ft, $k_x = 270$ md, and $k_y = 220$ md. The FVF and viscosity of the flowing fluid are 1.0 RB/STB and 2 cP, respectively. The reservoir south boundary is maintained at 3000 psia, the reservoir west boundary is sealed off to flow, the reservoir east boundary is kept at a constant pressure gradient of 0.1 psi/ft, and the reservoir loses fluid across its north boundary at a rate of 500 STB/D. Write the flow equations for boundary gridblocks 2, 5, 8, and 11.

Solution

The general flow equation for a 2-D horizontal reservoir is obtained from Eq. (4.2a) by neglecting the gravity term, resulting in

$$\sum_{l \in \psi_n} T_{l,n}^m \left(p_l^m - p_n^m\right) + \sum_{l \in \xi_n} q_{sc_{l,n}}^m + q_{sc_n}^m = \frac{V_{b_n}}{\alpha_c \Delta t}\left[\left(\frac{\phi}{B}\right)_n^{n+1} - \left(\frac{\phi}{B}\right)_n^n\right] \quad (4.14)$$

Note that $\Delta x = 250$ ft, $\Delta y = 300$ ft, k_x, k_y, μ, and B are constant. Therefore,

$$T_x^m = \beta_c \frac{k_x A_x}{\mu B \Delta x} = 0.001127 \times \frac{270 \times (300 \times 100)}{2 \times 1 \times 250} = 18.2574 \text{ STB/D-psi} \quad (4.50)$$

and

$$T_y^m = \beta_c \frac{k_y A_y}{\mu B \Delta y} = 0.001127 \times \frac{220 \times (250 \times 100)}{2 \times 1 \times 300} = 10.3308 \text{ STB/D-psi} \quad (4.51)$$

For boundary gridblock 2, $n = 2$, $\psi_2 = \{1, 3, 6\}$, $\xi_2 = \{b_S\}$, and $q_{sc_2}^m = 0$. $\sum_{l \in \xi_2} q_{sc_{l,2}}^m = q_{sc_{b_S,2}}^m$, where $q_{sc_{b_S,2}}^m$ is obtained from Eq. (4.37c) after discarding the gravity term, resulting in

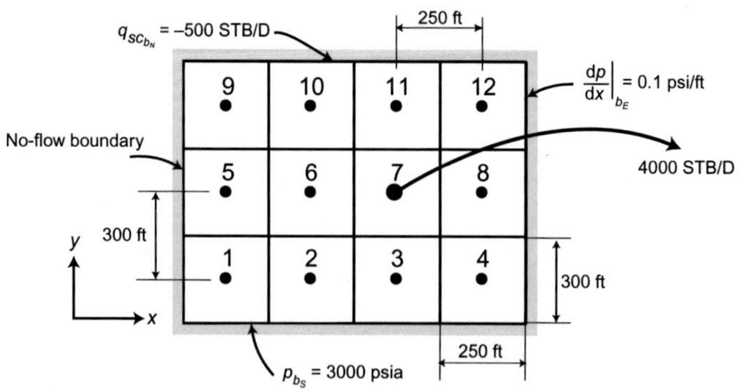

FIG. 4.12 Discretized 2-D reservoir in Examples 4.5 and 4.6.

$$q^m_{sc_{b_S},2} = \left[\beta_c \frac{k_y A_y}{\mu B(\Delta y/2)}\right]^m_2 (p_{b_S} - p^m_2)$$
$$= \left[0.001127 \times \frac{220 \times (250 \times 100)}{2 \times 1 \times (300/2)}\right] (3000 - p^m_2) \tag{4.52}$$

or

$$q^m_{sc_{b_S},2} = (20.6617)(3000 - p^m_2) \tag{4.53}$$

Substitution into Eq. (4.14) results in the flow equation for boundary gridblock 2,

$$(18.2574)(p^m_1 - p^m_2) + (18.2574)(p^m_3 - p^m_2) + (10.3308)(p^m_6 - p^m_2)$$
$$+(20.6617)(3000 - p^m_2) = \frac{V_{b_2}}{\alpha_c \Delta t}\left[\left(\frac{\phi}{B}\right)^{n+1}_2 - \left(\frac{\phi}{B}\right)^n_2\right] \tag{4.54}$$

For boundary gridblock 5, $n=5$, $\psi_5=\{1,6,9\}$, $\xi_5=\{b_W\}$, and $q^m_{sc_5}=0$. $\sum\limits_{l \in \xi_5} q^m_{sc_{l,5}} = q^m_{sc_{b_W},5}$, where $q^m_{sc_{b_W},5}$ is obtained from Eq. (4.32) for a no-flow boundary; that is, $q^m_{sc_{b_W},5}=0$.

Substitution into Eq. (4.14) results in the flow equation for boundary gridblock 5,

$$(10.3308)(p^m_1 - p^m_5) + (18.2574)(p^m_6 - p^m_5) + (10.3308)(p^m_9 - p^m_5) + 0$$
$$= \frac{V_{b_5}}{\alpha_c \Delta t}\left[\left(\frac{\phi}{B}\right)^{n+1}_5 - \left(\frac{\phi}{B}\right)^n_5\right] \tag{4.55}$$

For boundary gridblock 8, $n=8$, $\psi_8=\{4,7,12\}$, $\xi_8=\{b_E\}$, and $q^m_{sc_8}=0$. $\sum\limits_{l \in \xi_8} q^m_{sc_{l,8}} = q^m_{sc_{b_E},8}$, where $q^m_{sc_{b_E},8}$ is estimated using Eq. (4.24b) for the reservoir east boundary,

$$q^m_{sc_{b_E},8} = \left[\beta_c \frac{k_x A_x}{\mu B}\right]^m_8 \left[\frac{\partial p}{\partial x}\bigg|^m_{b_E} - \gamma^m_8 \frac{\partial Z}{\partial x}\bigg|_{b_E}\right] = \left[0.001127 \times \frac{270 \times (300 \times 100)}{2 \times 1}\right][0.1 - 0]$$
$$= 4564.35 \times (0.1) = 456.435 \tag{4.56}$$

Substitution into Eq. (4.14) results in the flow equation for boundary gridblock 8,

$$(10.3308)(p^m_4 - p^m_8) + (18.2574)(p^m_7 - p^m_8) + (10.3308)(p^m_{12} - p^m_8)$$
$$+456.435 = \frac{V_{b_8}}{\alpha_c \Delta t}\left[\left(\frac{\phi}{B}\right)^{n+1}_8 - \left(\frac{\phi}{B}\right)^n_8\right] \tag{4.57}$$

For boundary gridblock 11, $n = 11$, $\psi_{11} = \{7, 10, 12\}$, $\xi_{11} = \{b_N\}$, and $q^m_{sc_{11}} = 0$. $\sum_{l \in \xi_{11}} q^m_{sc_{l,11}} = q^m_{sc_{b_N,11}}$, where $q^m_{sc_{b_N,11}}$ is estimated using Eq. (4.28) because $q_{spsc} = -500$ STB/D is specified for the whole reservoir north boundary. This rate has to be prorated among all gridblocks sharing that boundary. Therefore,

$$q^m_{sc_{b_N,11}} = \frac{T^m_{b_N,11}}{\sum_{l \in \psi_{b_N}} T^m_{b_N,l}} q_{spsc} \tag{4.58}$$

where $\psi_{b_N} = \{9, 10, 11, 12\}$.

Using Eq. (4.29),

$$T^m_{b_N,l} = T^m_{b_N,11} = \left[\beta_c \frac{k_y A_y}{\mu B(\Delta y/2)}\right]^m_{11} = \left[0.001127 \times \frac{220 \times (250 \times 100)}{2 \times 1 \times (300/2)}\right] = 20.6616 \tag{4.59}$$

for all values of $l \in \psi_{b_N}$.

Substitution of Eq. (4.59) into Eq. (4.58) yields

$$q^m_{sc_{b_N,11}} = \frac{20.6616}{4 \times 20.6616} \times (-500) = -125 \text{ STB/D} \tag{4.60}$$

Substitution into Eq. (4.14) results in the flow equation for boundary gridblock 11:

$$(10.3308)\left(p^m_7 - p^m_{11}\right) + (18.2574)\left(p^m_{10} - p^m_{11}\right) + (18.2574)\left(p^m_{12} - p^m_{11}\right) - 125 = \frac{V_{b_{11}}}{\alpha_c \Delta t}\left[\left(\frac{\phi}{B}\right)^{n+1}_{11} - \left(\frac{\phi}{B}\right)^n_{11}\right] \tag{4.61}$$

Example 4.6 Consider single-phase fluid flow in the 2-D horizontal reservoir described in Example 4.5. Write the flow equations for gridblocks 1, 4, 9, and 12, where each gridblock falls on two reservoir boundaries.

Solution

The general flow equation for a 2-D horizontal reservoir is obtained from Eq. (4.2a) by neglecting the gravity term, resulting in Eq. (4.14):

$$\sum_{l \in \psi_n} T^m_{l,n}\left(p^m_l - p^m_n\right) + \sum_{l \in \xi_n} q^m_{sc_{l,n}} + q^m_{sc_n} = \frac{V_{b_n}}{\alpha_c \Delta t}\left[\left(\frac{\phi}{B}\right)^{n+1}_n - \left(\frac{\phi}{B}\right)^n_n\right] \tag{4.14}$$

The data necessary to write flow equations for any boundary gridblock were calculated in Example 4.5. The following is a summary:

$$T^m_x = 18.2574 \text{ STB/D-psi}$$

$$T^m_y = 10.3308 \text{ STB/D-psi}$$

$$q^m_{sc_{b_S,bB}} = (20.6617)\left(3000 - p^m_{bB}\right) \text{STB/D} \quad \text{for } bB = 1, 2, 3, 4 \tag{4.62}$$

$$q^m_{sc_{b_W},bB} = 0 \text{ STB/D} \quad \text{for } bB = 1,5,9$$

$$q^m_{sc_{b_E},bB} = 456.435 \text{ STB/D} \quad \text{for } bB = 4,8,12$$

and

$$q^m_{sc_{b_N},bB} = -125 \text{ STB/D} \quad \text{for } bB = 9,10,11,12$$

For boundary gridblock 1, $n=1$, $\psi_1=\{2,5\}$, $\xi_1=\{b_S,b_W\}$, $q^m_{sc_1}=0$, and

$$\sum_{l\in\xi_1} q^m_{sc_{l,1}} = q^m_{sc_{b_S},1} + q^m_{sc_{b_W},1} = (20.6617)\left(3000 - p^m_1\right) + 0$$
$$= (20.6617)\left(3000 - p^m_1\right) \text{ STB/D}$$

Substitution into Eq. (4.14) results in the flow equation for boundary grid-block 1,

$$(18.2574)\left(p^m_2 - p^m_1\right) + (10.3308)\left(p^m_5 - p^m_1\right) + (20.6617)\left(3000 - p^m_1\right)$$
$$= \frac{V_{b_1}}{\alpha_c \Delta t}\left[\left(\frac{\phi}{B}\right)^{n+1}_1 - \left(\frac{\phi}{B}\right)^n_1\right] \tag{4.63}$$

For boundary gridblock 4, $n=4$, $\psi_4=\{3,8\}$, $\xi_4=\{b_S,b_E\}$, $q^m_{sc_4}=0$, and

$$\sum_{l\in\xi_4} q^m_{sc_{l,4}} = q^m_{sc_{b_S},4} + q^m_{sc_{b_E},4} = (20.6617)\left(3000 - p^m_4\right) + 456.435 \text{ STB/D}$$

Substitution into Eq. (4.14) results in the flow equation for boundary grid-block 4,

$$(18.2574)\left(p^m_3 - p^m_4\right) + (10.3308)\left(p^m_8 - p^m_4\right) + (20.6617)\left(3000 - p^m_4\right) + 456.435$$
$$= \frac{V_{b_4}}{\alpha_c \Delta t}\left[\left(\frac{\phi}{B}\right)^{n+1}_4 - \left(\frac{\phi}{B}\right)^n_4\right] \tag{4.64}$$

For boundary gridblock 9, $n=9$, $\psi_9=\{5,10\}$, $\xi_9=\{b_W,b_N\}$, $q^m_{sc_9}=0$, and

$$\sum_{l\in\xi_9} q^m_{sc_{l,9}} = q^m_{sc_{b_W},9} + q^m_{sc_{b_N},9} = 0 - 125 = -125 \text{ STB/D}$$

Substitution into Eq. (4.14) results in the flow equation for boundary grid-block 9,

$$(10.3308)\left(p^m_5 - p^m_9\right) + (18.2574)\left(p^m_{10} - p^m_9\right) - 125 = \frac{V_{b_9}}{\alpha_c \Delta t}\left[\left(\frac{\phi}{B}\right)^{n+1}_9 - \left(\frac{\phi}{B}\right)^n_9\right] \tag{4.65}$$

For boundary gridblock 12, $n=12$, $\psi_{12}=\{8,11\}$, $\xi_{12}=\{b_E,b_N\}$, $q^m_{sc_{12}}=0$, and

$$\sum_{l\in\xi_{12}} q^m_{sc_{l,12}} = q^m_{sc_{b_E},12} + q^m_{sc_{b_N},12} = 456.435 - 125 = 331.435 \text{ STB/D}$$

Substitution into Eq. (4.14) results in the flow equation for boundary gridblock 12:

$$(10.3308)\left(p_8^m - p_{12}^m\right) + (18.2574)\left(p_{11}^m - p_{12}^m\right) + 331.435 = \frac{V_{b_{12}}}{\alpha_c \Delta t}\left[\left(\frac{\phi}{B}\right)^{n+1}_{12} - \left(\frac{\phi}{B}\right)^{n}_{12}\right] \quad (4.66)$$

4.5 Calculation of transmissibilities

Eq. (2.39) in Chapter 2 defines the transmissibilities in the flow equations in Cartesian coordinates. The definitions of transmissibility in the x-, y-, and z-directions are expressed as:

$$T_{x_{i\mp1/2,j,k}} = G_{x_{i\mp1/2,j,k}}\left(\frac{1}{\mu B}\right)_{x_{i\mp1/2,j,k}} \quad (4.67a)$$

$$T_{y_{i,j\mp1/2,k}} = G_{y_{i,j\mp1/2,k}}\left(\frac{1}{\mu B}\right)_{y_{i,j\mp1/2,k}} \quad (4.67b)$$

and

$$T_{z_{i,j,k\mp1/2}} = G_{z_{i,j,k\mp1/2}}\left(\frac{1}{\mu B}\right)_{z_{i,j,k\mp1/2}} \quad (4.67c)$$

where the geometric factors G for anisotropic porous media and irregular gridblock distribution are given in Table 4.1 (Ertekin et al., 2001). The treatment of

TABLE 4.1 Geometric factors in rectangular grids (Ertekin et al., 2001)

Direction	Geometric factor
x	$G_{x_{i\mp1/2,j,k}} = \dfrac{2\beta_c}{\Delta x_{i,j,k}/\left(A_{x_{i,j,k}} k_{x_{i,j,k}}\right) + \Delta x_{i\mp1,j,k}/\left(A_{x_{i\mp1,j,k}} k_{x_{i\mp1,j,k}}\right)}$
y	$G_{y_{i,j\mp1/2,k}} = \dfrac{2\beta_c}{\Delta y_{i,j,k}/\left(A_{y_{i,j,k}} k_{y_{i,j,k}}\right) + \Delta y_{i,j\mp1,k}/\left(A_{y_{i,j\mp1,k}} k_{y_{i,j\mp1,k}}\right)}$
z	$G_{z_{i,j,k\mp1/2}} = \dfrac{2\beta_c}{\Delta z_{i,j,k}/\left(A_{z_{i,j,k}} k_{z_{i,j,k}}\right) + \Delta z_{i,j,k\mp1}/\left(A_{z_{i,j,k\mp1}} k_{z_{i,j,k\mp1}}\right)}$

the pressure-dependent term (μB) in Eq. (4.67) is discussed in detail under linearization in Chapter 8 (Section 8.4.1).

Example 4.7 Derive the equation for the geometric factor of transmissibility in the x-direction between gridblocks i and $i+1$ in 1D flow using the following:

(1) Table 4.1
(2) Darcy's law.

Solution

1. The geometric factor of transmissibility in the x-direction is given as

$$G_{x_{i\mp1/2,j,k}} = \frac{2\beta_c}{\Delta x_{i,j,k}/\left(A_{x_{i,j,k}} k_{x_{i,j,k}}\right) + \Delta x_{i\mp1,j,k}/\left(A_{x_{i\mp1,j,k}} k_{x_{i\mp1,j,k}}\right)} \quad (4.68)$$

For flow between gridblocks i and $i+1$ in a 1-D reservoir, $j=1$, and $k=1$. Discarding these subscripts and the negative sign in Eq. (4.68) that yields the sought geometric factor,

$$G_{x_{i+1/2}} = \frac{2\beta_c}{\Delta x_i/\left(A_{x_i} k_{x_i}\right) + \Delta x_{i+1}/\left(A_{x_{i+1}} k_{x_{i+1}}\right)} \quad (4.69)$$

2. Consider the steady-state flow of incompressible fluid ($B=1$ and $\mu=$ constant) in incompressible porous media between gridblocks i and $i+1$. Gridblock i has cross-sectional area A_{x_i} and permeability k_{x_i}, and gridblock $i+1$ has cross-sectional area $A_{x_{i+1}}$ and permeability $k_{x_{i+1}}$. Boundary $i+½$ between the two blocks is δx_{i^+} away from point i and δx_{i+1^-} away from point $i+1$ as shown in Fig. 4.13. Fluid flows from gridblock i to block boundary $i+½$ and then from block boundary $i+½$ to gridblock $i+1$.

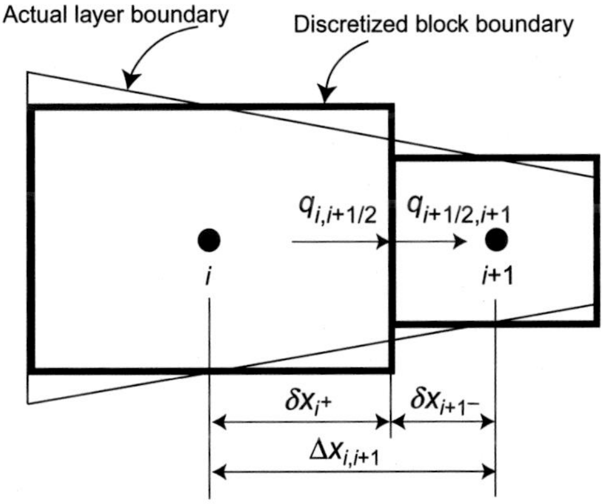

FIG. 4.13 Transmissibility between two adjacent blocks.

The rate of fluid flow from the center of gridblock i to block boundary $i+½$ is given by Darcy's law as

$$q_{i,i+1/2}=\frac{\beta_c k_{x_i} A_{x_i}}{B\mu\delta x_{i^+}}\left(p_i-p_{i+1/2}\right) \tag{4.70}$$

Similarly, the rate of fluid flow from block boundary $i+½$ to the center of gridblock $i+1$ is given by Darcy's law as

$$q_{i+1/2,i+1}=\frac{\beta_c k_{x_{i+1}} A_{x_{i+1}}}{B\mu\delta x_{i+1^-}}\left(p_{i+1/2}-p_{i+1}\right) \tag{4.71}$$

In this flow system, there is neither fluid accumulation nor fluid depletion. Therefore, the rate of fluid leaving gridblock i ($q_{i,i+1/2}$) has to be equal to the rate of fluid entering gridblock $i+1$ ($q_{i+1/2,i+1}$); that is,

$$q_{i,i+1/2}=q_{i+1/2,i+1}=q_{i,i+1} \tag{4.72}$$

The rate of fluid flow between the centers of gridblocks i and $i+1$ is given by Darcy's law as

$$q_{i,i+1}=\frac{G_{x_{i+1/2}}}{B\mu}\left(p_i-p_{i+1}\right) \tag{4.73}$$

The pressure drop between the centers of gridblocks i and $i+1$ is equal to the sum of the pressure drops between the block centers and the block boundary between them; that is,

$$\left(p_i-p_{i+1}\right)=\left(p_i-p_{i+1/2}\right)+\left(p_{i+1/2}-p_{i+1}\right) \tag{4.74}$$

Substituting for pressure drops in Eq. (4.74) using Eqs. (4.70), (4.71), and (4.73) yields

$$\frac{q_{i,i+1}B\mu}{G_{x_{i+1/2}}}=\frac{q_{i,i+1/2}B\mu\delta x_{i^+}}{\beta_c k_{x_i} A_{x_i}}+\frac{q_{i+1/2,i+1}B\mu\delta x_{i+1^-}}{\beta_c k_{x_{i+1}} A_{x_{i+1}}} \tag{4.75}$$

Combining Eqs. (4.75) and (4.72) and dividing by flow rate, FVF, and viscosity yields

$$\frac{1}{G_{x_{i+1/2}}}=\frac{\delta x_{i^+}}{\beta_c k_{x_i} A_{x_i}}+\frac{\delta x_{i+1^-}}{\beta_c k_{x_{i+1}} A_{x_{i+1}}} \tag{4.76}$$

Eq. (4.76) can be solved for $G_{x_{i+1/2}}$. The resulting equation is

$$G_{x_{i+1/2}}=\frac{\beta_c}{\delta x_{i^+}/\left(A_{x_i}k_{x_i}\right)+\delta x_{i+1^-}/\left(A_{x_{i+1}}k_{x_{i+1}}\right)} \tag{4.77}$$

Observing that $\delta x_{i^+}={}^1/_2\Delta x_i$ and $\delta x_{i+1^-}={}^1/_2\Delta x_{i+1}$ for a block-centered grid, Eq. (4.77) becomes

$$G_{x_{i+1/2}} = \frac{2\beta_c}{\Delta x_i/(A_{x_i}k_{x_i}) + \Delta x_{i+1}/(A_{x_{i+1}}k_{x_{i+1}})} \tag{4.78}$$

Eqs. (4.69) and (4.78) are identical.

Eq. (2.69) in Chapter 2 defines the transmissibilities in the flow equations in radial-cylindrical coordinates. The definitions of transmissibility in the r-, θ-, and z-directions are expressed as

$$T_{r_{i\mp1/2,j,k}} = G_{r_{i\mp1/2,j,k}} \left(\frac{1}{\mu B}\right)_{r_{i\mp1/2,j,k}} \tag{4.79a}$$

$$T_{\theta_{i,j\mp1/2,k}} = G_{\theta_{i,j\mp1/2,k}} \left(\frac{1}{\mu B}\right)_{\theta_{i,j\mp1/2,k}} \tag{4.79b}$$

and

$$T_{z_{i,j,k\mp1/2}} = G_{z_{i,j,k\mp1/2}} \left(\frac{1}{\mu B}\right)_{z_{i,j,k\mp1/2}} \tag{4.79c}$$

where the geometric factors G for anisotropic porous media and irregular grid-block distribution are given in Table 4.2 (Farouq Ali, 1986). Note that in this table, r_i and $r_{i\mp1/2}$ depend on the value of subscript i only for $j=1, 2, 3\ldots n_\theta$ and $k=1, 2, 3\ldots n_z$, $\Delta\theta_j$ and $\Delta\theta_{j\mp1/2}$ depend on the value of subscript j only for $i=1, 2, 3\ldots n_r$ and $k=1, 2, 3\ldots n_z$, and Δz_k $\Delta z_{k\mp1/2}$ depend on the value of subscript k only for $i=1, 2, 3\ldots n_r$ and $j=1, 2, 3\ldots n_\theta$. The treatment of the pressure-dependent term (μB) in Eq. (4.79) is discussed in detail under linearization in Chapter 8 (Section 8.4.1).

TABLE 4.2 Geometric factors in cylindrical grids (Farouq Ali, 1986)

Direction	Geometric factor
r	$G_{r_{i-1/2,j,k}} = \dfrac{\beta_c \Delta\theta_j}{\log_e\left(r_i/r^L_{i-1/2}\right)/\left(\Delta z_{i,j,k} k_{r_{i,j,k}}\right) + \log_e\left(r^L_{i-1/2}/r_{i-1}\right)/\left(\Delta z_{i-1,j,k} k_{r_{i-1,j,k}}\right)}$
	$G_{r_{i+1/2,j,k}} = \dfrac{\beta_c \Delta\theta_j}{\log_e\left(r^L_{i+1/2}/r_i\right)/\left(\Delta z_{i,j,k} k_{r_{i,j,k}}\right) + \log_e\left(r_{i+1}/r^L_{i+1/2}\right)/\left(\Delta z_{i+1,j,k} k_{r_{i+1,j,k}}\right)}$
θ	$G_{\theta_{i,j\mp1/2,k}} = \dfrac{2\beta_c \log_e\left(r^L_{i+1/2}/r^L_{i-1/2}\right)}{\Delta\theta_j/\left(\Delta z_{i,j,k} k_{\theta_{i,j,k}}\right) + \Delta\theta_{j\mp1}/\left(\Delta z_{i,j\mp1,k} k_{\theta_{i,j\mp1,k}}\right)}$
z	$G_{z_{i,j,k\mp1/2}} = \dfrac{2\beta_c\left(^1/_2\Delta\theta_j\right)\left(r^2_{i+1/2} - r^2_{i-1/2}\right)}{\Delta z_{i,j,k}/k_{z_{i,j,k}} + \Delta z_{i,j,k\mp1}/k_{z_{i,j,k\mp1}}}$

Table 4.2 uses gridblock dimensions and block boundaries in the z-direction as defined in Eq. (4.1), with z replacing x. Those in the θ-direction are defined in a similar way. Specifically,

$$\sum_{j=1}^{n_\theta} \Delta\theta_j = 2\pi$$

$$\Delta\theta_{j+1/2} = {}^1/_2\left(\Delta\theta_{j+1} + \Delta\theta_j\right),\ j = 1,2,3\ldots n_\theta - 1$$

$$\theta_{j+1} = \theta_j + \Delta\theta_{j+1/2},\ j = 1,2,3\ldots n_\theta - 1,\ \theta_1 = {}^1/_2\Delta\theta_1 \quad (4.80)$$

and

$$\theta_{j\mp1/2} = \theta_j \mp {}^1/_2\Delta\theta_j,\ \ i = 1,2,3\ldots n_\theta$$

In the r-direction, however, the points representing gridblocks are spaced such that the pressure drops between neighboring points are equal (see Example 4.8). Block boundaries for transmissibility calculations are spaced logarithmically in r to warrant that the radial flow rates between neighboring points using the integrated continuous and discretized forms of Darcy's law are identical (see Example 4.9). Block boundaries for bulk volume calculations are spaced logarithmically in r^2 to warrant that the actual and discretized bulk volumes of gridblocks are equal. Therefore, the radii for the pressure points ($r_{i\mp1}$), transmissibility calculations ($r^L_{i\mp1/2}$), and bulk volume calculations ($r_{i\mp1/2}$) are as follows (Aziz and Settari, 1979; Ertekin et al., 2001):

$$r_{i+1} = \alpha_{lg} r_i \ \text{ for } i = 1,2,3\ldots n_r - 1 \quad (4.81)$$

$$r^L_{i+1/2} = \frac{r_{i+1} - r_i}{\log_e(r_{i+1}/r_i)} \ \text{ for } i = 1,2,3\ldots n_r - 1 \quad (4.82a)$$

$$r^L_{i-1/2} = \frac{r_i - r_{i-1}}{\log_e(r_i/r_{i-1})} \ \text{ for } i = 2,3\ldots n_r \quad (4.83a)$$

$$r^2_{i+1/2} = \frac{r^2_{i+1} - r^2_i}{\log_e\left(r^2_{i+1}/r^2_i\right)} \ \text{ for } i = 1,2,3\ldots n_r - 1 \quad (4.84a)$$

$$r^2_{i-1/2} = \frac{r^2_i - r^2_{i-1}}{\log_e\left(r^2_i/r^2_{i-1}\right)} \ \text{ for } i = 2,3\ldots n_r \quad (4.85a)$$

where

$$\alpha_{lg} = \left(\frac{r_e}{r_w}\right)^{1/n_r} \quad (4.86)$$

and

$$r_1 = \left[\alpha_{lg}\log_e(\alpha_{lg})/(\alpha_{lg} - 1)\right] r_w \quad (4.87)$$

Note that the reservoir internal boundary (r_w) and the reservoir external boundary (r_e) through which fluid may enter or leave the reservoir are, respectively, the internal boundary of gridblock 1 and the external boundary of gridblock n_r that are used to calculate transmissibility. That is to say, $r^L_{1/2}=r_w$ and $r^L_{n_r+1/2}=r_e$ by definition for block-centered grid (Ertekin et al., 2001).

The bulk volume of gridblock (i,j,k) is calculated from

$$V_{b_{i,j,k}} = \left(r^2_{i+1/2} - r^2_{i-1/2}\right)\left({}^1/{}_2\Delta\theta_j\right)\Delta z_{i,j,k} \tag{4.88a}$$

for $i=1, 2, 3\ldots n_r-1, j=1, 2, 3\ldots n_\theta, k=1, 2, 3\ldots n_z$; and

$$V_{b_{n_r,j,k}} = \left(r^2_e - r^2_{n_r-1/2}\right)\left({}^1/{}_2\Delta\theta_j\right)\Delta z_{n_r,j,k} \tag{4.88c}$$

for $j=1, 2, 3\ldots n_\theta,\ k=1, 2, 3\ldots n_z$.

Example 4.8 Prove that the grid spacing in the radial direction defined by Eqs. (4.81) and (4.86) satisfies the condition of constant and equal pressure drops between successive points in steady-state radial flow of incompressible fluid.

Solution

The steady-state flow of incompressible fluid toward a well with radius r_w in a horizontal reservoir with an external radius r_e is expressed by Darcy's law:

$$q = \frac{-2\pi\beta_c k_H h}{B\mu \log_e\left(\frac{r_e}{r_w}\right)}(p_e - p_w) \tag{4.89}$$

The pressure drop across the reservoir is obtained from Eq. (4.89) as

$$(p_e - p_w) = \frac{-qB\mu \log_e\left(\frac{r_e}{r_w}\right)}{2\pi\beta_c k_H h} \tag{4.90}$$

Let the reservoir be divided into n_r radial segments that are represented by points $i=1, 2, 3\ldots n_r$ placed at $r_1, r_2, r_3, \ldots r_{i-1}, r_i, r_{i+1}, \ldots r_{n_r}$. The location of these points will be determined later (Eq. 4.81). For steady-state radial flow between points $i+1$ and i,

$$q = \frac{-2\pi\beta_c k_H h}{B\mu \log_e\left(\frac{r_{i+1}}{r_i}\right)}(p_{i+1} - p_i) \tag{4.91}$$

The pressure drop between points $i+1$ and i is obtained from Eq. (4.91) as

$$(p_{i+1} - p_i) = \frac{-qB\mu \log_e\left(\frac{r_{i+1}}{r_i}\right)}{2\pi\beta_c k_H h} \tag{4.92}$$

If the pressure drop over each of the radial distances $(r_{i+1} - r_i)$ for $i=1, 2, 3\ldots n_r-1$ is chosen to be constant and equal, then

$$(p_{i+1}-p_i)=\frac{(p_e-p_w)}{n_r} \tag{4.93}$$

for $i=1, 2, 3 \ldots n_r-1$.

Substituting Eqs. (4.90) and (4.92) into Eq. (4.93) yields

$$\log_e\left(\frac{r_{i+1}}{r_i}\right)=\frac{1}{n_r}\log_e\left(\frac{r_e}{r_w}\right) \tag{4.94}$$

or

$$\left(\frac{r_{i+1}}{r_i}\right)=\left(\frac{r_e}{r_w}\right)^{1/n_r} \tag{4.95a}$$

for $i=1, 2, 3 \ldots n_r-1$.

For the convenience of manipulation, define

$$\alpha_{lg}=\left(\frac{r_e}{r_w}\right)^{1/n_r} \tag{4.86}$$

then Eq. (4.95a) becomes

$$\left(\frac{r_{i+1}}{r_i}\right)=\alpha_{lg} \tag{4.95b}$$

or

$$r_{i+1}=\alpha_{lg}r_i \tag{4.81}$$

for $i=1, 2, 3 \ldots n_r-1$.

Eq. (4.81) defines the locations of the points in the r-direction that result in equal pressure drops between any two successive points.

Example 4.9 Show that the block boundaries defined by Eq. (4.82a) ensure that the flow rate across a block boundary is identical to that obtained from Darcy's law.

Solution

From Example 4.8, for steady-state radial flow of incompressible fluid between points $i+1$ and i,

$$q=\frac{-2\pi\beta_c k_H h}{B\mu\log_e\left(\frac{r_{i+1}}{r_i}\right)}(p_{i+1}-p_i) \tag{4.91}$$

The steady-state fluid flow rate across a block boundary is also expressed by the differential form of Darcy's law at block boundary $r^L_{i+1/2}$,

$$q_{r^L_{i+1/2}}=\frac{-2\pi\beta_c k_H h r^L_{i+1/2}}{B\mu}\left.\frac{\mathrm{d}p}{\mathrm{d}r}\right|_{r^L_{i+1/2}} \tag{4.96}$$

The pressure gradient at a block boundary can be approximated, using central differencing, as

$$\left.\frac{\mathrm{d}p}{\mathrm{d}r}\right|_{r^L_{i+1/2}} \cong \frac{p_{i+1}-p_i}{r_{i+1}-r_i} \tag{4.97}$$

Substitution of Eq. (4.97) into Eq. (4.96) results in

$$q_{r^L_{i+1/2}} = \frac{-2\pi\beta_c k_H h r^L_{i+1/2}}{B\mu}\frac{(p_{i+1}-p_i)}{r_{i+1}-r_i} \tag{4.98}$$

If the flow rate calculated from Darcy's law (Eq. 4.91) is identical to the flow rate calculated from the discretized Darcy's law (Eq. 4.98), then

$$\frac{-2\pi\beta_c k_H h}{B\mu \log_e\left(\frac{r_{i+1}}{r_i}\right)}(p_{i+1}-p_i) = \frac{-2\pi\beta_c k_H h r^L_{i+1/2}}{B\mu}\frac{(p_{i+1}-p_i)}{r_{i+1}-r_i} \tag{4.99}$$

which simplifies to give

$$r^L_{i+1/2} = \frac{r_{i+1}-r_i}{\log_e\left(\frac{r_{i+1}}{r_i}\right)} \tag{4.82a}$$

Eqs. (4.82a), (4.83a), (4.84a), (4.85a), (4.88a), and (4.88c) can be expressed in terms of r_i and α_{lg} as:

$$r^L_{i+1/2} = \{(\alpha_{lg}-1)/[\log_e(\alpha_{lg})]\}r_i \tag{4.82b}$$

for $i=1, 2, 3\ldots n_r-1$;

$$r^L_{i-1/2} = \{(\alpha_{lg}-1)/[\alpha_{lg}\log_e(\alpha_{lg})]\}r_i = \left({}^1/_{\alpha_{lg}}\right)r^L_{i+1/2} \tag{4.83b}$$

for $i=2, 3\ldots n_r$;

$$r^2_{i+1/2} = \left\{\left(\alpha^2_{lg}-1\right)/\left[\log_e\left(\alpha^2_{lg}\right)\right]\right\}r^2_i \tag{4.84b}$$

for $i=1, 2, 3\ldots n_r-1$;

$$r^2_{i-1/2} = \left\{\left(\alpha^2_{lg}-1\right)/\left[\alpha^2_{lg}\log_e\left(\alpha^2_{lg}\right)\right]\right\}r^2_i = \left({}^1/_{\alpha^2_{lg}}\right)r^2_{i+1/2} \tag{4.85b}$$

for $i=2, 3\ldots n_r$;

$$V_{b_{i,j,k}} = \left\{\left(\alpha^2_{lg}-1\right)^2/\left[\alpha^2_{lg}\log_e\left(\alpha^2_{lg}\right)\right]\right\}r^2_i\left({}^1/_2\Delta\theta_j\right)\Delta z_{i,j,k} \tag{4.88b}$$

for $i=1, 2, 3, \ldots n_r-1$; and

$$V_{b_{n_r,j,k}} = \left\{1 - \left[\log_e(\alpha_{lg})/(\alpha_{lg}-1)\right]^2 \left(\alpha_{lg}^2 - 1\right)/\left[\alpha_{lg}^2 \log_e\left(\alpha_{lg}^2\right)\right]\right\} r_e^2\left({}^1/_2\Delta\theta_j\right)\Delta z_{n_r,j,k} \tag{4.88d}$$

for $i = n_r$.

Example 4.10 Prove that Eqs. (4.82b), (4.83b), (4.84b), (4.85b), and (4.88b) are equivalent to Eqs. (4.82a), (4.83a), (4.84a), (4.85a), and (4.88a), respectively. In addition, express the arguments of the log terms that appear in Table 4.2 and the gridblock bulk volume in terms of α_{lg}.

Solution

Using Eq. (4.81), we obtain

$$r_{i+1} - r_i = \alpha_{lg} r_i - r_i = (\alpha_{lg} - 1) r_i \tag{4.100}$$

and

$$r_{i+1}/r_i = \alpha_{lg} \tag{4.101}$$

Substitution of Eqs. (4.100) and (4.101) into Eq. (4.82a) yields

$$r_{i+1/2}^L = \frac{r_{i+1} - r_i}{\log_e(r_{i+1}/r_i)} = \frac{(\alpha_{lg} - 1) r_i}{\log_e(\alpha_{lg})_i} = \left\{(\alpha_{lg} - 1)/\log_e(\alpha_{lg})\right\} r_i \tag{4.102}$$

Eq. (4.102) can be rearranged to give

$$r_{i+1/2}^L / r_i = (\alpha_{lg} - 1)/\log_e(\alpha_{lg}) \tag{4.103}$$

from which

$$\log_e\left(r_{i+1/2}^L / r_i\right) = \log_e\left[(\alpha_{lg} - 1)/\log_e(\alpha_{lg})\right] \tag{4.104}$$

Eqs. (4.101) and (4.102) can be combined by eliminating r_i, yielding

$$r_{i+1/2}^L = \frac{1}{\log_e(\alpha_{lg})}(\alpha_{lg} - 1)(r_{i+1}/\alpha_{lg}) = \left\{(\alpha_{lg} - 1)/\left[\alpha_{lg}\log_e(\alpha_{lg})\right]\right\} r_{i+1} \tag{4.105}$$

Eq. (4.105) can be rearranged to give

$$r_{i+1}/r_{i+1/2}^L = \left[\alpha_{lg}\log_e(\alpha_{lg})\right]/(\alpha_{lg} - 1) \tag{4.106}$$

from which

$$\log_e\left(r_{i+1}/r_{i+1/2}^L\right) = \log_e\left\{\left[\alpha_{lg}\log_e(\alpha_{lg})\right]/(\alpha_{lg} - 1)\right\} \tag{4.107}$$

Using Eq. (4.81) and replacing subscript i with $i-1$ yields

$$r_i = \alpha_{lg} r_{i-1} \tag{4.108}$$

and

$$r_i/r_{i-1} = \alpha_{lg} \tag{4.109}$$

Substitution of Eqs. (4.108) and (4.109) into Eq. (4.83a) yields

$$r^L_{i-1/2} = \frac{r_i - r_{i-1}}{\log_e(r_i/r_{i-1})} = \frac{r_i - r_i/\alpha_{lg}}{\log_e(\alpha_{lg})} = \{(\alpha_{lg}-1)/[\alpha_{lg}\log_e(\alpha_{lg})]\}r_i \quad (4.110)$$

Eq. (4.110) can be rearranged to give

$$r_i/r^L_{i-1/2} = [\alpha_{lg}\log_e(\alpha_{lg})]/(\alpha_{lg}-1) \quad (4.111)$$

from which

$$\log_e\left(r_i/r^L_{i-1/2}\right) = \log_e\{[\alpha_{lg}\log_e(\alpha_{lg})]/(\alpha_{lg}-1)\} \quad (4.112)$$

Eqs. (4.108) and (4.110) can be combined by eliminating r_i, yielding

$$r^L_{i-1/2} = \frac{1}{\log_e(\alpha_{lg})}[(\alpha_{lg}-1)/\alpha_{lg}](\alpha_{lg}r_{i-1}) = [(\alpha_{lg}-1)/\log_e(\alpha_{lg})]r_{i-1} \quad (4.113)$$

Eq. (4.113) can be rearranged to give

$$r^L_{i-1/2}/r_{i-1} = (\alpha_{lg}-1)/\log_e(\alpha_{lg}) \quad (4.114)$$

from which

$$\log_e\left(r^L_{i-1/2}/r_{i-1}\right) = \log_e[(\alpha_{lg}-1)/\log_e(\alpha_{lg})] \quad (4.115)$$

Eqs. (4.102) and (4.110) are combined to get

$$r^L_{i+1/2}/r^L_{i-1/2} = \frac{\{(\alpha_{lg}-1)/\log_e(\alpha_{lg})\}r_i}{\{(\alpha_{lg}-1)/[\alpha_{lg}\log_e(\alpha_{lg})]\}r_i} = \alpha_{lg} \quad (4.116)$$

from which

$$\log_e\left(r^L_{i+1/2}/r^L_{i-1/2}\right) = \log_e(\alpha_{lg}) \quad (4.117)$$

Substitution of Eqs. (4.81) and (4.101) into Eq. (4.84a) yields

$$r^2_{i+1/2} = \frac{r^2_{i+1} - r^2_i}{\log_e(r^2_{i+1}/r^2_i)} = \frac{\left(\alpha^2_{lg}-1\right)r^2_i}{\log_e\left(\alpha^2_{lg}\right)} = \left[\left(\alpha^2_{lg}-1\right)/\log_e\left(\alpha^2_{lg}\right)\right]r^2_i \quad (4.118)$$

Substitution of Eqs. (4.108) and (4.109) into Eq. (4.85a) yields

$$r^2_{i-1/2} = \frac{r^2_i - r^2_{i-1}}{\log_e(r^2_i/r^2_{i-1})} = \frac{\left(1-1/\alpha^2_{lg}\right)r^2_i}{\log_e\left(\alpha^2_{lg}\right)} = \left\{\left(\alpha^2_{lg}-1\right)/\left[\alpha^2_{lg}\log_e\left(\alpha^2_{lg}\right)\right]\right\}r^2_i \quad (4.119)$$

Subtraction of Eq. (4.119) from Eq. (4.118) yields

$$r_{i+1/2}^2 - r_{i-1/2}^2 = \frac{\left(\alpha_{lg}^2 - 1\right)}{\log_e\left(\alpha_{lg}^2\right)} r_i^2 - \frac{\left[\left(\alpha_{lg}^2 - 1\right)/\alpha_{lg}^2\right]}{\log_e\left(\alpha_{lg}^2\right)} r_i^2$$

$$= \frac{\left(\alpha_{lg}^2 - 1\right)\left(1 - 1/\alpha_{lg}^2\right)}{\log_e\left(\alpha_{lg}^2\right)} r_i^2 = \left\{\left(\alpha_{lg}^2 - 1\right)^2 / \left[\alpha_{lg}^2 \log_e\left(\alpha_{lg}^2\right)\right]\right\} r_i^2 \tag{4.120}$$

Combining Eqs. (4.88a) and (4.120) yields

$$V_{b_{i,j,k}} = \left\{\left(\alpha_{lg}^2 - 1\right)^2 / \left[\alpha_{lg}^2 \log_e\left(\alpha_{lg}^2\right)\right]\right\} r_i^2 \left({}^1/{}_2\Delta\theta_j\right) \Delta z_{i,j,k} \tag{4.121}$$

Eq. (4.121) can be used to calculate bulk volumes of gridblocks other than those that fall on the reservoir external boundary in the r-direction. For blocks with $i = n_r$, Eq. (4.88d) is used and the proof is left as an exercise (Exercise 4.13).

Example 4.10 demonstrates that quotients $r_i/r_{i-1/2}^L$, $r_{i-1/2}^L/r_{i-1}$, $r_{i+1/2}^L/r_i$, $r_{i+1}/r_{i+1/2}^L$, and $r_{i+1/2}^L/r_{i-1/2}^L$ are functions of the logarithmic spacing constant α_{lg} only as expressed in the following equations:

$$r_i/r_{i-1/2}^L = \left[\alpha_{lg} \log_e\left(\alpha_{lg}\right)\right] / \left(\alpha_{lg} - 1\right) \tag{4.111}$$

$$r_{i-1/2}^L/r_{i-1} = \left(\alpha_{lg} - 1\right) / \log_e\left(\alpha_{lg}\right) \tag{4.114}$$

$$r_{i+1/2}^L/r_i = \left(\alpha_{lg} - 1\right) / \log_e\left(\alpha_{lg}\right) \tag{4.103}$$

$$r_{i+1}/r_{i+1/2}^L = \left[\alpha_{lg} \log_e\left(\alpha_{lg}\right)\right] / \left(\alpha_{lg} - 1\right) \tag{4.106}$$

$$r_{i+1/2}^L/r_{i-1/2}^L = \alpha_{lg} \tag{4.116}$$

By substituting the above five equations into the equations in Table 4.2 and observing that $\left({}^1/{}_2\Delta\theta_j\right)\left(r_{i+1/2}^2 - r_{i-1/2}^2\right) = V_{b_{i,j,k}}/\Delta z_{i,j,k}$ using Eq. (4.88a), Table 4.3 is obtained.

Now, the calculation of geometric factors and pore volumes can be simplified using the following algorithm.

1. Define

$$\alpha_{lg} = \left(\frac{r_e}{r_w}\right)^{1/n_r} \tag{4.86}$$

2. Let

$$r_1 = \left[\alpha_{lg} \log_e\left(\alpha_{lg}\right) / \left(\alpha_{lg} - 1\right)\right] r_w \tag{4.87}$$

TABLE 4.3 Geometric factors in cylindrical grids

Direction	Geometric factor
r	$G_{r_{i-1/2,j,k}} = \dfrac{\beta_c \Delta\theta_j}{\left\{ \log_e[\alpha_{lg} \log_e(\alpha_{lg})/(\alpha_{lg}-1)]/\left(\Delta z_{i,j,k} k_{r_{i,j,k}}\right) + \log_e[(\alpha_{lg}-1)/\log_e(\alpha_{lg})]/\left(\Delta z_{i-1,j,k} k_{r_{i-1,j,k}}\right) \right\}}$
	$G_{r_{i+1/2,j,k}} = \dfrac{\beta_c \Delta\theta_j}{\left\{ \log_e[(\alpha_{lg}-1)/\log_e(\alpha_{lg})]/\left(\Delta z_{i,j,k} k_{r_{i,j,k}}\right) + \log_e[\alpha_{lg} \log_e(\alpha_{lg})/(\alpha_{lg}-1)]/\left(\Delta z_{i+1,j,k} k_{r_{i+1,j,k}}\right) \right\}}$
θ	$G_{\theta_{i,j\mp1/2,k}} = \dfrac{2\beta_c \log_e(\alpha_{lg})}{\Delta\theta_j/\left(\Delta z_{i,j,k} k_{\theta_{i,j,k}}\right) + \Delta\theta_{j\mp1}/\left(\Delta z_{i,j\mp1,k} k_{\theta_{i,j\mp1,k}}\right)}$
z	$G_{z_{i,j,k\mp1/2}} = \dfrac{2\beta_c \left(V_{b_{i,j,k}}/\Delta z_{i,j,k}\right)}{\Delta z_{i,j,k}/k_{z_{i,j,k}} + \Delta z_{i,j,k\mp1}/k_{z_{i,j,k\mp1}}}$

3. Set

$$r_i = \alpha_{lg}^{i-1} r_1 \tag{4.122}$$

where $i = 1, 2, 3, \ldots n_r$.

4. For $j = 1, 2, 3, \ldots n_\theta$ and $k = 1, 2, 3, \ldots n_z$; set

$$V_{b_{i,j,k}} = \left\{ \left(\alpha_{lg}^2 - 1\right)^2 / \left[\alpha_{lg}^2 \log_e\left(\alpha_{lg}^2\right)\right] \right\} r_i^2 \left({}^1/_2 \Delta\theta_j\right) \Delta z_{i,j,k} \tag{4.88b}$$

for $i = 1, 2, 3, \ldots n_r - 1$, and

$$V_{b_{n_r,j,k}} = \left\{ 1 - [\log_e(\alpha_{lg})/(\alpha_{lg}-1)]^2 \left(\alpha_{lg}^2 - 1\right) / \left[\alpha_{lg}^2 \log_e\left(\alpha_{lg}^2\right)\right] \right\} r_e^2 \left({}^1/_2 \Delta\theta_j\right) \Delta z_{n_r,j,k} \tag{4.88d}$$

for $i = n_r$.

5. Estimate the geometric factors using the equations in Table 4.3. Note that in the calculation of $G_{r_{1/2,j,k}}$, $G_{r_{n_r+1/2,j,k}}$, $G_{z_{i,j,1/2}}$, or $G_{z_{i,j,n_z+1/2}}$, terms that describe properties of blocks that fall outside the reservoir ($i = 0$, $i = n_r + 1$, $k = 0$, and $k = n_z + 1$) are discarded.

Examples 4.11 and 4.12 show that reservoir discretization in the radial direction can be accomplished using either the traditional equations reported in the previous literature (Eqs. 4.81, 4.82a, 4.83a, 4.84a, 4.85a, 4.86, 4.87, 4.88a, and 4.88c) or those reported in this book (Eqs. 4.81, 4.82b, 4.83b, 4.84b, 4.85b, 4.86, 4.87, 4.88b, and 4.88d) that led to Table 4.3. The equations reported in this book, however, are easier and less confusing because they only use r_i and α_{lg}.

In Example 4.13, we demonstrate how to use Eq. (4.2a) and the appropriate expressions for $q^m_{sc_{b,bB}}$, along with Table 4.3, to write the flow equations for boundary and interior gridblocks in a 2-D single-well simulation problem.

Example 4.11 Consider the simulation of a single well in 40-acre spacing. Wellbore diameter is 0.5 ft. The reservoir thickness is 100 ft. The reservoir can be simulated using a single layer discretized into five gridblocks in the radial direction.

1. Find the gridblock spacing in the r-direction.
2. Find the gridblock boundaries in the r-direction for transmissibility calculations.
3. Calculate the arguments of the $\log_e$ terms in Table 4.2.
4. Find the gridblock boundaries in the r-direction for bulk volume calculations and calculate bulk volumes.

Solution

1. The reservoir external radius can be estimated from well spacing

$r_e = \sqrt{43{,}560 \times 40/\pi} = 744.73$ ft, and well radius is given as $r_w = 0.25$ ft.
First, estimate α_{lg} using Eq. (4.86):

$$\alpha_{lg} = \left(\frac{r_e}{r_w}\right)^{1/n_r} = \left(\frac{744.73}{0.25}\right)^{1/5} = 4.9524$$

Second, let $r_1 = [(4.9524)\log_e(4.9524)/(4.9524 - 1)](0.25) = 0.5012$ ft according to Eq. (4.87). Third, calculate the location of the gridblocks in the r-direction using Eq. (4.122), $r_i = \alpha_{lg}^{i-1} r_1$. For example, for $i = 2$, $r_2 = (4.9524)^{2-1} \times 0.5012 = 2.4819$ ft. Table 4.4 shows the location of the other gridblocks along the r-direction.

2. Block boundaries for transmissibility calculations ($r^L_{i-1/2}$, $r^L_{i+1/2}$) are estimated using Eqs. (4.82a) and (4.83a).

For $i = 2$,

$$r^L_{2+1/2} = \frac{r_3 - r_2}{\log_e(r_3/r_2)} = \frac{12.2914 - 2.4819}{\log_e(12.2914/2.4819)} = 6.1315\,\text{ft} \tag{4.123}$$

and

$$r^L_{2-1/2} = \frac{r_2 - r_1}{\log_e(r_2/r_1)} = \frac{2.4819 - 0.5012}{\log_e(2.4819/0.5012)} = 1.2381\,\text{ft} \tag{4.124}$$

Table 4.4 shows the boundaries for transmissibility calculations for other gridblocks.

3. Table 4.4 Shows the calculated values for $r_i/r^L_{i-1/2}$, $r_{i+1}/r^L_{i+1/2}$, $r^L_{i-1/2}/r_{i-1}$, $r^L_{i+1/2}/r_i$, and $r^L_{i+1/2}/r^L_{i-1/2}$, which appear in the argument of $\log_e$ terms in Table 4.2
4. The block boundaries for bulk volume calculations ($r_{i-1/2}$, $r_{i+1/2}$) are estimated using Eqs. (4.84a) and (4.85a).

TABLE 4.4 r_i, $r^L_{i\mp 1/2}$, and $\log_e$ arguments in Table 4.2 for Example 4.11

i	r_i	$r^L_{i-1/2}$	$r^L_{i+1/2}$	$r_i / r^L_{i-1/2}$	$r_{i+1} / r^L_{i+1/2}$	$r^L_{i-1/2} / r_{i-1}$	$r^L_{i+1/2} / r_i$	$r^L_{i+1/2} / r^L_{i-1/2}$
1	0.5012	0.25[a]	1.2381	2.005	2.005	2.47	2.47	4.9528
2	2.4819	1.2381	6.1315	2.005	2.005	2.47	2.47	4.9524
3	12.2914	6.1315	30.3651	2.005	2.005	2.47	2.47	4.9524
4	60.8715	30.3651	150.379	2.005	2.005	2.47	2.47	4.9524
5	301.457	150.379	744.73[b]	2.005	2.005	2.47	2.47	–

[a] $r^L_{1-1/2} = r_w = 0.25$.

[b] $r^L_{5+1/2} = r_e = 744.73$.

For $i=2$,

$$r_{2+1/2}^2=\frac{r_3^2-r_2^2}{\log_e\left(r_3^2/r_2^2\right)}=\frac{(12.2914)^2-(2.4819)^2}{\log_e\left[(12.2914)^2/(2.4819)^2\right]}=45.2906\,\text{ft}^2 \quad (4.125)$$

and

$$r_{2-1/2}^2=\frac{r_2^2-r_1^2}{\log_e\left(r_2^2/r_1^2\right)}=\frac{(2.4819)^2-(0.5012)^2}{\log_e\left[(2.4819)^2/(0.5012)^2\right]}=1.8467\,\text{ft}^2 \quad (4.126)$$

Therefore, the gridblock boundaries for bulk volume calculations are

$$r_{2+1/2}=\sqrt{45.2906}=6.7298\,\text{ft}$$

and

$$r_{2-1/2}=\sqrt{1.8467}=1.3589\,\text{ft}$$

The bulk volume for the gridblocks can be calculated using Eqs.(4.88a), and (4.88c).

For $i=2$,

$$\begin{aligned}V_{b_2}&=\left(r_{2+1/2}^2-r_{2-1/2}^2\right)\left({}^1/_2\Delta\theta\right)\Delta z_2\\&=\left[(6.7299)^2-(1.3589)^2\right]\left({}^1/_2\times 2\pi\right)\times 100=13648.47\ \text{ft}^3\end{aligned} \quad (4.127)$$

For $i=5$,

$$\begin{aligned}V_{b_5}&=\left(r_e^2-r_{5-1/2}^2\right)\left({}^1/_2\Delta\theta\right)\Delta z_5\\&=\left[(744.73)^2-(165.056)^2\right]\left({}^1/_2\times 2\pi\right)\times 100=165.68114\times 10^6\,\text{ft}^3\end{aligned} \quad (4.128)$$

Table 4.5 shows the gridblock boundaries and the bulk volumes for other gridblocks.

Example 4.12 Solve Example 4.11 again, this time using Eqs. (4.82b), (4.83b), (4.84b), (4.85b), and (4.88d), which make use of r_i, α_{lg}, and Eq. (4.88d).

Solution

1. From Example 4.11, $r_e=744.73$ ft, $r_w=0.25$ ft, $r_1=0.5012$ ft, and $\alpha_{lg}=4.9524$. In addition, Table 4.4 reports radii of points representing gridblocks (r_i) calculated using Eq. (4.122).
2. Block boundaries for transmissibility calculations ($r_{i-1/2}^L$, $r_{i+1/2}^L$) are estimated using Eqs. (4.82b) and (4.83b), yielding

$$\begin{aligned}r_{i+1/2}^L&=\left\{\left(\alpha_{lg}-1\right)/\left[\log_e\left(\alpha_{lg}\right)\right]\right\}r_i=\{(4.9524-1)/[\log_e(4.9524)]\}r_i\\&=2.47045r_i\end{aligned} \quad (4.129)$$

TABLE 4.5 Gridblock boundaries and bulk volumes for gridblocks in Example 4.11

i	r_i	$r_{i-1/2}$	$r_{i+1/2}$	V_{b_i}
1	0.5012	0.2744	1.3589	556.4939
2	2.4819	1.3589	6.7299	13,648.47
3	12.2914	6.7299	33.3287	334,739.9
4	60.8715	33.3287	165.056	8,209,770
5	301.4573	165.056	744.73[a]	165,681,140

[a] $r_{5+1/2}=r_e=744.73$.

and

$$r^L_{i-1/2}=\{(\alpha_{lg}-1)/[\alpha_{lg}\log_e(\alpha_{lg})]\}r_i=\{(4.9524-1)/[4.9524\log_e(4.9524)]\}r_i$$
$$=0.49884r_i \tag{4.130}$$

Substitution of the values of r_i into Eqs. (4.129) and (4.130) produces the results reported in Table 4.4.

3. The ratios $r_i/r^L_{i-1/2}$, $r_{i+1}/r^L_{i+1/2}$, $r^L_{i-1/2}/r_{i-1}$, $r^L_{i+1/2}/r_i$, and $r^L_{i+1/2}/r^L_{i-1/2}$ as functions of α_{lg} were derived in Example 4.10 as Eqs. (4.111), (4.106), (4.114), (4.103), and (4.116), respectively. Substitution of $\alpha_{lg}=4.9524$ in these equations, we obtain:

$$r_i/r^L_{i-1/2}=[\alpha_{lg}\log_e(\alpha_{lg})]/(\alpha_{lg}-1)=[4.9524\log_e(4.9524)]/(4.9524-1)$$
$$=2.005 \tag{4.131}$$

$$r_{i+1}/r^L_{i+1/2}=[\alpha_{lg}\log_e(\alpha_{lg})]/(\alpha_{lg}-1)=2.005 \tag{4.132}$$

$$r^L_{i-1/2}/r_{i-1}=(\alpha_{lg}-1)/\log_e(\alpha_{lg})=(4.9524-1)/\log_e(4.9524)=2.470 \tag{4.133}$$

$$r^L_{i+1/2}/r_i=(\alpha_{lg}-1)/\log_e(\alpha_{lg})=2.470 \tag{4.134}$$

$$r^L_{i+1/2}/r^L_{i-1/2}=\alpha_{lg}=4.9524 \tag{4.135}$$

Note that the values of the above ratios are the same as those reported in Table 4.4.

4. Block boundaries for bulk volume calculations ($r_{i-1/2}$, $r_{i+1/2}$) are estimated using Eqs. (4.84b) and (4.85b):

$$r_{i+1/2}^2 = \left\{ \left(\alpha_{lg}^2 - 1\right) / \left[\log_e \left(\alpha_{lg}^2\right) \right] \right\} r_i^2 = \left\{ \left((4.9524)^2 - 1\right) / \left[\log_e \left((4.9524)^2\right) \right] \right\} r_i^2$$
$$= (7.3525) r_i^2 \quad (4.136)$$

and

$$r_{i-1/2}^2 = \left\{ \left(\alpha_{lg}^2 - 1\right) / \left[\alpha_{lg}^2 \log_e \left(\alpha_{lg}^2\right) \right] \right\} r_i^2 = \left\{ 7.3525 / (4.9524)^2 \right\} r_i^2$$
$$= (0.29978) r_i^2 \quad (4.137)$$

Therefore,

$$r_{i+1/2} = \sqrt{(7.3525) r_i^2} = (2.7116) r_i \quad (4.138)$$

and

$$r_{i-1/2} = \sqrt{(0.29978) r_i^2} = (0.54752) r_i \quad (4.139)$$

The bulk volume associated with each gridblock can be calculated using Eqs. (4.88b) and (4.88d).

For $i = 1, 2, 3, 4$;

$$V_{b_i} = \left\{ \left(\alpha_{lg}^2 - 1\right)^2 / \left[\alpha_{lg}^2 \log_e \left(\alpha_{lg}^2\right) \right] \right\} r_i^2 \left[{}^1/_2 (2\pi) \right] \Delta z$$
$$= \left\{ \left[(4.9524)^2 - 1 \right]^2 / \left[(4.9524)^2 \log_e (4.9524)^2 \right] \right\} r_i^2 \left[{}^1/_2 (2\pi) \right] \times 100 = 2215.7 r_i^2$$
$$(4.140)$$

For $i = 5$,

$$V_{b_5} = \left\{ 1 - \left[\log_e(4.9524) / (4.9524 - 1) \right]^2 \times \left[(4.9524)^2 - 1 \right] / \left[(4.9524)^2 \times \log_e \left((4.9524)^2 \right) \right] \right\} \times (744.73)^2 \left({}^1/_2 \times 2\pi \right) \times 100$$
$$= 165.681284 \times 10^6 \quad (4.141)$$

Note that the values of estimated bulk volumes slightly differ from those reported in Table 4.5 due to roundoff errors resulting from approximations in the various stages of calculations.

Example 4.13 A 0.5-ft diameter water well is located in 20-acre spacing. The reservoir thickness, horizontal permeability, and porosity are 30 ft, 150 md, and 0.23, respectively. The (k_V/k_H) for this reservoir is estimated from core data as 0.30. The flowing fluid has a density, FVF, and viscosity of 62.4 lbm/ft^3, 1 RB/B, and 0.5 cP, respectively. The reservoir external boundary in the radial direction is a no-flow boundary, and the well is completed in the top 20 ft only

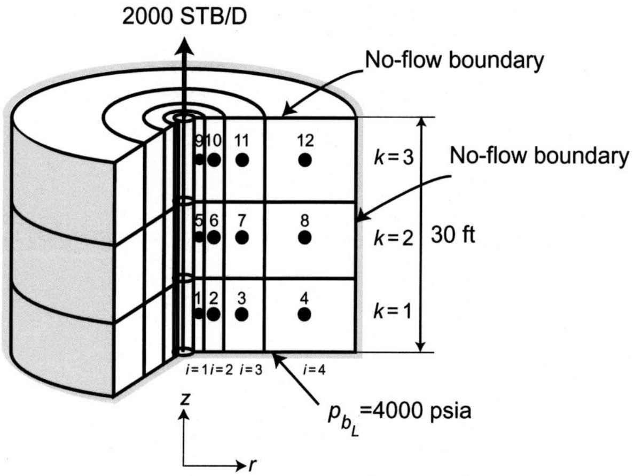

FIG. 4.14 Discretized 2-D radial-cylindrical reservoir in Example 4.13.

and produces at a rate of 2000 B/D. The reservoir bottom boundary is subject to influx such that the boundary is kept at 4000 psia. The reservoir top boundary is sealed to flow. Assuming the reservoir can be simulated using three equal gridblocks in the vertical direction and four gridblocks in the radial direction, as shown in Fig. 4.14, write the flow equations for gridblocks 1, 3, 5, 7, and 11.

Solution

To write the flow equations, the gridblocks are first ordered using natural ordering ($n=1, 2, 3, \ldots 10, 11, 12$) as shown in Fig. 4.14, in addition to being identified using the engineering notation along the radial direction ($i=1, 2, 3, 4$) and the vertical direction ($k=1, 2, 3$). This is followed by the estimation of reservoir rock and fluid property data, the determination of the location of points representing gridblocks in the radial direction, and the calculation of gridblock sizes and elevation in the vertical direction. Next, bulk volumes and transmissibilities in the r- and z-directions are calculated and the contributions of the gridblocks to well rates and fictitious well rates resulting from reservoir boundary conditions are estimated.

Reservoir rock and fluid data are restated as follows, $h=30$ ft, $\phi=0.23$, $k_r=k_H=150$ md, $k_z=k_H(k_V/k_H)=150\times 0.30=45$ md, $B=1$ RB/B, $\mu=0.5$ cP, $\gamma=\gamma_c\rho g=0.21584\times 10^{-3}(62.4)(32.174)=0.4333$ psi/ft, $r_w=0.25$ ft, and the reservoir external radius is estimated from well spacing as $r_e=(20\times 43560/\pi)^{1/2}=526.60$ ft. The reservoir east (external) and upper (top) boundaries are no-flow boundaries, the lower (bottom) boundary has $p_{b_L}=4000$ psia, and the reservoir west (internal) boundary has $q_{spsc}=-2000$ B/D to reflect the effect of the production well (i.e., the well is treated as a boundary condition).

For the block-centered grid shown in Fig. 4.14, $n_r=4$, $n_z=3$, and $\Delta z_k=h/n_z=30/3=10$ ft for $k=1, 2, 3$; hence, $\Delta z_n=10$ ft for $n=1, 2, 3, 4, 5, 6, 7, 8, 9, 10, 11, 12$, and $\Delta z_{k+1/2}=10$ ft for $k=1, 2$. Assuming the top of the reservoir as the reference level for elevation, $Z_n = 5$ ft for $n = 9, 10, 11, 12$; $Z_n = 15$ ft for $n = 5, 6, 7, 8$; $Z_n = 25$ ft for $n = 1, 2, 3, 4$; and $Z_{b_L} = 30$ ft.

The locations of gridblocks in the radial direction are calculated using Eqs. (4.86), (4.87), and (4.122); that is,

$$\alpha_{lg} = (526.60/0.25)^{1/4} = 6.7746$$

$$r_1 = [(6.7746)\log_e(6.7746)/(6.7746-1)] \times 0.25 = 0.56112 \text{ ft}$$

and

$$r_i = (6.7746)^{(i-1)}(0.56112)$$

for $i = 2, 3, 4$ or $r_2 = 3.8014$ ft, $r_3 = 25.753$ ft, and $r_4 = 174.46$ ft.

Eq. (4.88b) is used to calculate bulk volume for gridblocks that have $i = 1, 2, 3$:

$$\begin{aligned} V_{b_{i,k}} &= \left\{ \left(\alpha_{lg}^2 - 1\right)^2 / \left[\alpha_{lg}^2 \log_e\left(\alpha_{lg}^2\right)\right] \right\} r_i^2 \left({}^1/_2 \Delta\theta\right) \Delta z_{i,k} \\ &= \left\{ \left[(6.7746)^2 - 1\right]^2 / \left[(6.7746)^2 \log_e\left((6.7746)^2\right)\right] \right\} r_i^2 \left({}^1/_2 \times 2\pi\right) \Delta z_k \\ &= (36.0576) r_i^2 \Delta z_k \end{aligned}$$

and Eq. (4.88d) for gridblocks that have $i = n_r = 4$,

$$\begin{aligned} V_{b_{n_r,k}} &= \left\{ 1 - \left[\log_e\left(\alpha_{lg}\right)/\left(\alpha_{lg} - 1\right)\right]^2 \left(\alpha_{lg}^2 - 1\right) / \left[\alpha_{lg}^2 \log_e\left(\alpha_{lg}^2\right)\right] \right\} r_e^2 \left({}^1/_2 \Delta\theta_j\right) \Delta z_{n_r,k} \\ &= \{1 - [\log_e(6.7746)/(6.7746-1)]^2 [(6.7746)^2 - 1] / \\ &\quad [(6.7746)^2 \log_e((6.7746)^2)]\} \times (526.60)^2 \left({}^1/_2 \times 2\pi\right) \Delta z_k \\ &= \left(0.846740 \times 10^6\right) \Delta z_k \end{aligned}$$

Eq. (4.79c) defines the transmissibility in the vertical direction, resulting in

$$T_{z_{i,k\mp1/2}} = G_{z_{i,k\mp1/2}} \left(\frac{1}{\mu B}\right) = G_{z_{i,k\mp1/2}} \left(\frac{1}{0.5 \times 1}\right) = (2) G_{z_{i,k\mp1/2}} \quad (4.142)$$

where $G_{z_{i,k\mp1/2}}$ is defined in Table 4.3 as

$$G_{z_{i,k\mp1/2}} = \frac{2\beta_c \left(V_{b_{i,k}}/\Delta z_k\right)}{\Delta z_k / k_{z_{i,k}} + \Delta z_{k\mp1} / k_{z_{i,k\mp1}}} \quad (4.143)$$

For this problem, gridblock spacing, thickness, and permeability in the vertical direction are constants. Therefore, Eq. (4.143) reduces to

$$G_{z_{i,k\mp1/2}} = \frac{\beta_c k_z \left(V_{b_{i,k}}/\Delta z_k\right)}{\Delta z_k}$$

or after substitution for values it becomes

$$G_{z_{i,k\mp1/2}} = \frac{(1.127 \times 10^{-3})(45)(36.0576 \times r_i^2)}{10} = (0.182866)r_i^2 \quad (4.144a)$$

for $i=1, 2, 3$ and $k=1, 2, 3$.

$$G_{z_{i,k\mp1/2}} = \frac{(1.127 \times 10^{-3})(45)(0.846740 \times 10^6)}{10} = 4294.242 \quad (4.144b)$$

for $i=4$ and $k=1, 2, 3$.

Substituting Eq. (4.144) into Eq. (4.142) results in

$$T_{z_{i,k\mp1/2}} = 2(0.182866)r_i^2 = (0.365732)r_i^2 \quad (4.145a)$$

for $i=1, 2, 3$ and $k=1, 2, 3$; and

$$T_{z_{i,k\mp1/2}} = 2(4294.242) = 8588.484 \quad (4.145b)$$

for $i=4$ and $k=1, 2, 3$.

Eq. (4.79a) defines the transmissibility in the r-direction, yielding

$$T_{r_{i\mp1/2,k}} = G_{r_{i\mp1/2,k}}\left(\frac{1}{\mu B}\right) = G_{r_{i\mp1/2,k}}\left(\frac{1}{0.5 \times 1}\right) = (2)G_{r_{i\mp1/2,k}} \quad (4.146)$$

where $G_{r_{i\mp1/2,k}}$ is defined in Table 4.3. With $\Delta\theta = 2\pi$ and constant radial permeability, the equation for the geometric factor reduces to

$$G_{r_{i\mp1/2,k}} = \frac{2\pi\beta_c k_r \Delta z_k}{\log_e\{[\alpha_{lg}\log_e(\alpha_{lg})/(\alpha_{lg}-1)] \times [(\alpha_{lg}-1)/\log_e(\alpha_{lg})]\}}$$
$$= \frac{2\pi\beta_c k_r \Delta z_k}{\log_e(\alpha_{lg})} = \frac{2\pi(0.001127)(150)\Delta z_k}{\log_e(6.7746)} = (0.5551868)\Delta z_k \quad (4.147)$$

Therefore, transmissibility in the radial direction can be estimated by substituting Eq. (4.147) into Eq. (4.146):

$$T_{r_{i\mp1/2,k}} = (2)G_{r_{i\mp1/2,k}} = (2)(0.5551868)\Delta z_k = (1.1103736)\Delta z_k \quad (4.148)$$

Table 4.6 lists the estimated transmissibilities in the radial and vertical directions and bulk volumes. Before writing the flow equation, the well production rate (the specified rate for the reservoir west boundary) must be prorated between gridblocks 5 and 9 using Eq. (4.28):

$$q^m_{sc_{b,bB}} = \frac{T^m_{b,bB}}{\sum\limits_{l\in\psi_b} T^m_{b,l}} q_{spsc} \quad (4.28)$$

where $T^m_{b,bB}$ = transmissibility in the radial direction between reservoir boundary b and gridblock bB with the well-being the reservoir internal boundary and

TABLE 4.6 Gridblock location, bulk volume, and radial and vertical transmissibilities for Example 4.13

n	i	k	r_i (ft)	Δz_n (ft)	Z_n (ft)	V_{b_n} (ft^3)	$T_{r_{i\pm1/2,k}}$ (B/D-psi)	$T_{z_{i,k\pm1/2}}$ (B/D-psi)
1	1	1	0.56112	10	25	113.5318	11.10374	0.115155
2	2	1	3.8014	10	25	5210.583	11.10374	5.285098
3	3	1	25.753	10	25	239,123.0	11.10374	242.5426
4	4	1	174.46	10	25	8,467,440	11.10374	8588.532
5	1	2	0.56112	10	15	113.5318	11.10374	0.115155
6	2	2	3.8014	10	15	5210.583	11.10374	5.285098
7	3	2	25.753	10	15	239,123.0	11.10374	242.5426
8	4	2	174.46	10	15	8,467,440	11.10374	8588.532
9	1	3	0.56112	10	5	113.5318	11.10374	0.115155
10	2	3	3.8014	10	5	5210.583	11.10374	5.285098
11	3	3	25.753	10	5	239,123.0	11.10374	242.5426
12	4	3	174.46	10	5	8,467,440	11.10374	8588.532

$\psi_b = \psi_w = \{5,9\}$. Note that gridblock 1 has a no-flow boundary because it is not penetrated by the well; that is, $q^m_{sc_{b_W},1} = 0$.

Applying the equation for $G_{r_{i-1/2,1,k}}$ in Table 4.3 for $i=1, j=1, k=2, 3$ (i.e., $n=5, 9$) gives

$$G_{r_{i-1/2,1,k}} = \frac{2\pi\beta_c k_r \Delta z_k}{\log_e\{[\alpha_{lg}\log_e(\alpha_{lg})/(\alpha_{lg}-1)]\}}$$

$$= \frac{2\pi(0.001127)(150)\times\Delta z_k}{\log_e[6.7746\times\log_e 6.7746/(6.7746-1)]} = 1.3138\times\Delta z_k$$

$$T^m_{b_w,5} = \frac{G_{r_{1/2,1,2}}}{\mu B} = \frac{1.3138\times 10}{0.5\times 1} = 26.276\text{B/D-psi}$$

and

$$T^m_{b_w,9} = \frac{G_{r_{1/2,1,3}}}{\mu B} = \frac{1.3138\times 10}{0.5\times 1} = 26.276\text{B/D-psi}$$

The application of Eq. (4.28) results in

$$q^m_{sc_{b_W},5} = \frac{26.276}{26.276+26.276} \times (-2000) = -1000\,\text{B/D}$$

and

$$q^m_{sc_{b_W},9} = \frac{26.276}{26.276+26.276} \times (-2000) = -1000\,\text{B/D}$$

Note that the well penetrating gridblocks 5 and 9 are treated as fictitious well.

For the reservoir lower boundary, $p_{b_L}=4000$ psia. The flow rates of the fictitious wells in gridblocks 1, 2, 3, and 4 are estimated using Eq. (4.37c), yielding

$$q^m_{sc_{b_L},n} = T^m_{b_L,n}[(4000-p_n)-(0.4333)(30-25)]\ \text{B/D} \tag{4.149}$$

where $T^n_{b_L,n}$ is estimated using Eq. (4.29) and $A_{z_n}=V_{b_n}/\Delta z_n$

$$\begin{aligned} T^m_{b_L,n} &= \beta_c \frac{k_{z_n}A_{z_n}}{\mu B(\Delta z_n/2)} = 0.001127 \times \frac{45\times(V_{b_n}/\Delta z_n)}{0.5\times 1\times(10/2)} \\ &= (0.0020286)V_{b_n} \end{aligned} \tag{4.150}$$

For the reservoir east and upper (no-flow) boundaries, $q^m_{sc_{b_E},n}=0$ for $n=4, 8,$ 12 and $q^m_{sc_{b_U},n}=0$ for $n=9, 10, 11, 12$. Table 4.7 summarizes the contributions of gridblocks to well rates and fictitious well rates.

The general form of the flow equation for gridblock n is obtained from Eq. (4.2a):

$$\sum_{l\in\psi_n} T^m_{l,n}\left[\left(p^m_l-p^m_n\right)-\gamma^m_{l,n}(Z_l-Z_n)\right] + \sum_{l\in\xi_n} q^m_{sc_{l,n}} + q^m_{sc_n} = \frac{V_{b_n}}{\alpha_c\Delta t}\left[\left(\frac{\phi}{B}\right)^{n+1}_n - \left(\frac{\phi}{B}\right)^n_n\right] \tag{4.2a}$$

For gridblock 1, $n=1$, $i=1$, $k=1$, $\psi_1=\{2,5\}$, $\xi_1=\{b_L,b_W\}$, and $\sum_{l\in\xi_1} q^m_{sc_{l,1}} = q^m_{sc_{b_L},1} + q^m_{sc_{b_W},1}$, where from Table 4.7, $q^m_{sc_{b_L},1}=(0.23031)$ $[(4000-p^m_1)-(0.4333)(30-25)]$ B/D and $q^m_{sc_{b_W},1}=0$ and $q^m_{sc_1}=0$. Therefore, substitution into Eq. (4.2a) yields

$$\begin{aligned} &(11.10374)\left[\left(p^m_2-p^m_1\right)-(0.4333)(25-25)\right] \\ &+(0.115155)\left[\left(p^m_5-p^m_1\right)-(0.4333)(15-25)\right] \\ &+(0.23031)\left[\left(4000-p^m_1\right)-(0.4333)(30-25)\right]+0+0=\frac{113.5318}{\alpha_c\Delta t}\left[\left(\frac{\phi}{B}\right)^{n+1}_1-\left(\frac{\phi}{B}\right)^n_1\right] \end{aligned} \tag{4.151}$$

For gridblock 3, $n=3$, $i=3$, $k=1$, $\psi_3=\{2,4,7\}$, $\xi_3=\{b_L\}$, and $\sum_{l\in\xi_3} q^m_{sc_{l,3}} = q^m_{sc_{b_L},3}$, where from Table 4.7, $q^m_{sc_{b_L},3}=(485.085)[(4000-p^m_3)$

TABLE 4.7 Contribution of gridblocks to well rates and fictitious well rates

n	i	k	$q^m_{sc_n}$ (B/D)	$q^m_{sc_{b_L,n}}$ (B/D)	$q^m_{sc_{b_W,n}}$ (B/D)	$q^m_{sc_{b_E,n}}$ (B/D)	$q^m_{sc_{b_U,n}}$ (B/D)
1	1	1	0	$(0.23031)[(4000-p^m_1)-(0.4333)(30-25)]$	0		
2	2	1	0	$(10.5702)[(4000-p^m_2)-(0.4333)(30-25)]$			
3	3	1	0	$(485.085)[(4000-p^m_3)-(0.4333)(30-25)]$			
4	4	1	0	$(17177.1)[(4000-p^m_4)-(0.4333)(30-25)]$		0	
5	1	2	0		−1000		
6	2	2	0				
7	3	2	0				
8	4	2	0			0	
9	1	3	0		−1000		0
10	2	3	0				0
11	3	3	0				0
12	4	3	0			0	0

$-(0.4333)(30-25)]$ B/D and $q^m_{sc_3}=0$ (no wells). Therefore, substitution into Eq. (4.2a) yields

$$
\begin{aligned}
&(11.10374)\left[\left(p_2^m-p_3^m\right)-(0.4333)(25-25)\right]\\
&\quad+(11.10374)\left[\left(p_4^m-p_3^m\right)-(0.4333)(25-25)\right]\\
&\quad+(242.5426)\left[\left(p_7^m-p_3^m\right)-(0.4333)(15-25)\right]\\
&\quad+(485.0852)\left[\left(4000-p_3^m\right)-(0.4333)(30-25)\right]+0=\frac{239123.0}{\alpha_c\Delta t}\left[\left(\frac{\phi}{B}\right)_3^{n+1}-\left(\frac{\phi}{B}\right)_3^{n}\right]
\end{aligned}
\tag{4.152}
$$

For gridblock 5, $n=5$, $i=1$, $k=2$, $\psi_5=\{1,6,9\}$, $\xi_5=\{b_W\}$, and $\sum\limits_{l\in\xi_5} q^m_{sc_{l,5}}=q^m_{sc_{b_W,5}}$, where from Table 4.7, $q^m_{sc_{b_W,5}}=-1000$ B/D and $q^m_{sc_5}=0$

(the well is treated as a boundary condition). Therefore, substitution into Eq. (4.2a) yields

$$\begin{aligned}&(0.115155)\left[\left(p_1^m-p_5^m\right)-(0.4333)(25-15)\right]\\&+(11.10374)\left[\left(p_6^m-p_5^m\right)-(0.4333)(15-15)\right]\\&+(0.115155)\left[\left(p_9^m-p_5^m\right)-(0.4333)(5-15)\right]-1000+0=\frac{113.5318}{\alpha_c\Delta t}\left[\left(\frac{\phi}{B}\right)_5^{n+1}-\left(\frac{\phi}{B}\right)_5^n\right]\end{aligned} \quad (4.153)$$

For gridblock 7, $n=7$, $i=3$, $k=2$, $\psi_7=\{3,6,8,11\}$, $\xi_7=\{\}$, $\sum_{l\in\xi_7} q_{sc_{l,7}}^m=0$ (interior gridblock), and $q_{sc_7}^m=0$ (no wells). Therefore, substitution into Eq. (4.2a) yields

$$\begin{aligned}&(242.5426)\left[\left(p_3^m-p_7^m\right)-(0.4333)(25-15)\right]\\&+(11.10374)\left[\left(p_6^m-p_7^m\right)-(0.4333)(15-15)\right]\\&+(11.10374)\left[\left(p_8^m-p_7^m\right)-(0.4333)(15-15)\right]\\&+(242.5426)\left[\left(p_{11}^m-p_7^m\right)-(0.4333)(5-15)\right]+0+0=\frac{239123.0}{\alpha_c\Delta t}\left[\left(\frac{\phi}{B}\right)_7^{n+1}-\left(\frac{\phi}{B}\right)_7^n\right]\end{aligned} \quad (4.154)$$

For gridblock 11, $n=11$, $i=3$, $k=3$, $\psi_{11}=\{7,10,12\}$, $\xi_{11}=\{b_U\}$, $\sum_{l\in\xi_{11}} q_{sc_{l,11}}^m=q_{sc_{b_U,11}}^m$, $q_{sc_{b_U,11}}^m=0$ (no-flow boundary), and $q_{sc_{11}}^m=0$ (no wells). Therefore, substitution into Eq. (4.2a) yields

$$\begin{aligned}&(242.5426)\left[\left(p_7^m-p_{11}^m\right)-(0.4333)(15-5)\right]\\&+(11.10374)\left[\left(p_{10}^m-p_{11}^m\right)-(0.4333)(5-5)\right]\\&+(11.10374)\left[\left(p_{12}^m-p_{11}^m\right)-(0.4333)(5-5)\right]+0+0=\frac{239123.0}{\alpha_c\Delta t}\left[\left(\frac{\phi}{B}\right)_{11}^{n+1}-\left(\frac{\phi}{B}\right)_{11}^n\right]\end{aligned} \quad (4.155)$$

4.6 Symmetry and its use in solving practical problems

Reservoir rock properties are heterogeneous, and reservoir fluids and fluid-rock properties vary from one region to another within the same reservoir. In other words, it is rare to find a petroleum reservoir that has constant properties. The literature, however, is rich in study cases in which homogeneous reservoirs were modeled to study flood patterns such as five-spot and nine-spot patterns. In teaching reservoir simulation, educators and textbooks in this area make use of homogeneous reservoirs most of the time. If reservoir properties vary spatially region wise, then symmetry may exist. The use of symmetry reduces the efforts to solve a problem by solving a modified problem for one element of symmetry in the reservoir, usually the smallest element of symmetry

(Abou-Kassem et al., 1991). The smallest element of symmetry is a segment of the reservoir that is a mirror image of the rest of reservoir segments. Before solving the modified problem for one element of symmetry, however, symmetry must first be established. For symmetry to exist about a plane, there must be symmetry with regard to (1) the number of gridblocks and gridblock dimensions, (2) reservoir rock properties, (3) physical wells, (4) reservoir boundaries, and (5) initial conditions. Gridblock dimensions deal with gridblock size (Δx, Δy, and Δz) and gridblock elevation (Z). Reservoir rock properties deal with gridblock porosity (ϕ) and permeability in the various directions (k_x, k_y, and k_z). Wells deal with well location, well type (injection or production), and well operating condition. Reservoir boundaries deal with the geometry of boundaries and boundary conditions. Initial conditions deal with initial pressure and fluid saturation distributions in the reservoir. Failing to satisfy symmetry with respect to any of the items mentioned earlier means there is no symmetry about that plane. The formulation of the modified problem for the smallest element of symmetry involves replacing each plane of symmetry with a no-flow boundary and determining the new interblock geometric factors, bulk volume, wellblock rate, and wellblock geometric factor for those gridblocks that share their boundaries with the planes of symmetry. To elaborate on this point, we present a few possible cases. In the following discussion, we use bold numbers to identify the gridblocks that require determining new values for their bulk volume, wellblock rate, wellblock geometric factor, and interblock geometric factors in the element of symmetry.

The first two examples show planes of symmetry that coincide with the boundaries between gridblocks. Fig. 4.15a presents a 1-D flow problem in which the plane of symmetry A-A, which is normal to the flow direction (x-direction) and coincides with the boundary between gridblocks 3 and 4, and divides the reservoir into two symmetrical elements. Consequently,

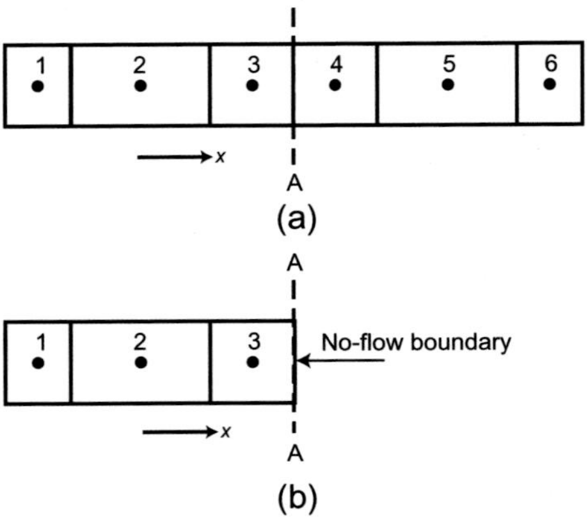

FIG. 4.15 Reservoir with even gridblocks exhibiting a vertical plane of symmetry. (a) Whole reservoir and plane of symmetry and (b) Boundary conditions at the plane of symmetry.

$p_1=p_6$, $p_2=p_5$, and $p_3=p_4$. The modified problem is represented by the element of symmetry shown in Fig. 4.15b, with the plane of symmetry being replaced with a no-flow boundary.

Fig. 4.16a presents a 2-D horizontal reservoir with two vertical planes of symmetry A-A and B-B. Plane of symmetry A-A is normal to the x-direction and coincides with the boundaries between gridblocks 2, 6, 10, and 14 on one side and gridblocks 3, 7, 11, and 15 on the other side. Plane of symmetry B-B is normal to the y-direction and coincides with the boundaries between gridblocks 5, 6, 7, and 8 on one side and gridblocks 9, 10, 11, and 12 on the other side. The two planes of symmetry divide the reservoir into four symmetrical elements. Consequently, $p_1=p_4=p_{13}=p_{16}$, $p_2=p_3=p_{14}=p_{15}$, $p_5=p_8=p_9=p_{12}$, and $p_6=p_7=p_{10}=p_{11}$. The modified problem is represented by the smallest element of symmetry shown in Fig. 4.16b, with each plane of symmetry being replaced with a no-flow boundary.

The second two examples show planes of symmetry that pass through the centers of gridblocks. Fig. 4.17a presents a 1-D flow problem where the plane of symmetry A-A, which is normal to the flow direction (x-direction) and passes through the center of gridblock 3, and divides the reservoir into two symmetrical elements. Consequently, $p_1=p_5$ and $p_2=p_4$. The modified problem is represented by the element of symmetry shown in Fig. 4.17b, with the plane

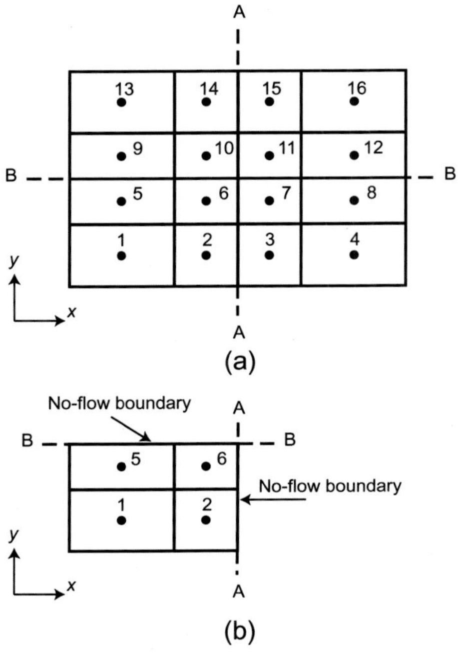

FIG. 4.16 Reservoir with even gridblocks in the x- and y-directions exhibiting two vertical planes of symmetry. (a) Whole reservoir and planes of symmetry and (b) Boundary conditions at the symmetry interface.

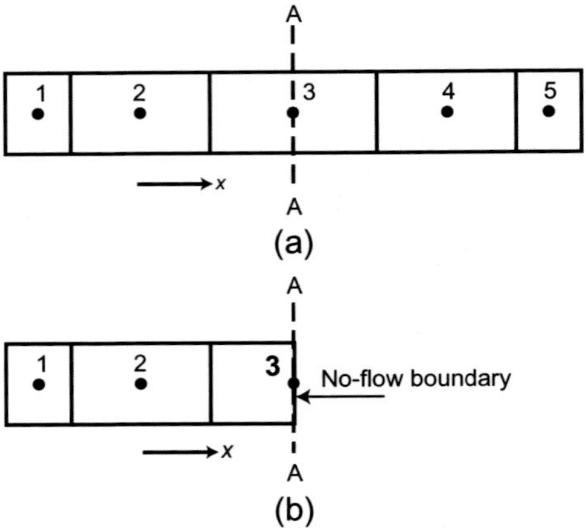

FIG. 4.17 Reservoir with odd gridblocks exhibiting a vertical plane of symmetry. (a) Whole reservoir and plane of symmetry and (b) Boundary conditions at the symmetry interface.

of symmetry being replaced with a no-flow boundary. This plane of symmetry bisects the gridblock bulk volume, wellblock rate, and wellblock geometric factor for gridblock 3 in Fig. 4.17a. Therefore, for gridblock **3** in Fig. 4.17b, $V_{b_{\mathbf{3}}}=\frac{1}{2}V_{b_3}$, $q_{sc_{\mathbf{3}}}=\frac{1}{2}q_{sc_3}$, and $G_{w_{\mathbf{3}}}=\frac{1}{2}G_{w_3}$. Note that the interblock geometric factor in the direction normal to the plane of symmetry ($G_{x_{2,3}}$) is not affected.

Fig. 4.18a presents a 2-D horizontal reservoir with two vertical planes of symmetry A-A and B-B. Plane A-A is a vertical plane of symmetry that is parallel to the *y-z* plane (normal to the *x*-direction) and passes through the centers of gridblocks 2, 5, and 8. Note that gridblocks 1, 4, and 7 are mirror images of gridblocks 3, 6, and 9. Plane B-B is a vertical plane of symmetry that is parallel to the *x-z* plane (normal to the *y*-direction) and passes through the centers of gridblocks 4, 5, and 6. Note that gridblocks 1, 2, and 3 are mirror images of gridblocks 7, 8, and 9. The two planes of symmetry divide the reservoir into four symmetrical elements. Consequently, $p_1=p_3=p_7=p_9$, $p_4=p_6$, and $p_2=p_8$. The modified problem is represented by the smallest element of symmetry shown in Fig. 4.18b, with each plane of symmetry being replaced with a no-flow boundary. Each plane of symmetry bisects the block bulk volume, wellblock rate, and wellblock geometric factor of the gridblock it passes through and bisects the interblock geometric factors in the directions that are parallel to the plane of symmetry. Therefore, $V_{b_{\mathbf{2}}}=\frac{1}{2}V_{b_2}$, $q_{sc_{\mathbf{2}}}=\frac{1}{2}q_{sc_2}$, and $G_{w_{\mathbf{2}}}=\frac{1}{2}G_{w_2}$; $V_{b_{\mathbf{4}}}=\frac{1}{2}V_{b_4}$, $q_{sc_{\mathbf{4}}}=\frac{1}{2}q_{sc_4}$, and $G_{w_{\mathbf{4}}}=\frac{1}{2}G_{w_4}$; $V_{b_{\mathbf{5}}}=\frac{1}{4}V_{b_5}$, $q_{sc_{\mathbf{5}}}=\frac{1}{4}q_{sc_5}$, and $G_{w_{\mathbf{5}}}=\frac{1}{4}G_{w_5}$; $G_{y_{\mathbf{2,5}}}=\frac{1}{2}G_{y_{2,5}}$; and $G_{x_{\mathbf{4,5}}}=\frac{1}{2}G_{x_{4,5}}$. Because gridblocks **2**, **4**, and **5** fall on the boundaries of the element of symmetry, they can be looked at as if they were gridpoints as in Chapter 5, and the same bulk volumes, wellblock rates, wellblock geometric factors, and interblock geometric factors will be calculated as those reported earlier. Note also that a plane of symmetry passing through the center of a gridblock results in a factor of $\frac{1}{2}$, as in gridblocks 2 and 4. Two planes of symmetry passing through the center of a gridblock result in a factor of $\frac{1}{2}\times\frac{1}{2}=\frac{1}{4}$, as in gridblock 5.

The third example presents two planes of symmetry, one coinciding with the boundaries between the gridblocks and the other passing through the centers of the gridblocks. Fig. 4.19a presents a 2-D horizontal reservoir with two vertical planes of symmetry A-A and B-B.

Plane A-A is a vertical plane of symmetry that is parallel to the y-z plane (normal to the x-direction) and passes through the centers of gridblocks 2, 5, 8, and 11. Note that gridblocks 1, 4, 7, and 10 are mirror images of gridblocks 3, 6, 9, and 12. Plane B-B is a vertical plane of symmetry that is parallel to the x-z plane (normal to the y-direction) and coincides with the boundaries between gridblocks 4, 5, and 6 on one side and gridblocks 7, 8, and 9 on the other side. Note that gridblocks 1, 2, and 3 are mirror images of gridblocks 10, 11, and 12. Additionally, gridblocks 4, 5, and 6 are mirror images of gridblocks 7, 8, and 9. The two planes of symmetry divide the reservoir into four symmetrical elements. Consequently, $p_1=p_3=p_{10}=p_{12}$, $p_4=p_6=p_7=p_9$, $p_2=p_{11}$, and $p_5=p_8$. The modified problem is represented by the smallest element of symmetry shown in Fig. 4.19b, with each plane of symmetry being replaced with a no-flow boundary. Plane of symmetry A-A bisects the block bulk volume, wellblock rate, and wellblock geometric factor of the gridblocks it passes through and bisects the interblock geometric factors in the directions that are parallel to the plane of symmetry (y-direction in this case). Therefore, $V_{b_2}=\frac{1}{2}V_{b_2}$, $q_{sc_2}=\frac{1}{2}q_{sc_2}$, and $G_{w_2}=\frac{1}{2}G_{w_2}$; $V_{b_5}=\frac{1}{2}V_{b_5}$, $q_{sc_5}=\frac{1}{2}q_{sc_5}$, and $G_{w_5}=\frac{1}{2}G_{w_5}$; $V_{b_8}=\frac{1}{2}V_{b_8}$, $q_{sc_8}=\frac{1}{2}q_{sc_8}$, and $G_{w_8}=\frac{1}{2}G_{w_8}$; $V_{b_{11}}=\frac{1}{2}V_{b_{11}}$, $q_{sc_{11}}=\frac{1}{2}q_{sc_{11}}$, and $G_{w_{11}}=\frac{1}{2}G_{w_{11}}$; $G_{y_{2,5}}=\frac{1}{2}G_{y_{2,5}}$; $G_{y_{5,8}}=\frac{1}{2}G_{y_{5,8}}$; and $G_{y_{8,11}}=\frac{1}{2}G_{y_{8,11}}$. Because gridblocks 2, 5, 8, and 11 fall on the boundaries of the element of symmetry, they can be looked at as if they were gridpoints as in Chapter 5, and the same bulk volumes, wellblock rates, wellblock geometric factors, and interblock geometric factors will be calculated as those reported earlier. Note also that a plane of symmetry passing through the center of a gridblock results in a factor of $\frac{1}{2}$, as in gridblocks 2, 5, 8, and 11 in Fig. 4.19a.

The fourth set of examples show oblique planes of symmetry. Fig. 4.20a shows a reservoir similar to that depicted in Fig. 4.16a, but the present reservoir has two additional planes of symmetry C-C and D-D. The four planes of symmetry divide the reservoir into eight symmetrical elements, each with a triangular shape as shown in Fig. 4.20b. Consequently, $p_1=p_4=p_{13}=p_{16}$, $p_6=p_7=p_{10}=p_{11}$, and $p_2=p_3=p_{14}=p_{15}=p_5=p_8=p_9=p_{12}$. The modified problem is represented by the smallest element of symmetry shown in Fig. 4.20b, with each plane of symmetry being replaced with a no-flow boundary.

Fig. 4.21a shows a reservoir similar to that depicted in Fig. 4.18a, but the present reservoir has two additional planes of symmetry C-C and D-D. The four planes of symmetry divide the reservoir into eight symmetrical elements, each

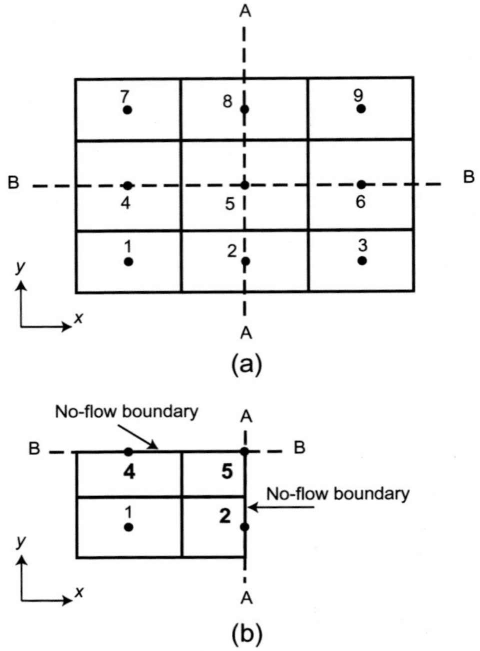

FIG. 4.18 Reservoir with odd gridblocks in the x- and y-directions exhibiting two vertical planes of symmetry. (a) Whole reservoir and planes of symmetry and (b) Boundary conditions at the symmetry interfaces.

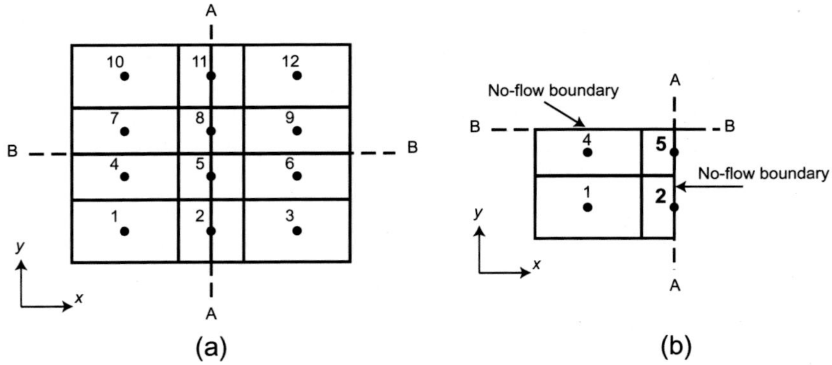

FIG. 4.19 Reservoir with even gridblocks in the y-direction and odd gridblocks in the x-direction exhibiting two vertical planes of symmetry. (a) Whole reservoir and planes of symmetry and (b) Boundary conditions at the symmetry interfaces.

with a triangular shape as shown in Fig. 4.21b. Consequently, $p_1=p_3=p_7=p_9$ and $p_4=p_6=p_2=p_8$. The modified problem is represented by the smallest element of symmetry shown in Fig. 4.21b, with each plane of symmetry being replaced with a no-flow boundary. A vertical plane of symmetry C-C or D-D that passes through the center of a gridblock but is neither parallel to the x-axis

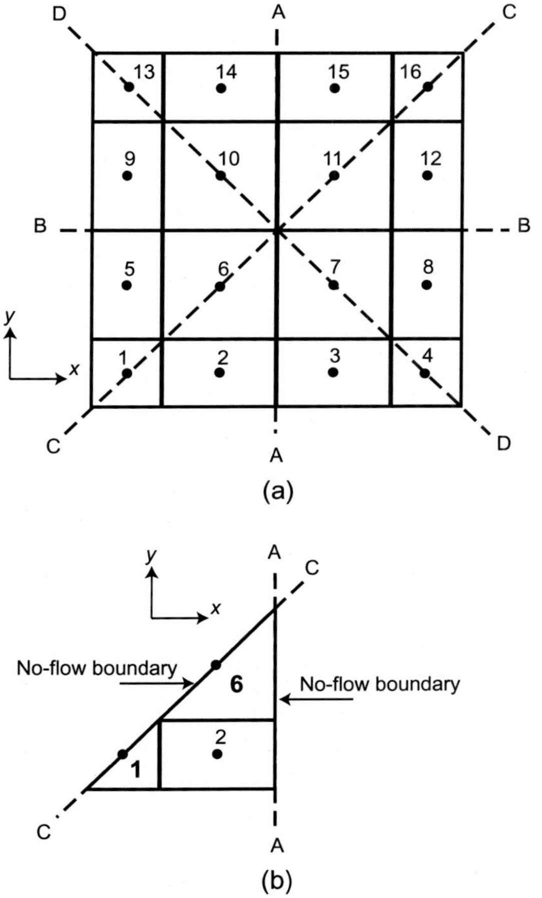

FIG. 4.20 Reservoir with even gridblocks in the x- and y-directions exhibiting four vertical planes of symmetry. (a) Whole reservoir and planes of symmetry and (b) Boundary conditions at the symmetry interfaces.

nor the y-axis (oblique plane), as shown in Figs. 4.20a and 4.21a, bisects the gridblock bulk volume, wellblock rate, and wellblock geometric factor of the gridblock it passes through. An oblique plane does not affect the interblock geometric factors in the x-axis or the y-axis. In reference to gridblocks **1**, **6**, and **5** in Figs. 4.20b and 4.21b, $V_{b_\mathbf{1}} = \frac{1}{2}V_{b_1}$, $q_{sc_\mathbf{1}} = \frac{1}{2}q_{sc_1}$, and $G_{w_\mathbf{1}} = \frac{1}{2}G_{w_1}$; $V_{b_\mathbf{6}} = \frac{1}{2}V_{b_6}$, $q_{sc_\mathbf{6}} = \frac{1}{2}q_{sc_6}$, and $G_{w_\mathbf{6}} = \frac{1}{2}G_{w_6}$; $V_{b_\mathbf{5}} = \frac{1}{8}V_{b_5}$, $q_{sc_\mathbf{5}} = \frac{1}{8}q_{sc_5}$, and $G_{w_\mathbf{5}} = \frac{1}{8}G_{w_5}$; $G_{y_{\mathbf{2},\mathbf{5}}} = \frac{1}{2}G_{y_{2,5}}$; and $G_{x_{\mathbf{2},\mathbf{6}}} = G_{x_{2,6}}$. Note that the four planes of symmetry (A-A, B-B, C-C, and D-D) passing through the center of gridblock 5 in Fig. 4.21a result in the factor of $\frac{1}{4} \times \frac{1}{2} = \frac{1}{8}$ used to calculate the actual bulk volume, wellblock rate, and wellblock geometric factor for gridblock **5** in Fig. 4.21b. That is to say, the modifying factor equals $\frac{1}{n_{vsp}} \times \frac{1}{2}$, where n_{vsp} is the number of vertical planes of symmetry passing through the center of a gridblock.

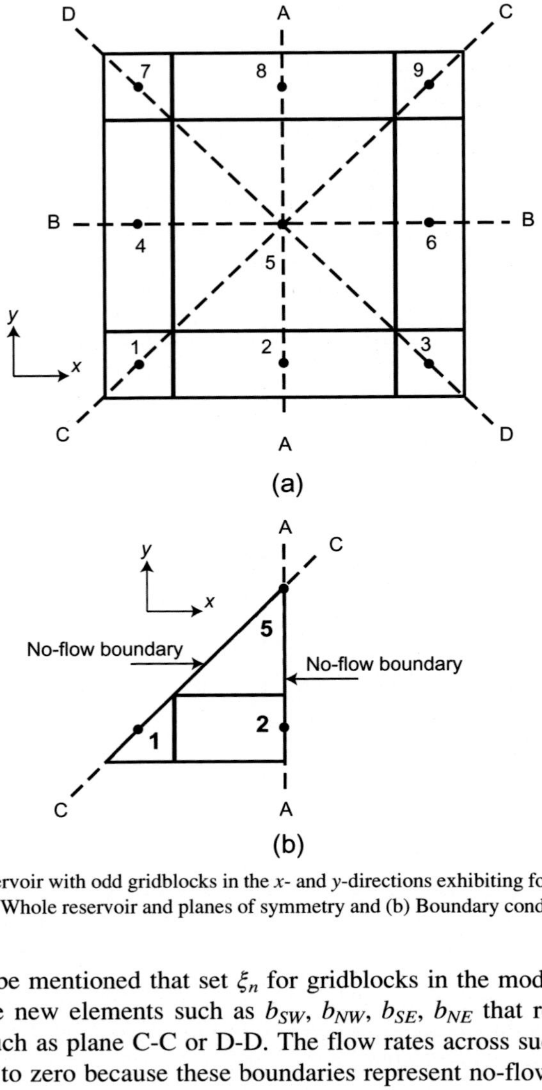

FIG. 4.21 Reservoir with odd gridblocks in the x- and y-directions exhibiting four vertical planes of symmetry. (a) Whole reservoir and planes of symmetry and (b) Boundary conditions at the symmetry interfaces.

It should be mentioned that set ξ_n for gridblocks in the modified problem might include new elements such as b_{SW}, b_{NW}, b_{SE}, b_{NE} that reflect oblique boundaries such as plane C-C or D-D. The flow rates across such boundaries ($q^m_{sc_{l,n}}$) are set to zero because these boundaries represent no-flow boundaries.

4.7 Summary

This chapter presents reservoir discretization in Cartesian and radial–cylindrical coordinates using a block-centered grid. For the Cartesian coordinate system, equations similar to those represented by Eq. (4.1) define gridblock locations and the relationships between gridblock sizes, gridblock boundaries, and distances between points representing gridblocks in the x-, y-, and z-directions, and Table 4.1 presents equations for the calculation of the transmissibility geometric factors in the three directions. For the radial-cylindrical coordinate system used for single-well simulation, the equations that define block locations and the relationships between gridblock sizes, gridblock boundaries, and distances between points representing blocks in the r-direction are given by Eqs. (4.81) through (4.88), Eq. (4.80) in the θ-direction, and an equation similar to Eq. (4.1) for the z-direction. The equations in either Table 4.2 or Table 4.3 can be used to calculate transmissibility geometric factors in the r-, θ-, and z-directions. Eq. (4.2) expresses the general form of the flow equation that applies to boundary gridblocks and interior gridblocks in 1-D, 2-D, or 3-D flow in both Cartesian and radial-cylindrical coordinates. The flow equation for any gridblock has flow terms equal to the number of existing neighboring gridblocks and fictitious wells equal to the number of boundary conditions. Each fictitious well represents a boundary condition. The flow rate of a fictitious well is given by Eq. (4.24b), (4.27), (4.32), or (4.37b) for a specified pressure gradient, specified flow rate, no-flow, or specified pressure boundary condition, respectively.

If reservoir symmetry exists, it can be exploited to define the smallest element of symmetry. Planes of symmetry may pass along gridblock boundaries or through gridblock centers. To simulate the smallest element of symmetry, planes of symmetry are replaced with no-flow boundaries and new interblock geometric factors, bulk volume, wellblock rate, and wellblock geometric factors for boundary gridblocks are calculated prior to simulation.

4.8 Exercises

4.1. What is the meaning of reservoir discretization into gridblocks?

4.2. Using your own words, describe how you discretize a reservoir of length L_x along the x-direction using n gridblocks.

4.3. Fig. 4.5 shows a reservoir with regular boundaries.
- **a.** How many boundaries does this reservoir have along the x-direction? Identify and name these boundaries.
- **b.** How many boundaries does this reservoir have along the y-direction? Identify and name these boundaries.
- **c.** How many boundaries does this reservoir have along the z-direction? Identify and name these boundaries.
- **d.** How many boundaries does this reservoir have along all directions?

4.4. Consider the 2-D reservoir described in Example 4.5 and shown in Fig. 4.12.
- **a.** Identify the interior and boundary gridblocks in the reservoir.
- **b.** Write the set of neighboring gridblocks (ψ_n) for each gridblock in the reservoir.
- **c.** Write the set of reservoir boundaries (ξ_n) for each gridblock in the reservoir.
- **d.** How many boundary conditions does each boundary gridblock have? How many fictitious wells does each boundary gridblock have? Write the terminology for the flow rate of each fictitious well.

e. How many flow terms does each boundary gridblock have?
f. Add the number of flow terms and number of fictitious wells for each boundary gridblock. Do they add up to four for each boundary gridblock?
g. How many flow terms does each interior gridblock have?
h. What can you conclude from your results of (f) and (g) earlier?

4.5. Consider fluid flow in the 1-D horizontal reservoir shown in Fig. 4.22.
a. Write the appropriate flow equation for gridblock n in this reservoir.
b. Write the flow equation for gridblock 1 by finding ψ_1 and ξ_1 and then use them to expand the equation in (a).
c. Write the flow equation for gridblock 2 by finding ψ_2 and ξ_2 and then use them to expand the equation in (a).
d. Write the flow equation for gridblock 3 by finding ψ_3 and ξ_3 and then use them to expand the equation in (a).

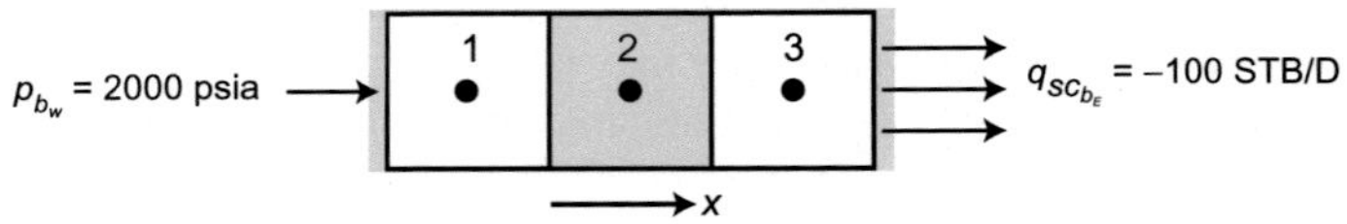

FIG. 4.22 1-D reservoir in Exercise 4.5.

4.6. Consider fluid flow in the 2-D horizontal reservoir shown in Fig. 4.23.
a. Write the appropriate flow equation for gridblock n in this reservoir.
b. Write the flow equation for gridblock 1 by finding ψ_1 and ξ_1 and then use them to expand the equation in (a).
c. Write the flow equation for gridblock 3 by finding ψ_3 and ξ_3 and then use them to expand the equation in (a).
d. Write the flow equation for gridblock 5 by finding ψ_5 and ξ_5 and then use them to expand the equation in (a).
e. Write the flow equation for gridblock 9 by finding ψ_9 and ξ_9 and then use them to expand the equation in (a).

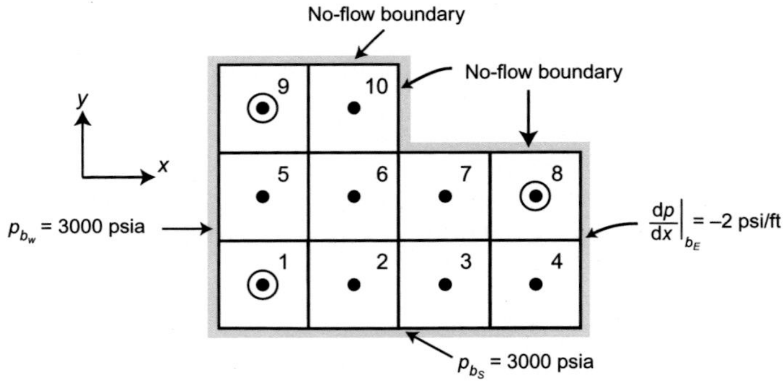

FIG. 4.23 2-D reservoir for Exercise 4.6.

4.7. Consider single-phase flow in a homogeneous, 1-D reservoir with constant pressure specification at the reservoir left boundary. The reservoir is discretized using a regular grid. Write the flow equation for gridblock 1, which shares its left boundary with the reservoir, and prove that $p_b = {}^1/_2(3p_1 - p_2)$. Aziz and Settari (1979) claim that the earlier equation represents a second-order correct approximation for boundary pressure.

4.8. A single-phase oil reservoir is described by four equal gridblocks as shown in Fig. 4.24. The reservoir is horizontal and has $k=25$ md. Gridblock dimensions are $\Delta x=500$, $\Delta y=700$, and $h=60$ ft. Oil properties are $B=1$ RB/STB and $\mu=0.5$ cP. The reservoir left boundary is maintained at constant pressure of 2500 psia, and the reservoir right boundary is sealed off to flow. A well in gridblock 3 produces 80 STB/D of oil. Assuming that the reservoir rock and oil are incompressible, calculate the pressure distribution in the reservoir.

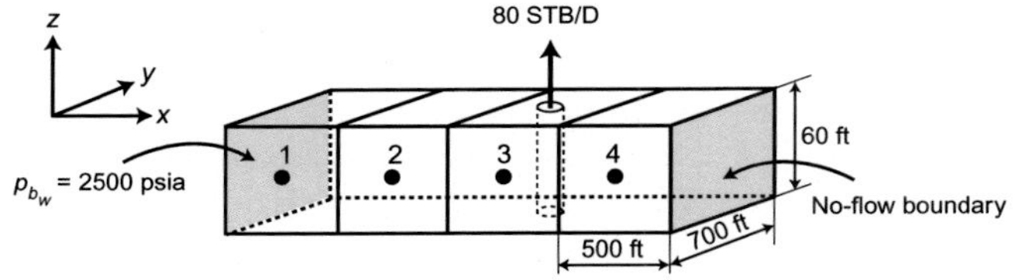

FIG. 4.24 Discretized 1D reservoir in Exercise 4.8.

4.9. A 1-D horizontal oil reservoir shown in Fig. 4.25 is described by four equal gridblocks. Reservoir blocks have $k=90$ md, $\Delta x=300$ ft, $\Delta y=250$ ft, and $h=45$ ft. Oil FVF and viscosity are 1 RB/STB and 2 cP, respectively. The reservoir left boundary is maintained at constant pressure of 2000 psia, and the reservoir right boundary has a constant influx of oil at a rate of 80 STB/D. A well in gridblock 3 produces 175 STB/D of oil. Assuming that the reservoir rock and oil are incompressible, calculate the pressure distribution in the reservoir.

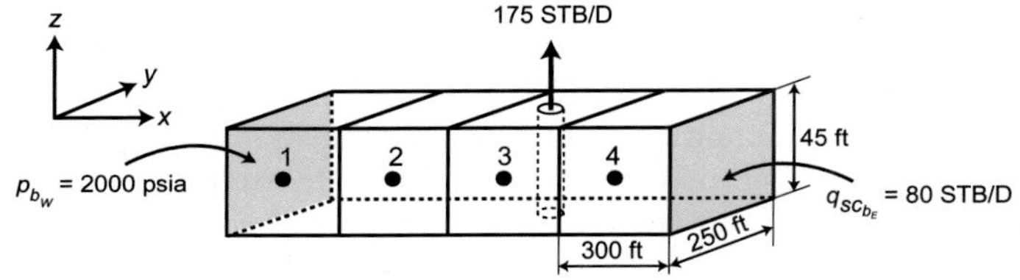

FIG. 4.25 Discretized 1D reservoir in Exercise 4.9.

4.10. A 1-D horizontal oil reservoir shown in Fig. 4.26 is described by four equal gridblocks. Reservoir blocks have $k=120$ md, $\Delta x=500$ ft,

$\Delta y = 450$ ft, and $h = 30$ ft. Oil FVF and viscosity are 1 RB/STB and 3.7 cP, respectively. The reservoir left boundary is subject to a constant pressure gradient of −0.2 psi/ft, and the reservoir right boundary is a no-flow boundary. A well in gridblock 3 produces oil at a rate such that the pressure of gridblock 3 is maintained at 1500 psia. Assuming that the reservoir rock and oil are incompressible, calculate the pressure distribution in the reservoir. Then, estimate the well production rate.

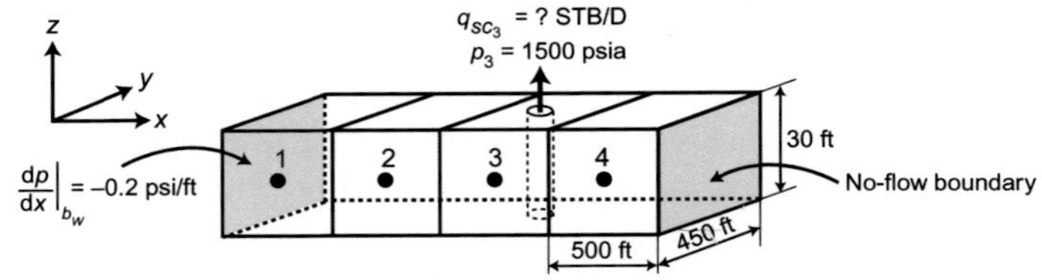

FIG. 4.26 Discretized 1-D reservoir in Exercise 4.10.

4.11. A 1-D horizontal oil reservoir shown in Fig. 4.27 is described by four equal gridblocks. Reservoir blocks have $k = 70$ md, $\Delta x = 400$ ft, $\Delta y = 660$ ft, and $h = 10$ ft. Oil FVF and viscosity are 1 RB/STB and 1.5 cP, respectively. The reservoir left boundary is maintained at constant pressure of 2700, while the boundary condition at the reservoir right boundary is not known, the pressure of gridblock 4 is maintained at 1900 psia. A well in gridblock 3 produces 150 STB/D of oil. Assuming that the reservoir rock and oil are incompressible, calculate the pressure distribution in the reservoir. Estimate the rate of oil that crosses the reservoir right boundary.

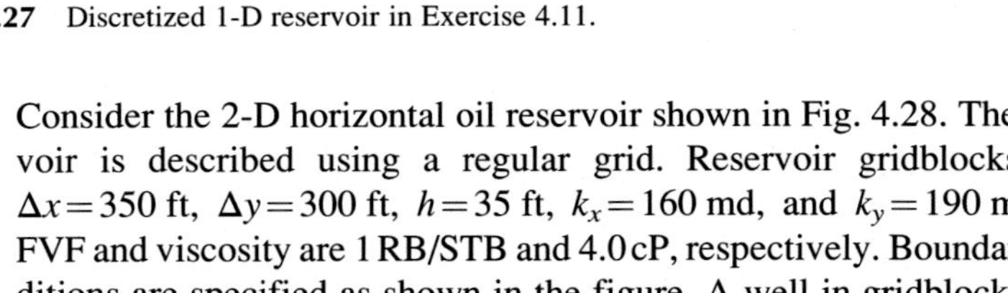

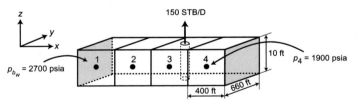

FIG. 4.27 Discretized 1-D reservoir in Exercise 4.11.

4.12. Consider the 2-D horizontal oil reservoir shown in Fig. 4.28. The reservoir is described using a regular grid. Reservoir gridblocks have $\Delta x = 350$ ft, $\Delta y = 300$ ft, $h = 35$ ft, $k_x = 160$ md, and $k_y = 190$ md. Oil FVF and viscosity are 1 RB/STB and 4.0 cP, respectively. Boundary conditions are specified as shown in the figure. A well in gridblock 5 produces oil at a rate of 2000 STB/D. Assume that the reservoir rock and oil are incompressible. Write the flow equations for all gridblocks. Do not solve the equations.

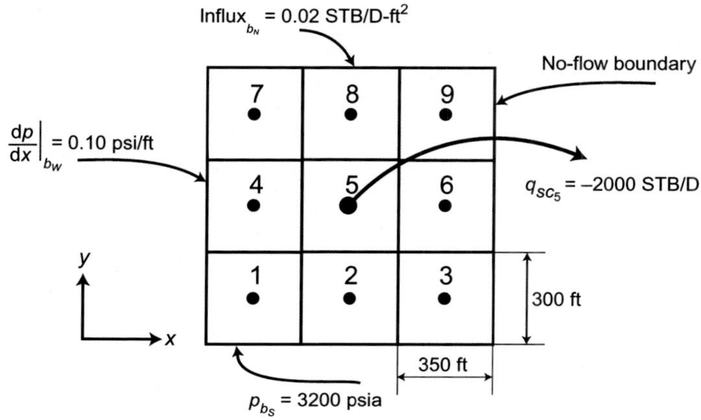

FIG. 4.28 Discretized 2-D reservoir in Exercise 4.12.

4.13. Starting with Eq. (4.88c), which expresses the bulk volume of gridblock (n_r, j, k) in terms of r_e and $r_{n_r-1/2}$, derive Eq. (4.88d), which expresses the bulk volume in terms of α_{lg} and r_e.

4.14. A 6-in. vertical well producing 500 STB/D of oil is located in 16-acre spacing. The reservoir is 30 ft thick and has horizontal permeability of 50 md. The oil FVF and viscosity are 1 RB/B and 3.5 cP, respectively. The reservoir external boundaries are no-flow boundaries. The reservoir is simulated using four gridblocks in the radial direction as shown in Fig. 4.29. Write the flow equations for all gridblocks. Do not substitute for values on the RHS of equations.

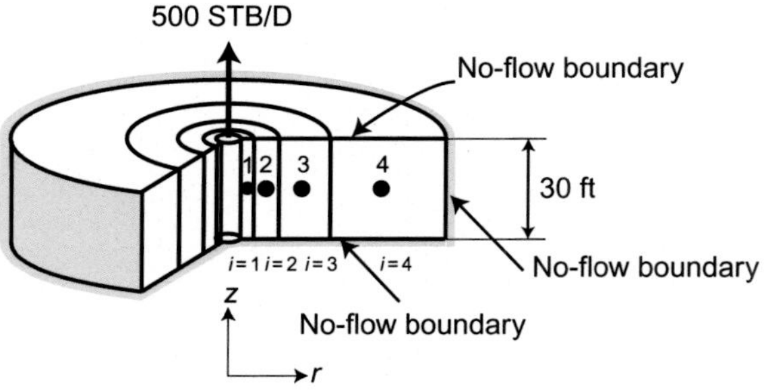

FIG. 4.29 Discretized reservoir in Exercise 4.14.

4.15. A 9⅝-in. vertical well is located in 12-acre spacing. The reservoir thickness is 50 ft. Horizontal and vertical reservoir permeabilities are 70 md and 40 md, respectively. The flowing fluid has a density, FVF, and viscosity of 62.4 lbm/ft^3, 1 RB/B, and 0.7 cP, respectively. The reservoir external boundary in the radial direction is a no-flow boundary, and the well is completed in the top 20 ft only and produces at a rate of 1000 B/D. The reservoir bottom boundary is subject to influx such that the boundary is maintained at 3000 psia. The reservoir top boundary is sealed to flow. Assuming the reservoir can be simulated using two gridblocks in the vertical direction and four gridblocks in the radial direction as shown in Fig. 4.30, write the flow equations for all gridblocks in this reservoir.

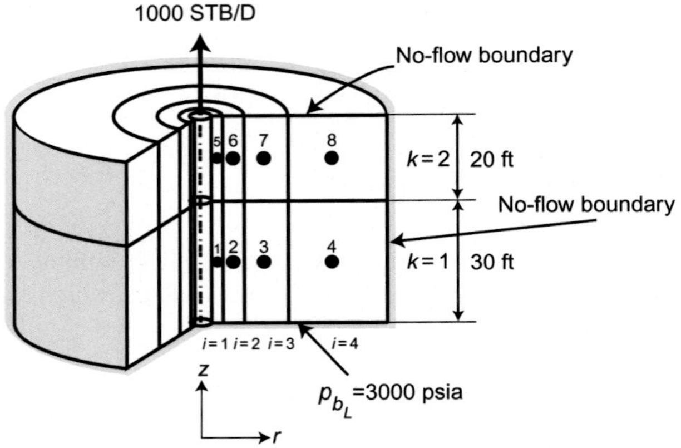

FIG. 4.30 Discretized 2-D radial-cylindrical reservoir in Exercise 4.15.

Chapter 5

Simulation with a point-distributed grid

Chapter outline

5.1 Introduction

Discretization process creates inherent challenges involving proper representation of natural processes. The problem is accentuated by boundaries, which create discontinuities—an absurd condition for natural systems. Historically, the petroleum engineers have identified these problems and have attempted to address many problems that emerge from discretization and boundary conditions, which must be addressed separately. Few, however, have recognized that the engineering approach keeps the process transparent and enables modelers to remedy with physically realistic solutions. This chapter presents discretization of 1-D, 2-D, and 3-D reservoirs using point-distributed grids in Cartesian and radial-cylindrical coordinate systems. This chapter describes the construction of a point-distributed grid for a reservoir and the relationships between the distances separating gridpoints, block boundaries, and sizes of the blocks represented by the gridpoints. The resulting gridpoints can be classified into interior and boundary gridpoints. While Chapter 2 derives the flow equations for interior gridpoints, the boundary gridpoints are subject to boundary conditions and thus require special treatment. This chapter presents the treatment

Petroleum Reservoir Simulation. https://doi.org/10.1016/B978-0-12-819150-7.00005-0

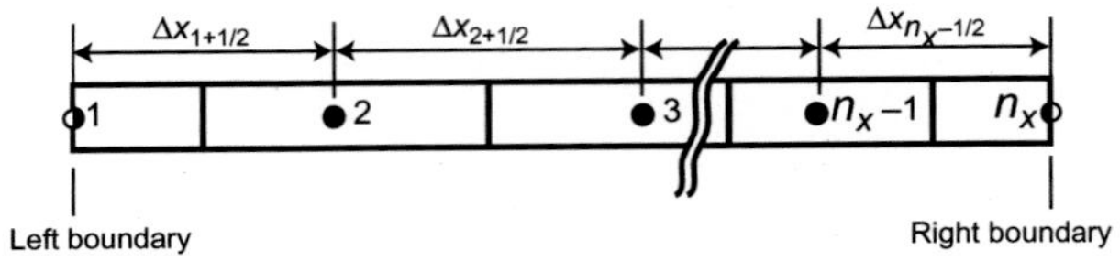

FIG. 5.1 Discretization of a 1-D reservoir using a point-distributed grid.

of various boundary conditions and introduces a general flow equation that is applicable to interior gridpoints and boundary gridpoints. This chapter also presents the equations for directional transmissibilities in both Cartesian and radial-cylindrical coordinate systems and discusses the use of symmetry in reservoir simulation.

There are three important differences between the block-centered grid discussed in Chapter 4 and the point-distributed grid discussed in this chapter. First, the boundary gridpoints for a point-distributed grid fall on reservoir boundaries, not inside reservoir boundaries as in the case of a block-centered grid. Second, the actual bulk volume and actual well rate of boundary gridpoints are a half, a quarter, or an eighth of those of whole blocks if they fall on one, two, or three reservoir boundaries, respectively. Third, the transmissibility parallel to the reservoir boundary for a boundary gridpoint has half of that of the whole block. These points are taken into consideration in developing the general flow equation for a point-distributed grid.

5.2 Reservoir discretization

As described in Chapter 4, reservoir discretization involves the assigning a set of gridpoints that represent blocks that are well defined in terms of properties, dimensions, boundaries, and locations in the reservoir. Fig. 5.1 shows a point-distributed grid for a 1-D reservoir in the direction of the x-axis. The point-distributed grid is constructed by choosing n_x gridpoints that span the entire reservoir length in the x-direction. In other words, the first gridpoint is placed at one reservoir boundary, and the last gridpoint is placed at the other reservoir boundary. The distances between gridpoints are assigned predetermined values ($\Delta x_{i+1/2}$, $i = 1, 2, 3 \ldots n_x - 1$) that are not necessarily equal. Each gridpoint represents a block whose boundaries are placed halfway between the gridpoint and its neighboring gridpoints.

Fig. 5.2 focuses on gridpoint i and its neighboring gridpoints. It shows how these gridpoints are related to each other. In addition, the figure shows block dimensions (Δx_{i-1}, Δx_i, Δx_{i+1}), block boundaries ($x_{i-1/2}$, $x_{i+1/2}$), distances between gridpoint i and block boundaries (δx_{i^-}, δx_{i^+}), and distances between gridpoints ($\Delta x_{i-1/2}$, $\Delta x_{i+1/2}$). Block dimensions, block boundaries, and gridpoint locations satisfy the following relationships:

$$
\begin{aligned}
&x_1 = 0,\ x_{n_x} = L_x,\ (\text{i.e.}, x_{n_x} - x_1 = L_x),\\
&\delta x_{i^-} = {}^1/_2 \Delta x_{i-1/2},\ i = 2,3 \ldots n_x,\\
&\delta x_{i^+} = {}^1/_2 \Delta x_{i+1/2},\ i = 1,2,3 \ldots n_x - 1,\\
&x_{i+1} = x_i + \Delta x_{i+1/2},\ i = 1,2,3 \ldots n_x - 1,\\
&x_{i-1/2} = x_i - \delta x_{i^-} = x_i - {}^1/_2 \Delta x_{i-1/2},\ i = 2,3 \ldots n_x,\\
&x_{i+1/2} = x_i + \delta x_{i^+} = x_i + {}^1/_2 \Delta x_{i+1/2},\ i = 1,2,3 \ldots n_x - 1,\\
&\Delta x_i = \delta x_{i^-} + \delta x_{i^+} = {}^1/_2 \left(\Delta x_{i-1/2} + \Delta x_{i+1/2}\right),\ i = 2,3 \ldots n_x - 1,\\
&\Delta x_1 = \delta x_{1^+} = {}^1/_2 \Delta x_{1+1/2},\\
&\text{and}\\
&\Delta x_{n_x} = \delta x_{n_x^-} = {}^1/_2 \Delta x_{n_x - 1/2}.
\end{aligned}
\tag{5.1}
$$

Fig. 5.3 shows the discretization of a 2-D reservoir into a 5×4 irregular grid. An irregular grid implies that the distances between the gridpoints in the direction of the x-axis ($\Delta x_{i \mp 1/2}$) and the y-axis ($\Delta y_{j \mp 1/2}$) are neither equal nor constant. Discretization using a regular grid means that distances between gridpoints in the x-direction and those in the y-direction are constant but not necessarily equal in both directions. The discretization in the x-direction uses the procedure just mentioned and the relationships presented in Eq. (5.1). Discretization in the y-direction uses a procedure and relationships similar to those for the x-direction, and the same can be said of the z-direction for a 3-D

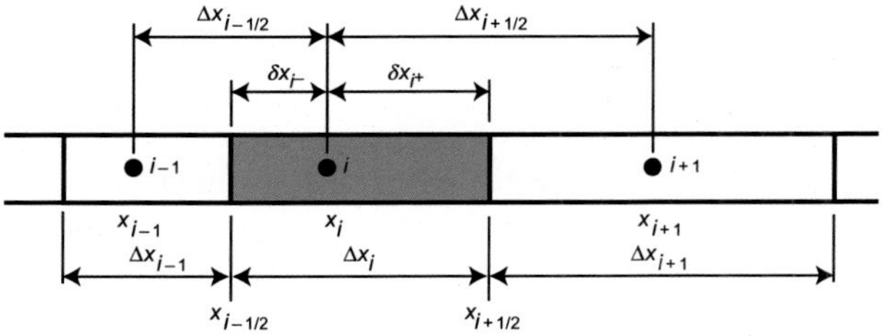

FIG. 5.2 Gridpoint i and its neighboring gridpoints in the x-direction.

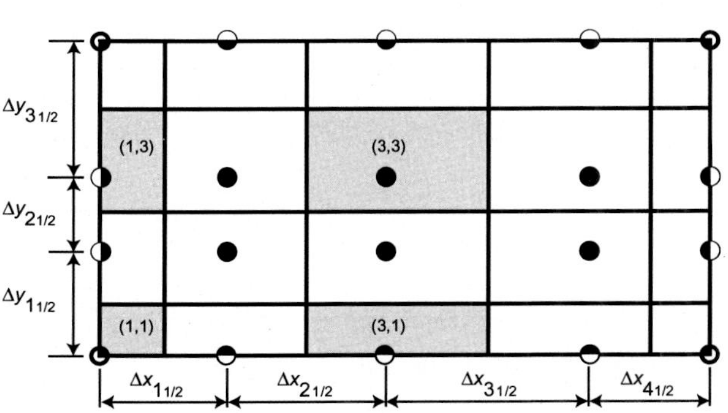

FIG. 5.3 Discretization of a 2-D reservoir using a point-distributed grid.

reservoir. Inspection of Figs. 5.1 and 5.3 shows that the boundary gridpoints fall on the boundaries of the reservoir. In addition, they are not completely enclosed by the blocks they represent.

Example 5.1 A 5000 ft × 1200 ft × 75 ft horizontal reservoir contains oil that flows along its length. The reservoir rock porosity and permeability are 0.18 and 15 md, respectively. The oil FVF and viscosity are 1 RB/STB and 10 cP, respectively. The reservoir has a well located at 4000 ft from the reservoir left boundary and produces oil at a rate of 150 STB/D. Discretize the reservoir into six equally spaced gridpoints using a point-distributed grid and assign properties to the gridpoints comprising this reservoir.

Solution

Using a point-distributed grid, the reservoir is divided along its length into six equally spaced gridpoints with gridpoints 1 and 6 being placed on the reservoir left and right boundaries, respectively. Each gridpoint represents a block whose boundaries are placed halfway between gridpoints. Therefore, $n_x=6$ and $\Delta x_{i\mp1/2}=L_x/(n_x-1)=5000/5=1000$ ft. Gridpoints are numbered from 1 to 6 as shown in Fig. 5.4.

Now, the reservoir is described through assigning properties to its six gridpoints ($i=1, 2, 3, 4, 5, 6$). All gridpoints have the same elevation because the reservoir is horizontal. The blocks that are represented by the gridpoints have the dimensions of $\Delta y=1200$ ft and $\Delta z=75$ ft and properties of $k_x=15$ md and $\phi=0.18$. The blocks for gridpoints 2, 3, 4, and 5 have $\Delta x=1000$ ft, whereas those for gridpoints 1 and 6 have $\Delta x=500$ ft. The distances between neighboring gridpoints are equal; that is, $\Delta x_{i\mp1/2}=1000$ ft and $A_{x_{i\mp1/2}}=A_x=\Delta y\times\Delta z=1200\times75=90,000$ ft^2. Gridpoint 1 falls on the reservoir west boundary, gridpoint 6 falls on the reservoir east boundary, and gridpoints 2, 3, 4, and 5 are interior gridpoints. In addition, the block enclosing gridpoint 5 hosts a well with $q_{sc_5}=-150$ STB/D. Fluid properties are $B=1$ RB/STB and $\mu=10$ cP.

5.3 Flow equation for boundary gridpoints

In this section, we present a form of the flow equation that applies to interior gridpoints and boundary gridpoints. That is to say, the proposed flow equation reduces to the flow equations presented in Chapters 2 and 3 for interior gridpoints, but it also includes the effects of boundary conditions for boundary

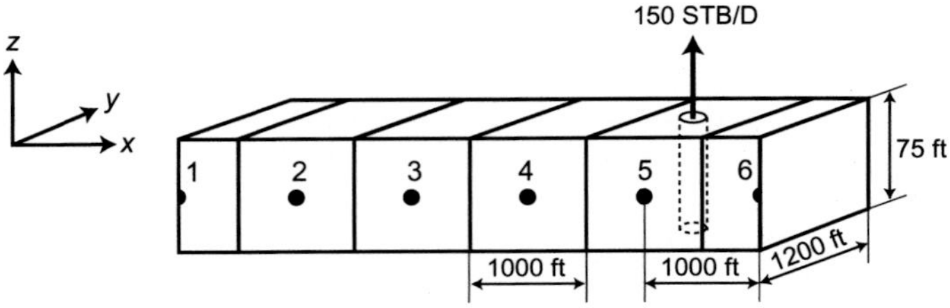

FIG. 5.4 Discretized 1-D reservoir in Example 5.1.

gridpoints. Fig. 5.1 shows a discretized 1-D reservoir in the direction of the x-axis. Gridpoints 2, 3, … $n_x - 1$ are interior gridpoints, whereas gridpoints 1 and n_x are boundary gridpoints that each falls on one reservoir boundary. Fig. 5.3 shows a discretized 2-D reservoir. The figure highlights an interior gridpoint, gridpoint (3,3); two boundary gridpoints that each falls on one reservoir boundary, gridpoints (1,3) and (3,1); and a gridpoint that falls at the intersection of two reservoir boundaries, gridpoint (1,1). Therefore, one can conclude that not all gridpoints fall inside reservoir boundaries, and the boundary gridpoints have incomplete blocks. As discussed in the previous chapter, there are interior gridpoints and boundary gridpoints, which may fall on one, two, or three reservoir boundaries. The terminology in this discussion has been presented in Chapter 4. This terminology is repeated in Fig. 5.5. Reservoir boundaries along the x-axis are termed reservoir west boundary (b_W) and reservoir east boundary (b_E), and those along the y-axis are termed reservoir south boundary (b_S) and reservoir north boundary (b_N). Reservoir boundaries along the z-axis are termed reservoir lower boundary (b_L) and reservoir upper boundary (b_U).

The flow equations for both interior and boundary gridpoints have a production (injection) term and an accumulation term. The treatment of a boundary condition by the engineering approach involves replacing the boundary condition with a no-flow boundary plus a fictitious well having flow rate of $q_{sc_{b,bP}}$ that reflects fluid transfer between the gridpoint that is exterior to the reservoir and the reservoir boundary itself (b) or the boundary gridpoint (bP). The flow equation for an interior gridpoint has a number of flow terms that equals the number of neighboring gridpoints (two, four, or six terms for a 1-D, 2-D, or 3-D reservoir, respectively). The flow equation for a boundary gridpoint has a number of flow terms that equals the number of existing neighboring gridpoints in the reservoir and a number of fictitious wells that equals the number of reservoir boundaries the boundary gridpoint falls on.

A general form of the flow equation that applies to boundary gridpoints and interior gridpoints in 1-D, 2-D, or 3-D flow in both Cartesian and radial-cylindrical coordinates can be expressed best using CVFD terminology. The use of summation operators in CVFD terminology makes it flexible and suitable for describing flow terms in the equation of any gridpoint that may or may not

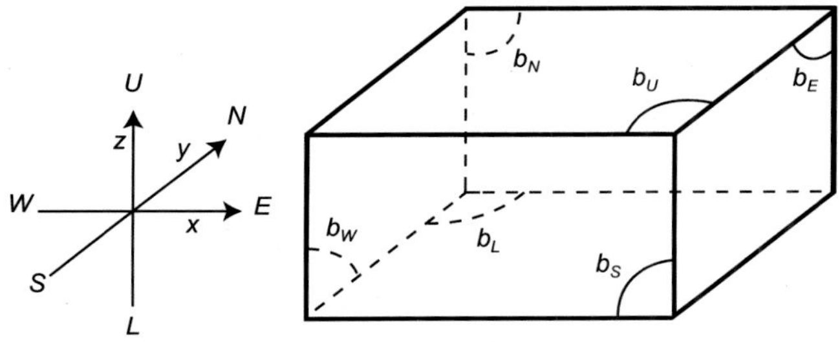

FIG. 5.5 Definition of the reservoir left and right boundaries in 3-D reservoirs.

be on a reservoir boundary. The general form of the flow equation for gridpoint n can be written as

$$\sum_{l\in\psi_n} T_{l,n}^m\left[\left(p_l^m - p_n^m\right) - \gamma_{l,n}^m (Z_l - Z_n)\right] + \sum_{l\in\xi_n} q_{sc_{l,n}}^m + q_{sc_n}^m = \frac{V_{b_n}}{\alpha_c \Delta t}\left[\left(\frac{\phi}{B}\right)_n^{n+1} - \left(\frac{\phi}{B}\right)_n^n\right] \quad (5.2a)$$

or, in terms of potentials, as

$$\sum_{l\in\psi_n} T_{l,n}^m\left(\Phi_l^m - \Phi_n^m\right) + \sum_{l\in\xi_n} q_{sc_{l,n}}^m + q_{sc_n}^m = \frac{V_{b_n}}{\alpha_c \Delta t}\left[\left(\frac{\phi}{B}\right)_n^{n+1} - \left(\frac{\phi}{B}\right)_n^n\right] \quad (5.2b)$$

where ψ_n= the set whose elements are the existing neighboring gridpoints in the reservoir, ξ_n= the set whose elements are the reservoir boundaries (b_L, b_S, b_W, b_E, b_N, b_U) that are shared by gridpoint n, and $q_{sc_{l,n}}^m$= flow rate of the fictitious well representing fluid transfer between reservoir boundary l and gridpoint n as a result of a boundary condition. For a 3-D reservoir, ξ_n is either an empty set for interior gridpoints or a set that contains one element for boundary gridpoints that fall on one reservoir boundary, two elements for boundary gridpoints that fall on two reservoir boundaries, or three elements for boundary gridpoints that fall on three reservoir boundaries. An empty set implies that the gridpoint does not fall on any reservoir boundary; that is, gridpoint n is an interior gridpoint, and hence, $\sum_{l\in\xi_n} q_{sc_{l,n}}^m = 0$. In engineering notation, $n \equiv (i,j,k)$, and Eq. (5.2a) becomes

$$\sum_{l\in\psi_{i,j,k}} T_{l,(i,j,k)}^m\left[\left(p_l^m - p_{i,j,k}^m\right) - \gamma_{l,(i,j,k)}^m\left(Z_l - Z_{i,j,k}\right)\right] + \sum_{l\in\xi_{i,j,k}} q_{sc_{l,(i,j,k)}}^m + q_{sc_{i,j,k}}^m = \frac{V_{b_{i,j,k}}}{\alpha_c \Delta t}\left[\left(\frac{\phi}{B}\right)_{i,j,k}^{n+1} - \left(\frac{\phi}{B}\right)_{i,j,k}^n\right] \quad (5.2c)$$

It is important to recognize that the flow equations for interior gridpoints in a point-distributed grid and those for interior gridblocks in a block-centered grid are the same because interior gridpoints represent the whole blocks. The flow equations for boundary blocks and boundary gridpoints, however, are different because of the way the two grids are constructed. To incorporate boundary conditions appropriately in the flow equation of a boundary gridpoint, we must write the flow equation for the whole block, which completely encloses the boundary gridpoint, in terms of the properties of the actual block and note that the whole block and the actual block are represented by the same boundary gridpoint.

It must be mentioned that reservoir blocks have a three-dimensional shape whether fluid flow is 1-D, 2-D, or 3-D. The number of existing neighboring gridpoints and the number of reservoir boundaries shared by a reservoir gridpoint add up to six as the case in 3-D flow. Existing neighboring gridpoints contribute to flow to or from the gridpoint, whereas reservoir boundaries may or may not contribute to flow depending on the dimensionality of flow and the prevailing boundary conditions. The dimensionality of flow implicitly defines those reservoir boundaries that do not contribute to flow at all. In 1-D flow problems, all reservoir gridpoints have four reservoir boundaries that do not contribute to flow. In 1-D flow in the x-direction, the reservoir south, north, lower, and upper boundaries do not contribute to flow to any reservoir gridpoint, including boundary gridpoints. These four reservoir boundaries (b_L, b_S, b_N, b_U) are discarded as if they did not exist. As a result, an interior reservoir gridpoint has two neighboring gridpoints and no reservoir boundaries, whereas a boundary gridpoint has one neighboring gridpoint and one reservoir boundary. In 2-D flow problems, all reservoir gridpoints have two reservoir boundaries that do not contribute to flow. For example, in 2-D flow in the x-y plane, the reservoir lower and upper boundaries do not contribute to flow to any reservoir gridpoint, including boundary gridpoints. These two reservoir boundaries (b_L, b_U) are discarded as if they did not exist. As a result, an interior reservoir gridpoint has four neighboring gridpoints and no reservoir boundaries, a reservoir gridpoint that falls on one reservoir boundary has three neighboring gridpoints and one reservoir boundary, and a reservoir gridpoint that falls on two reservoir boundaries has two neighboring gridpoints and two reservoir boundaries. In 3-D flow problems, any of the six reservoir boundaries may contribute to flow depending on the specified boundary condition. An interior gridpoint has six neighboring gridpoints. It does not share any of its boundaries with any of the reservoir boundaries. A boundary gridpoint may fall on one, two, or three of the reservoir boundaries. Therefore, a boundary gridpoint that falls on one, two, or three reservoir boundaries has five, four, or three neighboring gridpoints, respectively. The earlier discussion leads to a few conclusions related to the number of elements contained in sets ψ and ξ.

(1) For an interior reservoir gridpoint, set ψ contains two, four, or six elements for a 1-D, 2-D, or 3-D flow problem, respectively, and set ξ contains no elements or, in other words, is empty.

(2) For a boundary reservoir gridpoint, set ψ contains less than two, four, or six elements for a 1-D, 2-D, or 3-D flow problem, respectively, and set ξ is not empty.

(3) The sum of the number of elements in sets ψ and ξ for any reservoir gridpoint is a constant that depends on the dimensionality of flow. This sum is two, four, or six for a 1-D, 2-D, or 3-D flow problem, respectively.

For 1-D reservoirs, the flow equation for interior gridpoint i in Fig. 5.6 is given by Eq. (2.32):

$$T^m_{x_{i-1/2}}\left(\Phi^m_{i-1}-\Phi^m_i\right)+T^m_{x_{i+1/2}}\left(\Phi^m_{i+1}-\Phi^m_i\right)+q^m_{sc_i}=\frac{V_{b_i}}{\alpha_c\Delta t}\left[\left(\frac{\phi}{B}\right)^{n+1}_i-\left(\frac{\phi}{B}\right)^n_i\right] \tag{5.3}$$

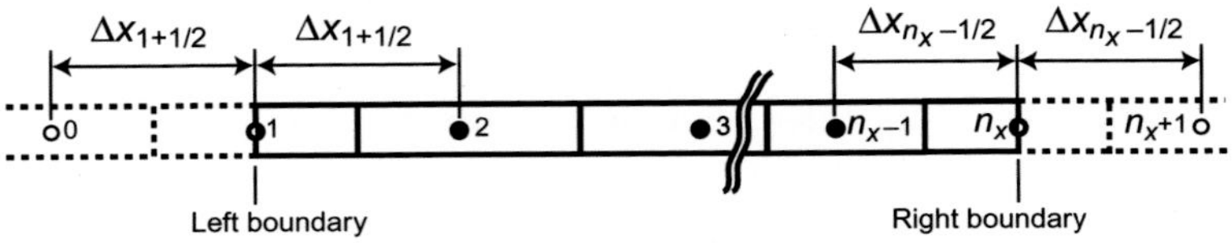

FIG. 5.6 Boundary gridpoints at the left and right boundaries of a 1-D reservoir.

The aforementioned flow equation can be obtained from Eq. (5.2b) for $n=i$, $\psi_i=\{i-1,i+1\}$, and $\xi_i=\{\}$ and by observing that $\sum_{l\in\xi_i} q_{sc_{l,i}}^m=0$ for an interior gridpoint and $T_{i\mp1,i}^m=T_{x_{i\mp1/2}}^m$.

To write the flow equation for boundary gridpoint 1, which falls on the reservoir west boundary in Fig. 5.6, we write the flow equation for the whole block of boundary gridpoint 1:

$$T_{x_{1-1/2}}^m\left(\Phi_0^m-\Phi_1^m\right)+T_{x_{1+1/2}}^m\left(\Phi_2^m-\Phi_1^m\right)+2q_{sc_1}^m=\frac{2V_{b_1}}{\alpha_c\Delta t}\left[\left(\frac{\phi}{B}\right)_1^{n+1}-\left(\frac{\phi}{B}\right)_1^n\right] \tag{5.4}$$

Note that the properties of the whole block in Eq. (5.4) are expressed in terms of those of the actual block; that is, $V_b=2V_{b_1}$ and $q_{sc}=2q_{sc_1}$. Adding and subtracting the flow term $T_{x_{1+1/2}}^m(\Phi_2^m-\Phi_1^m)$ to the LHS of the aforementioned equation gives

$$T_{x_{1-1/2}}^m\left(\Phi_0^m-\Phi_1^m\right)-T_{x_{1+1/2}}^m\left(\Phi_2^m-\Phi_1^m\right)+2T_{x_{1+1/2}}^m\left(\Phi_2^m-\Phi_1^m\right)+2q_{sc_1}^m$$
$$=\frac{2V_{b_1}}{\alpha_c\Delta t}\left[\left(\frac{\phi}{B}\right)_1^{n+1}-\left(\frac{\phi}{B}\right)_1^n\right] \tag{5.5}$$

Multiplying Eq. (5.5) by half results in the flow equation for the actual block represented by boundary gridpoint 1,

$$^1/_2\left[T_{x_{1-1/2}}^m\left(\Phi_0^m-\Phi_1^m\right)-T_{x_{1+1/2}}^m\left(\Phi_2^m-\Phi_1^m\right)\right]+T_{x_{1+1/2}}^m\left(\Phi_2^m-\Phi_1^m\right)+q_{sc_1}^m$$
$$=\frac{V_{b_1}}{\alpha_c\Delta t}\left[\left(\frac{\phi}{B}\right)_1^{n+1}-\left(\frac{\phi}{B}\right)_1^n\right] \tag{5.6a}$$

The first term on the LHS of Eq. (5.6a) represents the rate of fluid flow across the reservoir west boundary (b_W). This term can be replaced with the flow rate of a fictitious well ($q_{sc_{b_W,1}}^m$) that transfers fluid through the reservoir west boundary to gridpoint 1; that is,

$$q_{sc_{b_W,1}}^m={}^1/_2\left[T_{x_{1-1/2}}^m\left(\Phi_0^m-\Phi_1^m\right)-T_{x_{1+1/2}}^m\left(\Phi_2^m-\Phi_1^m\right)\right] \tag{5.7}$$

Substitution of Eq. (5.7) into Eq. (5.6a) yields

$$q_{sc_{b_W,1}}^m+T_{x_{1+1/2}}^m\left(\Phi_2^m-\Phi_1^m\right)+q_{sc_1}^m=\frac{V_{b_1}}{\alpha_c\Delta t}\left[\left(\frac{\phi}{B}\right)_1^{n+1}-\left(\frac{\phi}{B}\right)_1^n\right] \tag{5.6b}$$

The aforementioned flow equation can be obtained from Eq. (5.2b) for $n=1$, $\psi_1=\{2\}$, and $\xi_1=\{b_W\}$ and by observing that $\sum_{l\in\xi_1} q_{sc_{l,1}}^m = q_{sc_{b_W,1}}^m$ and $T_{2,1}^m = T_{x_{1+1/2}}^m$.

To write the flow equation for boundary gridpoint n_x, which falls on the reservoir east boundary in Fig. 5.6, we write the flow equation for the whole block of boundary gridpoint n_x:

$$T_{x_{n_x-1/2}}^m\left(\Phi_{n_x-1}^m-\Phi_{n_x}^m\right)+T_{x_{n_x+1/2}}^m\left(\Phi_{n_x+1}^m-\Phi_{n_x}^m\right)+2q_{sc_{n_x}}^m = \frac{2V_{b_{n_x}}}{\alpha_c\Delta t}\left[\left(\frac{\phi}{B}\right)_{n_x}^{n+1}-\left(\frac{\phi}{B}\right)_{n_x}^{n}\right] \tag{5.8}$$

Here again, note that the properties of the whole block in Eq. (5.8) are expressed in terms of those of the actual block; that is, $V_b=2V_{b_{n_x}}$ and $q_{sc}=2q_{sc_{n_x}}$. The aforementioned equation can be manipulated as was done for gridpoint 1 to obtain the flow equation for the actual block represented by boundary gridpoint n_x:

$$T_{x_{n_x-1/2}}^m\left(\Phi_{n_x-1}^m-\Phi_{n_x}^m\right)+{}^1/_2\left[T_{x_{n_x+1/2}}^m\left(\Phi_{n_x+1}^m-\Phi_{n_x}^m\right)-T_{x_{n_x-1/2}}^m\left(\Phi_{n_x-1}^m-\Phi_{n_x}^m\right)\right] + q_{sc_{n_x}}^m = \frac{V_{b_{n_x}}}{\alpha_c\Delta t}\left[\left(\frac{\phi}{B}\right)_{n_x}^{n+1}-\left(\frac{\phi}{B}\right)_{n_x}^{n}\right] \tag{5.9a}$$

The second term on the LHS of Eq. (5.9a) represents the rate of fluid flow across the reservoir east boundary (b_E). This term can be replaced with the flow rate of a fictitious well $\left(q_{sc_{b_E,n_x}}^m\right)$ that transfers fluid through the reservoir east boundary to gridpoint n_x; that is,

$$q_{sc_{b_E,n_x}}^m = {}^1/_2\left[T_{x_{n_x+1/2}}^m\left(\Phi_{n_x+1}^m-\Phi_{n_x}^m\right)-T_{x_{n_x-1/2}}^m\left(\Phi_{n_x-1}^m-\Phi_{n_x}^m\right)\right] \tag{5.10}$$

Substitution of Eq. (5.10) into Eq. (5.9a) yields

$$T_{x_{n_x-1/2}}^m\left(\Phi_{n_x-1}^m-\Phi_{n_x}^m\right)+q_{sc_{b_E,n_x}}^m+q_{sc_{n_x}}^m = \frac{V_{b_{n_x}}}{\alpha_c\Delta t}\left[\left(\frac{\phi}{B}\right)_{n_x}^{n+1}-\left(\frac{\phi}{B}\right)_{n_x}^{n}\right] \tag{5.9b}$$

The aforementioned flow equation can be obtained from Eq. (5.2b) for $n=n_x$, $\psi_{n_x}=\{n_x-1\}$, and $\xi_{n_x}=\{b_E\}$ and by observing that $\sum_{l\in\xi_{n_x}} q_{sc_{l,n_x}}^m = q_{sc_{b_E,n_x}}^m$ and $T_{n_x-1,n_x}^m = T_{x_{n_x-1/2}}^m$.

For 2-D reservoirs, the flow equation for interior gridpoint (i,j) is given by Eq. (2.37):

$$T_{y_{i,j-1/2}}^m\left(\Phi_{i,j-1}^m-\Phi_{i,j}^m\right)+T_{x_{i-1/2,j}}^m\left(\Phi_{i-1,j}^m-\Phi_{i,j}^m\right)+T_{x_{i+1/2,j}}^m\left(\Phi_{i+1,j}^m-\Phi_{i,j}^m\right) + T_{y_{i,j+1/2}}^m\left(\Phi_{i,j+1}^m-\Phi_{i,j}^m\right)+q_{sc_{i,j}}^m = \frac{V_{b_{i,j}}}{\alpha_c\Delta t}\left[\left(\frac{\phi}{B}\right)_{i,j}^{n+1}-\left(\frac{\phi}{B}\right)_{i,j}^{n}\right] \tag{5.11}$$

The aforementioned flow equation can be obtained from Eq. (5.2b) for $n \equiv (i,j)$, $\psi_{i,j} = \{(i,j-1),(i-1,j),(i+1,j),(i,j+1)\}$, and $\xi_{i,j} = \{\}$ and by observing that $\sum_{l \in \xi_{i,j}} q^m_{sc_{l,(i,j)}} = 0$ for an interior gridpoint, $T^m_{(i,j\mp 1),(i,j)} = T^m_{y_{i,j\mp 1/2}}$, and $T^m_{(i\mp 1,j),(i,j)} = T^m_{x_{i\mp 1/2,j}}$.

For a gridpoint that falls on one reservoir boundary, like gridpoint (3,1), which falls on the reservoir south boundary in Fig. 5.3, the bulk volume, well rate, and transmissibility in the x-direction for the whole block are twice the bulk volume, well rate, and transmissibility in the x-direction for the actual block represented by gridpoint (3,1). However, the transmissibility in the y-direction is the same for the whole and actual blocks. Therefore, the flow equation for the whole block expressed in terms of the properties of the actual block can be written as

$$\begin{aligned} &T^m_{y_{3,1-1/2}}\left(\Phi^m_{3,0} - \Phi^m_{3,1}\right) + 2T^m_{x_{3-1/2,1}}\left(\Phi^m_{2,1} - \Phi^m_{3,1}\right) + 2T^m_{x_{3+1/2,1}}\left(\Phi^m_{4,1} - \Phi^m_{3,1}\right) \\ &+ T^m_{y_{3,1+1/2}}\left(\Phi^m_{3,2} - \Phi^m_{3,1}\right) + 2q^m_{sc_{3,1}} = \frac{2V_{b_{3,1}}}{\alpha_c \Delta t}\left[\left(\frac{\phi}{B}\right)^{n+1}_{3,1} - \left(\frac{\phi}{B}\right)^{n}_{3,1}\right] \end{aligned} \tag{5.12}$$

Adding and subtracting the flow term $T^m_{y_{3,1+1/2}}(\Phi^m_{3,2} - \Phi^m_{3,1})$ to the LHS of the aforementioned equation gives

$$\begin{aligned} &\left[T^m_{y_{3,1-1/2}}\left(\Phi^m_{3,0} - \Phi^m_{3,1}\right) - T^m_{y_{3,1+1/2}}\left(\Phi^m_{3,2} - \Phi^m_{3,1}\right)\right] + 2T^m_{x_{3-1/2,1}}\left(\Phi^m_{2,1} - \Phi^m_{3,1}\right) \\ &+ 2T^m_{x_{3+1/2,1}}\left(\Phi^m_{4,1} - \Phi^m_{3,1}\right) + 2T^m_{y_{3,1+1/2}}\left(\Phi^m_{3,2} - \Phi^m_{3,1}\right) \\ &+ 2q^m_{sc_{3,1}} = \frac{2V_{b_{3,1}}}{\alpha_c \Delta t}\left[\left(\frac{\phi}{B}\right)^{n+1}_{3,1} - \left(\frac{\phi}{B}\right)^{n}_{3,1}\right] \end{aligned} \tag{5.13}$$

Multiplying Eq. (5.13) by half results in the flow equation for the actual block represented by boundary gridpoint (3,1):

$$\begin{aligned} &{}^1/_2\left[T^m_{y_{3,1-1/2}}\left(\Phi^m_{3,0} - \Phi^m_{3,1}\right) - T^m_{y_{3,1+1/2}}\left(\Phi^m_{3,2} - \Phi^m_{3,1}\right)\right] + T^m_{x_{3-1/2,1}}\left(\Phi^m_{2,1} - \Phi^m_{3,1}\right) \\ &+ T^m_{x_{3+1/2,1}}\left(\Phi^m_{4,1} - \Phi^m_{3,1}\right) + T^m_{y_{3,1+1/2}}\left(\Phi^m_{3,2} - \Phi^m_{3,1}\right) + q^m_{sc_{3,1}} = \frac{V_{b_{3,1}}}{\alpha_c \Delta t}\left[\left(\frac{\phi}{B}\right)^{n+1}_{3,1} - \left(\frac{\phi}{B}\right)^{n}_{3,1}\right] \end{aligned} \tag{5.14a}$$

The first term on the LHS of Eq. (5.14a) represents the rate of fluid flow across the reservoir south boundary (b_S). This term can be replaced with the flow rate of a fictitious well $\left(q^m_{sc_{b_S,(3,1)}}\right)$ that transfers fluid through the reservoir south boundary to boundary gridpoint (3,1); that is,

$$q_{sc_{b_S,(3,1)}} = {}^1/_2\left[T^m_{y_{3,1-1/2}}\left(\Phi^m_{3,0} - \Phi^m_{3,1}\right) - T^m_{y_{3,1+1/2}}\left(\Phi^m_{3,2} - \Phi^m_{3,1}\right)\right] \tag{5.15}$$

Substitution of Eq. (5.15) into Eq. (5.14a) yields

$$q^m_{sc_{b_S},(3,1)} + T^m_{x_{3-1/2,1}}\left(\Phi^m_{2,1} - \Phi^m_{3,1}\right) + T^m_{x_{3+1/2,1}}\left(\Phi^m_{4,1} - \Phi^m_{3,1}\right)$$
$$+T^m_{y_{3,1+1/2}}\left(\Phi^m_{3,2} - \Phi^m_{3,1}\right) + q^m_{sc_{3,1}} = \frac{V_{b_{3,1}}}{\alpha_c \Delta t}\left[\left(\frac{\phi}{B}\right)^{n+1}_{3,1} - \left(\frac{\phi}{B}\right)^{n}_{3,1}\right] \tag{5.14b}$$

The aforementioned flow equation can be obtained from Eq. (5.2) for $n \equiv (3,1)$, $\psi_{3,1} = \{(2,1),(4,1),(3,2)\}$, and $\xi_{3,1} = \{b_S\}$ and by observing that $\sum_{l \in \xi_{3,1}} q^m_{sc_{l,(3,1)}} = q^m_{sc_{b_S},(3,1)}$, $T^m_{(2,1),(3,1)} = T^m_{x_{3-1/2,1}}$, $T^m_{(4,1),(3,1)} = T^m_{x_{3+1/2,1}}$, and $T^m_{(3,2),(3,1)} = T^m_{y_{3,1+1/2}}$.

Another example is gridpoint (1,3), which falls on the reservoir west boundary in the 2-D reservoir shown in Fig. 5.3. In this case, the bulk volume, well rate, and transmissibility in the y-direction for the whole block are twice the bulk volume, well rate, and transmissibility in the y-direction for the actual block represented by gridpoint (1,3). However, the transmissibility in the x-direction is the same for the whole and actual blocks. Similarly, the flow equation for the actual block represented by gridpoint (1,3) can be expressed as

$$T^m_{y_{1,3-1/2}}\left(\Phi^m_{1,2} - \Phi^m_{1,3}\right) + q^m_{sc_{b_W},(1,3)} + T^m_{x_{1+1/2,3}}\left(\Phi^m_{2,3} - \Phi^m_{1,3}\right)$$
$$+T^m_{y_{1,3+1/2}}\left(\Phi^m_{1,4} - \Phi^m_{1,3}\right) + q^m_{sc_{1,3}} = \frac{V_{b_{1,3}}}{\alpha_c \Delta t}\left[\left(\frac{\phi}{B}\right)^{n+1}_{1,3} - \left(\frac{\phi}{B}\right)^{n}_{1,3}\right] \tag{5.16}$$

where

$$q_{sc_{b_W},(1,3)} = {}^1/_2\left[T^m_{x_{1-1/2,3}}\left(\Phi^m_{0,3} - \Phi^m_{1,3}\right) - T^m_{x_{1+1/2,3}}\left(\Phi^m_{2,3} - \Phi^m_{1,3}\right)\right] \tag{5.17}$$

The flow equation given by Eq. (5.16) can be obtained from Eq. (5.2b) for $n \equiv (1,3)$, $\psi_{1,3} = \{(1,2),(2,3),(1,4)\}$, and $\xi_{1,3} = \{b_W\}$ and by observing that $\sum_{l \in \xi_{1,3}} q^m_{sc_{l,(1,3)}} = q^m_{sc_{b_W},(1,3)}$, $T^m_{(1,2),(1,3)} = T^m_{y_{1,3-1/2}}$, $T^m_{(1,4),(1,3)} = T^m_{y_{1,3+1/2}}$, and $T^m_{(2,3),(1,3)} = T^m_{x_{1+1/2,3}}$.

Now, consider a gridpoint that falls on two reservoir boundaries, like boundary gridpoint (1,1), which falls on the reservoir south and west boundaries in Fig. 5.3. In this case, the bulk volume and well rate for the whole block are four times the bulk volume and well rate for the actual block represented by gridpoint (1,1). However, the transmissibilities in the x- and y-directions for the whole block are only twice the transmissibilities in the x- and y-directions for the actual block represented by gridpoint (3,1). Therefore, the flow equation for the whole block in terms of the properties of the actual block can be written as

$$2T^m_{y_{1,1-1/2}}\left(\Phi^m_{1,0} - \Phi^m_{1,1}\right) + 2T^m_{x_{1-1/2,1}}\left(\Phi^m_{0,1} - \Phi^m_{1,1}\right) + 2T^m_{x_{1+1/2,1}}\left(\Phi^m_{2,1} - \Phi^m_{1,1}\right)$$
$$+2T^m_{y_{1,1+1/2}}\left(\Phi^m_{1,2} - \Phi^m_{1,1}\right) + 4q^m_{sc_{1,1}} = \frac{4V_{b_{1,1}}}{\alpha_c \Delta t}\left[\left(\frac{\phi}{B}\right)^{n+1}_{1,1} - \left(\frac{\phi}{B}\right)^{n}_{1,1}\right] \tag{5.18}$$

Adding and subtracting $2T^m_{x_{1+1/2,1}}(\Phi^m_{2,1} - \Phi^m_{1,1}) + 2T^m_{y_{1,1+1/2}}(\Phi^m_{1,2} - \Phi^m_{1,1})$ to the LHS of the aforementioned equation gives

$$\begin{aligned}
&2\left[T^m_{y_{1,1-1/2}}\left(\Phi^m_{1,0} - \Phi^m_{1,1}\right) - T^m_{y_{1,1+1/2}}\left(\Phi^m_{1,2} - \Phi^m_{1,1}\right)\right] \\
&+2\left[T^m_{x_{1-1/2,1}}\left(\Phi^m_{0,1} - \Phi^m_{1,1}\right) - T^m_{x_{1+1/2,1}}\left(\Phi^m_{2,1} - \Phi^m_{1,1}\right)\right] \\
&+4T^m_{x_{1+1/2,1}}\left(\Phi^m_{2,1} - \Phi^m_{1,1}\right) + 4T^m_{y_{1,1+1/2}}\left(\Phi^m_{1,2} - \Phi^m_{1,1}\right) \\
&+4q^m_{sc_{1,1}} = \frac{4V_{b_{1,1}}}{\alpha_c \Delta t}\left[\left(\frac{\phi}{B}\right)^{n+1}_{1,1} - \left(\frac{\phi}{B}\right)^{n}_{1,1}\right]
\end{aligned} \tag{5.19}$$

Dividing the aforementioned equation by four results in the flow equation for the actual block represented by boundary gridpoint (1,1) results in

$$\begin{aligned}
&{}^1\!/_2\left[T^m_{y_{1,1-1/2}}\left(\Phi^m_{1,0} - \Phi^m_{1,1}\right) - T^m_{y_{1,1+1/2}}\left(\Phi^m_{1,2} - \Phi^m_{1,1}\right)\right] \\
&+{}^1\!/_2\left[T^m_{x_{1-1/2,1}}\left(\Phi^m_{0,1} - \Phi^m_{1,1}\right) - T^m_{x_{1+1/2,1}}\left(\Phi^m_{2,1} - \Phi^m_{1,1}\right)\right] \\
&+T^m_{x_{1+1/2,1}}\left(\Phi^m_{2,1} - \Phi^m_{1,1}\right) + T^m_{y_{1,1+1/2}}\left(\Phi^m_{1,2} - \Phi^m_{1,1}\right) + q^m_{sc_{1,1}} = \frac{V_{b_{1,1}}}{\alpha_c \Delta t}\left[\left(\frac{\phi}{B}\right)^{n+1}_{1,1} - \left(\frac{\phi}{B}\right)^{n}_{1,1}\right]
\end{aligned} \tag{5.20a}$$

The aforementioned equation can be rewritten as

$$\begin{aligned}
&q^m_{sc_{b_S},(1,1)} + q^m_{sc_{b_W},(1,1)} + T^m_{x_{1+1/2,1}}\left(\Phi^m_{2,1} - \Phi^m_{1,1}\right) + T^m_{y_{1,1+1/2}}\left(\Phi^m_{1,2} - \Phi^m_{1,1}\right) + q^m_{sc_{1,1}} \\
&= \frac{V_{b_{1,1}}}{\alpha_c \Delta t}\left[\left(\frac{\phi}{B}\right)^{n+1}_{1,1} - \left(\frac{\phi}{B}\right)^{n}_{1,1}\right]
\end{aligned} \tag{5.20b}$$

where

$$q_{sc_{b_S},(1,1)} = {}^1\!/_2\left[T^m_{y_{1,1-1/2}}\left(\Phi^m_{1,0} - \Phi^m_{1,1}\right) - T^m_{y_{1,1+1/2}}\left(\Phi^m_{1,2} - \Phi^m_{1,1}\right)\right] \tag{5.21}$$

and

$$q_{sc_{b_W},(1,1)} = {}^1\!/_2\left[T^m_{x_{1-1/2,1}}\left(\Phi^m_{0,1} - \Phi^m_{1,1}\right) - T^m_{x_{1+1/2,1}}\left(\Phi^m_{2,1} - \Phi^m_{1,1}\right)\right] \tag{5.22}$$

Eq. (5.20b) can be obtained from Eq. (5.2b) for $n \equiv (1,1), \psi_{1,1} = \{(2,1),(1,2)\}$, and $\xi_{1,1} = \{b_S, b_W\}$ and by observing that $\sum_{l \in \xi_{1,1}} q^m_{sc_{l,(1,1)}} = q^m_{sc_{b_S},(1,1)} + q^m_{sc_{b_W},(1,1)}$, $T^m_{(2,1),(1,1)} = T^m_{x_{1+1/2,1}}$, and $T^m_{(1,2),(1,1)} = T^m_{y_{1,1+1/2}}$.

The following example demonstrates the use of the general equation, Eq. (5.2a), to write the flow equations for interior gridpoints in a 1-D reservoir.

Example 5.2 For the 1-D reservoir described in Example 5.1, write the flow equations for interior gridpoints 2, 3, 4, and 5.

Solution

The flow equation for gridpoint n in a 1-D horizontal reservoir is obtained from Eq. (5.2a) by discarding the gravity term,

$$\sum_{l\in\psi_n} T_{l,n}^m\left(p_l^m - p_n^m\right) + \sum_{l\in\xi_n} q_{sc_{l,n}}^m + q_{sc_n}^m = \frac{V_{b_n}}{\alpha_c \Delta t}\left[\left(\frac{\phi}{B}\right)_n^{n+1} - \left(\frac{\phi}{B}\right)_n^n\right] \tag{5.23}$$

For interior gridpoints, $\psi_n=\{n-1,n+1\}$, and $\xi_n=\{\}$. Therefore, $\sum_{l\in\xi_n} q_{sc_{l,n}}^m=0$. The gridpoints in this problem are equally spaced ($\Delta x_{i\mp1/2}=\Delta x=1000$ ft) and have the same cross-sectional area ($\Delta y\times h=1200\times 75$ ft^2), permeability ($k_x=15$ md), and constants μ and B. Therefore, $T_x^m=\beta_c\frac{k_x A_x}{\mu B\Delta x}=0.001127\times\frac{15\times(1200\times75)}{10\times1\times1000}=0.1521$ STB/D-psi. In addition, $T_{1,2}^m=T_{2,3}^m=T_{3,4}^m=T_{4,5}^m=T_{5,6}^m=T_x^m=0.1521$ STB/D-psi.

For gridpoint 2, $n=2$, $\psi_2=\{1,3\}$, $\xi_2=\{\}$, $\sum_{l\in\xi_2} q_{sc_{l,2}}^m=0$, and $q_{sc_2}^m=0$.

Therefore, substitution into Eq. (5.23) yields

$$(0.1521)\left(p_1^m - p_2^m\right) + (0.1521)\left(p_3^m - p_2^m\right) = \frac{V_{b_2}}{\alpha_c \Delta t}\left[\left(\frac{\phi}{B}\right)_2^{n+1} - \left(\frac{\phi}{B}\right)_2^n\right] \tag{5.24}$$

For gridpoint 3, $n=3$, $\psi_3=\{2,4\}$, $\xi_3=\{\}$, $\sum_{l\in\xi_3} q_{sc_{l,3}}^m=0$, and $q_{sc_3}^m=0$.

Therefore, substitution into Eq. (5.23) yields

$$(0.1521)\left(p_2^m - p_3^m\right) + (0.1521)\left(p_4^m - p_3^m\right) = \frac{V_{b_3}}{\alpha_c \Delta t}\left[\left(\frac{\phi}{B}\right)_3^{n+1} - \left(\frac{\phi}{B}\right)_3^n\right] \tag{5.25}$$

For gridpoint 4, $n=4$, $\psi_4=\{3,5\}$, $\xi_4=\{\}$, $\sum_{l\in\xi_4} q_{sc_{l,4}}^m=0$, and $q_{sc_4}^m=0$.

Therefore, substitution into Eq. (5.23) yields

$$(0.1521)\left(p_3^m - p_4^m\right) + (0.1521)\left(p_5^m - p_4^m\right) = \frac{V_{b_4}}{\alpha_c \Delta t}\left[\left(\frac{\phi}{B}\right)_4^{n+1} - \left(\frac{\phi}{B}\right)_4^n\right] \tag{5.26}$$

For gridpoint 5, $n=5$, $\psi_5=\{4,6\}$, $\xi_5=\{\}$, $\sum_{l\in\xi_5} q_{sc_{l,5}}^m=0$, and $q_{sc_5}^m=-150$ STB/D. Therefore, substitution into Eq. (5.23) yields

$$(0.1521)\left(p_4^m - p_5^m\right) + (0.1521)\left(p_6^m - p_5^m\right) - 150 = \frac{V_{b_5}}{\alpha_c \Delta t}\left[\left(\frac{\phi}{B}\right)_5^{n+1} - \left(\frac{\phi}{B}\right)_5^n\right] \tag{5.27}$$

5.4 Treatment of boundary conditions

A reservoir boundary can be subject to one of four conditions: (1) no-flow boundary, (2) constant flow boundary, (3) constant pressure gradient boundary, and (4) constant pressure boundary. They have been discussed in Chapter 4. Block-centered grid and point-distributed grid are the most widely used grids to describe a petroleum reservoir as units in reservoir simulation. In the point-distributed grid, the boundary grid point falls on the boundary, whereas the point that represents the boundary grid block is half a block away from the boundary. As a result, the point-distributed grid gives an accurate representation of constant pressure boundary condition. In the block-centered grid, the approximation of a constant pressure boundary is implemented by assuming the boundary pressure being displaced half a block coincides with the point that represents the boundary grid block and by assigning boundary pressure to boundary grid block pressure. This is a first-order approximation. A second-order approximation was suggested, but it has not been used because it requires the addition of an extra equation for each reservoir boundary of a boundary grid block. Furthermore, the extra equations do not have the form of a flow equation. Abou-Kassem and Osman (2008) presented the engineering approach for the representation of a constant pressure boundary condition in a block-centered grid. The new approach involves adding a fictitious well term per boundary to the flow equation of a boundary grid block. This treatment is valid in both rectangular and radial-cylindrical grids. The flow toward a fictitious well is linear in rectangular coordinates and radial in radial-cylindrical coordinates. The flow rate equations for fictitious wells were derived from the interblock flow rate term between a boundary grid block and the grid block that falls immediately outside reservoir boundary. With the new treatment, both block-centered grid and point-distributed grid produce pressure profiles with comparable accuracy. In other words, the use of the point-distributed grid does not offer any advantage over the block-centered grid in rectangular and radial-cylindrical coordinates for the case of constant pressure boundaries.

The general form for the flow rate of the fictitious wells presented by Eqs. (5.7), (5.10), (5.15), (5.17), (5.21), and (5.22) can be expressed as

$$q^m_{sc_{b,bP}} = {}^1/_2\left[T^m_{b,bP^{**}}\left(\Phi^m_{bP^{**}} - \Phi^m_{bP}\right) - T^m_{b,bP^*}\left(\Phi^m_{bP^*} - \Phi^m_{bP}\right)\right] \tag{5.28a}$$

where, as shown in Fig. 5.7, $q^m_{sc_{b,bP}}$ = flow rate of a fictitious well representing flow across reservoir boundary (b) into the actual block represented by boundary gridpoint bP, $T_{b,bP^{**}}$ = transmissibility between reservoir boundary b (or boundary gridpoint bP) and the gridpoint that is exterior to the reservoir and located immediately next to the reservoir boundary (gridpoint bP^{**}), and T_{b,bP^*} = transmissibility between reservoir boundary b (or boundary gridpoint bP) and the gridpoint that is in the reservoir and located immediately next to the reservoir boundary (gridpoint bP^*). Since there is no geologic control for areas outside the reservoir (e.g., aquifers), it is not uncommon to assign reservoir rock properties to those areas in the neighborhood of the reservoir under consideration. Similar to Chapter 4, we use the reflection technique about the reservoir boundary, shown in Fig. 5.7, with regard to transmissibility only (i.e., $T^m_{b,bP^{**}} = T^m_{b,bP^*}$):

$$T^m_{b,bP^{**}} = \left[\beta_c \frac{k_l A_l}{\mu B \Delta l}\right]^m_{bP,bP^{**}} = \left[\beta_c \frac{k_l A_l}{\mu B \Delta l}\right]^m_{bP,bP^*} = T^m_{b,bP^*} \tag{5.29a}$$

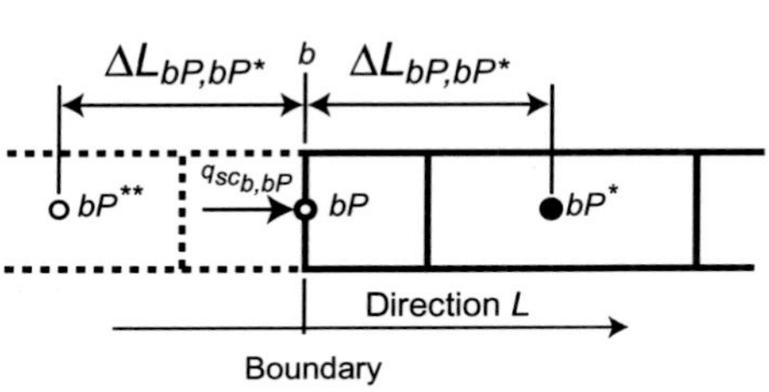

FIG. 5.7 Definition of the terminology used in Eq. (5.28).

where l is the direction normal to reservoir boundary (b). Substituting Eq. (5.29a) into Eq. (5.28a) results in

$$q^m_{sc_{b,bP}} = {}^1\!/_2 T^m_{b,bP^*}\left(\Phi^m_{bP^{**}} - \Phi^m_{bP*}\right) \tag{5.28b}$$

In the following sections, we derive expressions for $q^m_{sc_{b,bP}}$ under various boundary conditions for a point-distributed grid in Cartesian coordinates and stress that this rate must produce the same effects as the specified boundary condition. In Cartesian coordinates, real wells have radial flow, and fictitious wells have linear flow, whereas in radial-cylindrical coordinates in single-well simulation, both real wells and fictitious wells have radial flow. Therefore, in single-well simulation, (1) the equations for the flow rate of real wells presented in Sections 6.2.2 and 6.3.2 can be used to estimate the flow rate of fictitious wells representing boundary conditions in the radial direction only, (2) the flow rate equations of fictitious wells in the z-direction are similar to those presented next in this section because flow in the vertical direction is linear, and (3) there are no reservoir boundaries and hence fictitious wells in the θ-direction. The flow rate of a fictitious well is positive for fluid gain (injection) or negative for fluid loss (production) across a reservoir boundary.

5.4.1 Specified pressure gradient boundary condition

For the reservoir left (lower, south, or west) boundary, like boundary gridpoint 1 shown in Fig. 5.8, Eq. (5.28b) becomes

$$q^m_{sc_{bW},1} = {}^1\!/_2\left[T^m_{x_{1+1/2}}\left(\Phi^m_0 - \Phi^m_2\right)\right] = {}^1\!/_2\left[\left(\beta_c \frac{k_x A_x}{\mu B \Delta x}\right)^m_{1+1/2}\left(\Phi^m_0 - \Phi^m_2\right)\right]$$
$$= \left[\beta_c \frac{k_x A_x}{\mu B}\right]^m_{1+1/2} \frac{\left(\Phi^m_0 - \Phi^m_2\right)}{2\Delta x_{1+1/2}} = -\left[\beta_c \frac{k_x A_x}{\mu B}\right]^m_{1+1/2} \left.\frac{\partial \Phi}{\partial x}\right|^m_{bW} = -\left[\beta_c \frac{k_x A_x}{\mu B}\right]^m_{1,2} \left.\frac{\partial \Phi}{\partial x}\right|^m_{bW} \tag{5.30}$$

Note that to arrive at the aforementioned equation, the first-order derivative of potential was approximated by its central difference; that is, $-\left.\frac{\partial \Phi}{\partial x}\right|^m_{bW} \cong \frac{\left(\Phi^m_0 - \Phi^m_2\right)}{2\Delta x_{1+1/2}}$ (see Fig. 5.6). Substituting Eq. (2.10), which relates potential gradient to pressure gradient, into Eq. (5.30) gives:

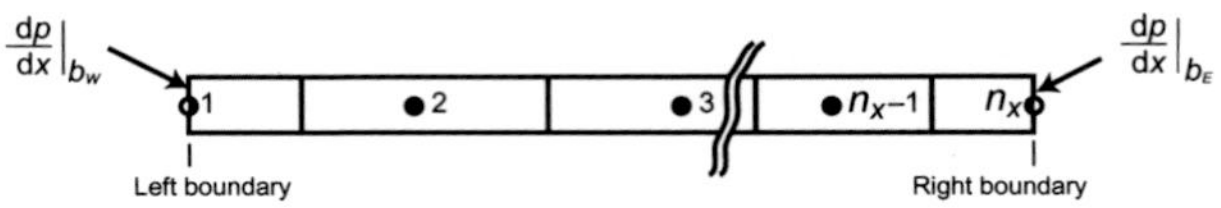

FIG. 5.8 Specified pressure gradient condition at reservoir boundaries in a point-distributed grid.

$$q^m_{sc_{b_W},1} = -\left[\beta_c \frac{k_x A_x}{\mu B}\right]^m_{1,2} \frac{\partial \Phi}{\partial x}\bigg|^m_{b_W} = -\left[\beta_c \frac{k_x A_x}{\mu B}\right]^m_{1,2} \left[\frac{\partial p}{\partial x}\bigg|^m_{b_W} - \gamma^m_{1,2} \frac{\partial Z}{\partial x}\bigg|_{b_W}\right] \tag{5.31}$$

Similar steps can be carried out for the reservoir right (east, north, or upper) boundary. For example, Eq. (5.28b) for boundary gridpoint n_x on the reservoir east boundary becomes

$$q^m_{sc_{b_E},n_x} = \left[\beta_c \frac{k_x A_x}{\mu B}\right]^m_{n_x,n_x-1} \frac{\partial \Phi}{\partial x}\bigg|^m_{b_E} = \left[\beta_c \frac{k_x A_x}{\mu B}\right]^m_{n_x,n_x-1} \left[\frac{\partial p}{\partial x}\bigg|^m_{b_E} - \gamma^m_{n_x,n_x-1} \frac{\partial Z}{\partial x}\bigg|_{b_E}\right] \tag{5.32}$$

In general, for a specified pressure gradient at the reservoir left (lower, south, or west) boundary,

$$q^m_{sc_{b,bP}} = -\left[\beta_c \frac{k_l A_l}{\mu B}\right]^m_{bP,bP^*} \left[\frac{\partial p}{\partial l}\bigg|^m_{b} - \gamma^m_{bP,bP^*} \frac{\partial Z}{\partial l}\bigg|_{b}\right] \tag{5.33a}$$

and at the reservoir right (east, north, or upper) boundary,

$$q^m_{sc_{b,bP}} = \left[\beta_c \frac{k_l A_l}{\mu B}\right]^m_{bP,bP^*} \left[\frac{\partial p}{\partial l}\bigg|^m_{b} - \gamma^m_{bP,bP^*} \frac{\partial Z}{\partial l}\bigg|_{b}\right] \tag{5.33b}$$

where l is the direction normal to the reservoir boundary.

5.4.2 Specified flow rate boundary condition

This condition arises when the reservoir near the boundary has higher or lower potential than that of a neighboring reservoir or aquifer. In this case, fluids move across the reservoir boundary. Methods such as water influx calculations and classical material balance in reservoir engineering can be used to estimate fluid flow rate, which we term here as specified flow rate (q_{spsc}). Therefore, for boundary gridpoint 1,

$$q^m_{sc_{b_W},1} = q_{spsc} \tag{5.34}$$

and for boundary gridpoint n_x,

$$q^m_{sc_{b_E},n_x} = q_{spsc} \tag{5.35}$$

In general, for a specified flow rate boundary condition, Eq. (5.28b) becomes

$$q_{sc_{b,bP}} = q_{spsc} \tag{5.36}$$

In multidimensional flow with q_{spsc} specified for the whole reservoir boundary, $q^m_{sc_{b,bP}}$ for each boundary gridpoint that falls on that boundary is obtained by prorating q_{spsc} among all boundary gridpoints that fall on that boundary; that is,

$$q^m_{sc_{b,bP}} = \frac{T^m_{bP,bP^*}}{\sum\limits_{l \in \psi_b} T^m_{l,l^*}} q_{spsc} \tag{5.37}$$

where ψ_b = the set that contains all boundary gridpoints that fall on the reservoir boundary in question and T^m_{l,l^*} = transmissibility between boundary gridpoint l (or reservoir boundary b) and gridpoint l^*, which falls inside the reservoir and is located immediately next to the reservoir boundary in a direction normal to it (see Fig. 5.7). T^m_{bP,bP^*} is defined as given in Eq. (5.29a):

$$T^m_{bP,bP^*} = \left[\beta_c \frac{k_l A_l}{\mu B \Delta l}\right]^m_{bP,bP^*} \tag{5.29b}$$

Subscript l in Eq. (4.29b) is replaced with x, y, or z depending on the boundary face of boundary block. It should be mentioned that Eq. (5.37) incorporates the assumption that the potential drops across the boundary for all gridpoints falling on the reservoir boundary are equal.

5.4.3 No-flow boundary condition

The no-flow boundary condition results from vanishing permeability at a reservoir boundary (e.g., $T^m_{x_{1/2}} = 0$ for the left boundary of gridpoint 1, and $T^m_{x_{nx+1/2}} = 0$ for the right boundary of gridpoint n_x) or because of symmetry about the reservoir boundary in Fig. 5.6 ($\Phi^m_0 = \Phi^m_2$ for gridpoint 1 and $\Phi^m_{n_x-1} = \Phi^m_{n_x+1}$ for gridpoint n_x). In either case, Eq. (5.28b) for boundary gridpoint 1 reduces to

$$q^m_{sc_{b_W},1} = {}^1/_2 T^m_{x_{1/2}} \left(\Phi^m_0 - \Phi^m_2\right) = {}^1/_2 (0) \left(\Phi^m_0 - \Phi^m_2\right) = {}^1/_2 T^m_{x_{1/2}} (0) = 0 \tag{5.38}$$

and for boundary gridpoint n_x, it reduces to

$$\begin{aligned} q^m_{sc_{b_E},n_x} &= {}^1/_2 T^m_{x_{n_x+1/2}} \left(\Phi^m_{n_x+1} - \Phi^m_{n_x-1}\right) = {}^1/_2 (0) \left(\Phi^m_{n_x+1} - \Phi^m_{n_x-1}\right) \\ &= {}^1/_2 T^m_{x_{n_x+1/2}} (0) = 0 \end{aligned} \tag{5.39}$$

In general, for a reservoir no-flow boundary, Eq. (5.28b) becomes

$$q^m_{sc_{b,bP}} = 0 \tag{5.40}$$

For multidimensional flow, $q^m_{sc_{b,bP}}$ for each boundary gridpoint that falls on a no-flow boundary in the x-, y-, or z-direction is set to zero as Eq. (5.40) implies.

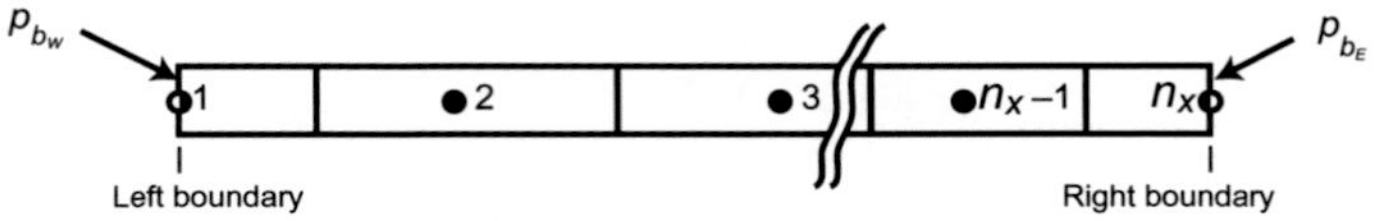

FIG. 5.9 Specified pressure condition at reservoir boundaries for a point-distributed grid.

5.4.4 Specified boundary pressure condition

The specified boundary pressure condition arises when the reservoir is in communication with a strong water aquifer or when wells on the other side of the reservoir boundary operate to maintain voidage replacement and as a result keep the reservoir boundary pressure (p_b) constant. Fig. 5.9 shows this boundary condition at the reservoir left and right boundaries.

For a point-distributed grid, boundary gridpoint 1 falls on the reservoir left boundary (b_W); therefore, $p_1 = p_{b_W}$, and $p_{n_x} = p_{b_E}$ for gridpoint n_x, which falls on the reservoir right boundary. The specified boundary pressure is used in the flow equation for gridpoint bP^* (e.g., gridpoint 2 and gridpoint $n_x - 1$ in Fig. 5.9). The flow equation for gridpoint 2 can be written as

$$T^m_{x_{1+1/2}}\left[\left(p_{b_W} - p_2^m\right) - \gamma^m_{1+1/2}(Z_1 - Z_2)\right] + T^m_{x_{2+1/2}}\left[\left(p_3^m - p_2^m\right) - \gamma^m_{2+1/2}(Z_3 - Z_2)\right] + q^m_{sc_2} = \frac{V_{b_2}}{\alpha_c \Delta t}\left[\left(\frac{\phi}{B}\right)^{n+1}_2 - \left(\frac{\phi}{B}\right)^n_2\right] \tag{5.41a}$$

Similarly, the flow equation for gridpoint $n_x - 1$ can be written as

$$T^m_{x_{n_x-3/2}}\left[\left(p^m_{n_x-2} - p^m_{n_x-1}\right) - \gamma^m_{n_x-3/2}(Z_{n_x-2} - Z_{n_x-1})\right] + T^m_{x_{n_x-1/2}}\left[\left(p_{b_E} - p^m_{n_x-1}\right) - \gamma^m_{n_x-1/2}(Z_{n_x} - Z_{n_x-1})\right] + q^m_{sc_{n_x-1}} = \frac{V_{b_{n_x-1}}}{\alpha_c \Delta t}\left[\left(\frac{\phi}{B}\right)^{n+1}_{n_x-1} - \left(\frac{\phi}{B}\right)^n_{n_x-1}\right] \tag{5.42a}$$

The condition that is responsible for maintaining the pressure of boundary gridpoint 1 constant at p_{b_W} can be obtained from Eq. (5.7):

$$q^m_{sc_{b_W,1}} = {}^1/_2\left[T^m_{x_{1-1/2}}\left(\Phi_0^m - \Phi_1^m\right) - T^m_{x_{1+1/2}}\left(\Phi_2^m - \Phi_1^m\right)\right] \tag{5.7}$$

To keep the pressure at the reservoir west boundary constant, the rate of fluid entering the boundary, $T^m_{x_{1-1/2}}(\Phi_0^m - \Phi_1^m)$, must equal the rate of fluid leaving the boundary, $T^m_{x_{1+1/2}}(\Phi_1^m - \Phi_2^m)$; that is,

$$T^m_{x_{1-1/2}}\left(\Phi_0^m - \Phi_1^m\right) = T^m_{x_{1+1/2}}\left(\Phi_1^m - \Phi_2^m\right) \tag{5.43}$$

Substituting Eq. (5.43) into Eq. (5.7) gives

$$q^m_{sc_{b_W,1}} = {}^1/_2\left[-T^m_{x_{1+1/2}}\left(\Phi_2^m - \Phi_1^m\right) - T^m_{x_{1+1/2}}\left(\Phi_2^m - \Phi_1^m\right)\right] = -T^m_{x_{1+1/2}}\left(\Phi_2^m - \Phi_1^m\right) = T^m_{x_{1+1/2}}\left(\Phi_1^m - \Phi_2^m\right) \tag{5.44a}$$

or

$$q^m_{sc_{b_W},1} = T^m_{x_{1+1/2}}\left[\left(p^m_1 - p^m_2\right) - \gamma^m_{1+1/2}(Z_1 - Z_2)\right] \tag{5.44b}$$

with $p_1 = p_{b_W}$.

Note that Eq. (5.44b) can be derived from Eq. (5.2a) for $n=1$, $\psi_1=\{2\}$, and $\xi_1=\{b_W\}$ and by observing that $\sum\limits_{l\in\xi_1} q^m_{sc_{l,1}} = q^m_{sc_{b_W},1}$, $T^m_{2,1}=T^m_{x_{1+1/2}}$, and the RHS of Eq. (5.2a) vanishes because $p_1 = p_{b_W}$ at all times.

Similarly, for boundary gridpoint n_x,

$$q^m_{sc_{b_E},n_x} = T^m_{x_{n_x-1/2}}\left(\Phi^m_{n_x} - \Phi^m_{n_x-1}\right) \tag{5.45a}$$

or

$$q^m_{sc_{b_E},n_x} = T^m_{x_{n_x-1/2}}\left[\left(p^m_{n_x} - p^m_{n_x-1}\right) - \gamma^m_{n_x-1/2}(Z_{n_x} - Z_{n_x-1})\right] \tag{5.45b}$$

with $p_{n_x} = p_{b_E}$.

The general equation becomes

$$q^m_{sc_{b,bP}} = T^m_{b,bP^*}\left(\Phi^m_{bP} - \Phi^m_{bP^*}\right) \tag{5.46a}$$

or

$$q^m_{sc_{b,bP}} = T^m_{b,bP^*}\left[\left(p^m_{bP} - p^m_{bP^*}\right) - \gamma^m_{b,bP^*}(Z_{bP} - Z_{bP^*})\right] \tag{5.46b}$$

where

$$T^m_{bP,bP^*} = \left[\beta_c \frac{k_l A_l}{\mu B \Delta l}\right]^m_{bP,bP^*} \tag{5.29b}$$

γ_{b,bP^*} = fluid gravity between boundary gridpoint bP and gridpoint bP^*, and $p_{bP} = p_b$.

Combining Eqs. (5.46b) and (5.29b) gives

$$q^m_{sc_{b,bP}} = \left[\beta_c \frac{k_l A_l}{\mu B \Delta l}\right]^m_{bP,bP^*}\left[\left(p^m_{bP} - p^m_{bP^*}\right) - \gamma^m_{b,bP^*}(Z_{bP} - Z_{bP^*})\right] \tag{5.46c}$$

In multidimensional flow, $q_{sc_{b,bP}}$ for a boundary gridpoint that falls on a specified pressure boundary in the x-, y-, or z-direction is estimated using Eq. (5.46c) with the corresponding x, y, or z replacing l.

5.4.5 Specified boundary gridpoint pressure

The specification of pressure at a reservoir boundary in a point-distributed grid results in the specification of the pressure of the boundary gridpoints that fall on that boundary as discussed in Section 5.4.4. This results in $p_1 \cong p_{b_W}$ for gridpoint 1 and $p_{n_x} \cong p_{b_E}$ for gridpoint n_x for the reservoir presented in Fig. 5.9. One way to implement this boundary condition is to write the flow equation for gridpoint bP^* (i.e., gridpoint 2 and gridpoint $n_x - 1$ in Fig. 5.9) and substitute for the pressure of boundary gridpoint bP (i.e., $p_1 \cong p_{b_W}$ and $p_{n_x} \cong p_{b_E}$) as has been mentioned in Section 5.4.4. The resulting flow equation is given for gridpoint 2 as

$$T^m_{x_{1+1/2}}\left[\left(p_{b_W} - p^m_2\right) - \gamma^m_{1+1/2}(Z_1 - Z_2)\right] + T^m_{x_{2+1/2}}\left[\left(p^m_3 - p^m_2\right) - \gamma^m_{2+1/2}(Z_3 - Z_2)\right] + q^m_{sc_2} = \frac{V_{b_2}}{\alpha_c \Delta t}\left[\left(\frac{\phi}{B}\right)^{n+1}_2 - \left(\frac{\phi}{B}\right)^n_2\right] \tag{5.41a}$$

and that for gridpoint $n_x - 1$ as

$$T^m_{x_{n_x-3/2}}\left[\left(p^m_{n_x-2} - p^m_{n_x-1}\right) - \gamma^m_{n_x-3/2}(Z_{n_x-2} - Z_{n_x-1})\right] + T^m_{x_{n_x-1/2}}\left[\left(p_{b_E} - p^m_{n_x-1}\right) - \gamma^m_{n_x-1/2}(Z_{n_x} - Z_{n_x-1})\right] + q^m_{sc_{n_x-1}} = \frac{V_{b_{n_x-1}}}{\alpha_c \Delta t}\left[\left(\frac{\phi}{B}\right)^{n+1}_{n_x-1} - \left(\frac{\phi}{B}\right)^n_{n_x-1}\right] \tag{5.42a}$$

Another way to implement this boundary condition is to assume that the block boundary between gridpoints bP^* and bP is a reservoir boundary with gridpoint bP falling outside the new reservoir description. Therefore, Eq. (5.41a) for gridpoint 2 becomes

$$T^m_{x_{2+1/2}}\left[\left(p^m_3 - p^m_2\right) - \gamma^m_{2+1/2}(Z_3 - Z_2)\right] + q^m_{sc_{b_W},2} + q^m_{sc_2} = \frac{V_{b_2}}{\alpha_c \Delta t}\left[\left(\frac{\phi}{B}\right)^{n+1}_2 - \left(\frac{\phi}{B}\right)^n_2\right] \tag{5.41b}$$

where $q^m_{sc_{b_W},2} = q^m_{sc_{b_W},1} = T^m_{x_{1+1/2}}[(p_{b_W} - p^m_2) - \gamma^m_{1+1/2}(Z_1 - Z_2)]$, and Eq. (5.42a) for gridpoint $n_x - 1$ becomes

$$T^m_{x_{n_x-3/2}}\left[\left(p^m_{n_x-2} - p^m_{n_x-1}\right) - \gamma^m_{n_x-3/2}(Z_{n_x-2} - Z_{n_x-1})\right] + q^m_{sc_{b_E},n_x-1} + q^m_{sc_{n_x-1}} = \frac{V_{b_{n_x-1}}}{\alpha_c \Delta t}\left[\left(\frac{\phi}{B}\right)^{n+1}_{n_x-1} - \left(\frac{\phi}{B}\right)^n_{n_x-1}\right] \tag{5.42b}$$

where $q^m_{sc_{b_E},n_x-1} = q^m_{sc_{b_E},n_x} = T^m_{x_{n_x-1/2}}[(p_{b_W} - p^m_{n_x-1}) - \gamma^m_{n_x-1/2}(Z_{n_x} - Z_{n_x-1})]$.

The following examples demonstrate the use of the general equation, Eq. (5.2a), and the appropriate expressions for $q^m_{sc_{b,bP}}$ to write the flow equations for boundary gridpoints in 1-D and 2-D reservoirs that are subject to various boundary conditions.

Example 5.3 For the 1-D reservoir described in Example 5.1, the reservoir left boundary is kept at a constant pressure of 5000 psia, and the reservoir right boundary is a no-flow (sealed) boundary as shown in Fig. 5.10. Write the flow equations for boundary gridpoints 1 and 6.

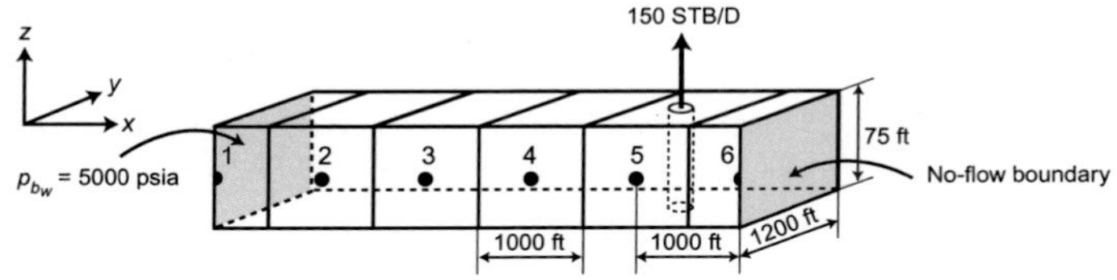

FIG. 5.10 Discretized 1-D reservoir in Example 5.3.

Solution

The flow equation for gridpoint n in a 1-D horizontal reservoir that is obtained from Eq. (5.2a) by discarding the gravity term yields:

$$\sum_{l \in \psi_n} T_{l,n}^m \left(p_l^m - p_n^m\right) + \sum_{l \in \xi_n} q_{sc_{l,n}}^m + q_{sc_n}^m = \frac{V_{b_n}}{\alpha_c \Delta t}\left[\left(\frac{\phi}{B}\right)_n^{n+1} - \left(\frac{\phi}{B}\right)_n^n\right] \tag{5.23}$$

For boundary gridpoint 1, $n=1$, and $p_1 = p_{b_W} = 5000$ psia because this gridpoint falls on the reservoir left boundary. Therefore, there is no need to write the flow equation for gridpoint 1. However, for the sake of generalization, let us proceed and write the flow equation. For $n=1$, $\psi_1 = \{2\}$, $\xi_1 = \{b_W\}$, $\sum_{l \in \xi_1} q_{sc_{l,1}}^m = q_{sc_{b_W,1}}^m$, and $q_{sc_1}^m = 0$. In addition, $T_{1,2}^m = 0.1521$ STB/D-psi from Example 5.2. Therefore, substitution into Eq. (5.23) yields

$$(0.1521)\left(p_2^m - p_1^m\right) + q_{sc_{b_W,1}}^m = \frac{V_{b_1}}{\alpha_c \Delta t}\left[\left(\frac{\phi}{B}\right)_1^{n+1} - \left(\frac{\phi}{B}\right)_1^n\right] \tag{5.47}$$

Furthermore, the RHS of Eq. (5.47) vanishes, resulting in

$$\frac{V_{b_1}}{\alpha_c \Delta t}\left[\left(\frac{\phi}{B}\right)_1^{n+1} - \left(\frac{\phi}{B}\right)_1^n\right] = 0 \tag{5.48}$$

because $p_1^{n+1} = p_1^n = p_{b_W} = 5000$ psia.

Combining Eqs. (5.47) and (5.48) and solving for $q_{sc_{b_W,1}}^m$ yields

$$q_{sc_{b_W,1}}^m = (0.1521)\left(5000 - p_2^m\right) \tag{5.49}$$

Note that Eq. (5.46c) also gives an estimate for $q_{sc_{b_W,1}}^m$:

$$q_{sc_{b_W,1}}^m = \left[\beta_c \frac{k_x A_x}{\mu B \Delta x}\right]_{1,2}^m \left(p_1^m - p_2^m\right) = 0.001127 \times \frac{15 \times (1200 \times 75)}{10 \times 1 \times (1000/2)}\left(5000 - p_2^m\right)$$
$$= (0.1521)\left(5000 - p_2^m\right) \tag{5.50}$$

Eqs. (5.49) and (5.50) give identical estimates for the flow rate of a fictitious well resulting from constant pressure boundary specification. Therefore, Eq. (5.46c) produces a result consistent with that obtained using the general flow equation for a boundary gridpoint.

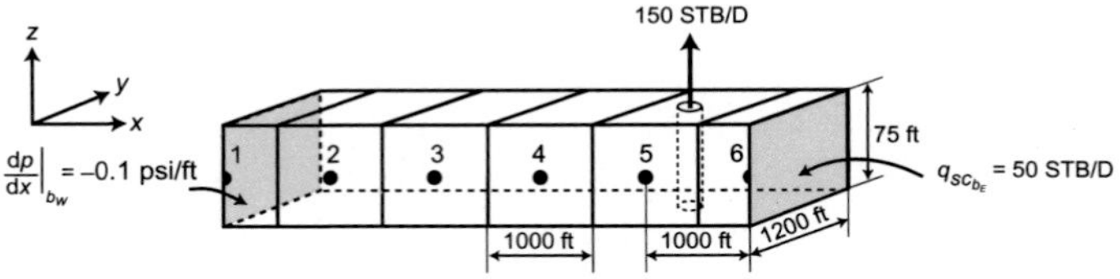

FIG. 5.11 Discretized 1-D reservoir in Example 5.4.

For boundary gridpoint 6, $n=6$, $\psi_6=\{5\}$, $\xi_6=\{b_E\}$, $\sum_{l\in\xi_6} q^m_{sc_{l,6}} = q^m_{sc_{b_E,6}}$, and $q^m_{sc_6}=0$. In addition, $T^m_{5,6}=0.1521$ STB/D-psi from Example 5.2. Therefore, substitution into Eq. (5.23) yields

$$(0.1521)\left(p_5^m - p_6^m\right) + q^m_{sc_{b_E,6}} = \frac{V_{b_6}}{\alpha_c \Delta t}\left[\left(\frac{\phi}{B}\right)_6^{n+1} - \left(\frac{\phi}{B}\right)_6^n\right] \tag{5.51}$$

where the flow rate of a fictitious well for a no-flow boundary is given by Eq. (5.40). For the reservoir east boundary, reservoir boundary $b\equiv b_E$, gridpoint $bP\equiv 6$, and $q^m_{sc_{b_E,6}}=0$.

Substitution into Eq. (5.51) results in the flow equation for boundary gridpoint 6,

$$(0.1521)\left(p_5^m - p_6^m\right) + 0 = \frac{V_{b_6}}{\alpha_c \Delta t}\left[\left(\frac{\phi}{B}\right)_6^{n+1} - \left(\frac{\phi}{B}\right)_6^n\right] \tag{5.52}$$

Example 5.4 For the 1-D reservoir described in Example 5.1, the reservoir left boundary is kept at a constant pressure gradient of −0.1 psi/ft, and the reservoir right boundary is supplied with fluid at a rate of 50 STB/D as shown in Fig. 5.11. Write the flow equations for boundary gridpoints 1 and 6.

Solution

The flow equation for gridpoint n in a 1-D horizontal reservoir is obtained from Eq. (5.2a) by discarding the gravity term, yielding

$$\sum_{l\in\psi_n} T^m_{l,n}\left(p_l^m - p_n^m\right) + \sum_{l\in\xi_n} q^m_{sc_{l,n}} + q^m_{sc_n} = \frac{V_{b_n}}{\alpha_c \Delta t}\left[\left(\frac{\phi}{B}\right)_n^{n+1} - \left(\frac{\phi}{B}\right)_n^n\right] \tag{5.23}$$

For boundary gridpoint 1, $n=1$, $\psi_1=\{2\}$, $\xi_1=\{b_W\}$, $\sum_{l\in\xi_1} q^m_{sc_{l,1}} = q^m_{sc_{b_W,1}}$, and $q^m_{sc_1}=0$. In addition, $T^m_{1,2}=0.1521$ STB/D-psi from Example 5.2. Therefore, substitution into Eq. (5.23) yields

$$(0.1521)\left(p_2^m - p_1^m\right) + q^m_{sc_{b_W,1}} = \frac{V_{b_1}}{\alpha_c \Delta t}\left[\left(\frac{\phi}{B}\right)_1^{n+1} - \left(\frac{\phi}{B}\right)_1^n\right] \tag{5.53}$$

where the flow rate of a fictitious well for specified pressure gradient at the reservoir left boundary is estimated using Eq. (5.33a):

$$q^m_{sc_{b_W,1}} = -\left[\beta_c \frac{k_x A_x}{\mu B}\right]^m_{1,2} \left[\frac{\partial p}{\partial x}\bigg|^m_{b_W} - \gamma^m_{1,2} \frac{\partial Z}{\partial x}\bigg|_{b_W}\right]$$
$$= -\left[0.001127 \times \frac{15 \times (1200 \times 75)}{10 \times 1}\right][-0.1 - 0] = -152.15 \times (-0.1) = 15.215 \tag{5.54}$$

Substitution into Eq. (5.53) results in the flow equation for boundary gridpoint 1:

$$(0.1521)\left(p^m_2 - p^m_1\right) + 15.215 = \frac{V_{b_1}}{\alpha_c \Delta t}\left[\left(\frac{\phi}{B}\right)^{n+1}_1 - \left(\frac{\phi}{B}\right)^n_1\right] \tag{5.55}$$

For boundary gridpoint 6, $n=6$, $\psi_6=\{5\}$, $\xi_6=\{b_E\}$, $\sum_{l\in\xi_6} q^m_{sc_{l,6}} = q^m_{sc_{b_E,6}}$, and $q^m_{sc_6}=0$. In addition, $T^m_{5,6}=0.1521$ STB/D-psi from Example 5.2. Therefore, substitution into Eq. (5.23) yields

$$(0.1521)\left(p^m_5 - p^m_6\right) + q^m_{sc_{b_E,6}} = \frac{V_{b_6}}{\alpha_c \Delta t}\left[\left(\frac{\phi}{B}\right)^{n+1}_6 - \left(\frac{\phi}{B}\right)^n_6\right] \tag{5.56}$$

where the flow rate of fictitious well for a specified rate boundary is estimated using Eq. (5.36); that is, $q^m_{sc_{b_E,6}}=50$ STB/D.

Substitution into Eq. (5.56) results in the flow equation for boundary gridpoint 6:

$$(0.1521)\left(p^m_5 - p^m_6\right) + 50 = \frac{V_{b_6}}{\alpha_c \Delta t}\left[\left(\frac{\phi}{B}\right)^{n+1}_6 - \left(\frac{\phi}{B}\right)^n_6\right] \tag{5.57}$$

Example 5.5 Consider single-phase fluid flow in the 2-D horizontal reservoir shown in Fig. 5.12.

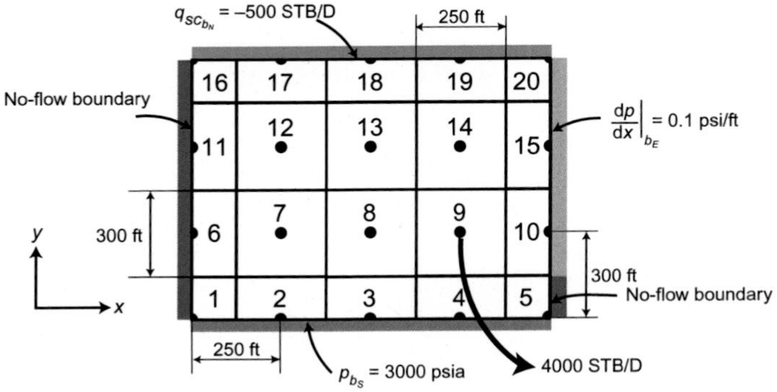

FIG. 5.12 Discretized 2-D reservoir in Examples 5.5 and 5.6.

A well located in gridpoint 9 produces at a rate of 4000 STB/D. All gridpoints have $\Delta x_{i\mp1/2}=250$ ft, $\Delta y_{j\mp1/2}=300$ ft, $h=100$ ft, $k_x=270$ md, and $k_y=220$ md. The FVF and viscosity of the flowing fluid are 1.0 RB/STB and 2 cP, respectively. The reservoir south boundary is maintained at 3000 psia, the reservoir west boundary is sealed off to flow, the reservoir east boundary is kept at constant pressure gradient of 0.1 psi/ft, and the reservoir loses fluid across its north boundary at a rate of 500 STB/D. Write the flow equations for gridpoints 2, 6, 10, and 18 that fall on one reservoir boundary.

Solution

The general flow equation for gridpoint n, in a 2-D horizontal reservoir that is obtained from Eq. (5.2a) by discarding the gravity term yields:

$$\sum_{l\in\psi_n} T^m_{l,n}\left(p^m_l - p^m_n\right) + \sum_{l\in\xi_n} q^m_{sc_{l,n}} + q^m_{sc_n} = \frac{V_{b_n}}{\alpha_c \Delta t}\left[\left(\frac{\phi}{B}\right)^{n+1}_n - \left(\frac{\phi}{B}\right)^n_n\right] \tag{5.23}$$

Before writing any flow equation, we calculate the transmissibilities in the x- and y-directions. The gridpoints in the x-direction are equally spaced ($\Delta x_{i\mp1/2,j}=\Delta x=250$ ft) and have the same cross-sectional area ($A_x=\Delta y \times h=300\times 100$ ft^2) and permeability in the x-direction ($k_x=270$ md), $\mu=2$ cP and $B=1$ RB/STB. Therefore, $T^m_x=\beta_c\frac{k_xA_x}{\mu B\Delta x}=0.001127\times\frac{270\times(300\times100)}{2\times1\times250}=18.2574$ STB/D-psi. The gridpoints in the y-direction are also equally spaced (i.e., $\Delta y_{i,j\mp1/2}=\Delta y=300$ ft) and have the same cross-sectional area of $A_y=\Delta x\times h=250\times100$ ft^2, permeability $k_y=220$ md, constant viscosity of 2 cP, and FVF of 1 RB/STB. Therefore,

$$T^m_y=\beta_c\frac{k_yA_y}{\mu B\Delta y}=0.001127\times\frac{220\times(250\times100)}{2\times1\times300}=10.3308\,\text{STB/D-psi}$$

In addition,

$$T^m_{6,7}=T^m_{7,8}=T^m_{8,9}=T^m_{9,10}=T^m_{11,12}=T^m_{12,13}=T^m_{13,14}=T^m_{14,15}=T^m_x \\ =18.2574\,\text{STB/D-psi}$$

because the gridpoints in the second and third rows have

$$A_x=\Delta y\times h=300\times100\,\text{ft}^2$$

However, $T^m_{1,2}=T^m_{2,3}=T^m_{3,4}=T^m_{4,5}=T^m_{16,17}=T^m_{17,18}=T^m_{18,19}=T^m_{19,20}={}^1/_2T^m_x=9.1287$ STB/D-psi because the gridpoints in the first and last rows have $A_x=(\Delta y/2)\times h=150\times100\,\text{ft}^2$. Similarly, $T^m_{2,7}=T^m_{7,12}=T^m_{12,17}=T^m_{3,8}=T^m_{8,13}=T^m_{13,18}=T^m_{4,9}=T^m_{9,14}=T^m_{14,19}=T^m_y=10.3308$ STB/D-psi because the gridpoints in the second, third, and fourth columns have

$A_y = \Delta x \times h = 250 \times 100\,\text{ft}^2$, but
$T^m_{1,6} = T^m_{6,11} = T^m_{11,16} = T^m_{5,10} = T^m_{10,15} = T^m_{15,20} = {}^1/_2 T^m_y = 5.1654\,\text{STB/D-psi}$
because the gridpoints in the first and last columns have

$$A_y = (\Delta x/2) \times h = 125 \times 100\,\text{ft}^2$$

For boundary gridpoint 2, $n=2$, $\psi_2=\{1,3,7\}$, $\xi_2=\{b_S\}$, and $q^m_{sc_2}=0$. $\sum_{l \in \xi_2} q^m_{sc_{l,2}} = q^m_{sc_{b_S,2}}$, where $q^m_{sc_{b_S,2}}$ is obtained using Eq. (5.46c) by discarding the gravity term, resulting in

$$q^m_{sc_{b_S,2}} = \left[\beta_c \frac{k_y A_y}{\mu B \Delta y}\right]^m_{2,7} \left(p_{b_S} - p^m_7\right) = \left[0.001127 \times \frac{220 \times (250 \times 100)}{2 \times 1 \times (300)}\right] \left(3000 - p^m_7\right)$$
$$= (10.3308)\left(3000 - p^m_7\right) \tag{5.58}$$

In addition, $T^m_{1,2} = T^m_{2,3} = ½T^m_x = 9.1287$ STB/D-psi, $T^m_{2,7} = T^m_y =$ 10.3308 STB/D-psi, and $V_{b_2} = 250 \times (300/2) \times 100\,\text{ft}^3$. Substitution into Eq. (5.23) results in the flow equation for boundary gridpoint 2:

$$(9.1287)\left(p^m_1 - p^m_2\right) + (9.1287)\left(p^m_3 - p^m_2\right) + (10.3308)\left(p^m_7 - p^m_2\right)$$
$$+(10.3308)\left(3000 - p^m_7\right) + 0 = \frac{V_{b_2}}{\alpha_c \Delta t}\left[\left(\frac{\phi}{B}\right)^{n+1}_2 - \left(\frac{\phi}{B}\right)^n_2\right] \tag{5.59}$$

The aforementioned equation reduces to identity equation because $p^m_1 = p^m_2 = p^m_3 = 3000$ psia, and the RHS vanishes because $p^n_2 = p^{n+1}_2 = p_{b_S} = 3000$ psia. In other words, Eq. (5.59) does not introduce new information, but it confirms that Eq. (5.46c) produces the correct fluid flow rate estimate across the constant pressure south boundary of gridpoint 2.

For boundary gridpoint 6, $n=6$, $\psi_6=\{1,7,11\}$, $\xi_6=\{b_W\}$, and $q^m_{sc_6}=0$. $\sum_{l \in \xi_6} q^m_{sc_{l,6}} = q^m_{sc_{b_W,6}}$, where $q^m_{sc_{b_W,6}}$ is obtained using Eq. (5.40) for the no-flow boundary; that is, $q^m_{sc_{b_W,6}} = 0$. In addition, $T^m_{7,6} = T^m_x = 18.2574$ STB/D-psi, $T^m_{1,6} = T^m_{11,6} = {}^1/_2 T^m_y = 5.1654$ STB/D-psi, and $V_{b_6} = (250/2) \times 300 \times 100\,\text{ft}^3$. Substitution into Eq. (5.23) results in the flow equation for boundary gridpoint 6:

$$(5.1654)\left(p^m_1 - p^m_6\right) + (18.2574)\left(p^m_7 - p^m_6\right) + (5.1654)\left(p^m_{11} - p^m_6\right)$$
$$+0 + 0 = \frac{V_{b_6}}{\alpha_c \Delta t}\left[\left(\frac{\phi}{B}\right)^{n+1}_6 - \left(\frac{\phi}{B}\right)^n_6\right] \tag{5.60}$$

For boundary gridpoint 10, $n=10$, $\psi_{10}=\{5,9,15\}$, $\xi_{10}=\{b_E\}$, and $q^m_{sc_{10}}=0$. $\sum_{l \in \xi_{10}} q^m_{sc_{l,10}} = q^m_{sc_{b_E,10}}$, where $q^m_{sc_{b_E,10}}$ is estimated using Eq. (5.33b) for the reservoir east boundary:

$$q^m_{sc_{b_E},10} = \left[\beta_c \frac{k_x A_x}{\mu B}\right]^m_{10,9} \left[\frac{\partial p}{\partial x}\bigg|^m_{b_E} - \gamma^m_{10,9} \frac{\partial Z}{\partial x}\bigg|_{b_E}\right]$$
$$= \left[0.001127 \times \frac{270 \times (300 \times 100)}{2 \times 1}\right][0.1 - 0] = 4564.35 \times (0.1)$$
$$= 456.435\,\mathrm{STB/D} \tag{5.61}$$

In addition, $T^m_{9,10} = T^m_x = 18.2574$ STB/D-psi, $T^m_{5,10} = T^m_{10,15} = {}^1/_2 T^m_y = 5.1654$ STB/D-psi, and $V_{b_{10}} = (250/2) \times 300 \times 100$ ft^3. Substitution into Eq. (5.23) results in the flow equation for boundary gridpoint 10,

$$(5.1654)\left(p^m_5 - p^m_{10}\right) + (18.2574)\left(p^m_9 - p^m_{10}\right) + (5.1654)\left(p^m_{15} - p^m_{10}\right)$$
$$+456.435 + 0 = \frac{V_{b_{10}}}{\alpha_c \Delta t}\left[\left(\frac{\phi}{B}\right)^{n+1}_{10} - \left(\frac{\phi}{B}\right)^n_{10}\right] \tag{5.62}$$

For boundary gridpoint 18, $n = 18$, $\psi_{18} = \{13, 17, 19\}$, $\xi_{18} = \{b_N\}$, and $q^m_{sc_{18}} = 0$. $\sum\limits_{l \in \xi_{18}} q^m_{sc_{l,18}} = q^m_{sc_{b_N},18}$, where $q^m_{sc_{b_N},18}$ is estimated using Eq. (5.37) because $q_{spsc} = -500$ STB/D is specified for the whole reservoir north boundary. This rate has to be prorated among all gridpoints falling on that boundary. Therefore, using Eq. (5.37),

$$q^m_{sc_{b_N},18} = \frac{T^m_{18,13}}{\sum\limits_{l \in \psi_{b_N}} T^m_{l,l^*}} q_{spsc} \tag{5.63}$$

where $\psi_{b_N} = \{16, 17, 18, 19, 20\}$. Note that, using Eq. (5.29b),

$$T^m_{18,13} = \left[\beta_c \frac{k_y A_y}{\mu B \Delta y}\right]^m_{18,13} = \left[0.001127 \times \frac{220 \times (250 \times 100)}{2 \times 1 \times 300}\right]$$
$$= 10.3308\,\mathrm{STB/D\text{-}psi} \tag{5.64}$$

Also, $T^m_{17,12} = T^m_{18,13} = T^m_{19,14} = 10.3308$ STB/D-psi, and $T^m_{16,11} = T^m_{20,15} = 5.1654$ STB/D-psi. Substitution into Eq. (5.37) yields.

$$q^m_{sc_{b_N},18} = \frac{10.3308}{5.1654 + 3 \times 10.3308 + 5.1654} \times (-500) = -125\,\mathrm{STB/D\text{-}psi} \tag{5.65}$$

In addition, $T^m_{17,18} = T^m_{19,18} = {}^1/_2 T^m_x = 9.1287$ STB/D-psi, $T^m_{13,18} = T^m_y = 10.3308$, and $V_{b_{18}} = 250 \times (300/2) \times 100$ ft^3. Substitution into Eq. (5.23) results in the flow equation for boundary gridpoint 18:

$$(10.3308)\left(p^m_{13} - p^m_{18}\right) + (9.1287)\left(p^m_{17} - p^m_{18}\right) + (9.1287)\left(p^m_{19} - p^m_{18}\right)$$
$$-125 + 0 = \frac{V_{b_{18}}}{\alpha_c \Delta t}\left[\left(\frac{\phi}{B}\right)^{n+1}_{18} - \left(\frac{\phi}{B}\right)^n_{18}\right] \tag{5.66}$$

Example 5.6 Consider single-phase fluid flow in the 2-D horizontal reservoir described in Example 5.5. Write the flow equations for gridpoints 1, 5, 16, and 20, which fall on two reservoir boundaries.

Solution

The general flow equation for gridpoint n in a 2-D horizontal reservoir that is obtained from Eq. (5.2a) by discarding the gravity term yields:

$$\sum_{l\in\psi_n} T_{l,n}^m (p_l^m - p_n^m) + \sum_{l\in\xi_n} q_{sc_{l,n}}^m + q_{sc_n}^m = \frac{V_{b_n}}{\alpha_c \Delta t}\left[\left(\frac{\phi}{B}\right)_n^{n+1} - \left(\frac{\phi}{B}\right)_n^n\right] \tag{5.23}$$

The data necessary to write flow equations for any boundary gridpoint were calculated in Example 5.5. The following is a summary:

$$T_x^m = 18.2574\,\text{STB/D-psi}$$

$$T_y^m = 10.3308\,\text{STB/D-psi}$$

$$T_{6,7}^m = T_{7,8}^m = T_{8,9}^m = T_{9,10}^m = T_{11,12}^m = T_{12,13}^m = T_{13,14}^m = T_{14,15}^m = T_x^m = 18.2574\,\text{STB/D-psi}$$

$$T_{1,2}^m = T_{2,3}^m = T_{3,4}^m = T_{4,5}^m = T_{16,17}^m = T_{17,18}^m = T_{18,19}^m = T_{19,20}^m = {}^1\!/_2 T_x^m = 9.1287\,\text{STB/D-psi}$$

$$T_{2,7}^m = T_{7,12}^m = T_{12,17}^m = T_{3,8}^m = T_{8,13}^m = T_{13,18}^m = T_{4,9}^m = T_{9,14}^m = T_{14,19}^m = T_y^m = 10.3308\,\text{STB/D-psi}$$

$$T_{1,6}^m = T_{6,11}^m = T_{11,16}^m = T_{5,10}^m = T_{10,15}^m = T_{15,20}^m = {}^1\!/_2 T_y^m = 5.1654\,\text{STB/D-psi}$$

$q_{sc_{b_S,bP}}^m = (10.3308)(3000 - p_{bP^*}^m)$ STB/D for $bP = 2$, 3, 4, where $2^* = 7$, $3^* = 8$, and $4^* = 9$; or more explicitly,

$$q_{sc_{b_S,2}}^m = (10.3308)(3000 - p_7^m)$$

$$q_{sc_{b_S,3}}^m = (10.3308)(3000 - p_8^m)$$

$$q_{sc_{b_S,4}}^m = (10.3308)(3000 - p_9^m)$$

$$q_{sc_{b_W,bP}}^m = 0\,\text{STB/D} \quad \text{for}\, bP = 6, 11$$

$$q_{sc_{b_E,bP}}^m = 456.435\,\text{STB/D} \quad \text{for}\, bP = 10, 15$$

and

$$q_{sc_{b_E,bP}}^m = -125\,\text{STB/D} \quad \text{for}\, bP = 17, 18, 19$$

For corner gridpoints, the areas open to flow in the x- and y-directions are half the size of those of the other gridpoints that fall on the same reservoir boundary; thus,

$q^m_{sc_{b_S,bP}} = (5.1654)(3000 - p^m_{bP^*})$ STB/D for $bP = 1$, 5, where $1^* = 6$ and $5^* = 10$ or more explicitly,

$$q^m_{sc_{b_S,1}} = (5.1654)\left(3000 - p^m_6\right)$$

and

$$q^m_{sc_{b_S,5}} = (5.1654)\left(3000 - p^m_{10}\right)$$

$$q^m_{sc_{b_W,bP}} = 0\,\text{STB/D} \quad \text{for}\, bP = 1,16$$

$$q^m_{sc_{b_E,5}} = 0\,\text{STB/D}$$

$$q^m_{sc_{b_E,20}} = 228.2175\,\text{STB/D}$$

and

$$q^m_{sc_{b_N,bP}} = -62.5\,\text{STB/D} \quad \text{for}\, bP = 16,20$$

For boundary gridpoint 1, $n = 1$, $\psi_1 = \{2,6\}$, $\xi_1 = \{b_S, b_W\}$, $q^m_{sc_1} = 0$, and $\sum_{l \in \xi_1} q^m_{sc_{l,1}} = q^m_{sc_{b_S,1}} + q^m_{sc_{b_W,1}} = (5.1654)\left(3000 - p^m_6\right) + 0$ STB/D.

In addition, $T^m_{1,2} = {}^1/{}_2 T^m_x = 9.1287$ STB/D-psi, $T^m_{1,6} = {}^1/{}_2 T^m_y = 5.1654$ STB/D-psi, and $V_{b_1} = (250/2) \times (300/2) \times 100$ ft^3. Substitution into Eq. (5.23) results in the flow equation for boundary gridpoint 1:

$$(9.1287)\left(p^m_2 - p^m_1\right) + (5.1654)\left(p^m_6 - p^m_1\right) + (5.1654)\left(3000 - p^m_6\right) + 0 + 0 = \frac{V_{b_1}}{\alpha_c \Delta t}\left[\left(\frac{\phi}{B}\right)^{n+1}_1 - \left(\frac{\phi}{B}\right)^n_1\right] \tag{5.67}$$

For boundary gridpoint 5, $n = 5$, $\psi_5 = \{4,10\}$, $\xi_5 = \{b_S, b_E\}$, $q^m_{sc_5} = 0$, and $\sum_{l \in \xi_5} q^m_{sc_{l,5}} = q^m_{sc_{b_S,5}} + q^m_{sc_{b_E,5}} = (5.1654)\left(3000 - p^m_{10}\right) + 0 = (5.1654)\left(3000 - p^m_{10}\right)$ STB/D

In addition, $T^m_{4,5} = {}^1/{}_2 T^m_x = 9.1287$ STB/D-psi, $T^m_{10,5} = {}^1/{}_2 T^m_y = 5.1654$ STB/D-psi, and $V_{b_5} = (250/2) \times (300/2) \times 100$ ft^3.

Substitution into Eq. (5.23) results in the flow equation for boundary gridpoint 5,

$$(9.1287)\left(p^m_4 - p^m_5\right) + (5.1654)\left(p^m_{10} - p^m_5\right) + (5.1654)\left(3000 - p^m_{10}\right) + 0 = \frac{V_{b_5}}{\alpha_c \Delta t}\left[\left(\frac{\phi}{B}\right)^{n+1}_5 - \left(\frac{\phi}{B}\right)^n_5\right] \tag{5.68}$$

For boundary gridpoint 16, $n = 16$, $\psi_{16} = \{11,17\}$, $\xi_{16} = \{b_W, b_N\}$, $q^m_{sc_{16}} = 0$, and $\sum_{l \in \xi_{16}} q^m_{sc_{l,16}} = q^m_{sc_{b_W,16}} + q^m_{sc_{b_N,16}} = 0 - 62.5$ STB/D.

In addition, $T^m_{17,16} = {}^1/{}_2 T^m_x = 9.1287$ STB/D-psi, $T^m_{11,16} = {}^1/{}_2 T^m_y = 5.1654$ STB/D-psi, and $V_{b_{16}} = (250/2) \times (300/2) \times 100$ ft^3.

Substitution into Eq. (5.23) results in the flow equation for boundary gridpoint 16:

$$(5.1654)\left(p_{11}^{m}-p_{16}^{m}\right)+(9.1287)\left(p_{17}^{m}-p_{16}^{m}\right)+0-62.5+0 \\ =\frac{V_{b_{16}}}{\alpha_c \Delta t}\left[\left(\frac{\phi}{B}\right)_{16}^{n+1}-\left(\frac{\phi}{B}\right)_{16}^{n}\right] \tag{5.69}$$

For boundary gridpoint 20, $n=20$, $\psi_{20}=\{15,19\}$, $\xi_{20}=\{b_E,b_N\}$, $q_{sc_{20}}^{m}=0$, and $\sum_{l\in\xi_{20}} q_{sc_{l,20}}^{m}=q_{sc_{b_E,20}}^{m}+q_{sc_{b_N,20}}^{m}=228.2175-62.5$ STB/D.

In addition, $T_{19,20}^{m}={}^{1}/_{2}T_x^{m}=9.1287$ STB/D-psi, $T_{15,20}^{m}={}^{1}/_{2}T_y^{m}=5.1654$ STB/D-psi, and $V_{b_{20}}=(250/2)\times(300/2)\times 100\ \text{ft}^3$.

Substitution into Eq. (5.23) results in the flow equation for boundary gridpoint 20:

$$(5.1654)\left(p_{15}^{m}-p_{20}^{m}\right)+(9.1287)\left(p_{19}^{m}-p_{20}^{m}\right)+228.2175-62.5+0 \\ =\frac{V_{b_{20}}}{\alpha_c \Delta t}\left[\left(\frac{\phi}{B}\right)_{20}^{n+1}-\left(\frac{\phi}{B}\right)_{20}^{n}\right] \tag{5.70}$$

5.5 Calculation of transmissibilities

The flow equations in Cartesian coordinates have transmissibilities in the x-, y-, and z-directions that are defined by Eq. (2.39) in Chapter 2:

$$T_{x_{i\mp1/2,j,k}}=G_{x_{i\mp1/2,j,k}}\left(\frac{1}{\mu B}\right)_{x_{i\mp1/2,j,k}} \tag{5.71a}$$

$$T_{y_{i,j\mp1/2,k}}=G_{y_{i,j\mp1/2,k}}\left(\frac{1}{\mu B}\right)_{y_{i,j\mp1/2,k}} \tag{5.71b}$$

and

$$T_{z_{i,j,k\mp1/2}}=G_{z_{i,j,k\mp1/2}}\left(\frac{1}{\mu B}\right)_{z_{i,j,k\mp1/2}} \tag{5.71c}$$

where the geometric factors G for anisotropic porous media and irregular gridpoint distribution are given in Table 5.1 (Ertekin et al., 2001). The treatment of the pressure-dependent term μB in Eq. (5.71) is discussed in detail under linearization in Chapter 8 (Section 8.4.1).The equations for geometric factors in Table 5.1 can be derived using the procedure followed in Example 4.7. For example, the derivation of $G_{x_{i+1/2}}$ for 1-D flow in the x-direction is the same as that presented in Example 4.7 except that $\delta x_{i^+}=\delta x_{i+1^-}={}^{1}/_{2}\Delta x_{i+1/2}$ for a point-distributed grid.

The flow equations in radial-cylindrical coordinates have transmissibility in the r-, θ-, and z-directions that are defined by Eq. (2.69) in Chapter 2:

TABLE 5.1 Geometric factors in rectangular grids (Ertekin et al., 2001).

Direction	Geometric factor
x	$G_{x_{i\mp1/2,j,k}} = \dfrac{2\beta_c}{\Delta x_{i\mp1/2,j,k}/\left(A_{x_{i,j,k}} k_{x_{i,j,k}}\right) + \Delta x_{i\mp1/2,j,k}/\left(A_{x_{i\mp1,j,k}} k_{x_{i\mp1,j,k}}\right)}$
y	$G_{y_{i,j\mp1/2,k}} = \dfrac{2\beta_c}{\Delta y_{i,j\mp1/2,k}/\left(A_{y_{i,j,k}} k_{y_{i,j,k}}\right) + \Delta y_{i,j\mp1/2,k}/\left(A_{y_{i,j\mp1,k}} k_{y_{i,j\mp1,k}}\right)}$
z	$G_{z_{i,j,k\mp1/2}} = \dfrac{2\beta_c}{\Delta z_{i,j,k\mp1/2}/\left(A_{z_{i,j,k}} k_{z_{i,j,k}}\right) + \Delta z_{i,j,k\mp1/2}/\left(A_{z_{i,j,k\mp1}} k_{z_{i,j,k\mp1}}\right)}$

$$T_{r_{i\mp1/2,j,k}} = G_{r_{i\mp1/2,j,k}} \left(\frac{1}{\mu B}\right)_{r_{i\mp1/2,j,k}} \tag{5.72a}$$

$$T_{\theta_{i,j\mp1/2,k}} = G_{\theta_{i,j\mp1/2,k}} \left(\frac{1}{\mu B}\right)_{\theta_{i,j\mp1/2,k}} \tag{5.72b}$$

and

$$T_{z_{i,j,k\mp1/2}} = G_{z_{i,j,k\mp1/2}} \left(\frac{1}{\mu B}\right)_{z_{i,j,k\mp1/2}} \tag{5.72c}$$

where the geometric factors G for anisotropic porous media and irregular gridpoint distribution are given in Table 5.2 (Pedrosa Jr. and Aziz, 1986). Note that for gridpoint (i,j,k), r_i and $r_{i\mp1/2}$ depend on the value of subscript i only, $\Delta\theta_j$ and $\Delta\theta_{j\mp1/2}$ depend on the value of subscript j only, and Δz_k and $\Delta z_{k\mp1/2}$ depend on the value of subscript k only. The treatment of the pressure-dependent term μB in Eq. (5.72) is discussed in detail under linearization in Chapter 8 (Section 8.4.1).

In Table 5.2, gridpoint spacing and block boundaries in the z-direction are defined as in Eq. (5.1), with z replacing x. Those in the θ-direction are defined in a similar way. Specifically,

$$\begin{aligned}
&\theta_1 = 0, \theta_{n_\theta} = 2\pi, \quad (\text{i.e.}, \theta_{n_\theta} - \theta_1 = 2\pi) \\
&\theta_{j+1} = \theta_j + \Delta\theta_{j+1/2}, \quad j = 1,2,3\ldots n_\theta - 1 \\
&\theta_{j+1/2} = \theta_j + {}^1/_2\Delta\theta_{j+1/2}, \quad j = 1,2,3\ldots n_\theta - 1 \\
&\Delta\theta_j = \theta_{j+1/2} - \theta_{j-1/2}, \quad j = 1,2,3\ldots n_\theta \\
&\theta_{1/2} = \theta_1, \text{ and } \theta_{n_\theta+1/2} = \theta_{n_\theta}
\end{aligned} \tag{5.73}$$

In the r-direction, however, gridpoints are spaced such that pressure drops between neighboring gridpoints are equal (see Example 4.8 and note that in this case, there are $n_r - 1$ spacings separating the n_r gridpoints). Additionally, block boundaries for transmissibility calculations are spaced logarithmically in r to

TABLE 5.2 Geometric factors in cylindrical grids (Pedrosa Jr. and Aziz, 1986).

Direction	Geometric factor
r	$G_{r_{i-1/2,j,k}} = \dfrac{\beta_c \Delta\theta_j \Delta z_k}{\log_e\left(r_i/r^L_{i-1/2}\right)/k_{r_{i,j,k}} + \log_e\left(r^L_{i-1/2}/r_{i-1}\right)/k_{r_{i-1,j,k}}}$
	$G_{r_{i+1/2,j,k}} = \dfrac{\beta_c \Delta\theta_j \Delta z_k}{\log_e\left(r^L_{i+1/2}/r_i\right)/k_{r_{i,j,k}} + \log_e\left(r_{i+1}/r^L_{i+1/2}\right)/k_{r_{i+1,j,k}}}$
θ	$G_{\theta_{i,j\mp 1/2,k}} = \dfrac{2\beta_c \log_e\left(r^L_{i+1/2}/r^L_{i-1/2}\right)\Delta z_k}{\Delta\theta_{j\mp 1/2}/k_{\theta_{i,j,k}} + \Delta\theta_{j\mp 1/2}/k_{\theta_{i,j\mp 1,k}}}$
z	$G_{z_{i,j,k\mp 1/2}} = \dfrac{2\beta_c (½\Delta\theta_j)\left(r^2_{i+1/2} - r^2_{i-1/2}\right)}{\Delta z_{k\mp 1/2}/k_{z_{i,j,k}} + \Delta z_{k\mp 1/2}/k_{z_{i,j,k\mp 1}}}$

warrant that the radial flow rates between neighboring gridpoints using the continuous and discretized forms of Darcy's law are identical (see Example 4.9), and block boundaries for bulk volume calculations are spaced logarithmically in r^2 to warrant that the actual and discretized bulk volumes of gridblocks are equal. Therefore, the radii for the pressure points ($r_{i\mp 1}$), transmissibility calculations ($r^L_{i\mp 1/2}$), and bulk volume calculations ($r_{i\mp 1/2}$) that appear in Table 5.2, are as follows (Aziz and Settari, 1979; Ertekin et al., 2001):

$$r_{i+1} = \alpha_{lg} r_i \tag{5.74}$$

$$r^L_{i+1/2} = \frac{r_{i+1} - r_i}{\log_e(r_{i+1}/r_i)} \tag{5.75a}$$

$$r^L_{i-1/2} = \frac{r_i - r_{i-1}}{\log_e(r_i/r_{i-1})} \tag{5.76a}$$

and

$$r^2_{i+1/2} = \frac{r^2_{i+1} - r^2_i}{\log_e\left(r^2_{i+1}/r^2_i\right)} \tag{5.77a}$$

$$r^2_{i-1/2} = \frac{r^2_i - r^2_{i-1}}{\log_e\left(r^2_i/r^2_{i-1}\right)} \tag{5.78a}$$

where

$$\alpha_{lg} = \left(\frac{r_e}{r_w}\right)^{1/(n_r - 1)} \tag{5.79}$$

and

$$r_1 = r_w \tag{5.80}$$

Note that gridpoint 1 falls on the reservoir internal boundary (r_w) and gridpoint n_r falls on the reservoir external boundary (r_e); therefore, $r_1 = r_w$ and $r_{n_r} = r_e$ by definition for a point-distributed grid. Furthermore, $r_{1-1/2} = r_w$ and $r_{n_r+1/2} = r_e$ define the internal boundary for gridpoint 1 and the external boundary for gridpoint n_r that are used to calculate block bulk volumes.

The bulk volume of gridpoint (i,j,k) is calculated from

$$V_{b_{i,j,k}} = \left(r^2_{i+1/2} - r^2_{i-1/2}\right)\left({}^1\!/_2\Delta\theta_j\right)\Delta z_k \tag{5.81a}$$

Note that $r^2_{i-1/2} = r^2_w$ for $i=1$ and $r^2_{i+1/2} = r^2_e$ for $i = n_r$.

It should be mentioned that the geometric factors in the r-direction given in Tables 4.2 and 5.2, $G_{r_{i\mp1/2,j,k}}$, differ only in the handling of block thickness. The block thickness in Table 5.2 is constant for all gridpoints in layer k, whereas in Table 4.2, it may assume different values for the gridblocks in layer k. This difference is a result of grid construction in block-centered and point-distributed grids.

Eqs. (5.75) through (5.78) and Eq. (5.81a) can be expressed in terms of r_i and α_{lg} (see Example 4.10), resulting in

$$r^L_{i+1/2} = \{(\alpha_{lg} - 1)/[\log_e(\alpha_{lg})]\}r_i \tag{5.75b}$$

$$r^L_{i-1/2} = \{(\alpha_{lg} - 1)/[\alpha_{lg}\log_e(\alpha_{lg})]\}r_i = \left({}^1\!/_{\alpha_{lg}}\right)r^L_{i+1/2} \tag{5.76b}$$

$$r^2_{i+1/2} = \left\{\left(\alpha^2_{lg} - 1\right)/\left[\log_e\left(\alpha^2_{lg}\right)\right]\right\}r^2_i \tag{5.77b}$$

$$r^2_{i-1/2} = \left\{\left(\alpha^2_{lg} - 1\right)/\left[\alpha^2_{lg}\log_e\left(\alpha^2_{lg}\right)\right]\right\}r^2_i = \left({}^1\!/_{\alpha^2_{lg}}\right)r^2_{i+1/2} \tag{5.78b}$$

and

$$V_{b_{i,j,k}} = \left\{\left(\alpha^2_{lg} - 1\right)^2/\left[\alpha^2_{lg}\log_e\left(\alpha^2_{lg}\right)\right]\right\}r^2_i\left({}^1\!/_2\Delta\theta_j\right)\Delta z_k \quad \text{for } i = 2,3,\ldots n_r - 1 \tag{5.81b}$$

Example 4.10 demonstrates that quotients $r_i/r^L_{i-1/2}$, $r^L_{i-1/2}/r_{i-1}$, $r^L_{i+1/2}/r_i$, $r_{i+1}/r^L_{i+1/2}$, and $r^L_{i+1/2}/r^L_{i-1/2}$ are functions of the logarithmic spacing constant α_{lg} only as given by Eqs. (4.111), (4.114), (4.103), (4.106), and (4.116), respectively. By substituting these equations, or Eqs. (5.82), (5.75b), (5.76b), (5.77b), and (5.78b), into Table 5.2 and observing that $\left({}^1\!/_2\Delta\theta_j\right)\left(r^2_{i+1/2} - r^2_{i-1/2}\right) = V_{b_{i,j,k}}/\Delta z_k$ using Eq. (5.81a), we obtain Table 5.3.

Now, the calculation of geometric factors and pore volumes can be simplified using the following algorithm:

1. Define

$$\alpha_{lg} = \left(\frac{r_e}{r_w}\right)^{1/(n_r-1)} \tag{5.79}$$

TABLE 5.3 Geometric factors in cylindrical grids.

Direction	Geometric factor
r	$G_{r_{i-1/2,j,k}} = \dfrac{\beta_c \Delta\theta_j \Delta z_k}{\log_e\left[\alpha_{lg}\log_e(\alpha_{lg})/(\alpha_{lg}-1)\right]/k_{r_{i,j,k}} + \log_e\left[(\alpha_{lg}-1)/\log_e(\alpha_{lg})\right]/k_{r_{i-1,j,k}}}$
	$G_{r_{i+1/2,j,k}} = \dfrac{\beta_c \Delta\theta_j \Delta z_k}{\log_e\left[(\alpha_{lg}-1)/\log_e(\alpha_{lg})\right]/k_{r_{i,j,k}} + \log_e\left[\alpha_{lg}\log_e(\alpha_{lg})/(\alpha_{lg}-1)\right]/k_{r_{i+1,j,k}}}$
θ	$G_{\theta_{i,j\mp1/2,k}} = \dfrac{2\beta_c \log_e(\alpha_{lg})\Delta z_k}{\Delta\theta_{j\mp1/2}/k_{\theta_{i,j,k}} + \Delta\theta_{j\mp1/2}/k_{\theta_{i,j\mp1,k}}}$
z	$G_{z_{i,j,k\mp1/2}} = \dfrac{2\beta_c (V_{b_{i,j,k}}/\Delta z_k)}{\Delta z_{k\mp1/2}/k_{z_{i,j,k}} + \Delta z_{k\mp1/2}/k_{z_{i,j,k\mp1}}}$

2. Let

$$r_1 = r_w \tag{5.80}$$

3. Set

$$r_i = \alpha_{lg}^{i-1} r_1 \tag{5.82}$$

where $i=1, 2, 3, \ldots n_r$

4. For $j=1, 2, 3, \ldots n_\theta$ and $k=1, 2, 3, \ldots n_z$, set

$$V_{b_{i,j,k}} = \left\{ \left(\alpha_{lg}^2 - 1\right)^2 / \left[\alpha_{lg}^2 \log_e\left(\alpha_{lg}^2\right)\right] \right\} r_i^2 \left(^1/_2 \Delta\theta_j\right) \Delta z_k \tag{5.81b}$$

for $i=2, 3, \ldots n_r - 1$

$$V_{b_{1,j,k}} = \left\{ \left[\left(\alpha_{lg}^2 - 1\right)/\log_e\left(\alpha_{lg}^2\right)\right] - 1 \right\} r_w^2 \left(^1/_2 \Delta\theta_j\right) \Delta z_k \tag{5.81c}$$

for $i=1$; and

$$V_{b_{n_r,j,k}} = \left\{ 1 - \left(\alpha_{lg}^2 - 1\right) / \left[\alpha_{lg}^2 \log_e\left(\alpha_{lg}^2\right)\right] \right\} r_e^2 \left(^1/_2 \Delta\theta_j\right) \Delta z_k \tag{5.81d}$$

for $i=n_r$. Note that Eq. (5.81b) is used to calculate bulk volumes of gridpoints other than those falling on the reservoir internal and external boundaries in the r-direction (see Example 5.7). For $i=1$ and $i=n_r$, Eqs. (5.81c) and (5.81d) are used.

5. Estimate the geometric factors using the equations in Table 5.3. Note that in the calculation of $G_{z_{i,j,1/2}}$ or $G_{z_{i,j,n_z+1/2}}$, terms that describe properties of blocks that fall outside the reservoir ($k=0$ and $k=n_z+1$) are discarded.

Examples 5.7 and 5.8 show that reservoir discretization in the radial direction can be accomplished using either the traditional equations reported in the previous literature (Eqs. 5.74, 5.75a, 5.76a, 5.77a, 5.78a, 5.79, 5.80, and 5.81a) or those reported in this book (Eqs. 7.74, 5.75b, 5.76b, 5.77b, 5.78b, 5.79, 5.80, 5.81b, 5.81c, and 5.81d) that led to Table 5.3. The equations reported in this book, however, are easier and less confusing because they use r_i and α_{lg} only. In Example 5.9, we demonstrate how to use Eq. (5.2a) and the appropriate expressions for $q^m_{sc_{b,bP}}$, along with Table 5.3, to write the flow equations for boundary and interior gridpoints in a 2-D single-well simulation problem.

Example 5.7 Consider the simulation of a single-well in 40-acre spacing. Wellbore diameter is 0.5 ft, and the reservoir thickness is 100 ft. The reservoir can be simulated using a single layer discretized into six gridpoints in the radial direction.

1. Find gridpoint spacing in the r-direction.
2. Find the gridpoint block boundaries in the r-direction for transmissibility calculations.
3. Calculate the arguments of the $\log_e$ terms in Table 5.2.
4. Find the gridpoint block boundaries in the r-direction for bulk volume calculations and calculate the bulk volumes.

Solution

1. The external reservoir radius can be estimated from well spacing, $r_e = \sqrt{43{,}560 \times 40/\pi} = 744.73$ ft, and well radius, $r_w = 0.25$ ft.
   First, estimate α_{lg}using Eq. (5.79):

$$\alpha_{lg} = \left(\frac{r_e}{r_w}\right)^{1/(n_r-1)} = \left(\frac{744.73}{0.25}\right)^{1/(6-1)} = 4.9524$$

   Second, according to Eq. (5.80), let $r_1 = r_w = 0.25$ ft. Third, calculate the location of the gridpoints in the r-direction using Eq. (5.82), $r_i = \alpha_{lg}^{i-1} r_1$. For example, for $i=2$, $r_2 = (4.9524)^{2-1} \times 0.25 = 1.2381$ ft. Table 5.4 shows the location of the other gridpoints along the r-direction.
2. Block boundaries for transmissibility calculations ($r^L_{i-1/2}$, $r^L_{i+1/2}$) are estimated using Eqs. (5.75a) and (5.76a).
   For $i=2$,

$$r^L_{2+1/2} = \frac{r_3 - r_2}{\log_e(r_3/r_2)} = \frac{6.1316 - 1.2381}{\log_e(6.1316/1.2381)} = 3.0587\,\text{ft}$$

   and

$$r^L_{2-1/2} = \frac{r_2 - r_1}{\log_e(r_2/r_1)} = \frac{1.2381 - 0.25}{\log_e(1.2381/0.25)} = 0.6176\,\text{ft}$$

   Table 5.4 shows the block boundaries for transmissibility calculations for the other gridpoints.
3. Table 5.4 Shows the calculated values for $r_i/r^L_{i-1/2}$, $r_{i+1}/r^L_{i+1/2}$, $r^L_{i-1/2}/r_{i-1}$, $r^L_{i+1/2}/r_i$, and $r^L_{i+1/2}/r^L_{i-1/2}$, which appear in the argument of $\log_e$ terms in Table 5.2

TABLE 5.4 r_i, $r^L_{i\mp1/2}$, and $\log_e$ arguments in Table 5.2 for Example 5.7.

i	r_i	$r^L_{i-1/2}$	$r^L_{i+1/2}$	$r_i / r^L_{i-1/2}$	$r_{i+1} / r^L_{i+1/2}$	$r^L_{i-1/2} / r_{i-1}$	$r^L_{i+1/2} / r_i$	$r^L_{i+1/2} / r^L_{i+1/2}$
1	0.25	–	0.6176	–	2.005	2.470	2.470	–
2	1.2381	0.6176	3.0587	2.005	2.005	2.470	2.470	4.9524
3	6.1316	3.0587	15.148	2.005	2.005	2.470	2.470	4.9524
4	30.366	15.148	75.018	2.005	2.005	2.470	2.470	4.9524
5	150.38	75.016	371.51	2.005	2.005	2.470	2.470	4.9524
6	744.73	371.51	–	2.005	2.005	–	–	–

4. The block boundaries for bulk volume calculations ($r_{i-1/2}$, $r_{i+1/2}$) are estimated using Eqs. (5.77a) and (5.78a).

For $i=2$,

$$r^2_{2+1/2}=\frac{r_3^2-r_2^2}{\log_e\left(r_3^2/r_2^2\right)}=\frac{(6.1316)^2-(1.2381)^2}{\log_e\left[(6.1316)^2/(1.2381)^2\right]}=11.2707\,\text{ft}^2$$

and

$$r^2_{2-1/2}=\frac{r_2^2-r_1^2}{\log_e\left(r_2^2/r_1^2\right)}=\frac{(1.2381)^2-(0.25)^2}{\log_e\left[(1.2381)^2/(0.25)^2\right]}=0.4595\,\text{ft}^2$$

Therefore, the block boundaries for bulk volume calculations are

$$r_{2+1/2}=\sqrt{11.2707}=3.3572\,\text{ft}$$

and

$$r_{1+1/2}=\sqrt{0.4595}=0.6779\,\text{ft}$$

The bulk volume for gridpoints can be calculated using Eq. (5.81a).

For $i=2$,

$$V_{b_2}=\left[(3.3572)^2-(0.6779)^2\right]\left({}^1/_2\times 2\pi\right)\times 100=3396.45\,\text{ft}^3$$

For $i=1$,

$$V_{b_1}=\left[(1.3558)^2-(0.25)^2\right]\left({}^1/_2\times 2\pi\right)\times 100=124.73\,\text{ft}^3$$

For $i=6$,

$$V_{b_6}=\left[(744.73)^2-(407.77)^2\right]\left({}^1/_2\times 2\pi\right)\times 100=122.003\times 10^6\,\text{ft}^3$$

Table 5.5 shows the block boundaries and bulk volumes of blocks for the other gridpoints.

Example 5.8 Solve Example 5.7 again, this time using Eqs. (5.75b), (5.76b), (5.77b), (5.78b), and (5.81b), which make use of r_i and α_{lg}, and Eqs. (5.81c) and (5.81d).

Solution

1. From Example 5.7, $r_e=744.73$ ft., $r_1=r_w=0.25$ ft, and $\alpha_{lg}=4.9524$. In addition, the radii of gridpoints are calculated using Eq. (5.82), $r_i=\alpha_{lg}^{i-1}r_1$, as shown in Example 5.7.

2. The block boundaries for transmissibility calculations ($r^L_{i-1/2}$, $r^L_{i+1/2}$), estimated using Eqs. (5.75b) and (5.76b), are

$$\begin{aligned}r^L_{i+1/2}&=\left\{\left(\alpha_{lg}-1\right)/\left[\log_e\left(\alpha_{lg}\right)\right]\right\}r_i=\{(4.9524-1)/[\log_e(4.9524)]\}r_i\\&=2.47045r_i\end{aligned}\tag{5.83}$$

TABLE 5.5 Gridpoint boundaries and bulk volumes for gridpoints in Example 5.7.

i	r_i	$r_{i-1/2}$	$r_{i+1/2}$	V_{b_i}
1	0.25	0.25^a	0.6779	124.73
2	1.2381	0.6779	3.3572	3396.5
3	6.1316	3.3572	16.626	83,300.3
4	30.366	16.626	82.337	2.04×10^6
5	150.38	82.337	407.77	50.1×10^6
6	744.73	407.77	744.73^b	122×10^6

[a] $r_{1-1/2}=r_w=0.25$.
[b] $r_{6+1/2}=r_e=744.73$.

and

$$r^L_{i-1/2}=\{(\alpha_{lg}-1)/[\alpha_{lg}\log_e(\alpha_{lg})]\}r_i$$
$$=\{(4.9524-1)/[4.9524\log_e(4.9524)]\}r_i=0.49884r_i \quad (5.84)$$

Substitution of values of r_i into Eqs. (5.83) and (5.84) produces the results reported in Table 5.4.

3. Example 4.10 derives the ratios $r_i/r^L_{i-1/2}$, $r_{i+1}/r^L_{i+1/2}$, $r^L_{i-1/2}/r_{i-1}$, $r^L_{i+1/2}/r_i$, and $r^L_{i+1/2}/r^L_{i-1/2}$ as functions of α_{lg} as Eqs. (4.111), (4.106), (4.114), (4.103), and (4.116), respectively. Substituting of $\alpha_{lg}=4.9524$ in these equations, one obtains

$$r_i/r^L_{i-1/2}=[\alpha_{lg}\log_e(\alpha_{lg})]/(\alpha_{lg}-1)=[4.9524\log_e(4.9524)]/(4.9524-1)$$
$$=2.005 \quad (5.85)$$

$$r_{i+1}/r^L_{i+1/2}=[\alpha_{lg}\log_e(\alpha_{lg})]/(\alpha_{lg}-1)=2.005 \quad (5.86)$$

$$r^L_{i-1/2}/r_{i-1}=(\alpha_{lg}-1)/\log_e(\alpha_{lg})$$
$$=(4.9524-1)/\log_e(4.9524)=2.470 \quad (5.87)$$

$$r^L_{i+1/2}/r_i=(\alpha_{lg}-1)/\log_e(\alpha_{lg})=2.470 \quad (5.88)$$

and

$$r^L_{i+1/2}/r^L_{i-1/2}=\alpha_{lg}=4.9524 \quad (5.89)$$

Note that the values of the aforementioned ratios are the same as those reported in Table 5.4.

4. The block boundaries for bulk volume calculations ($r_{i-1/2}$, $r_{i+1/2}$) are estimated using Eqs. (5.77b) and (5.78b), yielding

$$\begin{aligned} r_{i+1/2}^2 &= \left\{\left(\alpha_{lg}^2 - 1\right)/\left[\log_e\left(\alpha_{lg}^2\right)\right]\right\}r_i^2 \\ &= \left\{\left((4.9524)^2 - 1\right)/\left[\log_e\left((4.9524)^2\right)\right]\right\}r_i^2 = (7.3525)r_i^2 \end{aligned} \tag{5.90}$$

and

$$\begin{aligned} r_{i-1/2}^2 &= \left\{\left(\alpha_{lg}^2 - 1\right)/\left[\alpha_{lg}^2 \log_e\left(\alpha_{lg}^2\right)\right]\right\}r_i^2 = \left\{7.3525/(4.9524)^2\right\}r_i^2 \\ &= (0.29978)r_i^2 \end{aligned} \tag{5.91}$$

Therefore,

$$r_{i+1/2} = \sqrt{(7.3525)r_i^2} = (2.7116)r_i \tag{5.92}$$

and

$$r_{i-1/2} = \sqrt{(0.29978)r_i^2} = (0.54752)r_i \tag{5.93}$$

The bulk volume associated with each gridpoint can be calculated using Eqs. (5.81b), (5.81c), and (5.81d), yielding

$$\begin{aligned} V_{b_i} &= \left\{\left(\alpha_{lg}^2 - 1\right)^2/\left[\alpha_{lg}^2 \log_e\left(\alpha_{lg}^2\right)\right]\right\}r_i^2\left[{}^1/_2(2\pi)\right]\Delta z \\ &= \left\{\left((4.9524)^2 - 1\right)^2/\left[(4.9524)^2 \log_e\left((4.9524)^2\right)\right]\right\}r_i^2\left[{}^1/_2(2\pi)\right] \times 100 = 2215.7r_i^2 \end{aligned} \tag{5.94}$$

for $i=2, 3, 4, 5$.

$$\begin{aligned} V_{b_1} &= \left\{\left[\left(\alpha_{lg}^2 - 1\right)/\log_e\left(\alpha_{lg}^2\right)\right] - 1\right\}r_w^2\left({}^1/_2\Delta\theta\right)\Delta z \\ &= \left\{\left[\left((4.9524)^2 - 1\right)/\log_e\left((4.9524)^2\right)\right] - 1\right\}(0.25)^2\left[{}^1/_2(2\pi)\right] \times 100 = 124.73\,\text{ft}^3 \end{aligned} \tag{5.95}$$

and

$$\begin{aligned} V_{b_6} &= \left\{1 - \left(\alpha_{lg}^2 - 1\right)/\left[\alpha_{lg}^2 \log_e\left(\alpha_{lg}^2\right)\right]\right\}r_e^2\left({}^1/_2\Delta\theta\right)\Delta z \\ &= \left\{1 - \left((4.9524)^2 - 1\right)/\left[(4.9524)^2 \log_e\left((4.9524)^2\right)\right]\right\}(744.73)^2\left[{}^1/_2(2\pi)\right] \times 100 \\ &= 122.006 \times 10^6\,\text{ft}^3 \end{aligned} \tag{5.96}$$

Note that the values of the estimated bulk volumes slightly differ from those reported in Table 5.5 because of round-off errors resulting from approximations in the various stages of calculations.

Example 5.9 A 0.5-ft diameter water well is located in 20-acre spacing. The reservoir thickness, horizontal permeability, and porosity are 30 ft, 150 md, and 0.23, respectively. The (k_V/k_H) for this reservoir is estimated from core data as 0.30. The flowing fluid has a density, FVF, and viscosity of 62.4 lbm/ft^3, 1 RB/B, and 0.5 cP, respectively. The reservoir external boundary in the radial direction is a no-flow boundary, and the well is completed in the top 22.5 ft only and produces at a rate of 2000 B/D. The reservoir bottom boundary is subject to influx such that the boundary is kept at 4000 psia. The reservoir top boundary is sealed to flow. Assuming the reservoir can be simulated using three equispaced gridpoints in the vertical direction and four gridpoints in the radial direction, as shown in Fig. 5.13, write the flow equations for gridpoints 1, 3, 5, 7, and 11.

Solution

To write flow equations, the gridpoints are first ordered using natural ordering ($n=1, 2, 3, \ldots 10, 11, 12$), as shown in Fig. 5.13, in addition to being identified using the engineering notation along the radial direction ($i=1, 2, 3, 4$) and the vertical direction ($k=1, 2, 3$). This step is followed by the determination of the location of the gridpoints in the radial direction and the calculation of the gridpoints separation and elevation in the vertical direction, Next, the bulk volumes and transmissibilities in the r- and z-directions are calculated. We demonstrate in this example that block boundaries for transmissibility calculations and block boundaries for bulk volume calculations are not needed if we use Eqs. (5.81b), (5.81c), and (5.81d) for bulk volume calculations and Table 5.3. Making use of the aforementioned information, we estimate the contributions of the gridpoints to the well rates and the fictitious well rates resulting from reservoir boundary conditions.

The reservoir rock and fluid data are restated as follows: $h=30$ ft, $\phi=0.23$, $k_r=k_H=150$ md, $k_z=k_H(k_V/k_H)=150\times 0.30=45$ md, $B=1$ RB/B, $\mu=0.5$ cP, $\gamma=\gamma_c\rho g=0.21584\times 10^{-3}(62.4)(32.174)=0.4333$ psi/ft, $r_w=0.25$ ft, and the

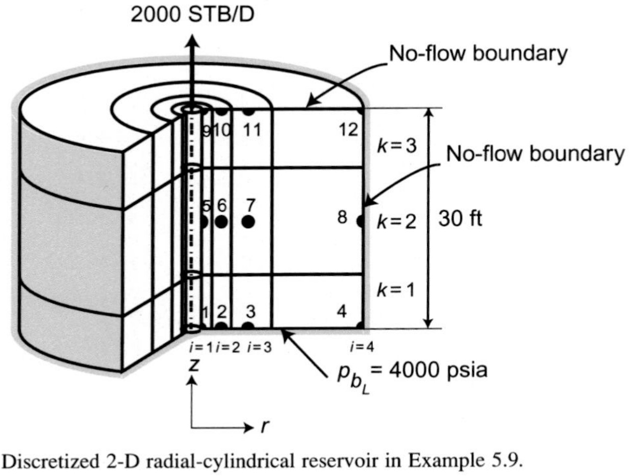

FIG. 5.13 Discretized 2-D radial-cylindrical reservoir in Example 5.9.

reservoir external radius is estimated from well spacing as $r_e = (20 \times 43560/\pi)^{1/2} =$ 526.60 ft. The reservoir east (external) and upper (top) boundaries are no-flow boundaries, the reservoir lower (bottom) boundary has $p_{b_L} = 4000$ psia, and the reservoir west (internal) boundary has $q_{spsc} = -2000$ B/D to reflect the effect of the production well (i.e., the well is treated as a boundary condition).

For the point-distributed grid shown in Fig. 5.13, $n_r = 4$, $n_z = 3$, and $\Delta z_{k+1/2} = h/(n_z - 1) = 30/(3-1) = 15$ ft for $k = 1, 2$; hence, $\Delta z_n = 15/2 = 7.5$ ft for $n = 1, 2, 3, 4$; $\Delta z_n = 15$ ft for $n = 5, 6, 7, 8$; and $\Delta z_n = 15/2 = 7.5$ ft for $n = 9, 10, 11, 12$. Assuming the top of the reservoir as the reference level for elevation, $Z_n = 0$ ft for $n = 9, 10, 11, 12$; $Z_n = 15$ ft for $n = 5, 6, 7, 8$; $Z_n = 30$ ft for $n = 1, 2, 3, 4$; and $Z_{b_L} = 30$ ft. The locations of the gridpoints in the radial direction are calculated using Eqs. (5.79), (5.80), and (5.82), yielding $\alpha_{lg} = (526.60/0.25)^{1/(4-1)} = 12.8188$; $r_1 = r_w = 0.25$ ft; and $r_i = (12.8188)^{(i-1)}(0.25)$ for $i = 2, 3, 4$ or $r_2 = 3.2047$ ft, $r_3 = 41.080$ ft, and $r_4 = 526.60$ ft.

The bulk volumes associated with the gridpoints are listed in Table 5.6. They are calculated using Eqs. (5.81b), (5.81c), and (5.81d). Note that subscript j is discarded and $\Delta\theta = 2\pi$.

$$
\begin{aligned}
V_{b_{1,k}} &= \left\{\left[\left(\alpha_{lg}^2 - 1\right)/\log_e\left(\alpha_{lg}^2\right)\right] - 1\right\} r_w^2 \left({}^1/_2 \Delta\theta\right) \Delta z_k \\
&= \left\{\left[\left((12.8188)^2 - 1\right)/\log_e\left((12.8188)^2\right)\right] - 1\right\} (0.25)^2 \left({}^1/_2 \times 2\pi\right) \Delta z_k \\
&= (6.0892685) \Delta z_k
\end{aligned}
$$

TABLE 5.6 Gridpoint locations, bulk volumes, and radial and vertical transmissibilities for Example 5.9.

n	i	k	r_i (ft)	Δz_n (ft)	Z_n (ft)	V_{b_n} (ft^3)	$T_{r_{i\pm1/2,k}}$ (B/D-psi)	$T_{z_{i,k\pm1/2}}$ (B/D-psi)
1	1	1	0.25	7.5	30	45.66941	6.245838	0.041176
2	2	1	3.2047	7.5	30	7699.337	6.245838	6.941719
3	3	1	41.080	7.5	30	1,265,140	6.245838	1140.650
4	4	1	526.60	7.5	30	5,261,005	6.245838	4743.320
5	1	2	0.25	15	15	91.33882	12.49168	0.041176
6	2	2	3.2047	15	15	15,398.67	12.49168	6.941719
7	3	2	41.080	15	15	2,530,280	12.49168	1140.650
8	4	2	526.60	15	15	10,522,011	12.49168	4743.320
9	1	3	0.25	7.5	0	45.66941	6.245838	0.041176
10	2	3	3.2047	7.5	0	7699.337	6.245838	6.941719
11	3	3	41.080	7.5	0	1,265,140	6.245838	1140.650
12	4	3	526.60	7.5	0	5,261,005	6.245838	4743.320

$$V_{b_{i,k}} = \left\{ \left(\alpha_{lg}^2 - 1\right)^2 / \left[\alpha_{lg}^2 \log_e\left(\alpha_{lg}^2\right)\right] \right\} r_i^2 \left({}^1/_2 \Delta\theta\right) \Delta z_k$$
$$= \left\{ \left((12.8188)^2 - 1\right)^2 / \left[(12.8188)^2 \log_e\left((12.8188)^2\right)\right] \right\} r_i^2 \left({}^1/_2 \times 2\pi\right) \Delta z_k$$
$$= (99.957858) r_i^2 \Delta z_k$$

for $i = 2$, 3, and

$$V_{b_{4,k}} = \left\{1 - \left(\alpha_{lg}^2 - 1\right) / \left[\alpha_{lg}^2 \log_e\left(\alpha_{lg}^2\right)\right]\right\} r_e^2 \left({}^1/_2 \Delta\theta\right) \Delta z_k$$
$$= \left\{1 - \left((12.8188)^2 - 1\right) / \left[(12.8188)^2 \log_e\left((12.8188)^2\right)\right]\right\} (526.60)^2 \left({}^1/_2 \times 2\pi\right) \Delta z_k$$
$$= (701466.65) \Delta z_k$$

The transmissibility in the r-direction is defined by Eq. (5.72a), yielding

$$T_{r_{i\mp1/2,k}} = G_{r_{i\mp1/2,k}} \left(\frac{1}{\mu B}\right) = G_{r_{i\mp1/2,k}} \left(\frac{1}{0.5 \times 1}\right) = (2) G_{r_{i\mp1/2,k}} \tag{5.97}$$

where $G_{r_{i\mp1/2,\,k}}$ is defined in Table 5.3. With $\Delta\theta = 2\pi$ and constant radial permeability, the equations for the geometric factor reduce to

$$\begin{aligned} G_{r_{i\mp1/2,k}} &= \frac{2\pi\beta_c k_r \Delta z_k}{\log_e\left[\alpha_{lg} \log_e(\alpha_{lg}) / (\alpha_{lg} - 1)\right] + \log_e\left[(\alpha_{lg} - 1) / \log_e(\alpha_{lg})\right]} \\ &= \frac{2\pi\beta_c k_r \Delta z_k}{\log_e(\alpha_{lg})} = \frac{2\pi(0.001127)(150)\Delta z_k}{\log_e(12.8188)} = (0.4163892)\Delta z_k \end{aligned} \tag{5.98}$$

Therefore, the transmissibility in the radial direction can be estimated by substituting Eq. (5.98) into Eq. (5.97), resulting in

$$T_{r_{i\mp1/2,k}} = (2) G_{r_{i\mp1/2,k}} = (2)(0.4163892)\Delta z_k = (0.8327784)\Delta z_k \tag{5.99}$$

The transmissibility in the vertical direction is defined by Eq. (5.72c), yielding

$$T_{z_{i,k\mp1/2}} = G_{z_{i,k\mp1/2}} \left(\frac{1}{\mu B}\right) = G_{z_{i,k\mp1/2}} \left(\frac{1}{0.5 \times 1}\right) = (2) G_{z_{i,k\mp1/2}} \tag{5.100}$$

where $G_{z_{i,k\mp1/2}}$ is defined in Table 5.3 as

$$G_{z_{i,k\mp1/2}} = \frac{2\beta_c \left(V_{b_{i,k}} / \Delta z_k\right)}{\Delta z_{k\mp1/2} / k_{z_{i,k}} + \Delta z_{k\mp1/2} / k_{z_{i,k\mp1}}} \tag{5.101}$$

For this problem, gridpoint spacing and vertical permeability are constants; therefore, the equation for the geometric factor reduces to

$$\begin{aligned} G_{z_{i,k\mp1/2}} &= \frac{2\beta_c k_z \left(V_{b_{i,k}} / \Delta z_k\right)}{2\Delta z_{k\mp1/2}} = \frac{\beta_c k_z \left(V_{b_{i,k}} / \Delta z_k\right)}{\Delta z_{k\mp1/2}} \\ &= \frac{(0.001127)(45)\left(V_{b_{i,k}} / \Delta z_k\right)}{15} = (0.003381)\left(V_{b_{i,k}} / \Delta z_k\right) \end{aligned} \tag{5.102}$$

Substituting Eq. (5.102) into Eq. (5.100) results in

$$T_{z_{i,k\mp1/2}} = (2)G_{z_{i,k\mp1/2}} = (2)(0.003381)\left(V_{b_{i,k}}/\Delta z_k\right) = (0.006762)\left(V_{b_{i,k}}/\Delta z_k\right) \tag{5.103}$$

The estimated transmissibilities in the radial and vertical directions are listed in Table 5.6.

Before writing the flow equations, the well production rate (specified rate for the reservoir west boundary) must be prorated between gridpoints 5 and 9 using

$$q^m_{sc_{b,bP}} = \frac{T^m_{bP,bP^*}}{\sum\limits_{l\in\psi_b} T^m_{l,l^*}} q_{spsc} \tag{5.37}$$

where T^m_{b,bP^*} = transmissibility in the radial direction between gridpoints bP and bP^* with the well being the reservoir internal boundary and $\psi_b = \psi_w = \{5, 9\}$. Note that gridpoint 1 has a no-flow boundary because it is not penetrated by the well; that is, $q^m_{sc_{b_W,1}} = 0$. Note also that $5^* = 6$ and $9^* = 10$ according to the terminology in Fig. 5.7. From Table 5.6,

$$T^m_{b_W,6} = T^m_{r_{5,6}} = 12.49168\,\text{B/D-psi}$$

and

$$T^m_{b_W,10} = T^m_{r_{9,10}} = 6.245838\,\text{B/D-psi}$$

The application of Eq. (5.37) results in

$$q^m_{sc_{b_W,9}} = \frac{6.245838}{6.245838 + 12.49168} \times (-2000) = -666.67\,\text{B/D}$$

and

$$q^m_{sc_{b_W,5}} = \frac{12.49168}{6.245838 + 12.49168} \times (-2000) = -1333.33\,\text{B/D}$$

With this treatment of the production well, $q^m_{sc_n} = 0$ for each gridpoint (including 1, 5, and 9).

For the reservoir lower boundary, $p^m_1 = p^m_2 = p^m_3 = p^m_4 = p_{b_L} = 4000$ psia. The flow rates of the fictitious wells in boundary gridpoints 1, 2, 3, and 4 are estimated using Eq. (5.46c), yielding.

$$q^m_{sc_{b_L,bP}} = T^m_{z_{i,k+1/2}}[(4000 - p_{bP^*}) - (0.4333)(30 - 15)]\,\text{B/D} \tag{5.104}$$

where according to Fig. 5.13 and our terminology in Fig. 5.7, $1^* = 5$, $2^* = 6$, $3^* = 7$, and $4^* = 8$. For the reservoir east and upper (no-flow) boundaries, $q^m_{sc_{b_E,n}} = 0$ for $n = 4$, 8, 12 and $q^m_{sc_{b_U,n}} = 0$ for $n = 9$, 10, 11, 12. The contributions of gridpoints to the well rates and the fictitious well rates are summarized in Table 5.7.

TABLE 5.7 Contribution of gridpoints to well rates and fictitious well rates for Example 5.9.

n	i	k	$q^m_{sc_n}$ (B/D)	$q^m_{sc_{b_L,n}}$ (B/D)	$q^m_{sc_{b_W,n}}$ (B/D)	$q^m_{sc_{b_E,n}}$ (B/D)	$q^m_{sc_{b_U,n}}$ (B/D)
1	1	1	0	(0.041176) $[(4000-p^m_5)-(0.4333)(30-15)]$	0		
2	2	1	0	(6.941719) $[(4000-p^m_6)-(0.4333)(30-15)]$			
3	3	1	0	(1140.650) $[(4000-p^m_7)-(0.4333)(30-15)]$			
4	4	1	0	(4743.320) $[(4000-p^m_8)-(0.4333)(30-15)]$		0	
5	1	2	0		−1333.33		
6	2	2	0				
7	3	2	0				
8	4	2	0			0	
9	1	3	0		−666.67		0
10	2	3	0				0
11	3	3	0				0
12	4	3	0			0	0

The general form of the flow equation for gridpoint n is written as:

$$\sum_{l\in\psi_n} T^m_{l,n}\left[(p^m_l - p^m_n) - \gamma^m_{l,n}(Z_l - Z_n)\right] + \sum_{l\in\xi_n} q^m_{sc_{l,n}} + q^m_{sc_n} = \frac{V_{b_n}}{\alpha_c \Delta t}\left[\left(\frac{\phi}{B}\right)^{n+1}_n - \left(\frac{\phi}{B}\right)^n_n\right] \tag{5.2a}$$

For gridpoint 1, $p^m_1 = 4000$ psia because gridpoint 1 falls on the constant pressure boundary. Let us write the flow equation for this gridpoint. For gridpoint 1, $n=1$, $i=1$, $k=1$, $\psi_1=\{2,5\}$, $\xi_1=\{b_L,b_W\}$, and $\sum_{l\in\xi_1} q^m_{sc_{l,1}} = q^m_{sc_{b_L,1}} + q^m_{sc_{b_W,1}}$, where from Table 5.7, $q^m_{sc_{b_L,1}} = (0.041176)[(4000 - p^m_5) - (0.4333)(30-15)]$ B/D, $q^m_{sc_{b_W,1}} = 0$, and $q^m_{sc_1} = 0$. Therefore, substitution into Eq. (5.2a) yields

$$\begin{aligned}&(6.245838)\left[\left(p_2^m-p_1^m\right)-(0.4333)(30-30)\right]\\&+(0.041176)\left[\left(p_5^m-p_1^m\right)-(0.4333)(15-30)\right]\\&+(0.041176)\left[\left(4000-p_5^m\right)-(0.4333)(30-15)\right]+0+0\\&=\frac{45.66941}{\alpha_c\Delta t}\left[\left(\frac{\phi}{B}\right)_1^{n+1}-\left(\frac{\phi}{B}\right)_1^n\right]\end{aligned} \tag{5.105}$$

where $p_1^m=4000$ psia. Note that the accumulation term vanishes because the gridpoint pressure is constant. Therefore, Eq. (5.105) after simplification becomes

$$(6.245838)\left[\left(p_1^m-p_2^m\right)-(0.4333)(30-30)\right]=0 \tag{5.106}$$

or

$$p_1^m=p_2^m \tag{5.107}$$

Eq. (5.107) does not introduce new knowledge because both gridpoints fall on the constant pressure bottom boundary, but it confirms that the flow equation for gridpoint 1, as expressed by Eq. (5.105), is correct.

For gridpoint 3, $p_3^m=4000$ psia because gridpoint 3 falls on the constant pressure boundary. Again, let us write the flow equation for this gridpoint. For gridpoint 3, $n=3$, $i=3$, $k=1$, $\psi_3=\{2,4,7\}$, $\xi_3=\{b_L\}$, and $\sum_{l\in\xi_3} q_{sc_{l,3}}^m=q_{sc_{b_L,3}}^m$, where from Table 5.7, $q_{sc_{b_L,3}}^m=(1140.650)[(4000-p_7^m)-(0.4333)(30-15)]$ B/D and $q_{sc_3}^m=0$ (no wells).

Therefore, substitution into Eq. (5.2a) yields

$$\begin{aligned}&(6.245838)\left[\left(p_2^m-p_3^m\right)-(0.4333)(30-30)\right]\\&+(6.245838)\left[\left(p_4^m-p_3^m\right)-(0.4333)(30-30)\right]\\&+(1140.650)\left[\left(p_7^m-p_3^m\right)-(0.4333)(15-30)\right]\\&+(1140.650)\left[\left(4000-p_7^m\right)-(0.4333)(30-15)\right]+0=\frac{1265140}{\alpha_c\Delta t}\left[\left(\frac{\phi}{B}\right)_3^{n+1}-\left(\frac{\phi}{B}\right)_3^n\right]\end{aligned} \tag{5.108}$$

where $p_3^m=4000$ psia. Note that the accumulation term vanishes because the gridpoint pressure is constant. Therefore, Eq. (5.108) after simplification becomes

$$\begin{aligned}&(6.245838)\left[\left(p_2^m-p_3^m\right)-(0.4333)(30-30)\right]\\&+(6.245838)\left[\left(p_4^m-p_3^m\right)-(0.4333)(30-30)\right]=0\end{aligned} \tag{5.109}$$

or

$$p_3^m={}^1/_2\left(p_2^m+p_4^m\right) \tag{5.110}$$

Eq. (5.110) does not introduce new knowledge because gridpoints 2, 3, and 4 fall on the constant pressure bottom boundary, but it confirms that the flow equation for gridpoint 3, as expressed by Eq. (5.108), is correct.

For gridpoint 5, $n=5$, $i=1$, $k=2$, $\psi_5=\{1,6,9\}$, $\xi_5=\{b_W\}$, $\sum_{l\in\xi_5} q_{sc_{l,5}}^m = q_{sc_{b_W,5}}^m = -1333.33$ B/D, and $q_{sc_5}^m=0$ (the well is treated as a boundary condition). Therefore, substitution into Eq. (5.2a) yields

$$\begin{aligned}&(0.041176)\left[(p_1^m-p_5^m)-(0.4333)(30-15)\right]\\&+(12.49168)\left[(p_6^m-p_5^m)-(0.4333)(15-15)\right]\\&+(0.041176)\left[(p_9^m-p_5^m)-(0.4333)(0-15)\right]-1333.33+0\\&=\frac{91.33882}{\alpha_c\Delta t}\left[\left(\frac{\phi}{B}\right)_5^{n+1}-\left(\frac{\phi}{B}\right)_5^n\right]\end{aligned}\tag{5.111}$$

In Eq. (5.111), the well is treated as a fictitious well. This treatment (or the substitution of well by a fictitious well and vice versa) is valid only in single-well simulation because, contrary to the situation in Cartesian coordinates, in cylindrical coordinates, both the well and the fictitious well have radial flow.

For gridpoint 7, $n=7$, $i=3$, $k=2$, $\psi_7=\{3,6,8,11\}$, $\xi_7=\{\}$, $\sum_{l\in\xi_7} q_{sc_{l,7}}^m=0$ (interior gridpoint), and $q_{sc_7}^m=0$ (no wells). Therefore, substitution into Eq. (5.2a) yields

$$\begin{aligned}&(1140.650)\left[(p_3^m-p_7^m)-(0.4333)(30-15)\right]\\&+(12.49168)\left[(p_6^m-p_7^m)-(0.4333)(15-15)\right]\\&+(12.49168)\left[(p_8^m-p_7^m)-(0.4333)(15-15)\right]\\&+(1140.650)\left[(p_{11}^m-p_7^m)-(0.4333)(0-15)\right]+0+0=\frac{2530280}{\alpha_c\Delta t}\left[\left(\frac{\phi}{B}\right)_7^{n+1}-\left(\frac{\phi}{B}\right)_7^n\right]\end{aligned}\tag{5.112}$$

For gridpoint 11, $n=11$, $i=3$, $k=3$, $\psi_{11}=\{7,10,12\}$, $\xi_{11}=\{b_U\}$, $\sum_{l\in\xi_{11}} q_{sc_{l,11}}^m = q_{sc_{b_U,11}}^m$ and $q_{sc_{b_U,11}}^m=0$ (no-flow boundary), and $q_{sc_{11}}^m=0$ (no wells). Therefore, substitution into Eq. (5.2a) yields

$$\begin{aligned}&(1140.650)\left[(p_7^m-p_{11}^m)-(0.4333)(15-0)\right]\\&+(6.245838)\left[(p_{10}^m-p_{11}^m)-(0.4333)(0-0)\right]\\&+(6.245838)\left[(p_{12}^m-p_{11}^m)-(0.4333)(0-0)\right]+0+0=\frac{1265140}{\alpha_c\Delta t}\left[\left(\frac{\phi}{B}\right)_{11}^{n+1}-\left(\frac{\phi}{B}\right)_{11}^n\right]\end{aligned}\tag{5.113}$$

5.6 Symmetry and its use in solving practical problems

The use of symmetry in solving practical problems has been discussed in Chapter 4. In most cases, the use of symmetry is justified if a pattern is found in the reservoir properties. The use of symmetry reduces the efforts to solve a problem by considering solving a modified problem for one element of symmetry in the reservoir, usually the smallest element of symmetry (Abou-Kassem et al., 1991). The smallest element of symmetry is a segment of the reservoir that is a mirror image of the rest of reservoir segments. Before solving the modified problem for one element of symmetry, however, symmetry must first be established. For symmetry to exist about a plane, there must be symmetry with regard to (1) the number of gridpoints and gridpoints spacing, (2) reservoir rock properties, (3) physical wells, (4) reservoir boundaries, and (5) initial conditions. Gridpoint spacing deals with the separation between gridpoints ($\Delta x_{i\mp 1/2}$, $\Delta y_{j\mp 1/2}$, $\Delta z_{k\mp 1/2}$) and gridpoint elevation (Z). Reservoir rock properties deal with gridpoint porosity (ϕ) and permeability in the various directions (k_x, k_y, k_z). Wells deal with well location, well type (injection or production), and well operating condition. Reservoir boundaries deal with the geometry of boundaries and boundary conditions. Initial conditions deal with initial pressure and fluid saturation distributions in the reservoir. Failing to satisfy symmetry with respect to any of the items mentioned earlier means there is no symmetry about that plane. The formulation of the modified problem for the smallest element of symmetry involves replacing each plane of symmetry with a no-flow boundary and determining the new interblock geometric factors, bulk volume, wellblock rate, and wellblock geometric factor for those gridpoints that share their boundaries with the planes of symmetry. To elaborate on this point, we present a few possible cases. In the following discussion, we use bold numbers to identify the gridpoints that require determining new values for their bulk volume, wellblock rate, wellblock geometric factor, and interblock geometric factors in the element of symmetry.

The first two examples show planes of symmetry that coincide with the boundaries between gridpoints. Fig. 5.14a presents a 1-D flow problem in which the plane of symmetry A-A, which is normal to the flow direction (x-direction) and coincides with the block boundary halfway between gridpoints 3 and 4, divides the reservoir into two symmetrical elements. Consequently, $p_1 = p_6$, $p_2 = p_5$, and $p_3 = p_4$. The modified problem is represented by the element of symmetry shown in Fig. 5.14b, with the plane of symmetry being replaced with a no-flow boundary. Fig. 5.15a presents a 2-D horizontal reservoir with two vertical planes of symmetry A-A and B-B. Plane of symmetry A-A is normal to the x-direction and coincides with the block boundaries halfway between gridpoints 2, 6, 10, and 14 on one side and gridpoints 3, 7, 11, and 15 on the other side. Plane of symmetry B-B is normal to the y-direction and coincides with the block boundaries halfway between gridpoints 5, 6, 7, and 8 on one side and gridpoints 9, 10, 11, and 12 on the other side. The two planes of symmetry divide the reservoir into four symmetrical elements. Consequently, $p_1 = p_4 = p_{13} = p_{16}$,

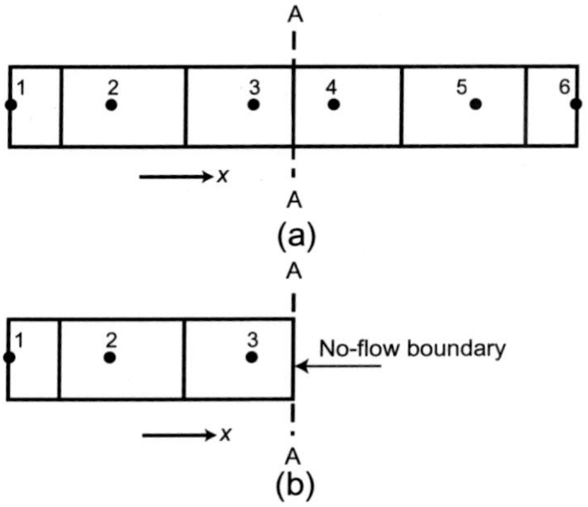

FIG. 5.14 1-D reservoir with even gridpoints exhibiting a vertical plane of symmetry. (a) Whole reservoir and planes of symmetry and (b) smallest element of symmetry.

$p_2=p_3=p_{14}=p_{15}$, $p_5=p_8=p_9=p_{12}$, and $p_6=p_7=p_{10}=p_{11}$. The modified problem is represented by the smallest element of symmetry shown in Fig. 5.15b, with each plane of symmetry being replaced with a no-flow boundary.

The second two examples show planes of symmetry that pass through gridpoints. Fig. 5.16a presents a 1-D flow problem in which the plane of symmetry

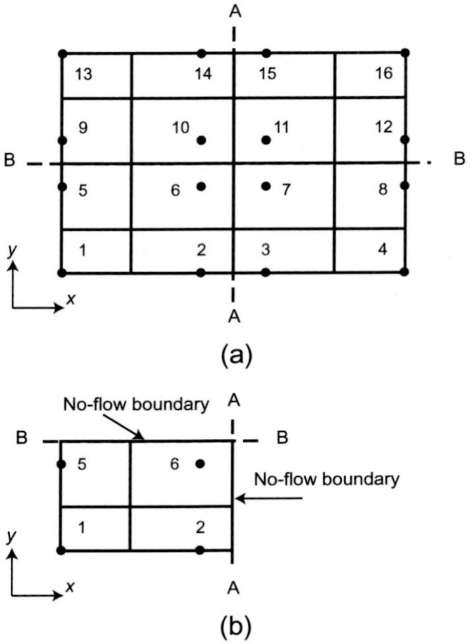

FIG. 5.15 2-D reservoir with even gridpoints in the x- and y-directions exhibiting two vertical planes of symmetry. (a) Whole reservoir and planes of symmetry and (b) smallest element of symmetry.

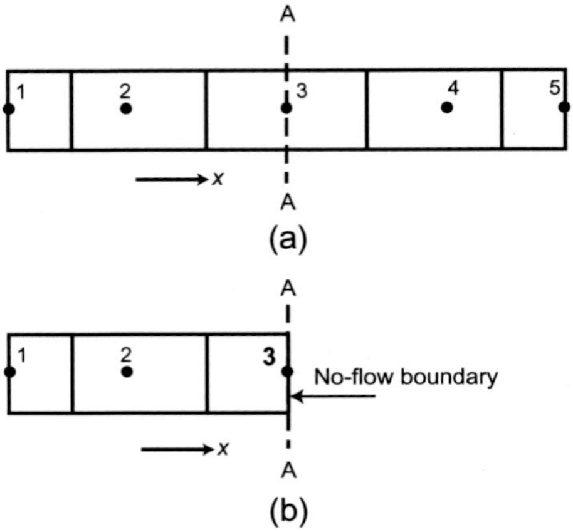

FIG. 5.16 1-D reservoir with odd gridpoints exhibiting a vertical plane of symmetry. (a) Whole reservoir and planes of symmetry and (b) smallest element of symmetry.

A-A, which is normal to the flow direction (x-direction) and passes through gridpoint 3, divides the reservoir into two symmetrical elements. Consequently, $p_1 = p_5$ and $p_2 = p_4$. The modified problem is represented by the element of symmetry shown in Fig. 5.16b, with the plane of symmetry being replaced with a no-flow boundary. This plane of symmetry bisects the gridpoint bulk volume, wellblock rate, and wellblock geometric factor for gridpoint 3 in Fig. 5.16a. Therefore, for gridpoint 3, $V_{b_3} = \frac{1}{2} V_{b_3}$, $q_{sc_3} = \frac{1}{2} q_{sc_3}$, and $G_{w_3} = \frac{1}{2} G_{w_3}$. Note that the interblock geometric factor in the direction normal to the plane of symmetry ($G_{x_{2,3}}$) is not affected. Fig. 5.17a presents a 2-D horizontal reservoir with two vertical planes of symmetry A-A and B-B. Plane A-A is a vertical plane of symmetry that is parallel to the y-z plane (normal to the x-direction) and passes through gridpoints 2, 5, and 8. Note that gridpoints 1, 4, and 7 are mirror images of gridpoints 3, 6, and 9. Plane B-B is a vertical plane of symmetry that is parallel to the x-z plane (normal to the y-direction) and passes through gridpoints 4, 5, and 6. Note that gridpoints 1, 2, and 3 are mirror images of gridpoints 7, 8, and 9. The two planes of symmetry divide the reservoir into four symmetrical elements. Consequently, $p_1 = p_3 = p_7 = p_9$, $p_4 = p_6$, and $p_2 = p_8$. The modified problem is represented by the smallest element of symmetry shown in Fig. 5.17b, with each plane of symmetry being replaced with a no-flow boundary. Each plane of symmetry bisects the gridpoint bulk volume, wellblock rate, and wellblock geometric factor of the gridpoint it passes through and bisects the interblock geometric factors in the directions that are parallel to the plane of symmetry. Therefore, $V_{b_2} = \frac{1}{2} V_{b_2}$, $q_{sc_2} = \frac{1}{2} q_{sc_2}$, $G_{w_2} = \frac{1}{2} G_{w_2}$; $V_{b_4} = \frac{1}{2} V_{b_4}$, $q_{sc_4} = \frac{1}{2} q_{sc_4}$, $G_{w_4} = \frac{1}{2} G_{w_4}$; $V_{b_5} = \frac{1}{4} V_{b_5}$, $q_{sc_5} = \frac{1}{4} q_{sc_5}$, $G_{w_5} = \frac{1}{4} G_{w_5}$; $G_{y_{2,5}} = \frac{1}{2} G_{y_{2,5}}$; and $G_{x_{4,5}} = \frac{1}{2} G_{x_{4,5}}$. Note that a plane of symmetry passing through a gridpoint results in a factor of $\frac{1}{2}$ as in gridpoints 2 and 4. Two planes of symmetry passing through a gridpoint result in a factor of $\frac{1}{2}\times\frac{1}{2}=\frac{1}{4}$ as in gridpoint 5.

The third example presents two vertical planes of symmetry, one coinciding with the boundaries between gridpoints and the other passing through the gridpoints. Fig. 5.18a presents a 2-D horizontal reservoir with two vertical planes of symmetry A-A and B-B. Plane A-A is a vertical plane of symmetry that is parallel to the *y-z* plane (normal to the *x*-direction) and passes through gridpoints 2, 5, 8, and 11. Note that gridpoints 1, 4, 7, and 10 are mirror images of gridpoints 3, 6, 9, and 12. Plane B-B is a vertical plane of symmetry that is parallel to the *x-z* plane (normal to the *y*-direction) and coincides with the boundaries between gridpoints 4, 5, and 6 on one side and gridpoints 7, 8, and 9 on the other side. Note that gridpoints 1, 2, and 3 are mirror images of gridpoints 10, 11, and 12 and gridpoints 4, 5, and 6 are mirror images of gridpoints 7, 8, and 9. The two planes of symmetry divide the reservoir into four symmetrical elements. Consequently, $p_1=p_3=p_{10}=p_{12}$, $p_4=p_6=p_7=p_9$, $p_2=p_{11}$, and $p_5=p_8$. The modified problem is represented by the smallest element of symmetry shown in Fig. 4.18b, with each plane of symmetry being replaced with a no-flow boundary. Plane of symmetry A-A bisects the block bulk volume, wellblock rate, and wellblock geometric factor of the gridpoints it passes through and bisects the interblock geometric factors in the directions that are parallel to the plane of symmetry (*y*-direction in this case). Therefore, $V_{\mathbf{b_2}}=\frac{1}{2}V_{b_2}$, $q_{sc\mathbf{2}}=\frac{1}{2}q_{sc_2}$, $G_{w\mathbf{2}}=\frac{1}{2}G_{w_2}$; $V_{\mathbf{b5}}=\frac{1}{2}V_{b_5}$, $q_{sc\mathbf{5}}=\frac{1}{2}q_{sc_5}$, $G_{w\mathbf{5}}=\frac{1}{2}G_{w_5}$; $V_{\mathbf{b8}}=\frac{1}{2}V_{b_8}$, $q_{sc\mathbf{8}}=\frac{1}{2}q_{sc_8}$, $G_{w\mathbf{8}}=\frac{1}{2}G_{w_8}$; $V_{\mathbf{b11}}=\frac{1}{2}V_{b_{11}}$, $q_{sc\mathbf{11}}=\frac{1}{2}q_{sc_{11}}$, $G_{w\mathbf{11}}=\frac{1}{2}G_{w_{11}}$; $G_{y\mathbf{2,5}}=\frac{1}{2}G_{y_{2,5}}$; $G_{y\mathbf{5,8}}=\frac{1}{2}G_{y_{5,8}}$; and $G_{y\mathbf{8,11}}=\frac{1}{2}G_{y_{8,11}}$. Note also a plane of symmetry passing through a gridpoint results in a factor of $\frac{1}{2}$ as in gridpoints 2, 5, 8, and 11 in Fig. 5.18a.

The fourth two examples show oblique planes of symmetry. Fig. 5.19a shows a reservoir similar to that depicted in Fig. 5.15a, but the present reservoir has two additional planes of symmetry C-C and D-D. The four planes of symmetry divide the reservoir into eight symmetrical elements, each with a triangular shape as shown in Fig. 5.19b. Consequently, $p_1=p_4=p_{13}=p_{16}$, $p_2=p_3=p_{14}=p_{15}=p_5=p_8=p_9=p_{12}$, $p_6=p_7=p_{10}=p_{11}$, and $p_2=p_3=p_{14}=p_{15}=p_5=p_8=p_9=p_{12}$. The modified problem is represented by the smallest element of symmetry shown in Fig. 5.19b, with each plane of symmetry being replaced with a no-flow boundary. Fig. 5.20a shows a reservoir similar to that depicted in Fig. 5.17a, but the present reservoir has two additional planes of symmetry C-C and D-D. The four planes of symmetry divide the reservoir into eight symmetrical elements, each with a triangular shape as shown in Fig. 5.20b. Consequently, $p_1=p_3=p_7=p_9$, and $p_4=p_6=p_2=p_8$. The modified problem is represented by the smallest element of symmetry shown in Fig. 5.20b, with each plane of symmetry being replaced with a no-flow boundary. A vertical plane of symmetry C-C or D-D that passes through a gridpoint but is neither

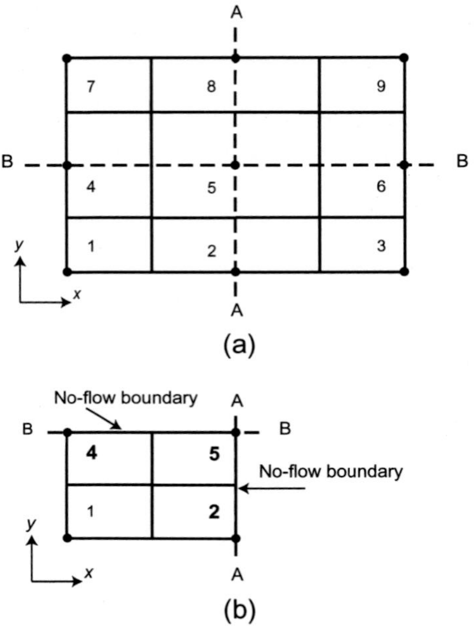

FIG. 5.17 2-D reservoir with odd gridpoints in the *x*- and *y*-directions exhibiting two vertical planes of symmetry. (a) Whole reservoir and planes of symmetry and (b) smallest element of symmetry.

parallel to the *x*-axis nor the *y*-axis (oblique plane), as shown in Figs. 5.19a and 5.20a, bisects the gridpoint bulk volume, wellblock rate, and wellblock geometric factor of the gridpoint it passes through. An oblique plane does not affect the interblock geometric factors in the *x*-axis or the *y*-axis. In reference to gridpoints 1, 6, and 5 in Figs. 5.19b and 5.20b, $\mathbf{V}_{\mathbf{b_1}} = \frac{1}{2} V_{b_1}$, $\mathbf{q}_{\mathbf{sc_1}} = \frac{1}{2} q_{sc_1}$, $\mathbf{G}_{\mathbf{w_1}} = \frac{1}{2} G_{w_1}$; $\mathbf{V}_{\mathbf{b_6}} = \frac{1}{2} V_{b_6}$, $\mathbf{q}_{\mathbf{sc_6}} = \frac{1}{2} q_{sc_6}$, $\mathbf{G}_{\mathbf{w_6}} = \frac{1}{2} G_{w_6}$; $\mathbf{V}_{\mathbf{b_5}} = \frac{1}{8} V_{b_5}$, $\mathbf{q}_{\mathbf{sc_5}} = \frac{1}{8} q_{sc_5}$, $\mathbf{G}_{\mathbf{w_5}} = \frac{1}{8} G_{w_5}$; $\mathbf{G}_{\mathbf{y_{1,2}}} = G_{y_{1,2}}$; $\mathbf{G}_{\mathbf{y_{2,5}}} = \frac{1}{2} G_{y_{2,5}}$; and $\mathbf{G}_{\mathbf{x_{2,6}}} = G_{x_{2,6}}$. Note that the four planes of symmetry (A-A, B-B, C-C, and D-D) passing through gridpoint 5 in Fig. 5.20a result in the factor of $\frac{1}{4} \times \frac{1}{2} = \frac{1}{8}$ used to calculate the actual gridpoint bulk volume, wellblock rate, and wellblock geometric factor for gridpoint 5 in Fig. 5.20b. That is to say, the modifying factor equals $\frac{1}{n_{vsp}} \times \frac{1}{2}$ where n_{vsp} is the number of vertical planes of symmetry passing through a gridpoint.

It should be mentioned that set ξ_n for gridblocks in the modified problem might include new elements such as b_{SW}, b_{NW}, b_{SE}, b_{NE} that reflect oblique boundaries such as plane C-C or D-D. The flow rates across such boundaries ($q^m_{sc_{l,n}}$) are set to zero because these boundaries represent no-flow boundaries.

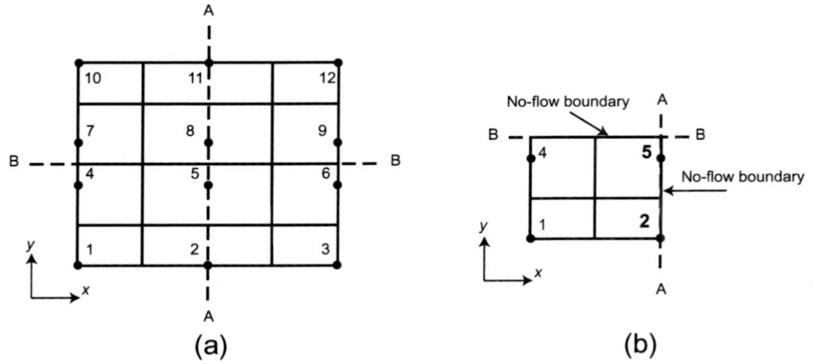

FIG. 5.18 Reservoir with even gridpoints in the y-direction and odd gridpoints in the x-direction exhibiting two vertical planes of symmetry. (a) Whole reservoir and planes of symmetry and (b) smallest element of symmetry.

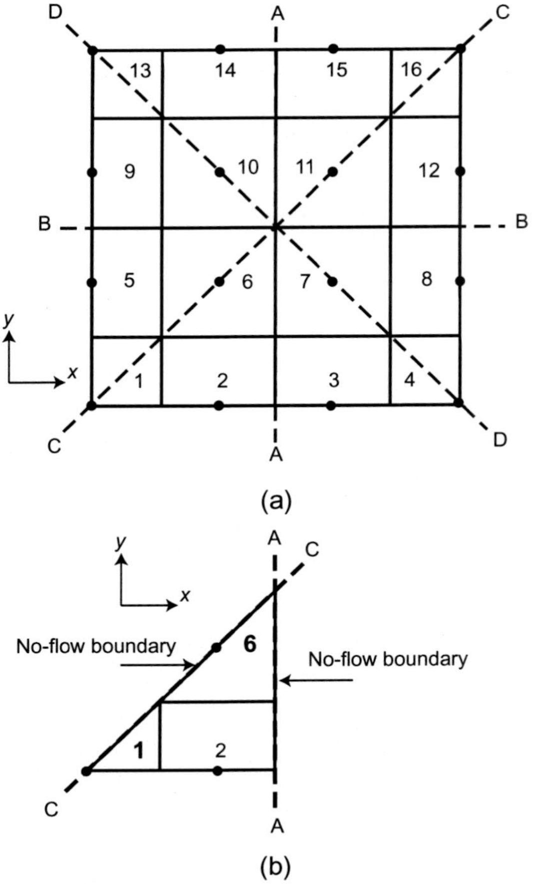

FIG. 5.19 Reservoir with even gridpoints in the x- and y-directions exhibiting four vertical planes of symmetry. (a) Whole reservoir and planes of symmetry and (b) smallest element of symmetry.

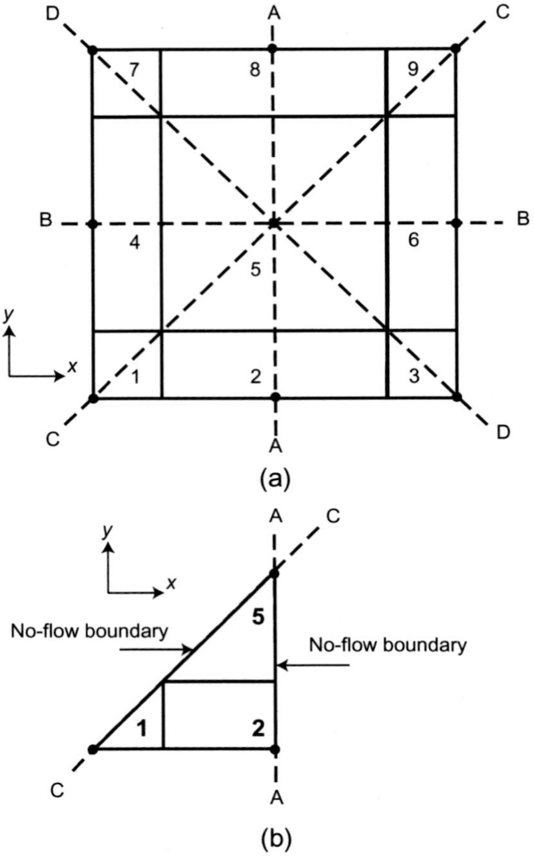

FIG. 5.20 Reservoir with odd gridpoints in the x- and y-directions exhibiting four vertical planes of symmetry. (a) Whole reservoir and planes of symmetry and (b) smallest element of symmetry.

5.7 Summary

This chapter presents reservoir discretization in Cartesian and radial-cylindrical coordinates using point-distributed grids. For the Cartesian coordinate system, equations similar to those represented by Eq. (5.1) define gridpoint locations and the relationships between the distances separating gridpoints, block boundaries, and sizes of the blocks represented by the gridpoints in the x-, y-, and z-directions. Table 5.1 presents equations for the calculation of the transmissibility geometric factors in the three directions. For the radial-cylindrical coordinate system used for single-well simulation, the equations that define gridpoint locations and the relationships between the distances separating gridpoints, block boundaries, and sizes of the blocks represented by the gridpoints in the r-direction are given by Eqs. (5.74) through (5.81), Eq. (5.73) in the θ-direction, and an equation similar to Eq. (5.1) for the z-direction. The equations in either Table 5.2 or 5.3 can be used to calculate the transmissibility geometric factors in the r-, θ-, and z-directions. Eq. (5.2) expresses the general form of the flow equation that applies to boundary gridpoints and interior gridpoints in 1-D, 2-D, or 3-D flow in both Cartesian and radial-cylindrical coordinates. The flow equation for any gridpoint has flow terms equal to the number of existing neighboring gridpoints and fictitious wells equal to the number of boundary conditions. Each fictitious well represents a boundary condition. The flow rate of a fictitious well is given by Eq. (5.33), (5.36), (5.40), or (5.46) for a specified pressure gradient, specified flow rate, no-flow, or specified pressure boundary condition, respectively.

If reservoir symmetry exists, it can be exploited to define the smallest element of symmetry. Planes of symmetry may pass through gridpoints or along block boundaries. To simulate the smallest element of symmetry, planes of symmetry are replaced with no-flow boundaries, and new interblock geometric factors, bulk volume, wellblock rate, and wellblock geometric factors for boundary gridpoints are calculated prior to simulation.

5.8 Exercises

5.1 What is the meaning of reservoir discretization into gridpoints?

5.2 Using your own words, describe how you discretize a reservoir of length L_x along the x-direction using n gridpoints.

5.3 Fig. 5.5 shows a reservoir with regular boundaries.

- **a.** How many boundaries does this reservoir have along the x-direction? Identify and name these boundaries.
- **b.** How many boundaries does this reservoir have along the y-direction? Identify and name these boundaries.
- **c.** How many boundaries does this reservoir have along the z-direction? Identify and name these boundaries.
- **d.** How many boundaries does this reservoir have along all directions?

5.4 Consider the 2-D reservoir described in Example 5.5 and shown in Fig. 5.12.

- **a.** Identify the interior and boundary gridpoints in the reservoir.
- **b.** Write the set of neighboring gridpoints (ψ_n) for each gridpoint in the reservoir.
- **c.** Write the set of reservoir boundaries (ξ_n) for each gridpoint in the reservoir.
- **d.** How many boundary conditions does each boundary gridpoint have? How many fictitious wells does each boundary gridpoint have? Write the terminology for the flow rate of each fictitious well.
- **e.** How many flow terms does each boundary gridpoint have?
- **f.** Add the number of flow terms and number of fictitious wells for each boundary gridpoint. Do they add up to four for each boundary gridpoint?
- **g.** How many flow terms does each interior gridpoint have?
- **h.** What can you conclude from your results of (f) and (g) earlier?

5.5 Consider fluid flow in the 1-D horizontal reservoir shown in Fig. 5.21.

a. Write the appropriate flow equation for gridpoint n in this reservoir.

b. Write the flow equation for gridpoint 1 by finding ψ_1 and ξ_1 and then use them to expand the equation in (a).

c. Write the flow equation for gridpoint 2 by finding ψ_2 and ξ_2 and then use them to expand the equation in (a).

d. Write the flow equation for gridpoint 3 by finding ψ_3 and ξ_3 and then use them to expand the equation in (a).

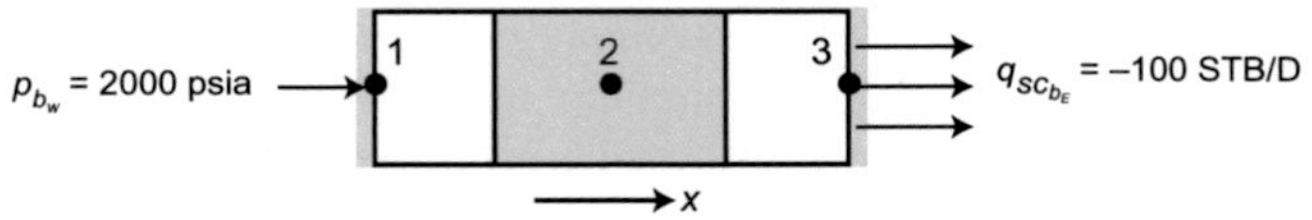

FIG. 5.21 1-D reservoir in Exercise 5.5.

5.6 Consider fluid flow in the 2-D horizontal reservoir shown in Fig. 5.22.

a. Write the appropriate flow equation for gridpoint n in this reservoir.

b. Write the flow equation for gridpoint 1 by finding ψ_1 and ξ_1 and then use them to expand the equation in (a).

c. Write the flow equation for gridpoint 7 by finding ψ_7 and ξ_7 and then use them to expand the equation in (a).

d. Write the flow equation for gridpoint 15 by finding ψ_{15} and ξ_{15} and then use them to expand the equation in (a).

e. Write the flow equation for gridpoint 19 by finding ψ_{19} and ξ_{19} and then use them to expand the equation in (a).

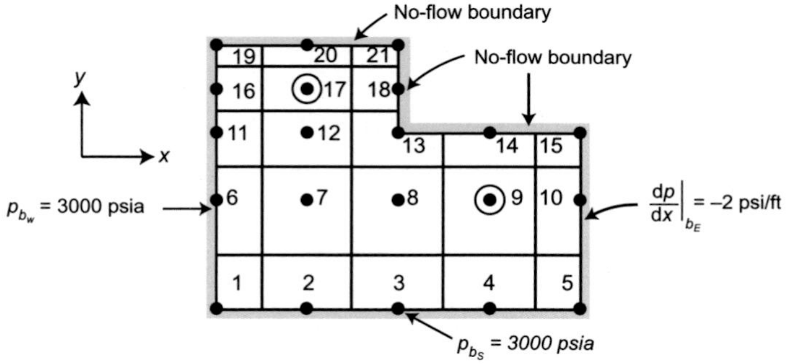

FIG. 5.22 1-D reservoir for Exercise 5.6.

5.7 Single-phase oil reservoir is described by four equally spaced gridpoints as shown in Fig. 5.23. The reservoir is horizontal and has $k=25$ md. Gridpoint spacing is $\Delta x=500$ ft, $\Delta y=700$ ft, and $h=60$ ft. Oil properties are $B=1$ RB/STB and $\mu=0.5$ cP. The reservoir left boundary is kept at constant pressure of 2500 psia, and the reservoir right boundary is sealed to flow. A well in gridpoint 3 produces 80 STB/D of oil. Assuming that the reservoir rock and oil are incompressible, calculate the pressure distribution in the reservoir.

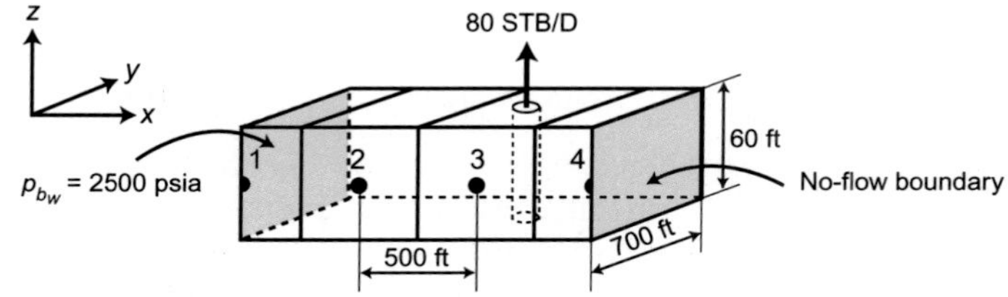

FIG. 5.23 Discretized 1-D reservoir in Exercise 5.7.

5.8 The 1-D horizontal oil reservoir shown in Fig. 5.24 is described by four equally spaced gridpoints. Reservoir gridpoints have $k=90$ md, $\Delta x=300$ ft, $\Delta y=250$ ft, and $h=45$ ft. Oil FVF and viscosity are 1 RB/STB and 2 cP, respectively. The reservoir left boundary is maintained at constant pressure of 2000 psia, and the reservoir right boundary has constant influx of oil at a rate of 80 STB/D. A well in gridpoint 3 produces 175 STB/D of oil. Assuming that the reservoir rock and oil are incompressible, calculate the pressure distribution in the reservoir.

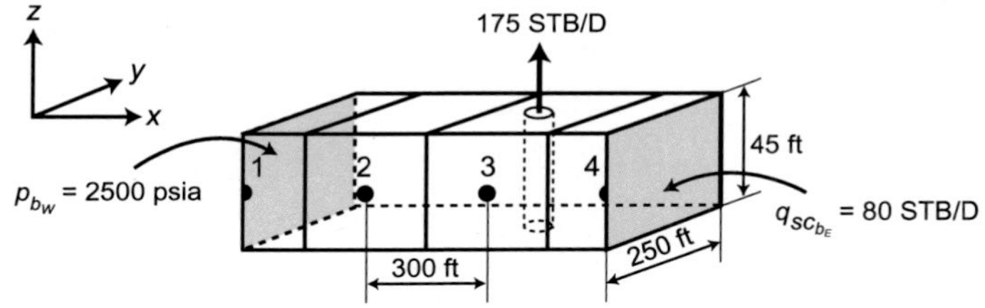

FIG. 5.24 Discretized 1-D reservoir in Exercise 5.8.

5.9 The 1-D horizontal oil reservoir shown in Fig. 5.25 is described by four equally spaced gridpoints. Reservoir gridpoints have $k=120$ md, $\Delta x=500$ ft, $\Delta y=450$ ft, and $h=30$ ft. Oil FVF and viscosity are 1 RB/STB and 3.7 cP, respectively. The reservoir left boundary is subject to constant pressure gradient of -0.2 psi/ft, and the reservoir right boundary is a no-flow boundary. A well in gridpoint 3 produces oil at a rate such

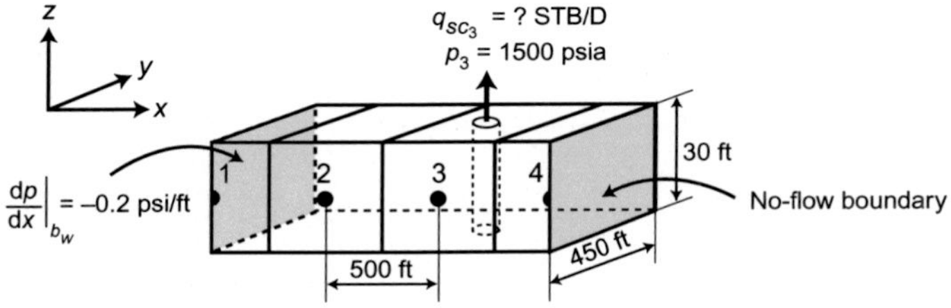

FIG. 5.25 Discretized 1-D reservoir in Exercise 5.9.

that the pressure of gridpoint 3 is maintained at 1500 psia. Assuming that the reservoir rock and oil are incompressible, calculate the pressure distribution in the reservoir. Then estimate well production rate.

5.10 The 1-D horizontal oil reservoir shown in Fig. 5.26 is described by four equally spaced gridpoints. Reservoir gridpoints have $k=70$ md, $\Delta x=400$ ft, $\Delta y=660$ ft, and $h=10$ ft. Oil FVF and viscosity are 1 RB/STB and 1.5 cP, respectively. The reservoir left boundary is maintained at constant pressure of 2700, and while the boundary condition at the reservoir right boundary is not known, the pressure of gridpoint 4 is maintained at 1900 psia. A well in gridpoint 3 produces 150 STB/D of oil. Assuming that the reservoir rock and oil are incompressible, calculate the pressure distribution in the reservoir. Estimate the rate of oil that crosses the reservoir right boundary.

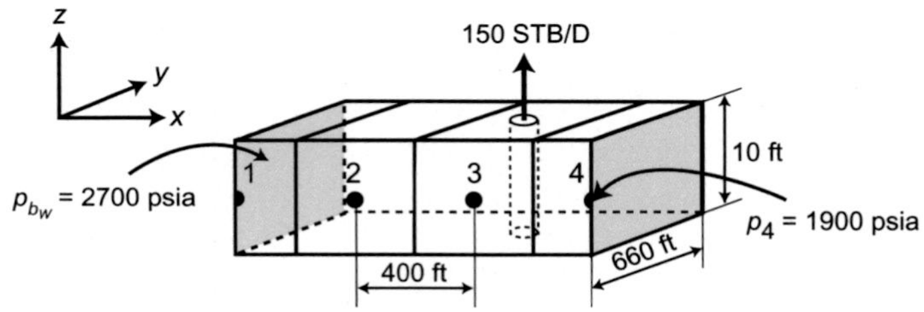

FIG. 5.26 Discretized 1-D reservoir in Exercise 5.10.

5.11 Consider the 2-D horizontal oil reservoir shown in Fig. 5.27. The reservoir is described using regular grid. Reservoir gridpoints have $\Delta x=350$ ft, $\Delta y=300$ ft, $h=35$ ft, $k_x=160$ md, and $k_y=190$ md. Oil FVF and viscosity are 1 RB/STB and 4.0 cP, respectively. Boundary

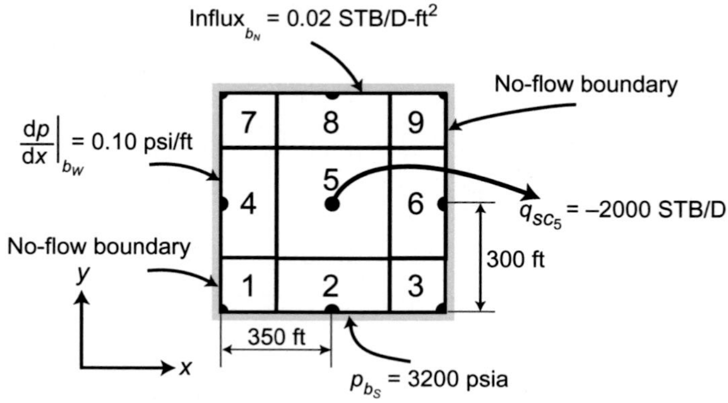

FIG. 5.27 Discretized 2-D reservoir in Exercise 5.11.

conditions are specified as shown in the figure. A well in gridpoint 5 produces oil at a rate of 2000 STB/D. Assume that the reservoir rock and oil are incompressible, and write the flow equations for gridpoints 4, 5, 6, 7, 8, and 9. Do not solve the equations.

5.12 Starting with Eq. (5.81a), which expresses the bulk volume of gridpoint (i, j, k), derive Eq. (5.81c) for gridpoint $(1, j, k)$ and Eq. (5.81d) for gridpoint (n_r, j, k).

5.13 A 6-in. vertical well producing 500 STB/D of oil is located in 16-acre spacing. The reservoir is 30-ft thick and has a horizontal permeability of 50 md. The oil FVF and viscosity are 1 RB/B and 3.5 cP, respectively. The reservoir external boundaries are no-flow boundaries. The reservoir is simulated using four gridpoints in the radial direction as shown in Fig. 5.28. Write the flow equations for all gridpoints. Do not substitute for values on the RHS of equations.

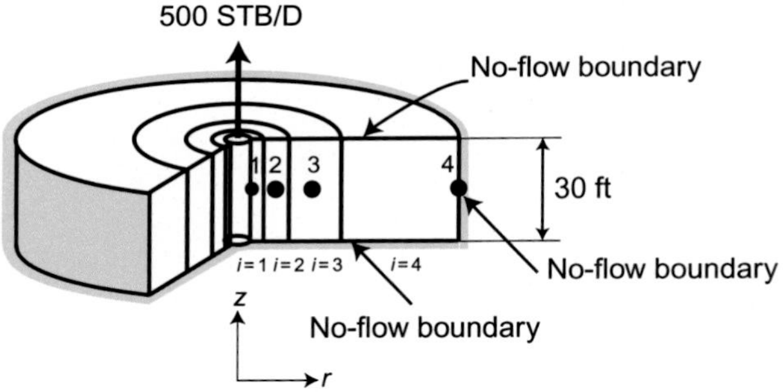

FIG. 5.28 Discretized reservoir in Exercise 5.13.

5.14 A 9⅝-in vertical well is located in 12-acre spacing. The reservoir thickness is 50 ft. The horizontal and vertical reservoir permeabilities are 70 and 40 md, respectively. The flowing fluid has a density, FVF, and viscosity of 62.4 lbm/ft^3, 1 RB/B, and 0.7 cP, respectively. The reservoir external boundary in the radial direction is no-flow boundary, and the well is completed in the top 25 ft only and produces at a rate of 1000 B/D. The reservoir bottom boundary is subject to influx such that the reservoir boundary is maintained at 3000 psia. The reservoir top boundary is sealed to flow. Assuming the reservoir can be simulated using two gridpoints in the vertical direction and four gridpoints in the radial direction, as shown in Fig. 5.29, write the flow equations for all gridpoints in this reservoir.

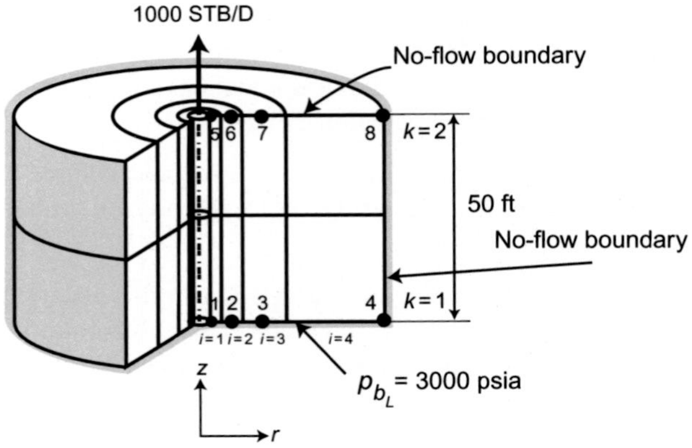

FIG. 5.29 Discretized 2-D radial-cylindrical reservoir in Exercise 5.14.

Chapter 6

Well representation in simulators

Chapter outline

6.1 Introduction

Wells in reservoir simulation are the most astute form of discontinuity. As such, the difficulties encountered due to boundary conditions are accentuated by the presence of wells. Yet, wells are paramount to reservoir evaluation because of the fact that engineering is all about optimizing well performance. In general, the contribution of any reservoir block penetrated by a well to the well flow rate is independent of the flow equation for that block. Such contribution has to be estimated separately from and then substituted into the flow equation for the wellblock. Fluid flow toward a well in a wellblock is radial regardless of the dimensionality of the flow problem. A well is modeled as a line source/sink term. In this chapter, the emphasis in 1-D and 2-D flow problems is on the estimation of the well geometric factor, while in 3-D flow problems, the focus is on the distribution of the well rate among the different blocks that are penetrated by the well. The estimation of the wellblock geometric factor is presented for a well hosted by one block and falling inside block boundaries and a well hosted by one block and falling on one or two of block boundaries (in 1-D and 2-D flow) that are reservoir boundaries. We present the production rate equation for a wellblock and the equations necessary for the estimation of the production rate or flowing bottom-hole pressure (FBHP) for wells operating under different

Petroleum Reservoir Simulation. https://doi.org/10.1016/B978-0-12-819150-7.00006-2

conditions, which include (1) a shut-in well, (2) a specified well production rate, (3) a specified well pressure gradient, and (4) a specified well FBHP.

The production rate equation for a wellblock has the form of

$$q_{sc_i} = -\frac{G_{w_i}}{B_i \mu_i}\left(p_i - p_{wf_i}\right) \tag{6.1}$$

where q_{sc_i}, G_{w_i}, and p_i = production rate, geometric factor, and pressure for wellblock i, respectively; p_{wf_i} = well pressure opposite wellblock i; and B_i and μ_i = fluid FVF and viscosity at the pressure of wellblock i. Eq. (6.1) is consistent with the sign convention of negative flow rate for production and positive flow rate for injection.

6.2 Single-block wells

In this section, we present the treatment of a well that penetrates a single block. Wells in 1-D linear flow, 1-D radial flow, and 2-D areal flow fall into this category.

6.2.1 Treatment of wells in 1-D linear flow

Fig. 6.1 depicts fluid flow in a 1-D linear flow problem. Fluid transfer into or out of a reservoir block has two components, global fluid transfer, and local fluid transfer. The global fluid transfer is linear and moves fluid from one block to another, and the local fluid transfer is radial and moves fluid within the block to a production well (or from an injection well). Although this treatment of wells is new for 1-D flow problems, it is consistent with and widely accepted in modeling fluid flow in 2-D, single-layer reservoirs. For a boundary gridblock (Fig. 6.2) or a boundary gridpoint (Fig. 6.3) in 1-D flow problems, it is important to differentiate between the source term that represents a real (or physical) well and the source term that represents a fictitious well (or boundary condition). This differentiation is crucial because flow resulting from a boundary condition is always linear, whereas flow to or from a real well is always radial (see Example 7.6). For example, the fluid that crosses the reservoir right boundary (gridblock 5 in Fig. 6.2 or gridpoint 5 in Fig. 6.3) is estimated from the specific

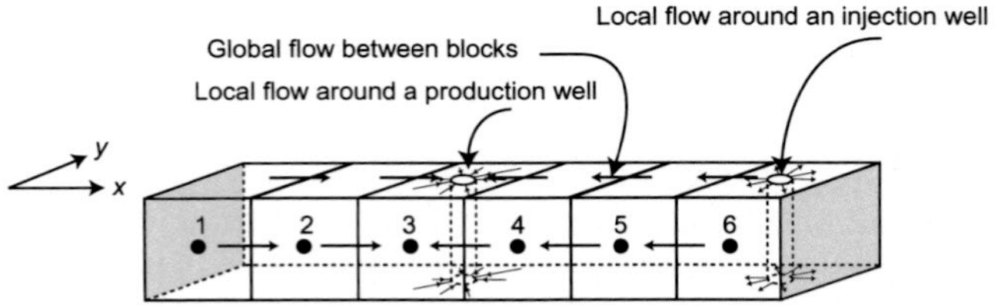

FIG. 6.1 Global flow and local flow around wells in 1-D reservoirs.

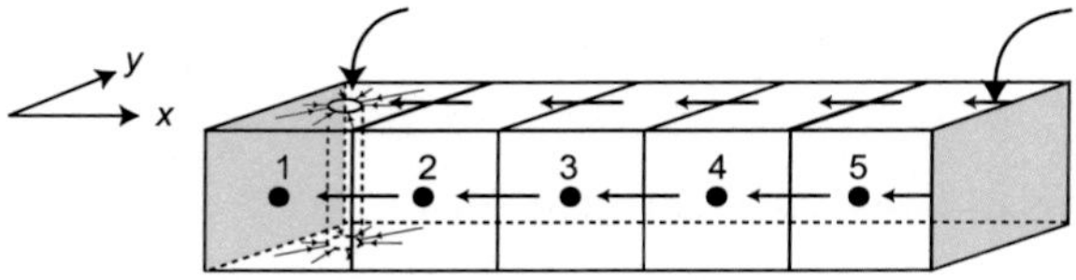

FIG. 6.2 Well at a boundary block in a block-centered grid.

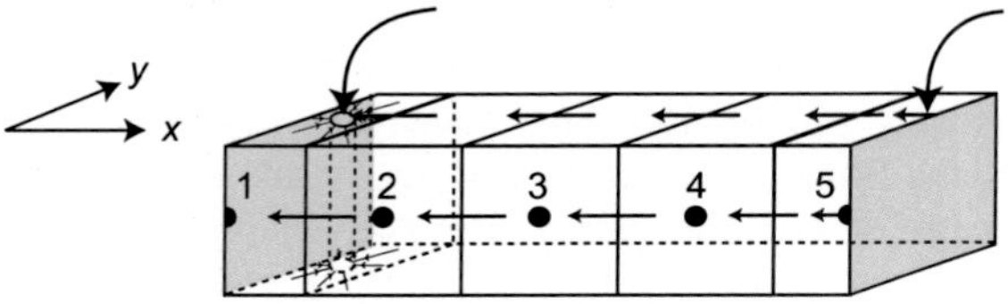

FIG. 6.3 Well at a boundary block in a point-distributed grid.

boundary conditions, the list of which was given in Chapters 4 and 5. However, the fluid that enters or leaves the block (gridblock 1 in Fig. 6.2 or gridpoint 1 in Fig. 6.3) at any point, including the boundary point, through a well is estimated from the radial flow equation of a real well given by Eq. (6.1). It must be mentioned; however, that modeling linear-flood experiments should use boundary conditions to represent injection and production at core end points. The logic behind this choice is that the injection and production ends of a core flood are designed such that the end effect is minimized and, consequently, linear flow near core end points is realized using end stems. An end stem (or end plug) is a thin cylinder that has a number of concentric grooves intersected by a number of radial grooves on the side adjacent to the core. The injected fluid enters through a hole at the center on the other side of the end stem and flows into the grooves making a uniform distribution of fluid across the face of the core adjacent to the grooves. This design of end stems results in linear flow of fluid along the axis of the core.

For a real well, the following equations apply.

Shut-in well

$$q_{sc_i} = 0 \tag{6.2}$$

Specified well production rate

$$q_{sc_i} = q_{spsc} \tag{6.3}$$

Specified well pressure gradient

$$q_{sc_i} = -\frac{2\pi\beta_c r_w k_{H_i} h_i}{B_i \mu_i} \frac{\partial p}{\partial r}\bigg|_{r_w} \tag{6.4}$$

Specified well FBHP

$$q_{sc_i} = -\frac{G_{w_i}}{B_i \mu_i}\left(p_i - p_{wf_i}\right) \tag{6.1}$$

where G_{w_i} is estimated using Eq. (6.12) in Section 6.2.3. The dimensions and rock properties of wellblock i are dealt with as explained for 2-D areal flow in Section 6.2.3.

6.2.2 Treatment of wells in 1-D radial flow

In 1-D radial flow in a single-well simulation, the well is hosted by the inner ringlike blocks termed here block 1 ($i=1$). Traditionally, wells in radial flow (single-well simulation) are treated as boundary conditions (Aziz and Settari, 1979; Ertekin et al., 2001). In the engineering approach, such wells can be treated as either source terms (real wells) or fictitious wells (boundary conditions) because in cylindrical coordinates, both real wells and fictitious wells have radial flow. Chapters 4 and 5 present equations for the flow rate of fictitious wells. In this section, we present equations for the flow rate of wells as a source term. The well production rate equations for block 1, under various well operating conditions, are given as follows:

Shut-in well

$$q_{sc_1}=0 \tag{6.5}$$

Specified well production rate

$$q_{sc_1}=q_{spsc} \tag{6.6}$$

Well FBHP can be estimated from Eq. (6.9), with q_{spsc} replacing q_{sc_1}.

Specified well pressure gradient

$$q_{sc_1}=-\frac{2\pi\beta_c r_w k_{H_1} h_1}{B_1\mu_1}\frac{\partial p}{\partial r}\bigg|_{r_w} \tag{6.7}$$

Specified well FBHP

Darcy's law for radial flow applies; that is,

$$q_{sc}=-\frac{2\pi\beta_c k_H h}{B\mu\log_e(r_e/r_w)}\left(p_e-p_{wf}\right) \tag{6.8}$$

For a block-centered grid, consider the flow of fluid in the radial segment enclosed between the external radius r_1 (the point representing gridblock 1) and the well radius r_w (the internal radius of gridblock 1). In this case, $r_e=r_1$, $p_e=p_1$, and $q_{sc}=q_{sc_1}$. Therefore, Eq. (6.8) becomes:

$$q_{sc_1}=-\frac{2\pi\beta_c k_{H_1} h_1}{B_1\mu_1\log_e(r_1/r_w)}\left(p_1-p_{wf}\right) \tag{6.9a}$$

from which

$$G_{w_1}=\frac{2\pi\beta_c k_{H_1} h_1}{\log_e(r_1/r_w)} \tag{6.10a}$$

Eq. (6.10a) can also be obtained by finding $G_{r_{i-1/2}}$ for $i=1$ in Table 4.2 or Table 4.3, discarding the second term in the denominator that corresponds to the nonexistent gridblock 0 and observing that, for a block-centered grid,

$r^L_{1/2} = r_w$ by definition if Table 4.2 is used or $(r_1/r_w) = [\alpha_{lg}\log_e(\alpha_{lg})/(\alpha_{lg} - 1)]$, as given by Eq. (4.87), if Table 4.3 is used.

For a point-distributed grid, consider the flow of fluid between gridpoints 1 and 2. These two gridpoints can be looked at as the internal and external boundaries of a radial reservoir segment. The application of Darcy's law for radial flow gives

$$q_{sc_1} = -\frac{2\pi\beta_c(k_H h/B\mu)_{1,2}}{\log_e(r_2/r_w)}(p_2 - p_1) \tag{6.9b}$$

because $p_e = p_2$, $p_{wf} = p_1$, $r_e = r_2$, and $r_1 = r_w$. Eq. (6.9b) is in the form of Eq. (6.8), where

$$G_{w_1} = \frac{2\pi\beta_c(k_H h)_{1,2}}{\log_e(r_2/r_w)} \tag{6.10b}$$

Eq. (6.10b) can also be obtained by finding $G_{r_{i+1/2}}$ for $i = 1$ in Table 5.2 or Table 5.3 and observing that for a point-distributed grid, $r_1 = r_w$ by definition and $(r_2/r_1) = \alpha_{lg}$ as given by Eq. (5.74). Note that for constant permeability $(k_1 = k_2 = k_H)$ and constant thickness $(h_1 = h_2 = h)$, $(k_H h)_{1,2} = k_H h = k_{H_1} h_1$.

You will notice that in a point-distributed grid, there is no need to write the flow equation for gridpoint 1 because the pressure of gridpoint 1 is known $(p_1 = p_{wf})$. In fact, this equation is nothing but Eq. (6.9b), which gives an estimate of the flow rate of wellblock 1 (refer to Exercise 6.7). The pressure of gridpoint 1 $(p_1 = p_{wf})$; however, is substituted in the flow equation for gridpoint 2.

6.2.3 Treatment of wells in 2-D areal flow

The wellblock pressure (p) and FBHP (p_{wf}) of a vertical well hosted by a wellblock in a single-layer reservoir are related through the inflow performance relationship (IPR) equation (Peaceman, 1983):

$$q_{sc} = -\frac{G_w}{B\mu}(p - p_{wf}) \tag{6.11}$$

where

$$G_w = \frac{2\pi\beta_c k_H h}{\left[\log_e(r_{eq}/r_w) + s\right]} \tag{6.12}$$

For anisotropic wellblock properties, k_H is estimated from the geometric mean permeability,

$$k_H = \left[k_x k_y\right]^{0.5} \tag{6.13}$$

The equivalent wellblock radius, for a well located at the center of a rectangular wellblock having anisotropic permeability as shown in Fig. 6.4, is given by

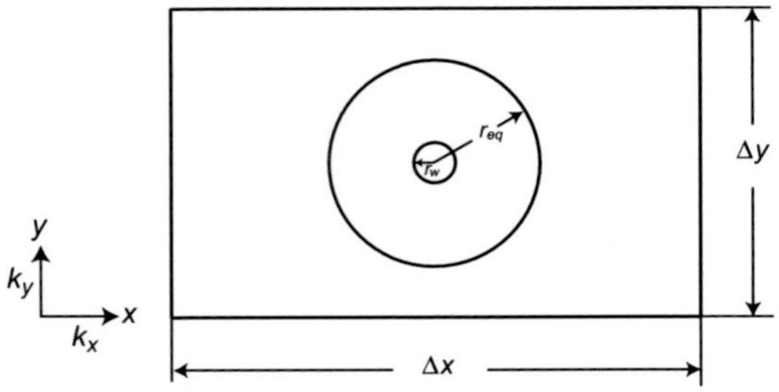

FIG. 6.4 Equivalent wellblock radius in a rectangular block showing anisotropy.

$$r_{eq} = 0.28 \frac{\left[(k_y/k_x)^{0.5} (\Delta x)^2 + (k_x/k_y)^{0.5} (\Delta y)^2 \right]^{0.5}}{\left[(k_y/k_x)^{0.25} + (k_x/k_y)^{0.25} \right]} \tag{6.14}$$

For isotropic permeability in the horizontal plane ($k_x = k_y$) and a rectangular wellblock, Eq. (6.14) reduces to

$$r_{eq} = 0.14 \left[(\Delta x)^2 + (\Delta y)^2 \right]^{0.5} \tag{6.15}$$

For isotropic permeability and a square wellblock ($\Delta x = \Delta y$), Eq. (6.15) becomes

$$r_{eq} = 0.198 \Delta x \tag{6.16}$$

Eqs. (6.14) through (6.16) apply to both block-centered and point-distributed grids. These equations, however, assume that the vertical well coincides with the center of the block hosting the well. They also have no provisions for the departure of the well axis from the block center. Therefore, the closer the well to the wellblock center, the better the representation of pressure distribution around the well. For centered wells in regularly distributed grids, the application of these equations is equally good for both grids, but the block-centered grid is preferred in an irregularly distributed grid because the wellblock center always coincides with the well. For wells that fall on reservoir boundaries (see Section 6.3.3); however, the point-distributed grid is preferred because the well and the gridpoint coincide.

For horizontal wells, Eq. (6.11) applies but with an appropriate definition of G_w. Further details on the estimation of G_w for horizontal wells can be found elsewhere (Babu and Odeh, 1989; Ertekin et al., 2001).

Examples 6.1 and 6.2 demonstrate the estimation of the wellblock geometric factor in square and rectangular blocks, isotropic and anisotropic permeability, and wells with and without skin. Examples 6.3 through 6.6 demonstrate the estimations of the well production rate and obtain the wellblock production rate equation under various well operating conditions.

TABLE 6.1 Dimensions, permeabilities, and skin factors of wellblocks.

Well ID	Wellblock					
	Δx (ft)	Δy (ft)	h (ft)	k_x (md)	k_y (md)	s
W-1	208	832	30	100	225	0
W-2	208	832	30	150	150	0
W-3	416	416	30	100	225	0
W-4	416	416	30	150	150	0

Example 6.1 A single-phase oil reservoir, consisting of a horizontal layer, has many vertical production wells. Table 6.1 identifies four of these wells and the dimensions, permeabilities, and skin factors of their wellblocks. Each well is located at the center of the wellblock and fully penetrates the layer. The oil FVF and viscosity are 1 RB/STB and 2 cP, respectively. Well diameter is 7 in. Calculate the wellblock geometric factors for the wells given in Table 6.1.

Solution

Well W-1

The wellblock has $k_x \neq k_y$ and $\Delta x \neq \Delta y$. Therefore, Eqs. (6.14) and (6.13) can be used to estimate the equivalent wellblock radius and horizontal permeability, respectively:

$$r_{eq} = 0.28 \frac{\left[(225/100)^{0.5}(208)^2 + (100/225)^{0.5}(832)^2\right]^{0.5}}{\left[(225/100)^{0.25} + (100/225)^{0.25}\right]} = 99.521 \text{ ft}$$

and

$$k_H = [100 \times 225]^{0.5} = 150 \text{ md}$$

The wellblock geometric factor is estimated using Eq. (6.12):

$$G_w = \frac{2\pi \times 0.001127 \times 150 \times 30}{\{\log_e[99.521/(3.5/12)] + 0\}} = 5.463 \text{ RB-cP/D-psi}$$

Well W-2

The wellblock has $k_x = k_y$, but $\Delta x \neq \Delta y$. Therefore, Eq. (6.15) can be used to estimate the equivalent wellblock radius:

$$r_{eq} = 0.14 \times \left[(208)^2 + (832)^2\right]^{0.5} = 120.065 \text{ ft}$$

and

$$k_H = k_x = k_y = 150 \text{ md}$$

The wellblock geometric factor is estimated by substituting values into Eq. (6.12), yielding

$$G_w = \frac{2\pi \times 0.001127 \times 150 \times 30}{\{\log_e[120.065/(3.5/12)] + 0\}} = 5.293 \text{ RB-cP/D-psi}$$

Well W-3

The wellblock has $k_x \neq k_y$, but $\Delta x = \Delta y$. Therefore, Eqs. (6.14) and (6.13) can be used to estimate the wellblock equivalent radius and horizontal permeability:

$$r_{eq} = 0.28 \frac{\left[(225/100)^{0.5}(416)^2 + (100/225)^{0.5}(416)^2\right]^{0.5}}{\left[(225/100)^{0.25} + (100/225)^{0.25}\right]} = 83.995 \text{ ft}$$

and

$$k_H = [100 \times 225]^{0.5} = 150 \text{ md}$$

The wellblock geometric factor is estimated by substituting values into Eq. (6.12), yielding

$$G_w = \frac{2\pi \times 0.001127 \times 150 \times 30}{\{\log_e[83.995/(3.5/12)] + 0\}} = 5.627 \text{ RB-cP/D-psi}$$

Well W-4

The wellblock has $k_x = k_y$ and $\Delta x = \Delta y$. Therefore, Eq. (6.16) can be used to estimate the equivalent wellblock radius:

$$r_{eq} = 0.198 \times 416 = 82.364 \text{ ft}$$

and

$$k_H = 150 \text{ md}$$

The wellblock geometric factor is estimated by substituting values into Eq. (6.12), yielding

$$G_w = \frac{2\pi \times 0.001127 \times 150 \times 30}{\{\log_e[82.364/(3.5/12)] + 0\}} = 5.647 \text{ RB-cP/D-psi}$$

It should be noted that even though all four wellblocks have the same thickness of 30 ft, area of 173,056 ft^2, and horizontal permeability of 150 md, the well geometric factors are different because of heterogeneity and/or wellblock dimensions.

Example 6.2 Consider well W-1 in Example 6.1 and estimate the well geometric factors for the following cases: (1) no mechanical well damage; that is, $s = 0$; (2) well damage resulting in $s = +1$; and (3) well stimulation resulting in $s = -1$.

Solution

The wellblock of well W-1 has $k_x \neq k_y$ and $\Delta x \neq \Delta y$. Therefore, Eqs. (6.14) and (6.13) can be used to estimate the equivalent wellblock radius and horizontal permeability:

$$r_{eq} = 0.28 \frac{\left[(225/100)^{0.5}(208)^2 + (100/225)^{0.5}(832)^2\right]^{0.5}}{\left[(225/100)^{0.25} + (100/225)^{0.25}\right]} = 99.521 \text{ ft}$$

and

$$k_H = [100 \times 225]^{0.5} = 150 \text{ md}$$

The wellblock geometric factor is estimated using Eq. (6.12):

1. For $s = 0$ (zero skin)

$$G_w = \frac{2\pi \times 0.001127 \times 150 \times 30}{\{\log_e[99.521/(3.5/12)] + 0\}} = 5.463 \text{ RB-cP/D-psi}$$

2. For $s = +1$ (positive skin)

$$G_w = \frac{2\pi \times 0.001127 \times 150 \times 30}{\{\log_e[99.521/(3.5/12)] + 1\}} = 4.664 \text{ RB-cP/D-psi}$$

3. For $s = -1$ (negative skin)

$$G_w = \frac{2\pi \times 0.001127 \times 150 \times 30}{\{\log_e[99.521/(3.5/12)] - 1\}} = 6.594 \text{ RB-cP/D-psi}$$

This example demonstrates the effect of well damage and stimulation on the well geometric factor and, in turn, on the well production rate. The reported damage in this well reduces the well geometric factor by 14.6%, where as the reported stimulation increases, the well geometric factor by 20.7%.

Example 6.3 Consider well W-1 in Example 6.1 and estimate the well production rate for the following possible operating conditions: (1) Well is closed, (2) has constant production rate of 3000 STB/D, and (3) has pressure gradient at sandface of 300 psi/ft, and (4) wellblock pressure is p_o, and FBHP is kept at 2000 psia.

Solution

For well W-1, $r_{eq} = 99.521$ ft, $k_H = 150$ md, and $G_w = 5.463$ RB-cP/D-psi from Example 6.1.

1. For a closed well, Eq. (6.2) applies. Therefore, $q_{sc_1} = 0$ STB/D.
2. For a specified production rate, Eq. (6.3) applies. Therefore, $q_{sc_1} = -3000$ STB/D.
3. For a specified pressure gradient, Eq. (6.4) applies. Therefore,

$$q_{sc_1} = -\frac{2\pi \times 0.001127 \times (3.5/12) \times 150 \times 30}{1 \times 2} \times 300 = -1394.1\ \text{STB/D}$$

4. For a specified FBHP, Eq. (6.1) applies. Therefore,

$$q_{sc_1} = -\frac{5.463}{1 \times 2}(p_o - 2000)$$

or

$$q_{sc_1} = -2.7315(p_o - 2000)\ \text{STB/D} \tag{6.17}$$

If, for example, the wellblock pressure is 3000 psia, then Eq. (6.17) predicts

$$q_{sc_1} = -2.7315(3000 - 2000) = -2731.5\ \text{STB/D}.$$

Example 6.4 Estimate the FBHP of the well hosted by gridblock 4 in Example 4.1. The wellbore diameter is 7 in., and the well has zero skin.

Solution

From Example 4.1, gridblock 4 has the following dimensions and properties: $\Delta x = 1000$ ft, $\Delta y = 1200$ ft, $h = \Delta z = 75$ ft, and $k_x = 15$ md; the flowing fluid has $B = 1$ RB/STB and $\mu = 10$ cP. The well in gridblock 4 has $q_{sc_4} = -150$ STB/D. The local flow of fluid toward this well is radial. The equivalent wellblock radius can be estimated using Eq. (6.15):

$$r_{eq} = 0.14 \times \left[(1000)^2 + (1200)^2\right]^{0.5} = 218.687\ \text{ft}$$

and

$$k_H = k_x = 15\ \text{md}$$

The wellblock geometric factor is estimated by substituting values into Eq. (6.12), yielding

$$G_w = \frac{2\pi \times 0.001127 \times 15 \times 75}{\{\log_e[218.687/(3.5/12)] + 0\}} = 1.203\ \text{RB-cP/D-psi}$$

Applying Eq. (6.1) gives

$$-150 = -\frac{1.203}{1 \times 10}\left(p_4 - p_{wf_4}\right)$$

from which the FBHP of the well in Example 4.1, where $q_{sc_4} = -150$ STB/D, can be estimated as a function of the pressure of gridblock 4 as

$$p_{wf_4} = p_4 - 1246.9\ \text{psia} \tag{6.18}$$

Example 6.5 Consider the single-well simulation in Example 4.11. Write the production rate equation for the well in gridblock 1 for each of the following well operating conditions: (1) The pressure gradient at sandface is specified at 200 psi/ft, and (2) the FBHP at the middle of formation is kept constant at 2000 psia. Rock and fluid properties are as follows: $k_H = 233$ md, $B = 1$ RB/STB, and $\mu = 1.5$ cP.

Solution

The following data are taken from Example 4.11: $r_e = 744.73$ ft, $r_w = 0.25$ ft, and $h = 100$ ft. In addition, discretization in the radial direction results in $r_1 = 0.5012$ ft, $r_2 = 2.4819$ ft, $r_3 = 12.2914$ ft, $r_4 = 60.8715$ ft, and $r_5 = 301.457$ ft.

1. For a specified pressure gradient, Eq. (6.7) applies. Therefore

$$q_{sc_1} = -\frac{2\pi \times 0.001127 \times 0.25 \times 233 \times 100}{1 \times 1.5} \times 200 = -5499.7\,\text{STB/D}$$

2. For a specified FBHP, Eq. (6.9a) applies. Therefore

$$q_{sc_1} = -\frac{2\pi \times 0.001127 \times 233 \times 100}{1 \times 1.5 \times \log_e(0.5012/0.25)}(p_1 - 2000)$$

or

$$q_{sc_1} = 158.1407(p_1 - 2000)\,\text{STB/D} \tag{6.19}$$

If, for example, the wellblock pressure is 2050 psia, then Eq. (6.19) predicts

$$q_{sc_1} = -158.1407(2050 - 2000) = -7907.0\,\text{STB/D}.$$

Example 6.6 Write the production rate equation for the well in gridpoint 9 in Example 5.5, and then estimate the FBHP of the well. The wellbore diameter is 7 in., and the wellblock has zero skin.

Solution

From Example 5.5, the block represented by gridpoint 9 has the following dimensions and properties: $\Delta x = 250$ ft, $\Delta y = 300$ ft, $h = 100$ ft, $k_x = 270$ md, and $k_y = 220$ md; the flowing fluid has $B = 1$ RB/STB and $\mu = 2$ cP. The well (or wellblock 9) production rate is specified at $q_{sc_9} = -4000$ STB/D.

The equivalent wellblock radius and horizontal permeability can be estimated using Eqs. (6.14) and (6.13), yielding

$$r_{eq} = 0.28\frac{\left[(220/270)^{0.5}(250)^2 + (270/220)^{0.5}(300)^2\right]^{0.5}}{\left[(220/270)^{0.25} + (270/220)^{0.25}\right]} = 55.245\,\text{ft}$$

and

$$k_H = [270 \times 220]^{0.5} = 243.72 \text{ md}$$

The wellblock geometric factor is estimated by substituting values into Eq. (6.12), resulting in

$$G_w = \frac{2\pi \times 0.001127 \times 243.72 \times 100}{\{\log_e[55.245/(3.5/12)] + 0\}} = 32.911 \text{ RB-cP/D-psi}$$

Applying Eq. (6.1) gives

$$q_{sc_9} = -\frac{32.911}{1 \times 2}\left(p_9 - p_{wf_9}\right) = -16.456\left(p_9 - p_{wf_9}\right) \text{ STB/D}$$

from which the FBHP of the well in Example 5.5, where $q_{sc_9} = -4000$ STB/D, can be estimated as a function of the pressure of gridpoint 9 as

$$p_{wf_9} = p_9 - 243.1 \text{ psia} \tag{6.20}$$

6.3 Multiblock wells

In this section, we present treatments of pressure variations within the wellbore, allocation of the well production rate among all layers penetrated by the well, and the treatment of the flow between hosting block and well, especially for wells that fall on reservoir boundaries sealed off to flow.

6.3.1 Vertical effects (flow within wellbore)

Pressures within the wellbore, opposite wellblocks, differ because of hydrostatic pressure, frictional loss due to flow, and kinetic energy. For vertical wells, the latter two factors can be neglected; therefore, pressure variation in the wellbore resulting from hydrostatic pressure can be expressed as.

$$p_{wf_i} = p_{wf_{ref}} + \overline{\gamma}_{wb}\left(Z_i - Z_{ref}\right) \tag{6.21}$$

where

$$\overline{\gamma}_{wb} = \gamma_c \overline{\rho}_{wb} g \tag{6.22}$$

and

$$\overline{\rho}_{wb} = \frac{\rho_{sc}}{\overline{B}} \tag{6.23}$$

Average FBHP can be used to obtain an estimate for $\overline{B}$.

6.3.2 Wellblock contribution to well rate

In this case, the vertical well penetrates several blocks. Fig. 6.5 shows a well that penetrates wellblocks located in different layers; that is, the wellblocks

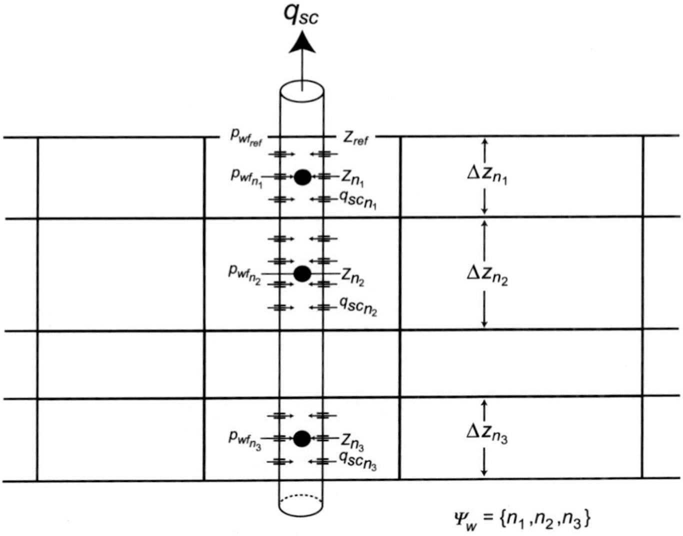

FIG. 6.5 Cross section showing pressures within vertical wellbore.

are vertically stacked. The concern here is to estimate the production rate of wellblock i, where wellblock i is a member of the set of all blocks that are penetrated by the well; that is, $i \in \psi_w$. The equations in this section also apply to the well in single-well simulation if the appropriate wellblock geometric factors are used.

Shut-in well

$$q_{sc_i} = 0 \tag{6.2}$$

Specified well production rate

The contribution of wellblock i to the well production rate is given by Eq. (6.1):

$$q_{sc_i} = -\frac{G_{w_i}}{B_i \mu_i}\left(p_i - p_{wf_i}\right) \tag{6.1}$$

where p_{wf_i} is given by Eq. (6.21).

Combining Eqs. (6.1) and (6.21) yields

$$q_{sc_i} = -\frac{G_{w_i}}{B_i \mu_i}\left[p_i - p_{wf_{ref}} - \overline{\gamma}_{wb}\left(Z_i - Z_{ref}\right)\right] \tag{6.24}$$

The sum of the production rates of all wellblocks must add up to the specified well production rate; that is,

$$q_{spsc} = \sum_{i \in \psi_w} q_{sc_i} \tag{6.25}$$

The FBHP of the well ($p_{wf_{ref}}$) can be estimated by combining Eqs. (6.24) and (6.25), yielding

$$p_{wf_{ref}} = \frac{\sum_{i \in \psi_w} \left\{ \left(\frac{G_w}{B\mu} \right)_i \left[p_i - \overline{\gamma}_{wb} \left(Z_i - Z_{ref} \right) \right] \right\} + q_{spsc}}{\sum_{i \in \psi_w} \left(\frac{G_w}{B\mu} \right)_i} \tag{6.26}$$

For a specified well production rate, Eq. (6.26) is used to estimate $p_{wf_{ref}}$, and this estimate is subsequently used in Eq. (6.24) to calculate the wellblock production rate. The use of Eq. (6.26), however, requires the knowledge of the unknown pressure values of all wellblocks. An implicit treatment of $p_{wf_{ref}}$ solves the problem, but such treatment leads to complications (e.g., construction and solution of the resulting matrix equation) that are beyond the scope of this introductory book. Ertekin et al. (2001) presented the details of the implicit treatment of $p_{wf_{ref}}$. One solution is to estimate $p_{wf_{ref}}$ at the beginning of each time step (old time level n); another solution is to assume that all vertically stacked wellblocks have the same pressure drop ($p_i - p_{wf_i} = \Delta p$). Solving Eq. (6.26) for Δp and substituting the result into Eq. (6.1) yields.

$$q_{sc_i} = \frac{\left(\frac{G_w}{B\mu} \right)_i}{\sum_{l \in \psi_w} \left(\frac{G_w}{B\mu} \right)_l} q_{spsc} \tag{6.27}$$

Furthermore, if fluid properties are not sensitive to small pressure variations and all vertically stacked wellblocks are assumed to have the same equivalent well radius and skin factor, the aforementioned equation can be simplified to.

$$q_{sc_i} = \frac{(k_H h)_i}{\sum_{l \in \psi_w} (k_H h)_l} q_{spsc} \tag{6.28}$$

Eq. (6.28) prorates the well production rate among vertically stacked wellblocks according to their capacities $(k_H h)_i$. In addition, if the horizontal permeability of various layers is the same, then the well production rate is prorated according to wellblock thickness:

$$q_{sc_i} = \frac{h_i}{\sum_{l \in \psi_w} h_l} q_{spsc} \tag{6.29}$$

Specified well pressure gradient

For a specified well pressure gradient, the contribution of wellblock i to the well production rate is given by.

$$q_{sc_i} = -\frac{2\pi F_i \beta_c r_w k_{H_i} h_i}{B_i \mu_i} \frac{\partial p}{\partial r}\bigg|_{r_w} \tag{6.30}$$

where F_i = ratio of wellblock i area to the theoretical area from which the well withdraws its fluid (see Section 6.3.3).

Specified well FBHP

The contribution of wellblock i to the well production rate is given by Eq. (6.24):

$$q_{sc_i} = -\frac{G_{w_i}}{B_i \mu_i}\left[p_i - p_{wf_{ref}} - \overline{\gamma}_{wb}\left(Z_i - Z_{ref}\right)\right] \quad (6.24)$$

The following example demonstrates the estimation of the production rate of individual wellblocks that are penetrated by the same well and the estimation of FBHP of the well.

Example 6.7 Consider the well in Example 5.9. The well production rate is specified at 2000 B/D of water. (1) Prorate the well production rate between wellblocks 5 and 9. (2) Estimate the FBHP of the well at the formation top if the pressure of gridpoints 5 and 9 are 3812.5 and 3789.7 psia, respectively. (3) Prorate the well production rate between wellblocks 5 and 9 if the pressure of gridpoints 5 and 9 is known a priori as given earlier. Assume that the well fully penetrates both wellblocks and uses open well completion.

Solution

The following data are taken from Example 5.9: $r_e = 526.6$ ft, $r_w = 0.25$ ft, $k_H = 150$ md, $B = 1$ RB/STB, $\mu = 0.5$ cP, and $\gamma = 0.4333$ psi/ft. In addition, discretization in the radial direction results in $r_1 = r_w = 0.25$ ft, $r_2 = 3.2047$ ft, $r_3 = 41.080$ ft, and $r_4 = 526.60$ ft; discretization in the vertical direction results in $h_5 = 15$, $Z_5 = 15$, $h_9 = 7.5$, and $Z_9 = 0$ ft. The FBHP is to be reported at the elevation of the formation top; that is, $Z_{ref} = 0$ ft.

1. The well in this problem is completed in wellblocks 5 and 9; that is, $\psi_w = \{5,9\}$. For a point-distributed grid, the wellblock geometric factors for wellblocks 5 and 9 are estimated using Eq. (6.10b), yielding

$$G_{w_5} = \frac{2\pi\beta_c k_{H_5} h_5}{\log_e(r_2/r_w)} = \frac{2\pi \times 0.001127 \times 150 \times 15}{\log_e(3.2047/0.25)} = 6.2458 \text{ B-cP/D-psi}$$

and

$$G_{w_9} = \frac{2\pi\beta_c k_{H_9} h_9}{\log_e(r_2/r_w)} = \frac{2\pi \times 0.001127 \times 150 \times 7.5}{\log_e(3.2047/0.25)} = 3.1229 \text{ B-cP/D-psi}$$

Eq. (6.27) can be used to prorate well rate among wellblocks, resulting in

$$q_{sc_5} = \frac{\left(\frac{G_w}{B\mu}\right)_5}{\left(\frac{G_w}{B\mu}\right)_5 + \left(\frac{G_w}{B\mu}\right)_9} q_{spsc} = \frac{\left(\frac{6.2458}{1 \times 0.5}\right)}{\left(\frac{6.2458}{1 \times 0.5}\right) + \left(\frac{3.1229}{1 \times 0.5}\right)} \times 2000 = 1333.33 \text{ B/D}$$

and

$$q_{sc_9} = \frac{\left(\frac{G_w}{B\mu}\right)_9}{\left(\frac{G_w}{B\mu}\right)_5 + \left(\frac{G_w}{B\mu}\right)_9} q_{spsc} = \frac{\left(\frac{3.12299}{1 \times 0.5}\right)}{\left(\frac{6.24588}{1 \times 0.5}\right) + \left(\frac{3.12299}{1 \times 0.5}\right)} \times 2000$$
$$= 666.67 \text{ B/D}$$

Note that in this case, the wellblock rates can be prorated according to thickness using Eq. (6.29) because the FVF, viscosity, and horizontal permeability are constant.

2. The FBHP at the reference depth can be estimated using Eq. (6.26):

$$p_{wf_{ref}} = \frac{\left(\frac{6.2458}{1 \times 0.5}\right)[3812.5 - 0.4333(15-0)] + \left(\frac{3.1229}{1 \times 0.5}\right)[3789.7 - 0.4333(0-0)] - 2000}{\frac{6.2458}{1 \times 0.5} + \frac{3.1229}{1 \times 0.5}}$$

or

$$p_{wf_{ref}} = 3693.8 \text{ psia} \tag{6.31}$$

3. The first step involves the estimation of the FBHP at the reference depth as shown in the previous step (2). The result is given by Eq. (6.31) as $p_{wf_{ref}} = 3693.8$ psia. The second step involves applying Eq. (6.24) for each wellblock, yielding

$$q_{sc_5} = \left(\frac{6.2458}{1 \times 0.5}\right)[3812.5 - 3693.8 - 0.4333(15-0)] = 1401.56 \text{ B/D}.$$

and

$$q_{sc_9} = \left(\frac{3.1229}{1 \times 0.5}\right)[3789.7 - 3693.8 - 0.4333(0-0)] = 598.97 \text{ B/D}$$

6.3.3 Estimation of the wellblock geometric factor

In general, the geometric factor for wellblock i (G_{w_i}) is a fraction of the theoretical well geometric factor ($G^*_{w_i}$):

$$G_{w_i} = F_i \times G^*_{w_i} \tag{6.32}$$

where F_i = ratio of wellblock area to the theoretical area from which the well withdraws its fluid. The geometric factor depends on well location in the wellblock and whether or not it falls on no-flow reservoir boundaries.

Fig. 6.6 shows a discretized reservoir surrounded by no-flow boundaries and penetrated by a few vertical wells. Two of these wells fall at the center of wellblocks (W-A and W-K), four fall on one reservoir boundary (wells

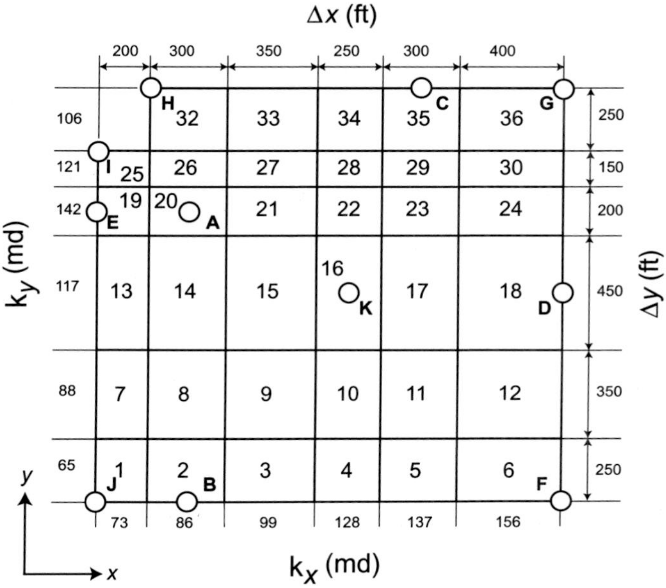

FIG. 6.6 Single-layer reservoir penetrated by vertical wells.

W-B, W-C, W-D, and W-E), and five fall at the intersection of two reservoir boundaries (W-F, W-G, W-H, W-I, and W-J). $F_i = 1$ if the well falls within the boundaries of a wellblock, $F_i = ½$ if the well falls on one reservoir boundary, and $F_i = ¼$ if the well falls at the intersection of two reservoir boundaries. The theoretical well geometric factor depends on well location, well radius, and the dimensions and permeabilities of the wellblock. To estimate the geometric factor for wellblock i, the dimensions of the area from which the well withdraws its fluid ($\Delta x \times \Delta y$) are first determined. This is followed by using Eq. (6.13) to estimate the horizontal permeability for wellblock i; Eq. (6.33), (6.34), or (6.35) to estimate the theoretical equivalent wellblock radius for block-centered grid; and Eq. (6.12) to estimate the theoretical well geometric factor ($G^*_{w_i}$). Finally, Eq. (6.32) is used to estimate the geometric factor for wellblock i (G_{w_i}).

For vertically stacked wellblocks such as those shown in Fig. 6.5, $F_i = 1$, $\Delta x = \Delta x_i$, and $\Delta y = \Delta y_i$. Therefore, the theoretical well geometric factor and the geometric factor for wellblock i are identical; that is, $G_{w_i} = G^*_{w_i}$. In this section, we present configurations in which the well is located on one and two reservoir boundaries sealed off to flow. We consider wells that are located at no-flow reservoir boundaries and that each produce from a single block

(W-B, W-C, W-D, W-E, and W-F). There are three possible configurations. The wellblock geometric factor in each configuration is estimated as follows (Peaceman, 1987).

Configuration 1 Fig. 6.7a presents a well located at the south boundary of a boundary wellblock that falls on the reservoir south boundary (W-B, hosted by block 2 in Fig. 6.6). Fig. 6.7b depicts the theoretical area from which the well withdraws fluid that is twice the area of the hosting wellblock. $F_i = ½$ as shown in Fig. 6.7c and r_{eq_i} and $G^*_{w_i}$ are calculated using Eqs. (6.33) and (6.12):

$$r_{eq_i} = 0.1403694\left[\Delta x^2 + \Delta y^2\right]^{0.5} \exp\left[(\Delta y/\Delta x)\tan^{-1}(\Delta x/\Delta y)\right] \tag{6.33}$$

A well that is located at the north boundary of a boundary wellblock (well W-C, hosted by block 35 in Fig. 6.6) receives similar treatment.

Configuration 2 Fig. 6.8a presents a well located at the east boundary of a boundary wellblock that falls on the reservoir east boundary (W-D, hosted by block 18 in Fig. 6.6). Fig. 6.8b depicts the theoretical area from which the well withdraws fluid that is twice the area of the hosting wellblock. $F_i = ½$ as shown in Fig. 6.8c and r_{eq_i} and $G^*_{w_i}$ are calculated using Eqs. (6.34) and (6.12):

$$r_{eq_i} = 0.1403694\left[\Delta x^2 + \Delta y^2\right]^{0.5} \exp\left[(\Delta x/\Delta y)\tan^{-1}(\Delta y/\Delta x)\right] \tag{6.34}$$

A well that is located at the west boundary of a boundary wellblock (W-E, hosted by block 19 in Fig. 6.6) receives similar treatment.

Configuration 3 Fig. 6.9a presents a well located at the intersection of the south and east boundaries of a wellblock that falls on the reservoir south and east boundaries (W-F, hosted by block 6 in Fig. 6.6). Fig. 6.9b depicts the theoretical area from which the well withdraws fluid that is four times the area of the

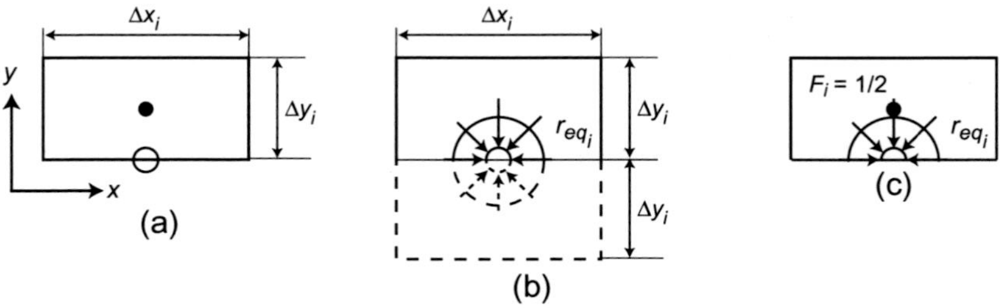

FIG. 6.7 Configuration 1 for a well on the reservoir south boundary.

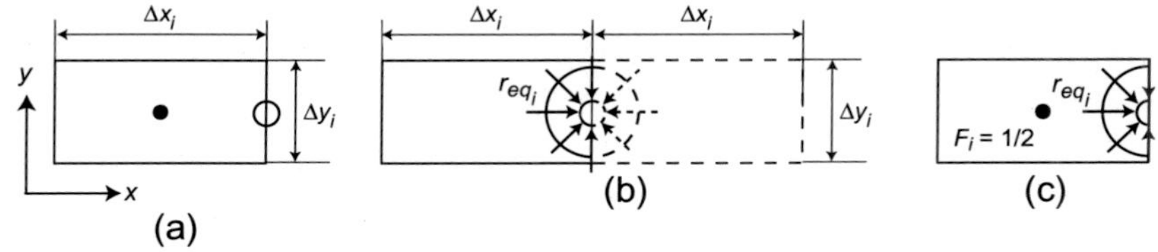

FIG. 6.8 Configuration 2 for a well on the reservoir east boundary.

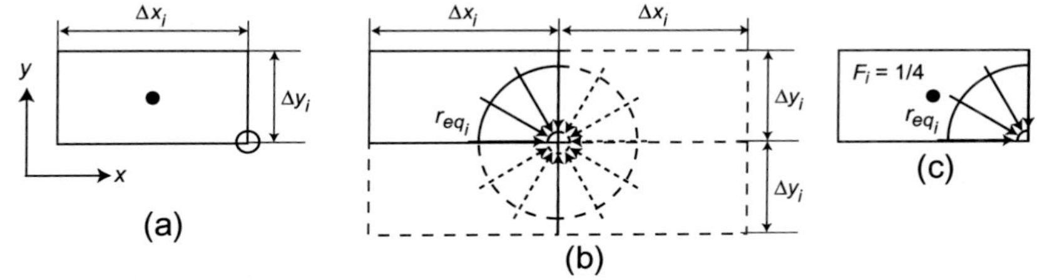

FIG. 6.9 Configuration 3 for a well on the reservoir south and east boundaries.

hosting wellblock. $F_i = {}^1/_4$ as shown in Fig. 6.9c and r_{eq_i} and $G^*_{w_i}$ are calculated using Eqs. (6.35) and (6.12):

$$r_{eq_i} = \left[\Delta x^2 + \Delta y^2\right]^{0.5}\left[0.3816 + \frac{0.2520}{(\Delta y/\Delta x)^{0.9401} + (\Delta x/\Delta y)^{0.9401}}\right] \tag{6.35}$$

A well that is located at the intersection of the south and west (W-J), west and north (W-H and W-I), or east and north (W-G) boundaries of the reservoir receives similar treatment.

Example 6.8 The single-phase oil, heterogeneous, anisotropic reservoir shown in Fig. 6.6 has many vertical production wells. The reservoir consists of a 40-ft-thick horizontal layer and has no-flow boundaries. Table 6.2 lists the identification of a few of these wells and the dimensions and permeabilities of the wellblocks. Each well fully penetrates the layer, and all wells were drilled with a 7-in. bit and have open-hole completion. Calculate the wellblock geometric factors for the wells given in Table 6.2. Assume zero skin factors.

Solution

Well W-A

TABLE 6.2 Wells and their wellblock dimensions and properties for Example 6.8.

Well ID	Wellblock order	Wellblock dimensions			Wellblock permeabilities	
		Δx (ft)	Δy (ft)	h (ft)	k_x (md)	k_y (md)
W-A	20	300	200	40	86	142
W-B	2	300	250	40	86	65
W-D	18	400	450	40	156	117
W-F	6	400	250	40	156	65

Wellblock 20 totally hosts well W-A, which falls at its center or inside the gridblock boundaries. Therefore, $F_i = 1$, $\Delta x = \Delta x_i = 300$ ft, and $\Delta y = \Delta y_i = 200$ ft. Wellblock 20 has $k_x = 86$ md and $k_y = 142$ md. Eqs. (6.14) and (6.13) are used to estimate the equivalent wellblock radius and horizontal permeability, yielding

$$r_{eq_{20}} = 0.28 \frac{\left[(142/86)^{0.5}(300)^2 + (86/142)^{0.5}(200)^2\right]^{0.5}}{\left[(142/86)^{0.25} + (86/142)^{0.25}\right]} = 53.217\,\text{ft}$$

and

$$k_{H_{20}} = [86 \times 142]^{0.5} = 110.51\,\text{md}$$

The well geometric factor specific to wellblock 20 is estimated by substituting values into Eq. (6.12), resulting in

$$G^*_{w_{20}} = \frac{2\pi \times 0.001127 \times 110.51 \times 40}{\{\log_e[53.217/(3.5/12)] + 0\}} = 6.012\,\text{RB-cP/D-psi}$$

The geometric factor for wellblock 20 is obtained using Eq. (6.32), yielding

$$G_{w_{20}} = 1 \times 6.012 = 6.012\,\text{RB-cP/D-psi}$$

Well W-B

Wellblock 2 hosts well W-B, which falls at the south gridblock boundary (Configuration 1). Therefore, $F_i = ½$, $\Delta x = \Delta x_i = 300$ ft, and $\Delta y = \Delta y_i = 250$ ft. Wellblock 2 has $k_x = 86$ md and $k_y = 65$ md. Eqs. (6.33) and (6.13) are used to estimate the equivalent wellblock radius and horizontal permeability, yielding

$$r_{eq_2} = 0.1403694\left[(300)^2 + (250)^2\right]^{0.5} \exp\left[(300/250)\tan^{-1}(250/300)\right]$$
$$= 126.175\,\text{ft}$$

and

$$k_{H_2} = [86 \times 65]^{0.5} = 74.766\,\text{md}$$

The well geometric factor specific to wellblock 2 is estimated by substituting values into Eq. (6.12), resulting in

$$G^*_{w_2} = \frac{2\pi \times 0.001127 \times 74.766 \times 40}{\{\log_e[126.175/(3.5/12)] + 0\}} = 3.489\,\text{RB-cP/D-psi}$$

The geometric factor for Wellblock 2 is obtained using Eq. (6.32), yielding

$$G_{w_2} = ½ \times 3.489 = 1.744\,\text{RB-cP/D-psi}$$

Well W-D

Wellblock 18 hosts well W-D, which falls at the east gridblock boundary (Configuration 2). Therefore, $F_i = {}^1/_2$, $\Delta x = \Delta x_i = 400$ ft, and $\Delta y = \Delta y_i =$ 450 ft. Wellblock 18 has $k_x = 156$ md and $k_y = 117$ md. Eqs. (6.34) and (6.13) are used to estimate the equivalent wellblock radius and horizontal permeability, yielding

$$r_{eq_{18}} = 0.1403694\left[(400)^2 + (450)^2\right]^{0.5} \exp\left[(400/450)\tan^{-1}(450/400)\right]$$
$$= 178.97\,\text{ft}$$

and

$$k_{H_{18}} = [156 \times 117]^{0.5} = 135.10\,\text{md}$$

The well geometric factor specific to wellblock 18 is estimated by substituting values into Eq. (6.12), resulting in

$$G^*_{w_{18}} = \frac{2\pi \times 0.001127 \times 135.10 \times 40}{\{\log_e[178.97/(3.5/12)] + 0\}} = 5.961\,\text{RB-cP/D-psi}$$

The geometric factor for wellblock 18 is obtained using Eq. (6.32), yielding

$$G_{w_{18}} = ½ \times 5.961 = 2.981\,\text{RB-cP/D-psi}$$

Well W-F

Wellblock 6 hosts well W-F, which falls at gridblock south and east boundaries (Configuration 3). Therefore, $F_i = ¼$, $\Delta x = \Delta x_i = 400$ ft, and $\Delta y = \Delta y_i =$ 250 ft. Wellblock 6 has $k_x = 156$ md and $k_y = 65$ md. Eqs. (6.35) and (6.13) are used to estimate the equivalent wellblock radius and horizontal permeability, yielding

$$r_{eq_6} = \left[(400)^2 + (250)^2\right]^{0.5}\left[0.3816 + \frac{0.2520}{(250/400)^{0.9401} + (400/250)^{0.9401}}\right]$$
$$= 234.1 \text{ ft}$$

and

$$k_{H_6} = [156 \times 65]^{0.5} = 100.70 \text{ md}$$

The well geometric factor specific to wellblock 6 is estimated by substituting values into Eq. (6.12), resulting in

$$G^*_{w_6} = \frac{2\pi \times 0.001127 \times 100.70 \times 40}{\{\log_e[234.1/(3.5/12)] + 0\}} = 4.265 \text{ RB-cP/D-psi}$$

The geometric factor for wellblock 6 is obtained using Eq. (6.32), yielding

$$G_{w_6} = ¼ \times 4.265 = 1.066 \text{ RB-cP/D-psi}$$

Table 6.3 shows the summary of intermediate and final results.

6.3.4 Estimation of well rate and FBHP

If the FBHP of a well ($p_{wf_{ref}}$) is specified, then the well production rate can be estimated as the sum of production rates from all wellblocks that are vertically penetrated by the well; that is,

$$q_{sc} = \sum_{i \in \psi_w} q_{sc_i} \tag{6.36}$$

If, on the other hand, the well production rate is specified, then the FBHP of the well ($p_{wf_{ref}}$) can be estimated using Eq. (6.26):

$$p_{wf_{ref}} = \frac{\sum_{i \in \psi_w} \left\{ \left(\frac{G_w}{B\mu}\right)_i \left[p_i - \overline{\gamma}_{wb}(Z_i - Z_{ref})\right] \right\} + q_{spsc}}{\sum_{i \in \psi_w} \left(\frac{G_w}{B\mu}\right)_i} \tag{6.26}$$

Eqs. (6.26) and (6.36) apply to vertical wells that are completed through vertically stacked wellblocks.

6.4 Practical considerations dealing with modeling well operating conditions

It is important for a reservoir model to represent the basic features of well performance. For example, a production well may not produce fluids at a constant rate indefinitely. We usually specify a desired constant rate for a well (q_{spsc}) and place a constraint on the FBHP of the well ($p_{wf_{sp}}$). The specified FBHP must be

TABLE 6.3 Estimated properties of theoretical wells and wellblock geometric factors.

Well ID	Wellblock i	Configuation	Theoretical well							Wellblock	
		#	Δx (ft)	Δy (ft)	k_x (md)	k_y (md)	k_{H_i} (md)	r_{eq_i} (ft)	$G^*_{w_i}$	F_i	G_{w_i}
W-A	20		300	200	86	142	110.51	53.220	6.012	1	6.012
W-B	2	1	300	250	86	65	74.766	126.17	3.489	1/2	1.744
W-D	18	2	400	450	156	117	135.10	178.97	5.961	1/2	2.981
W-F	6	3	400	250	156	65	100.70	234.1	4.265	1/4	1.066

sufficient to transport fluid from the bottom hole to the wellhead and maybe even to fluid treatment facilities. Additionally, an injection well may not inject fluid at a constant rate indefinitely. We usually specify a desired constant rate for a well (q_{spsc}) that is consistent with the availability of injected fluid and place a constraint on the FBHP of the well ($p_{wf_{sp}}$) that is consistent with the maximum pressure of the used pump or compressor (Abou-Kassem, 1996). The specified FBHP plus frictional loss in the injection well and the surface lines minus fluid head in the well must be less than or equal to the maximum pressure for the injection pump or compressor. To include the aforementioned practical features in a simulator, the following logic must be implemented in the developed simulator: (1) set $p_{wf_{ref}}=p_{wf_{sp}}$; (2) estimate the well FBHP ($p_{wf_{est}}$) that corresponds to the specified desired production (or injection) well rate using Eq. (6.26); and (3) use q_{spsc} for the well rate as long as $p_{wf_{est}} \geq p_{wf_{sp}}$ for a production well or $p_{wf_{est}} \leq p_{wf_{sp}}$ for an injection well, and distribute the well rate accordingly among the wellblocks (q_{sc_i}) as outlined in the text. Otherwise, (1) set $p_{wf_{est}}=p_{wf_{ref}}$, (2) estimate the wellblock rate (q_{sc_i}) for each wellblock in the well using Eq. (6.24), and (3) estimate the resulting well rate for multiblock wells using Eq. (6.36). These three steps are executed every iteration in every time step. A similar treatment is followed if the well pressure gradient at sandface is specified instead of the well rate. In this case, the desired wellblock rate is calculated using Eq. (6.4). If we neglect implementing provisions for the treatment of the aforementioned practical considerations in a simulator, the continuous withdrawal of fluids may result in negative simulated pressures, and the continuous injection of fluids may result in infinitely large simulated pressures. All reservoir simulators used by the petroleum industry, however, include logic for handling varying degrees of complicated well operating conditions.

6.5 Summary

Wells can be completed in a single block in 1-D and 2-D single-layer reservoirs or in multiblocks in multilayer reservoirs. Wells can be shut in or operated with a specified production rate, pressure gradient, or bottom-hole pressure. Shut-in wells have zero flow rates, and Eq. (6.2) defines the production rate of shut-in wells completed in wellblocks. Eq. (6.1) represents the IPR equation for a wellblock, and this equation can be used to estimate the production rate from the wellblock or the flowing bottom-hole pressure of the well in the wellblock. In single-well simulation, wells are incorporated in the flow equation as line source terms using Eq. (6.9). The wellblock geometric factor in a rectangular wellblock is estimated using Eq. (6.12). Eq. (6.4) can be used to estimate the wellblock production rate for a well operating with a specified pressure gradient, whereas Eq. (6.1) is used for a well operating with specified flowing bottom-hole pressure. In multiblock wells, proration of the well production rate among wellblocks can be achieved using Eq. (6.26) to estimate $p_{wf_{ref}}$ followed by Eq. (6.24) with wellblock geometric factor being estimated using Eq. (6.32).

6.6 Exercises

6.1 A well penetrates the whole thickness of a single layer. Does fluid flow toward (or away from) the well linearly, radially, or spherically?

6.2 In reservoir simulation, a well is represented as a source/sink line in the wellblock.

a. What is the fluid flow geometry within a wellblock in a 1-D reservoir?
b. What is the fluid flow geometry within a wellblock in a 2-D reservoir?
c. What is the fluid flow geometry within a wellblock in a 3-D reservoir?

6.3 You develop a model to simulate a 1-D, linear-flood experiment. Do you use fictitious wells or physical wells to reflect fluid input in the first block and fluid output out of the last block? Justify your answers.

6.4 You develop a single-well model. Justify why it is possible to use either a fictitious well or a physical well to describe the well rate in this case.

6.5 What are the different well operating conditions? Write the well production rate equation for each well operating condition.

6.6 Prove that Eq. (6.9a) is nothing but the flow rate of the fictitious well resulting from flow across the inner boundary of gridblock 1 in radial-cylindrical flow, which is equivalent to the flow term between the left boundary and the block center of gridblock 1; that is, $q_{sc_1} = \frac{G_{r_{1-1/2}}}{B_1 \mu_1}(p_0 - p_1)$, where $G_{r_{1-1/2}}$ is given in Table 4.3 in Chapter 4 and $p_0 = p_{wf}$.

6.7 Prove that Eq. (6.9b) can be derived from the steady-state flow equation for gridpoint 1 in radial-cylindrical flow and by using the definition of geometric factors given in Table 5.3 in Chapter 5.

6.8 Consider the reservoir presented in Example 6.8. Fig. 6.6 shows the block dimensions and permeabilities. Calculate the wellblock geometric factors for those penetrated by the wells identified as W-C, W-E, W-G, W-H, W-I, W-J, and W-K. All aforementioned wells have open-hole completion and were drilled with a 5-in. bit.

Chapter 7

Single-phase flow equation for various fluids

Chapter outline

7.1 Introduction

The single-phase, multidimensional flow equation for a reservoir block was derived in Chapter 2. In Chapter 3, this flow equation was rewritten using CVFD terminology for a reservoir block identified by engineering notation or block order. Chapters 4 and 5 presented the treatment of blocks that fall on reservoir boundaries using fictitious wells. In Chapter 6, the wellblock production rate equation was derived for various well operating conditions. In this chapter, the single-phase, multidimensional flow equation that incorporates the wellblock production rate and boundary conditions is presented for various fluids, including incompressible, slightly compressible, and compressible fluids. These fluids differ from each other by the pressure dependence of their densities, formation volume factors (FVFs), and viscosities. The presentation includes the flow equation for an incompressible system (rock and fluid) and the explicit, implicit, and Crank-Nicolson equations for slightly compressible and compressible fluids. The flow equations for block-centered grids and point-distributed grids have the same general form. The differences between the two grid systems lie in the construction of the grid, the treatment of boundary conditions, and the treatment of the wellblock production rate as was discussed in Chapters 4–6. The presentation in this chapter uses CVFD terminology to express the flow equation in a multidimensional domain.

Petroleum Reservoir Simulation. https://doi.org/10.1016/B978-0-12-819150-7.00007-4

7.2 Pressure dependence of fluid and rock properties

The pressure-dependent properties that are important in this chapter include those properties that appear in transmissibility, potential, production, and accumulation term, namely, fluid density, FVF, fluid viscosity, and rock porosity. Fluid density is needed for the estimation of fluid gravity using

$$\gamma = \gamma_c \rho g \tag{7.1}$$

The equations used for the estimation of these properties for various fluids and rock porosity are presented next.

7.2.1 Incompressible fluid

This type of fluid is an idealization of gas-free oil and water. An incompressible fluid has zero compressibility; therefore, regardless of pressure, it has a constant density, FVF, and viscosity. Mathematically,

$$\rho \neq \mathrm{f}(p) = \text{constant} \tag{7.2}$$

$$B \neq \mathrm{f}(p) = B^{\circ} \cong 1 \tag{7.3}$$

and

$$\mu \neq \mathrm{f}(p) = \text{constant} \tag{7.4}$$

7.2.2 Slightly compressible fluid

A slightly compressible fluid has a small but constant compressibility (c) that usually ranges from 10^{-5} to 10^{-6} psi^{-1}. Gas-free oil, water, and oil above bubble-point pressure are examples of slightly compressible fluids. The pressure dependence of the density, FVF, and viscosity for slightly compressible fluids is expressed as

$$\rho = \rho^{\circ}\left[1 + c\left(p - p^{\circ}\right)\right] \tag{7.5}$$

$$B = \frac{B^{\circ}}{[1 + c(p - p^{\circ})]} \tag{7.6}$$

and

$$\mu = \frac{\mu^{\circ}}{\left[1 - c_{\mu}(p - p^{\circ})\right]} \tag{7.7}$$

where ρ°, B°, and μ° are fluid density, FVF, and viscosity, respectively, at reference pressure (p°) and reservoir temperature and c_{μ} is the fractional change of viscosity with pressure change. Oil above its bubble-point pressure can be treated as a slightly compressible fluid with the reference pressure being the oil bubble-point pressure, and in this case, ρ°, B°, and μ° are the oil-saturated properties at the oil bubble-point pressure.

7.2.3 Compressible fluid

A compressible fluid has orders of magnitude higher compressibility than that of a slightly compressible fluid, usually 10^{-2} to 10^{-4} psi^{-1} depending on pressure. The density and viscosity of a compressible fluid increase as pressure increases but tend to level off at high pressures. The FVF decreases orders of magnitude as the pressure increases from atmospheric pressure to high pressure. Natural gas is a good example of a compressible fluid. The pressure dependencies of the density, FVF, and viscosity of natural gas are expressed as

$$\rho_g = \frac{pM}{zRT} \tag{7.8}$$

$$B_g = \frac{\rho_{g_{sc}}}{\alpha_c \rho_g} = \frac{p_{sc}}{\alpha_c T_{sc}} T \frac{z}{p} \tag{7.9}$$

and

$$\mu_g = \mathrm{f}(T, p, M) \tag{7.10}$$

The equations presented by Lee et al. (1966) and Dranchuk et al. (1986) are two forms of f(T,p,M) in Eq. (7.10). Although these gas properties can be estimated using Eqs. (7.8) through (7.10), these equations are used, external to a simulator, to calculate the density, FVF, and viscosity as functions of pressure over the pressure range of interest at reservoir temperature. The calculated FVF and viscosity are then supplied to the simulator in tabular form as functions of pressure. In addition, the gas density at standard conditions is supplied to calculate the gas density that corresponds to the gas FVF at any pressure.

7.2.4 Rock porosity

Porosity depends on reservoir pressure because of the combined compressibility of rock and pore. Porosity increases as reservoir pressure (pressure of the fluid contained in the pores) increases. This relationship can be expressed as

$$\phi = \phi^{\circ} \left[1 + c_{\phi}\left(p - p^{\circ}\right)\right] \tag{7.11}$$

where ϕ° is the porosity at the reference pressure (p°) and c_{ϕ} is the porosity compressibility. If the reference pressure is chosen as the initial reservoir pressure, then ϕ° may incorporate the effect of overburden on porosity.

7.3 General single-phase flow equation in multidimensions

The single-phase, multidimensional flow equation for Block (gridblock or gridpoint) n that incorporates boundary conditions is presented using CVFD terminology as in Eq. (4.2) (or Eq. 5.2)

$$\sum_{l\in\psi_n} T_{l,n}^m\left[\left(p_l^m - p_n^m\right) - \gamma_{l,n}^m (Z_l - Z_n)\right] + \sum_{l\in\xi_n} q_{sc_{l,n}}^m + q_{sc_n}^m = \frac{V_{b_n}}{\alpha_c \Delta t}\left[\left(\frac{\phi}{B}\right)_n^{n+1} - \left(\frac{\phi}{B}\right)_n^n\right] \tag{7.12}$$

where ψ_n = the set whose elements are the existing neighboring blocks in the reservoir, ξ_n = the set whose elements are the reservoir boundaries ($b_L, b_S, b_W, b_E, b_N, b_U$) that are shared by block n, and $q_{sc_{l,n}}^m$ = the flow rate of the fictitious well representing fluid transfer between reservoir boundary l and block n as a result of a boundary condition. For a 3-D reservoir, ξ_n is either an empty set for interior blocks or a set that contains one element for boundary blocks that fall on one reservoir boundary, two elements for boundary blocks that fall on two reservoir boundaries, or three elements for boundary blocks that fall on three reservoir boundaries. An empty set implies that the block does not fall on any reservoir boundary; that is, block n is an interior block, and hence $\sum_{l\in\xi_n} q_{sc_{l,n}}^m = 0$. Chapter 6 discusses the estimation of the production rate equation for a wellblock ($q_{sc_n}^m$) with the well producing (or injecting) fluid under a given operating condition. The accumulation term, represented by the RHS of Eq. (7.12), is presented for each type of fluid separately in Sections 7.3.1–7.3.3. In engineering notation, block order n is replaced with (i,j,k), and Eq. (7.12) becomes

$$\sum_{l\in\psi_{i,j,k}} T_{l,(i,j,k)}^m\left[\left(p_l^m - p_{i,j,k}^m\right) - \gamma_{l,(i,j,k)}^m \left(Z_l - Z_{i,j,k}\right)\right] + \sum_{l\in\xi_{i,j,k}} q_{sc_{l,(i,j,k)}}^m + q_{sc_{i,j,k}}^m = \frac{V_{b_{i,j,k}}}{\alpha_c \Delta t}\left[\left(\frac{\phi}{B}\right)_{i,j,k}^{n+1} - \left(\frac{\phi}{B}\right)_{i,j,k}^n\right] \tag{7.13}$$

7.3.1 Incompressible fluid flow equation

The density, FVF, and viscosity of an incompressible fluid are constant independent of pressure (Eqs. 7.2 through 7.4). Therefore, the accumulation term for an incompressible fluid ($c=0$) but a compressible porous medium reduces to

$$\frac{V_{b_n}}{\alpha_c \Delta t}\left[\left(\frac{\phi}{B}\right)_n^{n+1} - \left(\frac{\phi}{B}\right)_n^n\right] = \frac{V_{b_n}\phi_n^\circ c_\phi}{\alpha_c B^\circ \Delta t}\left[p_n^{n+1} - p_n^n\right] \tag{7.14}$$

with $B = B^\circ \cong 1$ for negligible fluid thermal expansion. If, in addition, the porous medium is treated as incompressible ($c_\phi = 0$), the accumulation term expressed by Eq. (7.14) becomes zero; that is,

$$\frac{V_{b_n}}{\alpha_c \Delta t}\left[\left(\frac{\phi}{B}\right)_n^{n+1} - \left(\frac{\phi}{B}\right)_n^n\right] = 0 \tag{7.15}$$

Substituting Eq. (7.15) into Eq. (7.12) yields the flow equation for incompressible systems:

$$\sum_{l\in\psi_n} T_{l,n}\left[(p_l - p_n) - \gamma_{l,n}(Z_l - Z_n)\right] + \sum_{l\in\xi_n} q_{sc_{l,n}} + q_{sc_n} = 0 \tag{7.16a}$$

or

$$\sum_{l\in\psi_{i,j,k}} T_{l,(i,j,k)}\left[\left(p_l - p_{i,j,k}\right) - \gamma_{l,(i,j,k)}\left(Z_l - Z_{i,j,k}\right)\right] + \sum_{l\in\xi_{i,j,k}} q_{sc_{l,(i,j,k)}} + q_{sc_{i,j,k}} = 0 \tag{7.16b}$$

The superscript m in Eq. (7.16) is dropped because none of the pressures depend on time, in addition to the condition that the wellblock production rate and boundary conditions do not change with time. Therefore, the pressure distribution for incompressible flow systems does not change with time.

7.3.1.1 Algorithm for obtaining the pressure solution

The pressure distribution for an incompressible flow problem is obtained using the following steps:

1. Calculate the interblock transmissibilities for all reservoir blocks.
2. Estimate the production rate (or write the production rate equation) for each wellblock in the reservoir as discussed in Chapter 6.
3. Estimate the flow rate (or write the flow rate equation) for each fictitious well in the reservoir as discussed in Chapter 4 (or Chapter 5); that is, estimate the flow rates resulting from the boundary conditions.
4. For every gridblock (or gridpoint) in the reservoir, define the set of existing neighboring reservoir blocks (ψ_n) and the set of reservoir boundaries that are block boundaries (ξ_n), expand the summation terms in the flow equation (Eq. 7.16 in this case), and substitute for the wellblock production rate obtained in (2) and the fictitious well rates obtained in (3).
5. Factorize, order, and place the unknown pressures on the LHS and place the known quantities on the RHS of each flow equation.
6. Solve the resulting set of equations for the unknown pressures using a linear equation solver such as those presented in Chapter 9.
7. Estimate the wellblock production rates and fictitious well rates if necessary using the flow rate equations obtained in (2) and (3).
8. Perform a material balance check.

7.3.1.2 Material balance check for an incompressible fluid flow problem

For an incompressible fluid flow problem (constant ϕ and B), there is no accumulation of mass in any reservoir block. Therefore, the sum of fluids entering and leaving the reservoir boundaries including wells must add up to zero (or a small number to account for round-off errors); that is,

$$\sum_{n=1}^{N}\left(q_{sc_n}+\sum_{l\in\xi_n}q_{sc_{l,n}}\right)=0 \tag{7.17a}$$

where N is the total number of blocks in the reservoir. The production (or injection) rate is set to zero for any reservoir block that is not penetrated by a well. The second term in the parentheses in Eq. (7.17a) takes care of fluid flow across reservoir boundaries resulting from boundary conditions. If reservoir blocks are identified using engineering notation, subscript n and summation $\sum_{n=1}^{N}$ in Eq. (7.17a) are replaced with subscripts (i,j,k) and $\sum_{i=1}^{n_x}\sum_{j=1}^{n_y}\sum_{k=1}^{n_z}$, respectively. The resulting equation is

$$\sum_{i=1}^{n_x}\sum_{j=1}^{n_y}\sum_{k=1}^{n_z}\left(q_{sc_{i,j,k}}+\sum_{l\in\xi_{i,j,k}}q_{sc_{l,(i,j,k)}}\right)=0 \tag{7.17b}$$

The material balance check that is expressed by Eq. (7.17) can be derived by writing Eq. (7.16) for each block in the system ($n=1, 2, 3\ldots N$) and then summing up all N equations. All interblock flow terms in the resulting equation cancel out, leading to Eq. (7.17). It is customary to perform a material balance check after solving any simulation problem. An unsatisfactory material balance check implies an incorrect pressure solution for the problem. A satisfactory material balance check; however, does not necessarily imply a correct pressure solution. If the material balance check is unsatisfactory, the flow equation and all of its elements (transmissibilities, well production rate, fictitious well rates, ψ_n, ξ_n, …etc.) for every gridblock (gridpoint) in the reservoir and the solution of the algebraic equations must be carefully investigated to find the cause of the error.

Examples 7.1 through 7.6 present the solutions for several variations of the 1-D flow problem. The variations include different boundary conditions, well operating conditions, and well location within the reservoir block. Example 7.1 demonstrates the application of the algorithm presented in this section to obtain the pressure solution. Example 7.2 presents an approximate solution method used by other reservoir simulation books when dealing with a constant pressure boundary in a block-centered grid. In Example 7.3, the well produces oil with a constant FBHP specification instead of a constant well production rate. In Example 7.4, the reservoir right boundary is specified as a constant pressure gradient boundary instead of a no-flow boundary. In Example 7.5, the reservoir is an inclined reservoir instead of horizontal. In Example 7.6, the well is relocated at a reservoir boundary, and the effect of treating it as a boundary condition is demonstrated. Example 7.7 presents a 2-D reservoir with anisotropic

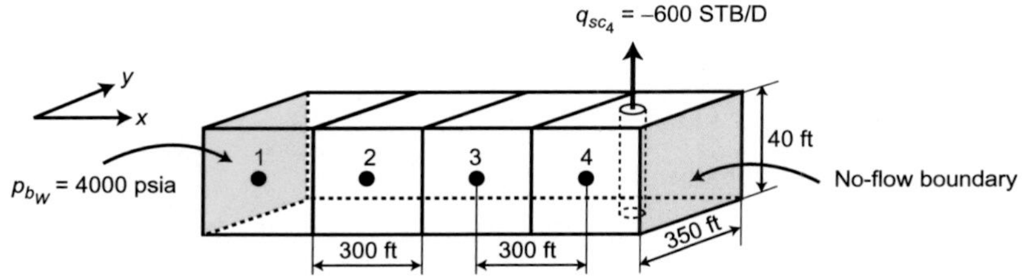

FIG. 7.1 Discretized 1-D reservoir in Example 7.1.

permeability. Example 7.8 presents a 2-D homogeneous and isotropic reservoir showing symmetry.

Example 7.1 A single-phase fluid reservoir is described by four equal blocks as shown in Fig. 7.1. The reservoir is horizontal and has homogeneous and isotropic rock properties, $k=270$ md and $\phi=0.27$. The gridblock dimensions are $\Delta x=300$ ft, $\Delta y=350$ ft, and $h=40$ ft. The reservoir fluid properties are $B=B^\circ=1$ RB/STB, $\rho=50$ lbm/ft^3, and $\mu=0.5$ cP. The reservoir left boundary is kept at constant pressure of 4000 psia, and the reservoir right boundary is sealed off to flow. A 7-in vertical well was drilled at the center of gridblock 4. The well produces 600 STB/D of fluid and has a skin factor of 1.5. Assuming that the reservoir rock and fluid are incompressible, find the pressure distribution in the reservoir and the FBHP of the well. Perform a material balance check.

Solution

The gridblocks have the same dimensions and rock properties. Therefore, $T_{1,2}=T_{2,3}=T_{3,4}=T_x$, where $T_x=\beta_c\frac{A_x k_x}{\mu B \Delta x}=0.001127\times\frac{(350\times40)\times270}{0.5\times1\times300}$ $=28.4004$ STB/D-psi. There is a production well in gridblock 4 only. Therefore, $q_{sc_4}=-600$ STB/D. In addition, for the other gridblocks, $q_{sc_1}=q_{sc_2}=q_{sc_3}=0$.

Gridblock 1 falls on the reservoir west boundary, which is kept at a constant pressure of 4000 psia. Therefore, Eq. (4.37c) can be used, yielding

$$q_{sc_{b_W,1}}=\left[\beta_c\frac{k_x A_x}{\mu B(\Delta x/2)}\right]_1[(p_{b_W}-p_1)-\gamma(Z_{b_W}-Z_1)]$$

$$=\left[0.001127\times\frac{270\times(350\times40)}{0.5\times1\times(300/2)}\right][(4000-p_1)-\gamma\times0]$$

or

$$q_{sc_{b_W,1}}=56.8008(4000-p_1)\,\text{STB/D} \tag{7.18}$$

Gridblock 4 falls on the reservoir east boundary, which is a no-flow boundary. Therefore, Eq. (4.32) applies giving $q_{sc_{b_E,4}}=0$ STB/D.

The general flow equation for this 1-D horizontal reservoir is obtained from Eq. (7.16a) by discarding the gravity term, yielding

$$\sum_{l\in\psi_n}T_{l,n}(p_l-p_n)+\sum_{l\in\xi_n}q_{sc_{l,n}}+q_{sc_n}=0 \tag{7.19}$$

For gridblock 1, $n=1$, $\psi_1=\{2\}$, and $\xi_1=\{b_W\}$. Therefore, $\sum_{l\in\xi_1} q_{sc_{l,1}} = q_{sc_{b_W,1}}$, and Eq. (7.19) becomes

$$T_{1,2}(p_2-p_1)+q_{sc_{b_W,1}}+q_{sc_1}=0 \tag{7.20}$$

Substitution of the values in this equation gives

$$28.4004(p_2-p_1)+56.8008(4000-p_1)+0=0$$

or after factorizing and ordering the unknowns,

$$-85.2012p_1+28.4004p_2=-227203.2 \tag{7.21}$$

For gridblock 2, $n=2$, $\psi_2=\{1,3\}$, and $\xi_2=\{\}$. Therefore, $\sum_{l\in\xi_2} q_{sc_{l,2}}=0$, and Eq. (7.19) becomes

$$T_{1,2}(p_1-p_2)+T_{2,3}(p_3-p_2)+q_{sc_2}=0 \tag{7.22}$$

Substitution of the values in this equation gives

$$28.4004(p_1-p_2)+28.4004(p_3-p_2)+0=0$$

or after factorizing and ordering the unknowns,

$$28.4004p_1-56.8008p_2+28.4004p_3=0 \tag{7.23}$$

For gridblock 3, $n=3$, $\psi_3=\{2,4\}$, and $\xi_3=\{\}$. Therefore, $\sum_{l\in\xi_3} q_{sc_{l,3}}=0$, and Eq. (7.19) becomes

$$T_{2,3}(p_2-p_3)+T_{3,4}(p_4-p_3)+q_{sc_3}=0 \tag{7.24}$$

Substitution of the values in this equation gives

$$28.4004(p_2-p_3)+28.4004(p_4-p_3)+0=0$$

or after factorizing and ordering the unknowns,

$$28.4004p_2-56.8008p_3+28.4004p_4=0 \tag{7.25}$$

For gridblock 4, $n=4$, $\psi_4=\{3\}$, and $\xi_4=\{b_E\}$. Therefore, $\sum_{l\in\xi_4} q_{sc_{l,4}}=q_{sc_{b_E,4}}$, and Eq. (7.19) becomes

$$T_{3,4}(p_3-p_4)+q_{sc_{b_E,4}}+q_{sc_4}=0 \tag{7.26}$$

Substitution of the values in this equation gives

$$28.4004(p_3-p_4)+0+(-600)=0$$

or after the ordering of the unknowns,

$$28.4004p_3-28.4004p_4=600 \tag{7.27}$$

The results of solving Eqs. (7.21), (7.23), (7.25), and (7.27) for the unknown pressures are $p_1=3989.44$ psia, $p_2=3968.31$ psia, $p_3=3947.18$ psia, and $p_4=3926.06$ psia.

Next, the flow rate across the reservoir left boundary ($q_{sc_{b_W,1}}$) is estimated using Eq. (7.18), yielding.

$$q_{sc_{b_W,1}} = 56.8008(4000 - p_1) = 56.8008(4000 - 3989.44) = 599.816\,\text{STB/D}$$

The FBHP of the well in gridblock 4 is estimated using Eq. (6.1). First, however, the equivalent wellblock radius using Eq. (6.15) followed by the wellblock geometric factor using Eq. (6.12) must be calculated, yielding

$$r_{eq} = 0.14\left[(300)^2 + (350)^2\right]^{0.5} = 64.537\,\text{ft}$$

$$G_w = \frac{2\pi \times 0.001127 \times 270 \times 40}{\log_e[64.537/(3.5/12)] + 1.5} = 11.0845\,\text{RB-cP/D-psi}$$

and

$$-600 = -\frac{11.0845}{1 \times 0.5}(3926.06 - p_{wf_4})$$

from which

$$p_{wf_4} = 3899.00\,\text{psia}$$

The material balance for an incompressible fluid and rock system is checked by substituting the values for the well production rates and fictitious well rates on the LHS of Eq. (7.17a), yielding

$$\begin{aligned}\sum_{n=1}^{N}\left(q_{sc_n} + \sum_{l \in \xi_n} q_{sc_{l,n}}\right) &= (q_{sc_1} + q_{sc_{b_W,1}}) + (q_{sc_2} + 0) + (q_{sc_3} + 0) + (q_{sc_4} + q_{sc_{b_E,4}}) \\ &= (0 + 599.816) + (0 + 0) + (0 + 0) + (-600 + 0) \\ &= -0.184\end{aligned}$$

Therefore, the material balance check is satisfied, and a small error of 0.184 STB/D is observed because of rounding off during calculations.

Example 7.2 Find the pressure distribution in the reservoir presented in Example 7.1, but this time, assume that the boundary pressure is displaced half a block to coincide with the center of boundary gridblock 1. In other words, the pressure of gridblock 1 is kept constant at 4000 psia as shown in Fig. 7.2.

Solution

For gridblock 1,

$$p_1 \cong p_{b_W} = 4000\,\text{psia} \tag{7.28}$$

What remains is to find the pressure of gridblocks 2, 3, and 4. The flow equations for these three blocks are obtained from Eqs. (7.23), (7.25), and (7.27) in Example 7.1.

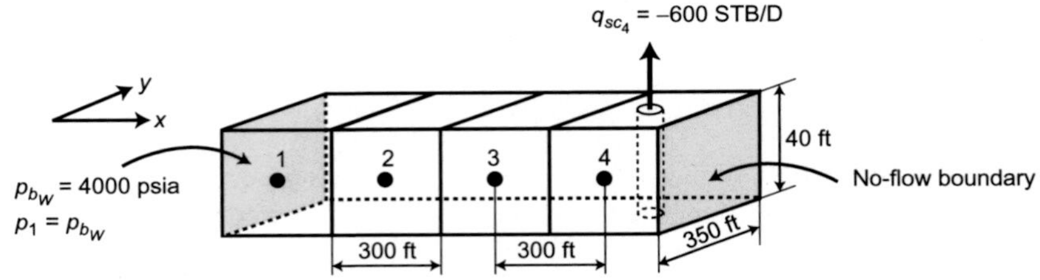

FIG. 7.2 Discretized 1-D reservoir in Example 7.2.

For gridblock 2,

$$28.4004p_1 - 56.8008p_2 + 28.4004p_3 = 0 \tag{7.23}$$

For gridblock 3,

$$28.4004p_2 - 56.8008p_3 + 28.4004p_4 = 0 \tag{7.25}$$

For gridblock 4,

$$28.4004p_3 - 28.4004p_4 = 600 \tag{7.27}$$

Substitution of Eq. (7.28) into Eq. (7.23) yields

$$28.4004 \times 4000 - 56.8008p_2 + 28.4004p_3 = 0$$

or the flow equation for gridblock 2 becomes

$$-56.8008p_2 + 28.4004p_3 = -113601.6 \tag{7.29}$$

The results of solving Eqs. (7.25), (7.27), and (7.29) for the unknown pressures are $p_2 = 3978.87$ psia, $p_3 = 3957.75$ psia, and $p_4 = 3936.62$ psia.

The flow rate across reservoir the left boundary ($q_{sc_{b_W,1}}$) can be estimated using the flow equation for gridblock 1, Eq. (7.20) obtained in Example 7.1, yielding

$$T_{1,2}(p_2 - p_1) + q_{sc_{b_W,1}} + q_{sc_1} = 0 \tag{7.20}$$

Substitution of the values of the gridblock pressures in this equation gives

$$28.4004(3978.87 - 4000) + q_{sc_{b_W,1}} + 0 = 0$$

or

$$q_{sc_{b_W,1}} = 600.100\,\text{STB/D}$$

The approximation presented by Eq. (7.28) results in $p_1 = 4000$ psia, compared with $p_1 = 3989.44$ psia using Eq. (4.37c) in Example 7.1. This approximation has been used in currently available books on reservoir simulation to obtain a solution for problems involving a specified pressure boundary condition in a block-centered grid. Such an approximation; however, is first-order correct and

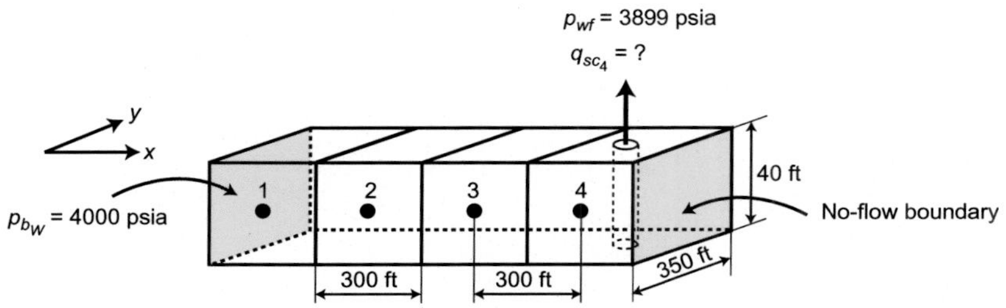

FIG. 7.3 Discretized 1-D reservoir in Example 7.3.

produces results that are less accurate than the treatment that uses Eq. (4.37c) and was demonstrated in Example 7.1.

Example 7.3 Consider the reservoir described in Example 7.1, but this time the well in gridblock 4 produces under a constant FBHP of 3899 psia as shown in Fig. 7.3. Find the pressure distribution in the reservoir. In addition, find the well production rate and flow rate across the reservoir west boundary.

Solution

From Example 7.1, the transmissibility and the flow rate across the reservoir left boundary are obtained as $T_x = 28.4004$ STB/D-psi and

$$q_{sc_{b_W,1}} = 56.8008(4000 - p_1)\ \text{STB/D} \tag{7.18}$$

respectively.

The flow equations for the first three gridblocks are obtained as in Example 7.1.

For gridblock 1,

$$-85.2012p_1 + 28.4004p_2 = -227203.2 \tag{7.21}$$

For gridblock 2,

$$28.4004p_1 - 56.8008p_2 + 28.4004p_3 = 0 \tag{7.23}$$

For gridblock 3,

$$28.4004p_2 - 56.8008p_3 + 28.4004p_4 = 0 \tag{7.25}$$

In addition, for the well in gridblock 4, $r_{eq} = 64.537$ ft, and $G_w = 11.0845$ RB-cP/D-psi.

The rate of production from the well in gridblock 4 can estimated using Eq. (6.1) for a constant FBHP specification, yielding

$$q_{sc_4} = -\frac{11.0845}{1 \times 0.5}(p_4 - 3899) = -22.1690(p_4 - 3899) \tag{7.30}$$

For gridblock 4, the flow equation is obtained from Eq. (7.26) in Example 7.1:

$$T_{3,4}(p_3 - p_4) + q_{sc_{b_E,4}} + q_{sc_4} = 0 \tag{7.26}$$

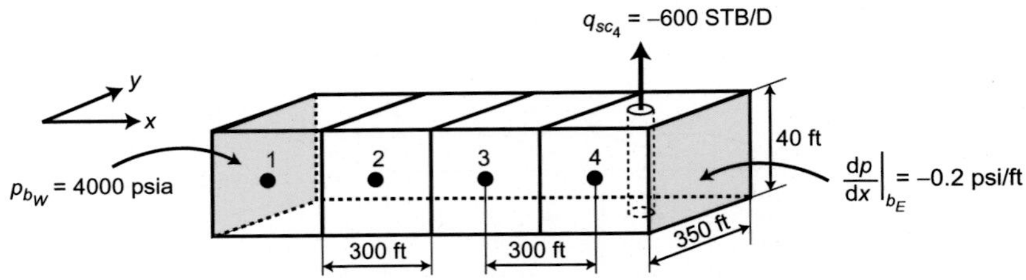

FIG. 7.4 Discretized 1-D reservoir in Example 7.4.

Substitution of transmissibility and Eq. (7.30) into Eq. (7.26) yields

$$28.4004(p_3 - p_4) + 0 + [-22.1690(p_4 - 3899)] = 0$$

or after factorizing and ordering the unknowns,

$$28.4004p_3 - 50.5694p_4 = -86436.93. \tag{7.31}$$

The results of solving Eqs. (7.21), (7.23), (7.25), and (7.31) for the unknown pressures are $p_1 = 3989.44$ psia, $p_2 = 3968.31$ psia, $p_3 = 3947.19$ psia, and $p_4 = 3926.06$ psia.

Substitution for pressures in the equations for q_{sc_4} (Eq. 7.30) and $q_{sc_{b_W,1}}$ (Eq. 7.18) yields

$$q_{sc_4} = -22.1690(p_4 - 3899) = -22.1690(3926.06 - 3899)$$
$$= -599.893 \text{ STB/D}$$

and

$$q_{sc_{b_W,1}} = 56.8008(4000 - p_1) = 56.8008(4000 - 3989.44) = 599.816 \text{ STB/D}$$

Example 7.4 Find the pressure distribution in the reservoir presented in Example 7.1, but this time, a pressure gradient of −0.2 psi/ft is specified at the reservoir right boundary as shown in Fig. 7.4.

Solution

From Example 7.1, the transmissibility and the flow rate across the reservoir west boundary are obtained as $T_x = 28.4004$ STB/D-psi and

$$q_{sc_{b_W,1}} = 56.8008(4000 - p_1) \text{ STB/D} \tag{7.18}$$

respectively.

The flow rate across the reservoir east boundary is estimated using Eq. (4.24b), yielding

$$q_{sc_{b_E,4}} = \left[\beta_c \frac{k_x A_x}{\mu B}\right]_4 \left[\left.\frac{\partial p}{\partial x}\right|_{b_E} - \gamma \left.\frac{\partial Z}{\partial x}\right|_{b_E}\right]$$
$$= 0.001127 \times \frac{270 \times (350 \times 40)}{0.5 \times 1}[-0.2 - \gamma \times 0]$$

or

$$q_{sc_{b_E,4}} = -1704.024\,\text{STB/D}$$

The flow equations for the first three gridblocks are obtained as in Example 7.1.

For gridblock 1,

$$-85.2012p_1 + 28.4004p_2 = -227203.2 \tag{7.21}$$

For gridblock 2,

$$28.4004p_1 - 56.8008p_2 + 28.4004p_3 = 0 \tag{7.23}$$

For gridblock 3,

$$28.4004p_2 - 56.8008p_3 + 28.4004p_4 = 0 \tag{7.25}$$

For gridblock 4, the flow equation is obtained from Eq. (7.26) in Example 7.1, yielding

$$T_{3,4}(p_3 - p_4) + q_{sc_{b_E,4}} + q_{sc_4} = 0 \tag{7.26}$$

Substitution of values in Eq. (7.26) gives

$$28.4004(p_3 - p_4) + (-1704.024) + (-600) = 0$$

or after the ordering of the unknowns,

$$28.4004p_3 - 28.4004p_4 = 2304.024 \tag{7.32}$$

The results of solving Eqs. (7.21), (7.23), (7.25), and (7.32) for the unknown pressures are $p_1 = 3959.44$ psia, $p_2 = 3878.31$ psia, $p_3 = 3797.18$ psia, and $p_4 = 3716.06$ psia.

Substitution for the pressures in the equation for $q_{sc_{b_W,1}}$(Eq. 7.18) yields

$$q_{sc_{b_W,1}} = 56.8008(4000 - p_1) = 56.8008(4000 - 3959.44) = 2304.024\,\text{STB/D}$$

Example 7.5 Consider the reservoir shown in Fig. 7.5. The reservoir has the same description as that presented in Example 7.1, with the exception that this reservoir is inclined along the formation dip. The elevations of the center of gridblocks 1, 2, 3, and 4 are, respectively, 3182.34, 3121.56, 3060.78, and 3000 ft below sea level. The centers of the reservoir west and east boundaries are, respectively, 3212.73 and 2969.62 ft below sea level. Assuming that the reservoir rock and fluid are incompressible, find the pressure distribution in the reservoir and the FBHP of the well in gridblock 4. Perform a material balance check.

Solution

The gridblocks have the same dimensions and rock properties. Therefore, $T_{1,2} = T_{2,3} = T_{3,4} = T_x = \beta_c \frac{A_x k_x}{\mu B \Delta x} = 0.001127 \times \frac{(350 \times 40) \times 270}{0.5 \times 1 \times 300} = 28.4004$ STB/D-psi. The fluid gravity is estimated using Eq. (7.1), yielding

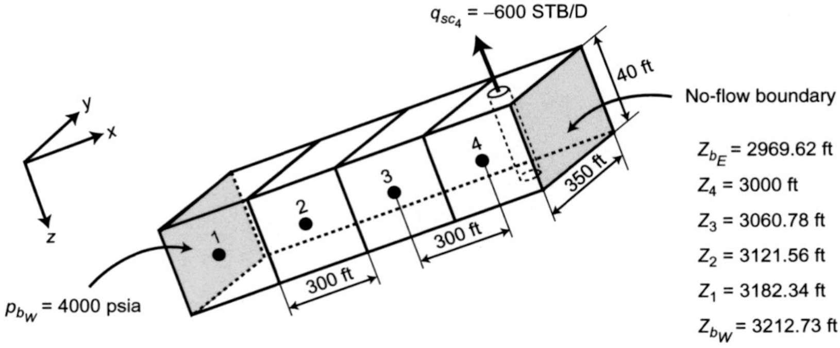

FIG. 7.5 Discretized 1-D reservoir in Example 7.5.

$$\gamma = \gamma_c \rho g = \left(0.21584 \times 10^{-3}\right) \times 50 \times 32.174 = 0.34722 \text{ psi/ft}$$

There is a production well in gridblock 4 only. Therefore, $q_{sc_4} = -600$ STB/D. In addition, for the other gridblocks, $q_{sc_1} = q_{sc_2} = q_{sc_3} = 0$.

Gridblock 1 falls on the reservoir west boundary, which is kept at a constant pressure of 4000 psia. Therefore, $q_{sc_{b_W,1}}$ can be estimated using Eq. (4.37c), which yields

$$q_{sc_{b_W,1}} = \left[\beta_c \frac{k_x A_x}{\mu B(\Delta x/2)}\right]_1 \left[(p_{b_W} - p_1) - \gamma(Z_{b_W} - Z_1)\right]$$

$$= \left[0.001127 \times \frac{270 \times (350 \times 40)}{0.5 \times 1 \times (300/2)}\right] [(4000 - p_1) - 0.34722 \times (3212.73 - 3182.34)]$$

or

$$q_{sc_{b_W,1}} = 56.8008(3989.448 - p_1) \text{ STB/D} \tag{7.33}$$

Gridblock 4 falls on the reservoir east boundary, which is a no-flow boundary. Therefore, Eq. (4.32) applies, giving $q_{sc_{b_E,4}} = 0$ STB/D.

The general flow equation for gridblock n in this 1-D inclined reservoir is expressed by Eq. (7.16a):

$$\sum_{l \in \psi_n} T_{l,n}\left[(p_l - p_n) - \gamma_{l,n}(Z_l - Z_n)\right] + \sum_{l \in \xi_n} q_{sc_{l,n}} + q_{sc_n} = 0 \tag{7.16a}$$

For gridblock 1, $n=1$, $\psi_1 = \{2\}$, and $\xi_1 = \{b_W\}$. Therefore, $\sum_{l \in \xi_1} q_{sc_{l,1}} = q_{sc_{b_W,1}}$, and Eq. (7.16a) becomes

$$T_{1,2}[(p_2 - p_1) - \gamma(Z_2 - Z_1)] + q_{sc_{b_W,1}} + q_{sc_1} = 0 \tag{7.34}$$

Substitution of Eq. (7.33) and the values into Eq. (7.34) gives $28.4004 \times [(p_2 - p_1) - 0.34722 \times (3121.56 - 3182.34)] + 56.8008(3989.448 - p_1) + 0 = 0$, or after factorizing and ordering the unknowns,

$$-85.2012 p_1 + 28.4004 p_2 = -227203.2 \tag{7.35}$$

For gridblock 2, $n=2$, $\psi_2 = \{1,3\}$, and $\xi_2 = \{\}$. Therefore, $\sum_{l \in \xi_2} q_{sc_{l,2}} = 0$, and Eq. (7.16a) becomes

$$T_{1,2}[(p_1 - p_2) - \gamma(Z_1 - Z_2)] + T_{2,3}[(p_3 - p_2) - \gamma(Z_3 - Z_2)] + q_{sc_2} = 0 \tag{7.36}$$

Substitution of the values in this equation gives

$$\begin{aligned} &28.4004[(p_1 - p_2) - 0.34722 \times (3182.34 - 3121.56)] \\ &+28.4004[(p_3 - p_2) - 0.34722 \times (3060.78 - 3121.56)] + 0 = 0 \end{aligned}$$

or after factorizing and ordering the unknowns,

$$28.4004 p_1 - 56.8008 p_2 + 28.4004 p_3 = 0 \tag{7.37}$$

For gridblock 3, $n=3$, $\psi_3 = \{2,4\}$, and $\xi_3 = \{\}$. Therefore, $\sum_{l \in \xi_3} q_{sc_{l,3}} = 0$, and Eq. (7.16a) becomes

$$T_{2,3}[(p_2 - p_3) - \gamma(Z_2 - Z_3)] + T_{3,4}[(p_4 - p_3) - \gamma(Z_4 - Z_3)] + q_{sc_3} = 0 \tag{7.38}$$

Substitution of the values in this equation gives

$$\begin{aligned} &28.4004[(p_2 - p_3) - 0.34722 \times (3121.56 - 3060.78)] \\ &+28.4004[(p_4 - p_3) - 0.34722 \times (3000 - 3060.78)] + 0 = 0 \end{aligned}$$

or after factorizing and ordering the unknowns,

$$28.4004 p_2 - 56.8008 p_3 + 28.4004 p_4 = 0 \tag{7.39}$$

For gridblock 4, $n=4$, $\psi_4 = \{3\}$, and $\xi_4 = \{b_E\}$. Therefore, $\sum_{l \in \xi_4} q_{sc_{l,4}} = q_{sc_{b_E,4}}$, and Eq. (7.16a) becomes

$$T_{3,4}[(p_3 - p_4) - \gamma(Z_3 - Z_4)] + q_{sc_{b_E,4}} + q_{sc_4} = 0 \tag{7.40}$$

Substitution of the values in this equation gives

$$28.4004[(p_3 - p_4) - 0.34722 \times (3060.78 - 3000)] + 0 + (-600) = 0$$

or after ordering the unknowns,

$$28.4004 p_3 - 28.4004 p_4 = 1199.366 \tag{7.41}$$

The results of solving Eqs. (7.35), (7.37), (7.39), and (7.41) for the unknown pressures are $p_1 = 3978.88$ psia, $p_2 = 3936.65$ psia, $p_3 = 3894.42$ psia, and $p_4 = 3852.19$ psia.

Next, the flow rate across the reservoir left boundary ($q_{sc_{b_W,1}}$) is estimated using Eq. (7.33), yielding

$$q_{sc_{b_W},1} = 56.8008(3989.448 - p_1) = 56.8008(3989.448 - 3978.88)$$
$$= 600.271\ \text{STB/D}$$

The FBHP of the well in gridblock 4 is estimated using Eq. (6.1), but first, the equivalent wellblock radius using Eq. (6.15) followed by the wellblock geometric factor using Eq. (6.12) must be calculated, giving

$$r_{eq} = 0.14\left[(300)^2 + (350)^2\right]^{0.5} = 64.537\ \text{ft}$$

$$G_w = \frac{2\pi \times 0.001127 \times 270 \times 40}{\log_e[64.537/(3.5/12)] + 1.5} = 11.0845\ \text{RB-cP/D-psi}$$

and

$$-600 = -\frac{11.0845}{1 \times 0.5}(3852.19 - p_{wf_4})$$

from which

$$p_{wf_4} = 3825.13\ \text{psia}$$

The material balance for an incompressible fluid and rock system is checked by substituting the values for the well production rates and the rates across reservoir boundaries on the LHS of Eq. (7.17a), which yields

$$\sum_{n=1}^{N}\left(q_{sc_n} + \sum_{l \in \xi_n} q_{sc_{l,n}}\right) = (q_{sc_1} + q_{sc_{b_W},1}) + (q_{sc_2} + 0) + (q_{sc_3} + 0) + (q_{sc_4} + q_{sc_{b_E},4})$$
$$= (0 + 600.271) + (0 + 0) + (0 + 0) + (-600 + 0)$$
$$= +0.271$$

Therefore, the material balance check is satisfied, and a small error of 0.271 STB/D is observed because of rounding off during calculations.

Example 7.6 Find the equation for the well production rate and pressure distribution in the reservoir presented in Example 7.1 if the vertical well is operated with a constant FBHP of 3850 psia for the following three cases:

1. The well is located at the center of gridblock 4.
2. The well is located at the east boundary of gridblock 4.
3. The well is treated as a boundary condition with the reservoir boundary pressure equal to 3850 psia.

Solution

From Example 7.1, the transmissibility and the flow rate across the reservoir west boundary are obtained as $T_x = 28.4004$ STB/D-psi and

$$q_{sc_{b_W},1} = 56.8008(4000 - p_1)\ \text{STB/D} \tag{7.18}$$

respectively.

The flow equations for the first three gridblocks are obtained as in Example 7.1.

For gridblock 1,

$$-85.2012p_1 + 28.4004p_2 = -227203.2 \tag{7.21}$$

For gridblock 2,

$$28.4004p_1 - 56.8008p_2 + 28.4004p_3 = 0 \tag{7.23}$$

For gridblock 3,

$$28.4004p_2 - 56.8008p_3 + 28.4004p_4 = 0 \tag{7.25}$$

For gridblock 4, $n=4$, $\psi_4=\{3\}$, and $\xi_4=\{b_E\}$. Therefore, $\sum_{l\in\xi_4} q_{sc_{l,4}} = q_{sc_{b_E,4}}$, and Eq. (7.16a) for a horizontal reservoir becomes

$$T_{3,4}(p_3 - p_4) + q_{sc_{b_E,4}} + q_{sc_4} = 0 \tag{7.26}$$

1. The well is located at center of gridblock 4. The equation for the well production rate is obtained using Eq. (6.15) for r_{eq}, Eq. (6.12) for G_w, and Eq. (6.1) for q_{sc_4} (see Fig. 7.6a), yielding

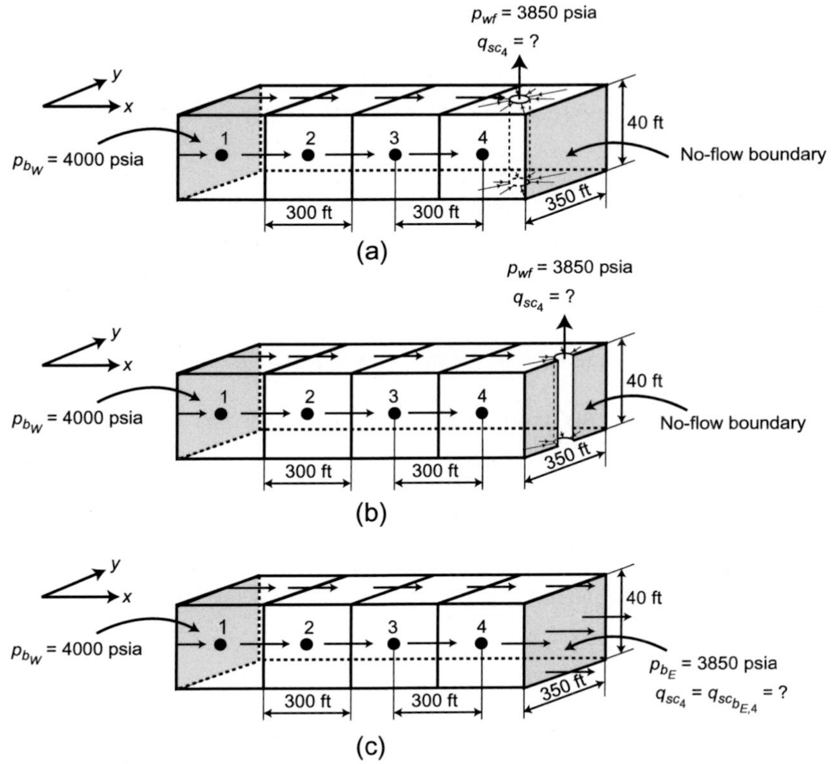

FIG. 7.6 Well location and treatment in Example 7.6. (a) Well is located at center of gridblock 4, (b) well is located at east boundary of gridblock 4, (c) well is replaced with a boundary condition at east boundary of gridblock 4.

$$r_{eq} = 0.14\left[(300)^2 + (350)^2\right]^{0.5} = 64.537\,\text{ft}$$

$$G_w = \frac{2\pi \times 0.001127 \times 270 \times 40}{\log_e[64.537/(3.5/12)] + 1.5} = 11.0845\,\text{RB-cP/D-psi}$$

and

$$q_{sc_4} = -\frac{11.0845}{1 \times 0.5}(p_4 - 3850)$$

or

$$q_{sc_4} = -22.1690(p_4 - 3850)\,\text{STB/D} \tag{7.42}$$

Substitution of the transmissibility and Eq. (7.42) into Eq. (7.26) gives

$$28.4004(p_3 - p_4) + 0 + [-22.1690(p_4 - 3850)] = 0$$

or after factorizing and ordering the unknowns,

$$28.4004p_3 - 50.5694p_4 = -85350.65 \tag{7.43}$$

The results of solving Eqs. (7.21), (7.23), (7.25), and (7.43) for the unknown pressures are $p_1 = 3984.31$ psia, $p_2 = 3952.94$ psia, $p_3 = 3921.56$ psia, and $p_4 = 3890.19$ psia.

Substitution for $p_4 = 3890.19$ into Eq. (7.42) yields

$$q_{sc_4} = -22.169(p_4 - 3850) = -22.169(3890.19 - 3850) = -890.972\,\text{STB/D} \tag{7.44}$$

2. The well is located at the east boundary of gridblock 4. The equation for the well production rate is obtained using Eq. (6.34) for r_{eq}, Eq. (6.32) for G_w, and Eq. (6.1) for q_{sc_4}. Note that the well at the block boundary withdraws only half of its fluid production potential from gridblock 4, as shown in Fig. 7.6b (i.e., $F_4 = {}^1/_2$, configuration 2 in Chapter 6). The geometric factor of wellblock 4 is half of that for the whole well. Therefore,

$$r_{eq_4} = 0.1403684\left[(300)^2 + (350)^2\right]^{0.5} \exp\left[(300/350)\tan^{-1}(350/300)\right]$$
$$= 135.487\,\text{ft}$$

$$G^*_{w_4} = \frac{2\pi \times 0.001127 \times 270 \times 40}{\log_e[135.487/(3.5/12)] + 1.5} = 10.009\,\text{RB-cp/D-psi}$$

$$G_{w_4} = {}^1/_2 G^*_{w_4} = {}^1/_2(10.009) = 5.0045\,\text{RB-cp/D-psi}$$

and

$$q_{sc_4} = -\frac{5.0045}{1 \times 0.5}(p_4 - 3850)$$

or

$$q_{sc_4} = -10.009(p_4 - 3850)\ \text{STB/D} \tag{7.45}$$

Substitution of the transmissibility and Eq. (7.45) into Eq. (7.26) gives

$$28.4004(p_3 - p_4) + 0 + [-10.009(p_4 - 3850)] = 0$$

or after factorizing and ordering the unknowns,

$$28.4004p_3 - 38.4094p_4 = -38534.65 \tag{7.46}$$

The results of solving Eqs. (7.21), (7.23), (7.25), and (7.46) for the unknown pressures are $p_1 = 3988.17$ psia, $p_2 = 3964.50$ psia, $p_3 = 3940.83$ psia, and $p_4 = 3917.16$ psia. Substitution for $p_4 = 3917.16$ into Eq. (7.45) yields

$$q_{sc_4} = -10.009(p_4 - 3850) = -10.009(3917.16 - 3850) = -672.20\ \text{STB/D} \tag{7.47}$$

3. The well is treated as a boundary condition with the reservoir east boundary pressure equal to 3850 psia as shown in Fig. 7.6c. Therefore, the flow rate of the fictitious well can be estimated using Eq. (4.37c) for a constant pressure boundary condition, whose application gives

$$q_{sc_{b_E,4}} = \left[\beta_c \frac{k_x A_x}{\mu B(\Delta x/2)}\right]_4 [(p_{b_E} - p_4) - \gamma(Z_{b_E} - Z_4)]$$

$$= \left[0.001127 \times \frac{270 \times (350 \times 40)}{0.5 \times 1 \times (300/2)}\right][(3850 - p_4) - \gamma \times 0]$$

or

$$q_{sc_{b_E,4}} = 56.8008(3850 - p_4)\ \text{STB/D} \tag{7.48}$$

Substitution of Eq. (7.48) and the values into Eq. (7.26) gives

$$28.4004(p_3 - p_4) + 56.8008(3850 - p_4) + 0 = 0$$

or after factorizing and ordering the unknowns,

$$28.4004p_3 - 85.2012p_4 = -218683.08 \tag{7.49}$$

The results of solving Eqs. (7.21), (7.23), (7.25), and (7.49) for the unknown pressures are $p_1 = 3981.25$ psia, $p_2 = 3943.75$ psia, $p_3 = 3906.25$ psia, and $p_4 = 3868.75$ psia.

Substitution for $p_4 = 3868.75$ into Eq. (7.48) yields the rate of flow across the reservoir east boundary. Therefore,

$$q_{sc_4} = q_{sc_{b_E,4}} = 56.8008(3850 - p_4) = 56.8008(3850 - 3868.75)$$

or

$$q_{sc_4} = -1065.015\ \text{STB/D} \tag{7.50}$$

The predicted well production rates given by Eqs. (7.44), (7.47), and (7.50) demonstrate that even for 1-D flow, it is not appropriate to treat wells at reservoir ends as boundary conditions; that it is important to differentiate between physical wells and fictitious wells as discussed in Chapter 6; and that well performance and pressure distribution are affected by well location (within a block or on a no-flow reservoir boundary).

Example 7.7 A 2-D oil reservoir is described by four equal blocks as shown in Fig. 7.7a. The reservoir is horizontal and has $\phi=0.27$ and anisotropic permeability, $k_x=150$ md and $k_y=100$ md. The gridblock dimensions are $\Delta x=350$ ft, $\Delta y=250$ ft, and $h=30$ ft. The reservoir fluid properties are $B=B^\circ=1$ RB/STB and $\mu=3.5$ cP. The reservoir boundaries are subject to the conditions shown in Fig. 7.7b. A vertical well in gridblock 2 produces oil with a constant FBHP of 2000 psia, and another vertical well in gridblock 3 produces 600 STB/D of oil. The wells have a 3-in radius. Assuming that the reservoir rock and fluid are incompressible, find the pressure distribution in the reservoir. Find the rate of production of the well in gridblock 2 and the FBHP of the well in gridblock 3. Find oil flow rates across the reservoir boundaries. Perform a material balance check.

Solution

The gridblocks have the same dimensions and rock properties. Therefore, $T_{1,2}=T_{3,4}=T_x=\beta_c\frac{A_x k_x}{\mu B\Delta x}=0.001127\times\frac{(250\times 30)\times 150}{3.5\times 1\times 350}=1.0350$ STB/D-psi.

$$T_{1,3}=T_{2,4}=T_y=\beta_c\frac{A_y k_y}{\mu B\Delta y}=0.001127\times\frac{(350\times 30)\times 100}{3.5\times 1\times 250}$$
$$=1.3524\,\text{STB/D-psi}$$

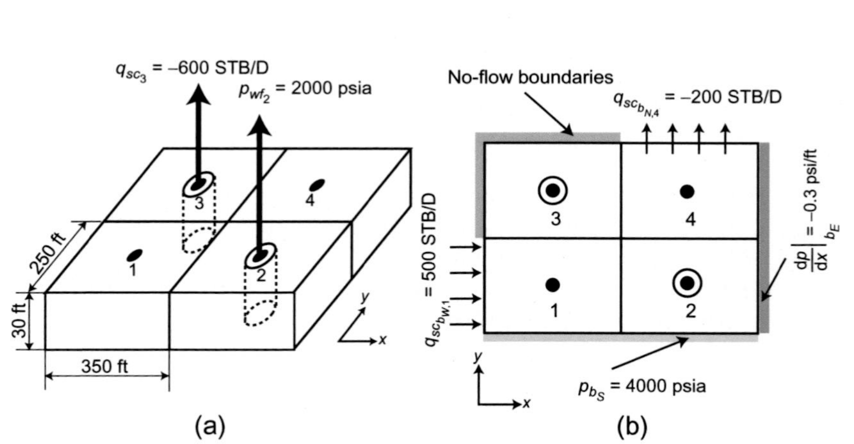

FIG. 7.7 Discretized 2-D reservoir in Example 7.7. (a) Gridblocks and wells and (b) boundary conditions.

There are two production wells in gridblock 2 and gridblock 3. For both wells, we use Eqs. (6.13), (6.14), and (6.12) to estimate k_H, r_{eq}, and G_w, respectively.

$$k_H = [150 \times 100]^{0.5} = 122.474\,\text{md}$$

$$r_{eq} = 0.28\frac{\left[(100/150)^{0.5}(350)^2 + (150/100)^{0.5}(250)^2\right]^{0.5}}{\left[(100/150)^{0.25} + (150/100)^{0.25}\right]} = 58.527\,\text{ft}$$

$$G_w = \frac{2\pi \times 0.001127 \times 122.474 \times 30}{\{\log_e[58.527/(3/12)] + 0\}} = 4.7688\,\text{RB-cp/D-psi}$$

For the well in gridblock 2, we apply Eq. (6.11), yielding

$$q_{sc_2} = -\frac{4.7688}{1 \times 3.5}(p_2 - 2000) = -1.3625(p_2 - 2000)\ \text{STB/D} \tag{7.51}$$

For the well in gridblock 3, $q_{sc_3} = -600$ STB/D. In addition, for the other gridblocks, $q_{sc_1} = q_{sc_4} = 0$.

Gridblock 1 falls on the reservoir south and west boundaries. The reservoir south boundary is kept at a constant pressure of 4000 psia. Therefore, the flow rate of the fictitious well can be estimated using Eq. (4.37c), whose application gives

$$q_{sc_{b_S,1}} = \left[\beta_c \frac{k_y A_y}{\mu B(\Delta y/2)}\right]_1 [(p_{b_S} - p_1) - \gamma(Z_{b_S} - Z_1)]$$
$$= \left[0.001127 \times \frac{100 \times (350 \times 30)}{3.5 \times 1 \times (250/2)}\right][(4000 - p_1) - \gamma \times 0]$$

or

$$q_{sc_{b_S,1}} = 2.7048(4000 - p_1)\ \text{STB/D} \tag{7.52}$$

The reservoir west boundary is a constant rate boundary supplying fluid to gridblock 1. Therefore, $q_{sc_{b_W,1}} = 500$ STB/D.

Gridblock 2 falls on the reservoir south and east boundaries. The reservoir south boundary is kept at a constant pressure of 4000 psia. Therefore, using Eq. (4.37c),

$$q_{sc_{b_S,2}} = \left[0.001127 \times \frac{100 \times (350 \times 30)}{3.5 \times 1 \times (250/2)}\right][(4000 - p_2) - \gamma \times 0]$$

or

$$q_{sc_{b_S,2}} = 2.7048(4000 - p_2)\ \text{STB/D} \tag{7.53}$$

The reservoir east boundary is a constant pressure gradient boundary. Therefore, using Eq. (4.24b),

$$q_{sc_{b_E,2}} = \left[\beta_c \frac{k_x A_x}{\mu B}\right]_2 \left[\frac{\partial p}{\partial x}\bigg|_{b_E} - \gamma \frac{\partial Z}{\partial x}\bigg|_{b_E}\right]$$
$$= 0.001127 \times \frac{150 \times (250 \times 30)}{3.5 \times 1}[-0.3 - \gamma \times 0]$$

or

$$q_{sc_{b_E,2}} = -108.675\ \mathrm{STB/D}$$

Gridblock 3 falls on the reservoir west and north boundaries. Both reservoir boundaries are no-flow boundaries. Therefore, $q_{sc_{b_W,3}} = q_{sc_{b_N,3}} = 0$.

Gridblock 4 falls on the reservoir east and north boundaries. The reservoir east boundary is a constant pressure gradient boundary. Therefore, using Eq. (4.24b),

$$q_{sc_{b_E,4}} = \left[\beta_c \frac{k_x A_x}{\mu B}\right]_4 \left[\frac{\partial p}{\partial x}\bigg|_{b_E} - \gamma \frac{\partial Z}{\partial x}\bigg|_{b_E}\right]$$
$$= 0.001127 \times \frac{150 \times (250 \times 30)}{3.5 \times 1}[-0.3 - \gamma \times 0]$$

or

$$q_{sc_{b_E,4}} = -108.675\ \mathrm{STB/D}$$

The reservoir north boundary is a constant rate boundary withdrawing fluid from gridblock 4. Therefore, $q_{sc_{b_N,4}} = -200$ STB/D.

The general flow equation for gridblock n in this 2-D horizontal reservoir can be obtained from Eq. (7.16a) by discarding the gravity term, yielding

$$\sum_{l \in \psi_n} T_{l,n}(p_l - p_n) + \sum_{l \in \xi_n} q_{sc_{l,n}} + q_{sc_n} = 0 \quad (7.19)$$

For gridblock 1, $n=1$, $\psi_1 = \{2,3\}$, and $\xi_1 = \{b_S, b_W\}$. Therefore, $\sum_{l \in \xi_1} q_{sc_{l,1}} = q_{sc_{b_S,1}} + q_{sc_{b_W,1}}$, and Eq. (7.19) becomes

$$T_{1,2}(p_2 - p_1) + T_{1,3}(p_3 - p_1) + q_{sc_{b_S,1}} + q_{sc_{b_W,1}} + q_{sc_1} = 0 \quad (7.54)$$

Upon substitution of the corresponding values, this equation becomes

$$1.0350(p_2 - p_1) + 1.3524(p_3 - p_1) + 2.7048(4000 - p_1) + 500 + 0 = 0$$

or after factorizing and ordering the unknowns,

$$-5.0922p_1 + 1.0350p_2 + 1.3524p_3 = -11319.20 \quad (7.55)$$

For gridblock 2, $n=2$, $\psi_2 = \{1,4\}$, and $\xi_2 = \{b_S, b_E\}$. Therefore, $\sum_{l \in \xi_2} q_{sc_{l,2}} = q_{sc_{b_S,2}} + q_{sc_{b_E,2}}$, and Eq. (7.19) becomes

$$T_{1,2}(p_1 - p_2) + T_{2,4}(p_4 - p_2) + q_{sc_{b_S,2}} + q_{sc_{b_E,2}} + q_{sc_2} = 0 \quad (7.56)$$

Upon substitution of the corresponding values, this equation becomes

$$1.0350(p_1 - p_2) + 1.3524(p_4 - p_2) + 2.7048(4000 - p_2) \\ -108.675 - 1.3625(p_2 - 2000) = 0$$

After factorizing and ordering the unknowns, the equation becomes

$$1.0350p_1 - 6.4547p_2 + 1.3524p_4 = -13435.554 \quad (7.57)$$

For gridblock 3, $n=3$, $\psi_3=\{1,4\}$, and $\xi_3=\{b_W, b_N\}$. Therefore, $\sum_{l\in\xi_3} q_{sc_{l,3}} = q_{sc_{b_W,3}} + q_{sc_{b_N,3}}$, and Eq. (7.19) becomes

$$T_{1,3}(p_1 - p_3) + T_{3,4}(p_4 - p_3) + q_{sc_{b_W,3}} + q_{sc_{b_N,3}} + q_{sc_3} = 0 \quad (7.58)$$

Upon substitution of the corresponding values, this equation becomes

$$1.3524(p_1 - p_3) + 1.0350(p_4 - p_3) + 0 + 0 - 600 = 0$$

After factorizing and ordering the unknowns, the equation becomes

$$1.3524p_1 - 2.3874p_3 + 1.0350p_4 = 600 \quad (7.59)$$

For gridblock 4, $n=4$, $\psi_4=\{2,3\}$, and $\xi_4=\{b_E, b_N\}$. Therefore, $\sum_{l\in\xi_4} q_{sc_{l,4}} = q_{sc_{b_E,4}} + q_{sc_{b_N,4}}$, and Eq. (7.19) becomes

$$T_{2,4}(p_2 - p_4) + T_{3,4}(p_3 - p_4) + q_{sc_{b_E,4}} + q_{sc_{b_N,4}} + q_{sc_4} = 0 \quad (7.60)$$

Upon substitution of the corresponding values, this equation becomes

$$1.3524(p_2 - p_4) + 1.0350(p_3 - p_4) - 108.675 - 200 + 0 = 0$$

After factorizing and ordering the unknowns, the equation becomes

$$1.3524p_2 + 1.0350p_3 - 2.3874p_4 = 308.675 \quad (7.61)$$

The results of solving Eqs. (7.55), (7.57), (7.59), and (7.61) for the unknown pressures are $p_1=3772.36$ psia, $p_2=3354.20$ psia, $p_3=3267.39$ psia, and $p_4=3187.27$ psia. The flow rates across the reservoir boundaries are estimated by substituting for the pressures in Eqs. (7.52) and (7.53), yielding

$$q_{sc_{b_S,1}} = 2.7048(4000 - p_1) = 2.7048(4000 - 3772.36) = 615.721 \text{ STB/D}$$

and

$$q_{sc_{b_S,2}} = 2.7048(4000 - p_2) = 2.7048(4000 - 3354.20) = 1746.787 \text{ STB/D}$$

The production rate for the well in gridblock 2 is obtained by substituting for gridblock pressure in Eq. (7.51), which gives

$$q_{sc_2} = -1.3625(p_2 - 2000) = -1.3625(3354.20 - 2000) = -1845.12 \text{ STB/D}$$

The FBHP of the well in gridblock 3 is estimated using Eq. (6.11), yielding

$$-600 = -\frac{4.7688}{1 \times 3.5}(3267.36 - p_{wf_3})$$

from which

$$p_{wf_3} = 2827.00\,\text{psia}$$

The material balance for an incompressible fluid and rock system is checked by substituting values for the well production rates and fictitious well rates on the LHS of Eq. (7.17a), resulting in

$$\sum_{n=1}^{N}\left(q_{sc_n} + \sum_{l\in\xi_n} q_{sc_{l,n}}\right) = \left[\begin{array}{l}\left(q_{sc_1} + q_{sc_{b_S,1}} + q_{sc_{b_W,1}}\right) + \left(q_{sc_2} + q_{sc_{b_S,2}} + q_{sc_{b_E,2}}\right)\\ +\left(q_{sc_3} + q_{sc_{b_W,3}} + q_{sc_{b_N,3}}\right) + \left(q_{sc_4} + q_{sc_{b_E,4}} + q_{sc_{b_N,4}}\right)\end{array}\right]$$

$$= \left[\begin{array}{l}(0 + 615.721 + 500) + (-1845.12 + 1746.787 - 108.675)\\ +(-600 + 0 + 0) + (0 - 108.675 - 200)\end{array}\right]$$

$$= 0.038$$

Therefore, the material balance check is satisfied, and a small error of 0.038 STB/D is observed because of rounding off during calculations.

Example 7.8 Find the pressure distribution in the 2-D horizontal reservoir shown in Fig. 7.8. The reservoir rock properties are homogeneous and isotropic: $\phi = 0.19$ and $k_x = k_y = 200$ md. Gridblocks have $\Delta x = \Delta y = 400$ ft and $h = 50$ ft, and fluid properties are $B \cong B^\circ = 1$ RB/STB, $\rho = 55$ lbm/ft^3, and $\mu = 3$ cP. The reservoir has no-flow boundaries, and there are three wells in this reservoir. The well in gridblock 7 produces fluid at a constant rate of 1000 STB/D. Each of the two wells in gridblocks 2 and 6 injects fluid with a constant FBHP of 3500 psia. The wells have a diameter of 6 in. Assume that reservoir rock and fluid are incompressible.

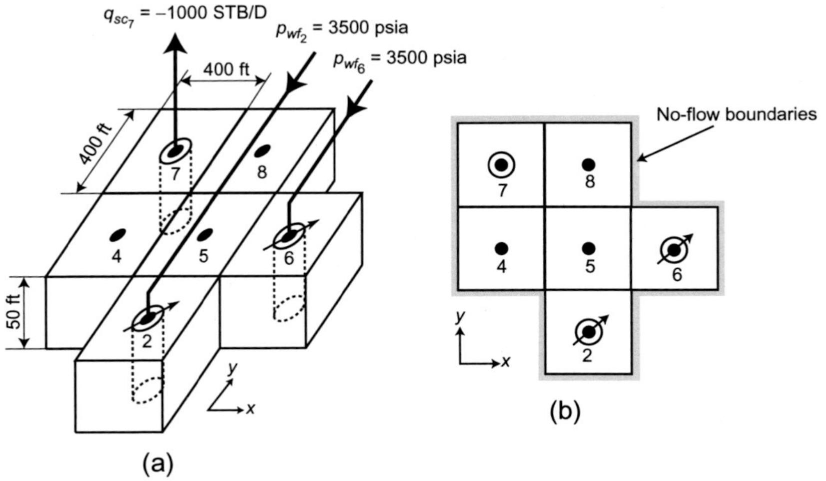

FIG. 7.8 Discretized 2-D reservoir in Example 7.8. (a) Gridblocks and wells and (b) boundary conditions.

Solution

The gridblocks have the same dimensions and rock properties. Therefore, $T_{4,5}=T_{5,6}=T_{7,8}=T_x=T_{2,5}=T_{4,7}=T_{5,8}=T_y=T$ where

$$T=\beta_c\frac{A_x k_x}{\mu B\Delta x}=0.001127\times\frac{(400\times 50)\times 200}{3\times 1\times 400}=3.7567\,\text{STB/D-psi}$$

For each of the three wells, we use Eqs. (6.13), (6.16), (6.12), and (6.1) for k_H, r_{eq}, G_w, and q_{sc}, respectively:

$$k_H=200\,\text{md}$$

$$r_{eq}=0.198\times 400=79.200\,\text{ft}$$

and

$$G_w=\frac{2\pi\times 0.001127\times 200\times 50}{\{\log_e[79.200/(3/12)]+0\}}=12.2974\,\text{RB-cp/D-psi}$$

The application of Eq. (6.1) gives

$$-1000=-\frac{12.2974}{1\times 3}\left(p_7-p_{wf_7}\right)$$

or

for wellblock 7,
$$p_{wf_7}=p_7-243.954\,\text{psia} \tag{7.62}$$

$$q_{sc_2}=-\frac{12.2974}{1\times 3}(p_2-3500)$$

or

for wellblock 2,
$$q_{sc_2}=-4.0991(p_2-3500)\,\text{STB/D} \tag{7.63}$$

and for wellblock 6, $q_{sc_6}=-4.0991(p_6-3500)\,\text{STB/D}$. (7.64)

In addition, $q_{sc_4}=q_{sc_5}=q_{sc_8}=0$.

For no-flow boundary conditions and interior blocks, $\sum_{l\in\xi_n} q_{sc_{l,n}}=0$ for $n=2, 4, 5, 6, 7, 8$.

The general flow equation for gridblock n in this 2-D horizontal reservoir can be obtained from Eq. (7.16a) by discarding the gravity term, yielding

$$\sum_{l\in\psi_n} T_{l,n}(p_l-p_n)+\sum_{l\in\xi_n} q_{sc_{l,n}}+q_{sc_n}=0 \tag{7.19}$$

For no-flow boundaries $\left(\sum_{l\in\xi_n} q_{sc_{l,n}}=0\right)$, Eq. (7.19) reduces to

$$\sum_{l\in\psi_n} T_{l,n}(p_l-p_n)+q_{sc_n}=0 \tag{7.65}$$

For gridblock 2, $n=2$, and $\psi_2=\{5\}$. Therefore, applying Eq. (7.65) gives

$$T_{2,5}(p_5-p_2)+q_{sc_2}=0 \tag{7.66}$$

Upon substitution of the corresponding values, this equation becomes

$$3.7567(p_5-p_2)-4.0991(p_2-3500)=0$$

After factorizing and ordering the unknowns, the equation becomes

$$-7.8558p_2+3.7567p_5=-14346.97 \tag{7.67}$$

For gridblock 4, $n=4$, and $\psi_4=\{5,7\}$. Therefore, applying Eq. (7.65) gives

$$T_{4,5}(p_5-p_4)+T_{4,7}(p_7-p_4)+q_{sc_4}=0 \tag{7.68}$$

Upon substitution of the corresponding values, this equation becomes

$$3.7567(p_5-p_4)+3.7567(p_7-p_4)+0=0$$

or after factorizing and ordering the unknowns,

$$-7.5134p_4+3.7567p_5+3.7567p_7=0 \tag{7.69}$$

For gridblock 5, $n=5$, and $\psi_5=\{2,4,6,8\}$. Therefore, applying Eq. (7.65) gives

$$T_{2,5}(p_2-p_5)+T_{4,5}(p_4-p_5)+T_{5,6}(p_6-p_5)+T_{5,8}(p_8-p_5)+q_{sc_5}=0 \tag{7.70}$$

Upon substitution of the corresponding values, this equation becomes

$$3.7567(p_2-p_5)+3.7567(p_4-p_5)+3.7567(p_6-p_5)+3.7567(p_8-p_5)+0=0$$

After factorizing and ordering the unknowns, the equation becomes

$$3.7567p_2+3.7567p_4-15.0268p_5+3.7567p_6+3.7567p_8=0 \tag{7.71}$$

For gridblock 6, $n=6$, and $\psi_6=\{5\}$. Therefore, applying Eq. (7.65) gives

$$T_{5,6}(p_5-p_6)+q_{sc_6}=0 \tag{7.72}$$

Upon substitution of the corresponding values, this equation becomes

$$3.7567(p_5-p_6)-4.0991(p_6-3500)=0$$

After factorizing and ordering the unknowns, the equation becomes

$$3.7567p_5-7.8558p_6=-14346.97 \tag{7.73}$$

For gridblock 7, $n=7$, and $\psi_7=\{4,8\}$. Therefore, applying Eq. (7.65) gives

$$T_{4,7}(p_4-p_7)+T_{7,8}(p_8-p_7)+q_{sc_7}=0 \tag{7.74}$$

Upon substitution of the corresponding values, this equation becomes

$$3.7567(p_4-p_7)+3.7567(p_8-p_7)-1000=0$$

After factorizing and ordering the unknowns, the equation becomes

$$3.7567p_4-7.5134p_7+3.7567p_8=1000 \tag{7.75}$$

For gridblock 8, $n=8$, and $\psi_8=\{5,7\}$. Therefore, applying Eq. (7.65) gives

$$T_{5,8}(p_5-p_8)+T_{7,8}(p_7-p_8)+q_{sc_8}=0 \tag{7.76}$$

Upon substitution of the corresponding values, this equation becomes

$$3.7567(p_5-p_8)+3.7567(p_7-p_8)+0=0$$

or after factorizing and ordering the unknowns,

$$3.7567p_5+3.7567p_7-7.5134p_8=0 \tag{7.77}$$

The results of solving Eqs. (7.67), (7.69), (7.71), (7.73), (7.75), and (7.77) for the unknown pressures are $p_2=3378.02$ psia, $p_4=3111.83$ psia, $p_5=3244.93$ psia, $p_6=3378.02$ psia, $p_7=2978.73$ psia, and $p_8=3111.83$ psia. Note the symmetry about the vertical plane that passes through the centers of gridblocks 5 and 7 (see Section 4.6). We could have made use of this symmetry and, accordingly, set $p_2=p_6$ and $p_4=p_8$; write the flow equations for gridblocks 2, 4, 5, and 7; and finally solve the resulting four equations for the unknowns p_2, p_4, p_5, and p_7.

Next, the production rate for the wells in gridblocks 2 and 6 are estimated by substituting for gridblock pressures in Eqs. (7.63) and (7.64), yielding

$$q_{sc_2}=-4.0991(p_2-3500)=-4.0991(3378.02-3500)=500.008\ \text{STB/D}$$

and

$$q_{sc_6}=-4.0991(p_6-3500)=-4.0991(3378.02-3500)=500.008\ \text{STB/D}$$

The FBHP of the well in gridblock 7 is estimated using Eq. (7.62), which gives

$$p_{wf_7}=p_7-243.954=2978.73-243.954=2734.8\ \text{psia}$$

The material balance for an incompressible fluid and rock system is checked by substituting the values for the well production rates and fictitious well rates on the LHS of Eq. (7.17a). For no-flow boundaries, the LHS of Eq. (7.17b) reduces to

$$\sum_{n=2}^{8}\left(q_{sc_n}+\sum_{l\in\xi_n}q_{sc_{l,n}}\right)=\sum_{n=2}^{8}(q_{sc_n}+0)=\sum_{n=2}^{8}q_{sc_n}$$
$$=500.008+0+0+500.008-1000+0=0.016$$

Therefore, the material balance check is satisfied, and a small error of 0.016 STB/D is observed because of rounding off during calculations.

7.3.2 Slightly compressible fluid flow equation

The density, FVF, and viscosity of slightly compressible fluids at reservoir temperature are functions of pressure. Such dependence; however, is weak. In this context, the FVF, viscosity, and density that appear on the LHS of a flow equation (Eq. 7.12) can be assumed constant. The accumulation term can be expressed in terms of pressure changes over a time step by substituting for B and ϕ (using Eqs. 7.6 and 7.11) into the RHS of Eq. (7.12). The resulting accumulation term is

$$\frac{V_{b_n}}{\alpha_c\Delta t}\left[\left(\frac{\phi}{B}\right)_n^{n+1}-\left(\frac{\phi}{B}\right)_n^{n}\right]\cong\frac{V_{b_n}\phi_n^{\circ}}{\alpha_c\Delta t B^{\circ}}(c+c_{\phi})\left[p_n^{n+1}-p_n^n\right] \tag{7.78}$$

Note that Eq. (7.78) reduces to Eq. (7.14) for an incompressible fluid where $c=0$. Substitution of Eq. (7.78) into Eq. (7.12) yields the flow equation for slightly compressible fluids:

$$\sum_{l\in\psi_n}T_{l,n}^m\left[(p_l^m-p_n^m)-\gamma_{l,n}^m(Z_l-Z_n)\right]+\sum_{l\in\xi_n}q_{sc_{l,n}}^m+q_{sc_n}^m$$
$$=\frac{V_{b_n}\phi_n^{\circ}(c+c_{\phi})}{\alpha_c B^{\circ}\Delta t}\left[p_n^{n+1}-p_n^n\right] \tag{7.79}$$

7.3.2.1 Formulations of the slightly compressible fluid flow equation

The time level m in Eq. (7.79) is approximated in reservoir simulation in one of three ways (t^n, t^{n+1}, or $t^{n+1/2}$) as mentioned in Chapter 2. The resulting equation is commonly known as the explicit formulation of the flow equation (or the forward-central-difference equation), the implicit formulation of the flow equation (or the backward-central-difference equation), and the Crank-Nicolson formulation of the flow equation (or the second-order-central-difference equation). The terminology in the parentheses above is usually used in the mathematical approach to reservoir simulation. It originates from the way the partial differential equation (PDE) describing the problem is approximated to give the finite-difference equation (or flow equation in algebraic form). The forward, backward, or second-order descriptor refers to the approximation of the time derivative (or accumulation) term with reference to the time level at which the PDE is written. The central-difference descriptor refers to using a second-order approximation of (interblock) flow terms in the PDE.

Explicit formulation of the flow equation

The explicit formulation of the flow equation can be obtained from Eq. (7.79) if the argument F^m (defined in Section 2.6.3) is dated at old time level t^n; that is, $t^m \cong t^n$, and as a result, $F^m \cong F^n$. Therefore, Eq. (7.79) becomes

$$\sum_{l\in\psi_n} T_{l,n}^n\left[\left(p_l^n - p_n^n\right) - \gamma_{l,n}^n (Z_l - Z_n)\right] + \sum_{l\in\xi_n} q_{sc_{l,n}}^n + q_{sc_n}^n \cong \frac{V_{b_n}\phi_n^\circ (c + c_\phi)}{\alpha_c B^\circ \Delta t}\left[p_n^{n+1} - p_n^n\right] \tag{7.80a}$$

or

$$\sum_{l\in\psi_{i,j,k}} T_{l,(i,j,k)}^n\left[\left(p_l^n - p_{i,j,k}^n\right) - \gamma_{l,(i,j,k)}^n\left(Z_l - Z_{i,j,k}\right)\right] + \sum_{l\in\xi_{i,j,k}} q_{sc_{l,(i,j,k)}}^n + q_{sc_{i,j,k}}^n \cong \frac{V_{b_{i,j,k}}\phi_{i,j,k}^\circ (c + c_\phi)}{\alpha_c B^\circ \Delta t}\left[p_{i,j,k}^{n+1} - p_{i,j,k}^n\right] \tag{7.80b}$$

Inspection of Eq. (7.80a) reveals that it has one unknown pressure, namely, p_n^{n+1}, and that all the neighboring blocks (nodes) have known pressures at the old time level. Therefore, the pressure solution at time level $n+1$ is obtained by solving Eq. (7.80a) for p_n^{n+1} for block n independent of the flow equations of the other blocks. Stability analysis performed in the mathematical approach (Ertekin et al., 2001) concludes that Eq. (7.80) is conditionally stable; that is, the use of Eq. (7.80) gives numerically stable pressure solutions for small time steps only (see Fig. 7.9). In other words, the allowable time step is quite small, and the amount of computational effort required to obtain the solution to practical problems at a given time level is prohibitive. Consequently, this formulation is not used in reservoir simulation. The explicit formulation is only of academic interest to mathematicians, and it is not pursued further in this book.

Implicit formulation of the flow equation

The implicit formulation of the flow equation can be obtained from Eq. (7.79) if the argument F^m (defined in Section 2.6.3) is dated at new time level t^{n+1}; that is, $t^m \cong t^{n+1}$, and as a result, $F^m \cong F^{n+1}$. Therefore, Eq. (7.79) becomes

$$\sum_{l\in\psi_n} T_{l,n}^{n+1}\left[\left(p_l^{n+1} - p_n^{n+1}\right) - \gamma_{l,n}^n (Z_l - Z_n)\right] + \sum_{l\in\xi_n} q_{sc_{l,n}}^{n+1} + q_{sc_n}^{n+1} \cong \frac{V_{b_n}\phi_n^\circ (c + c_\phi)}{\alpha_c B^\circ \Delta t}\left[p_n^{n+1} - p_n^n\right] \tag{7.81a}$$

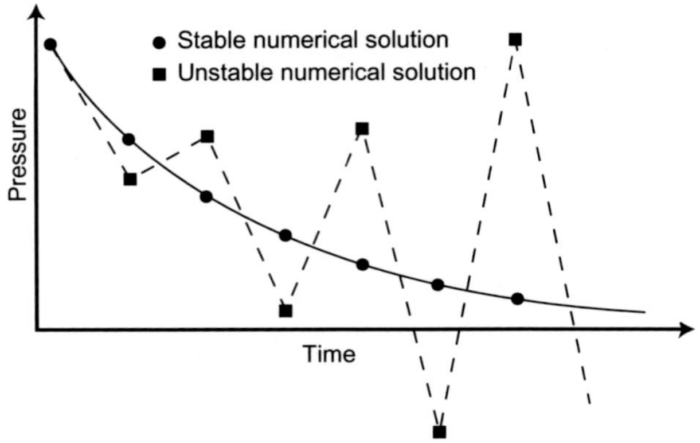

FIG. 7.9 Pressure behavior for a gridblock.

or

$$\sum_{l\in\psi_{i,j,k}} T^{n+1}_{l,(i,j,k)}\left[\left(p^{n+1}_{l}-p^{n+1}_{i,j,k}\right)-\gamma^{n}_{l,(i,j,k)}\left(Z_l-Z_{i,j,k}\right)\right] + \sum_{l\in\xi_{i,j,k}} q^{n+1}_{sc_{l,(i,j,k)}} + q^{n+1}_{sc_{i,j,k}} \cong \frac{V_{b_{i,j,k}}\phi^{\circ}_{i,j,k}\left(c+c_{\phi}\right)}{\alpha_c B^{\circ}\Delta t}\left[p^{n+1}_{i,j,k}-p^{n}_{i,j,k}\right] \tag{7.81b}$$

In this equation, dating fluid gravity at old time level n instead of new time level $n+1$ does not introduce any noticeable errors (Coats et al., 1974). This approximation will be used throughout this book. Inspection of Eq. (7.81a) reveals that block n and all its neighboring blocks (nodes) have unknown pressures at the current time level. Therefore, the pressure solution at current time level $n+1$ is obtained by solving simultaneously the system of equations, which result from writing Eq. (7.81a) for all blocks (nodes) in the reservoir. Stability analysis performed in the mathematical approach (Ertekin et al., 2001) concludes that Eq. (7.81) is unconditionally stable because of the linearity of this equation; that is, Eq. (7.81) gives numerically stable pressure solutions with no limits on the allowable time step. However, there has to be a limit on the time step to obtain an accurate solution, but this is not a stability consideration. The property of unconditional stability of the implicit formulation method makes it attractive in spite of the extra computational effort required per time step. The solution at a given simulation time can be obtained with much less computational effort by taking large time steps. The time step is limited only by accuracy requirements. Consequently, the implicit formulation method is commonly used in reservoir simulation.

Crank-Nicolson formulation of the flow equation

The Crank-Nicolson formulation of the flow equation can be obtained from Eq. (7.79) if the argument F^m (defined in Section 2.6.3) is dated at time $t^{n+1/2}$. In the mathematical approach, this time level was chosen to make the RHS of Eq. (7.79) a second-order approximation in time. In the engineering approach; however, the argument F^m can be approximated as $F^m \cong F^{n+1/2} = {}^1/_2(F^n + F^{n+1})$. Therefore, Eq. (7.79) becomes

$$
\begin{aligned}
&{}^1/_2 \sum_{l\in\psi_n} T_{l,n}^{n}\left[\left(p_l^n - p_n^n\right) - \gamma_{l,n}^n (Z_l - Z_n)\right] \\
&+{}^1/_2 \sum_{l\in\psi_n} T_{l,n}^{n+1}\left[\left(p_l^{n+1} - p_n^{n+1}\right) - \gamma_{l,n}^n (Z_l - Z_n)\right] + {}^1/_2\left(\sum_{l\in\xi_n} q_{sc_{l,n}}^n + \sum_{l\in\xi_n} q_{sc_{l,n}}^{n+1}\right) \\
&+{}^1/_2\left(q_{sc_n}^n + q_{sc_n}^{n+1}\right) \cong \frac{V_{b_n}\phi_n^{\circ}(c + c_\phi)}{\alpha_c B^{\circ} \Delta t}\left[p_n^{n+1} - p_n^n\right]
\end{aligned}
\tag{7.82a}
$$

Eq. (7.82a) can be rewritten in the form of Eq. (7.81a) as

$$
\begin{aligned}
&\sum_{l\in\psi_n} T_{l,n}^{n+1}\left[\left(p_l^{n+1} - p_n^{n+1}\right) - \gamma_{l,n}^n (Z_l - Z_n)\right] \\
&+\sum_{l\in\xi_n} q_{sc_{l,n}}^{n+1} + q_{sc_n}^{n+1} \cong \frac{V_{b_n}\phi_n^{\circ}(c + c_\phi)}{\alpha_c B^{\circ} (\Delta t/2)}\left[p_n^{n+1} - p_n^n\right] \\
&-\left\{\sum_{l\in\psi_n} T_{l,n}^{n}\left[\left(p_l^n - p_n^n\right) - \gamma_{l,n}^n (Z_l - Z_n)\right] + \sum_{l\in\xi_n} q_{sc_{l,n}}^n + q_{sc_n}^n\right\}
\end{aligned}
\tag{7.82b}
$$

Like Eq. (7.81), the pressure solution at current time level $n+1$ is obtained by solving simultaneously the system of equations, which result from writing Eq. (7.82b) for all blocks (nodes) in the reservoir. The Crank-Nicolson formulation is unconditionally stable, and the time step is limited only by accuracy requirements. The advantage of the Crank-Nicolson formulation over the implicit formulation is a more accurate solution for the same time step or larger time steps for the same accuracy (Hoffman, 1992). This gain in accuracy is obtained at no extra computational cost because the terms in the braces { } on the RHS of Eq. (7.82b) are calculated at the end of the previous time step. The drawback of the Crank-Nicolson formulation is that the numerical solution may exhibit overshoot and oscillations for some problems. Such oscillations are not due to instability but rather to an inherent feature of the Crank-Nicolson formulation (Hoffman, 1992). This formulation method finds infrequent use in reservoir simulation perhaps because of this drawback and the problems that may arise because of specifying a pressure gradient at reservoir boundaries (Keast and Mitchell, 1966).

7.3.2.2 Advancing the pressure solution in time

The pressure distribution in a slightly compressible flow problem changes with time. This means that the flow problem must be solved in its unsteady-state form. At time $t_0 = 0$, all reservoir block pressures (p_n^0, $n = 1, 2, 3 \ldots N$) must be specified. Initially, a fluid in the reservoir is in hydrodynamic equilibrium.

Therefore, it is sufficient to specify the pressure at one point in the reservoir, and the initial pressure of any block in the reservoir can be estimated from hydrostatic pressure considerations. Then, the procedure entails finding the pressure solution at discrete times (t_1, t_2, t_3, t_4, …etc.) by marching the latest value of pressure in time using time steps (Δt_1, Δt_2, Δt_3, Δt_4, …etc.). The pressure solution is advanced from initial conditions at $t_0=0$ (time level n) to $t_1=t_0+\Delta t_1$ (time level $n+1$). The solution then is advanced in time from t_1 (time level n) to $t_2=t_1+\Delta t_2$ (time level $n+1$), from t_2 to $t_3=t_2+\Delta t_3$, and from t_3 to $t_4=t_3+\Delta t_4$, and the process is repeated as many times as necessary until the desired simulation time is reached. To obtain the pressure solution at time level $n+1$, we assign the pressure solution just obtained as pressures at time level n, write the flow equation for every block (node) in the discretized reservoir, and solve the resulting set of linear equations for the set of unknown pressures.

For the explicit formulation, the calculation procedure within each time step follows:

1. Calculate the interblock transmissibilities and the coefficient of $(p_n^{n+1}-p_n^n)$, and define the pressure at the old time level for all reservoir blocks.
2. Estimate the production rate at time level n for each wellblock in the reservoir as discussed in Chapter 6.
3. Estimate the flow rate at time level n for each fictitious well in the reservoir as discussed in Chapter 4 (or Chapter 5); that is, estimate the flow rates resulting from boundary conditions.
4. For every gridblock (or gridpoint) in the reservoir, define the set of existing reservoir neighboring blocks (ψ_n) and the set of reservoir boundaries that are block boundaries (ξ_n), expand the summation terms in the flow equation (Eq. 7.80 in this case), and substitute for the wellblock production rate obtained in (2) and fictitious well rates obtained in (3).
5. Solve the flow equation of each reservoir block (node) for its unknown pressure independent of the other flow equations because each flow equation in the explicit formulation has only one unknown pressure.
6. Perform incremental and cumulative material balance checks.

For the implicit and the Crank-Nicolson formulations, the calculation procedure within each time step follows:

1. Calculate the interblock transmissibilities and the coefficient of $(p_n^{n+1}-p_n^n)$, and define pressure at the old time level for all reservoir blocks.
2. Estimate the production rate (or write the production rate equation) at time level $n+1$ for each wellblock in the reservoir as discussed in Chapter 6.
3. Estimate the flow rate (or write the flow rate equation) at time level $n+1$ for each fictitious well in the reservoir as discussed in Chapter 4 (or Chapter 5); that is, estimate the flow rates resulting from boundary conditions.
4. For every gridblock (or gridpoint) in the reservoir, define the set of existing reservoir neighboring blocks (ψ_n) and the set of reservoir boundaries that are

block boundaries (ξ_n), expand the summation terms in the flow equation (Eq. 7.81 or 7.82b), and substitute for the wellblock production rate obtained in (2) and fictitious well rates obtained in (3).

5. Factorize, order, and place the unknown pressures (at time level $n+1$) on the LHS and place known quantities on the RHS of each flow equaton.
6. Solve the resulting set of equations for the set of unknown pressures (at time level $n+1$) using a linear equation solver such as those presented in Chapter 9.
7. Estimate the wellblock production rates and fictitious well rates at time level $n+1$ if necessary by substituting the values of pressures obtained in (6) into the flow rate equations obtained in (2) and (3).
8. Perform incremental and cumulative material balance checks.

7.3.2.3 Material balance check for a slightly compressible fluid flow problem

For slightly compressible fluid flow problems, there are usually two material balance checks. The first is called the incremental material balance check (I_{MB}) and is used to check the material balance over a time step. The second is called the cumulative material balance check (C_{MB}) and is used to check the material balance from the initial conditions up to the current time step. The latter check tends to smooth errors that occur over all the previous time steps; therefore, it provides a less accurate check than the first check. In reservoir simulation, a material balance check is defined as the ratio of the accumulated mass to the net mass entering and leaving reservoir boundaries, including wells. If reservoir blocks are identified using block order and the implicit formulation is used, the equations for material balance checks are

$$I_{MB} = \frac{\sum_{n=1}^{N} \frac{V_{b_n}}{\alpha_c \Delta t} \left[\left(\frac{\phi}{B} \right)_n^{n+1} - \left(\frac{\phi}{B} \right)_n^{n} \right]}{\sum_{n=1}^{N} \left(q_{sc_n}^{n+1} + \sum_{l \in \xi_n} q_{sc_{l,n}}^{n+1} \right)} \tag{7.83}$$

and

$$C_{MB} = \frac{\sum_{n=1}^{N} \frac{V_{b_n}}{\alpha_c} \left[\left(\frac{\phi}{B} \right)_n^{n+1} - \left(\frac{\phi}{B} \right)_n^{0} \right]}{\sum_{m=1}^{n+1} \Delta t_m \sum_{n=1}^{N} \left(q_{sc_n}^{m} + \sum_{l \in \xi_n} q_{sc_{l,n}}^{m} \right)} \tag{7.84}$$

where N is the total number of blocks in the reservoir, subscript n is block number, and superscript n is old time level. In Eqs. (7.83) and (7.84), the production (or injection) rate is set to zero for any reservoir block that is not penetrated by a well. In addition, Eq. (7.11) defines rock porosity, and Eq. (7.6) defines FVF for slightly compressible fluid. Alternatively, we can substitute Eq. (7.78) for a slightly compressible fluid and porosity into Eqs. (7.83) and (7.84). The material balance checks become

$$I_{MB}=\frac{\sum_{n=1}^{N}\frac{V_{b_n}\phi_n^{\circ}(c+c_\phi)}{\alpha_c B^{\circ}\Delta t}\left[p_n^{n+1}-p_n^n\right]}{\sum_{n=1}^{N}\left(q_{sc_n}^{n+1}+\sum_{l\in\xi_n}q_{sc_{l,n}}^{n+1}\right)} \tag{7.85}$$

and

$$C_{MB}=\frac{\sum_{n=1}^{N}\frac{V_{b_n}\phi_n^{\circ}(c+c_\phi)}{\alpha_c B^{\circ}}\left[p_n^{n+1}-p_n^0\right]}{\sum_{m=1}^{n+1}\Delta t_m\sum_{n=1}^{N}\left(q_{sc_n}^{m}+\sum_{l\in\xi_n}q_{sc_{l,n}}^{m}\right)} \tag{7.86}$$

The second term in the parentheses in the denominator of Eqs. (7.85) and (7.86) takes care of fluid flow across reservoir boundaries. The numerical value of both I_{MB} and C_{MB} checks should be close to one. A value of 0.995–1.005 or better is acceptable for solving problems using handheld calculators, compared with 0.999995–1.000005 used in numerical simulators.

The incremental material balance check at time level $n+1$, which is expressed by Eq. (7.85), can be derived by writing Eq. (7.81a) for each block in the system ($n=1, 2, 3\ldots N$) and then summing up all n equations. The resulting equation is

$$\sum_{n=1}^{N}\left\{\sum_{l\in\psi_n}T_{l,n}^{n+1}\left[\left(p_l^{n+1}-p_n^{n+1}\right)-\gamma_{l,n}^{n}(Z_l-Z_n)\right]\right\}+\sum_{n=1}^{N}\left(\sum_{l\in\xi_n}q_{sc_{l,n}}^{n+1}+q_{sc_n}^{n+1}\right)$$
$$=\sum_{n=1}^{N}\frac{V_{b_n}\phi_n^{\circ}(c+c_\phi)}{\alpha_c B^{\circ}\Delta t}\left[p_n^{n+1}-p_n^n\right] \tag{7.87}$$

The sum of all interblock terms in the reservoir, which are expressed by the first term on the LHS of Eq. (7.87), adds up to zero, while the second term on the LHS represents the algebraic sum of all production rates through wells $\left(\sum_{n=1}^{N}q_{sc_n}^{n+1}\right)$ and those across reservoir boundaries $\left(\sum_{n=1}^{N}\sum_{l\in\xi_n}q_{sc_{l,n}}^{n+1}\right)$. The RHS of this equation represents the sum of the accumulation terms in all blocks in the reservoir. Therefore, Eq. (7.87) becomes

$$\sum_{n=1}^{N}\left(\sum_{l\in\xi_n}q_{sc_{l,n}}^{n+1}+q_{sc_n}^{n+1}\right)=\sum_{n=1}^{N}\frac{V_{b_n}\phi_n^{\circ}(c+c_\phi)}{\alpha_c B^{\circ}\Delta t}\left[p_n^{n+1}-p_n^n\right] \tag{7.88}$$

Dividing this equation by the term on the LHS yields

$$1 = \frac{\sum_{n=1}^{N} \frac{V_{b_n} \phi_n^{\circ}(c + c_\phi)}{\alpha_c B^{\circ} \Delta t} \left[p_n^{n+1} - p_n^n\right]}{\sum_{n=1}^{N} \left(q_{sc_n}^{n+1} + \sum_{l \in \xi_n} q_{sc_{l,n}}^{n+1} \right)} \tag{7.89}$$

Comparing Eqs. (7.85) and (7.89) dictates that I_{MB} must be equal or close to 1 to preserve the material balance. The equation for the cumulative material balance check is obtained by writing Eq. (7.88) for all time steps ($m=1, 2, 3\ldots n+1$), observing that $\Delta t_m = t^{m+1} - t^m$ replaces Δt, and summing up all resulting equations. It should be mentioned that for the explicit formulation, the denominator of Eq. (7.89) is replaced with $\sum_{n=1}^{N} \left(q_{sc_n}^n + \sum_{l \in \xi_n} q_{sc_{l,n}}^n \right)$. For the Crank-Nicolson formulation, the denominator of Eq. (7.89) becomes $\sum_{n=1}^{N} \left[{}^1/_2 \left(q_{sc_n}^{n+1} + q_{sc_n}^n \right) + \sum_{l \in \xi_n} {}^1/_2 \left(q_{sc_{l,n}}^{n+1} + q_{sc_{l,n}}^n \right) \right]$

Both Examples 7.9 and 7.10 demonstrate the application of the solution algorithm presented in this section to advance the pressure solution from one time step to another. The reservoir is discretized using a block-centered grid in Example 7.9, whereas a point-distributed grid is used in Example 7.10. Example 7.11 presents the simulation of a heterogeneous 1-D reservoir. Example 7.12 demonstrates the advancement of the pressure solution in time in single-well simulation.

Example 7.9 A single-phase fluid reservoir is described by four equal blocks as shown in Fig. 7.10. The reservoir is horizontal and has homogeneous rock properties, $k=270$ md, $\phi=0.27$, and $c_\phi = 1 \times 10^{-6}$ psi^{-1}. Initially, the reservoir pressure is 4000 psia. Gridblock dimensions are $\Delta x = 300$ ft, $\Delta y = 350$ ft, and $h=40$ ft. Reservoir fluid properties are $B = B^{\circ} = 1$ RB/STB, $\rho = 50$ lbm/ft^3, $\mu = 0.5$ cP, and $c = 1 \times 10^{-5}$ psi^{-1}. The reservoir left boundary is kept at a constant pressure of 4000 psia, and the reservoir right boundary is sealed off to flow. A 7-in vertical well was drilled at the center of gridblock 4. The well produces

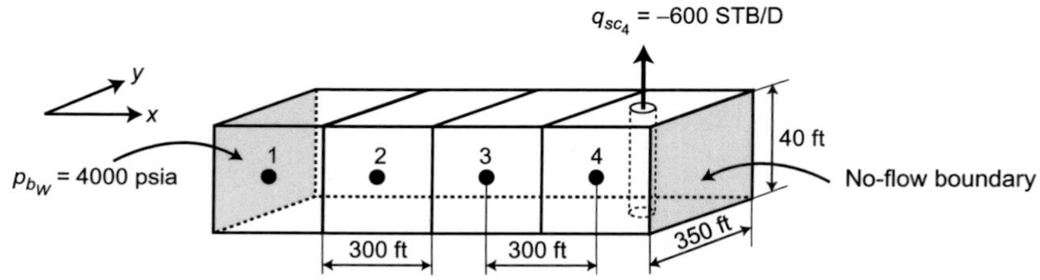

FIG. 7.10 Discretized 1-D reservoir in Example 7.9.

600 STB/D of fluid and has a skin factor of 1.5. Find the pressure distribution in the reservoir after 1 day and 2 days using the implicit formulation. Take time steps of 1 day. Perform a material balance check. This problem is the same as that presented in Example 7.1, except both the fluid and rock are slightly compressible.

Solution

The gridblocks have the same dimensions and rock properties. Therefore, $T_{1,2} = T_{2,3} = T_{3,4} = T_x = \beta_c \frac{A_x k_x}{\mu B \Delta x} = 0.001127 \times \frac{(350 \times 40) \times 270}{0.5 \times 1 \times 300} = 28.4004$ STB/D-psi and

$$\frac{V_{b_n} \phi_n^{\circ} (c + c_\phi)}{\alpha_c B^{\circ} \Delta t} = \frac{(300 \times 350 \times 40) \times 0.27 \times \left(1 \times 10^{-5} + 1 \times 10^{-6}\right)}{5.614583 \times 1 \times 1}$$
$$= 2.2217 \text{ for } n = 1,2,3,4$$

There is a production well in gridblock 4 only. Therefore, $q_{sc_1}^{n+1} = q_{sc_2}^{n+1} = q_{sc_3}^{n+1} = 0$ and $q_{sc_4}^{n+1} = -600$ STB/D.

Gridblock 1 falls on the reservoir west boundary, which is kept at a constant pressure of 4000 psia. Therefore, $q_{sc_{b_W,1}}^{n+1}$ can be estimated using Eq. (4.37c), whose application gives

$$q_{sc_{b_W,1}}^{n+1} = \left[\beta_c \frac{k_x A_x}{\mu B (\Delta x/2)}\right]_1 \left[\left(p_{b_W} - p_1^{n+1}\right) - \gamma (Z_{b_W} - Z_1)\right]$$
$$= \left[0.001127 \times \frac{270 \times (350 \times 40)}{0.5 \times 1 \times (300/2)}\right] \left[\left(4000 - p_1^{n+1}\right) - \gamma \times 0\right] \tag{7.90}$$

or

$$q_{sc_{b_W,1}}^{n+1} = 56.8008\left(4000 - p_1^{n+1}\right) \text{ STB/D} \tag{7.91}$$

Gridblock 4 falls on the reservoir east boundary, which is a no-flow boundary. Therefore, Eq. (4.32) applies, giving $q_{sc_{b_E,4}}^{n+1} = 0$ STB/D.

1. First time step calculations ($n=0$, $t_{n+1}=1$ day, and $\Delta t = 1$ day)

Assign $p_1^n = p_2^n = p_3^n = p_4^n = p_{in} = 4000$ psia.

The general flow equation for gridblock n in this 1-D horizontal reservoir is obtained from Eq. (7.81a) by discarding the gravity term, yielding

$$\sum_{l \in \psi_n} T_{l,n}^{n+1}\left(p_l^{n+1} - p_n^{n+1}\right) + \sum_{l \in \xi_n} q_{sc_{l,n}}^{n+1} + q_{sc_n}^{n+1} \cong \frac{V_{b_n} \phi_n^{\circ} (c + c_\phi)}{\alpha_c B^{\circ} \Delta t}\left[p_n^{n+1} - p_n^n\right] \tag{7.92}$$

For gridblock 1, $n=1$, $\psi_1 = \{2\}$, and $\xi_1 = \{b_W\}$. Therefore, $\sum_{l \in \xi_1} q_{sc_{l,1}}^{n+1} = q_{sc_{b_W,1}}^{n+1}$, and Eq. (7.92) becomes

$$T_{1,2}\left(p_2^{n+1} - p_1^{n+1}\right) + q_{sc_{b_W,1}}^{n+1} + q_{sc_1}^{n+1} = \frac{V_{b_1} \phi_1^{\circ} (c + c_\phi)}{\alpha_c B^{\circ} \Delta t}\left[p_1^{n+1} - p_1^n\right] \tag{7.93}$$

Substitution of the values in this equation gives

$$28.4004\left(p_2^{n+1}-p_1^{n+1}\right)+56.8008\left(4000-p_1^{n+1}\right)+0=2.2217\left[p_1^{n+1}-4000\right]$$

or after factorizing and ordering the unknowns,

$$-87.4229p_1^{n+1}+28.4004p_2^{n+1}=-236090.06 \tag{7.94}$$

For gridblock 2, $n=2$, $\psi_2=\{1,3\}$, and $\xi_2=\{\}$. Therefore, $\sum_{l\in\xi_2} q_{sc_{l,2}}^{n+1}=0$, and Eq. (7.92) becomes

$$T_{1,2}\left(p_1^{n+1}-p_2^{n+1}\right)+T_{2,3}\left(p_3^{n+1}-p_2^{n+1}\right)+0+q_{sc_2}^{n+1}=\frac{V_{b_2}\phi_2^{\circ}\left(c+c_\phi\right)}{\alpha_c B^{\circ}\Delta t}\left[p_2^{n+1}-p_2^{n}\right] \tag{7.95}$$

Substitution of the values in this equation gives

$$28.4004\left(p_1^{n+1}-p_2^{n+1}\right)+28.4004\left(p_3^{n+1}-p_2^{n+1}\right)+0+0=2.2217\left[p_2^{n+1}-4000\right]$$

or after factorizing and ordering the unknowns,

$$28.4004p_1^{n+1}-59.0225p_2^{n+1}+28.4004p_3^{n+1}=-8886.86 \tag{7.96}$$

For gridblock 3, $n=3$, $\psi_3=\{2,4\}$, and $\xi_3=\{\}$. Therefore, $\sum_{l\in\xi_3} q_{sc_{l,3}}^{n+1}=0$, and Eq. (7.92) becomes

$$T_{2,3}\left(p_2^{n+1}-p_3^{n+1}\right)+T_{3,4}\left(p_4^{n+1}-p_3^{n+1}\right)+0+q_{sc_3}^{n+1}=\frac{V_{b_3}\phi_3^{\circ}\left(c+c_\phi\right)}{\alpha_c B^{\circ}\Delta t}\left[p_3^{n+1}-p_3^{n}\right] \tag{7.97}$$

Substitution of the values in this equation gives

$$28.4004\left(p_2^{n+1}-p_3^{n+1}\right)+28.4004\left(p_4^{n+1}-p_3^{n+1}\right)+0+0=2.2217\left[p_3^{n+1}-4000\right]$$

or after factorizing and ordering the unknowns,

$$28.4004p_2^{n+1}-59.0225p_3^{n+1}+28.4004p_4^{n+1}=-8886.86 \tag{7.98}$$

For gridblock 4, $n=4$, $\psi_4=\{3\}$, and $\xi_4=\{b_E\}$. Therefore, $\sum_{l\in\xi_4} q_{sc_{l,4}}^{n+1}=q_{sc_{b_E,4}}^{n+1}$, and Eq. (7.92) becomes

$$T_{3,4}\left(p_3^{n+1}-p_4^{n+1}\right)+q_{sc_{b_E,4}}^{n+1}+q_{sc_4}^{n+1}=\frac{V_{b_4}\phi_4^{\circ}\left(c+c_\phi\right)}{\alpha_c B^{\circ}\Delta t}\left[p_4^{n+1}-p_4^{n}\right] \tag{7.99}$$

Substitution of the values in this equation gives

$$28.4004\left(p_3^{n+1}-p_4^{n+1}\right)+0-600=2.2217\left[p_4^{n+1}-4000\right]$$

or after factorizing and ordering the unknowns,

$$28.4004p_3^{n+1}-30.6221p_4^{n+1}=-8286.86 \tag{7.100}$$

The results of solving Eqs. (7.94), (7.96), (7.98), and (7.100) for the unknown pressures are $p_1^{n+1} = 3993.75$ psia, $p_2^{n+1} = 3980.75$ psia, $p_3^{n+1} = 3966.24$ psia, and $p_4^{n+1} = 3949.10$ psia

Next, the flow rate across the reservoir left boundary ($q_{sc_{b_W,1}}^{n+1}$) is estimated using Eq. (7.91), which gives

$$q_{sc_{b_W,1}}^{n+1} = 56.8008\left(4000 - p_1^{n+1}\right) = 56.8008(4000 - 3993.75)$$
$$= 355.005 \text{ STB/D}$$

The material balance for a slightly compressible fluid and rock system is checked using Eq. (7.85), yielding

$$I_{MB} = \frac{\sum_{n=1}^{N} \frac{V_{b_n}\phi_n^\circ (c + c_\phi)}{\alpha_c B^\circ \Delta t}\left[p_n^{n+1} - p_n^n\right]}{\sum_{n=1}^{N}\left(q_{sc_n}^{n+1} + \sum_{l \in \xi_n} q_{sc_{l,n}}^{n+1}\right)} = \frac{\frac{V_b \phi^\circ (c + c_\phi)}{\alpha_c B^\circ \Delta t}\sum_{n=1}^{4}\left[p_n^{n+1} - p_n^n\right]}{\sum_{n=1}^{4} q_{sc_n}^{n+1} + \sum_{n=1}^{4}\sum_{l \in \xi_n} q_{sc_{l,n}}^{n+1}}$$

$$= \frac{2.2217[(3993.75 - 4000) + (3980.75 - 4000) + (3966.24 - 4000) + (3949.10 - 4000)]}{[0 + 0 + 0 - 600] + [355.005 + 0 + 0 + 0]}$$

$$= \frac{-2.2217 \times 110.16}{-244.995} = 0.99897$$

Therefore, the material balance check is satisfied.

2. Second time step calculations ($n = 1$, $t_{n+1} = 2$ day, and $\Delta t = 1$ day)

Assign $p_1^n = 3993.75$ psia, $p_2^n = 3980.75$ psia, $p_3^n = 3966.24$ psia, and $p_4^n = 3949.10$ psia. Because Δt is constant, the flow equation for each gridblock in the second and succeeding time steps is obtained in a way similar to that used in the first time step, except the newly assigned p_n^n is used to replace the old p_n^n in the accumulation term. For example, p_n^n on the RHS of Eqs. (7.93), (7.95), (7.97), and (7.99) for this time step is replaced with 3993.75, 3980.75, 3966.24, and 3949.10, respectively.

For gridblock 1,

$$28.4004\left(p_2^{n+1} - p_1^{n+1}\right) + 56.8008\left(4000 - p_1^{n+1}\right) + 0 = 2.2217\left[p_1^{n+1} - 3993.75\right]$$

or after factorizing and ordering the unknowns,

$$-87.4229 p_1^{n+1} + 28.4004 p_2^{n+1} = -236076.16 \tag{7.101}$$

For gridblock 2,

$$28.4004\left(p_1^{n+1} - p_2^{n+1}\right) + 28.4004\left(p_3^{n+1} - p_2^{n+1}\right) + 0 + 0$$
$$= 2.2217\left[p_2^{n+1} - 3980.75\right]$$

or after factorizing and ordering the unknowns,

$$28.4004p_1^{n+1} - 59.0225p_2^{n+1} + 28.4004p_3^{n+1} = -8844.08 \tag{7.102}$$

For gridblock 3,

$$28.4004\left(p_2^{n+1} - p_3^{n+1}\right) + 28.4004\left(p_4^{n+1} - p_3^{n+1}\right) + 0 + 0$$
$$= 2.2217\left[p_3^{n+1} - 3966.24\right]$$

or after factorizing and ordering the unknowns,

$$28.4004p_2^{n+1} - 59.0225p_3^{n+1} + 28.4004p_4^{n+1} = -8811.86 \tag{7.103}$$

For gridblock 4,

$$28.4004\left(p_3^{n+1} - p_4^{n+1}\right) + 0 - 600 = 2.2217\left[p_4^{n+1} - 3949.10\right]$$

or after factorizing and ordering the unknowns,

$$28.4004p_3^{n+1} - 30.6221p_4^{n+1} = -8173.77 \tag{7.104}$$

The results of solving Eqs. (7.101) through (7.104) for the unknown pressures are $p_1^{n+1} = 3990.95$ psia, $p_2^{n+1} = 3972.64$ psia, $p_3^{n+1} = 3953.70$ psia, and $p_4^{n+1} = 3933.77$ psia.

Next, the flow rate across the reservoir left boundary ($q_{sc_{b_W,1}}^{n+1}$) is estimated using Eq. (7.91), which gives

$$q_{sc_{b_W,1}}^{n+1} = 56.8008\left(4000 - p_1^{n+1}\right) = 56.8008(4000 - 3990.95)$$
$$= 514.047\ \mathrm{STB/D}$$

The material balance is checked using Eq. (7.85), yielding

$$I_{MB} = \frac{\dfrac{V_b\phi^\circ\left(c + c_\phi\right)}{\alpha_c B^\circ \Delta t}\displaystyle\sum_{n=1}^{4}\left[p_n^{n+1} - p_n^n\right]}{\displaystyle\sum_{n=1}^{4} q_{sc_n}^{n+1} + \sum_{n=1}^{4}\sum_{l\in\xi_n} q_{sc_{l,n}}^{n+1}}$$

$$= \frac{\left\{2.2217\begin{bmatrix}(3990.95 - 3993.75) + (3972.64 - 3980.75) \\ +(3953.70 - 3966.24) + (3933.77 - 3949.10)\end{bmatrix}\right\}}{[0 + 0 + 0 - 600] + [514.047 + 0 + 0 + 0]}$$

$$= \frac{-2.2217 \times 38.78}{-85.953} = 1.00238$$

Example 7.10 Consider the problem presented in Example 7.9, but this time, the reservoir is described by five equally spaced gridpoints using a point-distributed grid as shown in Fig. 7.11.

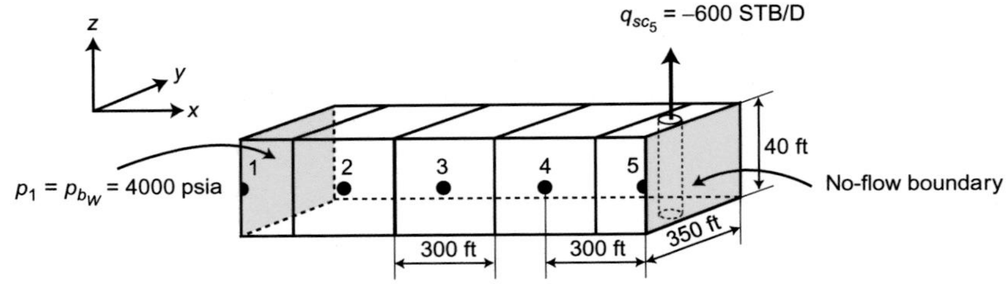

FIG. 7.11 Discretized 1-D reservoir in Example 7.10.

The problem is restated as follows. The reservoir length along the x-direction is 1200 ft, $\Delta y = 350$ ft, and $h = 40$ ft. The reservoir is horizontal and has homogeneous rock properties, $k = 270$ md, $\phi = 0.27$, and $c_\phi = 1 \times 10^{-6}\ \text{psi}^{-1}$. Initially, the reservoir pressure is 4000 psia. Reservoir fluid properties are $B = B^\circ = 1$ RB/STB, $\rho = 50\ \text{lbm/ft}^3$, $\mu = 0.5$ cP, and $c = 1 \times 10^{-5}\ \text{psi}^{-1}$. The reservoir left boundary is kept constant at 4000 psia, and the reservoir right boundary is sealed off to flow. A 7-in vertical well was drilled at the center of gridblock 5. The well produces 600 STB/D of fluid and has a skin factor of 1.5. Find the pressure distribution in the reservoir after 1 day and 2 days using the implicit formulation. Take time steps of 1 day. Perform a material balance check.

Solution

The reservoir is discretized into five gridpoints, $n_x = 5$. The distance between the gridpoints is $\Delta x_{i+1/2} = 1200/(5-1) = 300$ ft for $i = 1, 2, 3, 4$. Therefore, block sizes in the x-direction are $\Delta x_1 = \Delta x_5 = 300/2 = 150$ ft and $\Delta x_2 = \Delta x_3 = \Delta x_4 = 300$ ft. Blocks represented by the various gridpoints have the same rock properties. Therefore,

$$T_{1,2} = T_{2,3} = T_{3,4} = T_{4,5} = T_x = \beta_c \frac{A_x k_x}{\mu B \Delta x_{i+1/2}} = 0.001127 \times \frac{(350 \times 40) \times 270}{0.5 \times 1 \times 300}$$
$$= 28.4004\ \text{STB/D-psi}$$

$$\frac{V_{b_n} \phi_n^\circ (c + c_\phi)}{\alpha_c B^\circ \Delta t} = \frac{(300 \times 350 \times 40) \times 0.27 \times \left(1 \times 10^{-5} + 1 \times 10^{-6}\right)}{5.614583 \times 1 \times 1}$$
$$= 2.2217 \text{ for } n = 2, 3, 4$$

Additionally,

$$\frac{V_{b_n} \phi_n^\circ (c + c_\phi)}{\alpha_c B^\circ \Delta t} = \frac{(150 \times 350 \times 40) \times 0.27 \times \left(1 \times 10^{-5} + 1 \times 10^{-6}\right)}{5.614583 \times 1 \times 1}$$
$$= 1.11085 \text{ for } n = 1, 5$$

There is a production well in gridpoint 5 only. Therefore, $q_{sc_1}^{n+1} = q_{sc_2}^{n+1} = q_{sc_3}^{n+1} = q_{sc_4}^{n+1} = 0$, and $q_{sc_5}^{n+1} = -600$ STB/D.

Gridpoint 1 falls on the reservoir west boundary, which is kept at a constant pressure of 4000 psia. Therefore,

$$p_1^{n+1} = p_{b_W} = 4000\,\text{psia} \tag{7.105}$$

In addition, $q_{sc_{b_W,1}}^{n+1}$ can be estimated using Eq. (5.46c), whose application gives

$$\begin{aligned} q_{sc_{b_W},1}^{n+1} &= \left[\beta_c \frac{k_x A_x}{\mu B \Delta x}\right]_{1,2} \left[\left(p_{b_W} - p_2^{n+1}\right) - \gamma(Z_{b_W} - Z_2)\right] \\ &= \left[0.001127 \times \frac{270 \times (350 \times 40)}{0.5 \times 1 \times 300}\right] \left[\left(4000 - p_2^{n+1}\right) - \gamma \times 0\right] \end{aligned} \tag{7.106}$$

or

$$q_{sc_{b_W},1}^{n+1} = 28.4004\left(4000 - p_2^{n+1}\right)\,\text{STB/D} \tag{7.107}$$

Gridpoint 5 falls on the reservoir east boundary, which is a no-flow boundary. Therefore, Eq. (5.40) applies, giving $q_{sc_{b_E,5}}^{n+1} = 0$ STB/D.

1. First time step calculations ($n=0$, $t_{n+1}=1$ day, and $\Delta t = 1$ day)

Assign $p_1^n = p_2^n = p_3^n = p_4^n = p_5^n = p_{in} = 4000$ psia.

The general flow equation for gridpoint n in this 1-D horizontal reservoir is obtained from Eq. (7.81a) by discarding the gravity term, yielding

$$\sum_{l\in\psi_n} T_{l,n}^{n+1}\left(p_l^{n+1} - p_n^{n+1}\right) + \sum_{l\in\xi_n} q_{sc_{l,n}}^{n+1} + q_{sc_n}^{n+1} \cong \frac{V_{b_n}\phi_n^\circ(c + c_\phi)}{\alpha_c B^\circ \Delta t}\left[p_n^{n+1} - p_n^n\right] \tag{7.92}$$

For gridpoint 1, $n=1$, $\psi_1 = \{2\}$, and $\xi_1 = \{b_W\}$. Therefore, $\sum_{l\in\xi_1} q_{sc_{l,1}}^{n+1} = q_{sc_{b_W},1}^{n+1}$, and Eq. (7.92) becomes

$$T_{1,2}\left(p_2^{n+1} - p_1^{n+1}\right) + q_{sc_{b_W},1}^{n+1} + q_{sc_1}^{n+1} = \frac{V_{b_1}\phi_1^\circ(c + c_\phi)}{\alpha_c B^\circ \Delta t}\left[p_1^{n+1} - p_1^n\right] \tag{7.93}$$

In reality, we do not need to write or make use of the flow equation for gridpoint 1 because $p_1^{n+1} = 4000$ psia is defined by Eq. (7.105); however, Eq. (7.93) can be used to estimate $q_{sc_{b_W,1}}^{n+1}$. Substitution of values in Eq. (7.93) gives

$$28.4004\left(p_2^{n+1} - 4000\right) + q_{sc_{b_W},1}^{n+1} + 0 = 1.11085[4000 - 4000]$$

which when solved for $q_{sc_{b_W,1}}^{n+1}$ results in Eq. (7.107). Therefore, we may conclude for the case of a specified pressure boundary in a point-distributed grid that the rate of fluid flow across a reservoir boundary can be obtained by either using Eq. (5.46c) or writing the flow equation for the boundary gridpoint and making use of $p_{bP}^{n+1} = p_{bP}^{n} = p_b$.

For gridpoint 2, $n=2$, $\psi_2 = \{1,3\}$, and $\xi_2 = \{\}$. Therefore, $\sum_{l \in \xi_2} q_{sc_{l,2}}^{n+1} = 0$, and Eq. (7.92) becomes

$$T_{1,2}\left(p_1^{n+1} - p_2^{n+1}\right) + T_{2,3}\left(p_3^{n+1} - p_2^{n+1}\right) + 0 + q_{sc_2}^{n+1} = \frac{V_{b_2}\phi_2^{\circ}(c + c_\phi)}{\alpha_c B^{\circ} \Delta t}\left[p_2^{n+1} - p_2^{n}\right] \quad (7.95)$$

Substitution of the values in this equation gives

$$28.4004\left(4000 - p_2^{n+1}\right) + 28.4004\left(p_3^{n+1} - p_2^{n+1}\right) + 0 + 0 = 2.2217\left[p_2^{n+1} - 4000\right]$$

or after factorizing and ordering the unknowns,

$$-59.0225 p_2^{n+1} + 28.4004 p_3^{n+1} = -122488.46 \quad (7.108)$$

For gridblock 3, $n=3$, $\psi_3 = \{2,4\}$, and $\xi_3 = \{\}$. Therefore, $\sum_{l \in \xi_3} q_{sc_{l,3}}^{n+1} = 0$, and Eq. (7.92) becomes

$$T_{2,3}\left(p_2^{n+1} - p_3^{n+1}\right) + T_{3,4}\left(p_4^{n+1} - p_3^{n+1}\right) + 0 + q_{sc_3}^{n+1} = \frac{V_{b_3}\phi_3^{\circ}(c + c_\phi)}{\alpha_c B^{\circ} \Delta t}\left[p_3^{n+1} - p_3^{n}\right] \quad (7.97)$$

Substitution of the values in this equation gives

$$28.4004\left(p_2^{n+1} - p_3^{n+1}\right) + 28.4004\left(p_4^{n+1} - p_3^{n+1}\right) + 0 + 0 = 2.2217\left[p_3^{n+1} - 4000\right]$$

or after factorizing and ordering the unknowns,

$$28.4004 p_2^{n+1} - 59.0225 p_3^{n+1} + 28.4004 p_4^{n+1} = -8886.86 \quad (7.98)$$

For gridblock 4, $n=4$, $\psi_4 = \{3,5\}$, and $\xi_4 = \{\}$. Therefore, $\sum_{l \in \xi_4} q_{sc_{l,4}}^{n+1} = 0$, and Eq. (7.92) becomes

$$T_{3,4}\left(p_3^{n+1} - p_4^{n+1}\right) + T_{4,5}\left(p_5^{n+1} - p_4^{n+1}\right) + 0 + q_{sc_4}^{n+1} = \frac{V_{b_4}\phi_4^{\circ}(c + c_\phi)}{\alpha_c B^{\circ} \Delta t}\left[p_4^{n+1} - p_4^{n}\right] \quad (7.109)$$

Substitution of the values in this equation gives

$$28.4004\left(p_3^{n+1} - p_4^{n+1}\right) + 28.4004\left(p_5^{n+1} - p_4^{n+1}\right) + 0 + 0 = 2.2217\left[p_4^{n+1} - 4000\right]$$

or after factorizing and ordering the unknowns,

$$28.4004 p_3^{n+1} - 59.0225 p_4^{n+1} + 28.4004 p_5^{n+1} = -8886.86 \quad (7.110)$$

For gridblock 5, $n=5, \psi_5=\{4\}$, and $\xi_5=\{b_E\}$. Therefore, $\sum_{l\in\xi_5} q_{sc_{l,5}}^{n+1}=q_{sc_{b_E,5}}^{n+1}$, and Eq. (7.92) becomes

$$T_{4,5}\left(p_4^{n+1}-p_5^{n+1}\right)+q_{sc_{b_E,5}}^{n+1}+q_{sc_5}^{n+1}=\frac{V_{b_5}\phi_5^\circ\left(c+c_\phi\right)}{\alpha_c B^\circ \Delta t}\left[p_5^{n+1}-p_5^n\right] \tag{7.111}$$

Substitution of the values in this equation gives

$$28.4004\left(p_4^{n+1}-p_5^{n+1}\right)+0-600=1.11085\left[p_5^{n+1}-4000\right]$$

or after the factorizing and the ordering of unknowns,

$$28.4004p_4^{n+1}-29.51125p_5^{n+1}=-3843.4288 \tag{7.112}$$

The results of solving Eqs. (7.108), (7.98), (7.110), and (7.112) for the unknown pressures are $p_2^{n+1}=3987.49$ psia, $p_3^{n+1}=3974.00$ psia, $p_4^{n+1}=3958.48$ psia, and $p_5^{n+1}=3939.72$ psia.

Next, the flow rate across the reservoir left boundary ($q_{sc_{b_W,1}}^{n+1}$) is estimated using Eq. (7.107), which yields

$$\begin{aligned} q_{sc_{b_W,1}}^{n+1} &= 28.4004\left(4000-p_2^{n+1}\right)=28.4004(4000-3987.49) \\ &= 355.289\ \text{STB/D} \end{aligned}$$

The material balance for a slightly compressible fluid and rock system is checked using Eq. (7.85):

$$I_{MB}=\frac{\sum_{n=1}^{N}\frac{V_{b_n}\phi_n^\circ\left(c+c_\phi\right)}{\alpha_c B^\circ \Delta t}\left[p_n^{n+1}-p_n^n\right]}{\sum_{n=1}^{N}\left(q_{sc_n}^{n+1}+\sum_{l\in\xi_n}q_{sc_{l,n}}^{n+1}\right)}=\frac{\sum_{n=1}^{5}\frac{V_{b_n}\phi_n^\circ\left(c+c_\phi\right)}{\alpha_c B^\circ \Delta t}\left[p_n^{n+1}-p_n^n\right]}{\sum_{n=1}^{5}q_{sc_n}^{n+1}+\sum_{n=1}^{5}\sum_{l\in\xi_n}q_{sc_{l,n}}^{n+1}}$$

$$=\frac{\left[\begin{array}{l}1.11085\times(4000-4000)+2.2217\times(3987.49-4000)+2.2217\times(3974.00-4000)\\ +2.2217\times(3958.48-4000)+1.11085\times(3939.72-4000)\end{array}\right]}{[(0+0+0+0-600)+(355.289+0+0+0+0)]}$$

$$=\frac{-244.765}{-244.711}=1.00022$$

Therefore, the material balance check is satisfied.

2. Second time step calculations ($n=1$, $t_{n+1}=2$ day, and $\Delta t=1$ day)

Assign $p_2^n=3987.49$ psia, $p_3^n=3974.00$ psia, $p_4^n=3958.48$ psia, and $p_5^n=$ 3939.72 psia. Note that $p_1^{n+1}=4000$ psia.

Because Δt is constant, the flow equation for each gridblock in the second and succeeding time steps is obtained in a way similar to that used in the first time step, except the newly assigned p_n^n is used to replace the old p_n^n in the accumulation term, as mentioned in Example 7.9. For example, p_n^n on the RHS of Eqs. (7.95), (7.97), (7.109), and (7.111) for the present time step is replaced with 3987.49, 3974.00, 3958.48, and 3939.72, respectively.

For gridblock 2,

$$28.4004\left(4000-p_2^{n+1}\right)+28.4004\left(p_3^{n+1}-p_2^{n+1}\right)+0+0$$
$$=2.2217\left[p_2^{n+1}-3987.49\right]$$

or after factorizing and ordering the unknowns,

$$-59.0225p_2^{n+1}+28.4004p_3^{n+1}=-122460.667 \tag{7.113}$$

For gridblock 3,

$$28.4004\left(p_2^{n+1}-p_3^{n+1}\right)+28.4004\left(p_4^{n+1}-p_3^{n+1}\right)+0+0$$
$$=2.2217\left[p_3^{n+1}-3974.00\right]$$

or after factorizing and ordering the unknowns,

$$28.4004p_2^{n+1}-59.0225p_3^{n+1}+28.4004p_4^{n+1}=-8829.1026 \tag{7.114}$$

For gridblock 4,

$$28.4004\left(p_3^{n+1}-p_4^{n+1}\right)+28.4004\left(p_5^{n+1}-p_4^{n+1}\right)+0+0$$
$$=2.2217\left[p_4^{n+1}-3958.48\right]$$

or after factorizing and ordering the unknowns,

$$28.4004p_3^{n+1}-59.0225p_4^{n+1}+28.4004p_5^{n+1}=-8794.6200 \tag{7.115}$$

For gridblock 5,

$$28.4004\left(p_4^{n+1}-p_5^{n+1}\right)+0-600=1.11085\left[p_5^{n+1}-3939.72\right]$$

or after factorizing and ordering the unknowns,

$$28.4004p_4^{n+1}-29.51125p_5^{n+1}=-3776.4609 \tag{7.116}$$

The results of solving Eqs. (7.113), (7.114), (7.115), and (7.116) for the unknown pressures are $p_2^{n+1}=3981.91$ psia, $p_3^{n+1}=3963.38$ psia, $p_4^{n+1}=3944.02$ psia, and $p_5^{n+1}=3923.52$ psia.

The flow rate across the reservoir left boundary $(q_{sc_{b_W,1}}^{n+1})$ is estimated next using Eq. (7.107), yielding

$$q_{sc_{b_W,1}}^{n+1}=28.4004\left(4000-p_2^{n+1}\right)=28.4004(4000-3981.91)$$
$$=513.763\ \text{STB/D}$$

The application of Eq. (7.85) to check the material balance for the second time step gives

$$I_{MB} = \frac{\sum_{n=1}^{5} \frac{V_{b_n}\phi_n^\circ (c + c_\phi)}{\alpha_c B^\circ \Delta t}\left[p_n^{n+1} - p_n^n\right]}{\sum_{n=1}^{5} q_{sc_n}^{n+1} + \sum_{n=1}^{5}\sum_{l\in\xi_n} q_{sc_{l,n}}^{n+1}}$$

$$= \frac{\begin{bmatrix} 1.11085 \times (4000 - 4000) + 2.2217 \times (3981.91 - 3987.49) \\ +2.2217 \times (3963.38 - 3974.00) + 2.2217 \times (3944.02 - 3958.48) \\ +1.11085 \times (3923.52 - 3939.72) \end{bmatrix}}{[(0+0+0+0-600) + (513.763+0+0+0+0)]}$$

$$= \frac{-86.103}{-86.237} = 0.99845$$

Therefore, the material balance check is satisfied.

Example 7.11 A 1-D, horizontal, heterogeneous reservoir is discretized as shown in Fig. 7.12. The reservoir is described by five gridblocks whose dimensions and rock properties are shown in the figure. Reservoir fluid properties are $B = B^\circ = 1$ RB/STB, $\mu = 1.5$ cP, and $c = 2.5 \times 10^{-5}$ psi^{-1}. Initially, reservoir pressure is 3000 psia. The reservoir left and right boundaries are sealed off to flow. A 6-in vertical well was drilled at the center of gridblock 4. The well produces 400 STB/D of fluid and has zero skin. The well is switched to a constant FBHP of 1500 psia if the reservoir cannot sustain the specified production rate. Find the pressure distribution in the reservoir after 5 days and 10 days using the implicit formulation. Take time steps of 5 days. Tabulate reservoir pressure versus time until reservoir depletion.

Solution

The general flow equation for gridblock n in this 1-D horizontal reservoir is obtained from Eq. (7.81a) by discarding the gravity term, yielding

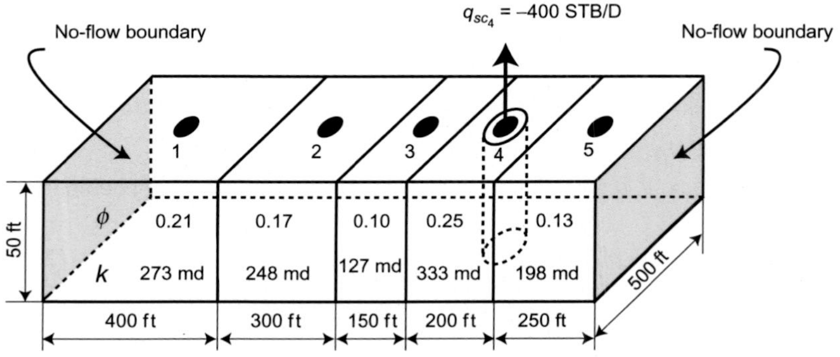

FIG. 7.12 Discretized 1-D reservoir in Example 7.11.

$$\sum_{l\in\psi_n} T_{l,n}^{n+1}\left(p_l^{n+1}-p_n^{n+1}\right)+\sum_{l\in\xi_n} q_{sc_{l,n}}^{n+1}+q_{sc_n}^{n+1}\cong\frac{V_{b_n}\phi_n^{\circ}\left(c+c_\phi\right)}{\alpha_c B^{\circ}\Delta t}\left[p_n^{n+1}-p_n^n\right] \quad (7.92)$$

Interblock transmissibilities can be calculated using Eq. (4.67a) with the geometric factors in the x-direction obtained from Table 4.1 because the gridblocks have an irregular grid size distribution and heterogeneous rock properties, resulting in

$$T_{n,n\pm1}=T_{x_{i\mp1/2}}=G_{x_{i\mp1/2}}\left(\frac{1}{\mu B}\right)_{x_{i\mp1/2}}=\frac{1}{\mu B}\times\frac{2\beta_c}{\Delta x_i/\left(A_{x_i}k_{x_i}\right)+\Delta x_{i\mp1}/\left(A_{x_{i\mp1}}k_{x_{i\mp1}}\right)} \quad (7.117)$$

Therefore,

$$T_{1,2}=\frac{1}{1.5\times1}\times\frac{2\times0.001127}{400/[(500\times50)\times273]+300/[(500\times50)\times248]}$$
$$=14.0442\ \text{STB/D-psi}.$$

Similarly, $T_{2,3}=15.7131$ STB/D-psi, $T_{3,4}=21.0847$ STB/D-psi, and $T_{4,5}=20.1622$ STB/D-psi.

$$\frac{V_{b_1}\phi_1^{\circ}\left(c+c_\phi\right)}{\alpha_c B^{\circ}\Delta t}=\frac{(400\times500\times50)\times0.21\times\left(2.5\times10^{-5}+0\right)}{5.614583\times1\times5}$$
$$=1.87013\ \text{STB/D-psi}$$

Similarly, $\frac{V_{b_2}\phi_2^{\circ}\left(c+c_\phi\right)}{\alpha_c B^{\circ}\Delta t}=1.13544$ STB/D-psi, $\frac{V_{b_3}\phi_3^{\circ}\left(c+c_\phi\right)}{\alpha_c B^{\circ}\Delta t}=0.333952$ STB/D-psi, $\frac{V_{b_4}\phi_4^{\circ}\left(c+c_\phi\right)}{\alpha_c B^{\circ}\Delta t}=1.11317$ STB/D-psi, and $\frac{V_{b_5}\phi_5^{\circ}\left(c+c_\phi\right)}{\alpha_c B^{\circ}\Delta t}=0.723562$ STB/D-psi.

There is a production well in gridblock 4 only. Therefore, $q_{sc_4}^{n+1}=-400$ STB/D and $q_{sc_1}^{n+1}=q_{sc_2}^{n+1}=q_{sc_3}^{n+1}=q_{sc_5}^{n+1}=0$. No-flow boundary conditions imply $q_{sc_{b_W,1}}^{n+1}=0$ and $q_{sc_{b_E,5}}^{n+1}=0$.

For no-flow boundaries, Eq. (7.92) reduces to

$$\sum_{l\in\psi_n} T_{l,n}^{n+1}\left(p_l^{n+1}-p_n^{n+1}\right)+q_{sc_n}^{n+1}\cong\frac{V_{b_n}\phi_n^{\circ}\left(c+c_\phi\right)}{\alpha_c B^{\circ}\Delta t}\left[p_n^{n+1}-p_n^n\right] \quad (7.118)$$

1. First time step calculations ($n=0$, $t_{n+1}=5$ days, and $\Delta t=5$ days)

Assign $p_1^n=p_2^n=p_3^n=p_4^n=p_5^n=p_{in}=3000$ psia.

For gridblock 1, $n=1$, and $\psi_1=\{2\}$. Therefore, Eq. (7.118) becomes

$$T_{1,2}\left(p_2^{n+1}-p_1^{n+1}\right)+q_{sc_1}^{n+1}=\frac{V_{b_1}\phi_1^{\circ}\left(c+c_\phi\right)}{\alpha_c B^{\circ}\Delta t}\left[p_1^{n+1}-p_1^n\right] \quad (7.119)$$

Substitution of the values in this equation gives

$$14.0442\left(p_2^{n+1}-p_1^{n+1}\right)+0=1.87013\left[p_1^{n+1}-3000\right]$$

or after factorizing and ordering the unknowns,

$$-15.9143p_1^{n+1}+14.0442p_2^{n+1}=-5610.39 \tag{7.120}$$

For gridblock 2, $n=2$, and $\psi_2=\{1,3\}$. Therefore, Eq. (7.118) becomes

$$T_{1,2}\left(p_1^{n+1}-p_2^{n+1}\right)+T_{2,3}\left(p_3^{n+1}-p_2^{n+1}\right)+q_{sc_2}^{n+1}=\frac{V_{b_2}\phi_2^{\circ}\left(c+c_\phi\right)}{\alpha_c B^{\circ}\Delta t}\left[p_2^{n+1}-p_2^n\right] \tag{7.121}$$

Substitution of the values in this equation gives

$$14.0442\left(p_1^{n+1}-p_2^{n+1}\right)+15.7131\left(p_3^{n+1}-p_2^{n+1}\right)+0=1.13544\left[p_2^{n+1}-3000\right]$$

or after factorizing and ordering the unknowns,

$$14.0442p_1^{n+1}-30.8927p_2^{n+1}+15.7131p_3^{n+1}=-3406.32 \tag{7.122}$$

For gridblock 3, $n=3$, and $\psi_3=\{2,4\}$. Therefore, Eq. (7.118) becomes

$$T_{2,3}\left(p_2^{n+1}-p_3^{n+1}\right)+T_{3,4}\left(p_4^{n+1}-p_3^{n+1}\right)+q_{sc_3}^{n+1}=\frac{V_{b_3}\phi_3^{\circ}\left(c+c_\phi\right)}{\alpha_c B^{\circ}\Delta t}\left[p_3^{n+1}-p_3^n\right] \tag{7.123}$$

Substitution of the values in this equation gives

$$15.7131\left(p_2^{n+1}-p_3^{n+1}\right)+21.0847\left(p_4^{n+1}-p_3^{n+1}\right)+0=0.333952\left[p_3^{n+1}-3000\right]$$

or after factorizing and ordering the unknowns,

$$15.7131p_2^{n+1}-37.1318p_3^{n+1}+21.0847p_4^{n+1}=-1001.856 \tag{7.124}$$

For gridblock 4, $n=4$, and $\psi_4=\{3,5\}$. Therefore, Eq. (7.118) becomes

$$T_{3,4}\left(p_3^{n+1}-p_4^{n+1}\right)+T_{4,5}\left(p_5^{n+1}-p_4^{n+1}\right)+q_{sc_4}^{n+1}=\frac{V_{b_4}\phi_4^{\circ}\left(c+c_\phi\right)}{\alpha_c B^{\circ}\Delta t}\left[p_4^{n+1}-p_4^n\right] \tag{7.125}$$

Substitution of the values in this equation gives

$$21.0847\left(p_3^{n+1}-p_4^{n+1}\right)+20.1622\left(p_5^{n+1}-p_4^{n+1}\right)-400=1.11317\left[p_4^{n+1}-3000\right]$$

or after factorizing and ordering the unknowns,

$$21.0847p_3^{n+1}-42.3601p_4^{n+1}+20.1622p_5^{n+1}=-2939.510 \tag{7.126}$$

For gridblock 5, $n=5$, and $\psi_5=\{4\}$. Therefore, Eq. (7.118) becomes

$$T_{4,5}\left(p_4^{n+1}-p_5^{n+1}\right)+q_{sc_5}^{n+1}=\frac{V_{b_5}\phi_5^{\circ}\left(c+c_\phi\right)}{\alpha_c B^{\circ}\Delta t}\left[p_5^{n+1}-p_5^n\right] \tag{7.127}$$

Substitution of the values in this equation gives

$$20.1622\left(p_4^{n+1}-p_5^{n+1}\right)+0=0.723562\left[p_5^{n+1}-3000\right]$$

or after factorizing and ordering the unknowns,

$$20.1622p_4^{n+1}-20.8857p_5^{n+1}=-2170.686 \tag{7.128}$$

The results of solving Eqs. (7.120), (7.122), (7.124), (7.126), and (7.128) for the unknown pressures are $p_1^{n+1}=2936.80$ psia, $p_2^{n+1}=2928.38$ psia, $p_3^{n+1}=2915.68$ psia, $p_4^{n+1}=2904.88$ psia, and $p_5^{n+1}=2908.18$ psia.

2. Second time step calculations ($n=1$, $t_{n+1}=10$ days, and $\Delta t=5$ days)

Assign $p_1^n=2936.80$ psia, $p_2^n=2928.38$ psia, $p_3^n=2915.68$ psia, $p_4^n=2904.88$ psia, and $p_5^n=2908.18$ psia. Because Δt is constant, the flow equation for each gridblock in the second and succeeding time steps is obtained in a way similar to that used in the first time step, except the newly assigned p_n^n is used to replace the old p_n^n in the accumulation term. In fact, for horizontal reservoirs having no-flow boundaries and constant production wells and simulated using a constant time step, only the RHSs of the final equations for the first time step change. The new value for the RHS of the equation for gridblock n is $[-q_{sc_n}^{n+1}-\frac{V_{b_n}\phi_n^{\circ}(c+c_\phi)}{\alpha_c B^{\circ}\Delta t}p_n^n]$.

For gridblock 1,

$$14.0442\left(p_2^{n+1}-p_1^{n+1}\right)+0=1.87013\left[p_1^{n+1}-2936.80\right] \tag{7.129}$$

or after factorizing and ordering the unknowns,

$$-15.9143p_1^{n+1}+14.0442p_2^{n+1}=-5492.20 \tag{7.130}$$

For gridblock 2,

$$14.0442\left(p_1^{n+1}-p_2^{n+1}\right)+15.7131\left(p_3^{n+1}-p_2^{n+1}\right)+0 \\ =1.13544\left[p_2^{n+1}-2928.38\right] \tag{7.131}$$

or after factorizing and ordering the unknowns,

$$14.0442p_1^{n+1}-30.8927p_2^{n+1}+15.7131p_3^{n+1}=-3325.00 \tag{7.132}$$

For gridblock 3,

$$15.7131\left(p_2^{n+1}-p_3^{n+1}\right)+21.0847\left(p_4^{n+1}-p_3^{n+1}\right)+0 \\ =0.333952\left[p_3^{n+1}-2915.68\right] \tag{7.133}$$

or after factorizing and ordering the unknowns,

$$15.7131p_2^{n+1}-37.1318p_3^{n+1}+21.0847p_4^{n+1}=-973.6972 \tag{7.134}$$

For gridblock 4,

$$21.0847\left(p_3^{n+1}-p_4^{n+1}\right)+20.1622\left(p_5^{n+1}-p_4^{n+1}\right)-400=1.11317\left[p_4^{n+1}-2904.88\right] \tag{7.135}$$

or after factorizing and ordering the unknowns,

$$21.0847p_3^{n+1}-42.3601p_4^{n+1}+20.1622p_5^{n+1}=-2833.63 \tag{7.136}$$

For gridblock 5,

$$20.1622\left(p_4^{n+1}-p_5^{n+1}\right)+0=0.723562\left[p_5^{n+1}-2908.180\right] \tag{7.137}$$

or after factorizing and ordering the unknowns,

$$20.1622p_4^{n+1}-20.8857p_5^{n+1}=-2104.248 \tag{7.138}$$

The results of solving Eqs. (7.130), (7.132), (7.134), (7.136), and (7.138) for the unknown pressures are $p_1^{n+1}=2861.76$ psia, $p_2^{n+1}=2851.77$ psia, ${p_3}^{n+1}=2837.30$ psia, $p_4^{n+1}=2825.28$ psia, and $p_5^{n+1}=2828.15$ psia.

Table 7.1 shows gridblock pressures, the well production rate, and the FBHP of the well as time progresses. Note that the reservoir produces at a constant rate for the first 90 days, after which the reservoir does not have the capacity to produce fluid at the specified rate and the well is switched to operation under a constant FBHP of 1500 psia. Observe also that reservoir pressure declines steadily from the initial condition of 3000 psia to ultimately 1500 psia at abandonment. The estimated p_{wf_4} reported in Table 7.1 used $k_H=333$ md, $r_e=75.392$ ft, and $G_{w_4}=20.652$ RB-cp/D-psi, which were based on the properties of wellblock 4 and the hosted well.

Example 7.12 A 0.5-ft-diameter water well is located in 20-acre spacing. The reservoir thickness, horizontal permeability, and porosity are 30 ft, 150 md, and 0.23, respectively. The flowing fluid has FVF, compressibility, and viscosity of 1 RB/B, 1×10^{-5} psi^{-1}, and 0.5 cP, respectively. The reservoir external boundaries are no-flow boundaries. The well has open-well completion and is placed on production at a rate of 2000 B/D. Initial reservoir pressure is 4000 psia. The reservoir can be simulated using five gridblocks in the radial direction as shown in Fig. 7.13. Find the pressure distribution in the reservoir after 1 day and 3 days, and check the material balance each time step. Use single time steps to advance the solution from one time to another.

Solution

The reservoir external radius is estimated from well spacing as $r_e=(20\times43560/\pi)^{1/2}=526.6040$ ft. The well in wellblock 1 has $r_w=0.25$ ft. Therefore, using Eq. (4.86) yields $\alpha_{lg}=(526.6040/0.25)^{1/5}=4.6207112$.

The location of gridblock 1 in the radial direction is calculated using Eq. (4.87), which yields

$$r_1=[(4.6207112)\log_e(4.6207112)/(4.6207112-1)]\times0.25=0.4883173\text{ ft}$$

TABLE 7.1 Performance of the reservoir described in Example 7.11.

Time (day)	p_1 (psia)	p_2 (psia)	p_3 (psia)	p_4 (psia)	p_5 (psia)	q_{sc_4} (STB/D)	p_{wf_4} (psia)
0	3000	3000	3000	3000	3000	0	3000
5	2936.80	2928.38	2915.68	2904.88	2908.18	−400	2875.83
10	2861.76	2851.77	2837.30	2825.28	2828.15	−400	2796.23
15	2784.83	2774.59	2759.86	2747.65	2750.44	−400	2718.60
20	2707.61	2697.33	2682.56	2670.32	2673.10	−400	2641.27
25	2630.34	2620.06	2605.28	2593.04	2595.81	−400	2563.98
30	2553.07	2542.78	2528.00	2515.76	2518.53	−400	2486.71
35	2475.79	2465.50	2450.72	2438.48	2441.26	−400	2409.43
40	2398.52	2388.23	2373.45	2361.21	2363.98	−400	2332.15
45	2321.24	2310.95	2296.17	2283.93	2286.71	−400	2254.88
50	2243.97	2233.68	2218.90	2206.66	2209.43	−400	2177.60
55	2166.69	2156.40	2141.62	2129.38	2132.15	−400	2100.33
60	2089.41	2079.12	2064.34	2052.10	2054.88	−400	2023.05
65	2012.14	2001.85	1987.07	1974.83	1977.60	−400	1945.78
70	1934.86	1924.57	1909.79	1897.55	1900.33	−400	1868.50
75	1857.59	1847.30	1832.52	1820.28	1823.05	−400	1791.22
80	1780.31	1770.02	1755.24	1743.00	1745.77	−400	1713.95
85	1703.03	1692.74	1677.96	1665.72	1668.50	−400	1636.67
90	1625.76	1615.47	1600.69	1588.45	1591.22	−400	1559.40
95	1557.58	1548.51	1535.55	1524.87	1527.17	−342.399	1500.00
100	1524.61	1520.22	1514.26	1509.47	1510.08	−130.389	1500.00
105	1510.35	1508.46	1505.91	1503.88	1504.09	−53.378	1500.00
110	1504.34	1503.54	1502.47	1501.61	1501.70	−22.229	1500.00
115	1501.82	1501.48	1501.03	1500.68	1500.71	−9.294	1500.00
120	1500.76	1500.62	1500.43	1500.28	1500.30	−3.890	1500.00
125	1500.32	1500.26	1500.18	1500.12	1500.12	−1.628	1500.00
130	1500.13	1500.11	1500.08	1500.05	1500.05	−0.682	1500.00
135	1500.06	1500.05	1500.03	1500.02	1500.02	−0.285	1500.00

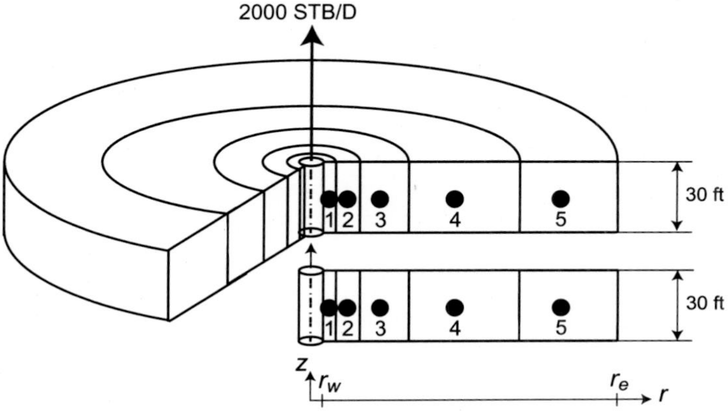

FIG. 7.13 Discretized 1-D reservoir in Example 7.12.

The locations of gridblocks 2, 3, 4, and 5 in the radial direction are calculated using Eq. (4.122), which gives

$$r_i = (4.6207112)^{(i-1)}(0.4883173) \tag{7.139}$$

for $i=2, 3, 4, 5$ or $r_2=2.2564$ ft, $r_3=10.4260$ ft, $r_4=48.1758$ ft, and $r_5=222.6063$ ft.

The gridblock bulk volumes are calculated using Eq. (4.88b), yielding

$$V_{b_i} = \left\{ \left(4.6207112^2 - 1\right)^2 / \left[4.6207112^2 \log_e\left(4.6207112^2\right)\right] \right\} r_i^2 \left({}^1/_2 \times 2\pi\right) \times 30$$
$$= (597.2369) r_i^2 \tag{7.140a}$$

for $i=1, 2, 3, 4$ and Eq. (4.88d) for $i=5$ is

$$V_{b_5} = \left\{ 1 - \left[\log_e(4.6207112)/(4.6207112-1)\right]^2 \left[4.6207112^2 - 1\right] / \left[(4.6207112)^2 \log_e\left(4.6207112^2\right)\right] \right\} (526.6040)^2 \left({}^1/_2 \times 2\pi\right) \times 30$$
$$= 0.24681778 \times 10^8 \tag{7.140b}$$

Table 7.2 lists gridblock bulk volumes.

The transmissibility in the r direction is defined by Eq. (4.79a), which gives

$$T_{r_{i\mp1/2}} = G_{r_{i\mp1/2}} \left(\frac{1}{\mu B}\right) = G_{r_{i\mp1/2}} \left(\frac{1}{0.5 \times 1}\right) = (2) G_{r_{i\mp1/2}} \tag{7.141}$$

TABLE 7.2 Gridblock bulk volumes and coefficients of the accumulation term.

				$\frac{V_{b_n}\phi^\circ_n(c+c_\phi)}{\alpha_c B^\circ \Delta t_n}$	
n	i	r_i (ft)	V_{b_n} (ft³)	$\Delta t_1 = 1$ *day*	$\Delta t_2 = 2$ *days*
1	1	0.4883	142.41339	58.339292×10^{-6}	29.169646×10^{-6}
2	2	2.2564	3040.6644	0.00124560063	$0.62280032 \times 10^{-3}$
3	3	10.4260	64,921.142	0.026594785	0.01329739
4	4	48.1758	1,386,129.5	0.56782451	0.28391226
5	5	222.6063	24,681,778	10.110829	5.0554145

where $G_{r_{i\mp1/2}}$ is defined in Table 4.3. With $\Delta\theta = 2\pi$ and constant radial permeability, the equation for the geometric factor reduces to

$$G_{r_{i\mp1/2}} = \frac{2\pi\beta_c k_r \Delta z}{\log_e\{[\alpha_{lg}\log_e(\alpha_{lg})/(\alpha_{lg}-1)] \times [(\alpha_{lg}-1)/\log_e(\alpha_{lg})]\}}$$
$$= \frac{2\pi\beta_c k_r \Delta z}{\log_e(\alpha_{lg})} = \frac{2\pi(0.001127)(150)(30)}{\log_e(4.6207112)} = 20.819446 \quad (7.142)$$

for all values of i.

Therefore, transmissibility in the radial direction can be estimated by substituting Eq. (7.142) into Eq. (7.141), yielding

$$T_{r_{i\mp1/2}} = (2)G_{r_{i\mp1/2}} = (2)(20.819446) = 41.6388914 \quad (7.143)$$

for all values of i, or

$$T_{1,2} = T_{2,3} = T_{3,4} = T_{4,5} = T = 41.6388914 \text{ B/D-psi} \quad (7.144)$$

Note that gridblocks 2, 3, and 4 are interior gridblocks and gridblocks 1 and 5 have no-flow boundaries; that is, $q^{n+1}_{sc_{b_W,1}} = 0$ and $q^{n+1}_{sc_{b_E,5}} = 0$. Therefore, $\sum_{l\in\xi_n} q^{n+1}_{sc_{l,n}} = 0$ for all gridblocks. There is a well in wellblock 1; that is, $q^{n+1}_{sc_1} = -2000$ B/D and $q^{n+1}_{sc_2} = q^{n+1}_{sc_3} = q^{n+1}_{sc_4} = q^{n+1}_{sc_5} = 0$.

The general form of the flow equation for gridblock n in this 1-D reservoir is obtained from Eq. (7.81a) by discarding the gravity term and noting that $\sum_{l\in\xi_n} q^{n+1}_{sc_{l,n}} = 0$ for all gridblocks, resulting in

$$\sum_{l\in\psi_n} T^{n+1}_{l,n}\left(p^{n+1}_l - p^{n+1}_n\right) + q^{n+1}_{sc_n} \cong \frac{V_{b_n}\phi^\circ_n(c+c_\phi)}{\alpha_c B^\circ \Delta t}\left[p^{n+1}_n - p^n_n\right] \quad (7.118)$$

1. First time step calculations ($n=0$, $t_{n+1}=1$ day, and $\Delta t=\Delta t_1=1$ day)

Assign $p_1^n=p_2^n=p_3^n=p_4^n=p_5^n=p_{in}=4000$ psia.

$$\frac{V_{b_1}\phi_1^{\circ}(c+c_\phi)}{\alpha_c B^{\circ}\Delta t_1}=\frac{142.41339\times 0.23\times\left(1\times 10^{-5}+0\right)}{5.614583\times 1\times 1}$$
$$=58.339292\times 10^{-6}\,\text{STB/D-psi}$$

The calculated values of $\frac{V_{b_n}\phi_n^{\circ}(c+c_\phi)}{\alpha_c B^{\circ}\Delta t_1}$ for $n=1$, 2, 3, 4, 5 are reported in Table 7.2.

For gridblock 1, $n=1$, and $\psi_1=\{2\}$. Therefore, Eq. (7.118) becomes

$$T_{1,2}\left(p_2^{n+1}-p_1^{n+1}\right)+q_{sc_1}^{n+1}=\frac{V_{b_1}\phi_1^{\circ}(c+c_\phi)}{\alpha_c B^{\circ}\Delta t}\left[p_1^{n+1}-p_1^n\right] \tag{7.119}$$

Substitution of the values in this equation gives

$$41.6388914\left(p_2^{n+1}-p_1^{n+1}\right)-2000=58.339292\times 10^{-6}\left[p_1^{n+1}-4000\right]$$

or after factorizing and ordering the unknowns,

$$-41.6389497p_1^{n+1}+41.6388914p_2^{n+1}=1999.76664 \tag{7.145}$$

For gridblock 2, $n=2$, and $\psi_2=\{1,3\}$. Therefore, Eq. (7.118) becomes

$$T_{1,2}\left(p_1^{n+1}-p_2^{n+1}\right)+T_{2,3}\left(p_3^{n+1}-p_2^{n+1}\right)+q_{sc_2}^{n+1}=\frac{V_{b_2}\phi_2^{\circ}(c+c_\phi)}{\alpha_c B^{\circ}\Delta t}\left[p_2^{n+1}-p_2^n\right] \tag{7.121}$$

Substitution of the values in this equation gives

$$41.6388914\left(p_1^{n+1}-p_2^{n+1}\right)+41.6388914\left(p_3^{n+1}-p_2^{n+1}\right)+0$$
$$=1.24560063\times 10^{-3}\left[p_2^{n+1}-4000\right]$$

or after factorizing and ordering the unknowns,

$$41.6388914p_1^{n+1}-83.2790283p_2^{n+1}+41.6388914p_3^{n+1}=-4.98240254 \tag{7.146}$$

For gridblock 3, $n=3$, and $\psi_3=\{2,4\}$. Therefore, Eq. (7.118) becomes

$$T_{2,3}\left(p_2^{n+1}-p_3^{n+1}\right)+T_{3,4}\left(p_4^{n+1}-p_3^{n+1}\right)+q_{sc_3}^{n+1}=\frac{V_{b_3}\phi_3^{\circ}(c+c_\phi)}{\alpha_c B^{\circ}\Delta t}\left[p_3^{n+1}-p_3^n\right] \tag{7.123}$$

Substitution of the values in this equation gives

$$41.6388914\left(p_2^{n+1}-p_3^{n+1}\right)+41.6388914\left(p_4^{n+1}-p_3^{n+1}\right)+0$$
$$=0.026594785\left[p_3^{n+1}-4000\right]$$

or after factorizing and ordering the unknowns,

$$41.6388914p_2^{n+1} - 83.3043775p_3^{n+1} + 41.6388914p_4^{n+1} = -106.379139 \tag{7.147}$$

For gridblock 4, $n=4$, and $\psi_4=\{3,5\}$. Therefore, Eq. (7.118) becomes

$$T_{3,4}\left(p_3^{n+1} - p_4^{n+1}\right) + T_{4,5}\left(p_5^{n+1} - p_4^{n+1}\right) + q_{sc_4}^{n+1} = \frac{V_{b_4}\phi_4^\circ (c + c_\phi)}{\alpha_c B^\circ \Delta t}\left[p_4^{n+1} - p_4^n\right] \tag{7.125}$$

Substitution of the values in this equation gives

$$41.6388914\left(p_3^{n+1} - p_4^{n+1}\right) + 41.6388914\left(p_5^{n+1} - p_4^{n+1}\right) + 0 = 0.56782451\left[p_4^{n+1} - 4000\right]$$

or after factorizing and ordering the unknowns,

$$41.6388914p_3^{n+1} - 83.8456072p_4^{n+1} + 41.6388914p_5^{n+1} = -2271.29805 \tag{7.148}$$

For gridblock 5, $n=5$, and $\psi_5=\{4\}$. Therefore, Eq. (7.118) becomes

$$T_{4,5}\left(p_4^{n+1} - p_5^{n+1}\right) + q_{sc_5}^{n+1} = \frac{V_{b_5}\phi_5^\circ (c + c_\phi)}{\alpha_c B^\circ \Delta t}\left[p_5^{n+1} - p_5^n\right] \tag{7.127}$$

Substitution of the values in this equation gives

$$41.6388914\left(p_4^{n+1} - p_5^{n+1}\right) + 0 = 10.110829\left[p_5^{n+1} - 4000\right]$$

or after factorizing and ordering the unknowns,

$$41.6388914p_4^{n+1} - 51.7497205p_5^{n+1} = -40443.3168 \tag{7.149}$$

The results of solving Eqs. (7.145), (7.146), (7.147), (7.148), and (7.149) for the unknown pressures are $p_1^{n+1}=3627.20$ psia, $p_2^{n+1}=3675.23$ psia, $p_3^{n+1}=3723.25$ psia, $p_4^{n+1}=3771.09$ psia, and $p_5^{n+1}=3815.82$ psia.

We apply Eq. (7.85) to check the material balance for the first time step:

$$I_{MB} = \frac{\sum_{n=1}^{5} \frac{V_{b_n}\phi_n^\circ (c + c_\phi)}{\alpha_c B^\circ \Delta t}\left[p_n^{n+1} - p_n^n\right]}{\sum_{n=1}^{5} q_{sc_n}^{n+1} + \sum_{n=1}^{5}\sum_{l\in\xi_n} q_{sc_{l,n}}^{n+1}}$$

$$= \frac{\begin{bmatrix} 58.339292 \times 10^{-6} \times (3627.20 - 4000) + 1.24560063 \times 10^{-3} \times (3675.23 - 4000) \\ +0.026594785 \times (3723.25 - 4000) + 0.56782451 \times (3771.09 - 4000) \\ +10.110829 \times (3815.82 - 4000) \end{bmatrix}}{[(0+0+0+0-2000) + (0+0+0+0+0)]}$$

$$= \frac{-1999.9796}{-2000} = 0.999990$$

Therefore, the material balance is satisfied.

2. Second time step calculations ($n=1$, $t_{n+1}=3$ days, and $\Delta t=\Delta t_2=2$ days)

Assign $p_1^n=3627.20$ psia, $p_2^n=3675.23$ psia, $p_3^n=3723.25$ psia, $p_4^n=3771.09$ psia, and $p_5^n=3815.82$ psia.

$$\frac{V_{b_1}\phi_1^\circ(c+c_\phi)}{\alpha_c B^\circ \Delta t_2}=\frac{142.41339\times 0.23\times(1\times 10^{-5}+0)}{5.614583\times 1\times 2}$$
$$=29.169646\times 10^{-6}\ \text{STB/D-psi}$$

The calculated values of $\frac{V_{b_i}\phi_i^\circ(c+c_\phi)}{\alpha_c B^\circ \Delta t_2}$ for $i=1, 2, 3, 4, 5$ are reported in Table 7.2.

The gridblock flow equations for the second time step are obtained by applying Eq. (7.118).

For gridblock 1,

$$41.6388914\left(p_2^{n+1}-p_1^{n+1}\right)-2000=29.169646\times 10^{-6}\left[p_1^{n+1}-3627.20\right]$$

or after factorizing and ordering the unknowns,

$$-41.6389205p_1^{n+1}+41.6388914p_2^{n+1}=1999.89420 \tag{7.150}$$

For gridblock 2,

$$41.6388914\left(p_1^{n+1}-p_2^{n+1}\right)+41.6388914\left(p_3^{n+1}-p_2^{n+1}\right)+0$$
$$=0.62280032\times 10^{-3}\left[p_2^{n+1}-3675.23\right]$$

or after factorizing and ordering the unknowns,

$$41.6388914p_1^{n+1}-83.2784055p_2^{n+1}+41.6388914p_3^{n+1}=-2.28893284 \tag{7.151}$$

For gridblock 3,

$$41.6388914\left(p_2^{n+1}-p_3^{n+1}\right)+41.6388914\left(p_4^{n+1}-p_3^{n+1}\right)+0$$
$$=0.01329739\left[p_3^{n+1}-3723.25\right]$$

or after factorizing and ordering the unknowns,

$$41.6388914p_2^{n+1}-83.2910801p_3^{n+1}+41.6388914p_4^{n+1}=-49.5095063 \tag{7.152}$$

For gridblock 4,

$$41.6388914\left(p_3^{n+1}-p_4^{n+1}\right)+41.6388914\left(p_5^{n+1}-p_4^{n+1}\right)+0$$
$$=0.28391226\left[p_4^{n+1}-3771.09\right]$$

or after factorizing and ordering the unknowns,

$$41.6388914p_3^{n+1}-83.561695p_4^{n+1}+41.6388914p_5^{n+1}=-1070.65989 \tag{7.153}$$

For gridblock 5,

$$41.6388914\left(p_4^{n+1}-p_5^{n+1}\right)+0=5.0554145\left[p_5^{n+1}-3815.82\right]$$

or after factorizing and ordering the unknowns,

$$41.6388914p_4^{n+1}-46.6943060p_5^{n+1}=-19290.5407 \tag{7.154}$$

The results of solving Eqs. (7.150), (7.151), (7.152), (7.153), and (7.154) for the unknown pressures are $p_1^{n+1}=3252.93$ psia, $p_2^{n+1}=3300.96$ psia, $p_3^{n+1}=3348.99$ psia, $p_4^{n+1}=3396.89$ psia, and $p_5^{n+1}=3442.25$ psia.

We apply Eq. (7.85) to check the material balance for the second time step:

$$I_{MB}=\frac{\sum_{n=1}^{5}\frac{V_{b_n}\phi_n^{\circ}\left(c+c_\phi\right)}{\alpha_c B^{\circ}\Delta t}\left[p_n^{n+1}-p_n^n\right]}{\sum_{n=1}^{5}q_{sc_n}^{n+1}+\sum_{n=1}^{5}\sum_{l\in\xi_n}q_{sc_{l,n}}^{n+1}}$$

$$=\frac{\left[\begin{array}{l}29.169646\times10^{-6}\times(3252.93-3627.20)+0.62280032\times10^{-3}\times(3300.96-3675.23)\\ +0.01329739\times(3348.99-3723.25)+0.28391226\times(3396.89-3771.09)\\ +5.0554145\times(3442.25-3815.82)\end{array}\right]}{[(0+0+0+0-2000)+(0+0+0+0+0)]}$$

$$=\frac{-2000.0119}{-2000}=1.000006$$

Therefore, the material balance is satisfied.

7.3.3 Compressible fluid flow equation

The density, FVF, and viscosity of compressible fluids at reservoir temperature are functions of pressure. Such dependence, however, is not as weak as the case in slightly compressible fluids. In this context, the FVF, viscosity, and density that appear on the LHS of the flow equation (Eq. 7.12) can be assumed constant but are updated at least once at the beginning of every time step. The accumulation term is expressed in terms of pressure change over a time step such that the material balance is preserved. The following expansion preserves material balance:

$$\frac{V_{b_n}}{\alpha_c\Delta t}\left[\left(\frac{\phi}{B}\right)_n^{n+1}-\left(\frac{\phi}{B}\right)_n^{n}\right]=\frac{V_{b_n}}{\alpha_c\Delta t}\left(\frac{\phi}{B_g}\right)_n'\left[p_n^{n+1}-p_n^n\right] \tag{7.155a}$$

where $\left(\frac{\phi}{B_g}\right)_n'$ is the chord slope of $\left(\frac{\phi}{B_g}\right)_n$ between the new pressure (p_n^{n+1}) and the old pressure (p_n^n). This chord slope is evaluated at the current time level but is one iteration lagging behind; that is,

$$\left(\frac{\phi}{B_g}\right)_n'=\left[\left(\frac{\phi}{B}\right)_n^{\overset{(\nu)}{n+1}}-\left(\frac{\phi}{B}\right)_n^{n}\right]/\left[p_n^{\overset{(\nu)}{n+1}}-p_n^n\right] \tag{7.156a}$$

As shown in Section 10.4.1, the RHS of Eq. (7.156a) can be expanded as

$$\left(\frac{\phi}{B_g}\right)'_n = \phi_n^{\overset{(\nu)}{n+1}}\left(\frac{1}{B_{g_n}}\right)' + \frac{1}{B_{g_n}^n}\phi'_n \tag{7.156b}$$

where again $\left(\frac{1}{B_{g_n}}\right)'$ and ϕ'_n are defined as the chord slopes estimated between values at the current time level at old iteration $\overset{(\nu)}{n+1}$ and old time level n,

$$\left(\frac{1}{B_{g_n}}\right)' = \left(\frac{1}{B_{g_n}^{\overset{(\nu)}{n+1}}} - \frac{1}{B_{g_n}^n}\right) / \left(p_n^{\overset{(\nu)}{n+1}} - p_n^n\right) \tag{7.157}$$

and

$$\phi'_n = \left(\phi_n^{\overset{(\nu)}{n+1}} - \phi_n^n\right) / \left(p_n^{\overset{(\nu)}{n+1}} - p_n^n\right) = \overset{\circ}{\phi}_n c_\phi \tag{7.158}$$

Alternatively, the accumulation term can be expressed in terms of pressure change over a time step using Eq. (7.9) and by observing that the contribution of rock compressibility is negligible compared with that of gas compressibility, resulting in

$$\frac{V_{b_n}}{\alpha_c \Delta t}\left[\left(\frac{\phi}{B}\right)_n^{n+1} - \left(\frac{\phi}{B}\right)_n^n\right] = \frac{V_{b_n}\overset{\circ}{\phi}_n}{\alpha_c \Delta t}\left[\frac{1}{B_{g_n}^{n+1}} - \frac{1}{B_{g_n}^n}\right]$$

$$= \frac{V_{b_n}\overset{\circ}{\phi}_n}{\alpha_c \Delta t}\left(\frac{\alpha_c T_{sc}}{p_{sc} T}\right)\left[\frac{p_n^{n+1}}{z_n^{n+1}} - \frac{p_n^n}{z_n^n}\right] = \frac{V_{b_n}\overset{\circ}{\phi}_n T_{sc}}{p_{sc} T \Delta t}\left[\frac{p_n^{n+1}}{z_n^{n+1}} - \frac{p_n^n}{z_n^n}\right] \tag{7.155b}$$

If we adopted the approximation given by Eq. (7.155b), then the flow equation for compressible fluids becomes

$$\sum_{l\in\psi_n} T_{l,n}^m\left[\left(p_l^m - p_n^m\right) - \gamma_{l,n}^n (Z_l - Z_n)\right] + \sum_{l\in\xi_n} q_{sc_{l,n}}^m + q_{sc_n}^m = \frac{V_{b_n}\overset{\circ}{\phi}_n T_{sc}}{p_{sc} T \Delta t}\left[\frac{p_n^{n+1}}{z_n^{n+1}} - \frac{p_n^n}{z_n^n}\right] \tag{7.159}$$

In this book, however, we adopt the approximation given by Eq. (7.155a), which is consistent with the treatment of multiphase flow in Chapter 10. The resulting flow equation for compressible fluids becomes

$$\sum_{l\in\psi_n} T_{l,n}^m\left[\left(p_l^m - p_n^m\right) - \gamma_{l,n}^n (Z_l - Z_n)\right] + \sum_{l\in\xi_n} q_{sc_{l,n}}^m + q_{sc_n}^m = \frac{V_{b_n}}{\alpha_c \Delta t}\left(\frac{\phi}{B_g}\right)'_n \left[p_n^{n+1} - p_n^n\right] \tag{7.160}$$

where $\left(\frac{\phi}{B_g}\right)'_n$ is defined by Eq. (7.156b).

7.3.3.1 Formulations of compressible fluid flow equation

The time level m in Eq. (7.160) is approximated in reservoir simulation in one of three ways, like in the case for slightly compressible fluids. The resulting equation is commonly known as the explicit formulation of the flow equation (or forward-central-difference equation), the implicit formulation of the flow equation (or backward-central-difference equation), and the Crank-Nicolson formulation of the flow equation (or second-order-central-difference equation).

Explicit formulation of the flow equation

The explicit formulation of the flow equation can be obtained from Eq. (7.160) if the argument F^m (defined in Section 2.6.3) is dated at old time level t^n; that is, $t^m \cong t^n$, and as a result, $F^m \cong F^n$. Therefore, Eq. (7.160) reduces to

$$\sum_{l\in\psi_n} T_{l,n}^n\left[\left(p_l^n - p_n^n\right) - \gamma_{l,n}^n(Z_l - Z_n)\right] + \sum_{l\in\xi_n} q_{sc_{l,n}}^n + q_{sc_n}^n \cong \frac{V_{b_n}}{\alpha_c \Delta t}\left(\frac{\phi}{B_g}\right)'_n \left[p_n^{n+1} - p_n^n\right] \tag{7.161a}$$

or

$$\sum_{l\in\psi_{i,j,k}} T_{l,(i,j,k)}^n\left[\left(p_l^n - p_{i,j,k}^n\right) - \gamma_{l,(i,j,k)}^n\left(Z_l - Z_{i,j,k}\right)\right] + \sum_{l\in\xi_{i,j,k}} q_{sc_{l,(i,j,k)}}^n + q_{sc_{i,j,k}}^n$$
$$\cong \frac{V_{b_{i,j,k}}}{\alpha_c \Delta t}\left(\frac{\phi}{B_g}\right)'_{i,j,k}\left[p_{i,j,k}^{n+1} - p_{i,j,k}^n\right] \tag{7.161b}$$

In addition to the remarks related to the explicit formulation method mentioned in Section 7.3.2.1, the solution of Eq. (7.161) requires iterations to remove the nonlinearity of the equation exhibited by $\overset{(v)}{B}{}_{g_n}^{n+1}$ in the definition of $\left(\frac{\phi}{B_g}\right)'_n$ on the RHS of the equation.

Implicit formulation of the flow equation

The implicit formulation of the flow equation can be obtained from Eq. (7.160) if the argument F^m (defined in Section 2.6.3) is dated at new time level t^{n+1}; that is, $t^m \cong t^{n+1}$, and as a result, $F^m \cong F^{n+1}$. Therefore, Eq. (7.160) reduces to

$$\sum_{l\in\psi_n} T_{l,n}^{n+1}\left[\left(p_l^{n+1} - p_n^{n+1}\right) - \gamma_{l,n}^n(Z_l - Z_n)\right] + \sum_{l\in\xi_n} q_{sc_{l,n}}^{n+1} + q_{sc_n}^{n+1}$$
$$\cong \frac{V_{b_n}}{\alpha_c \Delta t}\left(\frac{\phi}{B_g}\right)'_n \left[p_n^{n+1} - p_n^n\right] \tag{7.162a}$$

or

$$\sum_{l\in\psi_{i,j,k}} T_{l,(i,j,k)}^{n+1}\left[\left(p_l^{n+1}-p_{i,j,k}^{n+1}\right)-\gamma_{l,(i,j,k)}^{n}\left(Z_l-Z_{i,j,k}\right)\right]+\sum_{l\in\xi_{i,j,k}} q_{sc_{l,(i,j,k)}}^{n+1}+q_{sc_{i,j,k}}^{n+1}$$
$$\cong\frac{V_{b_{i,j,k}}}{\alpha_c\Delta t}\left(\frac{\phi}{B_g}\right)'_{i,j,k}\left[p_{i,j,k}^{n+1}-p_{i,j,k}^{n}\right] \tag{7.162b}$$

In this equation, dating fluid gravity at old time level n instead of time level $n+1$ does not introduce any noticeable errors (Coats et al., 1974). Unlike Eq. (7.81) for slightly compressible fluids, Eq. (7.162) is a nonlinear equation due to the dependence of transmissibility ($T_{l,n}^{n+1}$) and $\left(\frac{\phi}{B_g}\right)'_n$ on the pressure solution. These nonlinear terms present a serious numerical problem. Chapter 8 discusses the linearization of these terms in space and time. The time linearization; however, introduces additional truncation errors that depend on time steps. Thus, time linearization reduces the accuracy of solution and generally restricts time step. This leads to the erasing of the advantage of unconditional stability associated with the implicit formulation method mentioned in Section 7.3.2.1.

Crank-Nicolson formulation of the flow equation

The Crank-Nicolson formulation of the flow equation can be obtained from Eq. (7.160) if the argument F^m (defined in Section 2.6.3) is dated at time $t^{n+1/2}$. In the mathematical approach, this time level was selected to make the RHS of Eq. (7.160) a second-order approximation in time. In the engineering approach; however, the argument F^m can be approximated as $F^m\cong F^{n+1/2}={}^1/_2(F^n+F^{n+1})$. Therefore, Eq. (7.160) becomes

$$½\sum_{l\in\psi_n} T_{l,n}^{n}\left[\left(p_l^{n}-p_n^{n}\right)-\gamma_{l,n}^{n}(Z_l-Z_n)\right]+½\sum_{l\in\psi_n} T_{l,n}^{n+1}\left[\left(p_l^{n+1}-p_n^{n+1}\right)-\gamma_{l,n}^{n}(Z_l-Z_n)\right]$$
$$+½\left(\sum_{l\in\xi_n} q_{sc_{l,n}}^{n}+\sum_{l\in\xi_n} q_{sc_{l,n}}^{n+1}\right)+½\left(q_{sc_n}^{n}+q_{sc_n}^{n+1}\right)\cong\frac{V_{b_n}}{\alpha_c\Delta t}\left(\frac{\phi}{B_g}\right)'_n\left[p_n^{n+1}-p_n^{n}\right] \tag{7.163a}$$

Eq. (7.163a) can be rewritten in the form of Eq. (7.162) as

$$\sum_{l\in\psi_n} T_{l,n}^{n+1}\left[\left(p_l^{n+1}-p_n^{n+1}\right)-\gamma_{l,n}^{n}(Z_l-Z_n)\right]+\sum_{l\in\xi_n} q_{sc_{l,n}}^{n+1}+q_{sc_n}^{n+1}$$
$$\cong\frac{V_{b_n}}{\alpha_c(\Delta t/2)}\left(\frac{\phi}{B_g}\right)'_n\left[p_n^{n+1}-p_n^{n}\right] \tag{7.163b}$$
$$-\left\{\sum_{l\in\psi_n} T_{l,n}^{n}\left[\left(p_l^{n}-p_n^{n}\right)-\gamma_{l,n}^{n}(Z_l-Z_n)\right]+\sum_{l\in\xi_n} q_{sc_{l,n}}^{n}+q_{sc_n}^{n}\right\}$$

7.3.3.2 Advancing the pressure solution in time

The pressure distribution in a compressible flow problem changes with time, as is the case with slightly compressible fluid flow. Therefore, a compressible fluid flow problem has an unsteady-state solution, and the pressure solution is obtained in the same way as that for slightly compressible fluid flow discussed in Section 7.3.2.2, with a few exceptions. These include the following: (1) Initialization may require iteration because gas gravity is a function of pressure; (2) transmissibilities in step 1 are not kept constant but rather are calculated at the upstream blocks and updated at the beginning of each time step; (3) Eq. (7.161), (7.162), or (7.163) is used instead of Eq. (7.80), (7.81), or (7.82) in step 4; (4) an additional step immediately before step 5 is added to linearize the flow equations for compressible fluid (discussed in Chapter 8); and (5) obtaining the pressure solution may require iterations because the flow equation for compressible fluid is nonlinear compared with the almost-linear flow equation for slightly compressible fluid.

7.3.3.3 Material balance check for a compressible fluid flow problem

For the implicit formulation, the incremental and cumulative material balance checks for compressible fluid flow problems are given by Eqs. (7.83) and (7.84), where the rock porosity is defined by Eq. (7.11) and FVF is for natural gas, yielding

$$I_{MB} = \frac{\sum_{n=1}^{N} \frac{V_{b_n}}{\alpha_c \Delta t} \left[\left(\frac{\phi}{B_g} \right)_n^{n+1} - \left(\frac{\phi}{B_g} \right)_n^{n} \right]}{\sum_{n=1}^{N} \left(q_{sc_n}^{n+1} + \sum_{l \in \xi_n} q_{sc_{l,n}}^{n+1} \right)} \tag{7.164}$$

and

$$C_{MB} = \frac{\sum_{n=1}^{N} \frac{V_{b_n}}{\alpha_c} \left[\left(\frac{\phi}{B_g} \right)_n^{n+1} - \left(\frac{\phi}{B_g} \right)_n^{0} \right]}{\sum_{m=1}^{n+1} \Delta t_m \sum_{n=1}^{N} \left(q_{sc_n}^{m} + \sum_{l \in \xi_n} q_{sc_{l,n}}^{m} \right)} \tag{7.165}$$

where N is the total number of blocks in the reservoir.

The following example presents a single-well simulation of a natural gas reservoir. It demonstrates the iterative nature of the solution method within individual time steps and the progression of the solution in time.

Example 7.13 A vertical well is drilled on 20-acre spacing in a natural gas reservoir. The reservoir is described by four gridblocks in the radial direction as shown in Fig. 7.14. The reservoir is horizontal and has 30 ft net thickness

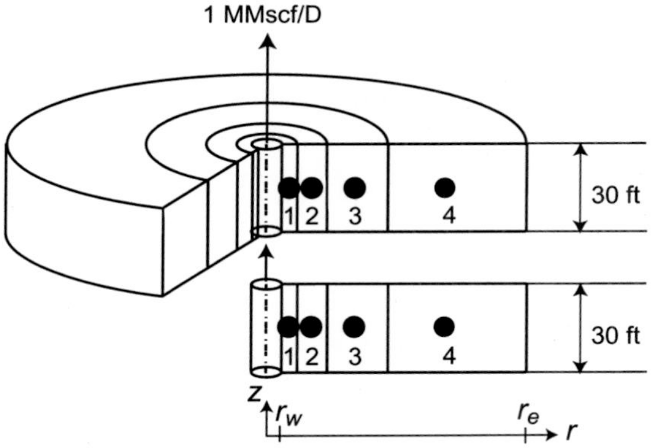

FIG. 7.14 Discretized 1-D reservoir in Example 7.13.

and homogeneous and isotropic rock properties with $k = 15$ md and $\phi = 0.13$. Initially, reservoir pressure is 4015 psia. Gas FVF and viscosity dependence on pressure are presented in Table 7.3. The reservoir external boundaries are sealed off to fluid flow. Well diameter is 6 in. The well produces 1 MMscf/D with a minimum FBHP of 515 psia. Find the pressure distribution in the reservoir every month (30.42 days) for 2 years. Take time steps of 30.42 days.

Solution

The interblock geometric factors in the radial direction and the gridblock bulk volumes can be calculated exactly as in Example 7.12. Alternatively, gridblock boundaries are estimated using Eqs. (4.82a), (4.83a), (4.84a), and (4.85a), followed by estimating interblock geometric factors using Table 4.2 and gridblock bulk volumes using Eqs. (4.88a) and (4.88c). The gridblock boundaries, bulk volume, and geometric factors are presented in Table 7.4.

For single-well simulation in a horizontal reservoir ($Z_n =$ constant) with no-flow boundaries $\left(\sum_{l \in \xi_n} q_{sc_{l,n}}^{n+1} = 0\right)$, Eq. (7.162a) reduces to

$$\sum_{l \in \psi_n} T_{l,n}^{n+1}\left[\left(p_l^{n+1} - p_n^{n+1}\right)\right] + q_{sc_n}^{n+1} = \frac{V_{b_n}}{\alpha_c \Delta t}\left(\frac{\phi}{B_g}\right)'_n \left[p_n^{n+1} - p_n^n\right] \tag{7.166a}$$

The gas in this reservoir flows toward the well in gridblock 1. Therefore, gridblock 4 is upstream to gridblock 3, gridblock 3 is upstream to gridblock 2, and gridblock 2 is upstream to gridblock 1. In solving this problem, we use the implicit formulation with simple iteration (Section 8.4.1.2) and upstream weighting (Section 8.4.1.1) of the pressure-dependent terms in transmissibility. Placing the iteration level, Eq. (6.166a) becomes

$$\sum_{l \in \psi_n} T_{l,n}^{\overset{(\nu)}{n+1}}\left[\left(p_l^{\overset{(\nu+1)}{n+1}} - p_n^{\overset{(\nu+1)}{n+1}}\right)\right] + q_{sc_n}^{\overset{(\nu+1)}{n+1}} = \frac{V_{b_n}}{\alpha_c \Delta t}\left(\frac{\phi}{B_g}\right)'_n \left[p_n^{\overset{(\nu+1)}{n+1}} - p_n^n\right] \tag{7.166b}$$

TABLE 7.3 Gas FVF and viscosity for Example 7.13.

Pressure (psia)	GFVF (RB/scf)	Gas viscosity (cP)
215.00	0.016654	0.0126
415.00	0.008141	0.0129
615.00	0.005371	0.0132
815.00	0.003956	0.0135
1015.00	0.003114	0.0138
1215.00	0.002544	0.0143
1415.00	0.002149	0.0147
1615.00	0.001857	0.0152
1815.00	0.001630	0.0156
2015.00	0.001459	0.0161
2215.00	0.001318	0.0167
2415.00	0.001201	0.0173
2615.00	0.001109	0.0180
2815.00	0.001032	0.0186
3015.00	0.000972	0.0192
3215.00	0.000922	0.0198
3415.00	0.000878	0.0204
3615.00	0.000840	0.0211
3815.00	0.000808	0.0217
4015.00	0.000779	0.0223

TABLE 7.4 Gridblock locations, boundaries, bulk volumes, and interblock geometric factors for Example 7.13.

i	n	r_i (ft)	$r^L_{i-1/2}$ (ft)	$r^L_{i+1/2}$ (ft)	$r_{i-1/2}$ (ft)	$r_{i+1/2}$ (ft)	$G_{r_{i+1/2}}$ (RB-cP/D-psi)	V_{b_n} (ft^3)
1	1	0.5611	0.2500	1.6937	0.2837	1.9221	1.6655557	340.59522
2	2	3.8014	1.6937	11.4739	1.9221	13.0213	1.6655557	15,631.859
3	3	25.7532	11.4739	77.7317	13.0213	88.2144	1.6655557	717,435.23
4	4	174.4683	77.7317	526.6040	88.2144	526.6040	1.6655557	25,402,604

1. First time step calculations ($n=0$, $t_{n+1}=30.42$ days, and $\Delta t=30.42$ days)

Assign $p_1^n=p_2^n=p_3^n=p_4^n=p_{in}=4015$ psia.

For the first iteration ($\nu=0$), assume $p_n^{\overset{(\nu)}{n+1}}=p_n^n=4015$ psia for $n=1, 2, 3, 4$. Table 7.5 presents the estimated values of FVF and viscosity using linear interpolation within table entries, chord slope $\left(\frac{\phi}{B_g}\right)_n'$, and $\frac{V_{b_n}}{\alpha_c \Delta t}\left(\frac{\phi}{B_g}\right)_n'$ for all grid blocks. It should be mentioned; however, that the calculation of $\left(\frac{\phi}{B_g}\right)_n'$ used a perturbed value of $p_n^{\overset{(\nu)}{n+1}}=p_n^n-\varepsilon=p_n^n-1=4015-1=4014$ psia for $n=1$, 2, 3, 4 only for the first iteration.

For example, for gridblock 1,

$$\left(\frac{\phi}{B_g}\right)_1'=\frac{\left(\frac{\phi}{B}\right)_1^{\overset{(\nu)}{n+1}}-\left(\frac{\phi}{B}\right)_1^n}{p_1^{\overset{(\nu)}{n+1}}-p_1^n}=\frac{\left(\frac{0.13}{0.00077914}\right)-\left(\frac{0.13}{0.000779}\right)}{4014-4015}=0.03105672$$

$$\frac{V_{b_1}}{\alpha_c \Delta t}\left(\frac{\phi}{B_g}\right)_1'=\frac{340.59522\times 0.03105672}{5.614583\times 30.42}=0.06193233$$

and

$$T_{r_{1,2}}^{\overset{(\nu)}{n+1}}=T_{r_{1,2}}^{\overset{(\nu)}{n+1}}\Big|_2=G_{r_{1+1/2}}\left(\frac{1}{\mu B}\right)_2^{\overset{(\nu)}{n+1}}=1.6655557\times\left(\frac{1}{0.0223000\times 0.00077900}\right)$$
$$=95877.5281$$

for upstream weighting of transmissibility.

Therefore, $T_{r_{1,2}}^{\overset{(\nu)}{n+1}}\Big|_2=T_{r_{2,3}}^{\overset{(\nu)}{n+1}}\Big|_3=T_{r_{3,4}}^{\overset{(\nu)}{n+1}}\Big|_4=95877.5281$ scf/D-psi. Note that upstream weighting is not evident for the first iteration in the first time step because all gridblock pressures are assumed equal.

TABLE 7.5 Estimated gridblock FVF, viscosity, and chord slope at old iteration $\nu=0$.

Block n	$p_n^{\overset{(0)}{n+1}}$ (psia)	B_g (RB/scf)	μ_g (cP)	$\left(\frac{\phi}{B_g}\right)_n'$	$\frac{V_{b_n}}{\alpha_c \Delta t}\left(\frac{\phi}{B_g}\right)_n'$
1	4015	0.00077900	0.0223000	0.03105672	0.06193233
2	4015	0.00077900	0.0223000	0.03105672	2.842428
3	4015	0.00077900	0.0223000	0.03105672	130.4553
4	4015	0.00077900	0.0223000	0.03105672	4619.097

For gridblock 1, $n=1$, and $\psi_1=\{2\}$. Therefore, Eq. (7.166b) becomes

$$T_{2,1}^{\overset{(\nu)}{n+1}}\Big|_2\left(\overset{(\nu+1)}{p_2^{n+1}}-\overset{(\nu+1)}{p_1^{n+1}}\right)+q_{sc_1}^{n+1}=\frac{V_{b_1}}{\alpha_c\Delta t}\left(\frac{\phi}{B_g}\right)'_1\left[\overset{(\nu+1)}{p_1^{n+1}}-p_1^n\right] \tag{7.167}$$

Substitution of the values in this equation gives

$$95877.5281\left(\overset{(\nu+1)}{p_2^{n+1}}-\overset{(\nu+1)}{p_1^{n+1}}\right)-10^6=0.06193233\left[\overset{(\nu+1)}{p_1^{n+1}}-4015\right]$$

or after factorizing and ordering the unknowns,

$$-95877.5900\overset{(\nu+1)}{p_1^{n+1}}+95877.5281\overset{(\nu+1)}{p_2^{n+1}}=999751.1342 \tag{7.168}$$

For gridblock 2, $n=2$, and $\psi_2=\{1,3\}$. Therefore, Eq. (7.166b) becomes

$$T_{1,2}^{\overset{(\nu)}{n+1}}\Big|_2\left(\overset{(\nu+1)}{p_1^{n+1}}-\overset{(\nu+1)}{p_2^{n+1}}\right)+T_{3,2}^{\overset{(\nu)}{n+1}}\Big|_3\left(\overset{(\nu+1)}{p_3^{n+1}}-\overset{(\nu+1)}{p_2^{n+1}}\right)$$
$$+q_{sc_2}^{n+1}=\frac{V_{b_2}}{\alpha_c\Delta t}\left(\frac{\phi}{B_g}\right)'_2\left[\overset{(\nu+1)}{p_2^{n+1}}-p_2^n\right] \tag{7.169}$$

Substitution of the values in this equation gives

$$95877.5281\left(\overset{(\nu+1)}{p_1^{n+1}}-\overset{(\nu+1)}{p_2^{n+1}}\right)+95877.5281\left(\overset{(\nu+1)}{p_3^{n+1}}-\overset{(\nu+1)}{p_2^{n+1}}\right)+0$$
$$=2.842428\left[\overset{(\nu+1)}{p_2^{n+1}}-4015\right]$$

or after factorizing and ordering the unknowns,

$$95877.5281\overset{(\nu+1)}{p_1^{n+1}}-191757.899\overset{(\nu+1)}{p_2^{n+1}}+95877.5281\overset{(\nu+1)}{p_3^{n+1}}=-11412.3496 \tag{7.170}$$

For gridblock 3, $n=3$, and $\psi_3=\{2,4\}$. Therefore, Eq. (7.166b) becomes

$$T_{2,3}^{\overset{(\nu)}{n+1}}\Big|_3\left(\overset{(\nu+1)}{p_2^{n+1}}-\overset{(\nu+1)}{p_3^{n+1}}\right)+T_{4,3}^{\overset{(\nu)}{n+1}}\Big|_4\left(\overset{(\nu+1)}{p_4^{n+1}}-\overset{(\nu+1)}{p_3^{n+1}}\right)+q_{sc_3}^{n+1}$$
$$=\frac{V_{b_3}}{\alpha_c\Delta t}\left(\frac{\phi}{B_g}\right)'_3\left[\overset{(\nu+1)}{p_3^{n+1}}-p_3^n\right] \tag{7.171}$$

Substitution of the values in this equation gives

$$95877.5281\left(\overset{(\nu+1)}{p_2^{n+1}}-\overset{(\nu+1)}{p_3^{n+1}}\right)+95877.5281\left(\overset{(\nu+1)}{p_4^{n+1}}-\overset{(\nu+1)}{p_3^{n+1}}\right)+0$$
$$=130.4553\left[\overset{(\nu+1)}{p_3^{n+1}}-4015\right]$$

or after factorizing and ordering the unknowns,

$$95877.5281 \overset{(\nu+1)}{p_2^{n+1}} - 191885.511 \overset{(\nu+1)}{p_3^{n+1}} + 95877.5281 \overset{(\nu+1)}{p_4^{n+1}} = -523777.862 \tag{7.172}$$

For gridblock 4, $n=4$, and $\psi_4=\{3\}$. Therefore, Eq. (7.166b) becomes

$$\overset{(\nu)}{T_{3,4}^{n+1}}\Big|_4 \left(\overset{(\nu+1)}{p_3^{n+1}} - \overset{(\nu+1)}{p_4^{n+1}}\right) + q_{sc_4}^{n+1} = \frac{V_{b_4}}{\alpha_c \Delta t}\left(\frac{\phi}{B_g}\right)_4' \left[\overset{(\nu+1)}{p_4^{n+1}} - p_4^n\right] \tag{7.173}$$

Substitution of the values in this equation gives

$$95877.5281\left(\overset{(\nu+1)}{p_3^{n+1}} - \overset{(\nu+1)}{p_4^{n+1}}\right) + 0 = 4619.097\left[\overset{(\nu+1)}{p_4^{n+1}} - 4015\right]$$

or after factorizing and ordering the unknowns,

$$95877.5281 \overset{(\nu+1)}{p_3^{n+1}} - 100496.626 \overset{(\nu+1)}{p_4^{n+1}} = -18545676.2 \tag{7.174}$$

The results of solving Eqs. (7.168), (7.170), (7.172), and (7.174) for the unknown pressures are $\overset{(1)}{p_1^{n+1}} = 3773.90$ psia, $\overset{(1)}{p_2^{n+1}} = 3784.33$ psia, $\overset{(1)}{p_3^{n+1}} = 3794.75$ psia, and $\overset{(1)}{p_4^{n+1}} = 3804.87$ psia.

For the second iteration ($\nu=1$), we use $\overset{(1)}{p_n^{n+1}}$ to estimate the values of FVF and viscosity using linear interpolation within table entries, chord slope $\left(\frac{\phi}{B_g}\right)_n'$, and $\frac{V_{b_n}}{\alpha_c \Delta t}\left(\frac{\phi}{B_g}\right)_n'$ for gridblock n. Table 7.6 lists these values. For example, for gridblock 1,

$$\left(\frac{\phi}{B_g}\right)_1' = \frac{\left(\frac{\phi}{B}\right)_1^{\overset{(\nu)}{n+1}} - \left(\frac{\phi}{B}\right)_1^n}{\overset{(\nu)}{p_1^{n+1}} - p_1^n} = \frac{\left(\frac{0.13}{0.00081458}\right) - \left(\frac{0.13}{0.000779}\right)}{3773.90 - 4015} = 0.03022975$$

TABLE 7.6 Estimated gridblock FVF, viscosity, and chord slope at old iteration $\nu=1$.

Block n	$\overset{(1)}{p_n^{n+1}}$ (psia)	B_g (RB/scf)	μ_g (cP)	$\left(\frac{\phi}{B_g}\right)_n'$	$\frac{V_{b_n}}{\alpha_c \Delta t}\left(\frac{\phi}{B_g}\right)_n'$
1	3773.90	0.00081458	0.0215767	0.03022975	0.0602832
2	3784.33	0.00081291	0.0216080	0.03017631	2.761849
3	3794.75	0.00081124	0.0216392	0.03011173	126.4858
4	3804.87	0.00080962	0.0216696	0.03003771	4467.390

$$\frac{V_{b_1}}{\alpha_c \Delta t}\left(\frac{\phi}{B_g}\right)'_1 = \frac{340.59522 \times 0.03022975}{5.614583 \times 30.42} = 0.0602832$$

and

$$\overset{(\nu)}{T^{n+1}_{r_{2,1}}} = \overset{(\nu)}{T^{n+1}_{r_{2,1}}}\bigg|_2 = G_{r_{1+1/2}}\left(\frac{1}{\mu B}\right)_2^{\overset{(\nu)}{n+1}}$$
$$= 1.6655557 \times \left(\frac{1}{0.0216080 \times 0.00081291}\right) = 94820.8191$$

for upstream weighting of transmissibility.

Similarly, $\overset{(\nu)}{T^{n+1}_{r_{3,2}}}\bigg|_3 = 94878.4477$ scf/D-psi and $\overset{(\nu)}{T^{n+1}_{r_{4,3}}}\bigg|_4 = 94935.0267$ scf/D-psi.

For gridblock 1, $n=1$. Substituting the values in Eq. (7.167) gives

$$94820.8191\left(\overset{(\nu+1)}{p_2^{n+1}} - \overset{(\nu+1)}{p_1^{n+1}}\right) - 10^6 = 0.0602832\left[\overset{(\nu+1)}{p_1^{n+1}} - 4015\right]$$

or after factorizing and ordering the unknowns,

$$-94820.8794\overset{(\nu+1)}{p_1^{n+1}} + 94820.8191\overset{(\nu+1)}{p_2^{n+1}} = 999757.963 \tag{7.175}$$

For gridblock 2, $n=2$. Substituting the values in Eq. (7.169) gives

$$94820.8191\left(\overset{(\nu+1)}{p_1^{n+1}} - \overset{(\nu+1)}{p_2^{n+1}}\right) + 94878.4477\left(\overset{(\nu+1)}{p_3^{n+1}} - \overset{(\nu+1)}{p_2^{n+1}}\right) + 0$$
$$= 2.761849\left[\overset{(\nu+1)}{p_2^{n+1}} - 4015\right]$$

or after factorizing and ordering the unknowns,

$$94820.8191\overset{(\nu+1)}{p_1^{n+1}} - 189702.029\overset{(\nu+1)}{p_2^{n+1}} + 94878.4477\overset{(\nu+1)}{p_3^{n+1}} = -11088.8252 \tag{7.176}$$

For gridblock 3, $n=3$. Substituting the values in Eq. (7.171) gives

$$94878.4477\left(\overset{(\nu+1)}{p_2^{n+1}} - \overset{(\nu+1)}{p_3^{n+1}}\right) + 94935.0267\left(\overset{(\nu+1)}{p_4^{n+1}} - \overset{(\nu+1)}{p_3^{n+1}}\right) + 0$$
$$= 126.4858\left[\overset{(\nu+1)}{p_3^{n+1}} - 4015\right]$$

or after factorizing and ordering the unknowns,

$$94878.4477\overset{(\nu+1)}{p_2^{n+1}} - 189939.960\overset{(\nu+1)}{p_3^{n+1}} + 94935.0267\overset{(\nu+1)}{p_4^{n+1}} = -507840.406 \tag{7.177}$$

TABLE 7.7 The pressure solution at $t_{n+1}=30.42$ days for successive iterations.

ν	p_1^{n+1} (psia)	p_2^{n+1} (psia)	p_3^{n+1} (psia)	p_4^{n+1} (psia)
1	3773.90	3784.33	3794.75	3804.87
2	3766.44	3776.99	3787.52	3797.75
3	3766.82	3777.37	3787.91	3798.14

For gridblock 4, $n=4$. Substituting the values in Eq. (7.173) gives

$$94935.0267\left(\overset{(\nu+1)}{p_3^{n+1}}-\overset{(\nu+1)}{p_4^{n+1}}\right)+0=4467.390\left[\overset{(\nu+1)}{p_4^{n+1}}-4015\right]$$

or after factorizing and ordering the unknowns,

$$94935.0267\overset{(\nu+1)}{p_3^{n+1}}-99402.4167\overset{(\nu+1)}{p_4^{n+1}}=-17936570.6 \tag{7.178}$$

The results of solving Eqs. (7.175), (7.176), (7.177), and (7.178) for the unknown pressures are $\overset{(2)}{p_1^{n+1}}=3766.44$ psia, $\overset{(2)}{p_2^{n+1}}=3776.99$ psia, $\overset{(2)}{p_3^{n+1}}=3787.52$ psia, and $\overset{(2)}{p_4^{n+1}}=3797.75$ psia.

The iterations continue until the convergence criterion is satisfied. The successive iterations for the first time step are shown in Table 7.7. It can be seen that it took three iterations to converge. The convergence criterion was set as

$$\max_{1\le n\le N}\left|\frac{\overset{(\nu+1)}{p_n^{n+1}}-\overset{(\nu)}{p_n^{n+1}}}{\overset{(\nu)}{p_n^{n+1}}}\right|\le 0.001 \tag{7.179}$$

After reaching convergence, time is incremented by $\Delta t=30.42$ days. and the above procedure is repeated. The converged solutions at various times up to 2 years of simulation time are shown in Table 7.8. Inspection of the simulation results reported in Table 7.8 reveals that the well switched to a constant FBHP of 500 psia after 21 months because the reservoir does not have the capacity to produce gas at the specified rate of 1 MMscf/D.

7.4 Summary

Reservoir fluids are incompressible, slightly compressible, or compressible. The flow equation for an incompressible fluid in incompressible porous media is described by Eq. (7.16). Reservoir pressure in this case has steady-state behavior and can be obtained using the algorithm presented in Section 7.3.1.1.

TABLE 7.8 The converged pressure solution and gas production at various times.

$n+1$	Time (day)	ν	p_1^{n+1} (psia)	p_2^{n+1} (psia)	p_3^{n+1} (psia)	p_4^{n+1} (psia)	$p_{wf_1}^{n+1}$ (psia)	q_{gsc}^{n+1} (MMscf/D)	Cumulative production (MMMscf)
1	30.42	3	3766.82	3777.37	3787.91	3798.14	3762.36	−1.000000	−0.0304200
2	60.84	3	3556.34	3567.01	3577.67	3588.02	3551.82	−1.000000	−0.0608400
3	91.26	3	3362.00	3372.80	3383.58	3394.05	3357.43	−1.000000	−0.0912600
4	121.68	3	3176.08	3187.08	3198.06	3208.72	3171.43	−1.000000	−0.121680
5	152.10	3	2995.56	3006.78	3017.97	3028.85	2990.81	−1.000000	−0.152100
6	182.52	3	2827.23	2838.72	2850.18	2861.32	2822.36	−1.000000	−0.182520
7	212.94	3	2673.43	2685.26	2697.06	2708.50	2668.42	−1.000000	−0.212940
8	243.36	2	2524.28	2536.47	2548.62	2560.41	2519.12	−1.000000	−0.243360
9	273.78	3	2375.01	2387.59	2400.12	2412.25	2369.67	−1.000000	−0.273780
10	304.20	3	2241.26	2254.33	2267.35	2279.97	2235.71	−1.000000	−0.304200
11	334.62	3	2103.68	2117.34	2130.93	2144.09	2097.88	−1.000000	−0.334620
12	365.04	3	1961.05	1975.39	1989.65	2003.42	1954.95	−1.000000	−0.365040
13	395.46	3	1821.72	1836.86	1851.91	1866.47	1815.29	−1.000000	−0.395460
14	425.88	3	1684.94	1701.18	1717.27	1732.78	1678.02	−1.000000	−0.425880
15	456.30	3	1543.26	1560.78	1578.11	1594.79	1535.78	−1.000000	−0.456300
16	486.72	4	1403.75	1422.64	1441.34	1459.36	1395.67	−1.000000	−0.486720
17	517.14	3	1263.19	1284.07	1304.65	1324.36	1254.24	−1.000000	−0.517140
18	547.56	3	1114.51	1137.93	1160.87	1182.74	1104.42	−1.000000	−0.547560
19	577.98	4	964.49	991.04	1016.79	1041.39	952.91	−1.000000	−0.577980
20	608.40	4	812.91	844.10	874.32	902.83	799.30	−1.000000	−0.608400
21	638.82	3	645.89	684.85	721.84	755.98	628.58	−1.000000	−0.638820
22	669.24	4	531.46	567.57	601.01	631.84	515.00	−0.759957	−0.661938
23	699.66	4	523.60	543.17	561.98	579.67	515.00	−0.391107	−0.673835
24	730.08	3	519.68	530.53	541.13	551.32	515.00	−0.211379	−0.680266

For a slightly compressible fluid, the flow equation can be expressed using explicit formulation (Eq. 7.80), implicit formulation (Eq. 7.81), or the Crank-Nicolson formulation (Eq. 7.82). For a compressible fluid (natural gas), the flow equation can be expressed using explicit formulation (Eq. 7.161), implicit formulation (Eq. 7.162), or the Crank-Nicolson formulation (Eq. 7.163). Reservoir pressure for slightly compressible and compressible fluids has unsteady-state behavior. The pressure solution is obtained by marching in time from the initial conditions to the desired time using time steps. Advancing the pressure solution one time step is obtained using the algorithms presented in Section 7.3.2.2 for a slightly compressible fluid or that presented in Section 7.3.3.2 for a compressible fluid. Material balance is checked every time a pressure solution is obtained. Eq. (7.17) applies to incompressible fluid flow. Eqs. (7.83) and (7.84) (or Eqs. 7.85 and 7.86) apply to slightly compressible fluid flow, and Eqs. (7.164) and (7.165) apply to compressible fluid flow. The incremental material balance checks, however, are more accurate than the cumulative material balance checks.

7.5 Exercises

7.1 Examine the various terms in Eq. (7.16a) for incompressible fluid flow and then give justification for describing it as a linear equation.

7.2 Examine Eq. (7.81a) for a slightly compressible fluid and then give justification and conditions under which it can be considered a linear equation.

7.3 Examine Eq. (7.162a) for a compressible fluid and then give justification for describing it as a nonlinear equation.

7.4 Explain why Eq. (7.80a) is explicit, whereas Eq. (7.81a) is implicit.

7.5 A 1-D reservoir consists of four gridblocks ($N=4$) and contains an incompressible fluid. The reservoir boundaries can be subject to any condition. Write the flow equation for each individual gridblock. Add up all flow equations and prove that the material balance for this reservoir is given by Eq. (7.17a) for $N=4$.

7.6 Repeat the procedure in Exercise 7.5 and prove that the incremental material balance for this reservoir is given by Eq. (7.85) for a slightly compressible fluid using the implicit formulation.

7.7 In order, list all the steps necessary to advance the pressure solution in time for compressible fluid flow in reservoirs.

7.8 Start with Eq. (7.88) and derive Eq. (7.86) as outlined in the text.

7.9 The estimation of transmissibility in incompressible fluid flow and in slightly compressible fluid flow, as presented in this chapter, does not mention or make use of upstream weighting. Explain why weighting (or upstream weighting) of transmissibility is not needed in these two cases.

7.10 Consider the single-phase flow of slightly compressible oil in the 2-D horizontal homogeneous reservoir shown in Fig. 7.15a. The reservoir is volumetric; that is, it has no-flow boundaries. Initial reservoir pressure is 4000 psia. Gridblock 5 houses a 7-in well at its center, which produces at a constant rate of 50 STB/D. Gridblock dimensions and properties are $\Delta x = \Delta y = 350$ ft, $h = 20$ ft, $k_x = k_y = 120$ md, and $\phi = 0.25$ Oil properties are $B_o = B_o^\circ = 1$ RB/STB, $c_o = 7 \times 10^{-6}$ psi^{-1}, $\mu_o = 6$ cP, and $c_\mu = 0$ psi^{-1}. Using the single-phase simulator, report the pressure distribution in the reservoir shown in Fig. 7.15a and the well FBHP at 10, 20, and 50 days. Use single time steps to advance the solution from one time to the next.

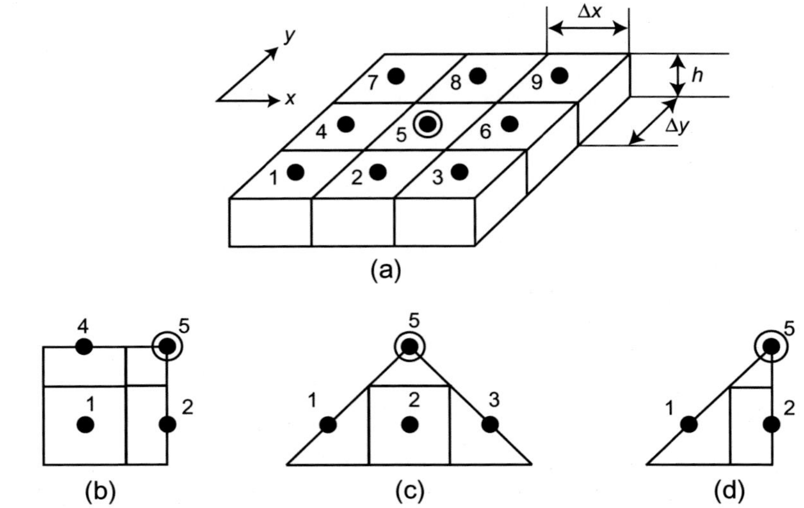

FIG. 7.15 2-D reservoir and elements of symmetry in Exercises 7.10 through 7.13. (a) Discretized 2-D reservoir in Exercise 7.10, (b) Element of symmetry in Exercise 7.11, (c) Element of symmetry in Exercise 7.12, and (d) Element of symmetry in Exercise 7.13.

7.11 Consider the flow problem presented in Exercise 7.10. In addition, consider symmetry about the two vertical planes passing through the center of gridblock 5 and perpendicular to either the x-axis or y-axis. Using the element of symmetry shown in Fig. 7.15b, estimate the pressure distribution in the reservoir and the well FBHP at 10, 20, and 50 days.

7.12 Consider the fluid flow problem described in Exercise 7.10. This time consider symmetry only about the two diagonal planes passing through the center of gridblock 5. Using the element of symmetry shown in Fig. 7.15c, estimate the pressure distribution in the reservoir and the well FBHP at 10, 20, and 50 days.

7.13 Consider the fluid flow problem described in Exercise 7.10. This time consider symmetry about all four planes passing through the center of gridblock 5. Using the smallest element of symmetry shown in Fig. 7.15d, estimate the pressure distribution in the reservoir and the well FBHP at 10, 20, and 50 days.

7.14 A single-phase fluid reservoir is described by three equal gridblocks as shown in Fig. 7.16. The reservoir is horizontal and has homogeneous and isotropic rock properties, $k=270$ md and $\phi=0.27$. Gridblock dimensions are $\Delta x=400$ ft, $\Delta y=650$ ft, and $h=60$ ft. Reservoir fluid properties are $B=1$ RB/STB and $\mu=1$ cP. The reservoir left boundary is kept constant at 3000 psia, and the reservoir right boundary is kept at a pressure gradient of -0.2 psi/ft. Two 7-in vertical wells were drilled at the centers of gridblocks 1 and 3. The well in gridblock 1 injects 300 STB/D of fluid, and the well in gridblock 3 produces 600 STB/D of fluid. Both wells have zero skin. Assume that the reservoir rock and fluid are incompressible. Find the pressure distribution in the reservoir.

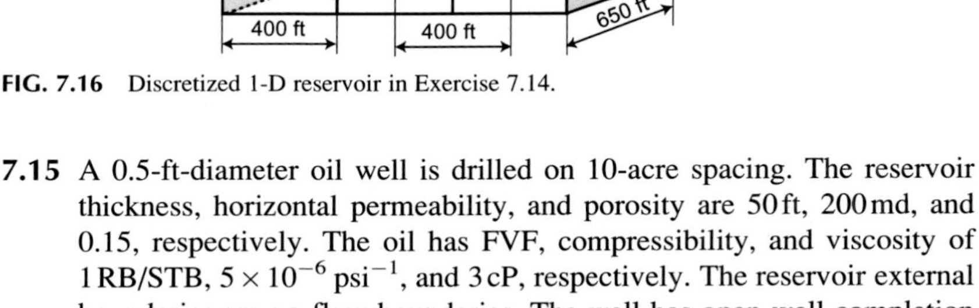

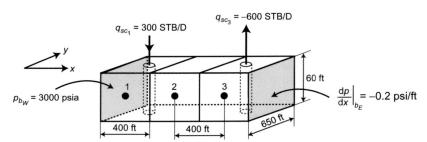

FIG. 7.16 Discretized 1-D reservoir in Exercise 7.14.

7.15 A 0.5-ft-diameter oil well is drilled on 10-acre spacing. The reservoir thickness, horizontal permeability, and porosity are 50 ft, 200 md, and 0.15, respectively. The oil has FVF, compressibility, and viscosity of 1 RB/STB, 5×10^{-6} psi^{-1}, and 3 cP, respectively. The reservoir external boundaries are no-flow boundaries. The well has open-well completion and is placed on production at a rate of 100 STB/D. Initial reservoir pressure is 4000 psia. The reservoir can be simulated using three gridblocks in the radial direction as shown in Fig. 7.17.

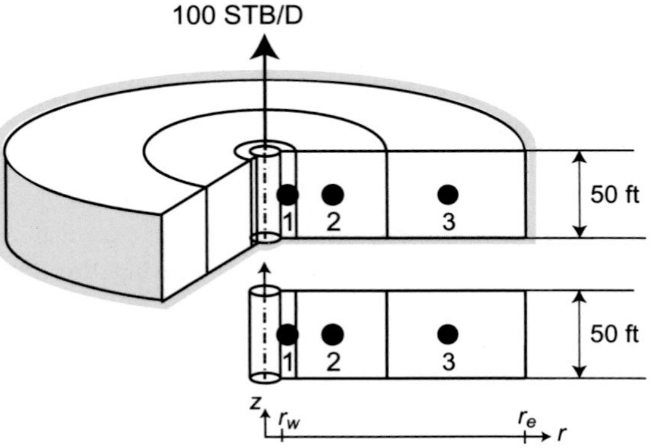

FIG. 7.17 Discretized 1-D reservoir in Exercise 7.15.

Use single time steps to advance the solution from one time to another. Find the pressure distribution in the reservoir and the FBHP of the well after 5 days. Check the material balance. Write the final form of the flow equations for this reservoir after 10 days.

7.16 A 0.5-ft-diameter oil well is drilled on 30-acre spacing. The reservoir thickness, horizontal permeability, and porosity are 50 ft, 210 md, and 0.17, respectively. The oil has FVF, compressibility, and viscosity of 1 RB/STB, 5×10^{-6} psi^{-1}, and 5 cP, respectively. The reservoir external boundaries are no-flow boundaries. The well has open-well completion and is placed on production at a rate of 1500 STB/D. Initial reservoir pressure is 3500 psia. The reservoir can be simulated using four gridblocks in the radial direction as shown in Fig. 7.18. Use single time steps to advance the solution from one time to another. Find the pressure distribution in the reservoir and the FBHP of the well after 1 day and 3 days. Check the material balance.

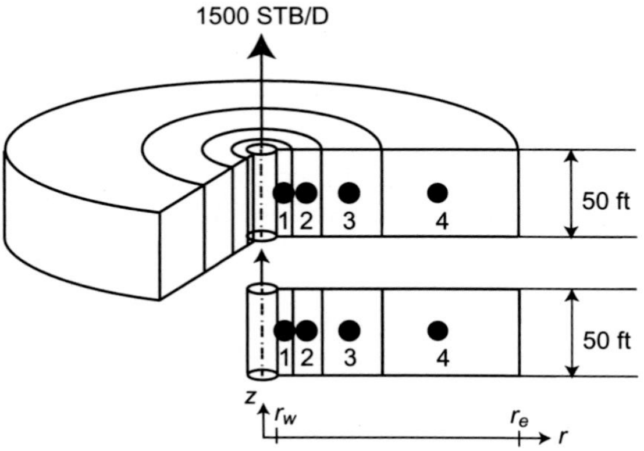

FIG. 7.18 Discretized 1-D reservoir in Exercise 7.16.

7.17 A single-phase fluid reservoir is discretized into four equal gridblocks as shown in Fig. 7.19. The reservoir is horizontal and has $k = 70$ md.

Gridblock dimensions are $\Delta x = 400$ ft, $\Delta y = 900$ ft, and $h = 25$ ft. Reservoir fluid properties are $B = 1$ RB/STB and $\mu = 1.5$ cP. The reservoir left boundary is kept at a constant pressure of 2600 psia, and the reservoir right boundary is kept at a constant pressure gradient of −0.2 psi/ft. A 6-in vertical well located at the center of gridblocks 3 produces fluid under a constant FBHP of 1000 psia. Assuming that the reservoir rock and fluid are incompressible, calculate the pressure distribution in the reservoir. Estimate the well production rate and the rates of fluid crossing the reservoir external boundaries. Perform a material balance check.

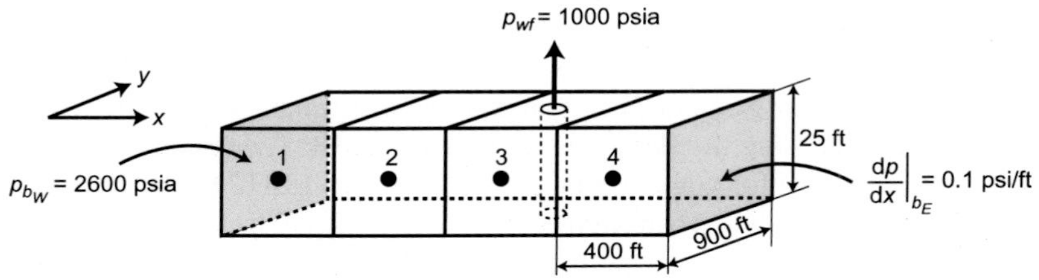

FIG. 7.19 Discretized 1-D reservoir in Exercise 7.17.

7.18 Consider the reservoir shown in Fig. 7.20. The reservoir is discretized into four equal gridblocks with $\Delta x = 300$ ft, $\Delta y = 600$ ft, $h = 30$ ft, and $k = 180$ md. The elevations of the center of gridblocks 1, 2, 3, and 4 are respectively 3532.34, 3471.56, 3410.78, and 3350.56 ft below sea level. The fluid FVF, viscosity, and density are 1 RB/STB, 2.4 cP, and 45 lbm/ft^3, respectively. The centers of the reservoir west and east boundaries are respectively 3562.73 and 3319.62 ft below sea level. The west boundary is sealed off to flow, and the east boundary is prescribed at a constant pressure gradient of 0.2 psi/ft. The reservoir has two 6-in wells. The first well is located at the center of gridblock 1 and injects fluid at a rate of 320 STB/D. The second well is located at the center of gridblocks 3 and produces fluid under a constant FBHP of 1200 psia. Assuming that the reservoir rock and fluid are incompressible, calculate the reservoir pressure distribution, the FBHP of the well in gridblock 1, and the production rate of the well in gridblock 3. Perform a material balance check.

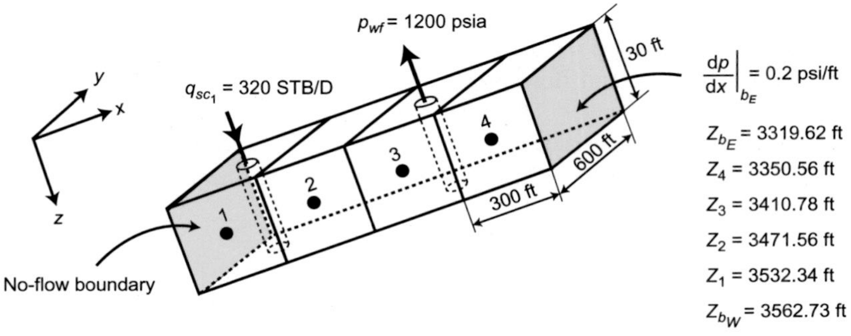

FIG. 7.20 Discretized 1-D reservoir in Exercise 7.18.

7.19 Perform a single-well simulation of the reservoir described in Exercise 7.16 assuming that the reservoir can be described using four gridpoints.

7.20 Consider the 2-D single-phase flow of incompressible oil taking place in the inclined, homogeneous reservoir shown in Fig. 7.21. The reservoir east and north external boundaries receive a constant influx of 0.02 STB/D-ft^2 from a neighboring reservoir. The reservoir west and south external boundaries are no-flow boundaries. The elevation below sea level of the center of gridblocks 1, 2, 3, and 4 are, respectively, 2000, 1700, 1700, and 1400 ft. The pressure of gridblock 1 is kept at 1000 psia. The gridblock properties are $\Delta x = \Delta y = 600$ ft, $h = 40$ ft, and $k_x = k_y = 500$ md. Oil density and viscosity are 37 lbm/ft^3 and 4 cP, respectively. Calculate the pressure of gridblocks 2, 3, and 4. Then, estimate the production rate of the well using the flow equation for gridblock 1, carry out a material balance check for your results, and estimate the FBHP of the well given that the well radius is 6 in. Consider symmetry about the vertical plane that passes through the centers of gridblocks 1 and 4.

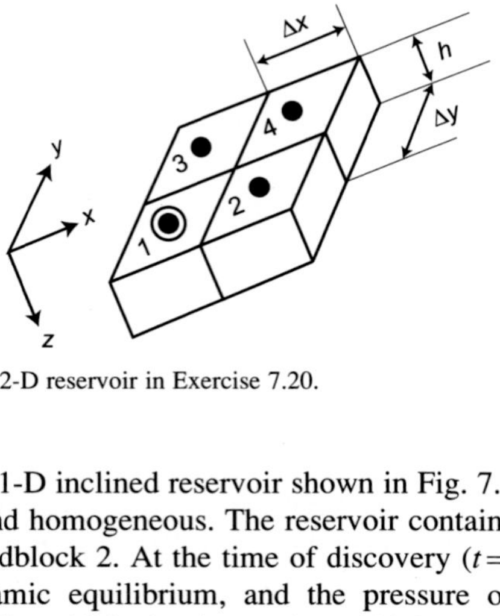

FIG. 7.21 Discretized 2-D reservoir in Exercise 7.20.

7.21 Consider the 1-D inclined reservoir shown in Fig. 7.22. The reservoir is volumetric and homogeneous. The reservoir contains a production well located in gridblock 2. At the time of discovery ($t = 0$), the fluids were in hydrodynamic equilibrium, and the pressure of gridblock 2 was 3000 psia. All gridblocks have $\Delta x = 400$ ft, $\Delta y = 200$ ft, $h = 80$ ft, $k = 222$ md, and $\phi = 0.20$. The well in gridblock 2 is produced at a rate of 200 STB/D, and fluid properties are $\mu_o = 2$ cP, $B_o = B_o^{\circ} = 1$ RB/STB, $\rho_o = 45$ lbm/ft^3, and $c_o = 5 \times 10^{-5}$ psi^{-1}. Estimate the initial pressure distribution in the reservoir. Find the well FBHP and pressure distribution in the system at 50 and 100 days using the implicit formulation. Check the material balance every time step.

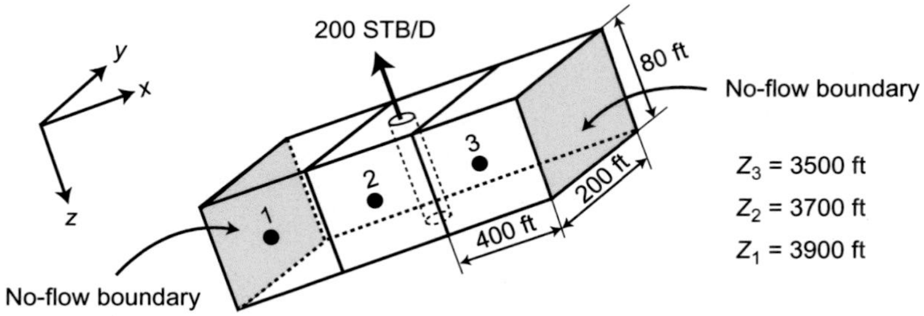

FIG. 7.22 Discretized 1-D reservoir in Exercise 7.21.

7.22 Consider the single-well simulation problem presented in Example 7.13. Solve the problem again, but this time, the reservoir is discretized into four gridpoints in the radial direction as shown in Fig. 7.23.

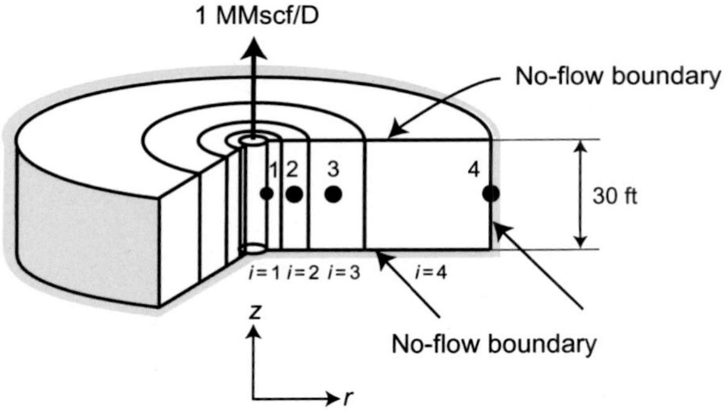

FIG. 7.23 Discretized reservoir in Exercise 7.22.

7.23 If the reservoir described in Exercise 7.21 is horizontal as shown in Fig. 7.24, observe and use the symmetry about the vertical plane that passes through the center of gridblock 2 and solve the problem.

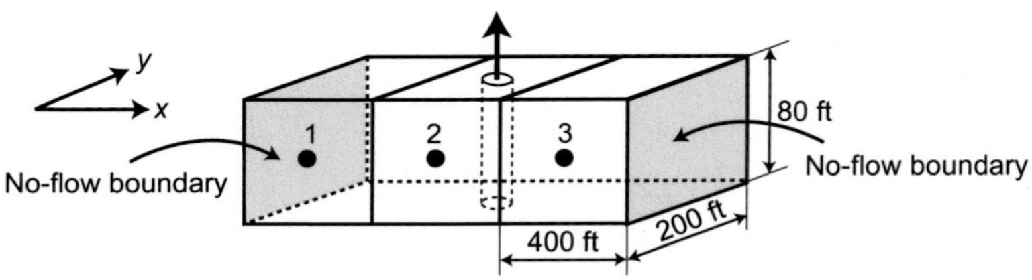

FIG. 7.24 Discretized 1-D reservoir in Exercise 7.23.

Chapter 8

Linearization of flow equations

Chapter outline

8.1 Introduction

The flow equations presented in Chapter 7 are generally nonlinear. Even if solved implicitly, the nonlinearity comes in boundary conditions and wells, which invoke discontinuities. Solving nonlinear algebraic equations is limited to trivial ones. All other forms have to be linearized before they are amenable to solutions. Only recently, some progress has been made for solving flow equations in their nonlinear forms (Mustafiz et al., 2008a,b). These solutions are extremely cumbersome to obtain and often result in hitting spurious solutions. Fortunately, such rigorous treatment is not necessary for most practical applications, for which a priori linearization suffices. To obtain the pressure distribution in the reservoir, these equations are linearized to use linear equation solvers. In this chapter, we aim at obtaining the linearized flow equation for an arbitrary gridblock (or gridpoint). To achieve this objective, we identify the nonlinear terms in the flow equations, present methods of linearizing these terms in space and time, and subsequently present the linearized

Petroleum Reservoir Simulation. https://doi.org/10.1016/B978-0-12-819150-7.00008-6

flow equation for single-phase flow problems. To simplify the presentation of concepts, we use the implicit formulation of the 1-D flow equation in the x-direction and use a block-centered grid in discretizing the reservoir. We first discuss the incompressible fluid flow equation that exhibits linearity, then the implicit formulation for the slightly compressible fluid flow equation that exhibits very weak nonlinearity, and finally the implicit formulation for the compressible fluid flow equation that exhibits a higher degree of nonlinearity. Although single-phase flow equations exhibit different degrees of nonlinearity, these equations are usually classified as having weak nonlinearities.

8.2 Nonlinear terms in flow equations

The terms composing any flow equation include interblock flow terms, the accumulation term, the well production rate term, and fictitious well rate terms reflecting flow across reservoir boundaries for boundary blocks. The number of interblock flow terms equals the number of all the existing neighboring blocks. The number of fictitious well rate terms equals the number of block boundaries that fall on reservoir boundaries. For any boundary block, the number of existing neighboring blocks and the number of fictitious wells always add up to two, four, or six for 1-D, 2-D, or 3-D flow, respectively. In single-phase flow problems, if the coefficients of unknown block pressures in the flow equation depend on block pressure, the algebraic equation is termed nonlinear; otherwise, the equation is linear. Therefore, the terms that may exhibit pressure dependence include transmissibilities, the well production rate, fictitious well rates, and the coefficient of block pressure difference in the accumulation term. This is true for equations in the mathematical approach. In the engineering approach; however, interblock flow terms, the well production rate, and fictitious well rates receive the same treatment; that is, block pressures contributing to flow potential (the pressure difference) in any term are treated implicitly as demonstrated in Chapter 7. Therefore, the nonlinear terms include transmissibilities in interblock flow terms and fictitious well rates, the coefficient of pressure drop in the well production rate term, and the coefficient of block pressure difference in the accumulation term.

8.3 Nonlinearity of flow equations for various fluids

In this section, we examine the nonlinearity of the flow equations for slightly compressible and compressible fluids. The flow equation for incompressible fluids is linear. We examine the pressure dependence of the various terms in a flow equation, namely, the interblock flow terms, the accumulation term, the well production rate term, and the fictitious well rate terms.

8.3.1 Linearity of the incompressible fluid flow equation

The 1-D flow equation in the x-direction for an incompressible fluid can be obtained from Eq. (7.16a), which states

$$\sum_{l\in\psi_n} T_{l,n}\left[(p_l - p_n) - \gamma_{l,n}(Z_l - Z_n)\right] + \sum_{l\in\xi_n} q_{sc_{l,n}} + q_{sc_n} = 0 \tag{8.1}$$

where $\psi_n = \{n-1, n+1\}$, $\xi_n = \{\ \}$, $\{b_W\}$, or $\{b_E\}$, and $n = 1, 2, 3, \ldots n_x$.

For gridblock 1,

$$T_{x_{1+1/2}}\left[(p_2 - p_1) - \gamma_{1+1/2}(Z_2 - Z_1)\right] + q_{sc_{b_W,1}} + q_{sc_1} = 0 \tag{8.2a}$$

For gridblock $i = 2, 3, \ldots n_x - 1$,

$$\begin{aligned} &T_{x_{i-1/2}}\left[(p_{i-1} - p_i) - \gamma_{i-1/2}(Z_{i-1} - Z_i)\right] \\ &\quad + T_{x_{i+1/2}}\left[(p_{i+1} - p_i) - \gamma_{i+1/2}(Z_{i+1} - Z_i)\right] + q_{sc_i} = 0 \end{aligned} \tag{8.2b}$$

For gridblock n_x,

$$T_{x_{n_x-1/2}}\left[(p_{n_x-1} - p_{n_x}) - \gamma_{n_x-1/2}(Z_{n_x-1} - Z_{n_x})\right] + q_{sc_{b_E,n_x}} + q_{sc_{n_x}} = 0 \tag{8.2c}$$

Transmissibility $T_{x_{i\mp1/2}}$ is expressed as Eq. (2.39a):

$$T_{x_{i\mp1/2}} = \left(\beta_c \frac{k_x A_x}{\mu B \Delta x}\right)\Bigg|_{x_{i\mp1/2}} = G_{x_{i\mp1/2}}\left(\frac{1}{\mu B}\right)_{x_{i\mp1/2}} \tag{8.3a}$$

Geometric factor $G_{x_{i\mp1/2}}$ is defined in Table 4.1 for a block-centered grid,

$$G_{x_{i\mp1/2}} = \frac{2\beta_c}{\Delta x_i/(A_{x_i}k_{x_i}) + \Delta x_{i\mp1}/\left(A_{x_{i\mp1}}k_{x_{i\mp1}}\right)} \tag{8.4}$$

The well production rate (q_{sc_i}) is estimated according to the well operating condition as discussed in Chapter 6, and fictitious well rates $\left(q_{sc_{b_W,1}}, q_{sc_{b_E,n_x}}\right)$ are estimated according to the type of boundary condition as discussed in Chapter 4. Note that $T_{x_{i\mp1/2}}$ and $G_{x_{i\mp1/2}}$ are functions of the space between gridblocks i and $i\mp1$ only. It should be mentioned that a numerical value for the well production rate could be calculated for well operating conditions other than a specified FBHP. Similarly, a numerical value for a fictitious well flow rate can be calculated for boundary conditions other than a specified pressure boundary. In such cases, both the well production rate and fictitious well rate are known quantities and, as a result, can be moved to the RHS of the flow equation (Eq. 8.2). Otherwise, the well production rate and fictitious well rate are functions of block pressure (p_i), and as a result, part of the rate equations appears in the coefficient of p_i, and the other part has to be moved to the RHS of the flow equation (Eq. 8.2). The FVF, viscosity, and gravity of an incompressible fluid are not functions of pressure. Therefore, transmissibilities and gravity are not functions of pressure; consequently, Eq. (8.2) represents a system of n_x linear algebraic equations. This system of linear equations can be solved for the unknown pressures $(p_1, p_2, p_3, \ldots p_{n_x})$ by the algorithm presented in Section 7.3.1.1.

8.3.2 Nonlinearity of the slightly compressible fluid flow equation

The implicit flow equation for a slightly compressible fluid is expressed as Eq. (7.81a):

$$\sum_{l\in\psi_n} T_{l,n}^{n+1}\left[\left(p_l^{n+1}-p_n^{n+1}\right)-\gamma_{l,n}^n(Z_l-Z_n)\right]+\sum_{l\in\xi_n} q_{sc_{l,n}}^{n+1}+q_{sc_n}^{n+1} = \frac{V_{b_n}\phi_n^\circ(c+c_\phi)}{\alpha_c B^\circ \Delta t}\left[p_n^{n+1}-p_n^n\right] \tag{8.5}$$

where the FVF, viscosity, and density are described by Eqs. (7.5) through (7.7):

$$B=\frac{B^\circ}{[1+c(p-p^\circ)]} \tag{8.6}$$

$$\mu=\frac{\mu^\circ}{[1-c_\mu(p-p^\circ)]} \tag{8.7}$$

and

$$\rho=\rho^\circ[1+c(p-p^\circ)] \tag{8.8}$$

The numerical values of c and c_μ for slightly compressible fluids are in the order of magnitude of 10^{-6} to 10^{-5}. Consequently, the effect of pressure variation on the FVF, viscosity, and gravity can be neglected without introducing noticeable errors. Simply stated $B\cong B^\circ$, $\mu\cong\mu^\circ$, and $\rho\cong\rho^\circ$, and in turn, transmissibilities and gravity are independent of pressure (i.e., $T_{l,n}^{n+1}\cong T_{l,n}$ and $\gamma_{l,n}^n\cong\gamma_{l,n}$). Therefore, Eq. (8.5) simplifies to

$$\sum_{l\in\psi_n} T_{l,n}\left[\left(p_l^{n+1}-p_n^{n+1}\right)-\gamma_{l,n}(Z_l-Z_n)\right]+\sum_{l\in\xi_n} q_{sc_{l,n}}^{n+1}+q_{sc_n}^{n+1} = \frac{V_{b_n}\phi_n^\circ(c+c_\phi)}{\alpha_c B^\circ \Delta t}\left[p_n^{n+1}-p_n^n\right] \tag{8.9}$$

Eq. (8.9) is a linear algebraic equation because the coefficients of the unknown pressures at time level $n+1$ are independent of pressure.

The 1-D flow equation in the x-direction for a slightly compressible fluid is obtained from Eq. (8.9) in the same way that was described in the previous section.

For gridblock 1,

$$T_{x_{1+1/2}}\left[\left(p_2^{n+1}-p_1^{n+1}\right)-\gamma_{1+1/2}(Z_2-Z_1)\right]+q_{sc_{b_W,1}}^{n+1}+q_{sc_1}^{n+1} = \frac{V_{b_1}\phi_1^{\circ}(c+c_\phi)}{\alpha_c B^{\circ}\Delta t}\left[p_1^{n+1}-p_1^n\right] \tag{8.10a}$$

For gridblock $i=2,3,\ldots n_x-1$,

$$T_{x_{i-1/2}}\left[\left(p_{i-1}^{n+1}-p_i^{n+1}\right)-\gamma_{i-1/2}(Z_{i-1}-Z_i)\right] + T_{x_{i+1/2}}\left[\left(p_{i+1}^{n+1}-p_i^{n+1}\right)-\gamma_{i+1/2}(Z_{i+1}-Z_i)\right]+q_{sc_i}^{n+1}=\frac{V_{b_i}\phi_i^{\circ}(c+c_\phi)}{\alpha_c B^{\circ}\Delta t}\left[p_i^{n+1}-p_i^n\right] \tag{8.10b}$$

For gridblock n_x,

$$T_{x_{n_x-1/2}}\left[\left(p_{n_x-1}^{n+1}-p_{n_x}^{n+1}\right)-\gamma_{n_x-1/2}(Z_{n_x-1}-Z_{n_x})\right]+q_{sc_{b_E,n_x}}^{n+1}+q_{sc_{n_x}}^{n+1} = \frac{V_{b_{n_x}}\phi_{n_x}^{\circ}(c+c_\phi)}{\alpha_c B^{\circ}\Delta t}\left[p_{n_x}^{n+1}-p_{n_x}^n\right] \tag{8.10c}$$

In the aforementioned equation, $T_{x_{i\mp1/2}}$ and $G_{x_{i\mp1/2}}$ for a block-centered grid are defined by Eqs. (8.3a) and (8.4):

$$T_{x_{i\mp1/2}}=\left(\beta_c\frac{k_xA_x}{\mu B\Delta x}\right)\bigg|_{x_{i\mp1/2}}=G_{x_{i\mp1/2}}\left(\frac{1}{\mu B}\right)_{x_{i\mp1/2}} \tag{8.3a}$$

and

$$G_{x_{i\mp1/2}}=\frac{2\beta_c}{\Delta x_i/\left(A_{x_i}k_{x_i}\right)+\Delta x_{i\mp1}/\left(A_{x_{i\mp1}}k_{x_{i\mp1}}\right)} \tag{8.4}$$

Here again, the well production rate ($q_{sc_i}^{n+1}$) and fictitious well rates $\left(q_{sc_{b_W,1}}^{n+1}, q_{sc_{b_E,n_x}}^{n+1}\right)$ are handled in exactly the same way as discussed in the previous section. The resulting set of n_x linear algebraic equations can be solved for the unknown pressures (p_1^{n+1}, p_2^{n+1}, p_3^{n+1}, … $p_{n_x}^{n+1}$) by the algorithm presented in Section 7.3.2.2.

Although each of Eqs. (8.2) and (8.10) represents a set of linear algebraic equations, there is a basic difference between them. In Eq. (8.2), the reservoir pressure depends on space (location) only, whereas in Eq. (8.10), reservoir pressure depends on both space and time. The implication of this difference is that the flow equation for an incompressible fluid (Eq. 8.2) has a steady-state solution (i.e., a solution that is independent of time), whereas the flow equation for a slightly compressible fluid (Eq. 8.10) has an unsteady-state solution (i.e., a solution that is dependent on time). It should be mentioned that the pressure solution for Eq. (8.10) at any time step is obtained without iteration because the equation is linear.

We must reiterate that the linearity of Eq. (8.9) is the result of neglecting the pressure dependence of FVF and viscosity in transmissibility, the well production rate, and the fictitious well rates on the LHS of Eq. (8.5). If Eqs. (8.6) and (8.7) are used to reflect such pressure dependence, the resulting flow equation becomes nonlinear. In conclusion, understanding the behavior of fluid properties has led to devising a practical way of linearizing the flow equation for a slightly compressible fluid.

8.3.3 Nonlinearity of the compressible fluid flow equation

The implicit flow equation for a compressible fluid is expressed as Eq. (7.162a):

$$\sum_{l\in\psi_n} T_{l,n}^{n+1}\left[\left(p_l^{n+1}-p_n^{n+1}\right)-\gamma_{l,n}^n(Z_l-Z_n)\right]+\sum_{l\in\xi_n} q_{sc_{l,n}}^{n+1}+q_{sc_n}^{n+1} = \frac{V_{b_n}}{\alpha_c\Delta t}\left(\frac{\phi}{B_g}\right)_n'\left[p_n^{n+1}-p_n^n\right] \tag{8.11}$$

The pressure dependence of density is expressed as Eq. (7.9):

$$\rho_g=\frac{\rho_{gsc}}{\alpha_c B_g} \tag{8.12}$$

In addition, gas FVF and viscosity are presented in a tabular form as functions of pressure at reservoir temperature:

$$B_g=\mathrm{f}(p) \tag{8.13}$$

and

$$\mu_g=\mathrm{f}(p) \tag{8.14}$$

As mentioned in Chapter 7, the density and viscosity of a compressible fluid increase as pressure increases but tend to level off at high pressures. The FVF decreases orders of magnitude as the pressure increases from low pressure to high pressure. Consequently, interblock transmissibilities, gas gravity, the coefficient of pressure difference in accumulation term, well production, and transmissibility in fictitious well terms are all functions of unknown block pressures. Therefore, Eq. (8.11) is nonlinear. The solution of this equation requires linearization of nonlinear terms in both space and time.

The 1-D flow equation in the x-direction for a compressible fluid can be obtained from Eq. (8.11) in the same way that was described in Section 8.3.1.

For gridblock 1,

$$T_{x_{1+1/2}}^{n+1}\left[\left(p_2^{n+1}-p_1^{n+1}\right)-\gamma_{1+1/2}^n(Z_2-Z_1)\right]+q_{sc_{b_W,1}}^{n+1}+q_{sc_1}^{n+1} = \frac{V_{b_1}}{\alpha_c\Delta t}\left(\frac{\phi}{B_g}\right)_1'\left[p_1^{n+1}-p_1^n\right] \tag{8.15a}$$

For gridblock $i = 2, 3, \ldots n_x - 1$,

$$T_{x_{i-1/2}}^{n+1}\left[\left(p_{i-1}^{n+1}-p_i^{n+1}\right)-\gamma_{i-1/2}^{n}(Z_{i-1}-Z_i)\right]$$
$$+T_{x_{i+1/2}}^{n+1}\left[\left(p_{i+1}^{n+1}-p_i^{n+1}\right)-\gamma_{i+1/2}^{n}(Z_{i+1}-Z_i)\right]+q_{sc_i}^{n+1}=\frac{V_{b_i}}{\alpha_c\Delta t}\left(\frac{\phi}{B_g}\right)_i^{\prime}\left[p_i^{n+1}-p_i^n\right] \tag{8.15b}$$

For gridblock n_x,

$$T_{x_{n_x-1/2}}^{n+1}\left[\left(p_{n_x-1}^{n+1}-p_{n_x}^{n+1}\right)-\gamma_{n_x-1/2}^{n}(Z_{n_x-1}-Z_{n_x})\right]+q_{sc_{b_E},n_x}^{n+1}+q_{sc_{n_x}}^{n+1}$$
$$=\frac{V_{b_{n_x}}}{\alpha_c\Delta t}\left(\frac{\phi}{B_g}\right)_{n_x}^{\prime}\left[p_{n_x}^{n+1}-p_{n_x}^{n}\right] \tag{8.15c}$$

In the aforementioned equation, $T_{x_{i\mp1/2}}^{n+1}$ and $G_{x_{i\mp1/2}}$ for a block-centered grid are defined by Eqs. (8.3b) and (8.4):

$$T_{x_{i\mp1/2}}^{n+1}=\left(\beta_c\frac{k_xA_x}{\mu B\Delta x}\right)\Bigg|_{x_{i\mp1/2}}^{n+1}=G_{x_{i\mp1/2}}\left(\frac{1}{\mu B}\right)_{x_{i\mp1/2}}^{n+1} \tag{8.3b}$$

and

$$G_{x_{i\mp1/2}}=\frac{2\beta_c}{\Delta x_i/\left(A_{x_i}k_{x_i}\right)+\Delta x_{i\mp1}/\left(A_{x_{i\mp1}}k_{x_{i\mp1}}\right)} \tag{8.4}$$

where B and μ stand for B_g and μ_g, respectively.

Here again, the well production rate ($q_{sc_i}^{n+1}$) and fictitious well rates $\left(q_{sc_{b_W},1}^{n+1}, q_{sc_{b_E},n_x}^{n+1}\right)$ are handled in exactly the same way as discussed in Section 8.3.1. In addition, interblock transmissibility (Eq. 8.3b) is a function of the space between gridblocks i and $i\mp1$ and time. The resulting set of n_x nonlinear algebraic equations has to be linearized prior to being solved for the unknown pressures ($p_1^{n+1}, p_2^{n+1}, p_3^{n+1}, \ldots p_{n_x}^{n+1}$). The algorithm outlined in Section 7.3.3.2 uses explicit transmissibility to linearize flow equations. This essentially involves transmissibility values being used from nth time step. The following section presents other methods of linearization. It should be mentioned that even though the solutions of Eqs. (8.10) and (8.15) are time dependent, the solution of Eq. (8.10) requires no iteration because of the linearity of the equation, while the solution of Eq. (8.15) requires iteration to remove the nonlinearity due to time. In addition, while the pressure coefficients in Eq. (8.10) are constant (i.e., they do not change from one time step to another), the pressure coefficients in Eq. (8.15) are not constant and need to be updated at least once at the beginning of each time step.

8.4 Linearization of nonlinear terms

In this section, we present the various methods used to treat nonlinearities. Although the methods of linearization presented here may not be required because nonlinearities in single-phase flow are weak, these linearization methods are needed for the simulation of multiphase flow in petroleum reservoirs that is presented in Chapter 11. Nonlinear terms have to be approximated in both space and time. Linearization in space defines the location where the nonlinearity is to be evaluated and which reservoir blocks should be used in its estimation. Linearization in time implies how the term is approximated to reflect its value at the current time level where the pressure solution is unknown. Fig. 8.1 sketches three commonly used linearization methods as they apply to a nonlinearity (f) that is a function of one variable (p): (a) the explicit method (Fig. 8.1a), (b) the simple iteration method (Fig. 8.1b), and (c) the fully implicit method (Fig. 8.1c).

Each figure shows the improvements in the linearized value of the nonlinearity as iteration progresses from the first iteration ($\nu=0$) to the second iteration ($\nu=1$) and so on until the pressure converges to p^{n+1}. Iteration on pressure in the case of a compressible fluid only is necessary to satisfy material balance and remove the nonlinearity of the accumulation term due to time. In Fig. 8.1, the value of the nonlinearity at time level n (the beginning of the time step) is represented by an empty circle, its value at time level $n+1$ (after reaching convergence) is represented by a solid circle, and its value at any iteration is represented by an empty square at that iteration. Note that the explicit method, sketched in Fig. 8.1a, does not provide for any improvement in the value of the nonlinearity as iteration progresses. The simple iteration method, sketched in Fig. 8.1b, provides for improvement in the value of the nonlinearity in stepwise fashion. In the fully implicit treatment, presented Fig. 8.1c, the improved value of the nonlinearity, as iteration progresses, falls on the tangent of the nonlinearity at the previous iteration. Other linearization methods, such as the linearized implicit method (MacDonald and Coats, 1970) and the semi-implicit method of Nolen and Berry (1972), are not applicable to single-phase flow. They are used in multiphase flow to deal with nonlinearities due to fluid saturation only. The treatments of the various nonlinear terms that appear in single-phase flow equations are presented in Sections 8.4.1–8.4.4.

8.4.1 Linearization of transmissibilities

Transmissibilities at time level $n+1$ are expressed by Eq. (8.3b):

$$T_{x_{i\mp1/2}}^{n+1} = \left(\beta_c \frac{k_x A_x}{\mu B \Delta x}\right)\Bigg|_{x_{i\mp1/2}}^{n+1} = G_{x_{i\mp1/2}} \left(\frac{1}{\mu B}\right)_{x_{i\mp1/2}}^{n+1} = G_{x_{i\mp1/2}} f_{p_{i\mp1/2}}^{n+1} \tag{8.16}$$

where $G_{x_{i\mp1/2}}$ is defined by Eq. (8.4) for a block-centered grid and $f_{p_{i\mp1/2}}^{n+1}$ is defined as

$$f_{p_{i\mp1/2}}^{n+1} = \left(\frac{1}{\mu B}\right)_{x_{i\mp1/2}}^{n+1} \tag{8.17}$$

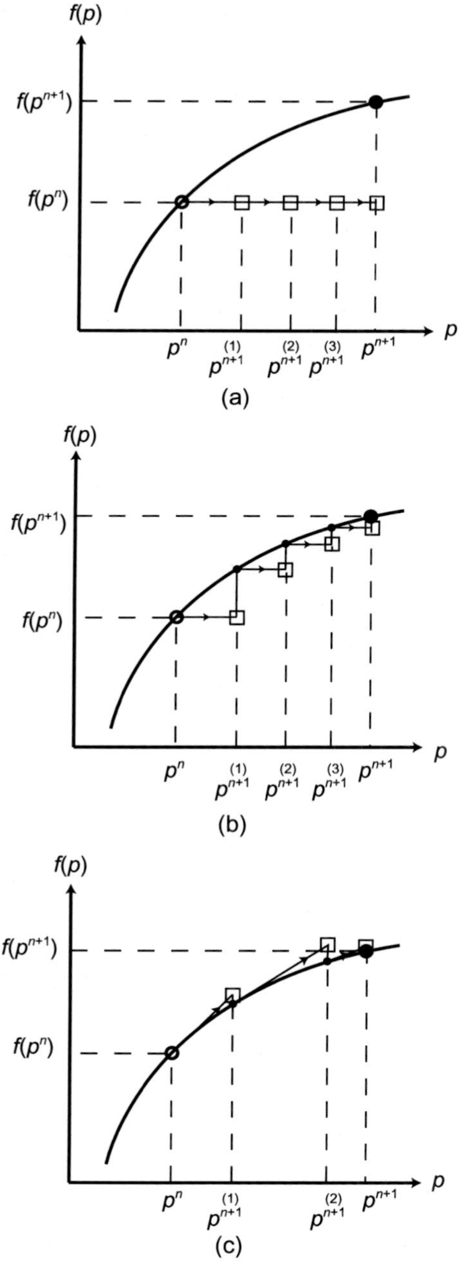

FIG. 8.1 Convergence of different methods of linearization. (a) Explicit linearization, (b) Simple-iteration linearization, and (c) Fully implicit linearization.

Therefore, linearization of transmissibility reduces to linearization of $f_{p_{i\mp1/2}}^{n+1}$. The function f_p is evaluated between the corresponding two blocks (termed here as block boundaries $x_{i\mp1/2}$) and at time level $n+1$, where the pressure solution is not known. Therefore, f_p needs to be expressed as a function of the pressure of the blocks on both sides of the specific block boundary and at some known time. These approximations are termed linearization in space and linearization in time.

8.4.1.1 Linearization of f_p in space

There are several methods used to approximate f_p in space.

With single-point upstream weighting,

$$f_{p_{i\mp1/2}} = f_{p_i} \tag{8.18a}$$

if block i is upstream to block $i\mp1$ or

$$f_{p_{i\mp1/2}} = f_{p_{i\mp1}} \tag{8.18b}$$

if block i is downstream to block $i\mp1$. The potential difference between blocks i and $i\mp1$ is used to determine the upstream and downstream blocks.

With average function value weighting,

$$f_{p_{i\mp1/2}} = \bar{f} = 1/2\left(f_{p_i} + f_{p_{i\mp1}}\right) \tag{8.19}$$

With average pressure value weighting,

$$f_{p_{i\mp1/2}} = f(\overline{p}) = 1/\mu(\overline{p})B(\overline{p}) \tag{8.20}$$

where

$$\overline{p} = 1/2(p_i + p_{i\mp1}) \tag{8.21}$$

With average function components value weighting,

$$f_{p_{i\mp1/2}} = f(\overline{p}) = 1/\overline{\mu}\overline{B} \tag{8.22}$$

where

$$\overline{\mu} = \frac{\mu(p_i) + \mu(p_{i\mp1})}{2} \tag{8.23}$$

and

$$\overline{B} = \frac{B(p_i) + B(p_{i\mp1})}{2} \tag{8.24}$$

Once f_p is linearized in space as in Eqs. (8.18) through (8.24), then the space-linearized transmissibility is obtained by applying Eq. (8.16):

$$T_{x_{i\mp1/2}} = G_{x_{i\mp1/2}} f_{p_{i\mp1/2}} \tag{8.25}$$

8.4.1.2 Linearization of f_p in time

The effect of the nonlinearity of f_p on the stability of the solution depends on the magnitude of the pressure change over a time step. The methods of time linearization presented earlier in Fig. 8.1 may be used to approximate f_p in time. Note that f_p is a function of the pressures of the blocks that surround a block boundary as mentioned in the previous section; that is, $f_p = \mathrm{f}(p_i, p_{i\mp1})$.

With the explicit method (see Fig. 8.1a), the nonlinearity is evaluated at the beginning of the time step (at time level n) as

$$f_{p_{i\mp1/2}}^{n+1} \cong f_{p_{i\mp1/2}}^{n} = \mathrm{f}\left(p_i^n, p_{i\mp1}^n\right) \tag{8.26}$$

With the simple iteration method (see Fig. 8.1b), the nonlinearity is evaluated one iteration behind the pressure solution

$$f_{p_{i\mp1/2}}^{n+1} \cong f_{p_{i\mp1/2}}^{\overset{(\nu)}{n+1}} = \mathrm{f}\left(p_i^{\overset{(\nu)}{n+1}}, p_{i\mp1}^{\overset{(\nu)}{n+1}}\right) \tag{8.27}$$

With the fully implicit method (see Fig. 8.1c), the nonlinearity is approximated by its value at iteration level (ν) plus a term that depends on the rate of change of pressure over iteration,

$$f_{p_{i\mp1/2}}^{n+1} \cong f_{p_{i\mp1/2}}^{\overset{(\nu+1)}{n+1}} \cong \mathrm{f}\left(p_i^{\overset{(\nu)}{n+1}}, p_{i\mp1}^{\overset{(\nu)}{n+1}}\right) + \frac{\partial \mathrm{f}(p_i, p_{i\mp1})}{\partial p_i}\bigg|^{\overset{(\nu)}{n+1}} \left(p_i^{\overset{(\nu+1)}{n+1}} - p_i^{\overset{(\nu)}{n+1}}\right)$$
$$+ \frac{\partial \mathrm{f}(p_i, p_{i\mp1})}{\partial p_{i\mp1}}\bigg|^{\overset{(\nu)}{n+1}} \left(p_{i\mp1}^{\overset{(\nu+1)}{n+1}} - p_{i\mp1}^{\overset{(\nu)}{n+1}}\right) \tag{8.28}$$

Once f_p is linearized in time as in Eq. (8.26), (8.27), or (8.28), then the time-linearized transmissibility is obtained by applying Eq. (8.16):

$$T_{x_{i\mp1/2}}^{n+1} = G_{x_{i\mp1/2}} f_{p_{i\mp1/2}}^{n+1} \tag{8.29}$$

8.4.2 Linearization of well rates

A wellblock production (injection) rate is evaluated in space at the gridblock (or gridpoint) for which the flow equation is written. Linearization in time of the wellblock production rate involves first linearizing the wellblock production (injection) rate equation and then substituting the result in the linearized flow equation for the wellblock. This method of linearization, which is usually used in reservoir simulation, parallels the linearization of interblock transmissibility. The following methods may be used to approximate a wellblock rate in time.

For wells operating with specified bottom-hole pressure condition, the nonlinearity involves the term $G_{w_i}\left(\frac{1}{B\mu}\right)_i^{n+1}$.

Explicit transmissibility method:

$$q_{sc_i}^{n+1} \cong -G_{w_i}\left(\frac{1}{B\mu}\right)_i^n \left(p_i^{n+1} - p_{wf_i}\right) \tag{8.30}$$

Simple iteration on transmissibility method:

$$q_{sc_i}^{n+1} \cong -G_{w_i}\left(\frac{1}{B\mu}\right)_i^{\overset{(v)}{n+1}} \left(p_i^{n+1} - p_{wf_i}\right) \tag{8.31}$$

Fully implicit method:

$$q_{sc_i}^{n+1} \cong q_{sc_i}^{\overset{(\nu+1)}{n+1}} \cong q_{sc_i}^{\overset{(\nu)}{n+1}} + \left.\frac{dq_{sc_i}}{dp_i}\right|^{\overset{(\nu)}{n+1}} \left(p_i^{\overset{(\nu+1)}{n+1}} - p_i^{\overset{(\nu)}{n+1}}\right) \tag{8.32}$$

For wells operating with specified pressure gradient condition, nonlinearity involves the term $2\pi\beta_c r_w(kh)_i\left(\frac{1}{B\mu}\right)_i^{n+1}$. Linearization in space involves evaluating this term for wellblock. Linearization of the term $\left(\frac{1}{B\mu}\right)_i^{n+1}$ in time parallels the time linearization of f_p in transmissibility. In this case $f_p = \left(\frac{1}{B\mu}\right)_i$.

Explicit transmissibility method:

$$q_{sc_i}^{n+1} \cong 2\pi\beta_c r_w(kh)_i\left(\frac{1}{B\mu}\right)_i^n \left.\frac{dp}{dr}\right|_{rw} \tag{8.33}$$

Simple iteration on transmissibility method:

$$q_{sc_i}^{n+1} \cong 2\pi\beta_c r_w(kh)_i\left(\frac{1}{B\mu}\right)_i^{\overset{(v)}{n+1}} \left.\frac{dp}{dr}\right|_{rw} \tag{8.34}$$

Fully implicit method:

$$q_{sc_i}^{n+1} \cong q_{sc_i}^{\overset{(\nu+1)}{n+1}} \cong q_{sc_i}^{\overset{(\nu)}{n+1}} + \left.\frac{dq_{sc_i}}{dp_i}\right|^{\overset{(\nu)}{n+1}} \left(p_i^{\overset{(\nu+1)}{n+1}} - p_i^{\overset{(\nu)}{n+1}}\right) \tag{8.32}$$

where

$$\left.\frac{dq_{sc_i}}{dp_i}\right|^{\overset{(\nu)}{n+1}} = 2\pi\beta_c r_w(kh)_i \left.\frac{dp}{dr}\right|_{rw} \left.\frac{d(1/B\mu)}{dp}\right|_i^{\overset{(v)}{n+1}} \tag{8.35}$$

8.4.3 Linearization of fictitious well rates

The fictitious well rate in point-distrusted grid, presented in Chapter 5, is the interblock flow term between the boundary gridpoint and the neighboring reservoir grid point. Therefore, the linearization, in space and time, of fictitious well rate is similar to the linearization of interblock flow terms. For a block-centered grid, presented in Chapter 4, the fictitious well rate is nothing but the flow term within the gridblock between the gridblock boundary and the point that represents the gridblock. Therefore, a fictitious well rate can be linearized, in space and time, the same way as that of a physical well rate.

8.4.4 Linearization of coefficients in accumulation term

The coefficient of pressure change in the accumulation term exhibits nonlinearity for a compressible fluid only (Eq. 8.11). This nonlinearity results from the pressure dependence of $B_{g_n}^{\overset{(v)}{n+1}}$ in Eq. (7.157) that is used in the definition of $\left(\frac{\phi}{B_g}\right)'_n$ given by Eq. (7.156a). Linearization in space involves evaluating $B_{g_n}^{\overset{(v)}{n+1}}$ and hence $\left(\frac{\phi}{B_g}\right)'_n$ at the pressure of the gridblock (or gridpoint) for which the flow equation is written (gridblock n). Linearization in time uses simple iteration; that is, $B_{g_n}^{\overset{(v)}{n+1}}$ is evaluated at the current block pressure with one iteration lagging behind.

8.5 Linearized flow equations in time

As mentioned earlier in this chapter, the flow equation for a compressible fluid exhibits the highest degree of nonlinearity among single-phase flow equations. Eq. (8.15b) for an interior block in 1-D flow having a well operating with specified bottom-hole pressure (Eq. 6.11) is used to demonstrate the various methods of linearizing flow equations. The flow equation considered here is

$$T_{x_{i-1/2}}^{n+1}\left[\left(p_{i-1}^{n+1}-p_i^{n+1}\right)-\gamma_{i-1/2}^{n}(Z_{i-1}-Z_i)\right]$$

$$+T_{x_{i+1/2}}^{n+1}\left[\left(p_{i+1}^{n+1}-p_i^{n+1}\right)-\gamma_{i+1/2}^{n}(Z_{i+1}-Z_i)\right] \; - \; G_{w_i}\left(\frac{1}{B\mu}\right)_i^{n+1}\left(p_i^{n+1}-p_{wf_i}\right)$$

$$=\frac{V_{b_i}}{\alpha_c\Delta t}\left(\frac{\phi}{B_g}\right)'_i\left[p_i^{n+1}-p_i^{n}\right] \tag{8.36}$$

where $q_{sc_i}^{n+1}=-G_{w_i}\left(\frac{1}{B\mu}\right)_i^{n+1}\left(p_i^{n+1}-p_{wf_i}\right)$

The final form of the linearized flow equation for a boundary block must be modified to include fictitious wells (boundary conditions).

8.5.1 Explicit transmissibility method

In the explicit transmissibility method, transmissibility of interblock flow and coefficient of pressure drop in well rate equation are dated at old time level (time level n). One still has to iterate on $\left(\frac{\phi}{B_g}\right)'_i$. Eq. (8.36) becomes

$$\begin{aligned} &T^n_{x_{i-1/2}}\left[\left(p^{n+1}_{i-1}-p^{n+1}_i\right)-\gamma^n_{i-1/2}(Z_{i-1}-Z_i)\right] \\ &+T^n_{x_{i+1/2}}\left[\left(p^{n+1}_{i+1}-p^{n+1}_i\right)-\gamma^n_{i+1/2}(Z_{i+1}-Z_i)\right]-G_{w_i}\left(\frac{1}{B\mu}\right)^n_i\left(p^{n+1}_i-p_{wf_i}\right) \\ &=\frac{V_{b_i}}{\alpha_c\Delta t}\left(\frac{\phi}{B_g}\right)'_i\left[p^{n+1}_i-p^n_i\right] \end{aligned} \tag{8.37}$$

By placing the iteration level and rearranging the terms, we obtain the final form of the flow equation for interior block i:

$$\begin{aligned} &T^n_{x_{i-1/2}}\overset{(\nu+1)}{p^{n+1}_{i-1}}-\left[T^n_{x_{i-1/2}}+T^n_{x_{i+1/2}}+\frac{V_{b_i}}{\alpha_c\Delta t}\left(\frac{\phi}{B_g}\right)'_i+G_{w_i}\left(\frac{1}{B\mu}\right)^n_i\right]\overset{(\nu+1)}{p^{n+1}_i} \\ &+T^n_{x_{i+1/2}}\overset{(\nu+1)}{p^{n+1}_{i+1}}=\left[T^n_{x_{i-1/2}}\gamma^n_{i-1/2}(Z_{i-1}-Z_i)+T^n_{x_{i+1/2}}\gamma^n_{i+1/2}(Z_{i+1}-Z_i)\right] \\ &-G_{w_i}\left(\frac{1}{B\mu}\right)^n_i p_{wf_i}-\frac{V_{b_i}}{\alpha_c\Delta t}\left(\frac{\phi}{B_g}\right)'_i p^n_i \end{aligned} \tag{8.38}$$

The unknowns in Eq. (8.38) are the pressures of blocks $i-1$, i, and $i+1$ at time level $n+1$ and current iteration $(\nu+1)$, $\overset{(\nu+1)}{p^{n+1}_{i-1}}$, $\overset{(\nu+1)}{p^{n+1}_i}$, and $\overset{(\nu+1)}{p^{n+1}_{i+1}}$.

The general flow equation for interior block n in multidimensional flow using explicit transmissibility can be expressed as

$$\begin{aligned} &\sum_{l\in\psi_n}T^n_{l,n}\overset{(\nu+1)}{p^{n+1}_l}-\left[\sum_{l\in\psi_n}T^n_{l,n}+\frac{V_{b_n}}{\alpha_c\Delta t}\left(\frac{\phi}{B_g}\right)'_n+G_{w_n}\left(\frac{1}{B\mu}\right)^n_n\right]\overset{(\nu+1)}{p^{n+1}_n} \\ &=\sum_{l\in\psi_n}T^n_{l,n}\gamma^n_{l,n}(Z_l-Z_n)-G_{w_n}\left(\frac{1}{B\mu}\right)^n_n p_{wf_n}-\frac{V_{b_n}}{\alpha_c\Delta t}\left(\frac{\phi}{B_g}\right)'_n p^n_n \end{aligned} \tag{8.39}$$

8.5.2 Simple iteration on transmissibility method

In the simple iteration on transmissibility method, transmissibilities and the coefficient of pressure drop in well flow rate are dated at the current time level $(n+1)$ with one iteration lagging behind (ν). Gravities are dated at the old time level as mentioned in Chapter 7. We still have to iterate on $\left(\frac{\phi}{B_g}\right)'_i$. Eq. (8.36) becomes

$$T^{\overset{(\nu)}{n+1}}_{x_{i-1/2}}\left[\left(p^{\overset{(\nu+1)}{n+1}}_{i-1}-p^{\overset{(\nu+1)}{n+1}}_{i}\right)-\gamma^n_{i-1/2}(Z_{i-1}-Z_i)\right]$$
$$+T^{\overset{(\nu)}{n+1}}_{x_{i+1/2}}\left[\left(p^{\overset{(\nu+1)}{n+1}}_{i+1}-p^{\overset{(\nu+1)}{n+1}}_{i}\right)-\gamma^n_{i+1/2}(Z_{i+1}-Z_i)\right] \quad (8.40)$$
$$-G_{w_i}\left(\frac{1}{B\mu}\right)^{\overset{(\nu)}{n+1}}_i\left(p^{\overset{(\nu+1)}{n+1}}_{i}-p_{wf_i}\right)=\frac{V_{b_i}}{\alpha_c\Delta t}\left(\frac{\phi}{B_g}\right)'_i\left[p^{\overset{(\nu+1)}{n+1}}_{i}-p^n_i\right]$$

The final form of the flow equation for interior block i is obtained by rearranging the terms, yielding

$$T^{\overset{(\nu)}{n+1}}_{x_{i-1/2}}p^{\overset{(\nu+1)}{n+1}}_{i-1}-\left[T^{\overset{(\nu)}{n+1}}_{x_{i-1/2}}+T^{\overset{(\nu)}{n+1}}_{x_{i+1/2}}+\frac{V_{b_i}}{\alpha_c\Delta t}\left(\frac{\phi}{B_g}\right)'_i+G_{w_i}\left(\frac{1}{B\mu}\right)^{\overset{(\nu)}{n+1}}_i\right]p^{\overset{(\nu+1)}{n+1}}_{i}$$
$$+T^{\overset{(\nu)}{n+1}}_{x_{i+1/2}}p^{\overset{(\nu+1)}{n+1}}_{i+1}=\left[T^{\overset{(\nu)}{n+1}}_{x_{i-1/2}}\gamma^n_{i-1/2}(Z_{i-1}-Z_i)+T^{\overset{(\nu)}{n+1}}_{x_{i+1/2}}\gamma^n_{i+1/2}(Z_{i+1}-Z_i)\right] \quad (8.41)$$
$$-G_{w_i}\left(\frac{1}{B\mu}\right)^{\overset{(\nu)}{n+1}}_i p_{wf_i}-\frac{V_{b_i}}{\alpha_c\Delta t}\left(\frac{\phi}{B_g}\right)'_i p^n_i$$

The unknowns in Eq. (8.41) are the pressures of blocks $i-1$, i, and $i+1$ at time level $n+1$ and current iteration $(\nu+1)$, $p^{\overset{(\nu+1)}{n+1}}_{i-1}$, $p^{\overset{(\nu+1)}{n+1}}_{i}$, and $p^{\overset{(\nu+1)}{n+1}}_{i+1}$.

The general flow equation for interior block n in multidimensional flow using simple iteration on transmissibility can be expressed as

$$\sum_{l\in\psi_n}T^{\overset{(\nu)}{n+1}}_{l,n}p^{\overset{(\nu+1)}{n+1}}_{l}-\left[\sum_{l\in\psi_n}T^{\overset{(\nu)}{n+1}}_{l,n}+\frac{V_{b_n}}{\alpha_c\Delta t}\left(\frac{\phi}{B_g}\right)'_n+G_{w_n}\left(\frac{1}{B\mu}\right)^{\overset{(\nu)}{n+1}}_n\right]p^{\overset{(\nu+1)}{n+1}}_{n}$$
$$=\sum_{l\in\psi_n}T^{\overset{(\nu)}{n+1}}_{l,n}\gamma^n_{l,n}(Z_l-Z_n)-G_{w_n}\left(\frac{1}{B\mu}\right)^{\overset{(\nu)}{n+1}}_n p_{wf_n}-\frac{V_{b_n}}{\alpha_c\Delta t}\left(\frac{\phi}{B_g}\right)'_n p^n_n \quad (8.42)$$

8.5.3 Fully implicit (Newton's iteration) method

In the fully implicit method, transmissibility, well production rate, and fictitious well rates if present are dated at the current time level $(n+1)$. Gravities are dated at the old time level as mentioned in Chapter 7. By dating nonlinear terms and unknown pressures at the current time level and current iteration and using the previous iteration in calculating $\left(\frac{\phi}{B_g}\right)'_i$, Eq. (8.36) becomes

$$T_{x_{i-1/2}}^{\overset{(\nu+1)}{n+1}}\left[\left(p_{i-1}^{\overset{(\nu+1)}{n+1}}-p_i^{\overset{(\nu+1)}{n+1}}\right)-\gamma_{i-1/2}^n(Z_{i-1}-Z_i)\right]$$
$$+T_{x_{i+1/2}}^{\overset{(\nu+1)}{n+1}}\left[\left(p_{i+1}^{\overset{(\nu+1)}{n+1}}-p_i^{\overset{(\nu+1)}{n+1}}\right)-\gamma_{i+1/2}^n(Z_{i+1}-Z_i)\right]+q_{sc_i}^{\overset{(\nu+1)}{n+1}}=\frac{V_{b_i}}{\alpha_c\Delta t}\left(\frac{\phi}{B_g}\right)_i'\left[p_i^{\overset{(\nu+1)}{n+1}}-p_i^n\right] \tag{8.43}$$

The first, second, and third terms on the LHS of Eq. (8.43) can be approximated using the fully implicit method as

$$T_{x_{i\mp1/2}}^{\overset{(\nu+1)}{n+1}}\left[\left(p_{i\mp1}^{\overset{(\nu+1)}{n+1}}-p_i^{\overset{(\nu+1)}{n+1}}\right)-\gamma_{i\mp1/2}^n(Z_{i\mp1}-Z_i)\right]$$
$$\cong T_{x_{i\mp1/2}}^{\overset{(\nu)}{n+1}}\left[\left(p_{i\mp1}^{\overset{(\nu)}{n+1}}-p_i^{\overset{(\nu)}{n+1}}\right)-\gamma_{i\mp1/2}^n(Z_{i\mp1}-Z_i)\right]+\left[\left(p_{i\mp1}^{\overset{(\nu)}{n+1}}-p_i^{\overset{(\nu)}{n+1}}\right)\right.$$
$$\left.-\gamma_{i\mp1/2}^n(Z_{i\mp1}-Z_i)\right]\left[\frac{\partial T_{x_{i\mp1/2}}}{\partial p_i}\bigg|^{\overset{(\nu)}{n+1}}\left(p_i^{\overset{(\nu+1)}{n+1}}-p_i^{\overset{(\nu)}{n+1}}\right)\right.$$
$$\left.+\frac{\partial T_{x_{i\mp1/2}}}{\partial p_{i\mp1}}\bigg|^{\overset{(\nu)}{n+1}}\left(p_{i\mp1}^{\overset{(\nu+1)}{n+1}}-p_{i\mp1}^{\overset{(\nu)}{n+1}}\right)\right]+T_{x_{i\mp1/2}}^{\overset{(\nu)}{n+1}}\left[\left(p_{i\mp1}^{\overset{(\nu+1)}{n+1}}-p_{i\mp1}^{\overset{(\nu)}{n+1}}\right)-\left(p_i^{\overset{(\nu+1)}{n+1}}-p_i^{\overset{(\nu)}{n+1}}\right)\right] \tag{8.44}$$

and

$$q_{sc_i}^{\overset{(\nu+1)}{n+1}}\cong q_{sc_i}^{\overset{(\nu)}{n+1}}+\frac{dq_{sc_i}}{dp_i}\bigg|^{\overset{(\nu)}{n+1}}\left(p_i^{\overset{(\nu+1)}{n+1}}-p_i^{\overset{(\nu)}{n+1}}\right) \tag{8.32}$$

The RHS of Eq. (8.43) can be rewritten as

$$\frac{V_{b_i}}{\alpha_c\Delta t}\left(\frac{\phi}{B_g}\right)_i'\left[p_i^{\overset{(\nu+1)}{n+1}}-p_i^n\right]=\frac{V_{b_i}}{\alpha_c\Delta t}\left(\frac{\phi}{B_g}\right)_i'\left[\left(p_i^{\overset{(\nu+1)}{n+1}}-p_i^{\overset{(\nu)}{n+1}}\right)+\left(p_i^{\overset{(\nu)}{n+1}}-p_i^n\right)\right] \tag{8.45}$$

Substitution of Eqs. (8.32), (8.44), and (8.45) into Eq. (8.43) and collecting terms yields the final form for the fully implicit flow equation for interior gridblock i,

$$\left\{T_{x_{i-1/2}}^{\overset{(\nu)}{n+1}}+\left[\left(p_{i-1}^{\overset{(\nu)}{n+1}}-p_i^{\overset{(\nu)}{n+1}}\right)-\gamma_{i-1/2}^n(Z_{i-1}-Z_i)\right]\frac{\partial T_{x_{i-1/2}}}{\partial p_{i-1}}\bigg|^{\overset{(\nu)}{n+1}}\right\}\delta p_{i-1}^{\overset{(\nu+1)}{n+1}}$$
$$-\left\{T_{x_{i-1/2}}^{\overset{(\nu)}{n+1}}-\left[\left(p_{i-1}^{\overset{(\nu)}{n+1}}-p_i^{\overset{(\nu)}{n+1}}\right)-\gamma_{i-1/2}^n(Z_{i-1}-Z_i)\right]\frac{\partial T_{x_{i-1/2}}}{\partial p_i}\bigg|^{\overset{(\nu)}{n+1}}+T_{x_{i+1/2}}^{\overset{(\nu)}{n+1}}\right.$$
$$-\left[\left(p_{i+1}^{\overset{(\nu)}{n+1}}-p_i^{\overset{(\nu)}{n+1}}\right)-\gamma_{i+1/2}^n(Z_{i+1}-Z_i)\right]\frac{\partial T_{x_{i+1/2}}}{\partial p_i}\bigg|^{\overset{(\nu)}{n+1}}$$

$$-\frac{dq_{sc_i}}{dp_i}\bigg|^{\overset{(\nu)}{n+1}} + \frac{V_{b_i}}{\alpha_c \Delta t}\left(\frac{\phi}{B_g}\right)'_i \Bigg\} \delta p_i^{\overset{(\nu+1)}{n+1}}$$

$$+\left\{ T_{x_{i+1/2}}^{\overset{(\nu)}{n+1}} + \left[\left(p_{i+1}^{\overset{(\nu)}{n+1}} - p_i^{\overset{(\nu)}{n+1}}\right) - \gamma_{i+1/2}^n (Z_{i+1} - Z_i)\right] \frac{\partial T_{x_{i+1/2}}}{\partial p_{i+1}}\bigg|^{\overset{(\nu)}{n+1}} \right\} \delta p_{i+1}^{\overset{(\nu+1)}{n+1}}$$

$$= -\left\{ T_{x_{i-1/2}}^{\overset{(\nu)}{n+1}} \left[\left(p_{i-1}^{\overset{(\nu)}{n+1}} - p_i^{\overset{(\nu)}{n+1}}\right) - \gamma_{i-1/2}^n (Z_{i-1} - Z_i)\right] + T_{x_{i+1/2}}^{\overset{(\nu)}{n+1}} \left[\left(p_{i+1}^{\overset{(\nu)}{n+1}} - p_i^{\overset{(\nu)}{n+1}}\right)\right.\right.$$

$$\left.\left. - \gamma_{i+1/2}^n (Z_{i+1} - Z_i)\right] + q_{sc_i}^{\overset{(\nu)}{n+1}} - \frac{V_{b_i}}{\alpha_c \Delta t}\left(\frac{\phi}{B_g}\right)'_i \left(p_i^{\overset{(\nu)}{n+1}} - p_i^n\right)\right\} \tag{8.46}$$

The unknowns in Eq. (8.46), which reflects the fully implicit treatment of nonlinearities in the flow equation for interior block i, are the pressure changes over an iteration in blocks $i-1$, i, and $i+1$, $\left(p_{i-1}^{\overset{(\nu+1)}{n+1}} - p_{i-1}^{\overset{(\nu)}{n+1}}\right)$, $\left(p_i^{\overset{(\nu+1)}{n+1}} - p_i^{\overset{(\nu)}{n+1}}\right)$, and $\left(p_{i+1}^{\overset{(\nu+1)}{n+1}} - p_{i+1}^{\overset{(\nu)}{n+1}}\right)$. Note that for the first iteration ($\nu=0$), $p_i^{\overset{(0)}{n+1}} = p_i^n$ for $i=1, 2, 3, \ldots n_x$ and the first-order derivatives are evaluated at old time level n.

The fully implicit method general equation for block n has the form

$$\sum_{l\in\psi_n}\left\{ T_{l,n}^{\overset{(\nu)}{n+1}} + \left[\left(p_l^{\overset{(\nu)}{n+1}} - p_n^{\overset{(\nu)}{n+1}}\right) - \gamma_{l,n}^n (Z_l - Z_n)\right] \frac{\partial T_{l,n}}{\partial p_l}\bigg|^{\overset{(\nu)}{n+1}} + \sum_{m\in\xi_n} \frac{\partial q_{sc_{m,n}}}{\partial p_l}\bigg|^{\overset{(\nu)}{n+1}} \right\} \delta p_l^{\overset{(\nu+1)}{n+1}}$$

$$-\left\{ \sum_{l\in\psi_n}\left(T_{l,n}^{\overset{(\nu)}{n+1}} - \left[\left(p_l^{\overset{(\nu)}{n+1}} - p_n^{\overset{(\nu)}{n+1}}\right) - \gamma_{l,n}^n (Z_l - Z_n)\right] \frac{\partial T_{l,n}}{\partial p_n}\bigg|^{\overset{(\nu)}{n+1}} \right) - \sum_{l\in\xi_n} \frac{\partial q_{sc_{l,n}}}{\partial p_n}\bigg|^{\overset{(\nu)}{n+1}}\right.$$

$$\left. - \frac{dq_{sc_n}}{dp_n}\bigg|^{\overset{(\nu)}{n+1}} + \frac{V_{b_n}}{\alpha_c \Delta t}\left(\frac{\phi}{B_g}\right)'_n \right\} \delta p_n^{\overset{(\nu+1)}{n+1}} = -\left\{ \sum_{l\in\psi_n} T_{l,n}^{\overset{(\nu)}{n+1}} \left[\left(p_l^{\overset{(\nu)}{n+1}} - p_n^{\overset{(\nu)}{n+1}}\right) - \gamma_{l,n}^n (Z_l - Z_n)\right]\right.$$

$$\left. + \sum_{l\in\xi_n} q_{sc_{l,n}}^{\overset{(\nu)}{n+1}} + q_{sc_n}^{\overset{(\nu)}{n+1}} - \frac{V_{b_n}}{\alpha_c \Delta t}\left(\frac{\phi}{B_g}\right)'_n \left(p_n^{\overset{(\nu)}{n+1}} - p_n^n\right)\right\} \tag{8.47a}$$

Note that the summation term $\sum_{m\in\xi_n} \frac{\partial q_{sc_{m,n}}}{\partial p_l}\bigg|^{\overset{(\nu)}{n+1}}$ in Eq. (8.47a) contributes a maximum of one term for neighboring block l if and only if block n is a boundary block and block l falls next to reservoir boundary m. In addition, $\frac{\partial q_{sc_{m,n}}}{\partial p_l}$ and $\frac{\partial q_{sc_{m,n}}}{\partial p_n}$ are obtained from the flow rate equation of the fictitious well, which depends on the prevailing boundary condition. Note also that Eq. (8.47a) does not produce a symmetric matrix because of the term

$$\left[\left(p_l^{\overset{(\nu)}{n+1}} - p_n^{\overset{(\nu)}{n+1}}\right) - \gamma_{l,n}^n (Z_l - Z_n)\right] \frac{\partial T_{l,n}}{\partial p_n}\bigg|^{\overset{(\nu)}{n+1}}$$

Coats et al. (1977) derived the fully implicit equations for their steam model without conservative expansions of the accumulation terms. Although their equations do not conserve the material balance during iterations, they preserve it at convergence. Their method of obtaining the fully implicit iterative equation is applied here for the compressible fluid described by the implicit form of Eq. (7.12). This equation is written in a residual from at time level $n+1$; that is, all terms are placed on one side of the equation and the other side is zero. Each term at time level $n+1$ in the resulting equation is approximated by its value at the current iteration level $(\nu+1)$, which in turn can be approximated by its value at the last iteration level (ν), plus a linear combination of the unknowns arising from partial differentiation with respect to all unknown pressures. The unknown quantities in the resulting equation are the changes over an iteration of all the unknown pressures in the original equation. The resulting fully implicit iterative equation for block n is

$$\sum_{l\in\psi_n}\left\{\overset{(\nu)}{T}{}^{n+1}_{l,n}+\left[\left(\overset{(\nu)}{p}{}^{n+1}_{l}-\overset{(\nu)}{p}{}^{n+1}_{n}\right)-\gamma^{n}_{l,n}(Z_l-Z_n)\right]\frac{\partial T_{l,n}}{\partial p_l}\Bigg|^{\overset{(\nu)}{n+1}}+\sum_{m\in\xi_n}\frac{\partial q_{sc_{m,n}}}{\partial p_l}\Bigg|^{\overset{(\nu)}{n+1}}\right\}\delta\overset{(\nu+1)}{p}{}^{n+1}_{l}$$
$$-\left\{\sum_{l\in\psi_n}\left(\overset{(\nu)}{T}{}^{n+1}_{l,n}-\left[\left(\overset{(\nu)}{p}{}^{n+1}_{l}-\overset{(\nu)}{p}{}^{n+1}_{n}\right)-\gamma^{n}_{l,n}(Z_l-Z_n)\right]\frac{\partial T_{l,n}}{\partial p_n}\Bigg|^{\overset{(\nu)}{n+1}}\right)-\sum_{l\in\xi_n}\frac{\partial q_{sc_{l,n}}}{\partial p_n}\Bigg|^{\overset{(\nu)}{n+1}}\right.$$
$$\left.-\frac{\mathrm{d}q_{sc_n}}{\mathrm{d}p_n}\Bigg|^{\overset{(\nu)}{n+1}}+\frac{V_{b_n}}{\alpha_c\Delta t}\left(\frac{\phi}{B_g}\right)'_n\Bigg|^{\overset{(\nu)}{n+1}}\right\}\delta\overset{(\nu+1)}{p}{}^{n+1}_{n}=-\left\{\sum_{l\in\psi_n}\overset{(\nu)}{T}{}^{n+1}_{l,n}\left[\left(\overset{(\nu)}{p}{}^{n+1}_{l}-\overset{(\nu)}{p}{}^{n+1}_{n}\right)-\gamma^{n}_{l,n}(Z_l-Z_n)\right]\right.$$
$$\left.+\sum_{l\in\xi_n}\overset{(\nu)}{q}{}^{n+1}_{sc_{l,n}}+\overset{(\nu)}{q}{}^{n+1}_{sc_n}-\frac{V_{b_n}}{\alpha_c\Delta t}\left[\left(\frac{\phi}{B_g}\right)^{\overset{(\nu)}{n+1}}_{n}-\left(\frac{\phi}{B_g}\right)^{n}_{n}\right]\right\} \tag{8.47b}$$

Eq. (8.47b) is similar to Eq. (8.47a) with three exceptions that are related to the accumulation term. First, while Eq. (8.47a) preserves material balance at every current iteration, Eq. (8.47b) preserves material balance only at convergence. Second, the term $\left(\frac{\phi}{B_g}\right)'_n$ in Eq. (8.47a) represents the chord slope that results from a conservative expansion, whereas the term $\left(\frac{\phi}{B_g}\right)'_n\Big|^{\overset{(\nu)}{n+1}}$ in Eq. (8.47b) represents the slope of $\left(\frac{\phi}{B_g}\right)_n$, both terms being evaluated at last iteration level ν. Third, the last term on the RHS of Eq. (8.47a), $\frac{V_{b_n}}{\alpha_c\Delta t}\left(\frac{\phi}{B_g}\right)'_n\left(\overset{(\nu)}{p}{}^{n+1}_{n}-p^n_n\right)$, is replaced with $\frac{V_{b_n}}{\alpha_c\Delta t}\left[\left(\frac{\phi}{B_g}\right)^{\overset{(\nu)}{n+1}}_{n}-\left(\frac{\phi}{B_g}\right)^{n}_{n}\right]$ in Eq. (8.47b). For single-phase flow, where the accumulation term is a function of pressure only, these two terms are equal because both represent the accumulation term evaluated at the last iteration.

The next set of examples demonstrates the mechanics of implementing the explicit transmissibility method, simple iteration on transmissibility method, and fully implicit method of linearization in solving the equations for single-well simulation. It should be noted that the simple iteration on transmissibility and fully implicit methods produce close results because, contrary to the explicit transmissibility method, the transmissibility in both methods is updated every iteration. All methods in this problem show the same convergence property for a time step of 1 month because, over the pressure range 1515–4015 psia, the product μB is approximately straight line having small slope (-4.5×10^{-6} cP-RB/scf-psi).

Example 8.1 Consider the reservoir described in Example 7.13, where a 6-in vertical well is drilled on 20-acre spacing in a natural gas reservoir. The reservoir is described by four gridblocks in the radial direction as shown in Fig. 8.2. The reservoir is horizontal and has 30-ft net thickness and homogeneous and isotropic rock properties with $k=15$ md and $\phi=0.13$.

Initially, reservoir pressure is 4015 psia. Table 8.1 presents gas FVF and viscosity dependence on pressure. The external reservoir boundaries are sealed to fluid flow. Let the well produce with a FBHP of 1515 psia. Find the pressure distribution in the reservoir after 1 month (30.42 days) using a single time step. Solve the problem using the implicit formulation with the explicit transmissibility method of linearization and present the simulation results up to 6 months.

Solution

Gridblock locations, bulk volumes, and geometric factors in the radial direction are calculated in exactly the same way as in Example 7.13. The results are presented in Table 8.2.

For single-well simulation in a horizontal reservoir ($Z_n=$ constant) with no-flow boundaries $\left(\sum_{l\in\xi_n} q_{sc_{l,n}}^{n+1}=0\right)$, the implicit flow equation with explicit transmissibility is obtained from Eq. (8.39). For gridblock n with a well operating under a specified FBHP,

$$\sum_{l\in\psi_n} T_{l,n}^{n} p_l^{\overset{(\nu+1)}{n+1}} - \left[\sum_{l\in\psi_n} T_{l,n}^{n} + \frac{V_{b_n}}{\alpha_c \Delta t}\left(\frac{\phi}{B_g}\right)_n^{\prime} + G_{w_n}\left(\frac{1}{B\mu}\right)_n^{n}\right] p_n^{\overset{(\nu+1)}{n+1}} = -G_{w_n}\left(\frac{1}{B\mu}\right)_n^{n} p_{wf_n} - \frac{V_{b_n}}{\alpha_c \Delta t}\left(\frac{\phi}{B_g}\right)_n^{\prime} p_n^n \quad (8.48a)$$

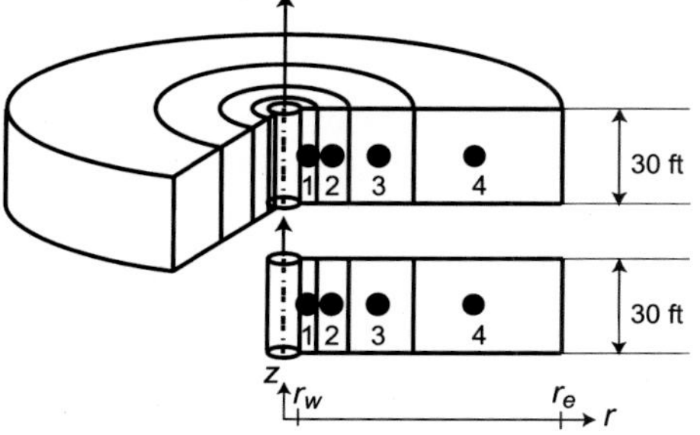

FIG. 8.2 Discretized 1-D reservoir in Example 8.1.

TABLE 8.1 Gas FVF and viscosity in Example 8.1.

Pressure (psia)	GFVF (RB/scf)	Gas viscosity (cP)
215.00	0.016654	0.0126
415.00	0.008141	0.0129
615.00	0.005371	0.0132
815.00	0.003956	0.0135
1015.00	0.003114	0.0138
1215.00	0.002544	0.0143
1415.00	0.002149	0.0147
1615.00	0.001857	0.0152
1815.00	0.001630	0.0156
2015.00	0.001459	0.0161
2215.00	0.001318	0.0167
2415.00	0.001201	0.0173
2615.00	0.001109	0.0180
2815.00	0.001032	0.0186
3015.00	0.000972	0.0192
3215.00	0.000922	0.0198
3415.00	0.000878	0.0204
3615.00	0.000840	0.0211
3815.00	0.000808	0.0217
4015.00	0.000779	0.0223

For gridblock n without a well,

$$\sum_{l\in\psi_n} T_{l,n}^{n} p_l^{n+1} - \left[\sum_{l\in\psi_n} T_{l,n}^{n} + \frac{V_{b_n}}{\alpha_c \Delta t}\left(\frac{\phi}{B_g}\right)'_n\right] p_n^{\overset{(\nu+1)}{n+1}} = -\frac{V_{b_n}}{\alpha_c \Delta t}\left(\frac{\phi}{B_g}\right)'_n p_n^n \quad (8.48b)$$

The gas in this reservoir flows toward the well in gridblock 1. Therefore, gridblock 4 is upstream to gridblock 3, gridblock 3 is upstream to gridblock 2, and gridblock 2 is upstream to gridblock 1. In solving this problem, we use upstream weighting (Section 8.4.1.1) of the pressure-dependent terms in transmissibility.

TABLE 8.2 Gridblock locations, bulk volumes, and geometric factors.

n	i	r_i (ft)	$G_{r_{i+1/2}}$ (RB-cP/D-psi)	V_{b_n} (ft^3)
1	1	0.5611	1.6655557	340.59522
2	2	3.8014	1.6655557	15,631.859
3	3	25.7532	1.6655557	717,435.23
4	4	174.4683	1.6655557	25,402,604

TABLE 8.3 Estimated gridblock FVF, viscosity, and chord slope at old iteration $\nu=0$.

Block n	$\overset{(0)}{p_n^{n+1}}$ (psia)	B_g (RB/scf)	μ_g (cP)	$(\phi/B_g)'_n$	$\frac{V_{b_n}}{\alpha_c \Delta t}\left(\frac{\phi}{B_g}\right)'_n$
1	4015	0.00077900	0.0223000	0.0310567	0.0619323
2	4015	0.00077900	0.0223000	0.0310567	2.84243
3	4015	0.00077900	0.0223000	0.0310567	130.455
4	4015	0.00077900	0.0223000	0.0310567	4619.10

First time step calculations ($n=0$, $t_{n+1}=30.42$ days, and $\Delta t=30.42$ days). Assign $p_1^n=p_2^n=p_3^n=p_4^n=p_{in}=4015$ psia.

For the first iteration ($\nu=0$), assume $\overset{(\nu)}{p_n^{n+1}}=p_n^n=4015$ psia for $n=1, 2, 3, 4$ In addition, we estimate $\left(\frac{\phi}{B_g}\right)'_n$ between p_n^n and $p_n^n-\varepsilon$ where $\varepsilon=1$ psi. Table 8.3 presents the estimated values of the FVF and viscosity using linear interpolation within table entries, chord slope $\left(\frac{\phi}{B_g}\right)'_n$, and $\frac{V_{b_n}}{\alpha_c \Delta t}\left(\frac{\phi}{B_g}\right)'_n$ for all grid blocks at the first iteration. Note that at $p=4014$ psia, $B_g=0.00077914$ RB/scf and $\mu_g=0.0222970$ cP. For example, for gridblock 1,

$$\left(\frac{\phi}{B_g}\right)'_1=\frac{\left(\frac{\phi}{B}\right)_1^{\overset{(\nu)}{n+1}}-\left(\frac{\phi}{B}\right)_1^{n}}{\overset{(\nu)}{p_1^{n+1}}-p_1^n}=\frac{\left(\frac{0.13}{0.00077914}\right)-\left(\frac{0.13}{0.000779}\right)}{4014-4015}=0.0310567$$

$$\frac{V_{b_1}}{\alpha_c \Delta t}\left(\frac{\phi}{B_g}\right)'_1=\frac{340.59522\times 0.0310567}{5.614583\times 30.42}=0.0619323$$

and

$$T^n_{r_{2,1}}\Big|_2 = T^n_{r_{1,2}}\Big|_2 = G_{r_{1+1/2}}\left(\frac{1}{\mu B}\right)^n_2 = 1.6655557 \times \left(\frac{1}{0.0223000 \times 0.00077900}\right)$$
$$= 95877.5281$$

for upstream weighting of transmissibility.

In addition, for the production well in wellblock 1, G_{w_1} is calculated using Eq. (6.10a), yielding

$$G_{w_1} = \frac{2 \times \pi \times 0.001127 \times 15 \times 30}{\log_e(0.5611/0.25)} = 3.941572$$

$$G_{w_1}\left(\frac{1}{\mu B}\right)^n_1 = 3.941572 \times \left(\frac{1}{0.0223000 \times 0.00077900}\right)$$
$$= 226896.16 \text{scf/D-psi}$$

Therefore, $T^n_{r_{1,2}}|_2 = T^n_{r_{2,3}}|_3 = T^n_{r_{3,4}}|_4 = 95877.5281$ scf/D-psi. Note also that $T^n_{r_{l,n}} = T^n_{r_{n,l}}$.

For gridblock 1, $n=1$ and $\psi_1 = \{2\}$. Therefore, Eq. (8.48a) becomes

$$-\left[T^n_{2,1}\big|_2 + \frac{V_{b_1}}{\alpha_c \Delta t}\left(\frac{\phi}{B_g}\right)'_1 + G_{w_1}\left(\frac{1}{B\mu}\right)^n_1\right] p_1^{\overset{(\nu+1)}{n+1}} + T^n_{2,1}\big|_2 p_2^{\overset{(\nu+1)}{n+1}}$$
$$= -G_{w_1}\left(\frac{1}{B\mu}\right)^n_1 p_{wf_1} - \frac{V_{b_1}}{\alpha_c \Delta t}\left(\frac{\phi}{B_g}\right)'_1 p_1^n \tag{8.49}$$

Substitution of the values in this equation gives

$$-[95877.5281 + 0.0619323 + 226896.16] p_1^{\overset{(\nu+1)}{n+1}} + 95877.5281 p_2^{\overset{(\nu+1)}{n+1}}$$
$$= -226896.16 \times 1515 - 0.0619323 \times 4015$$

or after simplification,

$$-322773.749 p_1^{\overset{(\nu+1)}{n+1}} + 95877.5281 p_2^{\overset{(\nu+1)}{n+1}} = -343747929 \tag{8.50}$$

For gridblock 2, $n=2$ and $\psi_2 = \{1,3\}$. Therefore, Eq. (8.48b) becomes

$$T^n_{1,2}\big|_2 p_1^{\overset{(\nu+1)}{n+1}} - \left[T^n_{1,2}\big|_2 + T^n_{3,2}\big|_3 + \frac{V_{b_2}}{\alpha_c \Delta t}\left(\frac{\phi}{B_g}\right)'_2\right] p_2^{\overset{(\nu+1)}{n+1}} + T^n_{3,2}\big|_3 p_3^{\overset{(\nu+1)}{n+1}}$$
$$= -\frac{V_{b_2}}{\alpha_c \Delta t}\left(\frac{\phi}{B_g}\right)'_2 p_2^n \tag{8.51}$$

Substitution of the values in this equation gives

$$95877.5281 p_1^{\overset{(\nu+1)}{n+1}} - [95877.5281 + 95877.5281 + 2.84243] p_2^{\overset{(\nu+1)}{n+1}}$$
$$+ 95877.5281 p_3^{\overset{(\nu+1)}{n+1}} = -2.84243 \times 4015$$

or after simplification,

$$95877.5281p_1^{\overset{(\nu+1)}{n+1}} - 191757.899p_2^{\overset{(\nu+1)}{n+1}} + 95877.5281p_3^{\overset{(\nu+1)}{n+1}} = -11412.3496 \tag{8.52}$$

For gridblock 3, $n=3$ and $\psi_3=\{2,4\}$. Therefore, Eq. (8.48b) becomes

$$T_{2,3}^n\big|_3 p_2^{\overset{(\nu+1)}{n+1}} - \left[T_{2,3}^n\big|_3 + T_{4,3}^n\big|_4 + \frac{V_{b_3}}{\alpha_c \Delta t}\left(\frac{\phi}{B_g}\right)_3'\right] p_3^{\overset{(\nu+1)}{n+1}} + T_{4,3}^n\big|_4 p_4^{\overset{(\nu+1)}{n+1}} = -\frac{V_{b_3}}{\alpha_c \Delta t}\left(\frac{\phi}{B_g}\right)_3' p_3^n \tag{8.53}$$

Substitution of the values in this equation gives

$$95877.5281p_2^{\overset{(\nu+1)}{n+1}} - [95877.5281 + 95877.5281 + 130.455]p_3^{\overset{(\nu+1)}{n+1}} + 95877.5281p_4^{\overset{(\nu+1)}{n+1}} = -130.455 \times 4015$$

or after simplification,

$$95877.5281p_2^{\overset{(\nu+1)}{n+1}} - 191885.511p_3^{\overset{(\nu+1)}{n+1}} + 95877.5281p_4^{\overset{(\nu+1)}{n+1}} = -523777.862 \tag{8.54}$$

For gridblock 4, $n=4$ and $\psi_4=\{3\}$. Therefore, Eq. (8.48b) becomes

$$T_{3,4}^n\big|_4 p_3^{\overset{(\nu+1)}{n+1}} - \left[T_{3,4}^n\big|_4 + \frac{V_{b_4}}{\alpha_c \Delta t}\left(\frac{\phi}{B_g}\right)_4'\right] p_4^{\overset{(\nu+1)}{n+1}} = -\frac{V_{b_4}}{\alpha_c \Delta t}\left(\frac{\phi}{B_g}\right)_4' p_4^n \tag{8.55}$$

Substitution of the values in this equation gives

$$95877.5281p_3^{\overset{(\nu+1)}{n+1}} - [95877.5281 + 4619.10]p_4^{\overset{(\nu+1)}{n+1}} = -4619.10 \times 4015$$

or after simplification,

$$95877.5281p_3^{\overset{(\nu+1)}{n+1}} - 100496.6251p_4^{\overset{(\nu+1)}{n+1}} = -18545676.2 \tag{8.56}$$

The results of solving Eqs. (8.50), (8.52), (8.54), and (8.56) for the unknown pressures are $p_1^{\overset{(1)}{n+1}} = 1559.88$ psia, $p_2^{\overset{(1)}{n+1}} = 1666.08$ psia, $p_3^{\overset{(1)}{n+1}} = 1772.22$ psia, and $p_4^{\overset{(1)}{n+1}} = 1875.30$ psia.

For the second iteration ($\nu=1$), we use $p_n^{\overset{(1)}{n+1}}$ to estimate the values of FVF to estimate chord slope $\left(\frac{\phi}{B_g}\right)_n'$ and $\frac{V_{b_n}}{\alpha_c \Delta t}\left(\frac{\phi}{B_g}\right)_n'$ for gridblock n. Table 8.4 lists these values. For example, for gridblock 1,

$$\left(\frac{\phi}{B_g}\right)_1' = \frac{\left(\frac{\phi}{B}\right)_1^{\overset{(\nu)}{n+1}} - \left(\frac{\phi}{B}\right)_1^n}{p_1^{\overset{(\nu)}{n+1}} - p_1^n} = \frac{\left(\frac{0.13}{0.0019375}\right) - \left(\frac{0.13}{0.000779000}\right)}{1559.88 - 4015} = 0.0406428$$

TABLE 8.4 Estimated gridblock FVF and chord slope at old iteration $\nu=1$.

Block n	$p_n^{\overset{(1)}{n+1}}$ (psia)	$B_{g_n}^{\overset{(1)}{n+1}}$ (RB/scf)	$(\phi/B_g)'_n$	$\frac{V_{b_n}}{\alpha_c \Delta t}\left(\frac{\phi}{B_g}\right)'_n$
1	1559.88	0.0019375	0.0406428	0.0810486
2	1666.08	0.0017990	0.0402820	3.68676
3	1772.22	0.0016786	0.0398760	167.501
4	1875.30	0.0015784	0.0395013	5875.07

$$\frac{V_{b_1}}{\alpha_c \Delta t}\left(\frac{\phi}{B_g}\right)'_1 = \frac{340.59522 \times 0.0406428}{5.614583 \times 30.42} = 0.0810486$$

Note that for the explicit transmissibility treatment, $T^n_{r_{1,2}}|_2 = T^n_{r_{2,3}}|_3 = T^n_{r_{3,4}}|_4 = 95877.5281$ scf/D-psi and $G_{w_1}\left(\frac{1}{\mu B}\right)^n_1 = 226896.16$ scf/D-psi for all iterations.

For gridblock 1, $n=1$. Substitution of the values in Eq. (8.49) gives

$$-[95877.5281 + 0.0810486 + 226896.16] p_1^{\overset{(\nu+1)}{n+1}} + 95877.5281 p_2^{\overset{(\nu+1)}{n+1}}$$
$$= -226896.16 \times 1515 - 0.0810486 \times 4015$$

or after simplification,

$$-322773.768 p_1^{\overset{(\nu+1)}{n+1}} + 95877.5281 p_2^{\overset{(\nu+1)}{n+1}} = -343748006 \tag{8.57}$$

For gridblock 2, $n=2$. Substitution of the values in Eq. (8.51) gives

$$95877.5281 p_1^{\overset{(\nu+1)}{n+1}} - [95877.5281 + 95877.5281 + 3.68676] p_2^{\overset{(\nu+1)}{n+1}}$$
$$+ 95877.5281 p_3^{\overset{(\nu+1)}{n+1}} = -3.68676 \times 4015$$

or after simplification,

$$95877.5281 p_1^{\overset{(\nu+1)}{n+1}} - 191758.743 p_2^{\overset{(\nu+1)}{n+1}} + 95877.5281 p_3^{\overset{(\nu+1)}{n+1}} = -14802.3438 \tag{8.58}$$

For gridblock 3, $n=3$. Substitution of the values in Eq. (8.53) gives

$$95877.5281 p_2^{\overset{(\nu+1)}{n+1}} - [95877.5281 + 95877.5281 + 167.501] p_3^{\overset{(\nu+1)}{n+1}}$$
$$+ 95877.5281 p_4^{\overset{(\nu+1)}{n+1}} = -167.501 \times 4015$$

or after simplification,

$$95877.5281 p_2^{\overset{(\nu+1)}{n+1}} - 191922.557 p_3^{\overset{(\nu+1)}{n+1}} + 95877.5281 p_4^{\overset{(\nu+1)}{n+1}} = -672516.495 \tag{8.59}$$

For gridblock 4, $n=4$. Substitution of the values in Eq. (8.55) gives

$$95877.5281 p_3^{\overset{(\nu+1)}{n+1}} - [95877.5281 + 5875.07] p_4^{\overset{(\nu+1)}{n+1}} = -5875.07 \times 4015$$

or after simplification,

$$95877.5281 p_3^{\overset{(\nu+1)}{n+1}} - 101752.599 p_4^{\overset{(\nu+1)}{n+1}} = -23588411.0 \tag{8.60}$$

The results of solving Eqs. (8.57), (8.58), (8.59), and (8.60) for the unknown pressures are $p_1^{\overset{(2)}{n+1}} = 1569.96$ psia, $p_2^{\overset{(2)}{n+1}} = 1700.03$ psia, $p_3^{\overset{(2)}{n+1}} = 1830.00$ psia, and $p_4^{\overset{(2)}{n+1}} = 1956.16$ psia. Iterations continue until the convergence criterion is satisfied. Table 8.5 shows the successive iterations for the first time step. Note that it took four iterations to converge. The convergence criterion was set as given by Eq. (7.179); that is,

$$\max_{1 \le n \le N} \left| \frac{p_n^{\overset{(\nu+1)}{n+1}} - p_n^{\overset{(\nu)}{n+1}}}{p_n^{\overset{(\nu)}{n+1}}} \right| \le 0.001 \tag{8.61}$$

After reaching convergence, the time is incremented by $\Delta t = 30.42$ days, and the earlier procedure is repeated. Table 8.6 shows the converged solutions at various times up to 6 months of simulation time.

Example 8.2 Consider the problem described in Example 8.1. Apply the simple iteration on transmissibility method to find the pressure distribution in the reservoir after 1 month (30.42 days) using a single time step. Present the simulation results up to 6 months.

TABLE 8.5 Pressure solution at $t_{n+1} = 30.42$ days for successive iterations.

$\nu+1$	$p_1^{\overset{(\nu+1)}{n+1}}$ (psia)	$p_2^{\overset{(\nu+1)}{n+1}}$ (psia)	$p_3^{\overset{(\nu+1)}{n+1}}$ (psia)	$p_4^{\overset{(\nu+1)}{n+1}}$ (psia)
0	4015.00	4015.00	4015.00	4015.00
1	1559.88	1666.08	1772.22	1875.30
2	1569.96	1700.03	1830.00	1956.16
3	1569.64	1698.94	1828.15	1953.57
4	1569.65	1698.98	1828.23	1953.68

TABLE 8.6 Converged pressure solution and gas production at various times.

$n+1$	Time (day)	ν	p_1^{n+1} (psia)	p_2^{n+1} (psia)	p_3^{n+1} (psia)	p_4^{n+1} (psia)	q_{gsc}^{n+1} (MMscf/D)	Cumulative production (MMMscf)
1	30.42	4	1569.65	1698.98	1828.23	1953.68	−12.4003	−0.377217
2	60.84	3	1531.85	1569.07	1603.85	1636.31	−2.28961	−0.446867
3	91.26	3	1519.81	1530.96	1541.87	1552.37	−0.639629	−0.466324
4	121.68	2	1516.45	1519.88	1523.27	1526.58	−0.191978	−0.472164
5	152.10	2	1515.44	1516.49	1517.53	1518.55	−0.058311	−0.473938
6	182.52	2	1515.13	1515.45	1515.77	1516.09	−0.017769	−0.474478

Solution

Table 8.2 reports the gridblock locations, bulk volumes, and geometric factors in the radial direction. For single-well simulation in a horizontal reservoir (Z_n= constant) with no-flow boundaries $\left(\sum_{l\in\xi_n} q_{sc_{l,n}}^{n+1}=0\right)$, the implicit flow equation with simple iteration on transmissibility is obtained from Eq. (8.42).

For gridblock n with a well operating under a specified FBHP,

$$\sum_{l\in\psi_n} T_{l,n}^{\overset{(\nu)}{n+1}} p_l^{\overset{(\nu+1)}{n+1}} - \left[\sum_{l\in\psi_n} T_{l,n}^{\overset{(\nu)}{n+1}} + \frac{V_{b_n}}{\alpha_c \Delta t}\left(\frac{\phi}{B_g}\right)_n' + G_{w_n}\left(\frac{1}{B\mu}\right)_n^{\overset{(\nu)}{n+1}}\right] p_n^{\overset{(\nu+1)}{n+1}}$$

$$= -G_{w_n}\left(\frac{1}{B\mu}\right)_n^{\overset{(\nu)}{n+1}} p_{wf_n} - \frac{V_{b_n}}{\alpha_c \Delta t}\left(\frac{\phi}{B_g}\right)_n' p_n^n \tag{8.62a}$$

For gridblock n without a well,

$$\sum_{l\in\psi_n} T_{l,n}^{\overset{(\nu)}{n+1}} p_l^{\overset{(\nu+1)}{n+1}} - \left[\sum_{l\in\psi_n} T_{l,n}^{\overset{(\nu)}{n+1}} + \frac{V_{b_n}}{\alpha_c \Delta t}\left(\frac{\phi}{B_g}\right)_n'\right] p_n^{\overset{(\nu+1)}{n+1}} = -\frac{V_{b_n}}{\alpha_c \Delta t}\left(\frac{\phi}{B_g}\right)_n' p_n^n \tag{8.62b}$$

As mentioned in Example 8.1, the gas in this reservoir flows toward the well in gridblock 1, gridblock 4 is upstream to gridblock 3, gridblock 3 is upstream to gridblock 2, and gridblock 2 is upstream to gridblock 1. In solving this problem, we use upstream weighting (Section 8.4.1.1) of the pressure-dependent terms in transmissibility.

First time step calculations ($n=0$, $t_{n+1}=30.42$ days, and $\Delta t=30.42$ days).

For the first iteration ($v=0$), assume $p_n^{\overset{(\nu)}{n+1}}=p_n^n=4015$ psia for $n=1, 2, 3, 4$.

Therefore, $G_{w_1}\left(\frac{1}{\mu B}\right)_1^{\overset{(0)}{n+1}}=G_{w_1}\left(\frac{1}{\mu B}\right)_1^n=226896.16$ scf/D-psi and $T_{r_{l,n}}^{\overset{(0)}{n+1}}=T_{r_{l,n}}^n$, or more explicitly,

$T_{r_{1,2}}^{\overset{(0)}{n+1}}\Big|_2=T_{r_{2,3}}^{\overset{(0)}{n+1}}\Big|_3=T_{r_{3,4}}^{\overset{(0)}{n+1}}\Big|_4=95877.5281$ scf/D-psi. Consequently, the equations for gridblocks 1, 2, 3, and 4 are given by Eqs. (8.50), (8.52), (8.54), and (8.56), respectively, and the unknown pressures are $p_1^{\overset{(1)}{n+1}}=1559.88$ psia, $p_2^{\overset{(1)}{n+1}}=1666.08$ psia, $p_3^{\overset{(1)}{n+1}}=1772.22$ psia, and $p_4^{\overset{(1)}{n+1}}=1875.30$ psia.

For the second iteration ($v=1$), we use $p_n^{\overset{(1)}{n+1}}$ to estimate the values of FVF, gas viscosity, and chord slope $\left(\frac{\phi}{B_g}\right)_n'$ and calculate $\frac{V_{b_n}}{\alpha_c\Delta t}\left(\frac{\phi}{B_g}\right)_n'$ for gridblock n. Table 8.7 lists these values in addition to the upstream value of interblock transmissibility $\left(T_{r_{n,n+1}}^{\overset{(\nu)}{n+1}}\right)$. For example, for gridblock 1,

$$\left(\frac{\phi}{B_g}\right)_1' = \frac{\left(\frac{\phi}{B}\right)_1^{\overset{(\nu)}{n+1}} - \left(\frac{\phi}{B}\right)_1^n}{p_1^{\overset{(\nu)}{n+1}} - p_1^n} = \frac{\left(\frac{0.13}{0.0019375}\right) - \left(\frac{0.13}{0.000779000}\right)}{1559.88 - 4015} = 0.0406428$$

TABLE 8.7 Estimated gridblock FVF and chord slope at old iteration $\nu = 1$.

Block n	$\overset{(1)}{p_n^{n+1}}$ (psia)	$\overset{(1)}{B_{g_n}^{n+1}}$ (RB/scf)	$\overset{(1)}{\mu_{g_n}^{n+1}}$ (cP)	$\overset{(\nu)}{T_{r_{n,n+1}}^{n+1}}\Big\|_{n+1}$	$(\phi/B_g)'_n$	$\frac{V_{b_n}}{\alpha_c \Delta t}\left(\frac{\phi}{B_g}\right)'_n$
1	1559.88	0.0019375	0.0150622	60,502.0907	0.0406428	0.0810486
2	1666.08	0.0017990	0.0153022	63,956.9105	0.0402820	3.68676
3	1772.22	0.0016786	0.0155144	66,993.0320	0.0398760	167.501
4	1875.30	0.0015784	0.0157508	–	0.0395013	5875.07

$$\frac{V_{b_1}}{\alpha_c \Delta t}\left(\frac{\phi}{B_g}\right)'_1 = \frac{340.59522 \times 0.0406428}{5.614583 \times 30.42} = 0.0810486$$

and

$$T_{r_{1,2}}^{\overset{(\nu)}{n+1}}\Big|_2 = T_{r_{2,1}}^{\overset{(\nu)}{n+1}}\Big|_2 = G_{r_{1+1/2}}\left(\frac{1}{\mu B}\right)_2^{\overset{(\nu)}{n+1}} = 1.6655557 \times \left(\frac{1}{0.0153022 \times 0.0017990}\right)$$
$$= 60502.0907$$

for upstream weighting of transmissibility. In addition, for the production well in wellblock 1,

$$G_{w_1}\left(\frac{1}{\mu B}\right)_1^{\overset{(\nu)}{n+1}} = 3.941572 \times \left(\frac{1}{0.01506220 \times 0.00193748}\right) = 135065.6$$

For gridblock 1, $n=1$ and $\psi_1=\{2\}$. Therefore, Eq. (8.62a) becomes

$$-\left[T_{2,1}^{\overset{(\nu)}{n+1}}\Big|_2 + \frac{V_{b_1}}{\alpha_c \Delta t}\left(\frac{\phi}{B_g}\right)'_1 + G_{w_1}\left(\frac{1}{B\mu}\right)_1^{\overset{(\nu)}{n+1}}\right] p_1^{\overset{(\nu+1)}{n+1}} + T_{2,1}^{\overset{(\nu)}{n+1}}\Big|_2 p_2^{\overset{(\nu+1)}{n+1}}$$
$$= -G_{w_1}\left(\frac{1}{B\mu}\right)_1^{\overset{(\nu)}{n+1}} p_{wf_1}^n - \frac{V_{b_1}}{\alpha_c \Delta t}\left(\frac{\phi}{B_g}\right)'_1 p_1^n \tag{8.63}$$

Substitution of the values in this equation gives

$$-[60502.0907 + 0.0810486 + 135065.6] p_1^{\overset{(\nu+1)}{n+1}} + 60502.0907 p_2^{\overset{(\nu+1)}{n+1}}$$
$$= -135065.6 \times 1515 - 0.0810486 \times 4015$$

or after simplification,

$$-195567.739 p_1^{\overset{(\nu+1)}{n+1}} + 60502.0907 p_2^{\overset{(\nu+1)}{n+1}} = -204624660 \tag{8.64}$$

For gridblock 2, $n=2$ and $\psi_2=\{1,3\}$. Therefore, Eq. (8.62b) becomes

$$T_{1,2}^{\overset{(\nu)}{n+1}}\Big|_2 p_1^{\overset{(\nu+1)}{n+1}} - \left[T_{1,2}^{\overset{(\nu)}{n+1}}\Big|_2 + T_{3,2}^{\overset{(\nu)}{n+1}}\Big|_3 + \frac{V_{b_2}}{\alpha_c \Delta t}\left(\frac{\phi}{B_g}\right)'_2\right] p_2^{\overset{(\nu+1)}{n+1}}$$
$$+ T_{3,2}^{\overset{(\nu)}{n+1}}\Big|_3 p_3^{\overset{(\nu+1)}{n+1}} = -\frac{V_{b_2}}{\alpha_c \Delta t}\left(\frac{\phi}{B_g}\right)'_2 p_2^n \tag{8.65}$$

Substitution of the values in this equation gives

$$60502.0907 p_1^{\overset{(\nu+1)}{n+1}} - [60502.0907 + 63956.9105 + 3.68676] p_2^{\overset{(\nu+1)}{n+1}}$$
$$+ 63956.9105 p_3^{\overset{(\nu+1)}{n+1}} = -3.68676 \times 4015$$

or after simplification,

$$60502.0907p_1^{\overset{(\nu+1)}{n+1}} - 124462.688p_2^{\overset{(\nu+1)}{n+1}} + 63956.9105p_3^{\overset{(\nu+1)}{n+1}} = -14802.3438 \tag{8.66}$$

For gridblock 3, $n=3$ and $\psi_3=\{2,4\}$. Therefore, Eq. (8.62b) becomes

$$T_{2,3}^{\overset{(\nu)}{n+1}}\Big|_3 p_2^{\overset{(\nu+1)}{n+1}} - \left[T_{2,3}^{\overset{(\nu)}{n+1}}\Big|_3 + T_{4,3}^{\overset{(\nu)}{n+1}}\Big|_4 + \frac{V_{b_3}}{\alpha_c \Delta t}\left(\frac{\phi}{B_g}\right)'_3\right] p_3^{\overset{(\nu+1)}{n+1}} + T_{4,3}^{\overset{(\nu)}{n+1}}\Big|_4 p_4^{\overset{(\nu+1)}{n+1}}$$
$$= -\frac{V_{b_3}}{\alpha_c \Delta t}\left(\frac{\phi}{B_g}\right)'_3 p_3^n \tag{8.67}$$

Substitution of the values in this equation gives

$$63956.9105p_2^{\overset{(\nu+1)}{n+1}} - [63956.9105 + 66993.0320 + 167.501]p_3^{\overset{(\nu+1)}{n+1}}$$
$$+ 66993.0320p_4^{\overset{(\nu+1)}{n+1}} = -167.501 \times 4015$$

or after simplification,

$$63956.9105p_2^{\overset{(\nu+1)}{n+1}} - 131117.443p_3^{\overset{(\nu+1)}{n+1}} + 66993.0320p_4^{\overset{(\nu+1)}{n+1}} = -672516.495 \tag{8.68}$$

For gridblock 4, $n=4$ and $\psi_4=\{3\}$. Therefore, Eq. (8.62b) becomes

$$T_{3,4}^{\overset{(\nu)}{n+1}}\Big|_4 p_3^{\overset{(\nu+1)}{n+1}} - \left[T_{3,4}^{\overset{(\nu)}{n+1}}\Big|_4 + \frac{V_{b_4}}{\alpha_c \Delta t}\left(\frac{\phi}{B_g}\right)'_4\right] p_4^{\overset{(\nu+1)}{n+1}} = -\frac{V_{b_4}}{\alpha_c \Delta t}\left(\frac{\phi}{B_g}\right)'_4 p_4^n \tag{8.69}$$

Substitution of the values in this equation gives

$$66993.0320p_3^{\overset{(\nu+1)}{n+1}} - [66993.0320 + 5875.07]p_4^{\overset{(\nu+1)}{n+1}} = -5875.07 \times 4015$$

or after simplification,

$$66993.0320p_3^{\overset{(\nu+1)}{n+1}} - 72868.1032p_4^{\overset{(\nu+1)}{n+1}} = -23588411.0 \tag{8.70}$$

The results of solving Eqs. (8.64), (8.66), (8.68), and (8.70) for the unknown pressures are $p_1^{\overset{(2)}{n+1}} = 1599.52$ psia, $p_2^{\overset{(2)}{n+1}} = 1788.20$ psia, $p_3^{\overset{(2)}{n+1}} = 1966.57$ psia, and $p_4^{\overset{(2)}{n+1}} = 2131.72$ psia.

Iterations continue until the convergence criterion is satisfied. Table 8.8 shows the successive iterations for the first time step. Note that it took five iterations to converge. The convergence criterion was set as given by Eq. (8.61). After reaching convergence, the time is incremented by $\Delta t = 30.42$ days, and

TABLE 8.8 Pressure solution at $t_{n+1}=30.42$ days for successive iterations.

$\nu+1$	$\overset{(\nu+1)}{p_1^{n+1}}$ (psia)	$\overset{(\nu+1)}{p_2^{n+1}}$ (psia)	$\overset{(\nu+1)}{p_3^{n+1}}$ (psia)	$\overset{(\nu+1)}{p_4^{n+1}}$ (psia)
0	4015.00	4015.00	4015.00	4015.00
1	1559.88	1666.08	1772.22	1875.30
2	1599.52	1788.20	1966.57	2131.72
3	1597.28	1773.65	1937.34	2087.32
4	1597.54	1775.64	1941.60	2094.01
5	1597.51	1775.38	1941.02	2093.08

the aforementioned procedure is repeated. Table 8.9 shows the converged solutions at various times up to 6 months of simulation time.

Example 8.3 Consider the problem described in Example 8.1. Apply Newton's iteration method to find the pressure distribution in the reservoir after 1 month (30.42 days) using a single time step, and present the simulation results up to 6 months.

Solution

Table 8.2 reports the gridblock locations, bulk volumes, and geometric factors in the radial direction. For single-well simulation in a horizontal reservoir ($Z_n=$ constant) with no-flow boundaries $\left(\sum_{l\in\xi_n} q_{sc_{l,n}}^{n+1}=0\right)$, the implicit flow equation with implicit transmissibility is obtained from Eq. (8.47a).

For gridblock n with a well operating under a specified FBHP,

$$\sum_{l\in\psi_n}\left\{\overset{(\nu)}{T_{l,n}^{n+1}}+\left(\overset{(\nu)}{p_l^{n+1}}-\overset{(\nu)}{p_n^{n+1}}\right)\left.\frac{\partial T_{l,n}}{\partial p_l}\right|^{\overset{(\nu)}{n+1}}\right\}\delta\overset{(\nu+1)}{p_l^{n+1}}$$
$$-\left\{\sum_{l\in\psi_n}\left[\overset{(\nu)}{T_{l,n}^{n+1}}-\left(\overset{(\nu)}{p_l^{n+1}}-\overset{(\nu)}{p_n^{n+1}}\right)\left.\frac{\partial T_{l,n}}{\partial p_n}\right|^{\overset{(\nu)}{n+1}}\right]-\left.\frac{dq_{sc_n}}{dp_n}\right|^{\overset{(\nu)}{n+1}}+\frac{V_{b_n}}{\alpha_c\Delta t}\left(\frac{\phi}{B_g}\right)_n'\right\}\delta\overset{(\nu+1)}{p_n^{n+1}}$$
$$=-\left\{\sum_{l\in\psi_n}\overset{(\nu)}{T_{l,n}^{n+1}}\left(\overset{(\nu)}{p_l^{n+1}}-\overset{(\nu)}{p_n^{n+1}}\right)+\overset{(\nu)}{q_{sc_n}^{n+1}}-\frac{V_{b_n}}{\alpha_c\Delta t}\left(\frac{\phi}{B_g}\right)_n'\left(\overset{(\nu)}{p_n^{n+1}}-p_n^n\right)\right\}$$

(8.71a)

TABLE 8.9 Converged pressure solution and gas production at various times.

$n+1$	Time (day)	ν	p_1^{n+1} (psia)	p_2^{n+1} (psia)	p_3^{n+1} (psia)	p_4^{n+1} (psia)	q_{gsc}^{n+1} (MMscf/D)	Cumulative production (MMMscf)
1	30.42	5	1597.51	1775.38	1941.02	2093.08	−11.3980	−0.346727
2	60.84	3	1537.18	1588.10	1637.63	1685.01	−2.95585	−0.436644
3	91.26	3	1521.54	1536.87	1552.07	1566.82	−0.863641	−0.462916
4	121.68	2	1517.03	1521.84	1526.63	1531.31	−0.268151	−0.471073
5	152.10	2	1515.62	1517.10	1518.58	1520.02	−0.082278	−0.473576
6	182.52	2	1515.19	1515.64	1516.09	1516.54	−0.025150	−0.474341

For gridblock n without a well,

$$\sum_{l\in\psi_n}\left\{\overset{(\nu)}{T_{l,n}^{n+1}}+\left(\overset{(\nu)}{p_l^{n+1}}-\overset{(\nu)}{p_n^{n+1}}\right)\frac{\partial T_{l,n}}{\partial p_l}\bigg|^{\overset{(\nu)}{n+1}}\right\}\delta\overset{(\nu+1)}{p_l^{n+1}}$$

$$-\left\{\sum_{l\in\psi_n}\left[\overset{(\nu)}{T_{l,n}^{n+1}}-\left(\overset{(\nu)}{p_l^{n+1}}-\overset{(\nu)}{p_n^{n+1}}\right)\frac{\partial T_{l,n}}{\partial p_n}\bigg|^{\overset{(\nu)}{n+1}}\right]+\frac{V_{b_n}}{\alpha_c\Delta t}\left(\frac{\phi}{B_g}\right)'_n\right\}\delta\overset{(\nu+1)}{p_n^{n+1}}$$

$$=-\left\{\sum_{l\in\psi_n}\overset{(\nu)}{T_{l,n}^{n+1}}\left(\overset{(\nu)}{p_l^{n+1}}-\overset{(\nu)}{p_n^{n+1}}\right)-\frac{V_{b_n}}{\alpha_c\Delta t}\left(\frac{\phi}{B_g}\right)'_n\left(\overset{(\nu)}{p_n^{n+1}}-p_n^n\right)\right\}$$

(8.71b)

As mentioned in Example 8.1, gridblock 4 is upstream to gridblock 3, gridblock 3 is upstream to gridblock 2, and gridblock 2 is upstream to gridblock 1. Upstream weighting of the pressure-dependent terms in transmissibility is used.

First time step calculations ($n=0$, $t_{n+1}=30.42$ days, and $\Delta t=30.42$ days).

For the first iteration ($v=0$), assume $\overset{(0)}{p_n^{n+1}}=p_n^n=4015$ psia for $n=1, 2, 3, 4$. Consequently, $T_{n,n+1}\big|_n^{\overset{(0)}{n+1}}=95877.5281$ for all gridblocks, $\left(\overset{(0)}{p_l^{n+1}}-p_n\right)\frac{\partial T_{l,n}}{\partial p_l}\Big|^{\overset{(\nu)}{n+1}}$ $=0$ for all values of l and n, and $\frac{V_{b_n}}{\alpha_c\Delta t}\left(\frac{\phi}{B_g}\right)'_n$ is obtained as shown in Table 8.3.

For wellblock 1, $\frac{d}{dp}\left(\frac{1}{\mu B}\right)\Big|_1^{\overset{(0)}{n+1}}=2.970747$

$$\overset{(0)}{q_{sc_1}^{n+1}}=-G_{w_1}\left(\frac{1}{\mu B}\right)_1^{\overset{(0)}{n+1}}\left(\overset{(0)}{p_1^{n+1}}-p_{wf_1}\right)$$
$$=-3.941572\times\left(\frac{1}{0.0223000\times0.0007790}\right)\times(4015-1515)$$
$$=-567240397$$

and

$$\frac{dq_{sc_1}}{dp_1}\bigg|^{\overset{(0)}{n+1}}=-G_{w_1}\left[\left(\frac{1}{\mu B}\right)_1^{\overset{(0)}{n+1}}+\frac{d}{dp}\left(\frac{1}{\mu B}\right)\bigg|_1^{\overset{(0)}{n+1}}\left(\overset{(0)}{p_1^{n+1}}-p_{wf_1}\right)\right]$$
$$=-3.941572\times\left[\left(\frac{1}{0.0223000\times0.0007790}\right)+2.970747\times(4015-1515)\right]$$
$$=-256169.692$$

In addition, the flow equation for gridblock n with a well (Eq. 8.71a) reduces to

$$\sum_{l\in\psi_n}\overset{(0)}{T_{l,n}^{n+1}}\delta\overset{(1)}{p_l^{n+1}}-\left\{\sum_{l\in\psi_n}\overset{(0)}{T_{l,n}^{n+1}}-\frac{dq_{sc_n}}{dp_n}\bigg|^{\overset{(0)}{n+1}}+\frac{V_{b_n}}{\alpha_c\Delta t}\left(\frac{\phi}{B_g}\right)'_n\right\}\delta\overset{(1)}{p_n^{n+1}}=-\overset{(0)}{q_{sc_n}^{n+1}}$$

(8.72a)

and that for gridblock n without a well (Eq. 8.71b) reduces to

$$\sum_{l\in\psi_n} \overset{(0)}{T}{}_{l,n}^{n+1} \delta \overset{(1)}{p}{}_l^{n+1} - \left\{ \sum_{l\in\psi_n} \overset{(0)}{T}{}_{l,n}^{n+1} + \frac{V_{b_n}}{\alpha_c \Delta t} \left(\frac{\phi}{B_g} \right)'_n \right\} \delta \overset{(1)}{p}{}_n^{n+1} = 0 \tag{8.72b}$$

For gridblock 1, $n=1$ and $\psi_1=\{2\}$. Substitution of the relevant values in Eq. (8.72a) yields

$$-\{95877.5281 - (-256169.692) + 0.06193233\} \times \delta \overset{(1)}{p}{}_1^{n+1} + 95877.5281 \times \delta \overset{(1)}{p}{}_2^{n+1}$$
$$= -(-567240397)$$

or

$$-352047.281 \times \delta \overset{(1)}{p}{}_1^{n+1} + 95877.5281 \times \delta \overset{(1)}{p}{}_2^{n+1} = 567240397 \tag{8.73}$$

For gridblock 2, $n=2$ and $\psi_2=\{1,3\}$. Substitution of the relevant values in Eq. (8.72b) results in

$$95877.5281 \times \delta \overset{(1)}{p}{}_1^{n+1} + 95877.5281 \times \delta \overset{(1)}{p}{}_3^{n+1}$$
$$-\{95877.5281 + 95877.5281 + 2.842428\} \times \delta \overset{(1)}{p}{}_2^{n+1} = 0$$

or

$$95877.5281 \times \delta \overset{(1)}{p}{}_1^{n+1} - 191757.899 \times \delta \overset{(1)}{p}{}_2^{n+1} + 95877.5281 \times \delta \overset{(1)}{p}{}_3^{n+1} = 0 \tag{8.74}$$

For gridblock 3, $n=3$ and $\psi_3=\{2,4\}$. Substitution of the relevant values in Eq. (8.72b) results in

$$95877.5281 \times \delta \overset{(1)}{p}{}_2^{n+1} + 95877.5281 \times \delta \overset{(1)}{p}{}_4^{n+1}$$
$$-\{95877.5281 + 95877.5281 + 130.4553\} \times \delta \overset{(1)}{p}{}_3^{n+1} = 0$$

or

$$95877.5281 \times \delta \overset{(1)}{p}{}_2^{n+1} - 191885.511 \times \delta \overset{(1)}{p}{}_3^{n+1} + 95877.5281 \times \delta \overset{(1)}{p}{}_4^{n+1} = 0 \tag{8.75}$$

For gridblock 4, $n=4$ and $\psi_4=\{3\}$. Substitution of the relevant values in Eq. (8.72b) results in

$$95877.5281 \times \delta \overset{(1)}{p}{}_3^{n+1} - \{95877.5281 + 4619.097\} \times \delta \overset{(1)}{p}{}_4^{n+1} = 0$$

or

$$95877.5281 \times \delta \overset{(1)}{p_3^{n+1}} - 100496.626 \times \delta \overset{(1)}{p_4^{n+1}} = 0 \qquad (8.76)$$

The results of solving Eqs. (8.73) through (8.76) for the pressure change over the first iteration are $\delta \overset{(1)}{p_1^{n+1}} = -2179.03$, $\delta \overset{(1)}{p_2^{n+1}} = -2084.77$, $\delta \overset{(1)}{p_3^{n+1}} = -1990.57$, and $\delta \overset{(1)}{p_4^{n+1}} = -1899.08$. Therefore, $\overset{(1)}{p_1^{n+1}} = 1835.97$ psia, $\overset{(1)}{p_2^{n+1}} = 1930.23$ psia, $\overset{(1)}{p_3^{n+1}} = 2024.43$ psia, and $\overset{(1)}{p_4^{n+1}} = 2115.92$ psia.

For second iteration ($v=1$), we use $\overset{(1)}{p_n^{n+1}}$ to estimate values of FVF, gas viscosity, $\left(\frac{\phi}{B_g}\right)'_n$, $\frac{V_{b_n}}{\alpha_c \Delta t}\left(\frac{\phi}{B_g}\right)'_n$, and transmissibility and its derivative with respect to block pressure. Table 8.10 lists these values. For example, for gridblock 1,

$$\left(\frac{\phi}{B_g}\right)'_1 = \frac{\left(\frac{\phi}{B}\right)_1^{\overset{(\nu)}{n+1}} - \left(\frac{\phi}{B}\right)_1^n}{\overset{(\nu)}{p_1^{n+1}} - p_1^n} = \frac{\left(\frac{0.13}{0.00161207}\right) - \left(\frac{0.13}{0.000779}\right)}{1835.97 - 4015} = 0.03957679$$

$$\frac{V_{b_1}}{\alpha_c \Delta t}\left(\frac{\phi}{B_g}\right)'_1 = \frac{340.59522 \times 0.03957679}{5.614583 \times 30.42} = 0.07892278$$

$$T_{r_{1,2}}^{\overset{(\nu)}{n+1}} = T_{1,2}\Big|_2^{\overset{(\nu)}{n+1}} = G_{r_{1+1/2}}\left(\frac{1}{\mu B}\right)_2^{\overset{(\nu)}{n+1}}$$
$$= 1.6655557 \times \left(\frac{1}{0.01588807 \times 0.00153148}\right) = 68450.4979$$

$$\frac{\partial T_{1,2}}{\partial p_1}\Big|_2^{\overset{(\nu)}{n+1}} = 0, \text{ and } \frac{\partial T_{1,2}}{\partial p_2}\Big|_2^{\overset{(\nu)}{n+1}} = G_{r_{1+1/2}} \frac{\mathrm{d}}{\mathrm{d}p}\left(\frac{1}{\mu B}\right)\Big|_2^{\overset{(\nu)}{n+1}} = 1.6655557 \times 16.47741$$
$$= 27.444044$$

for upstream weighting of transmissibility. In addition, for the production well in wellblock 1,

$$q_{sc_1}^{\overset{(\nu)}{n+1}} = -G_{w_1}\left(\frac{1}{\mu B}\right)_1^{\overset{(\nu)}{n+1}}\left(\overset{(\nu)}{p_1^{n+1}} - p_{wf_1}\right)$$
$$= -3.941572 \times \left(\frac{1}{0.01565241 \times 0.00161207}\right) \times (1835.97 - 1515)$$
$$= -50137330$$

$$\frac{\mathrm{d}q_{sc_1}}{\mathrm{d}p_1}\Big|^{\overset{(\nu)}{n+1}} = -G_{w_1}\left[\left(\frac{1}{\mu B}\right)_1^{\overset{(\nu)}{n+1}} + \frac{\mathrm{d}}{\mathrm{d}p}\left(\frac{1}{\mu B}\right)\Big|_1^{\overset{(\nu)}{n+1}}\left(\overset{(\nu)}{p_1^{n+1}} - p_{wf_1}\right)\right]$$

TABLE 8.10 Estimated gridblock functions at old iteration $\nu = 1$.

n	$\overset{(1)}{p_n^{n+1}}$ (psia)	$\overset{(1)}{B_{g_n}^{n+1}}$ (RB/scf)	$\overset{(1)}{\mu_{g_n}^{n+1}}$ (cP)	$(\phi/B_g)'_n$	$\frac{V_{b_n}}{\alpha_c \Delta t}\left(\frac{\phi}{B_g}\right)'_n$	$\frac{d}{dp}\left(\frac{1}{\mu B}\right)\Big\vert_n^{\overset{(\nu)}{n+1}}$	$\frac{\partial T_{n,n+1}}{\partial p_n}\Big\vert_n^{\overset{(\nu)}{n+1}}$	$T_{n,n+1}\Big\vert_n^{\overset{(\nu)}{n+1}}$
1	1835.97	0.00161207	0.01565241	0.03957679	0.07892278	14.68929	24.465831	66,007.6163
2	1930.23	0.00153148	0.01588807	0.03933064	3.599688	16.47741	27.444044	68,450.4979
3	2024.43	0.00145235	0.01612828	0.03886858	163.2694	12.78223	21.289516	71,104.7736
4	2115.92	0.00138785	0.01640276	0.03855058	5733.667	14.28023	23.784518	73,164.3131

or

$$\left.\frac{\mathrm{d}q_{sc_1}}{\mathrm{d}p_1}\right|^{\stackrel{(\nu)}{n+1}} = -3.941572 \times \left[\left(\frac{1}{0.01565241 \times 0.00161207}\right) + 14.68929 \times (1835.97 - 1515)\right] = -174791.4$$

For gridblock 1, $n=1$ and $\psi_1=\{2\}$. Therefore, Eq. (8.71a) becomes

$$-\left[T_{1,2}\Big|_2^{\stackrel{(\nu)}{n+1}} - \left(p_2^{\stackrel{(\nu)}{n+1}} - p_1^{\stackrel{(\nu)}{n+1}}\right)\left.\frac{\partial T_{1,2}}{\partial p_1}\right|_2^{\stackrel{(\nu)}{n+1}} - \left.\frac{\mathrm{d}q_{sc_1}}{\mathrm{d}p_1}\right|^{\stackrel{(\nu)}{n+1}} + \frac{V_{b_1}}{\alpha_c \Delta t}\left(\frac{\phi}{B_g}\right)'_1\right]\delta p_1^{\stackrel{(\nu+1)}{n+1}}$$
$$+\left[T_{1,2}\Big|_2^{\stackrel{(\nu)}{n+1}} + \left(p_2^{\stackrel{(\nu)}{n+1}} - p_1^{\stackrel{(\nu)}{n+1}}\right)\left.\frac{\partial T_{1,2}}{\partial p_2}\right|_2^{\stackrel{(\nu)}{n+1}}\right]\delta p_2^{\stackrel{(\nu+1)}{n+1}}$$
$$= -\left\{T_{1,2}\Big|_2^{\stackrel{(\nu)}{n+1}}\left(p_2^{\stackrel{(\nu)}{n+1}} - p_1^{\stackrel{(\nu)}{n+1}}\right) + q_{sc_1}^{\stackrel{(\nu)}{n+1}} - \frac{V_{b_1}}{\alpha_c \Delta t}\left(\frac{\phi}{B_g}\right)'_1\left(p_1^{\stackrel{(\nu)}{n+1}} - p_1^n\right)\right\} \tag{8.77}$$

Substitution of the values in Eq. (8.77) gives

$$-[68450.4979 - (1930.23 - 1835.97) \times 0 - (-174791.4) + 0.07892278]\delta p_1^{\stackrel{(\nu+1)}{n+1}}$$
$$+ [68450.4979 + (1930.23 - 1835.97) \times 27.444044]\delta p_2^{\stackrel{(\nu+1)}{n+1}}$$
$$= -\{68450.4979 \times (1930.23 - 1835.97) + (-50137330) - 0.07892278 \times (1835.97 - 4015)\}$$

After simplification, the equation becomes

$$-243242.024 \times \delta p_1^{\stackrel{(\nu+1)}{n+1}} + 71037.4371 \times \delta p_2^{\stackrel{(\nu+1)}{n+1}} = 43684856.7 \tag{8.78}$$

For gridblock 2, $n=2$ and $\psi_2=\{1,3\}$. Therefore, Eq. (8.71b) becomes

$$\left[T_{1,2}\Big|_2^{\stackrel{(\nu)}{n+1}} + \left(p_1^{\stackrel{(\nu)}{n+1}} - p_2^{\stackrel{(\nu)}{n+1}}\right)\left.\frac{\partial T_{1,2}}{\partial p_1}\right|_2^{\stackrel{(\nu)}{n+1}}\right]\delta p_1^{\stackrel{(\nu+1)}{n+1}}$$
$$-\left[T_{1,2}\Big|_2^{\stackrel{(\nu)}{n+1}} - \left(p_1^{\stackrel{(\nu)}{n+1}} - p_2^{\stackrel{(\nu)}{n+1}}\right)\left.\frac{\partial T_{1,2}}{\partial p_2}\right|_2^{\stackrel{(\nu)}{n+1}} + T_{3,2}\Big|_3^{\stackrel{(\nu)}{n+1}} - \left(p_3^{\stackrel{(\nu)}{n+1}} - p_2^{\stackrel{(\nu)}{n+1}}\right)\left.\frac{\partial T_{3,2}}{\partial p_2}\right|_3^{\stackrel{(\nu)}{n+1}}\right.$$
$$\left.+\frac{V_{b_2}}{\alpha_c \Delta t}\left(\frac{\phi}{B_g}\right)'_2\right]\delta p_2^{\stackrel{(\nu+1)}{n+1}} + \left[T_{3,2}\Big|_3^{\stackrel{(\nu)}{n+1}} + \left(p_3^{\stackrel{(\nu)}{n+1}} - p_2^{\stackrel{(\nu)}{n+1}}\right)\left.\frac{\partial T_{3,2}}{\partial p_3}\right|_3^{\stackrel{(\nu)}{n+1}}\right]\delta p_3^{\stackrel{(\nu+1)}{n+1}}$$
$$= -\left\{\left[T_{1,2}\Big|_2^{\stackrel{(\nu)}{n+1}}\left(p_1^{\stackrel{(\nu)}{n+1}} - p_2^{\stackrel{(\nu)}{n+1}}\right) + T_{3,2}\Big|_3^{\stackrel{(\nu)}{n+1}}\left(p_3^{\stackrel{(\nu)}{n+1}} - p_2^{\stackrel{(\nu)}{n+1}}\right)\right] - \frac{V_{b_2}}{\alpha_c \Delta t}\left(\frac{\phi}{B_g}\right)'_2\left(p_2^{\stackrel{(\nu)}{n+1}} - p_2^n\right)\right\} \tag{8.79}$$

In the earlier equation,

$$T_{r_{3,2}}^{\overset{(\nu)}{n+1}} = T_{3,2}\Big|_3^{\overset{(\nu)}{n+1}} = G_{r_{2+1/2}}\left(\frac{1}{\mu B}\right)_3^{\overset{(\nu)}{n+1}}$$
$$= 1.6655557 \times \left(\frac{1}{0.01612828 \times 0.00145235}\right) = 71104.7736$$

$$\frac{\partial T_{3,2}}{\partial p_2}\Big|_3^{\overset{(\nu)}{n+1}} = 0, \text{ and } \frac{\partial T_{3,2}}{\partial p_3}\Big|_3^{\overset{(\nu)}{n+1}} = G_{r_{2+1/2}}\frac{\mathrm{d}}{\mathrm{d}p}\left(\frac{1}{\mu B}\right)\Big|_3^{\overset{(\nu)}{n+1}}$$
$$= 1.6655557 \times 12.78223 = 21.289516$$

Substitution of these values in Eq. (8.79) gives

$$[68450.4979 + (1835.97 - 1930.23) \times 0]\delta p_1^{\overset{(\nu+1)}{n+1}}$$
$$-[68450.4979 - (1835.97 - 1930.23) \times 27.444044 + 71104.7736$$
$$-(2024.43 - 1930.23) \times 0 + 3.599688]\delta p_2^{\overset{(\nu+1)}{n+1}}$$
$$+ [71104.7736 + (2024.43 - 1930.23) \times 21.289516]\delta p_3^{\overset{(\nu+1)}{n+1}}$$
$$= -\{[68450.4979 \times (1835.97 - 1930.23)$$
$$+ 71104.7736 \times (2024.43 - 1930.23)] - 3.599688 \times (1930.23 - 4015)\}$$

or after simplification,

$$68450.4979 \times \delta p_1^{\overset{(\nu+1)}{n+1}} - 142145.810 \times \delta p_2^{\overset{(\nu+1)}{n+1}} + 73110.2577 \times \delta p_3^{\overset{(\nu+1)}{n+1}}$$
$$= -253308.066 \tag{8.80}$$

For gridblock 3, $n=3$ and $\psi_3=\{2,4\}$. Therefore, Eq. (8.71b) becomes

$$\left[T_{2,3}\Big|_3^{\overset{(\nu)}{n+1}} + \left(p_2^{\overset{(\nu)}{n+1}} - p_3^{\overset{(\nu)}{n+1}}\right)\frac{\partial T_{2,3}}{\partial p_2}\Big|_3^{\overset{(\nu)}{n+1}}\right]\delta p_2^{\overset{(\nu+1)}{n+1}} - \left[T_{2,3}\Big|_3^{\overset{(\nu)}{n+1}} - \left(p_2^{\overset{(\nu)}{n+1}} - p_3^{\overset{(\nu)}{n+1}}\right)\frac{\partial T_{2,3}}{\partial p_3}\Big|_3^{\overset{(\nu)}{n+1}}\right.$$
$$\left.+T_{4,3}\Big|_4^{\overset{(\nu)}{n+1}} - \left(p_4^{\overset{(\nu)}{n+1}} - p_3^{\overset{(\nu)}{n+1}}\right)\frac{\partial T_{4,3}}{\partial p_3}\Big|_4^{\overset{(\nu)}{n+1}} + \frac{V_{b_3}}{\alpha_c \Delta t}\left(\frac{\phi}{B_g}\right)_3'\right]\delta p_3^{\overset{(\nu+1)}{n+1}}$$
$$+\left[T_{4,3}\Big|_4^{\overset{(\nu)}{n+1}} + \left(p_4^{\overset{(\nu)}{n+1}} - p_3^{\overset{(\nu)}{n+1}}\right)\frac{\partial T_{4,3}}{\partial p_4}\Big|_4^{\overset{(\nu)}{n+1}}\right]\delta p_4^{\overset{(\nu+1)}{n+1}}$$
$$= -\left\{\left[T_{2,3}\Big|_3^{\overset{(\nu)}{n+1}}\left(p_2^{\overset{(\nu)}{n+1}} - p_3^{\overset{(\nu)}{n+1}}\right) + T_{4,3}\Big|_4^{\overset{(\nu)}{n+1}}\left(p_4^{\overset{(\nu)}{n+1}} - p_3^{\overset{(\nu)}{n+1}}\right)\right] - \frac{V_{b_3}}{\alpha_c \Delta t}\left(\frac{\phi}{B_g}\right)_3'\left(p_3^{\overset{(\nu)}{n+1}} - p_3^n\right)\right\} \tag{8.81}$$

where

$$T_{r_{4,3}}^{\overset{(\nu)}{n+1}} = T_{4,3}\Big|_4^{\overset{(\nu)}{n+1}} = G_{r_{3+1/2}}\left(\frac{1}{\mu B}\right)_4^{\overset{(\nu)}{n+1}}$$

$$= 1.6655557 \times \left(\frac{1}{0.01640276 \times 0.00138785}\right) = 73164.3131$$

$$\frac{\partial T_{4,3}}{\partial p_3}\Big|_4^{\overset{(\nu)}{n+1}} = 0, \text{ and } \frac{\partial T_{4,3}}{\partial p_4}\Big|_4^{\overset{(\nu)}{n+1}} = G_{r_{3+1/2}}\frac{\mathrm{d}}{\mathrm{d}p}\left(\frac{1}{\mu B}\right)\Big|_4^{\overset{(\nu)}{n+1}}$$

$$= 1.6655557 \times 14.28023 = 23.784518$$

Substitution of these values in Eq. (8.81) gives

$$[71104.7736 + (1930.23 - 2024.43) \times 0]\delta p_2^{\overset{(\nu+1)}{n+1}}$$
$$-[71104.7736 - (1930.23 - 2024.43) \times 21.289516 + 73164.3131$$
$$-(2115.92 - 2024.43) \times 0 + 163.2694]\delta p_3^{\overset{(\nu+1)}{n+1}}$$
$$+[73164.3131 + (2115.92 - 2024.43) \times 23.784518]\delta p_4^{\overset{(\nu+1)}{n+1}}$$
$$= -\{[71104.7736 \times (1930.23 - 2024.43) + 73164.3131$$
$$\times (2115.92 - 2024.43)] - 163.2694 \times (2024.43 - 4015)\}$$

After simplification, the equation becomes

$$71104.7736 \times \delta p_2^{\overset{(\nu+1)}{n+1}} - 146437.840 \times \delta p_3^{\overset{(\nu+1)}{n+1}} + 75340.4074 \times \delta p_4^{\overset{(\nu+1)}{n+1}} = -320846.394 \tag{8.82}$$

For gridblock 4, $n=4$ and $\psi_4=\{3\}$. Therefore, Eq. (8.71b) becomes

$$\left[T_{3,4}\Big|_4^{\overset{(\nu)}{n+1}} + \left(p_3^{\overset{(\nu)}{n+1}} - p_4^{\overset{(\nu)}{n+1}}\right)\frac{\partial T_{3,4}}{\partial p_3}\Big|_4^{\overset{(\nu)}{n+1}}\right]\delta p_3^{\overset{(\nu+1)}{n+1}}$$
$$-\left[T_{3,4}\Big|_4^{\overset{(\nu)}{n+1}} - \left(p_3^{\overset{(\nu)}{n+1}} - p_4^{\overset{(\nu)}{n+1}}\right)\frac{\partial T_{3,4}}{\partial p_4}\Big|_4^{\overset{(\nu)}{n+1}} + \frac{V_{b_4}}{\alpha_c \Delta t}\left(\frac{\phi}{B_g}\right)_4'\right]\delta p_4^{\overset{(\nu+1)}{n+1}} \tag{8.83}$$
$$= -\left\{\left[T_{3,4}\Big|_4^{\overset{(\nu)}{n+1}}\left(p_3^{\overset{(\nu)}{n+1}} - p_4^{\overset{(\nu)}{n+1}}\right)\right] - \frac{V_{b_4}}{\alpha_c \Delta t}\left(\frac{\phi}{B_g}\right)_4'\left(p_4^{\overset{(\nu)}{n+1}} - p_4^n\right)\right\}$$

Substitution of the values in Eq. (8.83) gives

$$[73164.3131 + (2024.43 - 2115.92) \times 0]\delta p_3^{\overset{(\nu+1)}{n+1}}$$
$$-[73164.3131 - (2024.43 - 2115.92) \times 23.784518 + 5733.667]\delta p_4^{\overset{(\nu+1)}{n+1}}$$
$$= -\{[73164.3131 \times (2024.43 - 2115.92)] - 5733.667 \times (2115.92 - 4015)\}$$

TABLE 8.11 Pressure solution at $t_{n+1} = 30.42$ days for successive iterations.

$\nu+1$	$\overset{(\nu+1)}{p_1^{n+1}}$ (psia)	$\overset{(\nu+1)}{p_2^{n+1}}$ (psia)	$\overset{(\nu+1)}{p_3^{n+1}}$ (psia)	$\overset{(\nu+1)}{p_4^{n+1}}$ (psia)
0	4015.00	4015.00	4015.00	4015.00
1	1835.97	1930.23	2024.43	2115.92
2	1614.00	1785.15	1946.71	2097.52
3	1597.65	1775.45	1941.04	2093.09
4	1597.51	1775.42	1941.09	2093.20

After simplification, the equation becomes

$$73164.3131 \times \overset{(\nu+1)}{\delta p_3^{n+1}} - 81074.0745 \times \overset{(\nu+1)}{\delta p_4^{n+1}} = -4194735.68 \qquad (8.84)$$

The results of solving Eqs. (8.78), (8.80), (8.82), and (8.84) for the pressure change over the second iteration are $\overset{(2)}{\delta p_1^{n+1}} = -221.97$, $\overset{(2)}{\delta p_2^{n+1}} = -145.08$, $\overset{(2)}{\delta p_3^{n+1}} = -77.72$, and $\overset{(2)}{\delta p_4^{n+1}} = -18.40$. Therefore, $\overset{(2)}{p_1^{n+1}} = 1614.00$ psia, $\overset{(2)}{p_2^{n+1}} = 1785.15$ psia, $\overset{(2)}{p_3^{n+1}} = 1946.71$ psia, and $\overset{(2)}{p_4^{n+1}} = 2097.52$ psia. Iterations continue until the convergence criterion is satisfied. Table 8.11 shows the successive iterations for the first time step. As can be seen, it took four iterations to converge. The convergence criterion was set as given by Eq. (8.61). After reaching convergence, time is incremented by $\Delta t = 30.42$ days, and the aforementioned procedure is repeated. Table 8.12 shows the converged solutions at various times up to 6 months of simulation time.

8.6 Summary

The flow equation for an incompressible fluid (Eq. 8.1) is linear. The flow equation for a slightly compressible fluid has very weak nonlinearity caused by the product μB that appears in the interblock flow terms, fictitious well flow rate, and well production rate. This product can be assumed constant without introducing noticeable errors; hence, the flow equation for a slightly compressible fluid becomes linear (Eq. 8.9). The flow equation for a compressible fluid has weak nonlinearity, but it needs to be linearized. Linearization involves treatment in both space and time of the transmissibilities, well production rate, fictitious well flow rate, and coefficient of pressure in the accumulation term. Linearization of transmissibility in space and time is accomplished by any of the methods mentioned in Section 8.4.1. In the engineering approach, the flow

TABLE 8.12 Converged pressure solution and gas production at various times.

$n+1$	Time (day)	ν	p_1^{n+1} (psia)	p_2^{n+1} (psia)	p_3^{n+1} (psia)	p_4^{n+1} (psia)	q_{gsc}^{n+1} (MMscf/D)	Cumulative production (MMMscf)
1	30.42	4	1597.51	1775.42	1941.09	2093.20	−11.3984	−0.346740
2	60.84	3	1537.18	1588.11	1637.66	1685.05	−2.95637	−0.436673
3	91.26	3	1521.54	1536.88	1552.08	1566.84	−0.863862	−0.462951
4	121.68	2	1517.04	1521.84	1526.63	1531.32	−0.268285	−0.471113
5	152.10	2	1515.63	1517.10	1518.58	1520.03	−0.082326	−0.473617
6	182.52	2	1515.19	1515.64	1516.10	1516.54	−0.025165	−0.474382

equation or any of its components (interblock flow term, well rate, fictitious well rate) can be linearized in time by the explicit transmissibility method, simple iteration on transmissibility method, or fully implicit method. Section 8.4.2 presented linearization of the physical well rates, Section 8.4.3 presented linearization of fictitious well rates, and Section 8.4.4 presented linearization of the coefficient of pressure change in the accumulation term. The linearized flow equation is obtained by substituting the linearized terms in the flow equation.

8.7 Exercises

8.1 Define the linearity of Eq. (8.1) by examining the various terms in the equation.

8.2 Define the linearity of Eq. (8.9) by examining the various terms in the equation.

8.3 Explain why Eq. (8.5) can be looked at as a nonlinear equation.

8.4 Explain why Eq. (8.11) is a nonlinear equation.

8.5 Examine Eq. (8.30), used for the linearization of the well production rate, and point out the differences between the explicit method and the explicit transmissibility method (Eq. 8.30).

8.6 Examine Eq. (8.31), used for the linearization of the well production rate, and point out the differences between the simple iteration method and the simple iteration on transmissibility method (Eq. 8.31).

8.7 Consider the 1-D, inclined reservoir shown in Fig. 8.3. The reservoir is volumetric and homogeneous. The reservoir contains a production well located in gridblock 2. At the time of discovery ($t=0$), fluids were in hydrodynamic equilibrium, and the pressure of gridblock 2 was 3000 psia. All gridblocks have $\Delta x=400$ ft, $w=200$ ft, $h=80$ ft, $k=222$ md, and $\phi=0.20$. The well in gridblock 2 produces fluid at a rate of 10^6 scf/D. Table 8.1 gives the gas FVF and viscosity. Gas density at standard

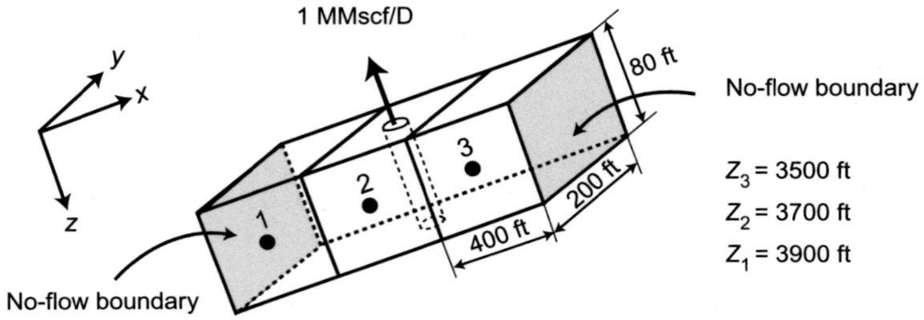

FIG. 8.3 Discretized 1-D reservoir in Exercise 8.7.

conditions is 0.05343 lbm/ft^3. Estimate the initial pressure distribution in the reservoir. Find the well FBHP and pressure distribution in the system at 50 and 100 days. Use the implicit formulation with the explicit transmissibility method.

8.8 Consider the 1-D flow problem described in Exercise 8.7. Find the pressure distribution in the reservoir at 50 and 100 days. Use the implicit formulation with the simple iteration on transmissibility method.

8.9 Consider the 1-D flow problem described in Exercise 8.7. Find the pressure distribution in the reservoir at 50 and 100 days. Use the implicit formulation with the fully implicit method.

8.10 A vertical well is drilled on 16-acre spacing in a natural gas reservoir. The reservoir is described by four gridpoints in the radial direction as shown in Fig. 8.4. The reservoir is horizontal and has 20-ft net thickness and homogeneous and isotropic rock properties with $k = 10$ md and $\phi = 0.13$. Initially, reservoir pressure is 3015 psia. Table 8.1 presents the gas FVF and viscosity dependence on pressure. The external reservoir boundaries are sealed to fluid flow. Well diameter is 6 in. The well produces under a constant FBHP of 2015 psia. Find the pressure distribution in the reservoir every month (30.42 days) for 2 months. Take time steps of 30.42 days. Use the implicit formulation with the explicit transmissibility method.

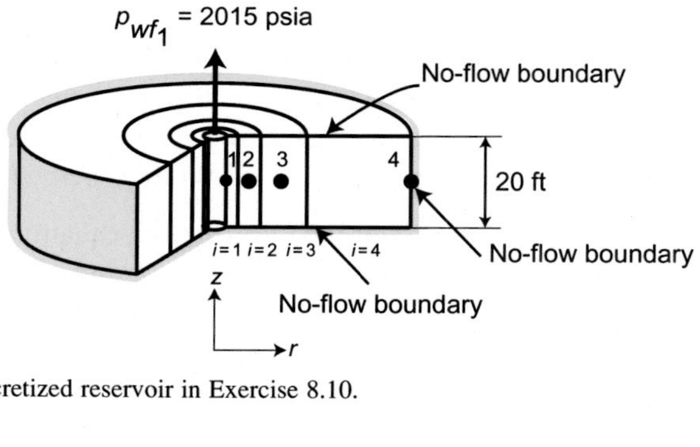

FIG. 8.4 Discretized reservoir in Exercise 8.10.

8.11 Consider the single-well simulation problem presented in Exercise 8.10. Find the pressure distribution in the reservoir at 1 and 2 months. Use the implicit formulation with the simple iteration on transmissibility method.

8.12 Consider the single-well simulation problem presented in Exercise 8.10. Find the pressure distribution in the reservoir at 1 and 2 months. Use the implicit formulation with the fully implicit transmissibility method.

8.13 Consider the 2-D single-phase flow of natural gas taking place in the horizontal, homogeneous reservoir shown in Fig. 8.5. The external

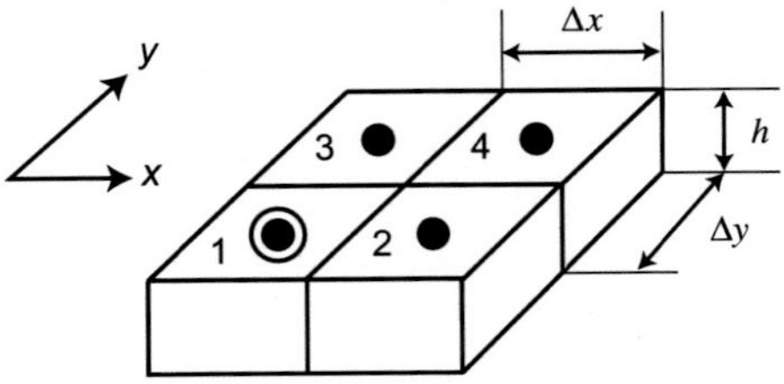

FIG. 8.5 Discretized 2-D reservoir in Exercise 8.13.

reservoir boundaries are sealed off to fluid flow. Gridblock properties are $\Delta x = \Delta y = 1000$ ft, $h = 25$ ft, $k_x = k_y = 20$ md, and $\phi = 0.12$. Initially, reservoir pressure is 4015 psia. Table 8.1 presents the gas FVF and viscosity dependence on pressure. The well in gridblock 1 produces gas at a rate of 10^6 scf/D. Well diameter is 6 in. Find the pressure distribution in the reservoir and the FBHP of the well every month (30.42 days) for 2 months. Check the material balance every time step. Use the implicit formulation with the explicit transmissibility method. Observe symmetry and take time steps of 30.42 days.

8.14 Consider the 2-D flow problem described in Exercise 8.13. Find the pressure distribution in the reservoir and the FBHP of the well at 1 and 2 months. Check the material balance every time step. Use the implicit formulation with the simple iteration on transmissibility method.

8.15 Consider the 2-D flow problem described in Exercise 8.13. Find the pressure distribution in the reservoir and the FBHP of the well at 1 and 2 months. Check the material balance every time step. Use the implicit formulation with the fully implicit transmissibility method.

8.16 Derive Eq. (8.47b) that represents the fully implicit equation without conservative expansion of accumulation term for compressible fluid, using the method of Coats et al. (1977) as outlined in the text.

8.17 What would be a rigorous treatment of nonlinear equations? What happens if multiple solutions emerge?

Chapter 9

Methods of solution of linear equations

Chapter outline

9.1 Introduction

Today, practically all aspects of reservoir engineering problems are solved with a reservoir simulator. The use of the simulators is so extensive that it will be no exaggeration to describe them as "the standard." The simulators enable us to predict reservoir performance, although this task becomes immensely difficult when dealing with complex reservoirs. The complexity can arise from variation in formation and fluid properties. The complexity of the reservoirs has always been handled with increasingly advanced approaches. Mustafiz and Islam (2008) reviewed latest advancements in petroleum reservoir simulation. Also, they discussed the framework of a futuristic reservoir simulator. They predicted that in the near future, the coupling of 3-D imaging with comprehensive reservoir models will enable one to use drilling data as input information for the simulator creating a real-time reservoir monitoring system. At the same time, coupling of ultrafast data acquisition system with digital/analog converters transforming signals into tangible sensations will make use of the capability of virtual reality incorporated into the state-of-the-art reservoir models. The basis of all these, however, is the formulation presented in this book. The reservoir was discretized into gridblocks in Chapter 4 and gridpoints in Chapter 5. These chapters demonstrated the flow equation for a general block while

Petroleum Reservoir Simulation. https://doi.org/10.1016/B978-0-12-819150-7.00009-8

incorporating the boundary conditions into the flow equation. Chapter 6 presented the well production rates. The resulting flow equation is either linear (incompressible fluid and slightly compressible fluid) or nonlinear (compressible fluid). Chapter 8 presented the linearization of a nonlinear flow equation. What remains is to write the linearized flow equation for each gridblock (or gridpoint) in the reservoir and solve the resulting set of linear equations. These tasks are the focus of this chapter. Linear equations can be solved using either direct or iterative methods. We restrict our discussion in this chapter to basic solution methods of both categories and present their application to 1-D, 2-D, and 3-D flow problems. The objective here is to introduce the reader to the mechanics of the basic methods of solution for linear equations of the form

$$[\mathbf{A}]\vec{x}=\vec{d} \tag{9.1}$$

where $[\mathbf{A}]$ = square coefficient matrix, $\vec{x}$ = vector of unknowns, and $\vec{d}$ = vector of known values.

9.2 Direct solution methods

The direct solution methods are characterized by their capacity to produce the solution vector for a given system of linear equations after a fixed number of operations. Direct solution methods not only require storing the information contained in the coefficient matrix $[\mathbf{A}]$ and the known vector $\vec{d}$ but also suffer from an accumulation of roundoff errors that occur during computations. In the following sections, we discuss methods such as Thomas' algorithm and Tang's algorithm, which are used for 1-D flow problems, and the g-band algorithm, which is used for 1-D, 2-D, or 3-D flow problems. These algorithms are based on the **LU** factorization of the coefficient matrix (i.e., $[\mathbf{A}]=[\mathbf{L}][\mathbf{U}]$).

9.2.1 1-D rectangular or radial flow problems (Thomas' algorithm)

This algorithm is applicable for a reservoir where flow takes place in the x-direction in rectangular flow problems, as shown in Fig. 9.1a, or in the r-direction in radial flow problems, as shown in Fig. 9.1b. In other words, there is one row of blocks arranged along a line (with $N=n_x$ or $N=n_r$).

The equation for the first block ($i=1$) has the form

$$c_1x_1+e_1x_2=d_1 \tag{9.2a}$$

because block 1 falls on the reservoir west boundary.

The equation for interior blocks $i=2, 3, \ldots, N-1$ has the form

$$w_ix_{i-1}+c_ix_i+e_ix_{i+1}=d_i \tag{9.2b}$$

The equation for the last block ($i=N$) has the form

$$w_Nx_{N-1}+c_Nx_N=d_N \tag{9.2c}$$

because block N falls on the reservoir east boundary.

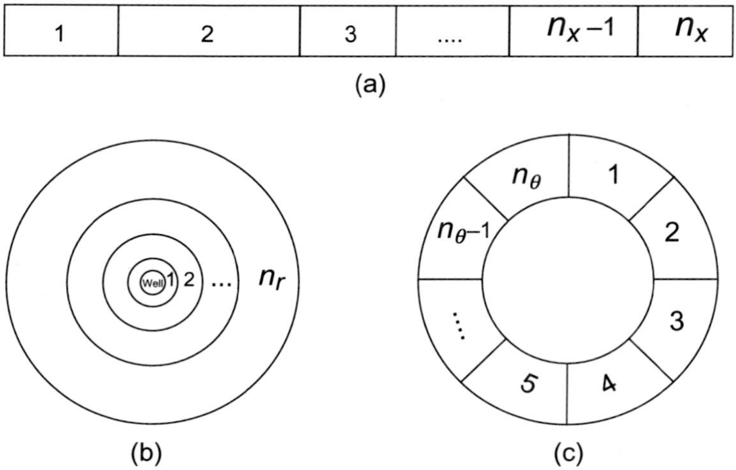

FIG. 9.1 Types of 1-D flow problems. (a) 1-D linear flow in x-direction, (b) 1-D radial flow in r-direction, and (c) 1-D tangential flow in θ-direction.

Inspection of Eq. (9.2) reveals that c_i is the coefficient of the unknown for block i (the center block), the block for which the flow equation is written, w_i is the coefficient of the unknown for neighboring block $i-1$ (the west block), and e_i is the coefficient of the unknown for neighboring block $i+1$ (the east block). The known RHS of the flow equation for block i is d_i. Consider Eq. (8.2b) for the flow of an incompressible fluid. This equation can be rewritten as

$$T_{x_{i-1/2}} p_{i-1} - \left[T_{x_{i-1/2}} + T_{x_{i+1/2}}\right] p_i + T_{x_{i+1/2}} p_{i+1} = \left[T_{x_{i-1/2}} \gamma_{i-1/2}(Z_{i-1} - Z_i) + T_{x_{i+1/2}} \gamma_{i+1/2}(Z_{i+1} - Z_i)\right] - q_{sc_i} \tag{9.3}$$

This equation has the form of Eq. (9.2b) with the unknowns p_{i-1}, p_i, and p_{i+1}; the coefficients $w_i = T_{x_{i-1/2}}$, $c_i = -[T_{x_{i-1/2}} + T_{x_{i+1/2}}]$, and $e_i = T_{x_{i+1/2}}$; and the known RHS $d_i = [T_{x_{i-1/2}} \gamma_{i-1/2}(Z_{i-1} - Z_i) + T_{x_{i+1/2}} \gamma_{i+1/2}(Z_{i+1} - Z_i)] - q_{sc_i}$. If we consider Eq. (8.10b) for the flow of a slightly compressible fluid and assume the well production rate is specified (say, $q_{sc_i}^{n+1} = q_{spsc_i}$), then we obtain

$$T_{x_{i-1/2}} p_{i-1}^{n+1} - \left[T_{x_{i-1/2}} + T_{x_{i+1/2}} + \frac{V_{b_i} \phi_i^{\circ} (c + c_\phi)}{\alpha_c B^{\circ} \Delta t}\right] p_i^{n+1} + T_{x_{i+1/2}} p_{i+1}^{n+1} = T_{x_{i-1/2}} \gamma_{i-1/2}(Z_{i-1} - Z_i) + T_{x_{i+1/2}} \gamma_{i+1/2}(Z_{i+1} - Z_i) - q_{spsc_i} - \frac{V_{b_i} \phi_i^{\circ} (c + c_\phi)}{\alpha_c B^{\circ} \Delta t} p_i^n \tag{9.4}$$

The unknowns in Eq. (9.4) are p_{i-1}^{n+1}, p_i^{n+1}, and p_{i+1}^{n+1}; the coefficients are $w_i = T_{x_{i-1/2}}$, $c_i = -\left[T_{x_{i-1/2}} + T_{x_{i+1/2}} + \frac{V_{b_i} \phi_i^{\circ} (c + c_\phi)}{\alpha_c B^{\circ} \Delta t}\right]$, and $e_i = T_{x_{i+1/2}}$; and the known RHS is $d_i = T_{x_{i-1/2}}\gamma_{i-1/2}(Z_{i-1} - Z_i) + T_{x_{i+1/2}}\gamma_{i+1/2}(Z_{i+1} - Z_i) - q_{spsc_i} - \frac{V_{b_i}\phi_i^\circ(c+c_\phi)}{\alpha_c B^\circ \Delta t} p_i^n$. In either case, block 1 does not have coefficient w_1, and block N does not have coefficient e_N because these blocks are boundary blocks. Depending on the boundary condition specification, its effects are embedded in d_i and c_i for boundary gridblocks (see Section 4.4) and in d_i, c_i, and w_i or e_i for boundary gridpoints (see Section 5.4).

The set of N equations expressed by Eq. (9.2) can be written in a matrix form as

$$\begin{bmatrix} c_1 & e_1 & & & & \\ w_2 & c_2 & e_2 & & & \\ & \cdots & \cdots & \cdots & & \\ & & \cdots & \cdots & \cdots & \\ & & & w_{N-1} & c_{N-1} & e_{N-1} \\ & & & & w_N & c_N \end{bmatrix} \begin{bmatrix} x_1 \\ x_2 \\ \cdots \\ \cdots \\ x_{N-1} \\ x_N \end{bmatrix} = \begin{bmatrix} d_1 \\ d_2 \\ \cdots \\ \cdots \\ d_{N-1} \\ d_N \end{bmatrix} \tag{9.5}$$

The matrix in Eq. (9.5) is called a tridiagonal matrix. This matrix equation can be solved using Thomas' algorithm. Thomas' algorithm is nothing more than an efficient procedure to solve a tridiagonal matrix equation (Eq. 9.5) through matrix factorization into lower **[L]** and upper **[U]** triangular matrices (Aziz and Settari, 1979). In addition, we do not have to store the whole matrix. Instead, it is sufficient to store four vectors ($\vec{w}$, $\vec{c}$, $\vec{e}$, and $\vec{d}$) of dimension N to store all information contained in Eq. (9.5). Thomas' algorithm is executed in two major steps that require the creation of two more vectors ($\vec{u}$ and $\vec{g}$) of dimension N. The two major steps are the forward solution and the backward solution.

9.2.1.1 *Forward solution*

$$\text{Set } u_1 = \frac{e_1}{c_1} \tag{9.6}$$

and

$$g_1 = \frac{d_1}{c_1} \tag{9.7}$$

For $i = 2, 3 \ldots N-1$,

$$u_i = \frac{e_i}{(c_i - w_i u_{i-1})} \tag{9.8}$$

and for $i = 2, 3 \ldots N$,

$$g_i = \frac{d_i - w_i g_{i-1}}{(c_i - w_i u_{i-1})} \tag{9.9}$$

9.2.1.2 Backward solution

$$\text{Set } x_N = g_N \tag{9.10}$$

For $i = N-1, N-2, \ldots, 3, 2, 1$;

$$x_i = g_i - u_i x_{i+1} \tag{9.11}$$

The following example demonstrates the application of Thomas' algorithm to the equations of a 1-D reservoir.

Example 9.1 The following equations were obtained for the 1-D reservoir in Example 7.1:

$$-85.2012p_1 + 28.4004p_2 = -227203.2 \tag{9.12}$$

$$28.4004p_1 - 56.8008p_2 + 28.4004p_3 = 0 \tag{9.13}$$

$$28.4004p_2 - 56.8008p_3 + 28.4004p_4 = 0 \tag{9.14}$$

and

$$28.4004p_3 - 28.4004p_4 = 600 \tag{9.15}$$

Solve these equations using Thomas' algorithm.

Solution

The first step is to calculate u_1 and g_1 using Eqs. (9.6) and (9.7), yielding

$$u_1 = e_1/c_1 = 28.4004/-85.2012 = -0.333333$$

and

$$g_1 == d_1/c_1 = -227203.2/-85.2012 = 2666.667$$

Then, u_2 and u_3 are calculated in that order using Eq. (9.8), which gives

$$u_2 = e_2/(c_2 - w_2u_1) = 28.4004/[-56.8008 - 28.4004 \times (-0.333333)] \\ = -0.600000$$

and

$$u_3 = e_3/(c_3 - w_3u_2) = 28.4004/[-56.8008 - 28.4004 \times (-0.600000)] \\ = -0.714286$$

This is followed by calculating g_2, g_3, and g_4 in that order using Eq. (9.9), resulting in

$$g_2 = \frac{(d_2 - w_2g_1)}{(c_2 - w_2u_1)} = \frac{(0 - 28.4004 \times 2666.667)}{[-56.8008 - 28.4004 \times (-0.333333)]} = 1600.000$$

$$g_3 = \frac{(d_3 - w_3g_2)}{(c_3 - w_3u_2)} = \frac{(0 - 28.4004 \times 1600.000)}{[-56.8008 - 28.4004 \times (-0.600000)]} = 1142.857$$

and

$$g_4 = \frac{(d_4 - w_4 g_3)}{(c_4 - w_4 u_3)} = \frac{(600 - 28.4004 \times 1142.857)}{[-28.4004 - 28.4004 \times (-0.714286)]} = 3926.06$$

Then $x_4 = g_4$ is set according to Eq. (9.10), yielding

$$x_4 = g_4 = 3929.06$$

This is followed by calculating x_3, x_2, and x_1 in that order using Eq. (9.11), which gives

$$x_3 = g_3 - u_3 x_4 = 1142.857 - (-0.714286) \times 3926.06 = 3947.18$$

$$x_2 = g_2 - u_2 x_3 = 1600.000 - (-0.600000) \times 3947.18 = 3968.31$$

and

$$x_1 = g_1 - u_1 x_2 = 2666.667 - (-0.333333) \times 3968.31 = 3989.44$$

Table 9.1 shows the results of the calculations as outlined here. The solution vector given in the last column in Table 9.1 is

$$\vec{x} = \begin{bmatrix} x_1 \\ x_2 \\ x_3 \\ x_4 \end{bmatrix} = \begin{bmatrix} 3989.44 \\ 3968.31 \\ 3947.18 \\ 3926.06 \end{bmatrix} \tag{9.16}$$

Therefore, the pressure solution of the set of equations in this example is $p_1 = 3989.44$ psia, $p_2 = 3968.31$ psia, $p_3 = 3947.18$ psia, and $p_4 = 3926.06$ psia.

9.2.2 1-D tangential flow problem (Tang's algorithm)

This algorithm is applicable when flow takes place only in the θ direction; that is, there is one row of blocks arranged in a circle as shown in Fig. 9.1c (with $N = n_\theta$). This is a 1-D flow problem that results in equations similar in form to those given by Eq. (9.2b) for a 1-D rectangular flow problem.

The equation for the first block ($i = 1$) has the form

$$w_1 x_N + c_1 x_1 + e_1 x_2 = d_1 \tag{9.17a}$$

The equation for blocks $i = 2, 3, \ldots, N-1$ has the form

$$w_i x_{i-1} + c_i x_i + e_i x_{i+1} = d_i \tag{9.17b}$$

The equation for the last block ($i = N$) has the form

$$w_N x_{N-1} + c_N x_N + e_N x_1 = d_N \tag{9.17c}$$

Note that Eqs. (9.17a) and (9.17c) have coefficients w_1 and e_N, respectively, because in this flow problem, blocks 1 and N are neighbors as shown in Fig. 9.1c.

TABLE 9.1 Use of Thomas' algorithm to solve the equations of Example 9.1.

i	w_i	c_i	e_i	d_i	u_i	g_i	x_i
1	–	−85.2012	28.4004	−227,203.2	−0.333333	2666.667	3989.44
2	28.4004	−56.8008	28.4004	0	−0.600000	1600.000	3968.31
3	28.4004	−56.8008	28.4004	0	−0.714286	1142.857	3947.18
4	28.4004	−28.4004	–	600	–	3926.057	3926.06

The set of N equations expressed by Eq. (9.17) can be written in a matrix form as

$$\begin{bmatrix} c_1 & e_1 & & & & w_1 \\ w_2 & c_2 & e_2 & & & \\ & \cdots & \cdots & \cdots & & \\ & & \cdots & \cdots & \cdots & \\ & & & w_{N-1} & c_{N-1} & e_{N-1} \\ e_N & & & & w_N & c_N \end{bmatrix} \begin{bmatrix} x_1 \\ x_2 \\ \cdots \\ \cdots \\ x_{N-1} \\ x_N \end{bmatrix} = \begin{bmatrix} d_1 \\ d_2 \\ \cdots \\ \cdots \\ d_{N-1} \\ d_N \end{bmatrix} \tag{9.18}$$

Tang (1969) presented the following algorithm for the solution of this matrix equation. As in Thomas' algorithm, this algorithm is based on **LU** matrix factorization. Here again, the solution is obtained in two major steps.

9.2.2.1 *Forward solution*

$$\text{Set } \zeta_1 = 0 \tag{9.19}$$

$$\beta_1 = -1 \tag{9.20}$$

and

$$\gamma_1 = 0 \tag{9.21}$$

$$\text{Set } \zeta_2 = \frac{d_1}{e_1} \tag{9.22}$$

$$\beta_2 = \frac{c_1}{e_1} \tag{9.23}$$

and

$$\gamma_2 = \frac{w_1}{e_1} \tag{9.24}$$

For $i = 2, 3 \ldots N-1$,

$$\zeta_{i+1} = -\frac{c_i\zeta_i + w_i\zeta_{i-1} - d_i}{e_i} \tag{9.25}$$

$$\beta_{i+1} = -\frac{c_i\beta_i + w_i\beta_{i-1}}{e_i} \tag{9.26}$$

and

$$\gamma_{i+1} = -\frac{c_i\gamma_i + w_i\gamma_{i-1}}{e_i} \tag{9.27}$$

9.2.2.2 Backward solution

First, calculate

$$A=\frac{\zeta_N}{1+\gamma_N} \quad (9.28)$$

$$B=\frac{\beta_N}{1+\gamma_N} \quad (9.29)$$

$$C=\frac{d_N-w_N\zeta_{N-1}}{c_N-w_N\gamma_{N-1}} \quad (9.30)$$

and

$$D=\frac{e_N-w_N\beta_{N-1}}{c_N-w_N\gamma_{N-1}} \quad (9.31)$$

Second, calculate the value of the first unknown (x_1) and the value of the last unknown (x_N) of the solution vector,

$$x_1=\frac{A-C}{B-D} \quad (9.32)$$

and

$$x_N=\frac{BC-AD}{B-D} \quad (9.33)$$

Third, calculate the value of the other unknowns of the solution vector. For $i=2, 3\ldots N-1$,

$$x_i=\zeta_i-\beta_i x_1-\gamma_i x_N \quad (9.34)$$

The next example demonstrates the application of Tang's algorithm to solve the equations of a ring-like 1-D reservoir.

Example 9.2 Using Tang's algorithm, solve the following set of equations:

$$2.84004x_4-5.68008x_1+2.84004x_2=0 \quad (9.35)$$

$$2.84004x_1-8.52012x_2+2.84004x_3=-22720.32 \quad (9.36)$$

$$2.84004x_2-5.68008x_3+2.84004x_4=0 \quad (9.37)$$

and

$$2.84004x_3-5.680084x_4+2.84004x_1=600 \quad (9.38)$$

Solution

The first step for the forward solution is to set $\zeta_1=0$, $\beta_1=-1$, and $\gamma_1=0$, according to Eqs. (9.19) through (9.21), and then to calculate ζ_2, β_2, and γ_2 using Eqs. (9.22), (9.23), and (9.24), which give

$$\zeta_2 = d_1/e_1 = 0/2.84004 = 0$$

$$\beta_2 = c_1/e_1 = -5.68008/2.84004 = -2$$

and

$$\gamma_2 = w_1/e_1 = 2.84004/2.84004 = 1$$

The next step is to calculate ζ_3 and ζ_4 using Eq. (9.25), β_3 and β_4 using Eq. (9.26) and γ_3 and γ_4 using Eq. (9.27), yielding

$$\zeta_3 = -\frac{c_2\zeta_2 + w_2\zeta_1 - d_2}{e_2} = -\frac{-8.52012 \times 0 + 2.84004 \times 0 - (-22720.32)}{2.84004}$$
$$= -8000$$

$$\zeta_4 = -\frac{c_3\zeta_3 + w_3\zeta_2 - d_3}{e_3} = -\frac{-5.68008 \times (-8000) + 2.84004 \times 0 - 0}{2.84004}$$
$$= -16000$$

$$\beta_3 = -\frac{c_2\beta_2 + w_2\beta_1}{e_2} = -\frac{-8.52012 \times (-2) + 2.84004 \times (-1)}{2.84004} = -5$$

$$\beta_4 = -\frac{c_3\beta_3 + w_3\beta_2}{e_3} = -\frac{-5.68008 \times (-5) + 2.84004 \times (-2)}{2.84004} = -8$$

$$\gamma_3 = -\frac{c_2\gamma_2 + w_2\gamma_1}{e_2} = -\frac{-8.52012 \times 1 + 2.84004 \times 0}{2.84004} = 3$$

and

$$\gamma_4 = -\frac{c_3\gamma_3 + w_3\gamma_2}{e_3} = -\frac{-5.68008 \times 3 + 2.84004 \times 1}{2.84004} = 5$$

Table 9.2 shows the results of the calculations as outlined here. The forward substitution step is followed by the backward substitution step, which involves calculating A, B, C, and D using Eqs. (9.28) through (9.31), resulting in

$$A = \frac{\zeta_4}{1+\gamma_4} = \frac{-16000}{1+5} = -2666.667$$

TABLE 9.2 Use of Tang's algorithm to solve the equations of Example 9.2.

i	w_i	c_i	e_i	d_i	ζ_i	β_i	γ_i
1	2.84004	−5.68008	2.84004	0	0	−1	0
2	2.84004	−8.52012	2.84004	−22,720.32	0	−2	1
3	2.84004	−5.68008	2.84004	0	−8000	−5	3
4	2.84004	−5.68008	2.84004	600	−16,000	−8	5

$$B=\frac{\beta_4}{1+\gamma_4}=\frac{-8}{1+5}=-1.33333$$

$$C=\frac{d_4-w_4\zeta_3}{c_4-w_4\gamma_3}=\frac{600-2.84004\times(-8000)}{-5.68008-2.84004\times 3}=-1642.253$$

and

$$D=\frac{e_4-w_4\beta_3}{c_4-w_4\gamma_3}=\frac{2.84004-2.84004\times(-5)}{-5.68008-2.84004\times 3}=-1.2$$

calculating x_1 and x_4 using Eqs. (9.32) and (9.33), yielding

$$x_1=\frac{A-C}{B-D}=\frac{-2666.667-(-1642.253)}{-1.33333-(-1.2)}=7683.30$$

and

$$x_4=\frac{BC-AD}{B-D}=\frac{-1.33333\times(-1642.253)-(-2666.667)\times(-1.2)}{-1.33333-(-1.2)}$$
$$=7577.70$$

and finally calculating x_2 and x_3 using Eq. (9.34) successively, which gives

$$x_2=\zeta_2-\beta_2x_1-\gamma_2x_4=0-(-2)(7683.30)-(1)(7577.70)=7788.90$$

and

$$x_3=\zeta_3-\beta_3x_1-\gamma_3x_4=-8000-(-5)(7683.30)-(3)(7577.70)=7683.40$$

Therefore, the solution vector is

$$\vec{x}=\begin{bmatrix}x_1\\x_2\\x_3\\x_4\end{bmatrix}=\begin{bmatrix}7683.30\\7788.90\\7683.40\\7577.70\end{bmatrix}\tag{9.39}$$

9.2.3 2-D and 3-D flow problems (sparse matrices)

The linear equations for 2-D and 3-D flow problems can be obtained by (1) writing the flow equation using the CVFD method, (2) writing the definition of set ψ_n for block n in 2-D or 3-D, using Fig. 3.1 for engineering notation of block identification or Fig. 3.3 for natural ordering of blocks, as explained in Sections 3.2.1 and 3.2.2, and the definition of set ξ_n for block n, and (3) writing the flow equation in an expanded form. For example, we use Eq. (8.1) in step 1 for 3-D flow of an incompressible fluid, yielding

$$\sum_{l\in\psi_n}T_{l,n}\left[(p_l-p_n)-\gamma_{l,n}(Z_l-Z_n)\right]+\sum_{l\in\xi_n}q_{sc_{l,n}}+q_{sc_n}=0\tag{9.40}$$

If the reservoir has no-flow boundaries ($\xi_n = \{\}$ and as a result $\sum_{l \in \xi_n} q_{sc_{l,n}} = 0$ for all values of n) and if the wells have specified flow rates, Eq. (9.40) can be rearranged as

$$\sum_{l \in \psi_n} T_{l,n} p_l - \left(\sum_{l \in \psi_n} T_{l,n} \right) p_n = \sum_{l \in \psi_n} T_{l,n} \gamma_{l,n} (Z_l - Z_n)] - q_{sc_n} \tag{9.41}$$

In step 2, we define block n as a block in 3-D space $[n \equiv (i,j,k)]$. Accordingly, ψ_n is given as in Fig. 3.3c:

$$\psi_n = \psi_{i,j,k} = \left\{ \left(n - n_x n_y\right), \left(n - n_x\right), \left(n - 1\right), \left(n + 1\right), \left(n + n_x\right), \left(n + n_x n_y\right) \right\} \tag{9.42}$$

provided that the reservoir blocks are ordered using natural ordering, with the blocks ordered in the *i* direction, the *j* direction, and finally the *k* direction. Now, Eq. (9.41) and the new definition of ψ_n given by Eq. (9.42) provide the sought equation.

In step 3, we expand Eq. (9.41) as

$$\begin{aligned}
& T_{n,n-n_x n_y} p_{n-n_x n_y} + T_{n,n-n_x} p_{n-n_x} + T_{n,n-1} p_{n-1} + T_{n,n+1} p_{n+1} + T_{n,n+n_x} p_{n+n_x} \\
& + T_{n,n+n_x n_y} p_{n+n_x n_y} - \left[T_{n,n-n_x n_y} + T_{n,n-n_x} + T_{n,n-1} + T_{n,n+1} + T_{n,n+n_x} + T_{n,n+n_x n_y} \right] p_n \\
& = [(T\gamma)_{n,n-n_x n_y} \left(Z_{n-n_x n_y} - Z_n \right) + (T\gamma)_{n,n-n_x} (Z_{n-n_x} - Z_n) + (T\gamma)_{n,n-1} (Z_{n-1} - Z_n) \\
& + (T\gamma)_{n,n+1} (Z_{n+1} - Z_n) + (T\gamma)_{n,n+n_x} (Z_{n+n_x} - Z_n) + (T\gamma)_{n,n+n_x n_y} \left(Z_{n+n_x n_y} - Z_n \right)] - q_{sc_n}
\end{aligned} \tag{9.43}$$

The unknown pressures in Eq. (9.43) are rearranged in the order shown in Fig. 9.2, yielding

$$\begin{aligned}
& T_{n,n-n_x n_y} p_{n-n_x n_y} + T_{n,n-n_x} p_{n-n_x} + T_{n,n-1} p_{n-1} \\
& - \left[T_{n,n-n_x n_y} + T_{n,n-n_x} + T_{n,n-1} + T_{n,n+1} + T_{n,n+n_x} + T_{n,n+n_x n_y} \right] p_n \\
& + T_{n,n+1} p_{n+1} + T_{n,n+n_x} p_{n+n_x} + T_{n,n+n_x n_y} p_{n+n_x n_y} \\
& = [(T\gamma)_{n,n-n_x n_y} \left(Z_{n-n_x n_y} - Z_n \right) + (T\gamma)_{n,n-n_x} (Z_{n-n_x} - Z_n) + (T\gamma)_{n,n-1} (Z_{n-1} - Z_n) \\
& + (T\gamma)_{n,n+1} (Z_{n+1} - Z_n) + (T\gamma)_{n,n+n_x} (Z_{n+n_x} - Z_n) + (T\gamma)_{n,n+n_x n_y} \left(Z_{n+n_x n_y} - Z_n \right)] - q_{sc_n}
\end{aligned} \tag{9.44}$$

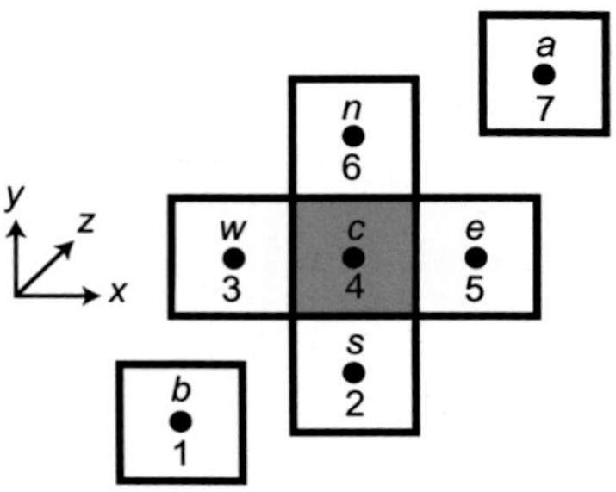

FIG. 9.2 Ordering of unknowns of neighboring blocks in flow equations.

Eq. (9.44) is the linear equation for 3-D flow of an incompressible fluid. The unknowns in this equation are $p_{n-n_x n_y}$, p_{n-n_x}, p_{n-1}, p_n, p_{n+1}, p_{n+n_x}, and $p_{n+n_x n_y}$. Eq. (9.44) can be expressed as

$$b_n x_{n-n_x n_y} + s_n x_{n-n_x} + w_n x_{n-1} + c_n x_n + e_n x_{n+1} + n_n x_{n+n_x} + a_n x_{n+n_x n_y} = d_n \quad (9.45)$$

where

$$b_n = T_{n,n-n_x n_y} = T_{z_{i,j,k-1/2}} \quad (9.46a)$$

$$s_n = T_{n,n-n_x} = T_{y_{i,j-1/2,k}} \quad (9.46b)$$

$$w_n = T_{n,n-1} = T_{x_{i-1/2,j,k}} \quad (9.46c)$$

$$e_n = T_{n,n+1} = T_{x_{i+1/2,j,k}} \quad (9.46d)$$

$$n_n = T_{n,n+n_x} = T_{y_{i,j+1/2,k}} \quad (9.46e)$$

$$a_n = T_{n,n+n_x n_y} = T_{z_{i,j,k+1/2}} \quad (9.46f)$$

$$c_n = -(b_n + s_n + w_n + e_n + n_n + a_n) \quad (9.46g)$$

and

$$\begin{aligned} d_n = {} & [(b\gamma)_n (Z_{n-n_x n_y} - Z_n) + (s\gamma)_n (Z_{n-n_x} - Z_n) + (w\gamma)_n (Z_{n-1} - Z_n) \\ & + (e\gamma)_n (Z_{n+1} - Z_n) + (n\gamma)_n (Z_{n+n_x} - Z_n) + (a\gamma)_n (Z_{n+n_x n_y} - Z_n)] - q_{sc_n} \end{aligned} \quad (9.46h)$$

If Eq. (9.45) is written for each block $n=1, 2, 3\ldots, N$ where $N = n_x \times n_y \times n_z$ in a rectangular reservoir, the matrix equation will have seven diagonals (a heptadiagonal coefficient matrix) as shown in Fig. 9.3c. Fluid flow in a 2-D reservoir ($b_n = a_n = 0$) with regular boundaries results in a matrix equation with five diagonals (a pentadiagonal coefficient matrix) as shown in Fig. 9.3b. Fluid flow in a 1-D reservoir ($b_n = s_n = n_n = a_n = 0$) results in a matrix equation with three diagonals (a tridiagonal coefficient matrix) as shown in Fig. 9.3a.

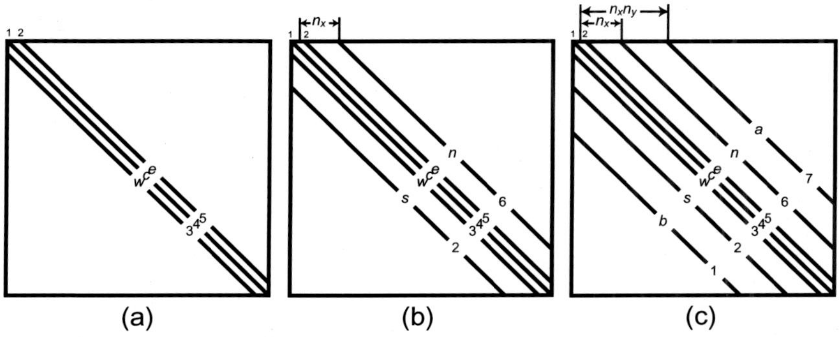

FIG. 9.3 Coefficient matrices in 1-D, 2-D, and 3-D flow problems. (a) Tridiagonal matrix, (b) Pentadiagonal matrix, and (c) Heptadiagonal matrix.

The solutions of these matrix equations can be obtained using a g-band matrix solver. Such a solver is nothing more than Gaussian elimination using **LU** factorization, which operates only on elements within the outermost bands of the sparse matrix. Zeros outside the outermost bands are not operated on. The number of row (or column) elements within the outermost bands is called the bandwidth $(2b_w+1)$, where $b_w=1$ for 1-D flow problems, $b_w=n_x$ for 2-D flow problems, and $b_w=n_x\times n_y$ for 3-D flow problems as shown in Fig. 9.3. The following algorithm is a g-band algorithm. The g-band algorithm is executed in three major steps: the initialization step, the forward elimination step, and the back substitution step.

9.2.3.1 *Initialization step*

For $i=1, 2, \ldots N$, set

$$d_i^{(0)}=d_i \tag{9.47}$$

$$j_{\min}=\max(1, i-b_w) \tag{9.48a}$$

$$j_{\max}=\min(i+b_w, N) \tag{9.48b}$$

and

$$a_{i,j}^{(0)}=a_{i,j} \tag{9.49}$$

for $j=j_{\min}, j_{\min}+1, \ldots j_{\max}$.

9.2.3.2 *Forward elimination step*

For $i=1, 2, \ldots N$, set

$$d_i^{(i)}=\frac{d_i^{(i-1)}}{a_{i,i}^{(i-1)}} \tag{9.50}$$

$$j_{\max}=\min(i+b_w, N) \tag{9.48b}$$

$$a_{i,j}^{(i)}=\frac{a_{i,j}^{(i-1)}}{a_{i,i}^{(i-1)}} \tag{9.51a}$$

for $j=i, i+1, \ldots j_{\max}$, and

$$a_{i,i}^{(i)}=1 \tag{9.51b}$$

For $k=i+1, i+2, \ldots j_{\max}$, set

$$d_k^{(i)}=d_k^{(i-1)}-d_i^{(i)}a_{k,i}^{(i-1)} \tag{9.52}$$

$$a_{k,j}^{(i)}=a_{k,j}^{(i-1)}-a_{i,j}^{(i)}a_{k,i}^{(i-1)} \tag{9.53a}$$

for $j = i, i+1, \ldots j_{\max}$, and

$$a_{k,i}^{(i)} = 0 \tag{9.53b}$$

9.2.3.3 *Back substitution step*

$$\text{Set } x_N = d_N^{(N)} \tag{9.54}$$

For $i = N-1, N-2, \ldots 2, 1$, set

$$j_{\max} = \min(i + b_w, N) \tag{9.48b}$$

and

$$x_i = d_i^{(N)} - \sum_{j=i+1}^{j_{\max}} a_{i,j}^{(N)} x_j \tag{9.55}$$

The FORTRAN computer codes that use this algorithm are available in the literature (Aziz and Settari, 1979; Abou-Kassem and Ertekin, 1992). Such programs require storing matrix elements within the outermost bands row-wise in a vector (a one-dimensional matrix).

9.3 Iterative solution methods

Iterative solution methods produce the solution vector for a given system of equations as the limit of a sequence of intermediate vectors that progressively converge toward the solution. Iterative solution methods do not require storing the coefficient matrix **[A]** as in the direct solution methods. In addition, these methods do not suffer from the accumulation of roundoff errors that occur during computations. In iterative methods, the reservoir blocks are usually ordered using natural ordering. In the following presentation, the blocks are ordered along the x-direction, then along the y-direction, and finally along the z-direction. We discuss basic iterative methods such as the point iterative methods [Jacobi, Gauss-Seidel, or point successive overrelaxation (PSOR)] that are most useful in solving equations for 1-D problems, line SOR (LSOR), and block SOR (BSOR) methods and alternating direction implicit procedure (ADIP) that are useful in solving equations for 2-D and 3-D problems. Although these methods are practically unused in today's simulators because of the development of advanced and more powerful iterative methods, they are sufficient for single-phase flow problems. We will use Eq. (9.45) to demonstrate the application of the various iterative solution methods for 1-D, 2-D, and 3-D problems. Initiation of the iterative methods requires the assignment of initial guesses for all the unknowns. For flow problems involving an incompressible fluid, the initial guess for unknown x_n is taken as zero; that is, $x_n^{(0)} = 0$. For flow problems involving slightly compressible and compressible fluids, the initial guess for unknown x_n for the first outer iteration ($k=1$) is taken as the value of the unknown at the old time level (p_n^n); that is, $x_n^{(0)}=p_n^n$. However, for the second ($k=2$), third ($k=3$), and higher ($k=4, 5, \ldots$) outer iterations, the initial guess for unknown x_n is taken as the value of the unknown at the latest outer iteration; that is, $x_n^{(0)}=p_n^{n+1^{(k-1)}}$. Outer iterations refer to the iterations used to linearize the equations in the process of advancing the pressure solution from old time level n to new time level $n+1$.

9.3.1 Point iterative methods

Point iterative methods include the point Jacobi, point Gauss-Seidel, and point successive overrelaxation (PSOR) methods. In these methods, the solution, at any iteration level ($\nu+1$), is obtained by solving for one unknown using one equation at a time. They start with the equation for block 1, followed by the equation for block 2, and proceed block by block (or point by point) to the last block (block N). Though these methods can be used for multidimensional problems, their use is recommended for 1-D problems because of their extremely slow convergence.

9.3.1.1 Point Jacobi method

To write the point Jacobi iterative equation for 1-D problems, we have to solve for the unknown of a general block n (x_n in this case) using the linear equation for the same block (Eq. 9.45 with $b_n=s_n=n_n=a_n=0$); that is,

$$x_n=\frac{1}{c_n}(d_n-w_nx_{n-1}-e_nx_{n+1}) \tag{9.56}$$

The unknown for block n on the LHS of the resulting equation (Eq. 9.56) is assigned current iteration level ($\nu+1$), whereas all other unknowns on the RHS of Eq. (9.56) are assigned old iteration level (ν). The point Jacobi iterative scheme becomes

$$x_n^{(\nu+1)}=\frac{1}{c_n}\left(d_n-w_nx_{n-1}^{(\nu)}-e_nx_{n+1}^{(\nu)}\right) \tag{9.57}$$

where $n=1, 2, \ldots N$ and $\nu=0, 1, 2, \ldots$

The iteration process starts from $\nu=0$ and uses initial guess values for all unknowns (say, $\vec{x}^{(0)}=\vec{0}$ for incompressible flow problems or the old time value, $\vec{x}^{(0)}=\vec{x}^n$ for slightly compressible and compressible flow problems as mentioned earlier in the introduction in Section 9.3). We start with block 1, then block 2, …, until block N and estimate the results of the first iteration ($\vec{x}^{(1)}$). The process is repeated for $\nu=1$, and second iteration estimates for all unknowns are obtained ($\vec{x}^{(2)}$). Iterations continue until a specified convergence criterion is satisfied. One form of convergence criterion is related to the maximum absolute difference between the successive iterations among all blocks; that is,

$$d_{\max}^{(\nu+1)} \leq \varepsilon \tag{9.58}$$

where

$$d_{\max}^{(\nu+1)} = \max_{1 \leq n \leq N} \left| x_n^{(\nu+1)} - x_n^{(\nu)} \right| \tag{9.59}$$

and ε is some acceptable tolerance.

A better convergence criterion is related to the residual (r_n) of the linear equation (Aziz and Settari, 1979):

$$\max_{1 \leq n \leq N} \left| r_n^{(\nu+1)} \right| \leq \varepsilon \tag{9.60}$$

The residual of Eq. (9.45) is defined as

$$r_n = b_n x_{n-n_x n_y} + s_n x_{n-n_x} + w_n x_{n-1} + c_n x_n + e_n x_{n+1} + n_n x_{n+n_x} + a_n x_{n+n_x n_y} - d_n \tag{9.61}$$

Fig. 9.4 shows the iteration level of the unknowns of the neighboring blocks that usually appear in iterative equations of the point Jacobi method in multi-dimensional problems. Fig. 9.5 illustrates the application of the method in a 2-D reservoir. It should be noted that the point Jacobi method requires storing the old iterate values of all unknowns. In addition, the convergence of this method is extremely slow. In Example 9.3, we apply the point Jacobi iterative method to solve the equations of a 1-D reservoir.

Example 9.3 The following equations were obtained for the 1-D reservoir in Example 7.1 and were solved in Example 9.1:

$$-85.2012 p_1 + 28.4004 p_2 = -227203.2 \tag{9.12}$$

$$28.4004 p_1 - 56.8008 p_2 + 28.4004 p_3 = 0 \tag{9.13}$$

$$28.4004 p_2 - 56.8008 p_3 + 28.4004 p_4 = 0 \tag{9.14}$$

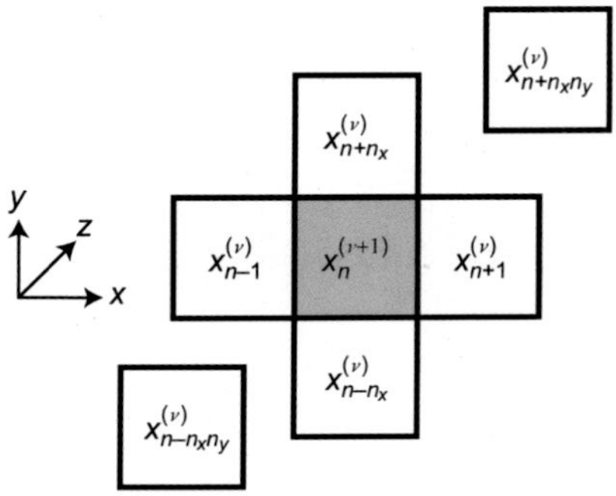

FIG. 9.4 Iteration level of the unknowns of the neighboring blocks in the point Jacobi method.

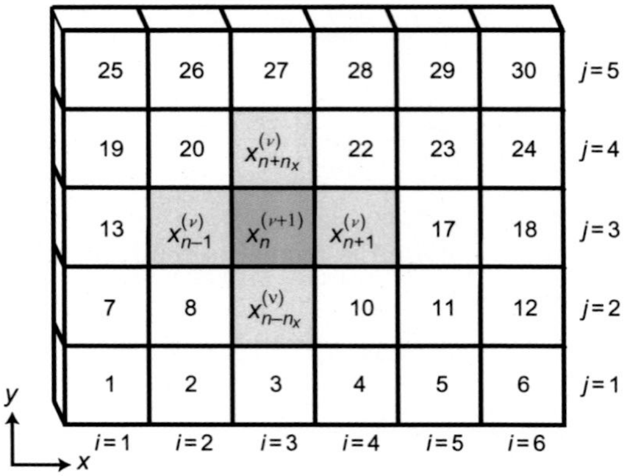

FIG. 9.5 Iteration level of the unknowns of the neighboring blocks in estimating values in the point Jacobi method in 2-D problems.

and

$$28.4004p_3 - 28.4004p_4 = 600 \tag{9.15}$$

Solve these equations using the point Jacobi iterative method.

Solution

First, we solve for p_1 using Eq. (9.12), p_2 using Eq. (9.13), p_3 using Eq. (9.14), and p_4 using Eq. (9.15):

$$p_1 = 2666.6667 + 0.33333333p_2 \tag{9.62}$$

$$p_2 = 0.5(p_1 + p_3) \tag{9.63}$$

$$p_3 = 0.5(p_2 + p_4) \tag{9.64}$$

and

$$p_4 = -21.126463 + p_3 \tag{9.65}$$

Second, the Jacobi iterative equations are obtained by placing levels of iteration according to Eq. (9.57),

$$p_1^{(\nu+1)} = 2666.6667 + 0.33333333p_2^{(\nu)} \tag{9.66}$$

$$p_2^{(\nu+1)} = 0.5\left(p_1^{(\nu)} + p_3^{(\nu)}\right) \tag{9.67}$$

$$p_3^{(\nu+1)} = 0.5\left(p_2^{(\nu)} + p_4^{(\nu)}\right) \tag{9.68}$$

and

$$p_4^{(\nu+1)} = -21.126463 + p_3^{(\nu)} \tag{9.69}$$

With an initial guess of 0 for all unknowns, the Jacobi iterative equations for the first iteration ($\nu = 0$) predict

$$p_1^{(1)} = 2666.6667 + 0.33333333 p_2^{(0)} = 2666.6667 + 0.33333333(0)$$
$$= 2666.6667$$

$$p_2^{(1)} = 0.5\left(p_1^{(0)} + p_3^{(0)}\right) = 0.5(0+0) = 0$$

$$p_3^{(1)} = 0.5\left(p_2^{(0)} + p_4^{(0)}\right) = 0.5(0+0) = 0$$

and

$$p_4^{(1)} = -21.126463 + p_3^{(0)} = -21.126463 + 0 = -21.126463$$

For the second iteration ($\nu = 1$), the Jacobi iterative equations predict

$$p_1^{(2)} = 2666.6667 + 0.33333333 p_2^{(1)} = 2666.6667 + 0.33333333(0)$$
$$= 2666.6667$$

$$p_2^{(2)} = 0.5\left(p_1^{(1)} + p_3^{(1)}\right) = 0.5(2666.6667 + 0) = 1333.33335$$

$$p_3^{(2)} = 0.5\left(p_2^{(1)} + p_4^{(1)}\right) = 0.5(0 - 21.126463) = -10.5632315$$

and

$$p_4^{(2)} = -21.126463 + p_3^{(1)} = -21.126463 + 0 = -21.126463$$

The procedure continues until the convergence criterion is satisfied. The convergence criterion set for this problem is $\varepsilon \leq 0.0001$. Table 9.3 presents the solution within the specified tolerance obtained after 159 iterations. At convergence, the maximum absolute difference calculated using Eq. (9.59) was 0.0000841.

9.3.1.2 *Point Gauss-Seidel method*

The point Gauss-Seidel method differs from the point Jacobi method in that it uses the latest available iterates of the unknowns in computing the unknown for block n at the current iteration ($x_n^{(\nu+1)}$). When we obtain the current iteration value for the unknown for block n, we already have obtained the current iteration values for the unknowns for blocks $1, 2, \ldots,$ and $n-1$ that precede block n. The unknowns for blocks $n+1$, $n+2$, …, and N still have their latest iteration value at iteration level v. Therefore, the point Gauss-Seidel iterative equation for block n in 1-D problems is

$$x_n^{(\nu+1)} = \frac{1}{c_n}\left(d_n - w_n x_{n-1}^{(\nu+1)} - e_n x_{n+1}^{(\nu)}\right) \tag{9.70}$$

TABLE 9.3 Jacobi iteration for Example 9.3.

$\nu+1$	p_1	p_2	p_3	p_4	$d_{max}^{(\nu+1)}$
	0	0	0	0	–
1	2666.67	0.00	0.00	−21.13	2666.6667
2	2666.67	1333.33	−10.56	−21.13	1333.3333
3	3111.11	1328.05	656.10	−31.69	666.6667
4	3109.35	1883.61	648.18	634.98	666.6667
5	3294.54	1878.77	1259.29	627.05	611.1111
6	3292.92	2276.91	1252.91	1238.17	611.1111
7	3425.64	2272.92	1757.54	1231.78	504.6296
…	…	…	…	…	…
21	3855.62	3565.91	3427.17	3285.97	120.0747
22	3855.30	3641.40	3425.94	3406.05	120.0747
23	3880.47	3640.62	3523.72	3404.81	97.7808
24	3880.21	3702.09	3522.72	3502.60	97.7808
…	…	…	…	…	…
45	3978.06	3934.09	3902.96	3871.62	10.2113
46	3978.03	3940.51	3902.86	3881.84	10.2113
47	3980.17	3940.44	3911.17	3881.73	8.3154
48	3980.15	3945.67	3911.09	3890.05	8.3154
…	…	…	…	…	…
67	3988.25	3964.74	3942.57	3920.37	1.0664
68	3988.25	3965.41	3942.55	3921.44	1.0664
69	3988.47	3965.40	3943.42	3921.43	0.8684
70	3988.47	3965.95	3943.41	3922.30	0.8684
…	…	…	…	…	…
90	3989.31	3968.01	3946.70	3925.58	0.1114
91	3989.34	3968.01	3946.79	3925.57	0.0907
92	3989.34	3968.06	3946.79	3925.66	0.0907
93	3989.35	3968.06	3946.86	3925.66	0.0738
…	…	…	…	…	…
112	3989.42	3968.28	3947.13	3926.01	0.0116
113	3989.43	3968.28	3947.14	3926.01	0.0095
…	…	…	…	…	…
158	3989.44	3968.31	3947.18	3926.06	0.0001
159	3989.44	3968.31	3947.18	3926.06	0.0001

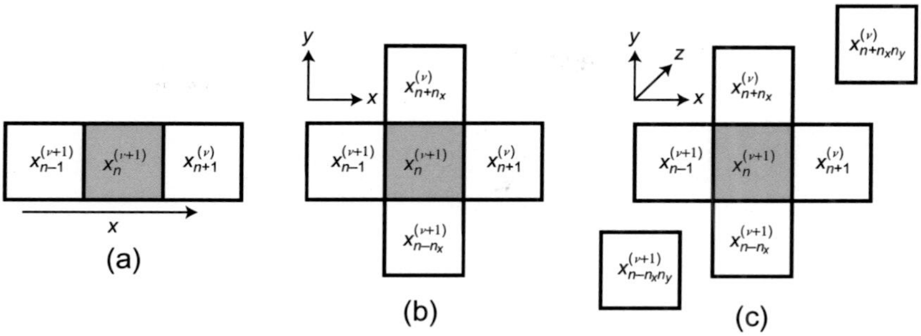

FIG. 9.6 Iteration level of the unknowns of the neighboring blocks in the point Gauss-Seidel method. (a) 1-D, (b) 2-D, and (c) 3-D.

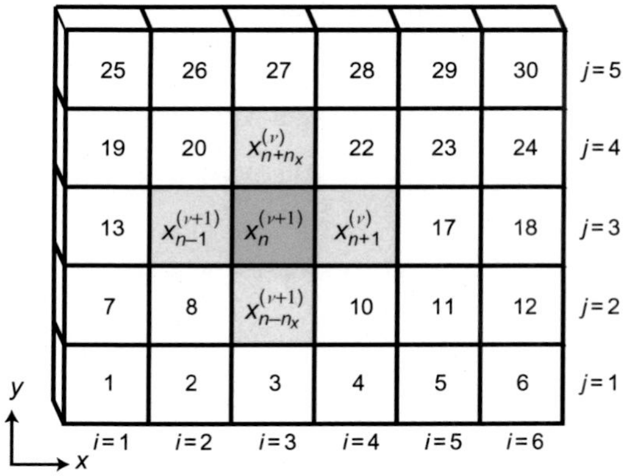

FIG. 9.7 Iteration level of the unknowns of the neighboring blocks in estimating values in the point Gauss-Seidel method in 2-D problems.

It should be mentioned that not only does the point Gauss-Seidel method not require storing the old iterate value of the unknowns but also it is easier to program and converges twice as fast as the point Jacobi method. Fig. 9.6 shows the iteration level of the unknowns of the neighboring blocks that usually appear in iterative equations in multidimensional problems. Fig. 9.7 illustrates the application of the method in a 2-D reservoir. Example 9.4 demonstrates the application of this iterative method to solve the equations presented in Example 9.3. Observe the improvement in the rate of convergence over that of the point Jacobi method.

Example 9.4 The following equations were obtained for the 1-D reservoir in Example 7.1:

$$-85.2012p_1 + 28.4004p_2 = -227203.2 \tag{9.12}$$

$$28.4004p_1 - 56.8008p_2 + 28.4004p_3 = 0 \quad (9.13)$$

$$28.4004p_2 - 56.8008p_3 + 28.4004p_4 = 0 \quad (9.14)$$

and

$$28.4004p_3 - 28.4004p_4 = 600 \quad (9.15)$$

Solve these equations using the point Gauss-Seidel iterative method.

Solution

First, we solve for p_1 using Eq. (9.12), p_2 using Eq. (9.13), p_3 using Eq. (9.14), and p_4 using Eq. (9.15) as in Example 9.3:

$$p_1 = 2666.6667 + 0.33333333p_2 \quad (9.62)$$

$$p_2 = 0.5(p_1 + p_3) \quad (9.63)$$

$$p_3 = 0.5(p_2 + p_4) \quad (9.64)$$

and

$$p_4 = -21.126463 + p_3 \quad (9.65)$$

Second, the Gauss-Seidel iterative equations are obtained by placing levels of iteration according to Eq. (9.70):

$$p_1^{(\nu+1)} = 2666.6667 + 0.33333333p_2^{(\nu)} \quad (9.71)$$

$$p_2^{(\nu+1)} = 0.5\left(p_1^{(\nu+1)} + p_3^{(\nu)}\right) \quad (9.72)$$

$$p_3^{(\nu+1)} = 0.5\left(p_2^{(\nu+1)} + p_4^{(\nu)}\right) \quad (9.73)$$

and

$$p_4^{(\nu+1)} = -21.126463 + p_3^{(\nu+1)} \quad (9.74)$$

With an initial guess of 0 for all unknowns, the Gauss-Seidel iterative equations for the first iteration ($\nu = 0$) predict

$$p_1^{(1)} = 2666.6667 + 0.33333333p_2^{(0)} = 2666.6667 + 0.33333333(0) = 2666.6667$$

$$p_2^{(1)} = 0.5\left(p_1^{(1)} + p_3^{(0)}\right) = 0.5(2666.6667 + 0) = 1333.33335$$

$$p_3^{(1)} = 0.5\left(p_2^{(1)} + p_4^{(0)}\right) = 0.5(1333.33335 + 0) = 666.66668$$

and

$$p_4^{(1)} = -21.126463 + p_3^{(1)} = -21.126463 + 666.66668 = 645.54021$$

For the second iteration ($\nu=1$), the Gauss-Seidel iterative equations predict

$$p_1^{(2)}=2666.6667+0.33333333\,p_2^{(1)}=2666.6667+0.33333333(1333.33335)$$
$$=3111.11115$$

$$p_2^{(2)}=0.5\left(p_1^{(2)}+p_3^{(1)}\right)=0.5(3111.11115+666.66668)=1888.88889$$

$$p_3^{(2)}=0.5\left(p_2^{(2)}+p_4^{(1)}\right)=0.5(1888.88889+645.54021)=1267.21455$$

and

$$p_4^{(2)}=-21.126463+p_3^{(2)}=-21.126463+1267.21455=1246.08809$$

The procedure continues until the convergence criterion is satisfied. The convergence criterion set for this problem is $\varepsilon\leq 0.0001$. Table 9.4 presents the solution within the specified tolerance obtained after 79 iterations. At convergence, the maximum absolute difference calculated using Eq. (9.59) was 0.0000828.

TABLE 9.4 Gauss-Seidel iteration for Example 9.4.

$\nu+1$	p_1	p_2	p_3	p_4	$d_{max}^{(\nu+1)}$
	0	0	0	0	–
1	2666.67	1333.33	666.67	645.54	2666.6667
2	3111.11	1888.89	1267.21	1246.09	600.5479
3	3296.30	2281.76	1763.92	1742.80	496.7072
4	3427.25	2595.59	2169.19	2148.06	405.2693
5	3531.86	2850.53	2499.30	2478.17	330.1046
6	3616.84	3058.07	2768.12	2746.99	268.8234
7	3686.02	3227.07	2987.03	2965.91	218.9128
8	3742.36	3364.69	3165.30	3144.17	178.2681
9	3788.23	3476.77	3310.47	3289.34	145.1697
10	3825.59	3568.03	3428.69	3407.56	118.2165
11	3856.01	3642.35	3524.95	3503.83	96.2677
12	3880.78	3702.87	3603.35	3582.22	78.3940
...	...	...	...	...	...
21	3972.33	3926.51	3893.04	3871.91	12.3454
22	3975.50	3934.27	3903.09	3881.96	10.0532
23	3978.09	3940.59	3911.28	3890.15	8.1867

Continued

TABLE 9.4 Gauss-Seidel iteration for Example 9.4.—cont'd

$\nu+1$	p_1	p_2	p_3	p_4	$d_{max}^{(\nu+1)}$
…	…	…	…	…	…
32	3987.65	3963.94	3941.53	3920.40	1.2892
33	3987.98	3964.76	3942.58	3921.45	1.0499
34	3988.25	3965.42	3943.43	3922.31	0.8549
…	…	…	…	…	…
42	3989.21	3967.75	3946.46	3925.33	0.1653
43	3989.25	3967.85	3946.59	3925.47	0.1346
44	3989.28	3967.94	3946.70	3925.58	0.1096
45	3989.31	3968.01	3946.79	3925.67	0.0893
46	3989.34	3968.06	3946.86	3925.74	0.0727
…	…	…	…	…	…
78	3989.44	3968.31	3947.18	3926.06	0.0001
79	3989.44	3968.31	3947.18	3926.06	0.0001

9.3.1.3 *Point SOR method*

The point SOR (PSOR) method offers improvements in convergence over the point Gauss-Seidel method by making use of the latest iterate value of the unknown ($x_n^{(\nu)}$) and introducing a parameter (ω) that accelerates convergence. Starting with the Gauss-Seidel method for 1-D problems, an intermediated value is estimated:

$$x_n^{*(\nu+1)} = \frac{1}{c_n}\left(d_n - w_n x_{n-1}^{(\nu+1)} - e_n x_{n+1}^{(\nu)}\right) \tag{9.75}$$

Fig. 9.8 shows the iteration level of the unknowns of the neighboring blocks that are used to estimate the intermediate value of the unknown of block n ($x_n^{*(\nu+1)}$). Fig. 9.9 illustrates the application of this step of the method in a 2-D reservoir. This intermediate value is improved and accelerated to obtain the current iterate value of the unknown before moving on to the next block:

$$x_n^{(\nu+1)} = (1-\omega)x_n^{(\nu)} + \omega x_n^{*(\nu+1)} \tag{9.76}$$

where $1 \le \omega \le 2$. The acceleration parameter has an optimum value that is called the optimum overrelaxation parameter (ω_{opt}). The use of this optimum value improves the convergence of the PSOR method that is roughly twice the

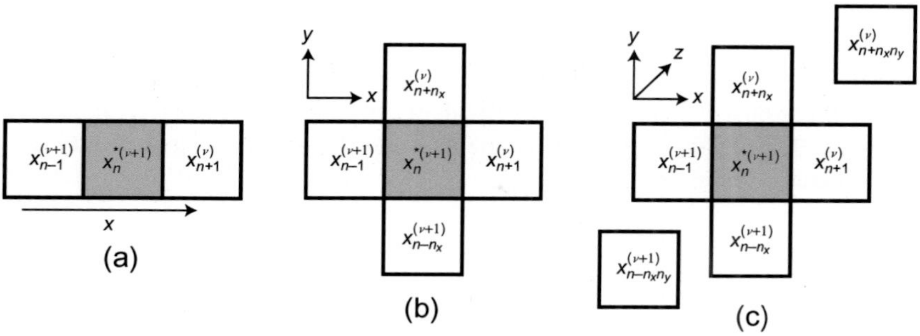

FIG. 9.8 Iteration level of the unknowns of the neighboring blocks in the PSOR method prior to acceleration. (a) 1-D, (b) 2-D, and (c) 3-D.

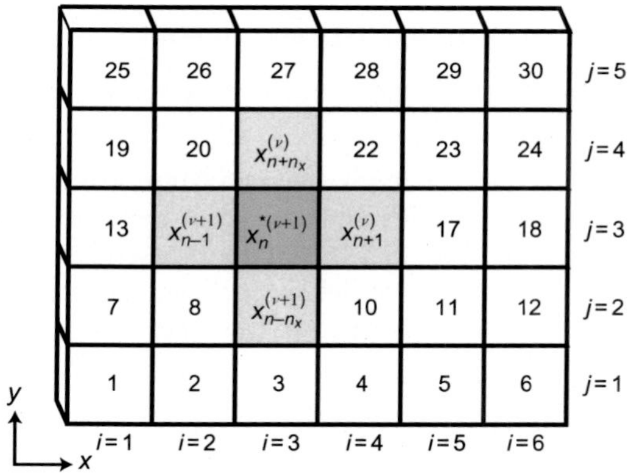

FIG. 9.9 Iteration level of the unknowns of the neighboring blocks in estimating the values in the PSOR method prior to acceleration in 2-D problems.

convergence rate of the point Gauss-Seidel method. The optimum overrelaxation parameter is estimated using

$$\omega_{opt} = \frac{2}{1+\sqrt{1-\rho_{GS}}} \tag{9.77}$$

where

$$\rho_{GS} = \frac{d_{\max}^{(\nu+1)}}{d_{\max}^{(\nu)}} \tag{9.78}$$

is obtained from the Gauss-Seidel method for sufficiently large values of ν. This means that the overrelaxation parameter (ω_{opt}) is estimated by solving

Eqs. (9.75) and (9.76) with $\omega=1$ until ρ_{GS}, estimated using Eq. (9.78), stabilizes (converges within 0.2%); then Eq. (9.77) is used. For 2-D and 3-D problems, Eq. (9.75) is replaced with the appropriate equation. Example 9.5 demonstrates the application of this iterative method to solve the equations presented in Example 9.3. Observe the improvement in the rate of convergence over that of the Gauss-Seidel method.

Example 9.5 The following equations were obtained for the 1-D reservoir in Example 7.1:

$$-85.2012p_1+28.4004p_2=-227203.2 \quad (9.12)$$

$$28.4004p_1-56.8008p_2+28.4004p_3=0 \quad (9.13)$$

$$28.4004p_2-56.8008p_3+28.4004p_4=0 \quad (9.14)$$

and

$$28.4004p_3-28.4004p_4=600 \quad (9.15)$$

Solve these equations using the PSOR iterative method.

Solution

First, we estimate the optimum overrelaxation parameter (ω_{opt}) using Eq. (9.77). This equation requires an estimate of the spectral radius that can be obtained by applying Eq. (9.78) and the Gauss-Seidel iteration as in Example 9.4. Table 9.5 shows that the spectral radius converges to 0.814531 within 0.15% after five iterations. Now, we can estimate ω_{opt} from Eq. (9.77) as

$$\omega_{opt}=\frac{2}{1+\sqrt{1-0.814531}}=1.397955$$

To write the PSOR iterative equations, we first write the Gauss-Seidel iterative equations as in Example 9.4:

$$p_1^{(\nu+1)}=2666.6667+0.33333333p_2^{(\nu)} \quad (9.71)$$

TABLE 9.5 Determination of spectral radius for Example 9.5.

$\nu+1$	p_1	p_2	p_3	p_4	$d_{max}^{(\nu+1)}$	ρ_{GS}
	0	0	0	0	–	–
1	2666.67	1333.33	666.67	645.54	2666.6667	–
2	3111.11	1888.89	1267.21	1246.09	600.5479	0.225205
3	3296.30	2281.76	1763.92	1742.80	496.7072	0.827090
4	3427.25	2595.59	2169.19	2148.06	405.2693	0.815912
5	3531.86	2850.53	2499.30	2478.17	330.1046	0.814531

$$p_2^{(v+1)} = 0.5\left(p_1^{(v+1)} + p_3^{(v)}\right) \tag{9.72}$$

$$p_3^{(v+1)} = 0.5\left(p_2^{(v+1)} + p_4^{(v)}\right) \tag{9.73}$$

and

$$p_4^{(v+1)} = -21.126463 + p_3^{(v+1)} \tag{9.74}$$

Then, applying Eq. (9.76), the PSOR iterative equations become

$$p_1^{(v+1)} = (1 - \omega_{opt})p_1^{(v)} + \omega_{opt}\left[2666.6667 + 0.33333333p_2^{(v)}\right] \tag{9.79}$$

$$p_2^{(v+1)} = (1 - \omega_{opt})p_2^{(v)} + \omega_{opt}\left[0.5\left(p_1^{(v+1)} + p_3^{(v)}\right)\right] \tag{9.80}$$

$$p_3^{(v+1)} = (1 - \omega_{opt})p_3^{(v)} + \omega_{opt}\left[0.5\left(p_2^{(v+1)} + p_4^{(v)}\right)\right] \tag{9.81}$$

and

$$p_4^{(v+1)} = (1 - \omega_{opt})p_4^{(v)} + \omega_{opt}\left[-21.126463 + p_3^{(v+1)}\right] \tag{9.82}$$

We continue the solution process with the PSOR iterative equations using $\omega_{opt} = 1.397955$ and starting with the results of the last Gauss-Seidel iteration, shown as the fifth iteration in Table 9.5, as an initial guess. The PSOR iterative equations for the first iteration ($\nu = 0$) predict

$$\begin{aligned} p_1^{(1)} &= (1 - \omega_{opt})p_1^{(0)} + \omega_{opt}\left[2666.6667 + 0.33333333p_2^{(0)}\right] \\ &= (1 - 1.397955)(3531.86) + (1.397955)[2666.6667 + 0.33333333 \times 2850.53] \\ &= 3650.66047 \end{aligned}$$

$$\begin{aligned} p_2^{(1)} &= (1 - \omega_{opt})p_2^{(0)} + \omega_{opt}\left[0.5\left(p_1^{(1)} + p_3^{(0)}\right)\right] \\ &= (1 - 1.397955)(2850.53) + (1.397955)[0.5(3650.66047 + 2499.30)] \\ &= 3164.29973 \end{aligned}$$

$$\begin{aligned} p_3^{(1)} &= (1 - \omega_{opt})p_3^{(0)} + \omega_{opt}\left[0.5\left(p_2^{(1)} + p_4^{(0)}\right)\right] \\ &= (1 - 1.397955)(2499.30) + (1.397955)[0.5(3164.29973 + 2478.17)] \\ &= 2949.35180 \end{aligned}$$

and

$$\begin{aligned} p_4^{(1)} &= (1 - \omega_{opt})p_4^{(0)} + \omega_{opt}\left[-21.126463 + p_3^{(1)}\right] \\ &= (1 - 1.397955)(2478.17) + (1.397955)[-21.126463 + 2949.35180] \\ &= 3107.32771 \end{aligned}$$

We continue with the second iteration using Eqs. (9.79) through (9.82), followed by the third iteration, and the iteration process is repeated until the

TABLE 9.6 PSOR iteration for Example 9.5.

$\nu+1$	p_1	p_2	p_3	p_4	$d_{max}^{(\nu+1)}$
	3531.86	2850.53	2499.30	2478.17	–
1	3650.66	3164.30	2949.35	3107.33	629.1586
2	3749.60	3423.17	3390.96	3474.30	441.6074
3	3830.85	3685.62	3655.17	3697.62	264.2122
4	3920.82	3828.73	3806.16	3819.82	150.9853
5	3951.70	3898.91	3880.53	3875.16	74.3782
6	3972.11	3937.23	3916.41	3903.29	38.3272
7	3981.85	3953.87	3933.42	3915.88	17.0095
8	3985.72	3961.84	3941.03	3921.50	7.97870
9	3987.90	3965.51	3944.49	3924.10	3.66090
10	3988.74	3967.06	3946.01	3925.20	1.5505
11	3989.13	3967.78	3946.68	3925.69	0.7205
12	3989.31	3968.08	3946.97	3925.90	0.3034
13	3989.38	3968.21	3947.09	3925.99	0.1311
14	3989.41	3968.27	3947.14	3926.03	0.0578
15	3989.43	3968.29	3947.17	3926.05	0.0237
16	3989.43	3968.30	3947.18	3926.05	0.0104
17	3989.44	3968.31	3947.18	3926.06	0.0043
18	3989.44	3968.31	3947.18	3926.06	0.0018
19	3989.44	3968.31	3947.18	3926.06	0.0008
20	3989.44	3968.31	3947.18	3926.06	0.0003
21	3989.44	3968.31	3947.18	3926.06	0.0001
22	3989.44	3968.31	3947.18	3926.06	0.0001

convergence criterion is satisfied. The convergence criterion set for this problem is $\varepsilon \leq 0.0001$. Table 9.6 presents the solution within the specified tolerance obtained after 22 iterations. The total number of iterations, including the Gauss-Seidel iterations necessary to estimate the optimum relaxation parameter, is 27. At convergence, the maximum absolute difference calculated using Eq. (9.59) was 0.00006.

9.3.2 Line and block SOR methods

Although point iterative methods can be used to solve equations for 2-D and 3-D problems, they are inefficient because of their extremely slow convergence. The line SOR (LSOR) and block SOR (BSOR) methods are more efficient in solving equations for these problems. The overrelaxation parameter (ω_{opt}) is estimated using the point Gauss-Seidel method until ρ_{GS} stabilizes and then using Eq. (9.77) as mentioned in Section 9.3.1.3.

9.3.2.1 Line SOR method

In the LSOR method, the reservoir is looked at as consisting of group of lines. These lines are usually aligned with the direction of highest transmissibility (Aziz and Settari, 1979) and are taken in order, one line at a time. For example, for a 2-D reservoir having the highest transmissibility along the x-direction, the lines are chosen parallel to the x-axis. Then, the lines are taken in order, one at a time for $j=1, 2, 3, \ldots, n_y$. In other words, the lines are swept in the y-direction. First, the equations for all blocks in a given line (line j) are written. In writing the equations for the current line (line j), the unknowns for the preceding line (line $j-1$) are assigned current iteration level $\nu+1$, and those for the succeeding line (line $j+1$) are assigned old iteration level ν as shown in Fig. 9.10a. In addition, the unknowns for the current line are assigned current iteration level $\nu+1$.

First, the equations for line j are written

$$w_n x_{n-1}^{(\nu+1)} + c_n x_n^{(\nu+1)} + e_n x_{n+1}^{(\nu+1)} = d_n - s_n x_{n-n_x}^{(\nu+1)} - n_n x_{n+n_x}^{(\nu)} \tag{9.83}$$

for $n=i+(j-1)\times n_x$; $i=1, 2, \ldots, n_x$.

Second, the resulting n_x equations for the current line (line j) are solved simultaneously, using Thomas' algorithm, for the intermediate values of the unknowns for the current line (line j) at current iteration level $\nu+1$ (e.g., line $j=3$ in Fig. 9.10a):

$$w_n x_{n-1}^{*(\nu+1)} + c_n x_n^{*(\nu+1)} + e_n x_{n+1}^{*(\nu+1)} = d_n - s_n x_{n-n_x}^{(\nu+1)} - n_n x_{n+n_x}^{(\nu)} \tag{9.84}$$

for $n=i+(j-1)\times n_x$; $i=1, 2, \ldots, n_x$.

Third, the intermediate solution for the current line (line j) is accelerated, using the acceleration parameter, to obtain the current iterate values of the unknowns for line j:

$$x_n^{(\nu+1)} = (1-\omega)x_n^{(\nu)} + \omega x_n^{*(\nu+1)} \tag{9.85}$$

for $n=i+(j-1)\times n_x$; $i=1, 2, \ldots, n_x$.

It should be mentioned that the improvement in the convergence of the LSOR method over the PSOR method is achieved because more unknowns are solved simultaneously at current iteration level $\nu+1$.

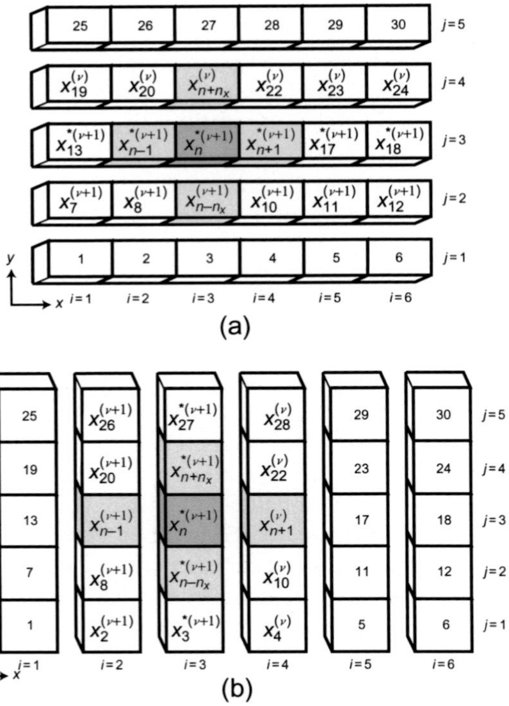

FIG. 9.10 Iteration level of the unknowns of the blocks in estimating the intermediate values in the LSOR method. (a) y-direction sweep and (b) x-direction sweep.

If lines are swept in the x-direction as shown in Fig. 9.10b, they are taken in order one at a time for $i = 1, 2, 3, \ldots, n_x$, and Eqs. (9.84) and (9.85) are replaced with Eqs. (9.86) and (9.87):

$$s_n x^{*(\nu+1)}_{n-n_x} + + c_n x^{*(\nu+1)}_n + n_n x^{*(\nu+1)}_{n+n_x} = d_n - w_n x^{(\nu+1)}_{n-1} + - e_n x^{(\nu)}_{n+1} \tag{9.86}$$

for $n = i + (j-1) \times n_x$; $j = 1, 2, \ldots, n_y$ and

$$x^{(\nu+1)}_n = (1-\omega) x^{(\nu)}_n + \omega x^{*(\nu+1)}_n \tag{9.87}$$

for $n = i + (j-1) \times n_x$; $j = 1, 2, \ldots, n_y$.

Eq. (9.86) assumes that the block ordering has not changed, that is, the blocks are ordered along the ith direction followed by the jth direction. The application of the LSOR method is presented in the next example.

Example 9.6 The following equations were obtained for the 2-D reservoir in Example 7.8 and shown in Fig. 9.11:

$$-7.8558 p_2 + 3.7567 p_5 = -14346.97 \tag{9.88}$$

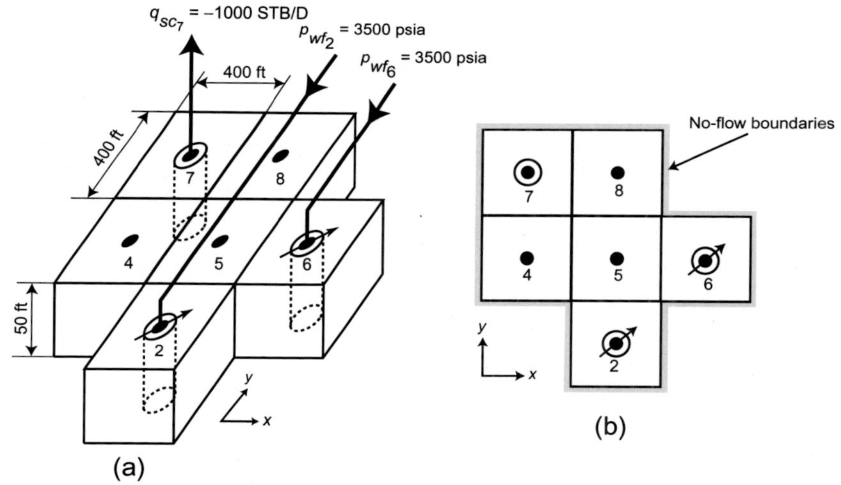

FIG. 9.11 Discretized 2-D reservoir in Example 7.8 (and Example 9.6). (a) Gridblocks and wells and (b) Boundary conditions.

$$-7.5134p_4 + 3.7567p_5 + 3.7567p_7 = 0 \tag{9.89}$$

$$3.7567p_2 + 3.7567p_4 - 15.0268p_5 + 3.7567p_6 + 3.7567p_8 = 0 \tag{9.90}$$

$$3.7567p_5 - 7.8558p_6 = -14346.97 \tag{9.91}$$

$$3.7567p_4 - 7.5134p_7 + 3.7567p_8 = 1000 \tag{9.92}$$

and

$$3.7567p_5 + 3.7567p_7 - 7.5134p_8 = 0 \tag{9.93}$$

Solve these equations using the LSOR iterative method by sweeping the lines in the y-direction.

Solution

For the y-direction sweep, Eq. (9.84) is applied to lines $j = 1, 2, \ldots n_y$. To obtain the LSOR equations for line j, the equation for each gridblock on that line is rearranged as follows. The unknowns on line j are assigned iteration level $*(\nu+1)$ and kept on the LHS of equation, the unknowns on line $j-1$ are assigned iteration level $\nu+1$ and moved to the RHS of equation, and those on line $j+1$ are assigned iteration level ν and moved to the RHS of equation. For the problem at hand, only gridblock 2 falls on line $j=1$; gridblocks 4, 5, and 6 fall on line $j=2$; and gridblocks 7 and 8 fall on line $j=3$.

The LSOR equations for line $j=1$ are obtained by considering Eq. (9.88):

$$-7.8558p_2^{*(\nu+1)} = -14346.97 - 3.7567p_5^{(\nu)} \tag{9.94}$$

After solving Eq. (9.94) for $p_2^{*(\nu+1)}$, Eq. (9.85) is applied to accelerate the solution, yielding

$$p_2^{(\nu+1)} = (1 - \omega_{opt})p_2^{(\nu)} + \omega_{opt} p_2^{*(\nu+1)} \tag{9.95}$$

The LSOR equations for line $j=2$ are obtained by considering Eqs. (9.89), (9.90), and (9.91), which state

$$-7.5134 p_4^{*(\nu+1)} + 3.7567 p_5^{*(\nu+1)} = -3.7567 p_7^{(\nu)} \tag{9.96}$$

$$3.7567 p_4^{*(\nu+1)} - 15.0268 p_5^{*(\nu+1)} + 3.7567 p_6^{*(\nu+1)} = -3.7567 p_2^{(\nu+1)} - 3.7567 p_8^{(\nu)} \tag{9.97}$$

and

$$3.7567 p_5^{*(\nu+1)} - 7.8558 p_6^{*(\nu+1)} = -14346.97 \tag{9.98}$$

After solving Eqs. (9.96), (9.97), and (9.98) for $p_4^{*(\nu+1)}$, $p_5^{*(\nu+1)}$, and $p_6^{*(\nu+1)}$ using Thomas' algorithm, Eq. (9.85) is applied to accelerate the solution, yielding

$$p_n^{(\nu+1)} = (1 - \omega_{opt})p_n^{(\nu)} + \omega_{opt} p_n^{*(\nu+1)} \tag{9.99}$$

for $n=4$, 5, 6.

The LSOR equations for line $j=3$ are obtained by considering Eqs. (9.92) and (9.93), which state

$$-7.5134 p_7^{*(\nu+1)} + 3.7567 p_8^{*(\nu+1)} = 1000 - 3.7567 p_4^{(\nu+1)} \tag{9.100}$$

and

$$3.7567 p_7^{*(\nu+1)} - 7.5134 p_8^{*(\nu+1)} = -3.7567 p_5^{(\nu+1)} \tag{9.101}$$

After solving Eqs. (9.100) and (9.101) for $p_7^{*(\nu+1)}$ and $p_8^{*(\nu+1)}$, Eq. (9.85) is applied to accelerate the solution, yielding

$$p_n^{(\nu+1)} = (1 - \omega_{opt})p_n^{(\nu)} + \omega_{opt} p_n^{*(\nu+1)} \tag{9.102}$$

for $n=7$ and 8.

Before applying the procedure given in Eqs. (9.94) through (9.102), we need to estimate the value of the optimum overrelaxation parameter ω_{opt} that must be estimated. The spectral radius for the system of Eqs. (9.88) through (9.93) is estimated using the point Gauss-Seidel iterative method, as in Example 9.5. Table 9.7 displays the results, which show that the spectral radius converges to 0.848526 within 0.22% after seven iterations. Now, we can calculate ω_{opt} from Eq. (9.77) as

$$\omega_{opt} = \frac{2}{1 + \sqrt{1 - 0.848526}} = 1.439681$$

TABLE 9.7 Determination of spectral radius for Example 9.6.

$\nu+1$	p_2	p_4	p_5	p_6	p_7	p_8	$d_{max}^{(\nu+1)}$	ρ_{GS}
	0	0	0	0	0	0	–	–
1	1826.26	0.00	456.57	2044.60	−133.10	161.73	2044.5959	–
2	2044.60	161.73	1103.17	2353.81	28.64	565.90	646.5998	0.316248
3	2353.81	565.90	1459.85	2524.38	432.80	946.33	404.1670	0.625065
4	2524.38	946.33	1735.35	2656.13	813.23	1274.29	380.4281	0.941265
5	2656.13	1274.29	1965.21	2766.05	1141.20	1553.20	327.9642	0.862092
6	2766.05	1553.20	2159.63	2859.02	1420.11	1789.87	278.9100	0.850428
7	2859.02	1789.87	2324.44	2937.84	1656.77	1990.61	236.6624	0.848526

For the first iteration ($\nu = 0$) of the LSOR method, the initial guess used is the pressure solution obtained at the seventh Gauss-Seidel iteration, which is shown in Table 9.7.

For $\nu = 0$, the LSOR equations for line $j = 1$ become

$$-7.8558 p_2^{*(1)} = -14346.97 - 3.7567 p_5^{(0)} \tag{9.103a}$$

After substitution for $p_5^{(0)} = 2324.44$, this equation becomes

$$-7.8558 p_2^{*(1)} = -14346.97 - 3.7567 \times 2324.44 \tag{9.103b}$$

The solution is $p_2^{*(1)} = 2937.8351$.

The accelerated solution is

$$\begin{aligned} p_2^{(1)} &= \left(1 - \omega_{opt}\right) p_2^{(0)} + \omega_{opt} p_2^{*(1)} \\ &= (1 - 1.439681) \times 2859.02 + 1.439681 \times 2937.835 \\ &= 2972.4896 \end{aligned} \tag{9.104}$$

For $\nu = 0$, the LSOR equations for line $j = 2$ become

$$-7.5134 p_4^{*(1)} + 3.7567 p_5^{*(1)} = -3.7567 p_7^{(0)} \tag{9.105a}$$

$$3.7567 p_4^{*(1)} - 15.0268 p_5^{*(1)} + 3.7567 p_6^{*(1)} = -3.7567 p_2^{(1)} - 3.7567 p_8^{(0)} \tag{9.106a}$$

and

$$3.7567 p_5^{*(1)} - 7.8558 p_6^{*(1)} = -14346.97 \tag{9.107a}$$

After substitution for $p_7^{(0)} = 1656.77$, $p_2^{(1)} = 2972.4896$, and $p_8^{(0)} = 1990.61$, these three equations become

$$-7.5134 p_4^{*(1)} + 3.7567 p_5^{*(1)} = -6223.9300 \tag{9.105b}$$

$$3.7567 p_4^{*(1)} - 15.0268 p_5^{*(1)} + 3.7567 p_6^{*(1)} = -18644.694 \tag{9.106b}$$

and

$$3.7567 p_5^{*(1)} - 7.8558 p_6^{*(1)} = -14346.97 \tag{9.107b}$$

The solution for these three equations is $p_4^{*(1)} = 2088.8534$, $p_5^{*(1)} = 2520.9375$, and $p_6^{*(1)} = 3031.8015$.

Next, the solution is accelerated, giving

$$\begin{aligned} p_4^{(1)} &= \left(1 - \omega_{opt}\right) p_4^{(0)} + \omega_{opt} p_4^{*(1)} \\ &= (1 - 1.439681) \times 1789.87 + 1.439681 \times 2088.8534 \\ &= 2220.3125 \end{aligned} \tag{9.108}$$

$$p_5^{(1)} = (1-\omega_{opt})p_5^{(0)} + \omega_{opt}p_5^{*(1)}$$
$$= (1-1.439681)\times 2324.44 + 1.439681\times 2520.9375 \tag{9.109}$$
$$= 2607.3329$$

and

$$p_6^{(1)} = (1-\omega_{opt})p_6^{(0)} + \omega_{opt}p_6^{*(1)}$$
$$= (1-1.439681)\times 2937.84 + 1.439681\times 3031.8015 \tag{9.110}$$
$$= 3073.1167$$

For $\nu=0$, the LSOR equations for line $j=3$ become

$$-7.5134p_7^{*(1)} + 3.7567p_8^{*(1)} = 1000 - 3.7567p_4^{(1)} \tag{9.111a}$$

and

$$3.7567p_7^{*(1)} - 7.5134p_8^{*(1)} = -3.7567p_5^{(1)} \tag{9.112a}$$

After substitution for $p_4^{(1)}=2220.3125$ and $p_5^{(1)}=2607.3329$, these two equations become

$$-7.5134p_7^{*(1)} + 3.7567p_8^{*(1)} = -7340.9740 \qquad 9.111b)$$

and

$$3.7567p_7^{*(1)} - 7.5134p_8^{*(1)} = -9794.8807 \tag{9.112b}$$

The solution for these two equations is $p_7^{*(1)}=2171.8570$ and $p_8^{*(1)}=2389.5950$.

Next, the solution is accelerated, giving

$$p_7^{(1)} = (1-\omega_{opt})p_7^{(0)} + \omega_{opt}p_7^{*(1)}$$
$$= (1-1.439681)\times 1656.77 + 1.439681\times 2171.8570 \tag{9.113}$$
$$= 2398.3313$$

and

$$p_8^{(1)} = (1-\omega_{opt})p_8^{(0)} + \omega_{opt}p_8^{*(1)}$$
$$= (1-1.439681)\times 1990.61 + 1.439681\times 2389.5950 \tag{9.114}$$
$$= 2565.0230$$

This completes the first LSOR iteration. Table 9.8 shows the results of this iteration. We perform calculations for the second iteration ($\nu=1$) and so on until convergence is reached. Table 9.8 shows the results of all LSOR iterations until the converged solution is obtained. The convergence criterion for this problem is set at a tolerance of $\varepsilon \le 0.0001$. The solution to the given system of equations is reached after 20 iterations. The results of solving

TABLE 9.8 LSOR iteration for Example 9.6.

$\nu+1$	p_2	p_4	p_5	p_6	p_7	p_8	$d_{max}^{(\nu+1)}$
	2859.02	1789.87	2324.44	2937.84	1656.77	1990.61	–
1	2972.49	2220.31	2607.33	3073.12	2398.33	2565.02	741.5620
2	3117.36	2824.53	3002.30	3261.99	2841.74	2981.51	604.2200
3	3325.58	3079.68	3231.90	3371.79	3001.86	3141.20	255.1517
4	3392.11	3155.72	3276.87	3393.29	3026.01	3150.63	76.0318
5	3393.82	3145.21	3268.16	3389.13	3001.13	3133.08	24.8878
6	3387.07	3123.16	3254.48	3382.59	2984.35	3117.09	22.0462
7	3380.62	3113.43	3245.81	3378.44	2978.22	3111.13	9.7293
8	3377.48	3110.40	3243.83	3377.49	2977.06	3110.40	3.1368
9	3377.50	3110.59	3244.08	3377.62	2977.87	3111.05	0.8096
10	3377.67	3111.38	3244.55	3377.84	2978.50	3111.60	0.7935
11	3377.92	3111.75	3244.87	3378.00	2978.74	3111.84	0.3718
12	3378.03	3111.87	3244.96	3378.04	2978.79	3111.87	0.1189
13	3378.04	3111.87	3244.95	3378.03	2978.76	3111.86	0.0253
14	3378.03	3111.84	3244.94	3378.03	2978.74	3111.83	0.0281
15	3378.02	3111.83	3244.92	3378.02	2978.73	3111.83	0.0142
16	3378.02	3111.83	3244.92	3378.02	2978.73	3111.82	0.0049
17	3378.02	3111.83	3244.92	3378.02	2978.73	3111.83	0.0007
18	3378.02	3111.83	3244.92	3378.02	2978.73	3111.83	0.0010
19	3378.02	3111.83	3244.92	3378.02	2978.73	3111.83	0.0005
20	3378.02	3111.83	3244.92	3378.02	2978.73	3111.83	0.0002

Eqs. (9.88) through (9.93) for the unknown pressures are $p_2=3378.02$ psia, $p_4=3111.83$ psia, $p_5=3244.92$ psia, $p_6=3378.02$ psia, $p_7=2978.73$ psia, and $p_8=3111.83$ psia.

9.3.2.2 Block SOR method

The block SOR (BSOR) method is a generalization of the LSOR method in that it treats any group of blocks instead of a line of blocks. The most commonly used group of blocks is a (horizontal) plane or a (vertical) slice. The following steps for obtaining the solution are similar to those for the LSOR method. Here again, planes (or slices) should be aligned with the direction of highest transmissibility and are taken in order, one plane (or slice) at a time. For example, for a 3-D reservoir having the highest transmissibility along the z-direction, slices are chosen parallel to the z-axis. Then, the slices are taken in order and one slice at a time for $i=1, 2, 3, \ldots, n_x$. In other words, the slices are swept in the x-direction.

First, the equations for slice i are written. In writing the equations for the current slice (slice i), the unknowns for the preceding slices (slice i-1) are assigned current iteration level $\nu+1$ and those for the succeeding slices (slice $i+1$) are assigned old iteration level ν. In addition, the unknowns in the current slice are assigned current iteration level $\nu+1$:

$$b_n x_{n-n_x n_y}^{(\nu+1)} + s_n x_{n-n_x}^{(\nu+1)} + c_n x_n^{(\nu+1)} + n_n x_{n+n_x}^{(\nu+1)} + a_n x_{n+n_x n_y}^{(\nu+1)} = d_n - w_n x_{n-1}^{(\nu+1)} + e_n x_{n+1}^{(\nu)} \tag{9.115}$$

for $n=i+(j-1)\times n_x+(k-1)\times n_x n_y$; $j=1, 2, \ldots, n_y$; $k=1, 2, \ldots, n_z$.

Second, the resulting $n_y n_z$ equations for the current slice (slice i) are solved simultaneously, using algorithms for sparse matrices, for the intermediate values of the unknowns of the current slice (slice i) at iteration level $*(\nu+1)$:

$$b_n x_{n-n_x n_y}^{*(\nu+1)} + s_n x_{n-n_x}^{*(\nu+1)} + c_n x_n^{*(\nu+1)} + n_n x_{n+n_x}^{*(\nu+1)} + a_n x_{n+n_x n_y}^{*(\nu+1)} = d_n - w_n x_{n-1}^{(\nu+1)} + e_n x_{n+1}^{(\nu)} \tag{9.116}$$

for $n=i+(j-1)\times n_x+(k-1)\times n_x n_y$; $j=1, 2, \ldots, n_y$; $k=1, 2, \ldots, n_z$.

Fig. 9.12a schematically shows slice SOR for slice $i=2$ and the iteration level for the unknowns of the blocks in preceding and succeeding slices.

Third, the intermediate solution for the current slice (slice i) is accelerated using acceleration parameter:

$$x_n^{(\nu+1)} = (1-\omega)x_n^{(\nu)} + \omega x_n^{*(\nu+1)} \tag{9.117}$$

for $n=i+(j-1)\times n_x+(k-1)\times n_x n_y$; $j=1, 2, \ldots, n_y$; $k=1, 2, \ldots, n_z$.

It should be mentioned that the improvement in the convergence of the BSOR method over the LSOR method is achieved because more unknowns are solved simultaneously at iteration level $\nu+1$.

If the blocks are swept in the z-direction (i.e., plane SOR) as shown in Fig. 9.12b, the planes are taken in order, one at a time for $k=1, 2, 3, \ldots, n_z$, and Eqs. (9.116) and (9.117) are replaced with Eqs. (9.118) and (9.119), which state

$$\begin{aligned} s_n x_{n-n_x}^{*(\nu+1)} + w_n x_{n-1}^{*(\nu+1)} + c_n x_n^{*(\nu+1)} + e_n x_{n+1}^{*(\nu+1)} + n_n x_{n+n_x}^{*(\nu+1)} \\ = d_n - b_n x_{n-n_x n_y}^{(\nu+1)} - a_n x_{n+n_x n_y}^{(\nu)} \end{aligned} \tag{9.118}$$

for $n=i+(j-1)\times n_x+(k-1)\times n_x n_y$; $i=1, 2, \ldots, n_x$; $j=1, 2, \ldots, n_y$; and

$$x_n^{(\nu+1)} = (1-\omega)x_n^{(\nu)} + \omega x_n^{*(\nu+1)} \tag{9.119}$$

for $n=i+(j-1)\times n_x+(k-1)\times n_x n_y$; $i=1, 2, \ldots, n_x$; $j=1, 2, \ldots, n_y$.

Eqs. (9.118) and (9.119) assume that the block ordering has not changed, that is, blocks are ordered along the ith direction, followed by the jth direction, and finally along the kth direction.

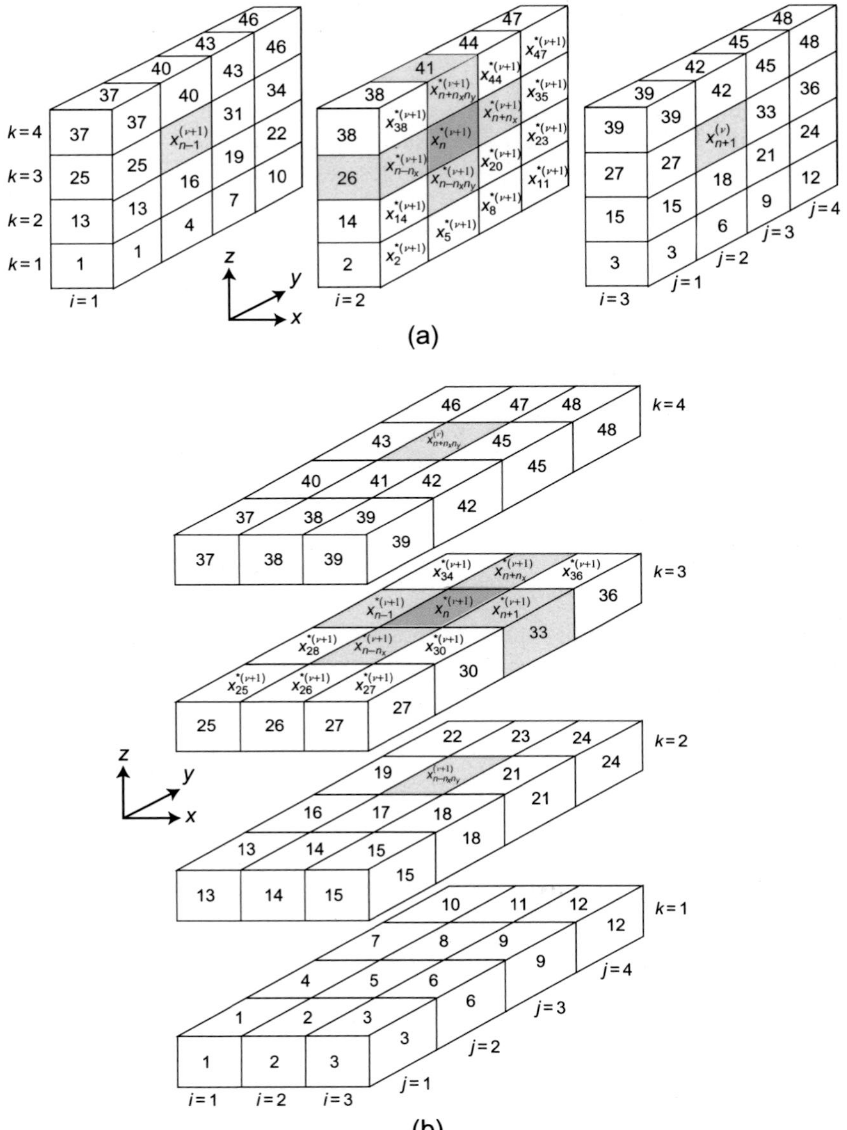

FIG. 9.12 Slice and plane sweeps in the BSOR method. (a) Slice sweep in BSOR and (b) Plane sweep in BSOR.

9.3.3 Alternating-direction implicit procedure

The alternating direction implicit procedure(ADIP) aims at replacing a 2-D or 3-D problem with two or three sets of 1-D problems in the x-, y-, and z-directions that are solved consecutively. This method was introduced by Peaceman and

Rachford (1955). In this section, we apply the method to a slightly compressible fluid flow problem in a 2-D parallelepiped reservoir ($n_x \times n_y$). The equation for block n in a 2-D problem is obtained from Eq. (9.45) as

$$s_n x_{n-n_x} + w_n x_{n-1} + c_n x_n + e_n x_{n+1} + n_n x_{n+n_x} = d_n \tag{9.120a}$$

for $n = i + (j-1) \times n_x$; $i = 1, 2, \ldots, n_x$; and $j = 1, 2, \ldots, n_y$, where s_n, w_n, e_n, and n_n are given by Eqs. (9.46b) through (9.46e), yielding

$$c_n = -\left[s_n + w_n + e_n + n_n + \frac{V_{b_n}\phi_n^{\circ}(c + c_\phi)}{\alpha_c B^{\circ}\Delta t}\right] \tag{9.120b}$$

$$\begin{aligned} D_n = {} & [(s\gamma)_n(Z_{n-n_x} - Z_n) + (w\gamma)_n(Z_{n-1} - Z_n) \\ & + (e\gamma)_n(Z_{n+1} - Z_n) + (n\gamma)_n(Z_{n+n_x} - Z_n)] - q_{sc_n} \end{aligned} \tag{9.120c}$$

and

$$d_n = D_n - \frac{V_{b_n}\phi_n^{\circ}(c + c_\phi)}{\alpha_c B^{\circ}\Delta t} x_n^n \tag{9.120d}$$

and the unknown x stands for pressure. The solution of this equation is obtained by finding the solutions of two sets of 1-D problems, one in the x-direction and the other in the y-direction as outlined in the succeeding text.

9.3.3.1 *Set of 1-D problems in the x-direction*

For each line $j = 1, 2, \ldots, n_y$ solve

$$w_n x_{n-1}^{*} + c_n^{*} x_n^{*} + e_n x_{n+1}^{*} = d_n^{*} \tag{9.121a}$$

for $n = i + (j-1) \times n_x$, $i = 1, 2, \ldots, n_x$, where

$$c_n^{*} = -\left[w_n + e_n + \frac{V_{b_n}\phi_n^{\circ}(c + c_\phi)}{\alpha_c B^{\circ}(\Delta t/2)}\right] \tag{9.121b}$$

and

$$d_n^{*} = D_n - \frac{V_{b_n}\phi_n^{\circ}(c + c_\phi)}{\alpha_c B^{\circ}(\Delta t/2)} x_n^n - \left[s_n\left(x_{n-n_x}^n - x_n^n\right) + n_n\left(x_{n+n_x}^n - x_n^n\right)\right] \tag{9.121c}$$

Each set of equations represented by Eq. (9.121a) consists of n_x linear equations that can be solved simultaneously using Thomas' algorithm or iteratively using the PSOR method.

9.3.3.2 *Set of 1-D problems in the y-direction*

For each line $i = 1, 2, \ldots, n_x$ solve

$$s_n x_{n-n_x}^{n+1} + c_n^{**} x_n^{n+1} + n_n x_{n+n_x}^{n+1} = d_n^{**} \tag{9.122a}$$

for $n=i+(j-1)\times n_x$, $j=1, 2, \ldots, n_y$, where

$$c_n^{**}=-\left[s_n+n_n+\frac{V_{b_n}\phi_n^{\circ}(c+c_\phi)}{\alpha_c B^{\circ}(\Delta t/2)}\right] \tag{9.122b}$$

and

$$d_n^{**}=D_n-\frac{V_{b_n}\phi_n^{\circ}(c+c_\phi)}{\alpha_c B^{\circ}(\Delta t/2)}x_n^*-\left[w_n\left(x_{n-1}^*-x_n^*\right)+e_n\left(x_{n+1}^*-x_n^*\right)\right] \tag{9.122c}$$

Each set of equations represented by Eq. (9.122a) consists of n_y linear equations that can be solved simultaneously using Thomas' algorithm or iteratively using the PSOR method.

While the ADIP just presented is a noniterative version of the method, other literature presents an iterative version that has better convergence (Ertekin et al., 2001). For 2-D problems, the ADIP is unconditionally stable. However, a direct extension of the ADIP presented here to 3-D problems is conditionally stable. Aziz and Settari (1979) reviewed unconditionally stable extensions of ADIP for 3-D problems.

9.3.4 Advanced iterative methods

As mentioned in the introduction, we restricted our discussion in this chapter to basic solution methods. The objective in this chapter was to introduce the mechanics of the basic methods of solution, although many of these iterative methods are not used in today's simulators. However, the algorithms for advanced iterative methods of solving systems of linear equations, such as conjugate gradient methods, the block iterative method, the nested factorization method, and Orthomin are beyond the scope of this book and can be found elsewhere (Ertekin et al., 2001; Behie and Vinsome, 1982; Appleyard and Cheshire, 1983; Vinsome, 1976). Such methods are very efficient for solving systems of linear equations for multiphase flow, compositional, and thermal simulation.

9.4 Summary

Systems of linear equations can be solved using direct solvers or iterative solvers. The methods presented in this chapter are basic methods that aim at introducing the mechanics of solving sets of linear equations resulting from reservoir simulation. Direct solvers include methods that use variations of **LU** factorization of the coefficient matrix **[A]**. These include Thomas' algorithm and Tang's algorithm for 1-D flow problems and the g-band matrix solver for 2-D and 3-D flow problems. Iterative solvers include point Jacobi, point Gauss-Seidel, and PSOR methods mainly for 1-D flow problems, the LSOR and BSOR methods for 2-D and 3-D flow problems, and the ADIP method for 2-D flow problems. The important issue in this chapter is how to relate the coefficients of matrix **[A]** to the linearized flow equation. The unknowns in the linearized equation for a general block n are placed on the LHS of equation, factorized, and ordered in ascending order; that is, they are ordered as shown in Fig. 9.2. Subsequently, the coefficients b_n, s_n, w_n, c_n, e_n, n_n, and a_n correspond to locations 1, 2, 3, 4, 5, 6, and 7 in Fig. 9.2. The RHS of equation corresponds to d_n.

9.5 Exercises

9.1 Define a direct solution method. Name any two methods under this category.

9.2 Define an iterative solution method. Name any two methods under this category.

9.3 What is the difference between the iteration level and the time level? When do you use each?

9.4 The following equations were obtained for the 1-D reservoir problem described in Example 7.2 and Fig. 7.2:

$$-56.8008p_2 + 28.4004p_3 = -113601.6$$

$$28.4004p_2 - 56.8008p_3 + 28.4004p_4 = 0$$

and

$$28.4004p_3 - 28.4004p_4 = 600$$

Solve these three equations for the unknowns p_2, p_3, and p_4 using the following:

a. Thomas' algorithm
b. Jacobi iterative method
c. Gauss-Seidel iterative method
d. PSOR method

For iterative methods, start with an initial guess of zero for all the unknowns and use a convergence tolerance of 1 psi (for hand calculations).

9.5 The following equations were obtained for the 1-D reservoir problem described in Example 7.5 and Fig. 7.5:

$$-85.2012p_1 + 28.4004p_2 = -227203.2$$

$$28.4004p_1 - 56.8008p_2 + 28.4004p_3 = 0$$

$$28.4004p_2 - 56.8008p_3 + 28.4004p_4 = 0$$

and

$$28.4004p_3 - 28.4004p_4 = 1199.366$$

Solve these four equations for the unknowns p_1, p_2, p_3, and p_4 using the following:

a. Thomas' algorithm
b. Jacobi iterative method

c. Gauss-Seidel iterative method
d. PSOR method

For iterative methods, start with an initial guess of zero for all the unknowns and use a convergence tolerance of 5 psi (for hand calculations).

9.6 The following equations were obtained for the 2-D reservoir problem described in Example 7.7 and Fig. 7.7:

$$-5.0922p_1 + 1.0350p_2 + 1.3524p_3 = -11319.20$$

$$1.0350p_1 - 6.4547p_2 + 1.3524p_4 = -13435.554$$

$$1.3524p_1 - 2.3874p_3 + 1.0350p_4 = 600$$

and

$$1.3524p_2 + 1.0350p_3 - 2.3874p_4 = 308.675$$

Solve these four equations for the unknowns p_1, p_2, p_3, and p_4 using the following:
a. Gaussian elimination
b. Jacobi iterative method
c. Gauss-Seidel iterative method
d. PSOR method
e. LSOR method

For iterative methods, start with an initial guess of zero for all the unknowns and use a convergence tolerance of 1 psi (for hand calculations).

9.7 The following equations were obtained for the 2-D reservoir problem described in Example 7.8 and Fig. 7.8:

$$-7.8558p_2 + 3.7567p_5 = -14346.97$$

$$-7.5134p_4 + 3.7567p_5 + 3.7567p_7 = 0$$

$$3.7567p_2 + 3.7567p_4 - 15.0268p_5 + 3.7567p_6 + 3.7567p_8 = 0$$

$$3.7567p_5 - 7.8558p_6 = -14346.97$$

$$3.7567p_4 - 7.5134p_7 + 3.7567p_8 = 1000$$

and

$$3.7567p_5 + 3.7567p_7 - 7.5134p_8 = 0$$

Solve these six equations for the unknowns p_2, p_4, p_5, p_6, p_7, and p_8 using the following:
a. Jacobi iterative method
b. Gauss-Seidel iterative method
c. PSOR iterative method
d. LSOR iterative method by sweeping lines in the x-direction

For iterative methods, start with an initial guess of zero for all the unknowns and use a convergence tolerance of 10 psi (for hand calculations).

9.8 Consider the 1-D flow problem presented in Example 7.11 and Fig. 7.12. Solve this problem for the first two time steps using the following:
a. Thomas' algorithm
b. Jacobi iterative method
c. Gauss-Seidel iterative method
d. PSOR iterative method
For iterative methods, take the pressures at the old time level as the initial guess and use a convergence tolerance of 1 psi (for hand calculations).

9.9 Consider the 1-D flow problem presented in Example 7.10 and Fig. 7.11. Solve this problem for the first two time steps using the following:
a. Thomas' algorithm
b. Jacobi iterative method
c. Gauss-Seidel iterative method
d. PSOR iterative method
For iterative methods, take the pressures at the old time level as the initial guess and use a convergence tolerance of 0.1 psi (for hand calculations).

9.10 Consider the 1-D single-well simulation problem presented in Exercise 7.16 and Fig. 7.18. Solve this problem using the following:
a. Thomas' algorithm
b. Jacobi iterative method
c. Gauss-Seidel iterative method
d. PSOR iterative method
For iterative methods, take the pressures at the old time level as the initial guess and use a convergence tolerance of 1 psi (for hand calculations).

9.11 Consider the 2-D flow problem presented in Exercise 7.10 and Fig. 7.15a. Solve this problem for the first time step using
a. Gauss-Seidel iterative method
b. PSOR iterative method
c. LSOR iterative method by sweeping lines in the y-direction
d. LSOR iterative method by sweeping lines in the x-direction
e. ADIP
For iterative methods, take the pressures at the old time level as the initial guess and use a convergence tolerance of 1 psi (for hand calculations).

Chapter 10

The engineering approach versus the mathematical approach in developing reservoir simulators

Chapter outline

10.1 Introduction

Traditionally, the steps involved in the development of a simulator include the following: (1) derivation of the partial differential equations (PDEs) describing the recovery process through formulation, (2) discretization of the PDEs in space and time to obtain the nonlinear algebraic equations, (3) linearization of resulting algebraic equations, (4) solving the linearized algebraic equations numerically, and (5) validation of the simulator. The mathematical approach refers to the first three steps. The engineering approach independently derives the same finite-difference equations, as special cases of approximating the integral equation in the engineering approach, without going through the rigor of PDEs and discretization. The two approaches, however, have a few differences in treating nonlinearities and boundary conditions. The objective in this chapter is to highlight the similarities and differences between the two approaches.

Petroleum Reservoir Simulation. https://doi.org/10.1016/B978-0-12-819150-7.00010-4

10.2 Derivation of fluid flow equations in discretized form

The fluid flow equations in discretized form (nonlinear algebraic equations) can be obtained by either the traditional mathematical approach or the engineering approach. Both of these approaches make use of the same basic principles and both approaches discretize the reservoir into gridblocks (or gridpoints). Both approaches yield the same discretized flow equations for modeling any reservoir fluid system (multiphase, multicomponent, thermal, and heterogeneous reservoir) using any coordinate system (Cartesian, cylindrical, and spherical) in one-dimensional (1-D), two-dimensional (2-D), or three-dimensional (3-D) reservoirs (Abou-Kassem, 2006). Therefore, the presentation here will be for modeling the flow of single-phase, compressible fluid in horizontal, 1-D reservoir using irregular block size distribution in rectangular coordinates. We will take advantage of this simple case to demonstrate the capacity of the engineering approach to give independent verification for the discretization methods used in the mathematical approach.

10.2.1 Basic principles

The basic principles include mass conservation, equation of state, and constitutive equation. The principle of mass conservation states that the total mass of fluid entering and leaving a volume element of the reservoir must equal the net increase in the mass of the fluid in the reservoir element:

$$m_i - m_o + m_s = m_a \tag{10.1}$$

An equation of state describes the density of fluid as a function of pressure and temperature:

$$B = \rho_{sc}/\rho \tag{10.2}$$

A constitutive equation describes the rate of fluid movement into (or out of) the reservoir element. In reservoir simulation, Darcy's law is used to relate fluid flow rate to potential gradient. The differential form of Darcy's law for a horizontal reservoir is

$$u_x = q_x/A_x = -\beta_c \frac{k_x}{\mu} \frac{\partial p}{\partial x} \tag{10.3}$$

10.2.2 Reservoir discretization

Reservoir discretization means that the reservoir is described by a set of gridblocks (or gridpoints) whose properties, dimensions, boundaries, and locations in the reservoir are well defined. Fig. 10.1 shows reservoir discretization in the x-direction for both block-centered and point-distributed grids in rectangular

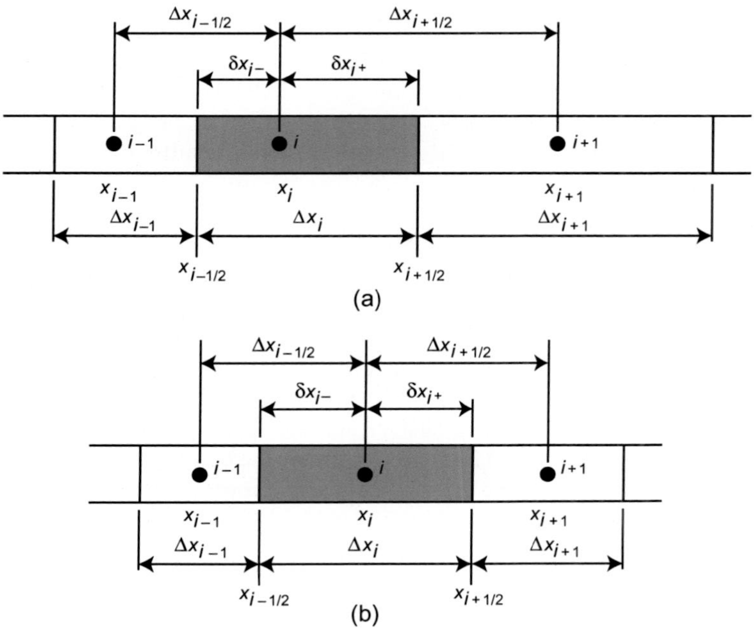

FIG. 10.1 Reservoir discretization. (a) Point-distributed grid and (b) block-centered grid.

coordinates as one focuses on gridblock i or gridpoint i. The figure shows how the blocks are related to each other [block i and its neighboring blocks (blocks $i-1$ and $i+1$)], block dimensions (Δx_i, Δx_{i-1}, Δx_{i+1}), block boundaries ($x_{i-1/2}$, $x_{i+1/2}$), distances between the point that represents the block and block boundaries (δx_{i-}, δx_{i+}), and distances between the gridpoints or points representing the blocks ($\Delta x_{i-1/2}$, $\Delta x_{i+1/2}$). In addition, each gridblock or gridpoint is assigned elevation and rock properties such as porosity and permeability.

In block-centered grid system, the grid is constructed by choosing n_x gridblocks that span the entire reservoir length in the x-direction. The gridblocks are assigned predetermined dimensions (Δx_i, $i=1, 2, 3\ldots\ n_x$) that are not necessarily equal. Then, the point that represents a gridblock is consequently located at the center of the gridblock. In point-distributed grid system, the grid is constructed by choosing n_x gridpoints that span the entire reservoir length in the x-direction. In other words, the first gridpoint is placed at one reservoir boundary and the last gridpoint is placed at the other reservoir boundary. The distances between gridpoints are assigned predetermined values ($\Delta x_{i+1/2}$, $i=1, 2, 3\ldots\ n_x-1$) that are not necessarily equal. Each gridpoint represents a gridblock whose boundaries are placed halfway between the gridpoint and its neighboring gridpoints.

10.2.3 The mathematical approach

In the mathematical approach, the algebraic flow equations are derived in three consecutive steps: (1) derivation of the PDE describing fluid flow in reservoir using basic principles, (2) discretization of reservoir into gridblocks or gridpoints, and (3) discretization of the resulting PDE in space and time.

10.2.3.1 Derivation of PDE

Fig. 10.2 shows a finite control volume with a cross-sectional area A_x perpendicular to the direction of flow, length Δx in the direction of flow, and volume $V_b = A_x \Delta x$. Point x represents control volume and falls at its center for block-centered grid. The fluid enters the control volume across its surface at $x - \Delta x/2$ and leaves across its surface at $x + \Delta x/2$ at mass rates of $w_x|_{x-\Delta x/2}$ and $w_x|_{x+\Delta x/2}$, respectively. The fluid also enters the control volume through a well at a mass rate of q_m. The mass of fluid in the control volume per unit volume of rock is m_v. Therefore, the material balance equation written over a time step Δt as expressed by Eq. (10.1) becomes

$$m_i|_{x-\Delta x/2} - m_o|_{x+\Delta x/2} + m_s = m_a \tag{10.4}$$

or

$$w_x|_{x-\Delta x/2}\Delta t - w_x|_{x+\Delta x/2}\Delta t + q_m \Delta t = m_a \tag{10.5}$$

where mass flow rate (w_x) and mass flux ($\dot{m}_x$) are related through

$$w_x = \dot{m}_x A_x \tag{10.6}$$

In addition, mass accumulation is defined as

$$m_a = \Delta_t(V_b m_v) = V_b\left(m_v|_{t+\Delta t} - m_v|_t\right) = V_b\left(m_v^{n+1} - m_v^n\right) \tag{10.7}$$

Substitution of Eqs. (10.6) and (10.7) into Eq. (10.5) yields

$$\left(\dot{m}_x A_x\right)|_{x-\Delta x/2}\Delta t - (\dot{m}_x A_x)|_{x+\Delta x/2}\Delta t + q_m \Delta t = V_b\left(m_v|_{t+\Delta t} - m_v|_t\right) \tag{10.8}$$

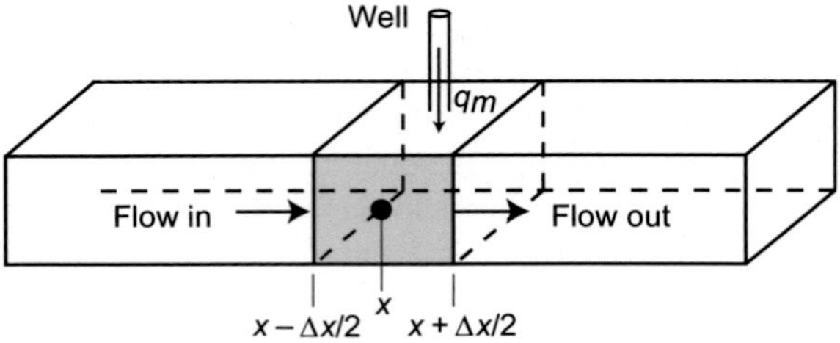

FIG. 10.2 Control volume in 1-D traditionally used for writing material balance.

Dividing Eq. (10.8) by $V_b \Delta t$, observing that $V_b = A_x \Delta x$, and rearranging results in

$$-\left[\left(\dot{m}_x|_{x+\Delta x/2} - \dot{m}_x|_{x-\Delta x/2}\right)/\Delta x\right] + \frac{q_m}{V_b} = \left[\left(m_v|_{t+\Delta t} - m_v|_t\right)/\Delta t\right] \tag{10.9}$$

The limits of the terms in brackets in Eq. (10.9) as Δx and Δt approach zero (i.e., as $\Delta x \rightarrow 0$ and $\Delta t \rightarrow 0$) become first-order partial derivatives and the resulting equation becomes

$$-\frac{\partial \dot{m}_x}{\partial x} + \frac{q_m}{V_b} = \frac{\partial m_v}{\partial t} \tag{10.10}$$

Mass flux ($\dot{m}_x$) can be stated in terms of fluid density (ρ) and volumetric velocity (u_x) as

$$\dot{m}_x = \alpha_c \rho u_x \tag{10.11}$$

m_v can be expressed in terms of fluid density and porosity (ϕ) as

$$m_v = \phi \rho \tag{10.12}$$

and q_m can be expressed in terms of well volumetric rate (q) and fluid density as

$$q_m = \alpha_c \rho q \tag{10.13}$$

Substituting Eqs. (10.11) through (10.13) into Eq. (10.10) results in the continuity equation:

$$-\frac{\partial(\rho u_x)}{\partial x} + \frac{\rho q}{V_b} = \frac{1}{\alpha_c}\frac{\partial(\rho \phi)}{\partial t} \tag{10.14}$$

The flow equation can be obtained by combining the continuity equation (Eq. 10.14), the equation of state (Eq. 10.2), and Darcy's law (Eq. 10.3), and noting that $q/B = q_{sc}$. The resulting flow equation for single-phase flow is

$$\frac{\partial}{\partial x}\left(\beta_c \frac{k_x}{\mu B}\frac{\partial p}{\partial x}\right) + \frac{q_{sc}}{V_b} = \frac{1}{\alpha_c}\frac{\partial}{\partial t}\left(\frac{\phi}{B}\right) \tag{10.15}$$

The above equation is the PDE that describes single-phase flow in 1-D rectangular coordinates.

10.2.3.2 Discretization of PDE in space and time

First, the reservoir is discretized as mentioned earlier. Second, Eq. (10.15) is rewritten in another form, to take care of variations of cross-sectional area through multiplying by $V_b = A_x \Delta x$, as

$$\frac{\partial}{\partial x}\left(\beta_c \frac{k_x A_x}{\mu B}\frac{\partial p}{\partial x}\right)\Delta x + q_{sc} = \frac{V_b}{\alpha_c}\frac{\partial}{\partial t}\left(\frac{\phi}{B}\right) \tag{10.16}$$

Eq. (10.16) is then written for gridblock i:

$$\frac{\partial}{\partial x}\left(\beta_c \frac{k_x A_x}{\mu B}\frac{\partial p}{\partial x}\right)_i \Delta x_i + q_{sc_i} = \frac{V_{b_i}}{\alpha_c}\frac{\partial}{\partial t}\left(\frac{\phi}{B}\right)_i \tag{10.17}$$

Space discretization

The second-order derivative w.r.t. x at Point i appearing on the LHS of Eq. (10.17) is approximated using second-order central differencing. The resulting approximation can be written as

$$\frac{\partial}{\partial x}\left(\beta_c \frac{k_x A_x}{\mu B}\frac{\partial p}{\partial x}\right)_i \Delta x_i \cong T_{x_{i-1/2}}(p_{i-1} - p_i) + T_{x_{i+1/2}}(p_{i+1} - p_i) \tag{10.18}$$

with transmissibility $T_{x_{i\mp1/2}}$ being defined as

$$T_{x_{i\mp1/2}} = \left(\beta_c \frac{k_x A_x}{\mu B \Delta x}\right)_{i\mp1/2} \tag{10.19}$$

The process of the approximation leading to Eq. (10.18) can be looked at as follows. Using the definition of central-difference approximation to the first-order derivative evaluated at Point i (see Fig. 10.1), one can write

$$\frac{\partial}{\partial x}\left(\beta_c \frac{k_x A_x}{\mu B}\frac{\partial p}{\partial x}\right)_i \cong \left[\left(\beta_c \frac{k_x A_x}{\mu B}\frac{\partial p}{\partial x}\right)_{i+1/2} - \left(\beta_c \frac{k_x A_x}{\mu B}\frac{\partial p}{\partial x}\right)_{i-1/2}\right] / \Delta x_i \tag{10.20}$$

Use of central differencing again to approximate $\left(\frac{\partial p}{\partial x}\right)_{i\mp1/2}$ yields

$$\left(\frac{\partial p}{\partial x}\right)_{i+1/2} \cong (p_{i+1} - p_i)/(x_{i+1} - x_i) = (p_{i+1} - p_i)/\Delta x_{i+1/2} \tag{10.21}$$

and

$$\left(\frac{\partial p}{\partial x}\right)_{i-1/2} \cong (p_i - p_{i-1})/(x_i - x_{i-1}) = (p_i - p_{i-1})/\Delta x_{i-1/2} \tag{10.22}$$

Substitution of Eqs. (10.21) and (10.22) into Eq. (10.20) and rearranging results in

$$\frac{\partial}{\partial x}\left(\beta_c \frac{k_x A_x}{\mu B}\frac{\partial p}{\partial x}\right)_i \Delta x_i \cong \left[\left(\beta_c \frac{k_x A_x}{\mu B \Delta x}\right)_{i+1/2}(p_{i+1} - p_i) - \left(\beta_c \frac{k_x A_x}{\mu B \Delta x}\right)_{i-1/2}(p_i - p_{i-1})\right] \tag{10.23}$$

or

$$\frac{\partial}{\partial x}\left(\beta_c \frac{k_x A_x}{\mu B}\frac{\partial p}{\partial x}\right)_i \Delta x_i \cong \left[\left(\beta_c \frac{k_x A_x}{\mu B \Delta x}\right)_{i+1/2}(p_{i+1} - p_i) + \left(\beta_c \frac{k_x A_x}{\mu B \Delta x}\right)_{i-1/2}(p_{i-1} - p_i)\right] \tag{10.24}$$

Eq. (10.18) results from the substitution of $T_{x_{i\mp 1/2}}$ given by Eq. (10.19) into Eq. (10.24).

Substitution of Eq. (10.18) into the PDE given by Eq. (10.17) yields an equation that is discrete in space but continuous in time:

$$T_{x_{i-1/2}}(p_{i-1}-p_i)+T_{x_{i+1/2}}(p_{i+1}-p_i)+q_{sc_i}\cong\frac{V_{b_i}}{\alpha_c}\frac{\partial}{\partial t}\left(\frac{\phi}{B}\right)_i \tag{10.25}$$

Time discretization

The discretization of Eq. (10.25) in time is accomplished by approximating the first-order derivative appearing on the RHS of the equation. We will consider here the forward-difference, backward-difference, and central-difference approximations. All three approximations can be written as

$$\frac{\partial}{\partial t}\left(\frac{\phi}{B}\right)_i\cong\frac{1}{\Delta t}\left[\left(\frac{\phi}{B}\right)_i^{n+1}-\left(\frac{\phi}{B}\right)_i^{n}\right] \tag{10.26}$$

Forward-difference discretization In the forward-difference discretization, one writes Eq. (10.25) at time level n (old time level t^n):

$$\left[T_{x_{i-1/2}}(p_{i-1}-p_i)+T_{x_{i+1/2}}(p_{i+1}-p_i)+q_{sc_i}\right]^n\cong\frac{V_{b_i}}{\alpha_c}\left[\frac{\partial}{\partial t}\left(\frac{\phi}{B}\right)_i\right]^n \tag{10.27}$$

In this case, it can be looked at Eq. (10.26) as forward difference of the first-order derivative w.r.t. time at time level n. The discretized flow equation is called a forward-difference equation:

$$T^n_{x_{i-1/2}}\left(p^n_{i-1}-p^n_i\right)+T^n_{x_{i+1/2}}\left(p^n_{i+1}-p^n_i\right)+q^n_{sc_i}\cong\frac{V_{b_i}}{\alpha_c\,\Delta t}\left[\left(\frac{\phi}{B}\right)_i^{n+1}-\left(\frac{\phi}{B}\right)_i^{n}\right] \tag{10.28}$$

The RHS of Eq. (10.28) can be expressed in terms of the pressure of gridblock i such that material balance is preserved. The resulting equation is

$$T^n_{x_{i-1/2}}\left(p^n_{i-1}-p^n_i\right)+T^n_{x_{i+1/2}}\left(p^n_{i+1}-p^n_i\right)+q^n_{sc_i}\cong\frac{V_{b_i}}{\alpha_c\,\Delta t}\left(\frac{\phi}{B}\right)_i^{\prime}\left[p_i^{n+1}-p_i^n\right] \tag{10.29}$$

where the derivative $\left(\frac{\phi}{B}\right)_i^{\prime}$ is defined as the chord; that is,

$$\left(\frac{\phi}{B}\right)_i^{\prime}=\left[\left(\frac{\phi}{B}\right)_i^{n+1}-\left(\frac{\phi}{B}\right)_i^{n}\right]/\left[p_i^{n+1}-p_i^n\right] \tag{10.30}$$

Backward-difference discretization In the backward-difference discretization, one writes Eq. (10.25) at time level $n+1$ (current time level t^{n+1}):

$$\left[T_{x_{i-1/2}}(p_{i-1}-p_i)+T_{x_{i+1/2}}(p_{i+1}-p_i)+q_{sc_i}\right]^{n+1}\cong\frac{V_{b_i}}{\alpha_c}\left[\frac{\partial}{\partial t}\left(\frac{\phi}{B}\right)_i\right]^{n+1} \tag{10.31}$$

In this case, it can be looked at Eq. (10.26) as backward-difference of the first-order derivative w.r.t. time at time level $n+1$. The discretized flow equation is called a backward-difference equation:

$$T^{n+1}_{x_{i-1/2}}\left(p^{n+1}_{i-1}-p^{n+1}_i\right)+T^{n+1}_{x_{i+1/2}}\left(p^{n+1}_{i+1}-p^{n+1}_i\right)+q^{n+1}_{sc_i}\cong\frac{V_{b_i}}{\alpha_c\Delta t}\left[\left(\frac{\phi}{B}\right)^{n+1}_i-\left(\frac{\phi}{B}\right)^{n}_i\right] \tag{10.32}$$

The equation that corresponds to Eq. (10.29) is

$$T^{n+1}_{x_{i-1/2}}\left(p^{n+1}_{i-1}-p^{n+1}_i\right)+T^{n+1}_{x_{i+1/2}}\left(p^{n+1}_{i+1}-p^{n+1}_i\right)+q^{n+1}_{sc_i}\cong\frac{V_{b_i}}{\alpha_c\Delta t}\left(\frac{\phi}{B}\right)'_i\left[p^{n+1}_i-p^n_i\right] \tag{10.33}$$

Central-difference discretization In the central-difference discretization, one writes Eq. (10.25) at time level $n+1/2$ (time level $t^{n+1/2}$):

$$\left[T_{x_{i-1/2}}(p_{i-1}-p_i)+T_{x_{i+1/2}}(p_{i+1}-p_i)+q_{sc_i}\right]^{n+1/2}\cong\frac{V_{b_i}}{\alpha_c}\left[\frac{\partial}{\partial t}\left(\frac{\phi}{B}\right)_i\right]^{n+1/2} \tag{10.34}$$

In this case, it can be looked at Eq. (10.26) as central-difference of the first-order derivative w.r.t. time at time level $n+1/2$. In addition, the flow terms at time level $n+1/2$ are approximated by the average values at time level $n+1$ and time level n. The discretized flow equation in this case is the Crank-Nicholson approximation:

$$\begin{aligned}&(1/2)\left[T^n_{x_{i-1/2}}\left(p^n_{i-1}-p^n_i\right)+T^n_{x_{i+1/2}}\left(p^n_{i+1}-p^n_i\right)\right]\\&+(1/2)\left[T^{n+1}_{x_{i-1/2}}\left(p^{n+1}_{i-1}-p^{n+1}_i\right)+T^{n+1}_{x_{i+1/2}}\left(p^{n+1}_{i+1}-p^{n+1}_i\right)\right]\\&+(1/2)\left[q^n_{sc_i}+q^{n+1}_{sc_i}\right]\cong\frac{V_{b_i}}{\alpha_c\Delta t}\left[\left(\tfrac{\phi}{B}\right)^{n+1}_i-\left(\tfrac{\phi}{B}\right)^{n}_i\right]\end{aligned} \tag{10.35}$$

The equation that corresponds to Eq. (10.29) is

$$\begin{aligned}&(1/2)\left[T^n_{x_{i-1/2}}\left(p^n_{i-1}-p^n_i\right)+T^n_{x_{i+1/2}}\left(p^n_{i+1}-p^n_i\right)\right]\\&+(1/2)\left[T^{n+1}_{x_{i-1/2}}\left(p^{n+1}_{i-1}-p^{n+1}_i\right)+T^{n+1}_{x_{i+1/2}}\left(p^{n+1}_{i+1}-p^{n+1}_i\right)\right]\\&+(1/2)\left[q^n_{sc_i}+q^{n+1}_{sc_i}\right]\cong\frac{V_{b_i}}{\alpha_c\Delta t}\left(\frac{\phi}{B}\right)'_i\left[p^{n+1}_i-p^n_i\right]\end{aligned} \tag{10.36}$$

10.2.3.3 Observations on the derivation of the mathematical approach

1. For heterogeneous block permeability distribution and irregular grid blocks (neither constant nor equal Δx), note that for a discretized reservoir, blocks have defined dimensions and permeabilities; therefore, interblock geometric factor $\left[\left(\beta_c \frac{k_x A_x}{\Delta x}\right)\Big|_{x_{i\mp1/2}}\right]$ is constant, independent of space and time. In addition, the pressure-dependent term $(\mu B)|_{x_{i\mp1/2}}$ of transmissibility uses some average viscosity and formation volume factor (FVF) of the fluid contained in block i and neighboring blocks $i\mp1$ or some weight (upstream weighting or average weighting) at any instant of time t. In other words, the term $(\mu B)|_{x_{i\mp1/2}}$ is not a function of space, but it is a function of time as block pressures change with time. Similarly, for multiphase flow, the relative permeability of phase $p=o, w, g$ between block i and neighboring blocks $i\mp1$ at any instant of time t ($k_{rp}|_{x_{i\mp1/2}}$) uses upstream value or two-point upstream value of block i and neighboring blocks $i\mp1$ that are already fixed in space. In other words, the term $k_{rp}|_{x_{i\mp1/2}}$ is not a function of space but it is a function of time as block saturations change with time. Hence, transmissibility $T_{x_{i\mp1/2}}$ between block i and its neighboring blocks $i\mp1$ is a function of time only; it does not depend on space at any instant of time.
2. A close inspection of the flow terms on the LHS of the discretized flow equation expressed by Eq. (10.25) reveals that these terms are nothing but Darcy's law describing volumetric flow rates at standard conditions ($q_{sc_{i\mp1/2}}$) between gridblock i and its neighboring gridblocks $i\mp1$ in the x-direction, that is,

$$T_{x_{i\mp1/2}}(p_{i\mp1}-p_i)=\left(\beta_c\frac{k_xA_x}{\mu B\Delta x}\right)_{i\mp1/2}(p_{i\mp1}-p_i)=q_{sc_{i\mp1/2}} \tag{10.37}$$

3. Interblock flow terms and production/injection rates that appear on the LHS of the discretized flow equations (Eqs. 10.29, 10.33, and 10.36) are dated at time level n for explicit flow equation, time level $n+1$ for implicit flow equation or time level $n+1/2$ for the Crank-Nicolson flow equation. In all cases, the RHS of the flow equations represent accumulation over a time step Δt. In other words, the accumulation term does not take into consideration the variation of interblock flow terms and production/injection rate (source/sink term) with time within a time step.

10.2.4 The engineering approach

In the engineering approach, the derivation of the algebraic flow equation is straightforward. It is accomplished in three consecutive steps: (1) discretization of reservoir into gridblocks (or gridpoints) as shown earlier to remove the effect of space variable as mentioned in Observation 1 earlier, (2) derivation of the algebraic flow equation for gridblock i (or gridpoint i) using the three basic principles mentioned earlier taking into consideration the variation of interblock flow terms and source/sink term with time within a time step, and (3) approximation of the time integrals in the resulting flow equation to produce the nonlinear algebraic flow equations.

10.2.4.1 Derivation of the algebraic flow equations

In the first step, the reservoir is discretized as mentioned earlier. Fig. 10.3 shows gridblock i (or gridpoint i) and its neighboring gridblocks in the x-direction (gridblock $i-1$ and gridblock $i+1$). At any instant in time, fluid enters gridblock i, coming from gridblock $i-1$, across its $x_{i-1/2}$ face at a mass rate of $w_x|_{x_{i-1/2}}$, and leaves to gridblock $i+1$ across its $x_{i+1/2}$ face at a mass rate of $w_x|_{x_{i+1/2}}$. The fluid also enters gridblock i through a well at a mass rate of q_{m_i}. The mass of fluid in gridblock i per unit volume of rock is m_{v_i}.

Therefore, the material balance equation written over a time step $\Delta t = t^{n+1} - t^n$ as expressed by Eq. (10.1) becomes

$$m_i|_{x_{i-1/2}} - m_o|_{x_{i+1/2}} + m_{s_i} = m_{a_i} \tag{10.38}$$

Terms like $w_x|_{x_{i-1/2}}$, $w_x|_{x_{i+1/2}}$ and q_{m_i} are functions of time only because space is not a variable for an already discretized reservoir (see Observation 1). Therefore,

$$m_i|_{x_{i-1/2}} = \int_{t^n}^{t^{n+1}} w_x|_{x_{i-1/2}} \mathrm{d}t \tag{10.39}$$

$$m_o|_{x_{i+1/2}} = \int_{t^n}^{t^{n+1}} w_x|_{x_{i+1/2}} \mathrm{d}t \tag{10.40}$$

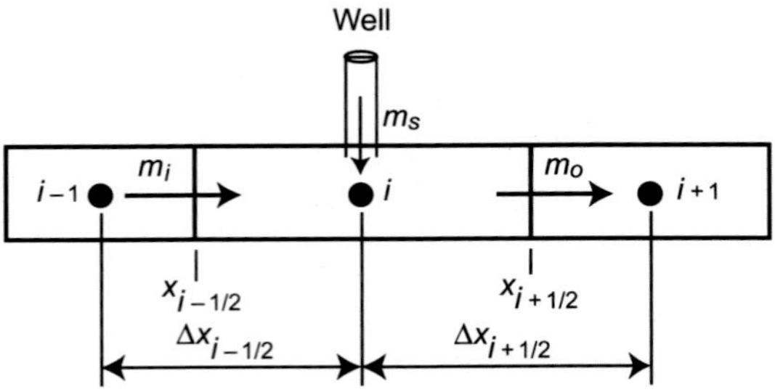

FIG. 10.3 Gridblock i (or gridpoint i) used for writing material balance in the engineering approach.

and

$$m_{s_i} = \int_{t^n}^{t^{n+1}} q_{m_i} \mathrm{d}t \tag{10.41}$$

Using Eqs. (10.39) through (10.41), Eq. (10.38) can be rewritten as

$$\int_{t^n}^{t^{n+1}} w_x|_{x_{i-1/2}} \mathrm{d}t - \int_{t^n}^{t^{n+1}} w_x|_{x_{i+1/2}} \mathrm{d}t + \int_{t^n}^{t^{n+1}} q_{m_i} \mathrm{d}t = m_{a_i} \tag{10.42}$$

Substitution of Eqs. (10.6) and (10.7) into Eq. (10.42) yields

$$\int_{t^n}^{t^{n+1}} (\dot{m}_x A_x)|_{x_{i-1/2}} \mathrm{d}t - \int_{t^n}^{t^{n+1}} (\dot{m}_x A_x)|_{x_{i+1/2}} \mathrm{d}t + \int_{t^n}^{t^{n+1}} q_{m_i} \mathrm{d}t = V_{b_i} \left(m_v^{n+1} - m_v^n\right)_i \tag{10.43}$$

Substitution of Eq. (10.11) through (10.13) into Eq. (10.43) yields

$$\int_{t^n}^{t^{n+1}} (\alpha_c \rho u_x A_x)|_{x_{i-1/2}} \mathrm{d}t - \int_{t^n}^{t^{n+1}} (\alpha_c \rho u_x A_x)|_{x_{i+1/2}} \mathrm{d}t + \int_{t^n}^{t^{n+1}} (\alpha_c \rho q)_i \mathrm{d}t = V_{b_i} \left[(\phi\rho)_i^{n+1} - (\phi\rho)_i^n\right]. \tag{10.44}$$

Substitution of Eq. (10.2) into Eq. (10.44), dividing through by $\alpha_c \rho_{sc}$ and noting that $q/B = q_{sc}$ yields

$$\int_{t^n}^{t^{n+1}} \left(\frac{u_x A_x}{B}\right)\bigg|_{x_{i-1/2}} \mathrm{d}t - \int_{t^n}^{t^{n+1}} \left(\frac{u_x A_x}{B}\right)\bigg|_{x_{i+1/2}} \mathrm{d}t + \int_{t^n}^{t^{n+1}} q_{sc_i} \mathrm{d}t = \frac{V_{b_i}}{\alpha_c}\left[\left(\frac{\phi}{B}\right)_i^{n+1} - \left(\frac{\phi}{B}\right)_i^n\right] \tag{10.45}$$

Fluid volumetric velocity (flow rate per unit cross-sectional area) from gridblock i-1 to gridblock i is given by the algebraic analog of Eq. (10.3),

$$u_x|_{x_{i-1/2}} = \beta_c \left(\frac{k_x}{\mu}\right)_{i-1/2} \frac{(p_{i-1} - p_i)}{\Delta x_{i-1/2}}. \tag{10.46}$$

Likewise, fluid flow rate per unit cross-sectional area from gridblock i to gridblock $i+1$ is

$$u_x|_{x_{i+1/2}} = \beta_c \left(\frac{k_x}{\mu}\right)_{i+1/2} \frac{(p_i - p_{i+1})}{\Delta x_{i+1/2}}. \tag{10.47}$$

Substitution of Eqs. (10.46) and (10.47) into Eq. (10.45) and rearranging results in

$$\int_{t^n}^{t^{n+1}} \left[\left(\beta_c \frac{k_x A_x}{\mu B \Delta x} \right) \Bigg|_{x_{i-1/2}} (p_{i-1} - p_i) \right] dt - \int_{t^n}^{t^{n+1}} \left[\left(\beta_c \frac{k_x A_x}{\mu B \Delta x} \right) \Bigg|_{x_{i+1/2}} (p_i - p_{i+1}) \right] dt + \int_{t^n}^{t^{n+1}} q_{sc_i} dt = \frac{V_{b_i}}{\alpha_c} \left[\left(\frac{\phi}{B} \right)_i^{n+1} - \left(\frac{\phi}{B} \right)_i^n \right] \tag{10.48}$$

or

$$\int_{t^n}^{t^{n+1}} \left[T_{x_{i-1/2}} (p_{i-1} - p_i) \right] dt + \int_{t^n}^{t^{n+1}} \left[T_{x_{i+1/2}} (p_{i+1} - p_i) \right] dt + \int_{t^n}^{t^{n+1}} q_{sc_i} dt = \frac{V_{b_i}}{\alpha_c} \left[\left(\frac{\phi}{B} \right)_i^{n+1} - \left(\frac{\phi}{B} \right)_i^n \right] \tag{10.49}$$

The derivation of Eq. (10.49) is rigorous and involves no assumptions other than the validity of Darcy's law (Eqs. 10.46 and 10.47) to estimate fluid volumetric velocity between gridblock i and its neighboring gridblocks i-1 and $i+1$. Such validity is not questionable by petroleum engineers.

Again, the accumulation term in the earlier equation can be expressed in terms of the pressure of gridblock i, and Eq. (10.49) becomes

$$\int_{t^n}^{t^{n+1}} \left[T_{x_{i-1/2}} (p_{i-1} - p_i) \right] dt + \int_{t^n}^{t^{n+1}} \left[T_{x_{i+1/2}} (p_{i+1} - p_i) \right] dt + \int_{t^n}^{t^{n+1}} q_{sc_i} dt = \frac{V_{b_i}}{\alpha_c} \left(\frac{\phi}{B} \right)_i' \left[p_i^{n+1} - p_i^n \right] \tag{10.50}$$

where $\left(\frac{\phi}{B} \right)_i'$ is chord slope defined by Eq. (10.30).

10.2.4.2 Approximation of time integrals

If the argument of an integral is an explicit function of time, the integral can be evaluated analytically. This is not the case for the integrals appearing on the LHS of either Eq. (10.49) or Eq. (10.50). The integration is schematically shown in Fig. 10.4. Performing the integrals on the LHS of Eq. (10.49) or (10.50) necessitates making certain assumptions. Such assumptions lead to deriving equations as those expressed by Eqs. (10.28), (10.32), and (10.35) or Eqs. (10.29), (10.33), and (10.36).

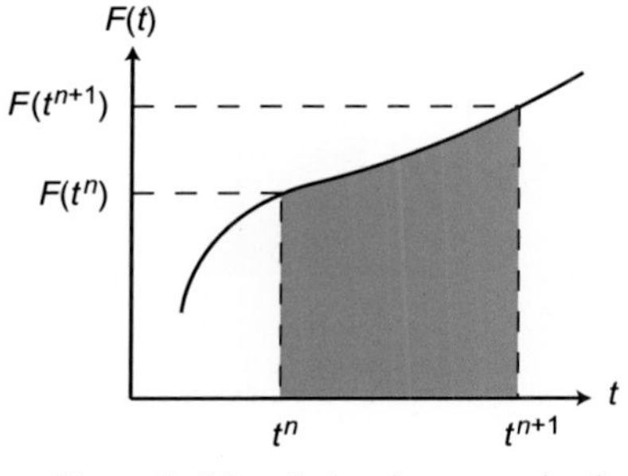

FIG. 10.4 Representation of integral of function as the area under the curve.

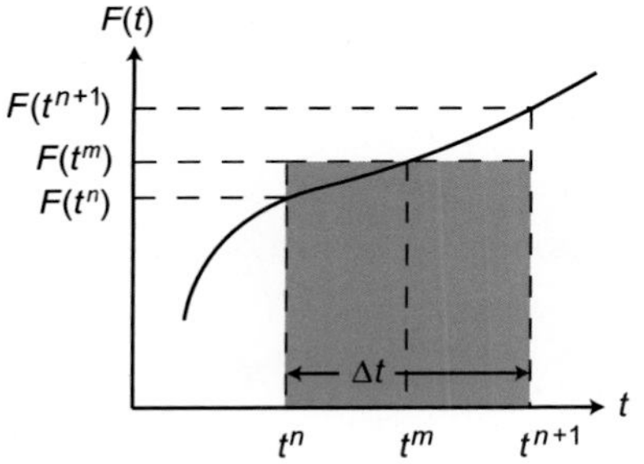

FIG. 10.5 Representation of integral of function as $F(t^m) \times \Delta t$.

Consider the integral $\int_{t^n}^{t^{n+1}} F(t)\mathrm{d}t$ shown in Fig. 10.5. This integral can be evaluated as follows

$$\int_{t^n}^{t^{n+1}} F(t)\mathrm{d}t \cong \int_{t^n}^{t^{n+1}} F(t^m)\mathrm{d}t = \int_{t^n}^{t^{n+1}} F^m\mathrm{d}t = F^m \int_{t^n}^{t^{n+1}} \mathrm{d}t = F^m t\Big|_{t^n}^{t^{n+1}} = F^m\left(t^{n+1} - t^n\right)$$
$$= F^m \Delta t \tag{10.51}$$

The argument F stands for $[T_{x_{i-1/2}}(p_{i-1} - p_i)]$, $[T_{x_{i+1/2}}(p_{i+1} - p_i)]$, or q_{sc_i} that appears on the LHS of Eq. (10.49) and F^m = approximation of F at time t^m = constant over the time interval Δt.

Forward-difference equation

The forward-difference equation given by Eq. (10.28) can be obtained from Eq. (10.49) if the argument F of integrals is dated at time t^n; that is, $F \cong F^m = F^n$

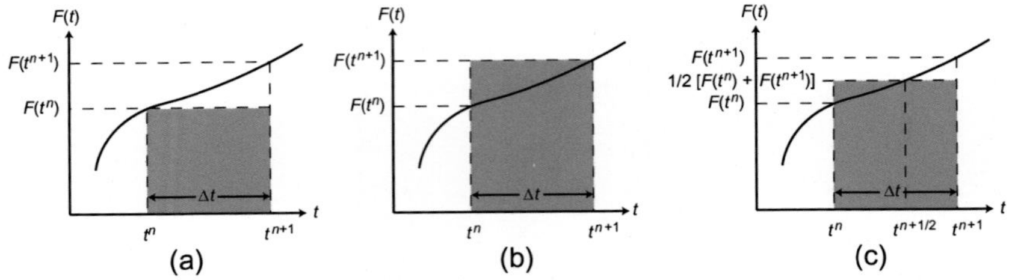

FIG. 10.6 Different methods of approximation of the integral of a function. (a) Forward difference; (b) Backward difference; (c) Central difference (Crank Nicholson).

as shown in Fig. 10.6a. Therefore, Eq. (10.51) becomes $\int_{t^n}^{t^{n+1}} F(t)\mathrm{d}t \cong F^n \Delta t$, and Eq. (10.49) reduces to

$$\left[T^n_{x_{i-1/2}}\left(p^n_{i-1} - p^n_i\right)\right]\Delta t + \left[T^n_{x_{i+1/2}}\left(p^n_{i+1} - p^n_i\right)\right]\Delta t + q^n_{sc_i}\Delta t \cong \frac{V_{b_i}}{\alpha_c}\left[\left(\frac{\phi}{B}\right)^{n+1}_i - \left(\frac{\phi}{B}\right)^n_i\right] \tag{10.52}$$

Dividing above equation by Δt gives Eq. (10.28):

$$T^n_{x_{i-1/2}}\left(p^n_{i-1} - p^n_i\right) + T^n_{x_{i+1/2}}\left(p^n_{i+1} - p^n_i\right) + q^n_{sc_i} \cong \frac{V_{b_i}}{\alpha_c \Delta t}\left[\left(\frac{\phi}{B}\right)^{n+1}_i - \left(\frac{\phi}{B}\right)^n_i\right] \tag{10.28}$$

If one starts with Eq. (10.50) instead of Eq. (10.49), he ends up with Eq. (10.29):

$$T^n_{x_{i-1/2}}\left(p^n_{i-1} - p^n_i\right) + T^n_{x_{i+1/2}}\left(p^n_{i+1} - p^n_i\right) + q^n_{sc_i} \cong \frac{V_{b_i}}{\alpha_c \Delta t}\left(\frac{\phi}{B}\right)'_i \left[p^{n+1}_i - p^n_i\right] \tag{10.29}$$

Backward-difference equation

The backward-difference equation given by Eq. (10.32) can be obtained from Eq. (10.49) if the argument F of integrals is dated at time t^{n+1}; that is, $F \cong F^m = F^{n+1}$ as shown in Fig. 10.6b. Therefore, Eq. (10.51) becomes $\int_{t^n}^{t^{n+1}} F(t)\mathrm{d}t \cong F^{n+1}\Delta t$, and Eq. (10.49) reduces to

$$\left[T^{n+1}_{x_{i-1/2}}\left(p^{n+1}_{i-1} - p^{n+1}_i\right)\right]\Delta t + \left[T^{n+1}_{x_{i+1/2}}\left(p^{n+1}_{i+1} - p^{n+1}_i\right)\right]\Delta t + q^{n+1}_{sc_i}\Delta t \cong \frac{V_{b_i}}{\alpha_c}\left[\left(\frac{\phi}{B}\right)^{n+1}_i - \left(\frac{\phi}{B}\right)^n_i\right] \tag{10.53}$$

Dividing above equation by Δt gives Eq. (10.32):

$$T^{n+1}_{x_{i-1/2}}\left(p^{n+1}_{i-1} - p^{n+1}_i\right) + T^{n+1}_{x_{i+1/2}}\left(p^{n+1}_{i+1} - p^{n+1}_i\right) + q^{n+1}_{sc_i} \cong \frac{V_{b_i}}{\alpha_c \Delta t}\left[\left(\frac{\phi}{B}\right)^{n+1}_i - \left(\frac{\phi}{B}\right)^n_i\right] \tag{10.32}$$

If one starts with Eq. (10.50) instead of Eq. (10.49), he ends up with Eq. (10.33)

$$T^{n+1}_{x_{i-1/2}}(p^{n+1}_{i-1}-p^{n+1}_{i})+T^{n+1}_{x_{i+1/2}}(p^{n+1}_{i+1}-p^{n+1}_{i})+q^{n+1}_{sc_i}\cong\frac{V_{b_i}}{\alpha_c\Delta t}\left(\frac{\phi}{B}\right)'_i\left[p^{n+1}_i-p^n_i\right] \tag{10.33}$$

Central-difference (Crank-Nicholson) equation

The second order in time Crank-Nicholson approximation given by Eq. (10.35) can be obtained from Eq. (10.49) if the argument F of integrals is dated at time $t^{n+1/2}$. This choice of time level was made to make the RHS of Eq. (10.26) to appear as second-order approximation in time in the mathematical approach. In this case, the argument F in the integrals can be approximated as $F\cong F^m=F^{n+1/2}=(F^n+F^{n+1})/2$ as shown in Fig. 10.6c. Therefore, $\int_{t^n}^{t^{n+1}} F(t)\mathrm{d}t\cong\frac{1}{2}(F^n+F^{n+1})\Delta t$, and Eq. (10.49) reduces to

$$\begin{aligned}&(1/2)\left[T^{n}_{x_{i-1/2}}(p^{n}_{i-1}-p^{n}_{i})+T^{n+1}_{x_{i-1/2}}(p^{n+1}_{i-1}-p^{n+1}_{i})\right]\Delta t\\&+(1/2)\left[T^{n}_{x_{i+1/2}}(p^{n}_{i+1}-p^{n}_{i})+T^{n+1}_{x_{i+1/2}}(p^{n+1}_{i+1}-p^{n+1}_{i})\right]\Delta t\\&+(1/2)\left[q^{n}_{sc_i}+q^{n+1}_{sc_i}\right]\Delta t\cong\frac{V_{b_i}}{\alpha_c}\left[\left(\frac{\phi}{B}\right)^{n+1}_i-\left(\frac{\phi}{B}\right)^{n}_i\right]\end{aligned} \tag{10.54}$$

Dividing above equation by Δt and rearranging terms give Eq. (10.35):

$$\begin{aligned}&(1/2)\left[T^{n}_{x_{i-1/2}}(p^{n}_{i-1}-p^{n}_{i})+T^{n}_{x_{i+1/2}}(p^{n}_{i+1}-p^{n}_{i})\right]\\&+(1/2)\left[T^{n+1}_{x_{i-1/2}}(p^{n+1}_{i-1}-p^{n+1}_{i})+T^{n+1}_{x_{i+1/2}}(p^{n+1}_{i+1}-p^{n+1}_{i})\right]\\&+(1/2)\left[q^{n}_{sc_i}+q^{n+1}_{sc_i}\right]\cong\frac{V_{b_i}}{\alpha_c\Delta t}\left[\left(\frac{\phi}{B}\right)^{n+1}_i-\left(\frac{\phi}{B}\right)^{n}_i\right]\end{aligned} \tag{10.35}$$

If one starts with Eq. (10.50) instead of Eq. (10.49), he ends up with Eq. (10.36):

$$\begin{aligned}&(1/2)\left[T^{n}_{x_{i-1/2}}(p^{n}_{i-1}-p^{n}_{i})+T^{n}_{x_{i+1/2}}(p^{n}_{i+1}-p^{n}_{i})\right]\\&+(1/2)\left[T^{n+1}_{x_{i-1/2}}(p^{n+1}_{i-1}-p^{n+1}_{i})+T^{n+1}_{x_{i+1/2}}(p^{n+1}_{i+1}-p^{n+1}_{i})\right]\\&+(1/2)\left[q^{n}_{sc_i}+q^{n+1}_{sc_i}\right]\cong\frac{V_{b_i}}{\alpha_c\Delta t}\left(\frac{\phi}{B}\right)'_i\left[p^{n+1}_i-p^{n}_i\right]\end{aligned} \tag{10.36}$$

Therefore, one can conclude that the same nonlinear algebraic equations can be derived by the mathematical and engineering approaches.

10.3 Treatment of initial and boundary conditions

Initial conditions receive the same treatment by both the mathematical and engineering approaches. Therefore, this section focuses on the treatment of boundary conditions by both approaches and highlights differences. An external (or internal) reservoir boundary can be subject to one of four conditions: no-flow boundary, constant-flow boundary, constant pressure-gradient boundary, or constant pressure boundary. In fact, the first three boundary conditions reduce to specified pressure-gradient condition (Neumann boundary condition) and the fourth boundary condition is the Dirichlet boundary condition. In the following presentation, we demonstrate the treatment of boundary conditions at $x=0$ only as an example.

10.3.1 Specified boundary pressure condition

10.3.1.1 The mathematical approach

For point-distributed grid (see Fig. 10.7a), $p_1=p_b$. Therefore, fictitious well rate across left boundary becomes

$$q_{sc_{b,bp}}=T_{1+1/2}(p_1-p_2)=T_{1+1/2}(p_b-p_2) \tag{10.55}$$

which is the interblock flow rate ($q_{sc_{1+1/2}}$) between gridpoints 1 and 2.

For block-centered grid, one sets $p_1 \cong p_b$ at reservoir left boundary (see Fig. 10.7b) and the flow equation for gridblock 1 is removed from the system of flow equations. This is a first-order approximation.

If a second-order approximation is used at reservoir left boundary (Settari and Aziz, 1975), the following pressure equation is added and solved with the system of flow equations.

$$p_b \cong \frac{\Delta x_{1/2}+\Delta x_{1+1/2}}{\Delta x_{1+1/2}}p_1-\frac{\Delta x_{1/2}}{\Delta x_{1+1/2}}p_2 \tag{10.56a}$$

or for equal size gridblocks,

$$p_b \cong \frac{1}{2}(3p_1-p_2) \tag{10.56b}$$

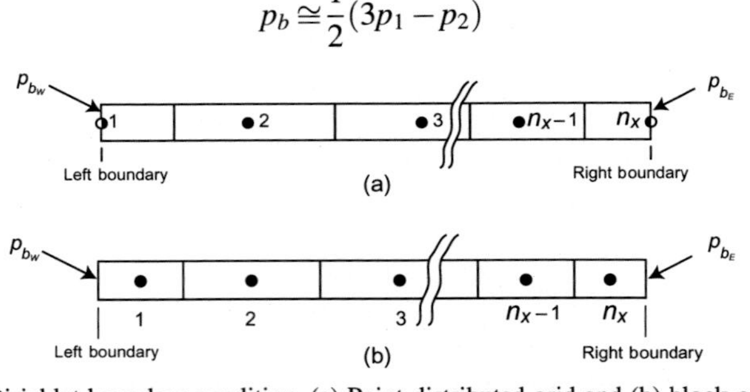

FIG. 10.7 Dirichlet boundary condition. (a) Point-distributed grid and (b) block-centered grid.

This treatment increases the number of equations to be solved by one equation for each boundary block having specified boundary pressure. Furthermore, this extra equation does not have the same form as that of the final pressure equation for a gridblock.

10.3.1.2 The engineering approach

For point-distributed grid, the fictitious well flow rate was derived earlier in Chapter 5 and expressed by Eq. (5.46c):

$$q^m_{sc_{b,bP}} = \left[\beta_c \frac{k_l A_l}{\mu B \Delta l}\right]^m_{bP,bP^*} \left[\left(p^m_{bP} - p^m_{bP^*}\right) - \gamma^m_{b,bP^*}\left(Z_{bP} - Z_{bP^*}\right)\right] \tag{5.46c}$$

where l is the direction normal to the boundary.

Replacing direction l by x and discarding time level m and gravity term, Eq. (5.46c) reduces to

$$q_{sc_{b,bP}} = \left[\beta_c \frac{k_x A_x}{\mu B \Delta x}\right]_{bP,bP^*} \left(p_{bP} - p_{bP^*}\right) \tag{10.57a}$$

or

$$q_{sc_{b,bP}} = T_{b,bP^*}\left(p_{bP} - p_{bP^*}\right) \tag{10.57b}$$

where

$$T_{b,bP^*} = \left(\beta_c \frac{k_x A_x}{\mu B\, x}\right)_{bP,bP^*} \tag{10.58}$$

For point-distributed grid (see Figs. 10.7a and 10.8a), $p_1 = p_{bP}, p_2 = p_{bP^*}$, and $T_{1+1/2} = T_{b,bP^*}$. Substitution of these relations into Eq. (10.57b) gives

$$q_{sc_{b,bP}} = T_{1+1/2}\left(p_1 - p_2\right) \tag{10.59}$$

which is the interblock flow rate ($q_{sc_{1+1/2}}$) between gridpoints 1 and 2 as given by Eq. (10.55) in the mathematical approach.

For block-centered grid the fictitious well flow rate was derived earlier in Chapter 4 and expressed by Eq. (4.37c):

$$q^m_{sc_{b,bB}} = \left[\beta_c \frac{k_l A_l}{\mu B(\Delta l/2)}\right]^m_{bB} \left[\left(p_b - p^m_{bB}\right) - \gamma^m_{b,bB}\left(Z_b - Z_{bB}\right)\right] \tag{4.37c}$$

Replacing direction l by x and discarding time level m and gravity term, Eq. (4.37c) reduces to

$$q_{sc_{b,bB}} = \left[\beta_c \frac{k_x A_x}{\mu B(\Delta x/2)}\right]_{bB} \left(p_b - p_{bB}\right) \tag{10.60a}$$

or

$$q_{sc_{b,bB}} = T_{b,bB}(p_b - p_{bB}) \tag{10.60b}$$

where

$$T_{b,bB} = \left[\beta_c \frac{k_x A_x}{\mu B(\Delta x/2)}\right]_{bB} \tag{10.61}$$

The application of Eq. (10.60b) for boundary gridblock 1 gives

$$q_{sc_{b,bB}} = T_{b,1}(p_b - p_1) = \left[\beta_c \frac{k_x A_x}{(\mu B \Delta x/2)}\right]_1 (p_b - p_1) \tag{10.62}$$

Note that the fictitious well rate presented by Eq. (10.62) is a second-order approximation and does not need the introduction of an extra equation as required by the mathematical approach.

10.3.2 Specified boundary pressure-gradient condition

10.3.2.1 The mathematical approach

For the mathematical approach, we will demonstrate the application of boundary pressure-gradient specification for gridblock 1 and gridpoint 1. A second-order approximation for the pressure gradient is possible using the "reflection technique" by introducing an auxiliary point (p_0) outside the reservoir on the other side of the boundary as shown in Fig. 10.8. Aziz and Settari (1979) reported the discretization of this boundary condition for both block-centered and point-distributed grids for regular grids. The discretization of this boundary condition is presented here for irregular grids.

For point-distributed grid (Fig. 10.8a),

$$\left.\frac{\partial p}{\partial x}\right|_b \cong \frac{p_2 - p_0}{2\Delta x_{1+1/2}} \tag{10.63}$$

The difference flow equation for the whole boundary block in terms of the original reservoir boundary block represented by gridpoint 1 is

$$T_{x_{1/2}}(p_0 - p_1) + T_{x_{1+1/2}}(p_2 - p_1) + 2q_{sc_1} = \frac{2V_{b_1}}{\alpha_c \Delta t}\left[\left(\frac{\phi}{B}\right)_1^{n+1} - \left(\frac{\phi}{B}\right)_1^n\right] \tag{10.64}$$

because $V_b = 2V_{b_1}$ and $q_{sc} = 2q_{sc_1}$. Using Eq. (10.63) to eliminate p_0 from Eq. (10.64), dividing the resulting equation by 2, and observing that $\Delta x_{1/2} = \Delta x_{1+1/2}$ and $T_{x_{1/2}} = T_{x_{1+1/2}}$ because of the reflection technique, one obtains

$$-T_{x_{1+1/2}} \Delta x_{1+1/2} \left.\frac{\partial p}{\partial x}\right|_b + T_{x_{1+1/2}}(p_2 - p_1) + q_{sc_1} = \frac{V_{b_1}}{\alpha_c \Delta t}\left[\left(\frac{\phi}{B}\right)_1^{n+1} - \left(\frac{\phi}{B}\right)_1^n\right] \tag{10.65a}$$

Noting that the first term on the LHS in above equation is nothing but $q_{sc_{b,1}}$ and that $T_{x_{1+1/2}}\Delta x_{1+1/2} = \left(\beta_c \frac{k_x A_x}{\mu B \Delta x}\right)_{1+1/2} \Delta x_{1+1/2} = \left(\beta_c \frac{k_x A_x}{\mu B}\right)_{1+1/2}$, then,

$$q_{sc_{b,1}} = -T_{x_{1/2}}\Delta x_{1/2} \left.\frac{\partial p}{\partial x}\right|_b = -\left(\beta_c \frac{k_x A_x}{\mu B}\right)_{1+1/2} \left.\frac{\partial p}{\partial x}\right|_b \tag{10.66}$$

Therefore, Eq. (10.65a) becomes

$$q_{sc_{b,1}} + T_{x_{1+1/2}}(p_2 - p_1) + q_{sc_1} = \frac{V_{b_1}}{\alpha_c \Delta t}\left[\left(\frac{\phi}{B}\right)_1^{n+1} - \left(\frac{\phi}{B}\right)_1^n\right] \tag{10.65b}$$

For specified pressure gradient at reservoir east boundary, the fictitious well flow rate for gridpoint n_x is defined by:

$$q_{sc_{b,nx}} = +T_{x_{n_x-1/2}}\Delta x_{n_x-1/2} \left.\frac{\partial p}{\partial x}\right|_b = +\left(\beta_c \frac{k_x A_x}{\mu B}\right)_{n_x-1/2} \left.\frac{\partial p}{\partial x}\right|_b \tag{10.67}$$

For block-centered grid (Fig. 10.8b),

$$\left.\frac{\partial p}{\partial x}\right|_b = \frac{p_1 - p_0}{\Delta x_1} \tag{10.68}$$

The difference equation for gridblock 1 is

$$T_{x_{1/2}}(p_0 - p_1) + T_{x_{1+1/2}}(p_2 - p_1) + q_{sc_1} = \frac{V_{b_1}}{\alpha_c \Delta t}\left[\left(\frac{\phi}{B}\right)_1^{n+1} - \left(\frac{\phi}{B}\right)_1^n\right] \tag{10.69}$$

Using Eq. (10.68) to eliminate p_0 from Eq. (10.69), one obtains

$$-T_{x_{1/2}}\Delta x_{1/2} \left.\frac{\partial p}{\partial x}\right|_b + T_{x_{1+1/2}}(p_2 - p_1) + q_{sc_1} = \frac{V_{b_1}}{\alpha_c \Delta t}\left[\left(\frac{\phi}{B}\right)_1^{n+1} - \left(\frac{\phi}{B}\right)_1^n\right] \tag{10.70a}$$

or

$$q_{sc_{b,1}} + T_{x_{1+1/2}}(p_2 - p_1) + q_{sc_1} = \frac{V_{b_1}}{\alpha_c \Delta t}\left[\left(\frac{\phi}{B}\right)_1^{n+1} - \left(\frac{\phi}{B}\right)_1^n\right] \tag{10.70b}$$

where

$$q_{sc_{b,1}} = -T_{x_{1/2}}\Delta x_{1/2} \left.\frac{\partial p}{\partial x}\right|_b = -\left(\beta_c \frac{k_x A_x}{\mu B}\right)_1 \left.\frac{\partial p}{\partial x}\right|_b \tag{10.71}$$

Because in this case $\Delta x_{1/2} = \Delta x_1$ and $T_{x_{1/2}} = \left(\beta_c \frac{k_x A_x}{\mu B \Delta x}\right)_1$ because properties and dimensions of gridblock 0 are the same as those of gridblock 1 (reflection technique about reservoir boundary [see Eq. 4.19]).

For specified pressure gradient at reservoir east boundary, the fictitious well flow rate for gridpoint n_x is defined by Eq. (10.72):

$$q_{sc_{b,nx}} = +T_{x_{n_x+1/2}} \Delta x_{n_x+1/2} \left.\frac{\partial p}{\partial x}\right|_b = +\left(\beta_c \frac{k_x A_x}{\mu B}\right)_{n_x} \left.\frac{\partial p}{\partial x}\right|_b \tag{10.72}$$

10.3.2.2 The engineering approach

For point-distributed grid, if pressure gradient at reservoir boundary is specified (see Fig. 10.9a), Chapter 5 defines the flow rate across the reservoir boundary by Eq. (5.31) for reservoir left boundary and Eq. (5.32) for reservoir right boundary. Discarding the time level m and gravity term in these equations, they reduce to

$$q_{sc_{b,1}} = -\left(\beta_c \frac{k_x A_x}{\mu B}\right)_{1+1/2} \left.\frac{\partial p}{\partial x}\right|_b \tag{10.73}$$

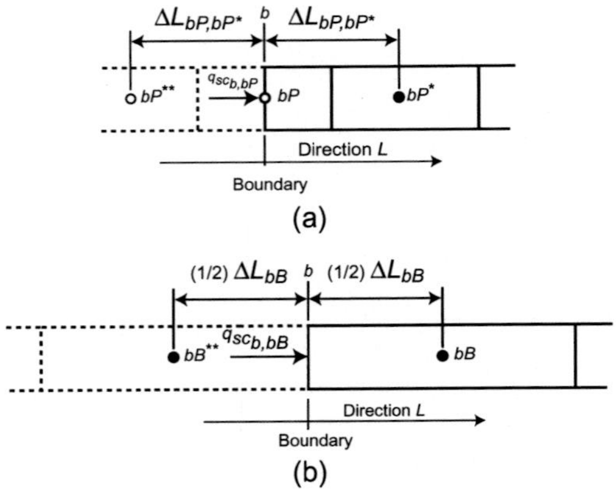

FIG. 10.8 Reflection technique. (a) Point-distributed grid and (b) block-centered grid. $bP^{**}=0$, $bP=1$, $bP^{*}=2$, $bB^{**}=0$, $bB=1$.

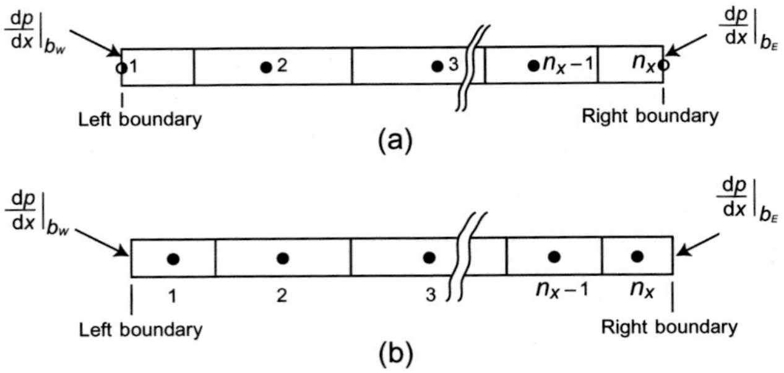

FIG. 10.9 Neumann boundary condition. (a) Point-distributed grid and (b) block-centered grid.

for boundary gridpoint 1, on reservoir left (west) boundary, and

$$q_{sc_{b,n_x}} = +\left(\beta_c \frac{k_x A_x}{\mu B}\right)_{n_x-1/2} \frac{\partial p}{\partial x}\bigg|_b \tag{10.74}$$

for boundary gridpoint n_x, on reservoir right (east) boundary.

For block-centered grid, if pressure gradient at reservoir boundary is specified (see Fig. 10.9b), Chapter 4 defines the flow rate across the reservoir boundary as Eq. (4.23b) for reservoir left boundary and Eq. (4.24b) for reservoir right boundary. Discarding the time level m and gravity term in these equations and replacing direction l by x, these two equation reduce to

$$q_{sc_{b,bB}} = -\left(\beta_c \frac{k_x A_x}{\mu B}\right)_{bB} \frac{\partial p}{\partial x}\bigg|_b \tag{10.75}$$

for reservoir left (west) boundary, and

$$q_{sc_{b,bB}} = \left(\beta_c \frac{k_x A_x}{\mu B}\right)_{bB} \frac{\partial p}{\partial x}\bigg|_b \tag{10.76}$$

for reservoir right (east) boundary.

Applying Eq. (10.75) for boundary gridblock 1 on reservoir west boundary results in

$$q_{sc_{b,1}} = -\left(\beta_c \frac{k_x A_x}{\mu B}\right)_1 \frac{\partial p}{\partial x}\bigg|_b \tag{10.77}$$

and Eq. (10.76) for boundary gridblock n_x on reservoir east results in

$$q_{sc_{b,n_x}} = +\left(\beta_c \frac{k_x A_x}{\mu B}\right)_{n_x} \frac{\partial p}{\partial x}\bigg|_b \tag{10.78}$$

10.3.3 Specified flow rate condition

10.3.3.1 The mathematical approach

In the mathematical approach, the specified flow rate boundary condition is expressed in terms of pressure-gradient condition using an equation similar to Eq. (10.66) for point-distributed grid or Eq. (10.71) for block-centered grid. This is followed by the treatment of specified pressure-gradient condition as presented in Section 10.3.2.1.

10.3.3.2 The engineering approach

In the engineering approach, the specified flow rate across reservoir boundary of a specific boundary gridblock ($q_{sc_{b,bB}} = q_{spsc}$) or boundary gridpoint ($q_{sc_{b,bP}} = q_{spsc}$) is substituted in the flow equation for that gridblock or gridpoint.

Note that $q_{spsc} \neq 0$ reflects constant flow rate boundary condition and $q_{spsc} = 0$ reflects no-flow boundary condition.

10.4 Linearization of well flow rates

A wellblock production (or injection) rate is evaluated in space at the gridblock (or gridpoint) for which the flow equation is written. Linearization in time of the wellblock flow rate equation involves first linearizing the wellblock production rate equation and then substituting the result in the linearized flow equation for the wellblock.

The well flow rate equation for wellblock i that needs linearization is

$$q_{sc_i}^{n+1} \cong -G_{w_i} \left(\frac{1}{B\mu} \right)_i^{n+1} \left(p_i^{n+1} - p_{wf_i} \right) \tag{10.79}$$

for specified bottom-hole pressure (Mode 1), or

$$q_{sc_i}^{n+1} \cong 2\pi\beta_c r_w (kh)_i \left(\frac{1}{B\mu} \right)_i^{n+1} \left. \frac{\mathrm{d}p}{\mathrm{d}r} \right|_{rw} \tag{10.80}$$

for specified pressure gradient at well radius (Mode 2).

In the mathematical approach, time nonlinearity in well rate includes both $\left(\frac{1}{B\mu} \right)_i^{n+1}$ and $(p_i^{n+1} - p_{wf_i})$, whereas in the engineering approach, it is limited to $\left(\frac{1}{B\mu} \right)_i^{n+1}$. This difference results from considering $q_{sc_i}^{n+1}$ as the average of the time integral of well flow rate over a time step. This difference leads to having different methods of linearization as shown in the next sections. The treatment of well rate nonlinearities in time is presented for wells operating with Mode 1.

10.4.1 The mathematical approach

The explicit method,

$$q_{sc_i}^{n+1} \cong q_{sc_i}^{n} = -G_{w_i} \left(\frac{1}{B\mu} \right)_i^{n} \left(p_i^{n} - p_{wf_i} \right) \tag{10.81}$$

The simple iteration method,

$$q_{sc_i}^{n+1} \cong q_i^{\overset{(v)}{n+1}} = -G_{w_i} \left(\frac{1}{B\mu} \right)_i^{\overset{(v)}{n+1}} \left(p_i^{\overset{(v)}{n+1}} - p_{wf_i} \right) \tag{10.82}$$

The explicit transmissibility method,

$$q_{sc_i}^{n+1} \cong -G_{w_i}\left(\frac{1}{B\mu}\right)_i^{n}\left(\overset{(\nu+1)}{p_i^{n+1}} - p_{wf_i}\right) \tag{10.83}$$

The simple iteration on transmissibility method,

$$q_{sc_i}^{n+1} \cong -G_{w_i}\left(\frac{1}{B\mu}\right)_i^{\overset{(\nu)}{n+1}}\left(\overset{(\nu+1)}{p_i^{n+1}} - p_{wf_i}\right) \tag{10.84}$$

The fully implicit method,

$$q_{sc_i}^{n+1} \cong \overset{(\nu+1)}{q_{sc_i}^{n+1}} \cong \overset{(\nu)}{q_{sc_i}^{n+1}} + \left.\frac{dq_{sc_i}}{dp_i}\right|^{\overset{(\nu)}{n+1}}\left(\overset{(\nu+1)}{p_i^{n+1}} - \overset{(\nu)}{p_i^{n+1}}\right) \tag{10.85}$$

10.4.2 The engineering approach

The explicit transmissibility method,

$$q_{sc_i}^{n+1} \cong -G_{w_i}\left(\frac{1}{B\mu}\right)_i^{n}\left(\overset{(\nu+1)}{p_i^{n+1}} - p_{wf_i}\right) \tag{10.83}$$

The simple iteration on transmissibility method,

$$q_{sc_i}^{n+1} \cong -G_{w_i}\left(\frac{1}{B\mu}\right)_i^{\overset{(\nu)}{n+1}}\left(\overset{(\nu+1)}{p_i^{n+1}} - p_{wf_i}\right) \tag{10.84}$$

The fully implicit method,

$$q_{sc_i}^{n+1} \cong \overset{(\nu+1)}{q_{sc_i}^{n+1}} \cong \overset{(\nu)}{q_{sc_i}^{n+1}} + \left.\frac{dq_{sc_i}}{dp_i}\right|^{\overset{(\nu)}{n+1}}\left(\overset{(\nu+1)}{p_i^{n+1}} - \overset{(\nu)}{p_i^{n+1}}\right) \tag{10.85}$$

The degree of implicitness increases with the equation selection from Eq. (10.81) to Eq. (10.85). Furthermore, the use of Eqs. (10.83)–(10.85) provides tremendous improvement in implicitness and hence stability over the linearization with Eqs. (10.81) and (10.82). This is the case because the primary nonlinearity in time of the production rate is due to $(p_i^{n+1} - p_{wf_i})$ term; the contribution of the $\left(\frac{1}{B\mu}\right)_i^{n+1}$ term to nonlinearity is secondary.

For wells operating with Mode 2, the explicit method is the same as explicit transmissibility method, and simple iteration method is the same as the simple iteration on transmissibility method because there is only one nonlinear term in the wellblock rate equation, namely, $\left(\frac{1}{B\mu}\right)_i^{n+1}$.

Another method of well rate linearization in time involves substituting the appropriate well rate equation into the flow equation for the wellblock prior to linearization and subsequently linearizing all terms in the resulting flow equation. That is to say, the well rate, fictitious well rates, and interblock flow rate terms receive identical linearization treatments. For a well operating with bottom-hole-pressure specification, this method results in the implicit treatment of wellblock pressure compared with the explicit treatments provided by the explicit transmissibility method, Eq. (10.83), and simple iteration on transmissibility method, Eq. (10.84). This method of linearization is identical to the linearization method used in the engineering approach because all terms in the flow equation except accumulation (well rate, fictitious rate, interblock flow rates) receive the same treatment of nonlinear terms in time.

10.5 Summary

The following conclusions can be drawn.

1. The discretized flow equations (nonlinear algebraic equations) in reservoir simulation of any process can be obtained in a rigorous way by the engineering approach without going through the rigor of obtaining the PDEs describing the process and space and time discretizations (mathematical approach).
2. The engineering approach rather than the mathematical approach is closer to engineer's thinking. While the mathematical approach derives the nonlinear algebraic equations by first deriving the PDEs, followed by discretizing the reservoir, and finally discretizing the PDEs, the engineering approach first discretizes the reservoir, then derives the algebraic flow equations with time integrals, and finally approximates the time integrals to obtain the same nonlinear algebraic flow equations.
3. Both the engineering and mathematical approaches treat boundary conditions with the same accuracy if second-order approximation is used. If discretization of specified boundary pressure condition in block-centered grid is first-order correct, then the engineering approach gives a representation that is more accurate. If a second-order approximation of boundary conditions in block-centered grid is used, then the engineering approach provides lesser number of equations.
4. The engineering approach is closer to the physical meaning of various terms in the algebraic flow equation. It also provides confirmation for using central-difference approximation of the second-order space derivative and gives interpretation of the forward-, backward-, and central-difference approximations of the first-order time derivative in the PDE. Analysis of local truncation errors, consistency, convergence, and stability; however, can be studied by the mathematical approach only. Therefore, one may conclude that the mathematical and engineering approaches complement each other.

Chapter 11

Introduction to modeling multiphase flow in petroleum reservoirs

Chapter Outline

11.1 Introduction

Nature is inherently multiphase and multicomponent. Water being ubiquitous in nature, any oil and gas formation is necessarily multiphase. In general, conditions pertaining to fluid, commonly designated as "black oil," show the presence of water, oil, and gas. For simplicity, previous chapters have dealt with single-phase fluid. This chapter presents the basics of modeling a black-oil reservoir. In this context, we present the necessary engineering concepts for multiphase flow in porous media, followed by the derivation of the flow equation for any component in the system in a 1-D rectangular reservoir. Then, using CVFD terminology, we present the component general flow equations in a multiphase, multidimensional system, which apply to interior and boundary reservoir blocks. From these component flow equations, the basic flow models of

Petroleum Reservoir Simulation. https://doi.org/10.1016/B978-0-12-819150-7.00011-6

two-phase oil/water, oil/gas, and gas/water and three-phase oil/water/gas are derived. The accumulation terms in flow equations are expressed in terms of changes in the reservoir block unknowns over a time step. We present the equations for phase production and injection rates from single-block and multiblock wells operating with different conditions. The treatment of boundary conditions as fictitious wells is presented and discussed in detail. Methods of linearization of nonlinear terms in multiphase flow are discussed. We introduce two of the basic methods for solving the linearized multiphase flow equations, the implicit pressure-explicit saturation (IMPES) and simultaneous solution (SS) methods. Because this chapter forms an introduction to the simulation of multiphase flow, we present the two solution methods (IMPES and SS) as they apply to the two-phase oil/water flow model only. The extensions of these methods to other flow models are straightforward, whereas the application of additional solution methods, such as the sequential (SEQ) and the fully implicit methods, is discussed elsewhere.

11.2 Reservoir engineering concepts in multiphase flow

The reservoir engineering concepts discussed in this chapter pertain to the simultaneous flow of oil, water, and gas. These three phases coexist and fill the pore volume of the reservoir; that is,

$$S_o + S_w + S_g = 1 \tag{11.1}$$

The properties of interest in modeling multiphase flow in petroleum reservoirs include the PVT and transport properties of oil phase, water phase, and gas phase; the relative permeabilities to oil phase, water phase, and gas phase; and oil/water capillary pressure and gas/oil capillary pressure. Such data are usually available and supplied to simulators in a tabular form.

11.2.1 Fluid properties

In a black-oil system, the oil, water, and gas phases coexist in equilibrium under isothermal conditions. To describe this behavior in a practical sense at reservoir temperature and any reservoir pressure, the oil and water phases can be assumed immiscible, neither the oil component nor the water component dissolves in the gas phase, and the gas-component miscibility may be large in the oil phase but is negligible in the water phase. Therefore, the water-phase and gas-phase properties that were discussed previously in single-phase flow are applicable for multiphase flow, whereas the oil-phase properties in multiphase flow are affected by pressure and solution-gas/oil ratio only. Fig. 11.1 demonstrates the dependence of the gas FVF and viscosity on pressure. Fig. 11.2 shows the pressure dependence of the water FVF and viscosity. Fig. 11.3 shows the oil FVF, oil viscosity, and solution-gas/oil ratio dependence on pressure. Fig. 11.3 highlights the effect of the solution-gas/oil ratio on oil FVF and viscosity below the oil bubble-point pressure. Above the oil bubble-point pressure,

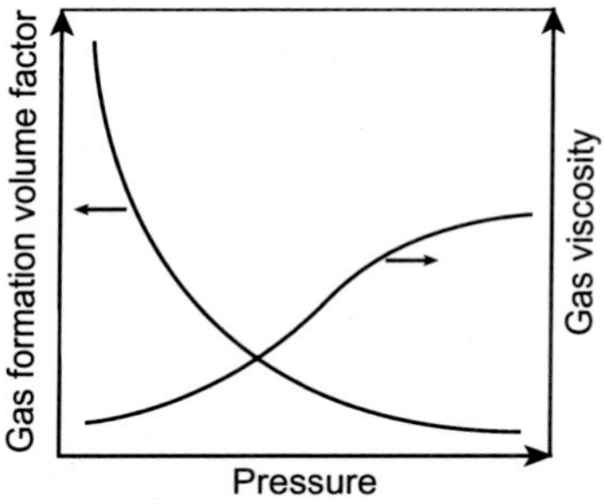

FIG. 11.1 Gas properties.

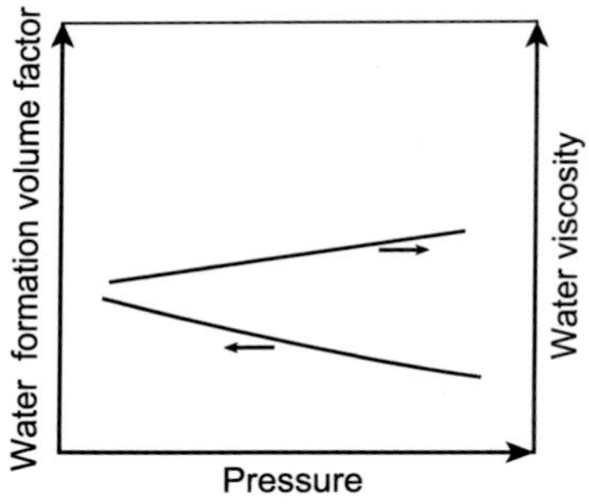

FIG. 11.2 Water properties.

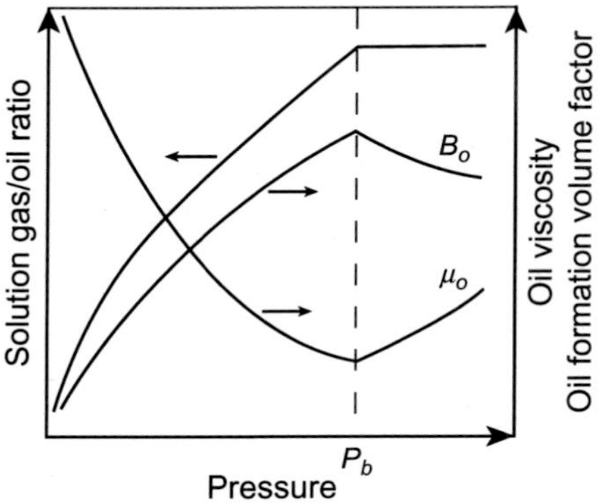

FIG. 11.3 Oil properties.

these properties are similar to those for a slightly compressible fluid and can be estimated from the values at the bubble-point pressure using

$$B_o = \frac{B_{ob}}{[1 + c_o(p - p_b)]} \tag{11.2}$$

and

$$\mu_o = \frac{\mu_{ob}}{\left[1 - c_\mu(p - p_b)\right]} \tag{11.3}$$

where c_o and c_μ are treated as constants although they, in general, depend on the solution-gas/oil ratio at the bubble-point pressure.

The densities of oil, water, and gas at standard conditions are usually supplied to simulators to aid in estimating the phase densities at reservoir temperature and any pressure using

$$\rho_w = \frac{\rho_{wsc}}{B_w} \tag{11.4}$$

for the water phase,

$$\rho_g = \frac{\rho_{gsc}}{\alpha_c B_g} \tag{11.5}$$

for the gas phase,

$$\rho_{osat} = \frac{\left(\rho_{osc} + \rho_{gsc} R_{sat}/\alpha_c\right)}{B_{osat}} \tag{11.6a}$$

for the saturated oil phase (oil at saturation pressures that are below or equal to the bubble-point pressure, $p = p_{sat}$ and $p_{sat} \le p_b$), and

$$\rho_o = \rho_{ob}\left[1 + c_o(p - p_b)\right] \tag{11.6b}$$

for the undersaturated oil phase (oil at pressures above the saturation pressure, $p > p_{sat}$).

Example 11.1 Table 11.1 lists the properties of gas, water, and saturated oil at reservoir temperature. Other pertinent data are $\rho_{osc} = 45$ lbm/ft^3, $\rho_{wsc} = 67$ lbm/ft^3, $\rho_{gsc} = 0.057922$ lbm/ft^3, $c_o = 21 \times 10^{-6}$ psi^{-1}, and $c_\mu = 40 \times 10^{-6}$ psi^{-1}. Estimate the oil-, water-, and gas-phase properties (B, μ, and ρ) at the following reservoir conditions:

1. $p = 4000$ psia and $R_s = 724.92$ scf/STB
2. $p = 4000$ psia and $R_s = 522.71$ scf/STB

Solution

1. $p = 4000$ psia and $R_s = 724.92$ scf/STB

   Water and gas properties are obtained from Table 11.1 at the reported reservoir pressure, $p = 4000$ psia. Therefore, $B_w = 1.01024$ RB/B, $\mu_w = 0.5200$ cP, and Eq. (11.4) is used to estimate the water density, $\rho_w = \frac{\rho_{wsc}}{B_w} = \frac{67}{1.01024} = 66.321$ lbm/ft^3; $B_g = 0.00069$ RB/scf, $\mu_g = 0.0241$ cP, and Eq. (11.5) is used to estimate the gas density, $\rho_g = \frac{\rho_{gsc}}{\alpha_c B_g} = \frac{0.057922}{5.614583 \times 0.00069} = 14.951$ lbm/ft^3. Note that if the sought entry value ($p = 4000$ in this example) is not listed in the table, linear interpolation

TABLE 11.1 Fluid PVT and viscosity data for Example 11.1.

Pressure (psia)	Oil			Water		Gas	
	R_s (scf/STB)	B_o (RB/STB)	μ_o (cP)	B_w (RB/B)	μ_w (cP)	B_g (RB/scf)	μ_g (cP)
1500	292.75	1.20413	1.7356	1.02527	0.5200	0.00180	0.0150
2000	368.00	1.23210	1.5562	1.02224	0.5200	0.00133	0.0167
2500	443.75	1.26054	1.4015	1.01921	0.5200	0.00105	0.0185
3000	522.71	1.29208	1.2516	1.01621	0.5200	0.00088	0.0204
3500	619.00	1.32933	1.1024	1.01321	0.5200	0.00077	0.0222
4000	724.92	1.37193	0.9647	1.01024	0.5200	0.00069	0.0241
4500	818.60	1.42596	0.9180	1.00731	0.5200	0.00064	0.0260

within table entries is used (linear interpolation is widely used in commercial reservoir simulators). For oil properties, we first determine if the oil, at the reported pressure conditions, falls into the saturated or undersaturated oil region using the saturated oil properties reported in Table 11.1. From the pressure entries in the table, $R_{sat}=724.92$ scf/STB at $p_{sat}=4000$ psia. Since $R_{sat}=724.92=R_s$, then $p=4000=p_{sat}$, the oil in the reservoir is saturated, and the oil properties at the reported pressure conditions are those of saturated oil at $p=p_{ast}=4000$ psia. Second, $R_s=R_{sat}=724.92$ scf/STB, $B_o=B_{osat}=1.37193$ RB/STB, $\mu_o=\mu_{osat}=0.9647$ cP, and the density of oil is estimated using Eq. (11.6a) at p_{sat}, which gives

$$\rho_{osat}=\frac{\left(\rho_{osc}+\rho_{gsc}R_{sat}/\alpha_c\right)}{B_{osat}}=\frac{(45+0.057922\times 724.92/5.614583)}{1.37193}$$
$$=32.943\,\text{lbm/ft}^3$$

Therefore, $\rho_o=\rho_{osat}=32.943$ lbm/ft^3 because $p=p_{ast}$.

2. $p=4000$ psia and $R_s=522.71$ scf/STB

Water and gas properties are obtained from Table 11.1 at the reported reservoir pressure, $p=4000$ psia as in part 1. Therefore, $B_w=1.01024$ RB/B, $\mu_w=0.5200$ cP, and Eq. (11.4) is used to estimate the water density, $\rho_w=\frac{\rho_{wsc}}{B_w}=\frac{67}{1.01024}=66.321$ lbm/ft^3; $B_g=0.00069$ RB/scf, $\mu_g=0.0241$ cP, and Eq. (11.5) is used to estimate the gas density, $\rho_g=\frac{\rho_{gsc}}{\alpha_c B_g}=\frac{0.057922}{5.614583\times 0.00069}=14.951$ lbm/ft^3. For oil properties, we first determine if the oil, at the reported pressure, falls into the saturated or undersaturated oil region using the saturated oil properties reported in Table 11.1. From the pressure entries in the table, $R_{sat}=724.92$ scf/STB at $p_{sat}=4000$ psia. Since $R_{sat}=724.92>522.71=R_s$, the oil in the reservoir is undersaturated. The oil bubble-point pressure is obtained by searching the table for the saturation pressure that corresponds to $R_{sb}=R_{sat}=R_s=522.71$ scf/STB. The search in Table 11.1 results in $p_b=p_{sat}=3000$ psia, $B_{ob}=B_{osat}=1.29208$ RB/STB, $\mu_{ob}=\mu_{osat}=1.2516$ cP, and $B_{gb}=0.00088$ RB/scf. The FVF, viscosity, and density of undersaturated oil at $p=4000$ psia are estimated using Eqs. (11.2), (11.3), and (11.6b), respectively. The use of Eq. (11.6b) requires the calculation of ρ_{ob} from Eq. (11.6a). Therefore,

$$B_o=\frac{B_{ob}}{[1+c_o(p-p_b)]}=\frac{1.29208}{\left[1+\left(21\times 10^{-6}\right)(4000-3000)\right]}=1.26550\,\text{RB/STB}$$

$$\mu_o=\frac{\mu_{ob}}{[1-c_\mu(p-p_b)]}=\frac{1.2516}{\left[1-\left(40\times 10^{-6}\right)(4000-3000)\right]}=1.3038\,\text{cP}$$

$$\rho_{ob} = \frac{\left(\rho_{osc} + \rho_{gsc} R_{sb}/\alpha_c\right)}{B_{ob}} = \frac{(45 + 0.057922 \times 522.71/5.614583)}{1.29208}$$
$$= 39.001\,\text{lbm/ft}^3$$

and

$$\rho_o = \rho_{ob}[1 + c_o(p - p_b)] = 39.001 \times \left[1 + \left(21 \times 10^{-6}\right)(4000 - 3000)\right]$$
$$= 39.820\text{lbm/ft}^3$$

11.2.2 Relative permeability

In multiphase flow, oil, water, and gas may coexist in any reservoir block at any time. The capacity of the rock to transmit any phase through its pores is described by the relative permeability to that phase. The flow rate of the same phase is described by Darcy's law in multiphase flow (Section 11.2.4). Figs. 11.4 and 11.5 show sketches of the phase relative permeability dependence on saturation in two-phase oil/water and gas/oil systems.

The relative permeability in three-phase oil/water/gas system can be estimated using data obtained from two-phase systems (Figs. 11.4 and 11.5). A widely used model for that purpose is Stone's Three-Phase Model II presented by Eqs. (11.7)–(11.9):

$$k_{rw} = \text{f}(S_w) \tag{11.7}$$

for the water phase,

$$k_{rg} = \text{f}\left(S_g\right) \tag{11.8}$$

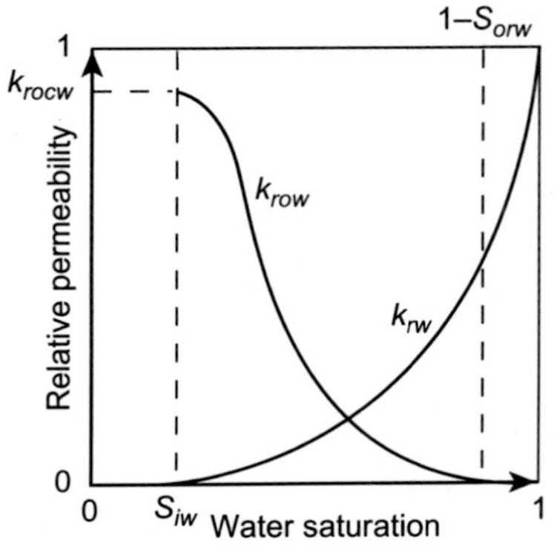

FIG. 11.4 O/W relative permeability.

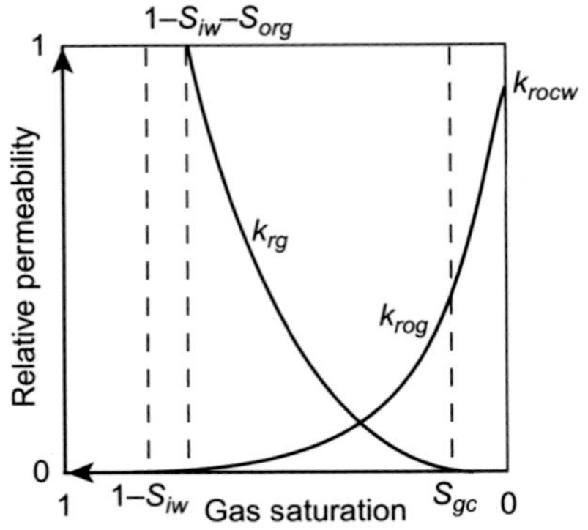

FIG. 11.5 G/O relative permeability.

for the gas phase, and

$$k_{ro} = k_{rocw}\left[\left(k_{row}/k_{rocw} + k_{rw}\right)\left(k_{rog}/k_{rocw} + k_{rg}\right) - \left(k_{rw} + k_{rg}\right)\right] \tag{11.9}$$

for the oil phase, where $k_{ro} \geq 0$, k_{row} and k_{rw} at a given S_w are obtained from two-phase oil/water data (Fig. 11.4), k_{rog} and k_{rg} at a given S_g are obtained from two-phase oil/gas data (Fig. 11.5), and k_{rocw} is the relative permeability to oil at irreducible water saturation ($k_{row}|_{S_w=S_{iw}}$ obtained from Fig. 11.4 or $k_{rog}|_{S_g=0}$ obtained from Fig. 11.5). It should be mentioned that the oil/gas relative permeability data in Fig. 11.5 must be obtained in the presence of irreducible water. Although Eq. (11.9) reduces to $k_{ro} = k_{row}$ at $S_g = 0$ (i.e., for a two-phase oil/water system) and to $k_{ro} = k_{rog}$ at $S_w = S_{iw}$ (i.e., for a two-phase oil/gas system), the estimation of relative permeabilities uses Fig. 11.4 for oil/water reservoirs and Fig. 11.5 for oil/gas reservoirs.

Example 11.2 Table 11.2 lists two-phase oil/water and oil/gas relative permeability data that will be used in three-phase relative permeability calculations. Estimate the relative permeability to oil, water, and gas using Stone's Three-Phase Model II for the following fluid saturation distributions:

1. $S_o = 0.315$, $S_w = 0.490$, and $S_g = 0.195$
2. $S_o = 0.510$, $S_w = 0.490$, and $S_g = 0.000$
3. $S_o = 0.675$, $S_w = 0.130$, and $S_g = 0.195$

Solution

1. $S_o = 0.315$, $S_w = 0.490$, and $S_g = 0.195$

At $S_w = 0.490$, $k_{rw} = 0.0665$ and $k_{row} = 0.3170$ using the two-phase oil/water relative permeability data. At $S_g = 0.195$, $k_{rg} = 0.0195$, and $k_{rog} = 0.2919$ using the two-phase oil/gas relative permeability data. According to Stone's Three-Phase Model II, the application of Eq. (11.7) gives relative permeability to the water phase, that is, $k_{rw} = 0.0665$; the application of Eq. (11.8) gives relative

TABLE 11.2 Two-phase relative permeability data (Coats et al., 1974).

Oil/water data			Oil/gas data		
S_w	k_{rw}	k_{row}	S_g	k_{rg}	k_{rog}
0.130	0.0000	1.0000	0.000	0.0000	1.0000
0.191	0.0051	0.9990	0.101	0.0026	0.5169
0.250	0.0102	0.8000	0.150	0.0121	0.3373
0.294	0.0168	0.7241	0.195	0.0195	0.2919
0.357	0.0275	0.6206	0.250	0.0285	0.2255
0.414	0.0424	0.5040	0.281	0.0372	0.2100
0.490	0.0665	0.3170	0.337	0.0500	0.1764
0.557	0.0970	0.3029	0.386	0.0654	0.1433
0.630	0.1148	0.1555	0.431	0.0761	0.1172
0.673	0.1259	0.0956	0.485	0.0855	0.0883
0.719	0.1381	0.0576	0.567	0.1022	0.0461
0.789	0.1636	0.0000	0.605	0.1120	0.0294
1.000	1.0000	0.0000	0.800	0.1700	0.0000

permeability to the gas phase, that is, $k_{rg}=0.0195$; and the application of Eq. (11.9) gives relative permeability to the oil phase, that is,

$$k_{ro}=1.0000[(0.3170/1.0000+0.0665)(0.2919/1.0000+0.0195) \\ -(0.0665+0.0195)]$$

or $k_{ro}=0.03342$. Note that $k_{rocw}=1.0000$ from the oil/water data at the irreducible water saturation of 0.13 or from oil/gas data at $S_g=0$.

2. $S_o=0.510$, $S_w=0.490$, and $S_g=0.000$

This is an example of two-phase flow of oil and water only because the gas saturation is zero. Therefore, at $S_w=0.490$, $k_{rw}=0.0665$, and $k_{ro}=k_{row}=0.3170$. Alternatively, the application of Stone's Three-Phase Model II gives $k_{rw}=0.0665$ and $k_{row}=0.3170$ at $S_w=0.490$ from the oil/water data and $k_{rg}=0.0000$, $k_{rog}=1.0000$ at $S_g=0.000$ from the oil/gas data. Therefore, $k_{rw}=0.0665$, $k_{rg}=0.0000$, and the application of Eq. (11.9) gives

$$k_{ro}=1.0000[(0.3170/1.0000+0.0665)(1.0000/1.0000+0.0000) \\ -(0.0665+0.0000)]$$

or $k_{ro}=0.3170$.

3. $S_o=0.675$, $S_w=0.130$, and $S_g=0.195$

This is a case of two-phase flow of oil and gas only because the water saturation is at the irreducible value of 0.130. Therefore, at $S_g=0.195$, $k_{rg}=0.0195$ and $k_{ro}=k_{rog}=0.2919$. Alternatively, the application of Stone's Three-Phase Model II gives $k_{rw}=0.0000$ and $k_{row}=1.0000$ at $S_w=0.130$ from the oil/water data and $k_{rg}=0.0195$, $k_{rog}=0.2919$ at $S_g=0.195$ from the oil/gas data. Therefore, $k_{rw}=0.0000$, $k_{rg}=0.0195$, and the application of Eq. (11.9) gives

$$k_{ro}=1.0000[(1.0000/1.0000+0.0000)(0.2919/1.0000+0.0195)-(0.0000+0.0195)]$$

or $k_{ro}=0.2919$.

The results of parts 2 and 3 confirm that Stone's Three-Phase Model II reduces to two-phase oil/water relative permeability data at zero gas saturation and to two-phase oil/gas relative permeability data at irreducible water saturation.

11.2.3 Capillary pressure

The coexistence of more than one phase in the capillary size pores of the reservoir rock is responsible for the creation of pressure difference between any two phases across the interface. This pressure difference is called capillary pressure, and it is a function of fluid saturation. Capillary pressure is defined as the pressure of the nonwetting phase minus the pressure of the wetting phase. Therefore,

$$P_{cow}=p_o-p_w=\mathrm{f}(S_w) \tag{11.10}$$

for a two-phase oil/water system in water-wet rock, and

$$P_{cgo}=p_g-p_o=\mathrm{f}(S_g) \tag{11.11}$$

for a two-phase gas/oil system. Note that in the presence of gas, liquid (oil or water) always wets the rock. Figs. 11.6 and 11.7 show sketches of the dependence of f(S_w) and f(S_g) on saturation.

Leverett and Lewis (1941) reported that the capillary pressures in a three-phase oil/water/gas system can be described by those obtained from two-phase systems.

Example 11.3 Table 11.3 lists two-phase oil/water and gas/oil capillary pressure data. Estimate the oil/water and gas/oil capillary pressures in a three-phase oil/water/gas reservoir at $S_o=0.26$, $S_w=0.50$, and $S_g=0.24$.

Solution

Using two-phase oil/water capillary pressure data, $P_{cow}=2.42$ psi at $S_w=0.50$. Also, using two-phase gas/oil capillary pressure data, $P_{cgo}=0.54$ psi at $S_g=0.24$. Now, the three-phase capillary pressure data at the given fluid saturations are those obtained from two-phase data; that is, $P_{cow}=2.42$ psi and $P_{cgo}=0.54$ psi.

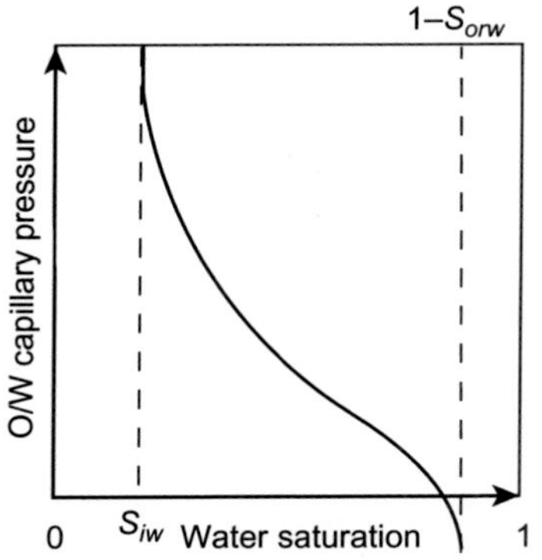

FIG. 11.6 O/W capillary pressure.

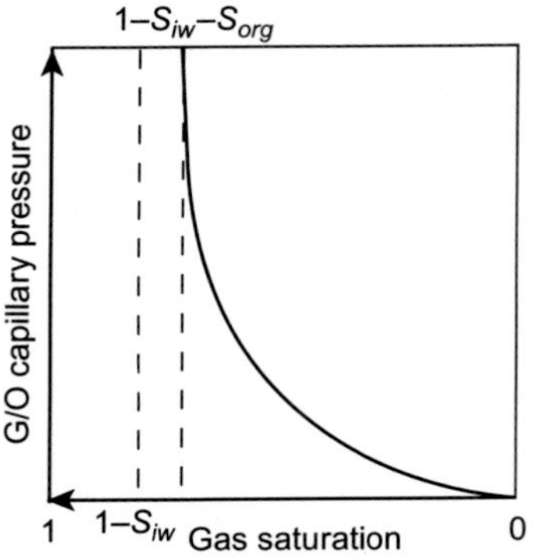

FIG. 11.7 G/O capillary pressure.

11.2.4 Darcy's law in multiphase flow

In multiphase flow in petroleum reservoirs, the fluid volumetric velocity (flow rate per unit cross-sectional area) of phase $p=o$, w, or g from block $i-1$ to block i is given by

$$u_{px}\big|_{x_{i-1/2}} = \beta_c \frac{(k_x k_{rp})\big|_{x_{i-1/2}}}{\mu_p\big|_{x_{i-1/2}}} \left[\frac{(\Phi_{p_{i-1}} - \Phi_{p_i})}{\Delta x_{i-1/2}} \right] \tag{11.12}$$

The potential difference between block $i-1$ and block i is

$$\Phi_{p_{i-1}} - \Phi_{p_i} = (p_{p_{i-1}} - p_{p_i}) - \gamma_{p_{i-1/2}}(Z_{i-1} - Z_i) \tag{11.13}$$

for $p=o$, w, or g.

TABLE 11.3 Two-phase capillary pressure data.

Oil/water data		Gas/oil data	
S_w	P_{cow} (psi)	S_g	P_{cgo} (psi)
0.20	16.00	0.04	0.02
0.25	8.60	0.24	0.54
0.30	6.00	0.34	1.02
0.40	3.56	0.49	2.08
0.50	2.42	0.59	2.98
0.60	1.58	0.69	4.44
0.70	0.86	0.74	5.88
0.80	0.20	0.79	9.52
0.90	0.00		

Substituting Eq. (11.13) into Eq. (11.12) yields

$$u_{px}\big|_{x_{i-1/2}} = \beta_c \frac{(k_x k_{rp})\big|_{x_{i-1/2}}}{\mu_p\big|_{x_{i-1/2}}} \left[\frac{(p_{p_{i-1}} - p_{p_i}) - \gamma_{p_{i-1/2}}(Z_{i-1} - Z_i)}{\Delta x_{i-1/2}} \right] \tag{11.14}$$

Eq. (11.14) can be rewritten as

$$u_{px}\big|_{x_{i-1/2}} = \beta_c \frac{k_x\big|_{x_{i-1/2}}}{\Delta x_{i-1/2}} \left(\frac{k_{rp}}{\mu_p} \right)\bigg|_{x_{i-1/2}} \left[(p_{p_{i-1}} - p_{p_i}) - \gamma_{p_{i-1/2}}(Z_{i-1} - Z_i) \right] \tag{11.15a}$$

for $p = o$, w, or g.

Likewise, the fluid volumetric velocity of phase p from block i to block $i+1$ is expressed as

$$u_{px}\big|_{x_{i+1/2}} = \beta_c \frac{k_x\big|_{x_{i+1/2}}}{\Delta x_{i+1/2}} \left(\frac{k_{rp}}{\mu_p} \right)\bigg|_{x_{i+1/2}} \left[(p_{p_i} - p_{p_{i+1}}) - \gamma_{p_{i+1/2}}(Z_i - Z_{i+1}) \right] \tag{11.15b}$$

for $p = o$, w, or g.

11.3 Multiphase flow models

In this section, we derive the equations for two-phase and three-phase flow models. As in the case for single-phase, the flow equations are obtained by first discretizing the reservoir into gridblocks as shown in Fig. 4.1 (or gridpoints as shown in Fig. 5.1), followed by writing the material balance for the component under consideration for block i and combining it with Darcy's law and FVF. We must clarify that once the reservoir is discretized and elevation and rock properties are assigned to gridblocks (or gridpoints), space is no longer a variable, and the functions that depend on space, such as interblock properties, become well defined. In other words, reservoir discretization removes space from being a variable in the formulation of the problem. In the black-oil model, we have three components in the system compared with one component in a single-phase flow model. The three components are oil, water, and gas at standard conditions ($c=o$, w, g). As implied in Section 11.2.1, the oil component ($c=o$) is contained in the oil phase ($p=o$), the water component ($c=w$) is contained in the water phase ($p=w$), and the gas component ($c=g$) is distributed between the oil phase ($p=o$) as solution gas and the gas phase ($p=fg=g$) as free gas. In deriving the flow equation for the gas component, we fictitiously split the gas component ($c=g$) into a free-gas component ($c=fg$) that is contained in the gas phase ($p=g$) and a solution-gas component ($c=sg$) that is contained in the oil phase ($p=o$); that is, $c=g=fg+sg$. In addition, the oil phase consists of the oil component and the solution-gas component. Close inspection of the density of the gas-saturated oil phase as given by Eq. (11.6a) gives the definition of the apparent density of the oil component and the solution-gas component at reservoir conditions (on the basis of the oil-phase volume) as ρ_{osc}/B_o and $\rho_{gsc}R_s/(\alpha_c B_o)$, respectively. It does not need mentioning that R_s and B_o are saturated oil properties (i.e., $R_s=R_{sat}$ and $B_o=B_{osat}$) and the density of the oil component and the solution-gas component at standard conditions are ρ_{osc} and ρ_{gsc}, respectively. The flow equations for the water and free-gas components ($c=w, fg$) are similar in form because each of these two components is the sole occupant of its phase. However, the flow equations for the oil component ($c=o$) and the solution-gas component ($c=sg$) (both occupy the oil phase) are obtained by considering the flow of the oil phase at reservoir conditions and the apparent densities of these two components.

Fig. 11.8 shows block i and its neighboring blocks in the x-direction (block $i-1$ and block $i+1$). At any instant in time, oil, water, free-gas, and solution-gas

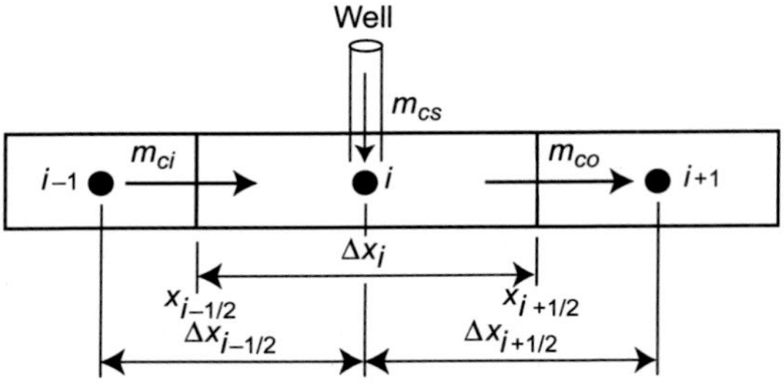

FIG. 11.8 Block i as a reservoir volume element in 1-D flow.

components enter block i, coming from block $i-1$ across its $x_{i-1/2}$ face at mass rates of $w_{cx}|_{x_{i-1/2}}$, and leave to block $i+1$ across its $x_{i+1/2}$ face at mass rates of $w_{cx}|_{x_{i+1/2}}$. Any of the components $c=o$, w, fg, and sg may also enter block i through a well at mass rates of q_{cm_i}. The mass of component $c=o$, w, fg, or sg contained in a unit volume of rock is m_{cv}. In the following steps, we derive the material balance equation for component $c=o, w, fg$, and sg for block i written over a time step $\Delta t = t^{n+1} - t^n$. For block i in Fig. 11.8, the mass balance equation for component c can be written as

$$m_{ci}|_{x_{i-1/2}} - m_{co}|_{x_{i+1/2}} + m_{cs_i} = m_{ca_i} \tag{11.16}$$

where

$$m_{ci}|_{x_{i-1/2}} = \int_{t^n}^{t^{n+1}} w_{cx}|_{x_{i-1/2}} \mathrm{d}t \tag{11.17}$$

$$m_{co}|_{x_{i+1/2}} = \int_{t^n}^{t^{n+1}} w_{cx}|_{x_{i+1/2}} \mathrm{d}t \tag{11.18}$$

and

$$m_{cs_i} = \int_{t^n}^{t^{n+1}} q_{cm_i} \mathrm{d}t \tag{11.19}$$

because terms like $w_{cx}|_{x_{i-1/2}}$, $w_{cx}|_{x_{i+1/2}}$, and q_{cm_i} for an already discretized reservoir are functions of time only, as discussed earlier. Further justification is presented later in this section.

Substitution of Eqs. (11.17) through (11.19) into Eq. (11.16) yields

$$\int_{t^n}^{t^{n+1}} w_{cx}|_{x_{i-1/2}} \mathrm{d}t - \int_{t^n}^{t^{n+1}} w_{cx}|_{x_{i+1/2}} \mathrm{d}t + \int_{t^n}^{t^{n+1}} q_{cm_i} \mathrm{d}t = m_{ca_i} \tag{11.20}$$

The mass accumulation of component c is defined as

$$m_{ca_i} = \Delta_t (V_b m_{cv})_i = V_{b_i} \Delta_t m_{cv_i} = V_{b_i} \left(m_{cv_i}^{n+1} - m_{cv_i}^{n} \right) \tag{11.21}$$

Note that the mass flow rate and mass flux for component c are related through

$$w_{cx} = \dot{m}_{cx} A_x \tag{11.22}$$

mass flux ($\dot{m}_{cx}$) can be expressed in terms of the component density (or apparent density) and phase volumetric velocity as

$$\dot{m}_{wx} = \alpha_c \rho_w u_{wx} \tag{11.23a}$$

$$\dot{m}_{fgx} = \alpha_c \rho_g u_{gx} \tag{11.23b}$$

$$\dot{m}_{ox} = \alpha_c \left(\frac{\rho_{osc}}{B_o} \right) u_{ox} \tag{11.23c}$$

and

$$\dot{m}_{sgx} = \alpha_c \left(\frac{\rho_{gsc} R_s}{\alpha_c B_o} \right) u_{ox} \tag{11.23d}$$

The mass of component c contained per unit rock volume (m_{cv}) can be expressed in terms of porosity, fluid saturation, and component density (or apparent density) as

$$m_{wv} = \phi \rho_w S_w \tag{11.24a}$$

$$m_{fgv} = \phi \rho_g S_g \tag{11.24b}$$

$$m_{ov} = \phi \left(\frac{\rho_{osc}}{B_o} \right) S_o \tag{11.24c}$$

and

$$m_{sgv} = \phi \left(\frac{\rho_{gsc} R_s}{\alpha_c B_o} \right) S_o \tag{11.24d}$$

The mass production rate of component c (q_{cm}) can be expressed in terms of the phase volumetric production rate (q_p) and component density (or apparent density) as

$$q_{wm} = \alpha_c \rho_w q_w \tag{11.25a}$$

$$q_{fgm} = \alpha_c \rho_g q_{fg} = \alpha_c \rho_g q_g \tag{11.25b}$$

$$q_{om} = \alpha_c \left(\frac{\rho_{osc}}{B_o} \right) q_o \tag{11.25c}$$

and

$$q_{sgm} = \alpha_c \left(\frac{\rho_{gsc} R_s}{\alpha_c B_o} \right) q_o \tag{11.25d}$$

It should be mentioned that, in Eqs. (11.23) through (11.25), u_{ox}, u_{wx}, u_{gx}, q_o, q_w, q_g, S_o, S_w, S_g, B_o, B_w, B_g, R_s, ρ_w, and ρ_g are all phase properties, whereas ρ_{osc}, ρ_{wsc}, and ρ_{gsc} are component properties.

Substitution of Eqs. (11.21) and (11.22) into Eq. (11.20) yields

$$\int_{t^n}^{t^{n+1}} (\dot{m}_{cx}A_x)|_{x_{i-1/2}} dt - \int_{t^n}^{t^{n+1}} (\dot{m}_{cx}A_x)|_{x_{i+1/2}} dt + \int_{t^n}^{t^{n+1}} q_{cm_i} dt = V_{b_i}\left(m_{cv_i}^{n+1} - m_{cv_i}^{n}\right) \tag{11.26}$$

Substituting Eqs. (11.23) through (11.25) into Eq. (11.26) after using $\rho_w = \rho_{wsc}/B_w$ in Eqs. (11.23a), (11.24a), and (11.25a) and $\rho_g = \rho_{gsc}/(\alpha_c B_g)$ in Eqs. (11.23b), (11.24b), and (11.25b); dividing by the appropriate $\alpha_c \rho_{psc}$ for $p = o, w, g$; and noting that $q_p/B_p = q_{psc}$ for $p = o, w, g$ yield

$$\int_{t^n}^{t^{n+1}} \left(\frac{u_{wx}A_x}{B_w}\right)\Bigg|_{x_{i-1/2}} dt - \int_{t^n}^{t^{n+1}} \left(\frac{u_{wx}A_x}{B_w}\right)\Bigg|_{x_{i+1/2}} dt + \int_{t^n}^{t^{n+1}} q_{wsc_i} dt = \frac{V_{b_i}}{\alpha_c}\left[\left(\frac{\phi S_w}{B_w}\right)_i^{n+1} - \left(\frac{\phi S_w}{B_w}\right)_i^{n}\right] \tag{11.27a}$$

for the water component,

$$\int_{t^n}^{t^{n+1}} \left(\frac{u_{gx}A_x}{B_g}\right)\Bigg|_{x_{i-1/2}} dt - \int_{t^n}^{t^{n+1}} \left(\frac{u_{gx}A_x}{B_g}\right)\Bigg|_{x_{i+1/2}} dt + \int_{t^n}^{t^{n+1}} q_{fgsc_i} dt = \frac{V_{b_i}}{\alpha_c}\left[\left(\frac{\phi S_g}{B_g}\right)_i^{n+1} - \left(\frac{\phi S_g}{B_g}\right)_i^{n}\right] \tag{11.27b}$$

for the free-gas component,

$$\int_{t^n}^{t^{n+1}} \left(\frac{u_{ox}A_x}{B_o}\right)\Bigg|_{x_{i-1/2}} dt - \int_{t^n}^{t^{n+1}} \left(\frac{u_{ox}A_x}{B_o}\right)\Bigg|_{x_{i+1/2}} dt + \int_{t^n}^{t^{n+1}} q_{osc_i} dt = \frac{V_{b_i}}{\alpha_c}\left[\left(\frac{\phi S_o}{B_o}\right)_i^{n+1} - \left(\frac{\phi S_o}{B_o}\right)_i^{n}\right] \tag{11.27c}$$

for the oil component, and

$$\int_{t^n}^{t^{n+1}} \left(\frac{R_s u_{ox}A_x}{B_o}\right)\Bigg|_{x_{i-1/2}} dt - \int_{t^n}^{t^{n+1}} \left(\frac{R_s u_{ox}A_x}{B_o}\right)\Bigg|_{x_{i+1/2}} dt + \int_{t^n}^{t^{n+1}} R_{s_i} q_{osc_i} dt = \frac{V_{b_i}}{\alpha_c}\left[\left(\frac{\phi R_s S_o}{B_o}\right)_i^{n+1} - \left(\frac{\phi R_s S_o}{B_o}\right)_i^{n}\right] \tag{11.27d}$$

for the solution-gas component.

Consider the equation for the water component. Water-phase volumetric velocities from block $i-1$ to block i and from block i to block $i+1$ are given by Eq. (11.15) for $p=w$. Substitution of Eq. (11.15) for $p=w$ into Eq. (11.27a) yields

$$\int_{t^n}^{t^{n+1}}\left\{\left(\beta_c\frac{k_xA_xk_{rw}}{\mu_wB_w\Delta x}\right)\bigg|_{x_{i-1/2}}\left[(p_{w_{i-1}}-p_{w_i})-\gamma_{w_{i-1/2}}(Z_{i-1}-Z_i)\right]\right\}dt$$
$$-\int_{t^n}^{t^{n+1}}\left\{\left(\beta_c\frac{k_xA_xk_{rw}}{\mu_wB_w\Delta x}\right)\bigg|_{x_{i+1/2}}\left[(p_{w_i}-p_{w_{i+1}})-\gamma_{w_{i+1/2}}(Z_i-Z_{i+1})\right]\right\}dt+\int_{t^n}^{t^{n+1}}q_{wsc_i}dt$$
$$=\frac{V_{b_i}}{\alpha_c}\left[\left(\frac{\phi S_w}{B_w}\right)_i^{n+1}-\left(\frac{\phi S_w}{B_w}\right)_i^n\right] \tag{11.28}$$

Define the transmissibility of phase w in the x-direction between block i and neighboring block $i\mp1$ as

$$T_{wx_{i\mp1/2}}=\left(\beta_c\frac{k_xA_xk_{rw}}{\mu_wB_w\Delta x}\right)\bigg|_{x_{i\mp1/2}} \tag{11.29}$$

Combining Eq. (11.29) and Eq. (11.28) and rearranging the terms result in

$$\int_{t^n}^{t^{n+1}}\left\{T_{wx_{i-1/2}}\left[(p_{w_{i-1}}-p_{w_i})-\gamma_{w_{i-1/2}}(Z_{i-1}-Z_i)\right]\right\}dt$$
$$+\int_{t^n}^{t^{n+1}}\left\{T_{wx_{i+1/2}}\left[(p_{w_{i+1}}-p_{w_i})-\gamma_{w_{i+1/2}}(Z_{i+1}-Z_i)\right]\right\}dt \tag{11.30}$$
$$+\int_{t^n}^{t^{n+1}}q_{wsc_i}dt=\frac{V_{b_i}}{\alpha_c}\left[\left(\frac{\phi S_w}{B_w}\right)_i^{n+1}-\left(\frac{\phi S_w}{B_w}\right)_i^n\right]$$

The derivation of Eq. (11.30) is rigorous and involves no assumptions other than the validity of Darcy's law for multiphase flow (Eq. 11.15) to estimate the water-phase volumetric velocities between block i and its neighboring blocks $i-1$ and $i+1$. Such validity is widely accepted by petroleum engineers. As discussed in Section 2.6.2 for single-phase flow, once the reservoir is discretized into blocks (or nodes), the interblock geometric factor between block i and its neighboring block $i \mp 1$ $\left[\left(\beta_c \frac{k_x A_x}{\Delta x}\right)\bigg|_{x_{i \mp 1/2}}\right]$ is constant, independent of space and time. In addition, the pressure-dependent term $(\mu_w B_w)|_{x_{i \mp 1/2}}$ of transmissibility of the water phase uses some average viscosity and FVF for block i and neighboring block $i \mp 1$, or some weight (upstream weighting, average weighting, etc.) at any instant of time t. In other words, the term $(\mu_w B_w)|_{x_{i \mp 1/2}}$ is not a function of space but a function of time as the block pressures change with time. Similarly, the relative permeability of the water phase between block i and neighboring block $i \mp 1$ at any instant of time t ($k_{rw}|_{x_{i \mp 1/2}}$) uses the upstream value or two-point upstream value of block i and neighboring block $i \mp 1$ that are already fixed in space. In other words, the term $k_{rw}|_{x_{i \mp 1/2}}$ is not a function of space but a function of time as the block saturations change with time. Hence, transmissibility $T_{wx_{i \mp 1/2}}$ between block i and its neighboring block $i \mp 1$ is a function of time only; it does not depend on space at any instant of time.

As discussed in Chapter 2, the integral $\int_{t^n}^{t^{n+1}} F(t)\mathrm{d}t$ is equal to the area under the curve $F(t)$ in the interval $t^n \le t \le t^{n+1}$. This area is also equal to the area of a rectangle with the dimensions of $F(t^m)$ and Δt where F^m is evaluated at time t^m and $t^n \le t^m \le t^{n+1}$. Therefore,

$$\int_{t^n}^{t^{n+1}} F(t)\mathrm{d}t = \int_{t^n}^{t^{n+1}} F(t^m)\mathrm{d}t = \int_{t^n}^{t^{n+1}} F^m \mathrm{d}t = F^m \int_{t^n}^{t^{n+1}} \mathrm{d}t = F^m t\big|_{t^n}^{t^{n+1}} = F^m\left(t^{n+1} - t^n\right)$$
$$= F^m \Delta t \tag{11.31}$$

Substituting Eq. (11.31) for the integrals into Eq. (11.30) and dividing by Δt result in the flow equation for the water component,

$$\begin{aligned} &T^m_{wx_{i-1/2}}\left[\left(p^m_{w_{i-1}} - p^m_{w_i}\right) - \gamma^m_{w_{i-1/2}}(Z_{i-1} - Z_i)\right] \\ &+ T^m_{wx_{i+1/2}}\left[\left(p^m_{w_{i+1}} - p^m_{w_i}\right) - \gamma^m_{w_{i+1/2}}(Z_{i+1} - Z_i)\right] \\ &+ q^m_{wsc_i} = \frac{V_{b_i}}{\alpha_c \Delta t}\left[\left(\frac{\phi S_w}{B_w}\right)^{n+1}_i - \left(\frac{\phi S_w}{B_w}\right)^n_i\right] \end{aligned} \tag{11.32a}$$

Steps similar to those that resulted in Eq. (11.32a) can be carried out on Eqs. (11.27b), (11.27c), and (11.27d) to derive the flow equations for the free-gas, oil, and solution-gas components, respectively.

For the free-gas component,

$$
\begin{aligned}
&T^m_{gx_{i-1/2}}\left[\left(p^m_{g_{i-1}}-p^m_{g_i}\right)-\gamma^m_{g_{i-1/2}}(Z_{i-1}-Z_i)\right] \\
&+T^m_{gx_{i+1/2}}\left[\left(p^m_{g_{i+1}}-p^m_{g_i}\right)-\gamma^m_{g_{i+1/2}}(Z_{i+1}-Z_i)\right] \\
&+q^m_{fgsc_i}=\frac{V_{b_i}}{\alpha_c\Delta t}\left[\left(\frac{\phi S_g}{B_g}\right)^{n+1}_i-\left(\frac{\phi S_g}{B_g}\right)^n_i\right]
\end{aligned}
\tag{11.32b}
$$

For the oil component,

$$
\begin{aligned}
&T^m_{ox_{i-1/2}}\left[\left(p^m_{o_{i-1}}-p^m_{o_i}\right)-\gamma^m_{o_{i-1/2}}(Z_{i-1}-Z_i)\right] \\
&+T^m_{ox_{i+1/2}}\left[\left(p^m_{o_{i+1}}-p^m_{o_i}\right)-\gamma^m_{o_{i+1/2}}(Z_{i+1}-Z_i)\right] \\
&+q^m_{osc_i}=\frac{V_{b_i}}{\alpha_c\Delta t}\left[\left(\frac{\phi S_o}{B_o}\right)^{n+1}_i-\left(\frac{\phi S_o}{B_o}\right)^n_i\right]
\end{aligned}
\tag{11.32c}
$$

For the solution-gas component,

$$
\begin{aligned}
&(T_{ox}R_s)^m_{i-1/2}\left[\left(p^m_{o_{i-1}}-p^m_{o_i}\right)-\gamma^m_{o_{i-1/2}}(Z_{i-1}-Z_i)\right] \\
&+(T_{ox}R_s)^m_{i+1/2}\left[\left(p^m_{o_{i+1}}-p^m_{o_i}\right)-\gamma^m_{o_{i+1/2}}(Z_{i+1}-Z_i)\right] \\
&+(R_s q_{osc})^m_i=\frac{V_{b_i}}{\alpha_c\Delta t}\left[\left(\frac{\phi R_s S_o}{B_o}\right)^{n+1}_i-\left(\frac{\phi R_s S_o}{B_o}\right)^n_i\right]
\end{aligned}
\tag{11.32d}
$$

The general flow equations for the various components present in block n, written in CVFD terminology, are now presented in Eq. (11.33).

For the water component,

$$
\begin{aligned}
&\sum_{l\in\psi_n}T^m_{w_{l,n}}\left[\left(p^m_{w_l}-p^m_{w_n}\right)-\gamma^m_{w_{l,n}}(Z_l-Z_n)\right]+\sum_{l\in\xi_n}q^m_{wsc_{l,n}}+q^m_{wsc_n} \\
&=\frac{V_{b_n}}{\alpha_c\Delta t}\left[\left(\frac{\phi S_w}{B_w}\right)^{n+1}_n-\left(\frac{\phi S_w}{B_w}\right)^n_n\right]
\end{aligned}
\tag{11.33a}
$$

For the free-gas component,

$$
\begin{aligned}
&\sum_{l\in\psi_n}T^m_{g_{l,n}}\left[\left(p^m_{g_l}-p^m_{g_n}\right)-\gamma^m_{g_{l,n}}(Z_l-Z_n)\right]+\sum_{l\in\xi_n}q^m_{fgsc_{l,n}}+q^m_{fgsc_n} \\
&=\frac{V_{b_n}}{\alpha_c\Delta t}\left[\left(\frac{\phi S_g}{B_g}\right)^{n+1}_n-\left(\frac{\phi S_g}{B_g}\right)^n_n\right]
\end{aligned}
\tag{11.33b}
$$

For the oil component,

$$\sum_{l\in\psi_n} T_{o_{l,n}}^m\left[\left(p_{o_l}^m-p_{o_n}^m\right)-\gamma_{o_{l,n}}^m(Z_l-Z_n)\right]+\sum_{l\in\xi_n} q_{osc_{l,n}}^m+q_{osc_n}^m$$
$$=\frac{V_{b_n}}{\alpha_c\Delta t}\left[\left(\frac{\phi S_o}{B_o}\right)_n^{n+1}-\left(\frac{\phi S_o}{B_o}\right)_n^n\right] \tag{11.33c}$$

For the solution-gas component,

$$\sum_{l\in\psi_n}(T_oR_s)_{l,n}^m\left[\left(p_{o_l}^m-p_{o_n}^m\right)-\gamma_{o_{l,n}}^m(Z_l-Z_n)\right]+\sum_{l\in\xi_n}(R_sq_{osc})_{l,n}^m+(R_sq_{osc})_n^m$$
$$=\frac{V_{b_n}}{\alpha_c\Delta t}\left[\left(\frac{\phi R_sS_o}{B_o}\right)_n^{n+1}-\left(\frac{\phi R_sS_o}{B_o}\right)_n^n\right] \tag{11.33d}$$

As defined in the previous chapters, $\psi_n=$ a set whose elements are the existing neighboring blocks to block n in the reservoir, $\xi_n=$ a set whose elements are the reservoir boundaries (b_L, b_S, b_W, b_E, b_N, b_U) that are shared by block n, and $q_{psc_{l,n}}^{n+1}=$ flow rate of the fictitious well that represents transfer of phase $p=o$, w, fg between reservoir boundary l and block n as a result of a boundary condition. As mentioned in Chapters 4 and 5, ξ_n is either an empty set for interior blocks or a set that contains one element for boundary blocks that fall on one reservoir boundary, two elements for boundary blocks that fall on two reservoir boundaries, or three elements for blocks that fall on three reservoir boundaries. An empty set implies that the block does not fall on any reservoir boundary; that is, block n is an interior block, and hence, $\sum_{l\in\xi_n} q_{psc_{l,n}}^{n+1}=0$ for $p=o$, w, fg.

The explicit, implicit, and Crank-Nicolson formulations are derived from Eq. (11.33) by specifying the approximation of time t^m as t^n, t^{n+1}, or $t^{n+1/2}$, which are equivalent to using the first, second, and third integral approximation methods referred to in Section 2.6.3. The explicit formulation, however, is not used in multiphase flow because of time step limitations, and the Crank-Nicolson formulation is not commonly used. Consequently, we limit our presentation to the implicit formulation. In the following equations, fluid gravity is dated at old time level n instead of new time level $n+1$, as this approximation does not introduce any noticeable errors (Coats et al. 1974).

For the water component,

$$\sum_{l\in\psi_n} T_{w_{l,n}}^{n+1}\left[\left(p_{w_l}^{n+1}-p_{w_n}^{n+1}\right)-\gamma_{w_{l,n}}^n(Z_l-Z_n)\right]+\sum_{l\in\xi_n} q_{wsc_{l,n}}^{n+1}+q_{wsc_n}^{n+1}$$
$$=\frac{V_{b_n}}{\alpha_c\Delta t}\left[\left(\frac{\phi S_w}{B_w}\right)_n^{n+1}-\left(\frac{\phi S_w}{B_w}\right)_n^n\right] \tag{11.34a}$$

For the free-gas component,

$$\sum_{l\in\psi_n} T_{g_{l,n}}^{n+1}\left[\left(p_{g_l}^{n+1}-p_{g_n}^{n+1}\right)-\gamma_{g_{l,n}}^{n}(Z_l-Z_n)\right]+\sum_{l\in\xi_n} q_{fgsc_{l,n}}^{n+1}+q_{fgsc_n}^{n+1}$$
$$=\frac{V_{b_n}}{\alpha_c\Delta t}\left[\left(\frac{\phi S_g}{B_g}\right)_n^{n+1}-\left(\frac{\phi S_g}{B_g}\right)_n^{n}\right] \tag{11.34b}$$

For the oil component,

$$\sum_{l\in\psi_n} T_{o_{l,n}}^{n+1}\left[\left(p_{o_l}^{n+1}-p_{o_n}^{n+1}\right)-\gamma_{o_{l,n}}^{n}(Z_l-Z_n)\right]+\sum_{l\in\xi_n} q_{osc_{l,n}}^{n+1}+q_{osc_n}^{n+1}$$
$$=\frac{V_{b_n}}{\alpha_c\Delta t}\left[\left(\frac{\phi S_o}{B_o}\right)_n^{n+1}-\left(\frac{\phi S_o}{B_o}\right)_n^{n}\right] \tag{11.34c}$$

For the solution-gas component,

$$\sum_{l\in\psi_n} (T_oR_s)_{l,n}^{n+1}\left[\left(p_{o_l}^{n+1}-p_{o_n}^{n+1}\right)-\gamma_{o_{l,n}}^{n}(Z_l-Z_n)\right]+\sum_{l\in\xi_n} (R_sq_{osc})_{l,n}^{n+1}+(R_sq_{osc})_n^{n+1}$$
$$=\frac{V_{b_n}}{\alpha_c\Delta t}\left[\left(\frac{\phi R_sS_o}{B_o}\right)_n^{n+1}-\left(\frac{\phi R_sS_o}{B_o}\right)_n^{n}\right] \tag{11.34d}$$

The transmissibility of phase $p=o$, w, or g between blocks l and n is defined as

$$T_{p_{l,n}}=G_{l,n}\left(\frac{1}{\mu_pB_p}\right)_{l,n}k_{rp_{l,n}} \tag{11.35}$$

where $G_{l,n}=$ the geometric factor between blocks n and l presented in Chapter 4 for a block-centered grid or Chapter 5 for a point-distributed grid.

We limit our presentation in this chapter to the $p_o-S_w-S_g$ formulation, that is, the formulation that uses p_o, S_w, and S_g as the primary unknowns in the reservoir. The secondary unknowns in this formulation are p_w, p_g, and S_o. Explicitly, the flow models of oil/water, oil/gas, and oil/water/gas use p_o-S_w, p_o-S_g, and $p_o-S_w-S_g$ formulations, respectively. Other formulations such as $p_o-p_w-p_g$, $p_o-P_{cow}-P_{cgo}$, or $p_o-P_{cow}-S_g$ break down for negligible or zero capillary pressures. To obtain the reduced set of equations for each block, we express the secondary unknowns in the flow equations in terms of the primary unknowns and thus eliminate the secondary unknowns from the flow equations. The equations used to eliminate the secondary unknowns are the saturation constraint equation (Eq. 11.1):

$$S_o=1-S_w-S_g \tag{11.36}$$

and the capillary pressure relationships (Eqs. 11.10 and 11.11),

$$p_w = p_o - P_{cow}(S_w) \tag{11.37}$$

and

$$p_g = p_o + P_{cgo}(S_g) \tag{11.38}$$

The gas/water flow model uses the $p_g - S_g$ formulation, and thus, the equations used to eliminate the secondary unknowns are

$$S_w = 1 - S_g \tag{11.39}$$

and

$$p_w = p_g - P_{cgw}(S_g) \tag{11.40}$$

Once the primary unknowns are solved for, the saturation and capillary pressure relationships (Eqs. 11.36 through 11.40) are used to solve for the secondary unknowns for each reservoir block.

11.3.1 Flow equations for oil/water flow model

The two components in the oil/water flow model are oil (or gas-free oil) and water at standard conditions. The oil phase in this case contains the oil component only. The flow equations for block n in the oil/water flow model are expressed by Eqs. (11.34a) and (11.34c). Combine these two equations with $S_o = 1 - S_w$ and $p_w = p_o - P_{cow}(S_w)$ to obtain the $p_o - S_w$ formulation.

For the oil component,

$$\sum_{l \in \psi_n} T_{o_{l,n}}^{n+1}\left[\left(p_{o_l}^{n+1} - p_{o_n}^{n+1}\right) - \gamma_{o_{l,n}}^{n}(Z_l - Z_n)\right] + \sum_{l \in \xi_n} q_{osc_{l,n}}^{n+1} + q_{osc_n}^{n+1} = \frac{V_{b_n}}{\alpha_c \Delta t}\left\{\left[\frac{\phi(1-S_w)}{B_o}\right]_n^{n+1} - \left[\frac{\phi(1-S_w)}{B_o}\right]_n^{n}\right\} \tag{11.41}$$

For the water component,

$$\sum_{l \in \psi_n} T_{w_{l,n}}^{n+1}\left[\left(p_{o_l}^{n+1} - p_{o_n}^{n+1}\right) - \left(P_{cow_l}^{n+1} - P_{cow_n}^{n+1}\right) - \gamma_{w_{l,n}}^{n}(Z_l - Z_n)\right] + \sum_{l \in \xi_n} q_{wsc_{l,n}}^{n+1} + q_{wsc_n}^{n+1} = \frac{V_{b_n}}{\alpha_c \Delta t}\left[\left(\frac{\phi S_w}{B_w}\right)_n^{n+1} - \left(\frac{\phi S_w}{B_w}\right)_n^{n}\right] \tag{11.42}$$

Eqs. (11.41) and (11.42) also model the flow of undersaturated oil and water as long as the reservoir is operated above the oil bubble-point pressure.

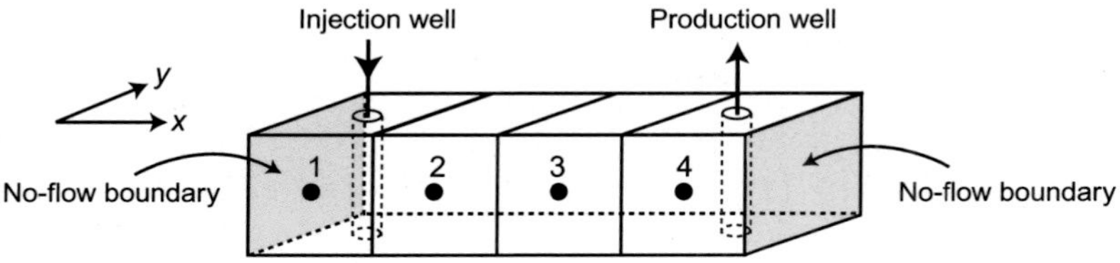

FIG. 11.9 1-D reservoir in Example 11.4.

Under such condition, the gas remains in the solution, and the undersaturated oil behaves as a slightly compressible fluid with $B_o^{\circ}=B_{ob}$ and a constant oil compressibility (c_o) whose value depends on the solution GOR at the bubble-point pressure (R_{sb}).

Example 11.4 A homogeneous, 1-D horizontal, two-phase oil/water reservoir is described by four equal blocks as shown in Fig. 11.9. Initial reservoir pressure and phase saturations are known. The reservoir left and right boundaries are sealed off to flow. The reservoir has a water injection well in gridblock 1 and a production well in gridblock 4. Write the flow equations for interior gridblock 3.

Solution

For gridblock 3, $n=3$, and $\psi_3=\{2,4\}$. Gridblock 3 is an interior block; therefore, $\xi_3=\{\ \}$, $\sum_{l\in\xi_3} q_{osc_{l,3}}^{n+1}=0$, and $\sum_{l\in\xi_3} q_{wsc_{l,3}}^{n+1}=0$. Gridblock 3 has no wells; therefore, $q_{osc_3}^{n+1}=0$ and $q_{wsc_3}^{n+1}=0$.

The oil equation is obtained by substituting the given values into Eq. (11.41) and expanding the summation terms, yielding

$$
\begin{aligned}
&T_{o_{2,3}}^{n+1}\left[\left(p_{o_2}^{n+1}-p_{o_3}^{n+1}\right)-\gamma_{o_{2,3}}^{n}(Z_2-Z_3)\right]\\
&+T_{o_{4,3}}^{n+1}\left[\left(p_{o_4}^{n+1}-p_{o_3}^{n+1}\right)-\gamma_{o_{4,3}}^{n}(Z_4-Z_3)\right]+0+0\\
&=\frac{V_{b_3}}{\alpha_c\Delta t}\left\{\left[\frac{\phi(1-S_w)}{B_o}\right]_3^{n+1}-\left[\frac{\phi(1-S_w)}{B_o}\right]_3^{n}\right\}
\end{aligned}
\tag{11.43a}
$$

Observing that $Z_2=Z_3=Z_4$ for a horizontal reservoir, the oil equation becomes

$$
\begin{aligned}
&T_{o_{2,3}}^{n+1}\left(p_{o_2}^{n+1}-p_{o_3}^{n+1}\right)+T_{o_{4,3}}^{n+1}\left(p_{o_4}^{n+1}-p_{o_3}^{n+1}\right)\\
&=\frac{V_{b_3}}{\alpha_c\Delta t}\left\{\left[\frac{\phi(1-S_w)}{B_o}\right]_3^{n+1}-\left[\frac{\phi(1-S_w)}{B_o}\right]_3^{n}\right\}
\end{aligned}
\tag{11.43b}
$$

The water equation is obtained by substituting the given values into Eq. (11.42) and expanding the summation terms, yielding

$$T_{w_{2,3}}^{n+1}\left[\left(p_{o_2}^{n+1}-p_{o_3}^{n+1}\right)-\left(P_{cow_2}^{n+1}-P_{cow_3}^{n+1}\right)-\gamma_{w_{2,3}}^{n}(Z_2-Z_3)\right]+T_{w_{4,3}}^{n+1}\left[\left(p_{o_4}^{n+1}-p_{o_3}^{n+1}\right)\right.$$
$$\left.-\left(P_{cow_4}^{n+1}-P_{cow_3}^{n+1}\right)-\gamma_{w_{4,3}}^{n}(Z_4-Z_3)\right]+0+0=\frac{V_{b_3}}{\alpha_c\Delta t}\left[\left(\frac{\phi S_w}{B_w}\right)_3^{n+1}-\left(\frac{\phi S_w}{B_w}\right)_3^{n}\right] \tag{11.44a}$$

Observing that $Z_2=Z_3=Z_4$ for a horizontal reservoir, the water equation becomes

$$\begin{aligned}&T_{w_{2,3}}^{n+1}\left[\left(p_{o_2}^{n+1}-p_{o_3}^{n+1}\right)-\left(P_{cow_2}^{n+1}-P_{cow_3}^{n+1}\right)\right]\\&+T_{w_{4,3}}^{n+1}\left[\left(p_{o_4}^{n+1}-p_{o_3}^{n+1}\right)-\left(P_{cow_4}^{n+1}-P_{cow_3}^{n+1}\right)\right]\\&=\frac{V_{b_3}}{\alpha_c\Delta t}\left[\left(\frac{\phi S_w}{B_w}\right)_3^{n+1}-\left(\frac{\phi S_w}{B_w}\right)_3^{n}\right]\end{aligned} \tag{11.44b}$$

Eqs. (11.43b) and (11.44b) are the two flow equations for gridblock 3 in this 1-D reservoir.

11.3.2 Flow equations for gas/water flow model

The two components in the gas/water flow model are water at standard conditions and the free-gas component at standard conditions. Gas solubility in the water phase is assumed negligible; hence, the gas phase contains all the gas that exists in this system. Therefore, Eqs. (11.34a) and (11.34b) express the gas/water flow equations for block n. Combine these two equations with $S_w=1-S_g$ and $p_w=p_g-P_{cgw}(S_g)$ to obtain the p_g-S_g formulation.

For the gas component,

$$\begin{aligned}&\sum_{l\in\psi_n}T_{g_{l,n}}^{n+1}\left[\left(p_{g_l}^{n+1}-p_{g_n}^{n+1}\right)-\gamma_{g_{l,n}}^{n}(Z_l-Z_n)\right]+\sum_{l\in\xi_n}q_{gsc_{l,n}}^{n+1}+q_{gsc_n}^{n+1}\\&=\frac{V_{b_n}}{\alpha_c\Delta t}\left[\left(\frac{\phi S_g}{B_g}\right)_n^{n+1}-\left(\frac{\phi S_g}{B_g}\right)_n^{n}\right]\end{aligned} \tag{11.45}$$

where $q_{gsc_i}^{n+1}=q_{fgsc_i}^{n+1}$ and $q_{gsc_{l,n}}^{n+1}=q_{fgsc_{l,n}}^{n+1}$.

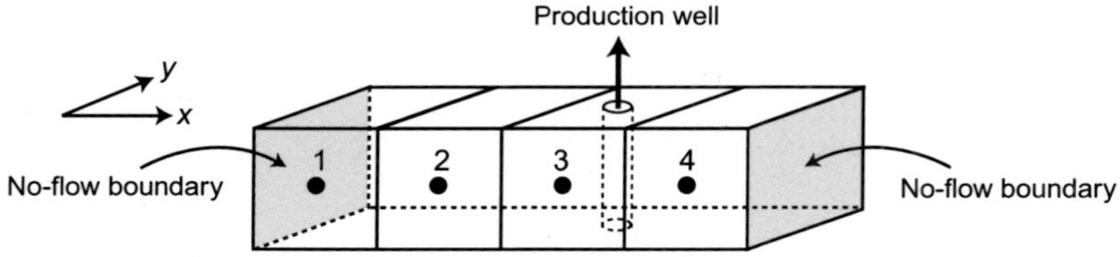

FIG. 11.10 1-D reservoir in Example 11.5.

For the water component,

$$\sum_{l\in\psi_n} T_{w_{l,n}}^{n+1}\left[\left(p_{g_l}^{n+1}-p_{g_n}^{n+1}\right)-\left(P_{cgw_l}^{n+1}-P_{cgw_n}^{n+1}\right)-\gamma_{w_{l,n}}^{n}(Z_l-Z_n)\right]+\sum_{l\in\xi_n} q_{wsc_{l,n}}^{n+1}+q_{wsc_n}^{n+1}$$
$$=\frac{V_{b_n}}{\alpha_c\Delta t}\left\{\left[\frac{\phi(1-S_g)}{B_w}\right]_n^{n+1}-\left[\frac{\phi(1-S_g)}{B_w}\right]_n^{n}\right\} \tag{11.46}$$

Example 11.5 A homogeneous, 1-D horizontal, two-phase gas/water reservoir is described by four equal blocks as shown in Fig. 11.10. Initial reservoir pressure and phase saturations are known. The reservoir left and right boundaries are sealed off to flow. The reservoir has a production well in gridblock 3. Write the flow equations for interior gridblock 2. Assume negligible gas/water capillary pressure.

Solution

For gridblock 2, $n=2$ and $\psi_3=\{1,3\}$. Gridblock 2 is an interior block; therefore, $\xi_2=\{\}$, $\sum_{l\in\xi_2} q_{gsc_{l,2}}^{n+1}=0$, and $\sum_{l\in\xi_2} q_{wsc_{l,2}}^{n+1}=0$. Gridblock 2 has no wells; therefore, $q_{gsc_2}^{n+1}=0$ and $q_{wsc_2}^{n+1}=0$.

The gas flow equation is obtained by substituting the given values into Eq. (11.45) and expanding the summation terms, yielding

$$T_{g_{1,2}}^{n+1}\left[\left(p_{g_1}^{n+1}-p_{g_2}^{n+1}\right)-\gamma_{g_{1,2}}^{n}(Z_1-Z_2)\right]$$
$$+T_{g_{3,2}}^{n+1}\left[\left(p_{g_3}^{n+1}-p_{g_2}^{n+1}\right)-\gamma_{g_{3,2}}^{n}(Z_3-Z_2)\right]+0+0 \tag{11.47a}$$
$$=\frac{V_{b_2}}{\alpha_c\Delta t}\left[\left(\frac{\phi S_g}{B_g}\right)_2^{n+1}-\left(\frac{\phi S_g}{B_g}\right)_2^{n}\right]$$

Observing that $Z_1=Z_2=Z_3$ for a horizontal reservoir, the gas flow equation becomes

$$T_{g_{1,2}}^{n+1}\left(p_{g_1}^{n+1}-p_{g_2}^{n+1}\right)+T_{g_{3,2}}^{n+1}\left(p_{g_3}^{n+1}-p_{g_2}^{n+1}\right)=\frac{V_{b_2}}{\alpha_c\Delta t}\left[\left(\frac{\phi S_g}{B_g}\right)_2^{n+1}-\left(\frac{\phi S_g}{B_g}\right)_2^{n}\right] \tag{11.47b}$$

The water flow equation is obtained by substituting the given values into Eq. (11.46) and expanding the summation terms, yielding

$$T_{w_{1,2}}^{n+1}\left[\left(p_{g_1}^{n+1}-p_{g_2}^{n+1}\right)-\left(P_{cgw_1}^{n+1}-P_{cgw_2}^{n+1}\right)-\gamma_{w_{1,2}}^{n}(Z_1-Z_2)\right]$$
$$+T_{w_{3,2}}^{n+1}\left[\left(p_{g_3}^{n+1}-p_{g_2}^{n+1}\right)-\left(P_{cgw_3}^{n+1}-P_{cgw_2}^{n+1}\right)-\gamma_{w_{3,2}}^{n}(Z_3-Z_2)\right]+0+0 \tag{11.48a}$$
$$=\frac{V_{b_2}}{\alpha_c\Delta t}\left\{\left[\frac{\phi(1-S_g)}{B_w}\right]_2^{n+1}-\left[\frac{\phi(1-S_g)}{B_w}\right]_2^{n}\right\}$$

Observing that $Z_1=Z_2=Z_3$ for a horizontal reservoir and for negligible gas/water capillary pressure, the water flow equation becomes

$$T_{w_{1,2}}^{n+1}\left(p_{g_1}^{n+1}-p_{g_2}^{n+1}\right)+T_{w_{3,2}}^{n+1}\left(p_{g_3}^{n+1}-p_{g_2}^{n+1}\right)$$
$$=\frac{V_{b_2}}{\alpha_c\Delta t}\left\{\left[\frac{\phi(1-S_g)}{B_w}\right]_2^{n+1}-\left[\frac{\phi(1-S_g)}{B_w}\right]_2^{n}\right\} \tag{11.48b}$$

Eqs. (11.47b) and (11.48b) are the two flow equations for gridblock 2 in this 1-D reservoir.

11.3.3 Flow equations for oil/gas flow model

The components in the oil/gas flow model are oil at standard conditions, gas at standard conditions, and irreducible water (immobile water). Gas consists of both free-gas and solution-gas components. The flow equation for gas is obtained by adding Eqs. (11.34b) and (11.34d):

$$\sum_{l\in\psi_n}\left\{T_{g_{l,n}}^{n+1}\left[\left(p_{g_l}^{n+1}-p_{g_n}^{n+1}\right)-\gamma_{g_{l,n}}^{n}(Z_l-Z_n)\right]+(T_oR_s)_{l,n}^{n+1}\left[\left(p_{o_l}^{n+1}-p_{o_n}^{n+1}\right)-\gamma_{o_{l,n}}^{n}(Z_l-Z_n)\right]\right\}$$
$$+\sum_{l\in\xi_n}\left[q_{fgsc_{l,n}}^{n+1}+(R_sq_{osc})_{l,n}^{n+1}\right]+\left[q_{fgsc_n}^{n+1}+(R_sq_{osc})_n^{n+1}\right]$$
$$=\frac{V_{b_n}}{\alpha_c\Delta t}\left\{\left[\left(\frac{\phi S_g}{B_g}\right)_n^{n+1}-\left(\frac{\phi S_g}{B_g}\right)_n^{n}\right]+\left[\left(\frac{\phi R_sS_o}{B_o}\right)_n^{n+1}-\left(\frac{\phi R_sS_o}{B_o}\right)_n^{n}\right]\right\} \tag{11.49}$$

Therefore, Eqs. (11.34c) and (11.49) express the oil/gas flow equations for block n.

Combine these two equations with $S_o=(1-S_{iw})-S_g$ and $p_g=p_o+P_{cgo}(S_g)$ to obtain the p_o-S_g formulation.

For the oil component,

$$\sum_{l\in\psi_n}T_{o_{l,n}}^{n+1}\left[\left(p_{o_l}^{n+1}-p_{o_n}^{n+1}\right)-\gamma_{o_{l,n}}^{n}(Z_l-Z_n)\right]+\sum_{l\in\xi_n}q_{osc_{l,n}}^{n+1}+q_{osc_n}^{n+1}$$
$$=\frac{V_{b_n}}{\alpha_c\Delta t}\left\{\left[\frac{\phi(1-S_{iw}-S_g)}{B_o}\right]_n^{n+1}-\left[\frac{\phi(1-S_{iw}-S_g)}{B_o}\right]_n^{n}\right\} \tag{11.50}$$

For the gas component,

$$\sum_{l\in\psi_n}\left\{T_{g_{l,n}}^{n+1}\left[\left(p_{o_l}^{n+1}-p_{o_n}^{n+1}\right)+\left(P_{cgo_l}^{n+1}-P_{cgo_n}^{n+1}\right)-\gamma_{g_{l,n}}^{n}(Z_l-Z_n)\right]\right.$$
$$\left.+(T_oR_s)_{l,n}^{n+1}\left[\left(p_{o_l}^{n+1}-p_{o_n}^{n+1}\right)-\gamma_{o_{l,n}}^{n}(Z_l-Z_n)\right]\right\}+\sum_{l\in\xi_n}\left[q_{fgsc_{l,n}}^{n+1}+R_{s_{l,n}}^{n+1}q_{osc_{l,n}}^{n+1}\right]$$
$$+\left[q_{fgsc_n}^{n+1}+R_{s_n}^{n+1}q_{osc_n}^{n+1}\right]$$
$$=\frac{V_{b_n}}{\alpha_c\Delta t}\left\{\left(\frac{\phi S_g}{B_g}\right)_n^{n+1}-\left(\frac{\phi S_g}{B_g}\right)_n^{n}+\left[\frac{\phi R_s(1-S_{iw}-S_g)}{B_o}\right]_n^{n+1}-\left[\frac{\phi R_s(1-S_{iw}-S_g)}{B_o}\right]_n^{n}\right\} \tag{11.51}$$

The irreducible water in this model is assumed to have the same compressibility as that of porosity. If the irreducible water is assumed incompressible, then $\phi_{HC}=\phi(1-S_{iw})$ replaces ϕ, and $(1-S_g)$ replaces $(1-S_{iw}-S_g)$ in Eqs. (11.50) and (11.51).

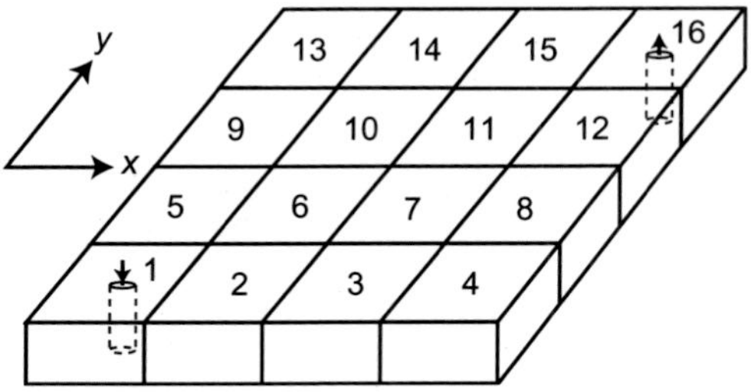

FIG. 11.11 2-D reservoir in Example 11.6.

Example 11.6 A homogeneous, 2-D horizontal, two-phase oil/gas reservoir is shown in Fig. 11.11. Initial reservoir pressure and phase saturations are known. The reservoir has no-flow boundaries. There is a gas injection well in gridblock 1 and a production well in gridblock 16. Write the flow equations for interior gridblock 10. Assume negligible gas/oil capillary pressure.

Solution

For gridblock 10, $n=10$ and $\psi_{10}=\{6,9,11,14\}$. Gridblock 10 is an interior block; therefore, $\xi_{10}=\{\}$, $\sum_{l\in\xi_{10}} q_{osc_{l,10}}^{n+1}=0$, and $\sum_{l\in\xi_{10}} q_{fgsc_{l,10}}^{n+1}=0$. Gridblock 10 has no wells; therefore, $q_{osc_{10}}^{n+1}=0$ and $q_{fgsc_{10}}^{n+1}=0$.

The oil equation is obtained by substituting the given values into Eq. (11.50) and expanding the summation terms, yielding

$$
\begin{aligned}
&T^{n+1}_{o6,10}\left[\left(p^{n+1}_{o6}-p^{n+1}_{o10}\right)-\gamma^{n}_{o6,10}(Z_6-Z_{10})\right]\\
&+T^{n+1}_{o9,10}\left[\left(p^{n+1}_{o9}-p^{n+1}_{o10}\right)-\gamma^{n}_{o9,10}(Z_9-Z_{10})\right]\\
&+T^{n+1}_{o11,10}\left[\left(p^{n+1}_{o11}-p^{n+1}_{o10}\right)-\gamma^{n}_{o11,10}(Z_{11}-Z_{10})\right]\\
&+T^{n+1}_{o14,10}\left[\left(p^{n+1}_{o14}-p^{n+1}_{o10}\right)-\gamma^{n}_{o14,10}(Z_{14}-Z_{10})\right]+0+0\\
&=\frac{V_{b_{10}}}{\alpha_c\Delta t}\left\{\left[\frac{\phi(1-S_{iw}-S_g)}{B_o}\right]^{n+1}_{10}-\left[\frac{\phi(1-S_{iw}-S_g)}{B_o}\right]^{n}_{10}\right\}
\end{aligned}
\tag{11.52a}
$$

For a horizontal reservoir, $Z_6=Z_9=Z_{10}=Z_{11}=Z_{14}$, and the oil equation becomes

$$
\begin{aligned}
&T^{n+1}_{o6,10}\left(p^{n+1}_{o6}-p^{n+1}_{o10}\right)+T^{n+1}_{o9,10}\left(p^{n+1}_{o9}-p^{n+1}_{o10}\right)\\
&+T^{n+1}_{o11,10}\left(p^{n+1}_{o11}-p^{n+1}_{o10}\right)+T^{n+1}_{o14,10}\left(p^{n+1}_{o14}-p^{n+1}_{o10}\right)\\
&=\frac{V_{b_{10}}}{\alpha_c\Delta t}\left\{\left[\frac{\phi(1-S_{iw}-S_g)}{B_o}\right]^{n+1}_{10}-\left[\frac{\phi(1-S_{iw}-S_g)}{B_o}\right]^{n}_{10}\right\}
\end{aligned}
\tag{11.52b}
$$

The gas equation is obtained by substituting the given values into Eq. (11.51) and expanding the summation terms, yielding

$$
\begin{aligned}
&T^{n+1}_{g6,10}\left[\left(p^{n+1}_{o6}-p^{n+1}_{o10}\right)+\left(P^{n+1}_{cgo_6}-P^{n+1}_{cgo_{10}}\right)-\gamma^{n}_{g6,10}(Z_6-Z_{10})\right]\\
&+(T_oR_s)^{n+1}_{6,10}\left[\left(p^{n+1}_{o6}-p^{n+1}_{o10}\right)-\gamma^{n}_{o6,10}(Z_6-Z_{10})\right]\\
&+T^{n+1}_{g9,10}\left[\left(p^{n+1}_{o9}-p^{n+1}_{o10}\right)+\left(P^{n+1}_{cgo_9}-P^{n+1}_{cgo_{10}}\right)-\gamma^{n}_{g9,10}(Z_9-Z_{10})\right]\\
&+(T_oR_s)^{n+1}_{9,10}\left[\left(p^{n+1}_{o9}-p^{n+1}_{o10}\right)-\gamma^{n}_{o9,10}(Z_9-Z_{10})\right]\\
&+T^{n+1}_{g11,10}\left[\left(p^{n+1}_{o11}-p^{n+1}_{o10}\right)+\left(P^{n+1}_{cgo_{11}}-P^{n+1}_{cgo_{10}}\right)-\gamma^{n}_{g11,10}(Z_{11}-Z_{10})\right]\\
&+(T_oR_s)^{n+1}_{11,10}\left[\left(p^{n+1}_{o11}-p^{n+1}_{o10}\right)-\gamma^{n}_{o11,10}(Z_{11}-Z_{10})\right]\\
&+T^{n+1}_{g14,10}\left[\left(p^{n+1}_{o14}-p^{n+1}_{o10}\right)+\left(P^{n+1}_{cgo_{14}}-P^{n+1}_{cgo_{10}}\right)-\gamma^{n}_{g14,10}(Z_{14}-Z_{10})\right]\\
&+(T_oR_s)^{n+1}_{14,10}\left[\left(p^{n+1}_{o14}-p^{n+1}_{o10}\right)-\gamma^{n}_{o14,10}(Z_{14}-Z_{10})\right]+0+\left[0+R^{n+1}_{s_{10}}\times 0\right]\\
&=\frac{V_{b_{10}}}{\alpha_c\Delta t}\left\{\left(\frac{\phi S_g}{B_g}\right)^{n+1}_{10}-\left(\frac{\phi S_g}{B_g}\right)^{n}_{10}+\left[\frac{\phi R_s(1-S_{iw}-S_g)}{B_o}\right]^{n+1}_{10}-\left[\frac{\phi R_s(1-S_{iw}-S_g)}{B_o}\right]^{n}_{10}\right\}
\end{aligned}
\tag{11.53a}
$$

Observing that $Z_6=Z_9=Z_{10}=Z_{11}=Z_{14}$ for a horizontal reservoir and for negligible gas/oil capillary pressure, the gas equation becomes

$$\left[T^{n+1}_{g_{6,10}}+(T_oR_s)^{n+1}_{6,10}\right]\left(p^{n+1}_{o_6}-p^{n+1}_{o_{10}}\right)+\left[T^{n+1}_{g_{9,10}}+(T_oR_s)^{n+1}_{9,10}\right]\left(p^{n+1}_{o_9}-p^{n+1}_{o_{10}}\right)$$
$$+\left[T^{n+1}_{g_{11,10}}+(T_oR_s)^{n+1}_{11,10}\right]\left(p^{n+1}_{o_{11}}-p^{n+1}_{o_{10}}\right)+\left[T^{n+1}_{g_{14,10}}+(T_oR_s)^{n+1}_{14,10}\right]\left(p^{n+1}_{o_{14}}-p^{n+1}_{o_{10}}\right)$$
$$=\frac{V_{b_{10}}}{\alpha_c\Delta t}\left\{\left(\frac{\phi S_g}{B_g}\right)^{n+1}_{10}-\left(\frac{\phi S_g}{B_g}\right)^{n}_{10}+\left[\frac{\phi R_s(1-S_{iw}-S_g)}{B_o}\right]^{n+1}_{10}-\left[\frac{\phi R_s(1-S_{iw}-S_g)}{B_o}\right]^{n}_{10}\right\} \tag{11.53b}$$

Eqs. (11.52b) and (11.53b) are the two flow equations for gridblock 10 in this 2-D reservoir.

11.3.4 Flow equations for black-oil model

The isothermal oil/water/gas flow model is known as the black-oil model. The oil component forms the bulk of the oil phase. The solution-gas component dissolves in it, and the remaining gas (the free-gas component) forms the gas phase. Oil and water are immiscible, and both do not dissolve in the gas phase. Therefore, the black-oil system consists of the water component, the oil component, and the gas component (solution gas plus free gas). Accordingly, a black-oil model consists of Eqs. (11.34a), (11.34c), and (11.49).

For the oil component,

$$\sum_{l\in\psi_n}T^{n+1}_{o_{l,n}}\left[\left(p^{n+1}_{o_l}-p^{n+1}_{o_n}\right)-\gamma^n_{o_{l,n}}(Z_l-Z_n)\right]+\sum_{l\in\xi_n}q^{n+1}_{osc_{l,n}}+q^{n+1}_{osc_n}$$
$$=\frac{V_{b_n}}{\alpha_c\Delta t}\left[\left(\frac{\phi S_o}{B_o}\right)^{n+1}_n-\left(\frac{\phi S_o}{B_o}\right)^{n}_n\right] \tag{11.34c}$$

For the gas component,

$$\sum_{l\in\psi_n}\left\{T^{n+1}_{g_{l,n}}\left[\left(p^{n+1}_{g_l}-p^{n+1}_{g_n}\right)-\gamma^n_{g_{l,n}}(Z_l-Z_n)\right]\right.$$
$$\left.+(T_oR_s)^{n+1}_{l,n}\left[\left(p^{n+1}_{o_l}-p^{n+1}_{o_n}\right)-\gamma^n_{o_{l,n}}(Z_l-Z_n)\right]\right\}$$
$$+\sum_{l\in\xi_n}\left[q^{n+1}_{fgsc_{l,n}}+(R_sq_{osc})^{n+1}_{l,n}\right]+\left[q^{n+1}_{fgsc_n}+(R_sq_{osc})^{n+1}_n\right]$$
$$=\frac{V_{b_n}}{\alpha_c\Delta t}\left\{\left[\left(\frac{\phi S_g}{B_g}\right)^{n+1}_n-\left(\frac{\phi S_g}{B_g}\right)^{n}_n\right]+\left[\left(\frac{\phi R_sS_o}{B_o}\right)^{n+1}_n-\left(\frac{\phi R_sS_o}{B_o}\right)^{n}_n\right]\right\} \tag{11.49}$$

For the water component,

$$\sum_{l\in\psi_n} T_{w_{l,n}}^{n+1}\left[\left(p_{w_l}^{n+1}-p_{w_n}^{n+1}\right)-\gamma_{w_{l,n}}^{n}(Z_l-Z_n)\right]+\sum_{l\in\xi_n} q_{wsc_{l,n}}^{n+1}+q_{wsc_n}^{n+1} = \frac{V_{b_n}}{\alpha_c\Delta t}\left[\left(\frac{\phi S_w}{B_w}\right)_n^{n+1}-\left(\frac{\phi S_w}{B_w}\right)_n^{n}\right] \tag{11.34a}$$

Combine these three equations with $S_o=1-S_w-S_g$, $p_w=p_o-P_{cow}(S_w)$, and $p_g=p_o+P_{cgo}(S_g)$ to obtain the $p_o-S_w-S_g$ formulation.

For the oil component,

$$\sum_{l\in\psi_n} T_{o_{l,n}}^{n+1}\left[\left(p_{o_l}^{n+1}-p_{o_n}^{n+1}\right)-\gamma_{o_{l,n}}^{n}(Z_l-Z_n)\right]+\sum_{l\in\xi_n} q_{osc_{l,n}}^{n+1}+q_{osc_n}^{n+1} = \frac{V_{b_n}}{\alpha_c\Delta t}\left\{\left[\frac{\phi(1-S_w-S_g)}{B_o}\right]_n^{n+1}-\left[\frac{\phi(1-S_w-S_g)}{B_o}\right]_n^{n}\right\} \tag{11.54}$$

For the gas component,

$$\begin{aligned}&\sum_{l\in\psi_n}\left\{T_{g_{l,n}}^{n+1}\left[\left(p_{o_l}^{n+1}-p_{o_n}^{n+1}\right)+\left(P_{cgo_l}^{n+1}-P_{cgo_n}^{n+1}\right)-\gamma_{g_{l,n}}^{n}(Z_l-Z_n)\right]\right.\\&\left.+(T_oR_s)_{l,n}^{n+1}\left[\left(p_{o_l}^{n+1}-p_{o_n}^{n+1}\right)-\gamma_{o_{l,n}}^{n}(Z_l-Z_n)\right]\right\}\\&+\sum_{l\in\xi_n}\left[q_{fgsc_{l,n}}^{n+1}+R_{s_{l,n}}^{n+1}q_{osc_{l,n}}^{n+1}\right]+\left[q_{fgsc_n}^{n+1}+R_{s_n}^{n+1}q_{osc_n}^{n+1}\right]\\&=\frac{V_{b_n}}{\alpha_c\Delta t}\left\{\left(\frac{\phi S_g}{B_g}\right)_n^{n+1}-\left(\frac{\phi S_g}{B_g}\right)_n^{n}+\left[\frac{\phi R_s(1-S_w-S_g)}{B_o}\right]_n^{n+1}-\left[\frac{\phi R_s(1-S_w-S_g)}{B_o}\right]_n^{n}\right\}\end{aligned} \tag{11.55}$$

For the water component,

$$\sum_{l\in\psi_n} T_{w_{l,n}}^{n+1}\left[\left(p_{o_l}^{n+1}-p_{o_n}^{n+1}\right)-\left(P_{cow_l}^{n+1}-P_{cow_n}^{n+1}\right)-\gamma_{w_{l,n}}^{n}(Z_l-Z_n)\right]+\sum_{l\in\xi_n} q_{wsc_{l,n}}^{n+1}+q_{wsc_n}^{n+1} = \frac{V_{b_n}}{\alpha_c\Delta t}\left[\left(\frac{\phi S_w}{B_w}\right)_n^{n+1}-\left(\frac{\phi S_w)}{B_w}\right)_n^{n}\right] \tag{11.56}$$

It is noteworthy to mention that the flow equations in a black-oil model (Eqs. 11.54, 11.55, and 11.56) can be reduced to any of the two-phase flow models already presented. This is accomplished by discarding the flow equation for the missing phase and setting the saturation of the missing phase to zero in the remaining flow equations. For example, the oil/water flow model is obtained from the black-oil model by discarding the gas flow equation (Eq. 11.55) and setting $S_g=0$ in Eq. (11.54). The oil/gas flow model is obtained by discarding

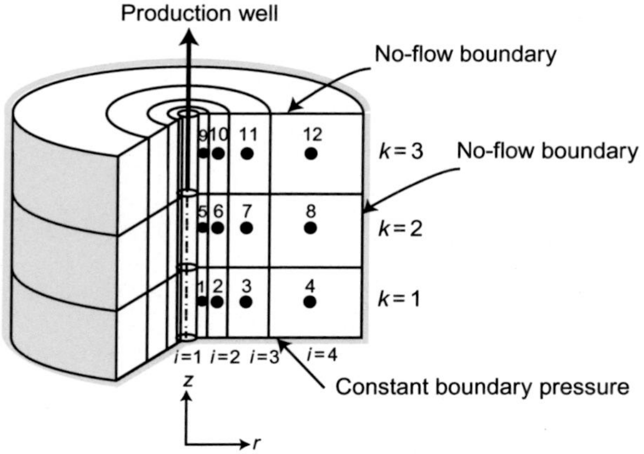

FIG. 11.12 2-D radial-cylindrical reservoir in Example 11.7.

the water flow equation (Eq. 11.56) and setting $S_w = S_{iw}$ in Eqs. (11.54) and (11.55).

Example 11.7 A single-well simulation problem is presented in Fig. 11.12. The reservoir is horizontal and contains oil, water, and gas. Initial reservoir pressure and phase saturations are known. The reservoir top and lateral boundaries are no-flow boundaries, whereas the reservoir bottom boundary represents a constant pressure WOC. The well is completed through the top layer. Write the flow equations for interior gridblock 7.

Solution

For gridblock 7, $n=7$ and $\psi_7 = \{3, 6, 8, 11\}$. Gridblock 7 is an interior block; therefore, $\xi_7 = \{\}$, $\sum\limits_{l \in \xi_7} q_{osc_{l,7}}^{n+1} = 0$, $\sum\limits_{l \in \xi_7} q_{wsc_{l,7}}^{n+1} = 0$, and $\sum\limits_{l \in \xi_7} q_{fgsc_{l,7}}^{n+1} = 0$. Gridblock 7 has no wells; therefore, $q_{osc_7}^{n+1} = 0$, $q_{wsc_7}^{n+1} = 0$, and $q_{fgsc_7}^{n+1} = 0$. Observe also that $Z_6 = Z_7 = Z_8$.

The oil equation is obtained by substituting the given values into Eq. (11.54) and expanding the summation terms, yielding

$$T_{o_{3,7}}^{n+1}\left[\left(p_{o_3}^{n+1} - p_{o_7}^{n+1}\right) - \gamma_{o_{3,7}}^{n}(Z_3 - Z_7)\right] + T_{o_{6,7}}^{n+1}\left(p_{o_6}^{n+1} - p_{o_7}^{n+1}\right)$$
$$+ T_{o_{8,7}}^{n+1}\left(p_{o_8}^{n+1} - p_{o_7}^{n+1}\right) + T_{o_{11,7}}^{n+1}\left[\left(p_{o_{11}}^{n+1} - p_{o_7}^{n+1}\right) - \gamma_{o_{11,7}}^{n}(Z_{11} - Z_7)\right] + 0 + 0$$
$$= \frac{V_{b_7}}{\alpha_c \Delta t}\left\{\left[\frac{\phi(1 - S_w - S_g)}{B_o}\right]_7^{n+1} - \left[\frac{\phi(1 - S_w - S_g)}{B_o}\right]_7^{n}\right\} \tag{11.57}$$

The gas equation is obtained by substituting the given values into Eq. (11.55) and expanding the summation terms, yielding

$$
\begin{aligned}
&T_{g_{3,7}}^{n+1}\left[\left(p_{o_3}^{n+1}-p_{o_7}^{n+1}\right)+\left(P_{cgo_3}^{n+1}-P_{cgo_7}^{n+1}\right)-\gamma_{g_{3,7}}^{n}(Z_3-Z_7)\right]\\
&+(T_oR_s)_{3,7}^{n+1}\left[\left(p_{o_3}^{n+1}-p_{o_7}^{n+1}\right)-\gamma_{o_{3,7}}^{n}(Z_3-Z_7)\right]\\
&+T_{g_{6,7}}^{n+1}\left[\left(p_{o_6}^{n+1}-p_{o_7}^{n+1}\right)+\left(P_{cgo_6}^{n+1}-P_{cgo_7}^{n+1}\right)\right]+(T_oR_s)_{6,7}^{n+1}\left[\left(p_{o_6}^{n+1}-p_{o_7}^{n+1}\right)\right]\\
&+T_{g_{8,7}}^{n+1}\left[\left(p_{o_8}^{n+1}-p_{o_7}^{n+1}\right)+\left(P_{cgo_8}^{n+1}-P_{cgo_7}^{n+1}\right)\right]+(T_oR_s)_{8,7}^{n+1}\left[\left(p_{o_8}^{n+1}-p_{o_7}^{n+1}\right)\right]\\
&+T_{g_{11,7}}^{n+1}\left[\left(p_{o_{11}}^{n+1}-p_{o_7}^{n+1}\right)+\left(P_{cgo_{11}}^{n+1}-P_{cgo_7}^{n+1}\right)-\gamma_{g_{11,7}}^{n}(Z_{11}-Z_7)\right]\\
&+(T_oR_s)_{11,7}^{n+1}\left[\left(p_{o_{11}}^{n+1}-p_{o_7}^{n+1}\right)-\gamma_{o_{11,7}}^{n}(Z_{11}-Z_7)\right]+0+\left[0+R_{s_7}^{n+1}\times 0\right]\\
&=\frac{V_{b_7}}{\alpha_c\Delta t}\left\{\left(\frac{\phi S_g}{B_g}\right)_7^{n+1}-\left(\frac{\phi S_g}{B_g}\right)_7^{n}+\left[\frac{\phi R_s(1-S_w-S_g)}{B_o}\right]_7^{n+1}-\left[\frac{\phi R_s(1-S_w-S_g)}{B_o}\right]_7^{n}\right\}
\end{aligned}
\tag{11.58}
$$

The water equation is obtained by substituting the given values into Eq. (11.56) and expanding the summation terms, yielding

$$
\begin{aligned}
&T_{w_{3,7}}^{n+1}\left[\left(p_{o_3}^{n+1}-p_{o_7}^{n+1}\right)-\left(P_{cow_3}^{n+1}-P_{cow_7}^{n+1}\right)-\gamma_{w_{3,7}}^{n}(Z_3-Z_7)\right]\\
&+T_{w_{6,7}}^{n+1}\left[\left(p_{o_6}^{n+1}-p_{o_7}^{n+1}\right)-\left(P_{cow_6}^{n+1}-P_{cow_7}^{n+1}\right)\right]\\
&+T_{w_{8,7}}^{n+1}\left[\left(p_{o_8}^{n+1}-p_{o_7}^{n+1}\right)-\left(P_{cow_8}^{n+1}-P_{cow_7}^{n+1}\right)\right]\\
&+T_{w_{11,7}}^{n+1}\left[\left(p_{o_{11}}^{n+1}-p_{o_7}^{n+1}\right)-\left(P_{cow_{11}}^{n+1}-P_{cow_7}^{n+1}\right)-\gamma_{w_{11,7}}^{n}(Z_{11}-Z_7)\right]+0+0\\
&=\frac{V_{b_7}}{\alpha_c\Delta t}\left[\left(\frac{\phi S_w}{B_w}\right)_7^{n+1}-\left(\frac{\phi S_w}{B_w}\right)_7^{n}\right]
\end{aligned}
\tag{11.59}
$$

Eqs. (11.57), (11.58), and (11.59) are the three flow equations for gridblock 7 in this 2-D radial flow reservoir.

11.4 Solution of multiphase flow equations

The equations for the whole reservoir consist of the flow equations contributed by all reservoir blocks. The unknowns in the system are the unknowns of the formulation for all reservoir blocks. To solve the flow equations of a reservoir model, several steps are taken. The accumulation terms in the flow equations are expanded in a conservative way and expressed in terms of the changes of the unknowns of the block over a time step, the boundary conditions are implemented (or the rates of fictitious wells are estimated), production and injection rates are included, and the nonlinear terms are linearized both in space and time. The treatments of boundary conditions, production (injection), and linearization are, to some extent, similar to those for single-flow models presented in Chapters 4, 5, 6, and 8. In this section, we present, in elaborate detail, the expansion of the accumulation terms, the treatments of production and injection wells, boundary conditions, and solution methods of the equations of multiphase flow models. In addition, we highlight differences in the treatment of nonlinear terms from single-phase flow.

11.4.1 Expansion of accumulation terms

The accumulation terms of the reduced set of equations for each reservoir block must be expanded and expressed in terms of the changes of the primary unknowns of formulation over a time step. These accumulation terms form the RHS of Eqs. (11.41) and (11.42) for the oil/water model, Eqs. (11.45) and (11.46) for the gas/water model, Eqs. (11.50) and (11.51) for the oil/gas model, and Eqs. (11.54) through (11.56) for the oil/water/gas model. The expansion scheme used must preserve material balance. For example, consider the expansion of the RHS of Eq. (11.42), $\frac{V_{b_n}}{\alpha_c \Delta t}\left[\left(\frac{\phi S_w}{B_w}\right)_n^{n+1} - \left(\frac{\phi S_w}{B_w}\right)_n^n\right]$, in terms of p_o and S_w. Add and subtract the term $S_{w_n}^n\left(\frac{\phi}{B_w}\right)_n^{n+1}$ and factorize the terms as follows:

$$\begin{aligned}
&\frac{V_{b_n}}{\alpha_c \Delta t}\left[\left(\frac{\phi S_w}{B_w}\right)_n^{n+1} - \left(\frac{\phi S_w}{B_w}\right)_n^n\right] \\
&= \frac{V_{b_n}}{\alpha_c \Delta t}\left[\left(\frac{\phi S_w}{B_w}\right)_n^{n+1} - S_{w_n}^n\left(\frac{\phi}{B_w}\right)_n^{n+1} + S_{w_n}^n\left(\frac{\phi}{B_w}\right)_n^{n+1} - \left(\frac{\phi S_w}{B_w}\right)_n^n\right] \\
&= \frac{V_{b_n}}{\alpha_c \Delta t}\left\{\left(\frac{\phi}{B_w}\right)_n^{n+1}\left(S_{w_n}^{n+1} - S_{w_n}^n\right) + S_{w_n}^n\left[\left(\frac{\phi}{B_w}\right)_n^{n+1} - \left(\frac{\phi}{B_w}\right)_n^n\right]\right\}
\end{aligned} \tag{11.60}$$

Again, add and subtract the term $\phi_n^{n+1}\frac{1}{B_{w_n}^n}$ in the square bracket on the RHS of Eq. (11.60) and factorize the terms as follows:

$$\begin{aligned}
&\frac{V_{b_n}}{\alpha_c \Delta t}\left[\left(\frac{\phi S_w}{B_w}\right)_n^{n+1} - \left(\frac{\phi S_w}{B_w}\right)_n^n\right] \\
&= \frac{V_{b_n}}{\alpha_c \Delta t}\left\{\left(\frac{\phi}{B_w}\right)_n^{n+1}\left(S_{w_n}^{n+1} - S_{w_n}^n\right) + S_{w_n}^n\left[\left(\frac{\phi}{B_w}\right)_n^{n+1} - \phi_n^{n+1}\frac{1}{B_{w_n}^n} + \phi_n^{n+1}\frac{1}{B_{w_n}^n} - \left(\frac{\phi}{B_w}\right)_n^n\right]\right\} \\
&= \frac{V_{b_n}}{\alpha_c \Delta t}\left\{\left(\frac{\phi}{B_w}\right)_n^{n+1}\left(S_{w_n}^{n+1} - S_{w_n}^n\right) + S_{w_n}^n\left[\phi_n^{n+1}\left(\frac{1}{B_{w_n}^{n+1}} - \frac{1}{B_{w_n}^n}\right) + \frac{1}{B_{w_n}^n}\left(\phi_n^{n+1} - \phi_n^n\right)\right]\right\}
\end{aligned} \tag{11.61}$$

Expressing the changes in $\frac{1}{B_{w_n}}$ and ϕ_n over a time step in terms of the changes in oil-phase pressure over the same time step, results in

$$\frac{V_{b_n}}{\alpha_c \Delta t}\left[\left(\frac{\phi S_w}{B_w}\right)_n^{n+1} - \left(\frac{\phi S_w}{B_w}\right)_n^n\right]$$
$$= \frac{V_{b_n}}{\alpha_c \Delta t}\left\{\left(\frac{\phi}{B_w}\right)_n^{n+1}\left(S_{w_n}^{n+1} - S_{w_n}^n\right) + S_{w_n}^n\left[\phi_n^{n+1}\left(\frac{1}{B_{w_n}}\right)' + \frac{1}{B_{w_n}^n}\phi_n'\right]\left(p_{o_n}^{n+1} - p_{o_n}^n\right)\right\} \tag{11.62}$$

where $\left(\frac{1}{B_{w_n}}\right)'$ and ϕ_n' are defined as the chord slopes estimated between values at current time level at old iteration $\overset{(\nu)}{n+1}$ and old time level n

$$\left(\frac{1}{B_{w_n}}\right)' = \left(\frac{1}{B_{w_n}^{\overset{(\nu)}{n+1}}} - \frac{1}{B_{w_n}^n}\right) \Big/ \left(p_{o_n}^{\overset{(\nu)}{n+1}} - p_{o_n}^n\right) \tag{11.63}$$

and

$$\phi_n' = \left(\phi_n^{\overset{(\nu)}{n+1}} - \phi_n^n\right) \Big/ \left(p_{o_n}^{\overset{(\nu)}{n+1}} - p_{o_n}^n\right) \tag{11.64}$$

The RHS of Eq. (11.62), along with the definitions of chord slopes given by Eqs. (11.63) and (11.64), is termed a conservative expansion of the accumulation term represented by the LHS of Eq. (11.62).

Other accumulation terms can be expanded using similar steps as those that led to Eq. (11.62). Ertekin et al. (2001) derived a generic equation for a conservative expansion of any accumulation term, which states

$$\frac{V_b}{\alpha_c \Delta t}\left[(UVXY)^{n+1} - (UVXY)^n\right] = \frac{V_b}{\alpha_c \Delta t}[(VXY)^n(U^{n+1} - U^n)$$
$$+U^{n+1}(XY)^n(V^{n+1} - V^n) + (UV)^{n+1}Y^n(X^{n+1} - X^n) + (UVX)^{n+1}(Y^{n+1} - Y^n)] \tag{11.65}$$

where U is the weakest nonlinear function, Y is the strongest nonlinear function, and the degree of nonlinearity of V and X increases in the direction from U to Y. Usually, $U \equiv \phi$, $V \equiv 1/B_p$, $X \equiv R_s$, and $Y \equiv S_p$. If U, V, X, or Y does not exist, then it is assigned a value of 1. Because ϕ, $1/B_p$, and R_s are functions of the oil-phase pressure that is a primary unknown and S_p is either a primary unknown as in case of S_w and S_g or a function of the saturations that are primary unknowns (S_w, S_g) as in the case of S_o, Eq. (11.65) can be developed further to give

$$\frac{V_b}{\alpha_c \Delta t}\left[(UVXY)^{n+1} - (UVXY)^n\right] = \frac{V_b}{\alpha_c \Delta t}\Big\{(VXY)^n U'(p^{n+1} - p^n)$$
$$+U^{n+1}(XY)^n V'(p^{n+1} - p^n) + (UV)^{n+1}Y^n X'(p^{n+1} - p^n) \tag{11.66}$$
$$+(UVX)^{n+1}\left[(\partial Y/\partial S_w)(S_w^{n+1} - S_w^n) + (\partial Y/\partial S_g)\left(S_g^{n+1} - S_g^n\right)\right]\Big\}$$

or

$$\frac{V_b}{\alpha_c \Delta t}\left[(UVXY)^{n+1}-(UVXY)^n\right]=\frac{V_b}{\alpha_c \Delta t}\left\{[(VXY)^n U'+U^{n+1}(XY)^n V' \right.$$
$$+(UV)^{n+1}Y^n X'](p^{n+1}-p^n)+(UVX)^{n+1}(\partial Y/\partial S_w)\left(S_w^{n+1}-S_w^n\right) \tag{11.67}$$
$$\left.+(UVX)^{n+1}\left(\partial Y/\partial S_g\right)\left(S_g^{n+1}-S_g^n\right)\right\}$$

where

$$U'=\left(U^{\overset{(\nu)}{n+1}}-U^n\right)\Big/\left(p^{\overset{(\nu)}{n+1}}-p^n\right) \tag{11.68a}$$

$$V'=\left(V^{\overset{(\nu)}{n+1}}-V^n\right)\Big/\left(p^{\overset{(\nu)}{n+1}}-p^n\right) \tag{11.68b}$$

and

$$X'=\left(X^{\overset{(\nu)}{n+1}}-X^n\right)\Big/\left(p^{\overset{(\nu)}{n+1}}-p^n\right) \tag{11.68c}$$

Moreover, for $Y \equiv S_w$, $\partial Y/\partial S_w = 1$ and $\partial Y/\partial S_g = 0$; for $Y \equiv S_g$, $\partial Y/\partial S_w = 0$ and $\partial Y/\partial S_g = 1$; and for $Y \equiv S_o$, $\partial Y/\partial S_w = -1$ and $\partial Y/\partial S_g = -1$.

Let us apply Eq. (11.67) to obtain the expansion given by Eq. (11.62). In this case, we have $U \equiv \phi$, $V \equiv 1/B_w$, $X \equiv 1$, and $Y \equiv S_w$. Note that $p \equiv p_o$. Substitution into Eq. (11.68) gives

$$\phi'=\left(\phi^{\overset{(v)}{n+1}}-\phi^n\right)\Big/\left(p_o^{\overset{(v)}{n+1}}-p_o^n\right) \tag{11.69}$$

$$\left(\frac{1}{B_w}\right)'=\left(\frac{1}{B_w^{\overset{(v)}{n+1}}}-\frac{1}{B_w^n}\right)\Big/\left(p_o^{\overset{(v)}{n+1}}-p_o^n\right) \tag{11.70}$$

and

$$X'=0 \tag{11.71}$$

In addition,

$$\partial Y/\partial S_w = \partial S_w/\partial S_w = 1 \tag{11.72a}$$

and

$$\partial Y/\partial S_g = \partial S_w/\partial S_g = 0 \tag{11.72b}$$

Substitution of Eqs. (11.69) and (11.70) and the definitions of U, V, X, and Y into Eq. (11.67) gives

$$\frac{V_b}{\alpha_c \Delta t}\left[\left(\phi \frac{1}{B_w} S_w\right)^{n+1} - \left(\phi \frac{1}{B_w} S_w\right)^{n}\right] = \frac{V_b}{\alpha_c \Delta t}\left\{\left[\left(\frac{1}{B_w} S_w\right)^{n} \phi' + \phi^{n+1} S_w^n \left(\frac{1}{B_w}\right)' + \left(\phi \frac{1}{B_w}\right)^{n+1} S_w^n X'\right]\left(p_o^{n+1} - p_o^n\right) + \left(\phi \frac{1}{B_w}\right)^{n+1} \left(\partial Y/\partial S_w\right)\left(S_w^{n+1} - S_w^n\right) + \left(\phi \frac{1}{B_w}\right)^{n+1} \left(\partial Y/\partial S_g\right)\left(S_g^{n+1} - S_g^n\right)\right\} \tag{11.73}$$

Substitution of Eqs. (11.71) and (11.72) into this equation yields

$$\frac{V_b}{\alpha_c \Delta t}\left[\left(\phi \frac{1}{B_w} S_w\right)^{n+1} - \left(\phi \frac{1}{B_w} S_w\right)^{n}\right] = \frac{V_b}{\alpha_c \Delta t}\left\{\left[\left(\frac{1}{B_w} S_w\right)^{n} \phi' + \phi^{n+1} S_w^n \left(\frac{1}{B_w}\right)' + \left(\phi \frac{1}{B_w}\right)^{n+1} S_w^n \times 0\right]\left(p_o^{n+1} - p_o^n\right) + \left(\phi \frac{1}{B_w}\right)^{n+1} \times 1 \times \left(S_w^{n+1} - S_w^n\right) + \left(\phi \frac{1}{B_w}\right)^{n+1} \times 0 \times \left(S_g^{n+1} - S_g^n\right)\right\} \tag{11.74}$$

which upon simplification, term factorization, and addition of subscript n to all functions to identify the block gives Eq. (11.62), which states

$$\frac{V_{b_n}}{\alpha_c \Delta t}\left[\left(\frac{\phi S_w}{B_w}\right)_n^{n+1} - \left(\frac{\phi S_w}{B_w}\right)_n^{n}\right] = \frac{V_{b_n}}{\alpha_c \Delta t}\left\{\left(\frac{\phi}{B_w}\right)_n^{n+1}\left(S_{w_n}^{n+1} - S_{w_n}^n\right) + S_{w_n}^n\left[\phi_n^{n+1}\left(\frac{1}{B_{w_n}}\right)' + \frac{1}{B_{w_n}^n}\phi_n'\right]\left(p_{o_n}^{n+1} - p_{o_n}^n\right)\right\} \tag{11.62}$$

11.4.2 Well rate terms

Production and injection wells are treated separately because injection usually involves one phase only, either water or gas, but production involves all phases present in wellblocks.

11.4.2.1 Production terms

Fluid production rates in multiphase flow are dependent on each other through at least relative permeabilities. In other words, the specification of the production rate of any phase implicitly dictates the production rates of the other phases. In this section, we emphasize the treatment of a vertical well that is completed in several blocks, as shown in Fig. 11.13, and produces fluids from a multiphase reservoir.

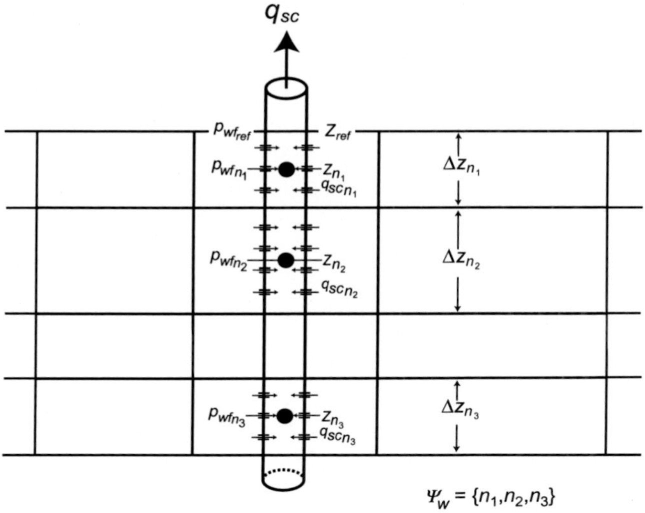

FIG. 11.13 Cross section showing pressures within a vertical production wellbore.

If the FBHP at reference depth ($p_{wf_{ref}}$) is assumed to be known, then well pressure opposite wellblock i can be estimated using the following equation:

$$p_{wf_i} = p_{wf_{ref}} + \overline{\gamma}_{wb}(Z_i - Z_{ref}) \tag{11.75}$$

where

$$\overline{\gamma}_{wb} = \gamma_c \overline{\rho}_{wb} g \tag{11.76}$$

In addition, the average fluid density in the wellbore opposite the producing formation is approximated as

$$\overline{\rho}_{wb} = \frac{\sum\limits_{p \in \{o, w, fg\}} \overline{\rho}_p \overline{B}_p q_{psc}}{\sum\limits_{p \in \{o, w, fg\}} \overline{B}_p q_{psc}} \tag{11.77a}$$

where average FBHP or $p_{wf_{ref}}$ can be used to obtain estimates for $\overline{B}_p$ and $\overline{\rho}_p$ for phase $p = o, w, g$.

The concern here is to estimate the production rate of phase $p = o, w, fg$ from wellblock i under different well operating conditions, where wellblock i is a member of the set of all blocks that contribute to well production; that is, $i \in \psi_w$.

Shut-in well

$$q_{psc_i} = 0 \tag{11.78}$$

where $p = o, w, fg$.

Specified well flow rate

TABLE 11.4 Well rate specification and definitions of set η_{prd} and M_p.

Well rate specification q_{sp}	Set of specified phases η_{prd}	Phase relative mobility M_p
q_{osp}	$\{o, w\}$	k_{rp}/μ_p
q_{Lsp}	$\{o, w\}$	k_{rp}/μ_p
q_{Tsp}	$\{o, w, g\}$	k_{rp}/μ_p
q_{ospsc}	$\{o, w\}$	$k_{rp}/(B_p\mu_p)$
q_{Lspsc}	$\{o, w\}$	$k_{rp}/(B_p\mu_p)$

The production rate of phase $p=o, w, fg$ from wellblock i is given by

$$q_{psc_i} = -G_{w_i}\left(\frac{k_{rp}}{B_p\mu_p}\right)_i \left(p_i - p_{wf_i}\right) \tag{11.79a}$$

This equation can be combined with Eq. (11.75) to give

$$q_{psc_i} = -G_{w_i}\left(\frac{k_{rp}}{B_p\mu_p}\right)_i \left[p_i - p_{wf_{ref}} - \overline{\gamma}_{wb}\left(Z_i - Z_{ref}\right)\right] \tag{11.80a}$$

For a multiblock well, $p_{wf_{ref}}$ is estimated from the well rate specification (q_{sp}) using

$$p_{wf_{ref}} = \frac{\sum\limits_{i\in\psi_w}\left\{G_{w_i}\left[p_i - \overline{\gamma}_{wb}\left(Z_i - Z_{ref}\right)\right]\sum\limits_{p\in\eta_{prd}} M_{p_i}\right\} + q_{sp}}{\sum\limits_{i\in\psi_w} G_{w_i}\sum\limits_{p\in\eta_{prd}} M_{p_i}} \tag{11.81a}$$

where η_{prd} and M_p depend on the type of well rate specification as listed in Table 11.4. The use of Eq. (11.81a) requires solving for $p_{wf_{ref}}$ implicitly along with the reservoir block pressures. An explicit treatment, however, uses Eq. (11.81a) at old time level n to estimate $p^n_{wf_{ref}}$, which is subsequently substituted into Eq. (11.80a) to estimate the production rate of phase $p=o, w, fg$ from wellblock i (q_{psc_i}). For a single-block well, the application of Eq. (11.81a) for $\psi_w = \{i\}$ followed by substitution for $p_{wf_{ref}}$ into Eq. (11.80a) yields.

$$p_{wf_{ref}} = \left[p_i - \overline{\gamma}_{wb}\left(Z_i - Z_{ref}\right)\right] + \frac{q_{sp}}{G_{w_i}\sum\limits_{p\in\eta_{prd}} M_{p_i}} \tag{11.81c}$$

and

$$q_{psc_i} = \left(\frac{k_{rp}}{B_p\mu_p}\right)_i \frac{q_{sp}}{\sum\limits_{p\in\eta_{prd}} M_{p_i}} \tag{11.80c}$$

for $p=o,\ w,fg$.

Specified well pressure gradient

For a specified well pressure gradient, the production rate of phase $p=o,\ w,\ fg$ from wellblock i is given by

$$q_{psc_i} = -2\pi\beta_c r_w k_{H_i} h_i \left(\frac{k_{rp}}{B_p\mu_p}\right)_i \left.\frac{\partial p}{\partial r}\right|_{r_w} \tag{11.82a}$$

Specified well FBHP

If the FBHP of a well ($p_{wf_{ref}}$) is specified, then the production rate of phase $p=o,\ w,fg$ from wellblock i can be estimated using Eq. (11.80a):

$$q_{psc_i} = -G_{w_i}\left(\frac{k_{rp}}{B_p\mu_p}\right)_i \left[p_i - p_{wf_{ref}} - \overline{\gamma}_{wb}\left(Z_i - Z_{ref}\right)\right] \tag{11.80a}$$

Example 11.8 Consider the single-well simulation problem presented in Example 11.7. Write the production rate equations for oil, water, and gas from the well in gridblock 9 given that the well is producing at a specified constant liquid rate of q_{Lspsc}.

Solution

The concern here is to find the production rate of the individual phases from the well in wellblock 9 given that $q_{sp}=q_{Lspsc}$. For single-block wells, Eq. (11.80c) is applicable, stating

$$q_{psc_i} = \left(\frac{k_{rp}}{B_p\mu_p}\right)_i \frac{q_{sp}}{\sum\limits_{p\in\eta_{prd}} M_{p_i}} \tag{11.80c}$$

where $p=o,\ w,\ fg$. For $q_{sp}=q_{Lspsc}$ in Table 11.4, we have $\eta_{prd}=\{o,w\}$ and $M_p=k_{rp}/B_p\mu_p$. Therefore, substitution into Eq. (11.80c) for wellblock 9 (i.e., $i=9$) gives

$$q_{osc_9} = \left(\frac{k_{ro}}{B_o\mu_o}\right)_9 \frac{q_{Lspsc}}{\left[\left(\frac{k_{ro}}{B_o\mu_o}\right)_9 + \left(\frac{k_{rw}}{B_w\mu_w}\right)_9\right]} \tag{11.83a}$$

$$q_{wsc_9} = \left(\frac{k_{rw}}{B_w\mu_w}\right)_9 \frac{q_{Lspsc}}{\left[\left(\frac{k_{ro}}{B_o\mu_o}\right)_9 + \left(\frac{k_{rw}}{B_w\mu_w}\right)_9\right]} \tag{11.83b}$$

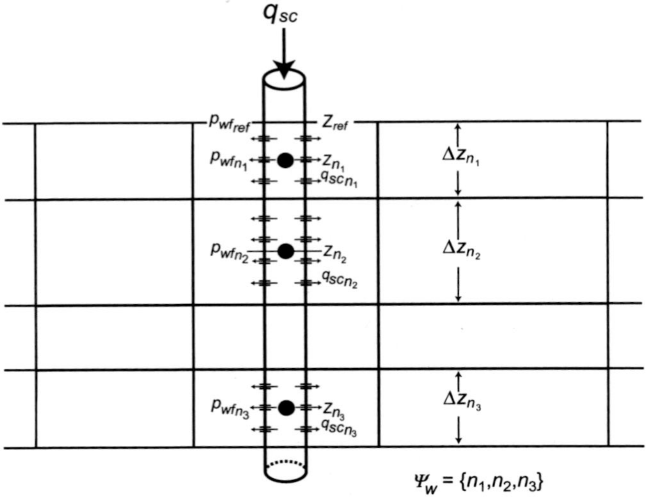

FIG. 11.14 Cross section showing pressures within a vertical injection wellbore.

and

$$q_{fgsc_9} = \left(\frac{k_{rg}}{B_g \mu_g}\right)_9 \frac{q_{Lspsc}}{\left[\left(\frac{k_{ro}}{B_o \mu_o}\right)_9 + \left(\frac{k_{rw}}{B_w \mu_w}\right)_9\right]} \tag{11.83c}$$

It should be noted that $q_{gsc_9} = q_{fgsc_9} + R_{s_9} q_{osc_9}$.

11.4.2.2 Injection terms

For injection wells, one phase (usually water or gas) is injected. The mobility of the injected fluid at reservoir conditions in a wellblock is equal to the sum of the mobilities of all phases present in the wellblock (Abou-Kassem, 1996); that is,

$$M_{inj} = \sum_{p \in \eta_{inj}} M_p \tag{11.84}$$

where

$$\eta_{inj} = \{o, w, g\} \tag{11.85}$$

$M_p = (k_{rp}/\mu_p)$, and $\beta_c k_H M_p =$ mobility of phase p at reservoir conditions of the wellblock.

In this section, we emphasize the treatment of a well that is completed in several blocks, as shown in Fig. 11.14, and injects either water or gas into a multiphase reservoir. If the FBHP at reference depth ($p_{wf_{ref}}$) is assumed to be known, then the well pressure opposite wellblock i can be estimated using Eq. (11.75):

$$p_{wf_i} = p_{wf_{ref}} + \overline{\gamma}_{wb}(Z_i - Z_{ref}) \tag{11.75}$$

where

$$\overline{\gamma}_{wb} = \gamma_c \overline{\rho}_{wb} g \tag{11.76}$$

and the average density of the injected fluid, opposite the formation, is estimated as

$$\overline{\rho}_{wb} = \frac{\rho_{psc}}{\overline{B}_p} \tag{11.77b}$$

Average FBHP or $p_{wf_{ref}}$ can be used to obtain an estimate for $\overline{B}_p$ of the injected phase p. The concern here is to estimate the injection rate of the injected phase (usually water or gas) into wellblock i under different well operating conditions, where wellblock i is a member of the set of all blocks that receive the injected fluid; that is, $i \in \psi_w$. Of course, the rates of injection of the remaining phases are set to zero.

Shut-in well

$$q_{psc_i} = 0 \tag{11.78}$$

where $p = w$ or fg.

Specified well flow rate

The injection rate of the injected fluid $p = w$ or fg into wellblock i is given by

$$q_{psc_i} = -G_{w_i}\left(\frac{M_{inj}}{B_p}\right)_i (p_i - p_{wf_i}) \tag{11.79b}$$

This equation can be combined with Eq. (11.75) to give

$$q_{psc_i} = -G_{w_i}\left(\frac{M_{inj}}{B_p}\right)_i \left[p_i - p_{wf_{ref}} - \overline{\gamma}_{wb}(Z_i - Z_{ref})\right] \tag{11.80b}$$

For a single-block well, $q_{psc_i} = q_{spsc}$ and Eq. (11.80b) is used to estimate $p_{wf_{ref}}$. For a multiblock well, however, $p_{wf_{ref}}$ is estimated from the well rate specification at standard conditions (q_{spsc}) using

$$p_{wf_{ref}} = \frac{\sum\limits_{i \in \psi_w} \left\{ G_{w_i}\left(\frac{M_{inj}}{B_p}\right)_i \left[p_i - \overline{\gamma}_{wb}(Z_i - Z_{ref})\right] \right\} + q_{spsc}}{\sum\limits_{i \in \psi_w} G_{w_i}\left(\frac{M_{inj}}{B_p}\right)_i} \tag{11.81b}$$

Then, the injection rate of the injected fluid $p = w$ or fg into wellblock i (q_{psc_i}) is estimated using Eq. (11.80b). The use of Eq. (11.81b) requires solving for $p_{wf_{ref}}$ implicitly along with the reservoir block pressures. An explicit treatment, however, uses Eq. (11.81b) at old time level n to estimate $p^n_{wf_{ref}}$, which is subsequently substituted into Eq. (11.80b) to estimate the injection rate of the injected phase $p = w$ or fg into wellblock i (q_{psc_i}).

Specified well pressure gradient

For a specified well pressure gradient, the injection rate of fluid $p=w$ or fg into wellblock i is given by

$$q_{psc_i} = -2\pi\beta_c r_w k_{H_i} h_i \left(\frac{M_{inj}}{B_p}\right)_i \frac{\partial p}{\partial r}\bigg|_{r_w} \tag{11.82b}$$

Specified well FBHP

If the FBHP of a well ($p_{wf_{ref}}$) is specified, then the injection rate of the injected fluid $p=w$ or fg into wellblock i can be estimated using Eq. (11.80b):

$$q_{psc_i} = -G_{w_i}\left(\frac{M_{inj}}{B_p}\right)_i \left[p_i - p_{wf_{ref}} - \overline{\gamma}_{wb}\left(Z_i - Z_{ref}\right)\right] \tag{11.80b}$$

11.4.3 Treatment of boundary conditions

A reservoir boundary can be subject to one of four conditions: (1) a no-flow boundary, (2) a constant flow boundary, (3) a constant pressure gradient boundary, and (4) a constant pressure boundary. As discussed in single-phase flow in Chapters 4 and 5, the first three boundary conditions reduce to a specified pressure gradient condition (the Neumann boundary condition), and the fourth boundary condition is the Dirichlet boundary condition. The treatment of boundary conditions for 1-D flow in the x-direction is similar to that presented in Section 4.4 for a block-centered grid and Section 5.4 for a point-distributed grid. In this section, we present the fictitious well rate equations as they apply to multiphase flow in reservoirs discretized using a block-centered grid only. The effect of capillary pressure is assumed negligible. The fictitious well rate of phase p ($q_{psc_{b,bB}}^{n+1}$) reflects fluid transfer of phase p between the boundary block (bB) and the reservoir boundary itself (b) or the block next to the reservoir boundary that falls outside the reservoir. In multiphase flow, a reservoir boundary may (1) separate two segments of one reservoir that has same fluids, (2) separate an oil reservoir from a water aquifer or a gas cap, or (3) seal off the reservoir from a neighboring reservoir. If the neighboring reservoir segment is an aquifer, then either water invades the reservoir across the reservoir boundary (WOC), or reservoir fluids leave the reservoir block to the aquifer. Similarly, if the neighboring reservoir segment is a gas cap, then either gas invades the reservoir across the reservoir boundary (GOC), or reservoir fluids leave the reservoir block to the gas cap.

11.4.3.1 Specified pressure gradient boundary condition

For a specified pressure gradient at the reservoir left (west) boundary,

$$q_{psc_{b,bB}}^{n+1} \cong -\left[\beta_c \frac{k_l k_{rp} A_l}{\mu_p B_p}\right]_{bB}^{n+1} \left[\frac{\partial p_p}{\partial l}\bigg|_b^{n+1} - \left(\gamma_p\right)_{bB}^n \frac{\partial Z}{\partial l}\bigg|_b\right] \tag{11.86a}$$

for $p = o, w, fg$, and at the reservoir right (east) boundary,

$$q_{psc_{b,bB}}^{n+1} \cong \left[\beta_c \frac{k_l k_{rp} A_l}{\mu_p B_p}\right]_{bB}^{n+1} \left[\frac{\partial p_p}{\partial l}\bigg|_b^{n+1} - (\gamma_p)_{bB}^n \frac{\partial Z}{\partial l}\bigg|_b\right] \tag{11.86b}$$

for $p = o, w, fg$, where the component physical properties and phase physical properties other than the flow rate for the gas phase and the free-gas component are the same. The flow rate at standard conditions of the gas component, however, equals the sum of flow rates at standard conditions of the free-gas and solution-gas components; that is,

$$q_{gsc_{b,bB}}^{n+1} = q_{fgsc_{b,bB}}^{n+1} + R_{s_{bB}}^{n+1} q_{osc_{b,bB}}^{n+1} \tag{11.87}$$

In Eq. (11.86), the specified pressure gradient may replace the phase pressure gradient at the boundary. Eq. (11.86) applies to fluid flow across a reservoir boundary that separates two segments of the same reservoir or across a reservoir boundary that represents WOC with fluids being lost to the water aquifer. If the reservoir boundary represents WOC and water invades the reservoir, then

$$q_{wsc_{b,bB}}^{n+1} \cong -\left[\beta_c \frac{k_l A_l}{\mu_w B_w}\right]_{bB}^{n+1} (k_{rw})_{aq}^{n+1} \left[\frac{\partial p}{\partial l}\bigg|_b^{n+1} - (\gamma_w)_{bB}^n \frac{\partial Z}{\partial l}\bigg|_b\right] \tag{11.88a}$$

for the reservoir left (west) boundary, and

$$q_{wsc_{b,bB}}^{n+1} \cong \left[\beta_c \frac{k_l A_l}{\mu_w B_w}\right]_{bB}^{n+1} (k_{rw})_{aq}^{n+1} \left[\frac{\partial p}{\partial l}\bigg|_b^{n+1} - (\gamma_w)_{bB}^n \frac{\partial Z}{\partial l}\bigg|_b\right] \tag{11.88b}$$

for the reservoir right (east) boundary.

Moreover,

$$q_{osc_{b,bB}}^{n+1} = q_{fgsc_{b,bB}}^{n+1} = q_{gsc_{b,bB}}^{n+1} = 0 \tag{11.89}$$

Note that, in Eq. (11.88), the rock and fluid properties in the aquifer are approximated by those of the boundary block properties because of the lack of geologic control in aquifers and because the effect of oil/water capillary pressure is neglected. In addition, $(k_{rw})_{aq}^{n+1} = 1$ because $S_w = 1$ in the aquifer.

11.4.3.2 Specified flow rate boundary condition

If the specified flow rate stands for water influx across a reservoir boundary, then

$$q_{wsc_{b,bB}}^{n+1} = q_{sp}/B_{w_{bB}} \tag{11.90}$$

In addition, Eq. (11.89) applies (i.e., $q_{osc_{b,bB}}^{n+1} = q_{fgsc_{b,bB}}^{n+1} = q_{gsc_{b,bB}}^{n+1} = 0$). If, however, the specified flow rate stands for fluid transfer between two segments of the same reservoir or fluid loss to an aquifer across WOC, then

$$q_{psc_{b,bB}}^{n+1} = \frac{\left(T_p^R\right)_{b,bB}^{n+1}}{B_p \sum_{l \in \{o,w,fg\}} \left(T_l^R\right)_{b,bB}^{n+1}} q_{sp} = \frac{\left(\frac{k_{rp}}{\mu_p}\right)_{bB}^{n+1}}{B_p \sum_{l \in \{o,w,fg\}} \left(\frac{k_{rl}}{\mu_l}\right)_{bB}^{n+1}} q_{sp} \quad (11.91)$$

for $p = o, w, fg$ because in this case,

$$\left(T_p^R\right)_{b,bB}^{n+1} = \left(T_p^R\right)_{bB}^{n+1} = \left[\beta_c \frac{k_l k_{rp} A_l}{\mu_p (\Delta l/2)}\right]_{bB}^{n+1} \quad (11.92)$$

Eq. (11.91) neglects the effects of gravity forces and capillary pressures.

11.4.3.3 No-flow boundary condition

This condition results from vanishing permeability at a reservoir boundary or because of symmetry about a reservoir boundary. In either case, for a reservoir no-flow boundary,

$$q_{psc_{b,bB}}^{n+1} = 0 \quad (11.93)$$

for $p = o, w, fg$.

11.4.3.4 Specified boundary pressure condition

This condition arises due to the presence of wells on the other side of a reservoir boundary that operate to maintain voidage replacement and as a result keep the boundary pressure (p_b) constant. The flow rate of phase p across a reservoir boundary that separates two segments of the same reservoir or across a reservoir boundary that represents WOC with fluid loss to an aquifer is estimated using

$$q_{psc_{b,bB}}^{n+1} = \left[\beta_c \frac{k_l k_{rp} A_l}{\mu_p B_p (\Delta l/2)}\right]_{bB}^{n+1} \left[\left(p_b - p_{bB}^{n+1}\right) - \left(\gamma_p\right)_{bB}^{n} (Z_b - Z_{bB})\right] \quad (11.94)$$

for $p = o, w, fg$.

If the reservoir boundary represents WOC with water influx, then

$$q_{wsc_{b,bB}}^{n+1} = \left[\beta_c \frac{k_l A_l}{\mu_w B_w (\Delta l/2)}\right]_{bB}^{n+1} (k_{rw})_{aq}^{n+1} \left[\left(p_b - p_{bB}^{n+1}\right) - (\gamma_w)_{bB}^{n} (Z_b - Z_{bB})\right] \quad (11.95)$$

In addition, Eq. (11.89) applies (i.e., $q_{osc_{b,bB}}^{n+1} = q_{fgsc_{b,bB}}^{n+1} = q_{gsc_{b,bB}}^{n+1} = 0$). Note that, in Eq. (11.95), the rock and fluid properties in the aquifer are approximated by those of the boundary block properties because of the lack of geologic control in aquifers. In addition, $(k_{rw})_{aq}^{n+1} = 1$ because $S_w = 1$ in the aquifer.

It is worth mentioning that when reservoir boundary b stands for WOC, the flow rate of phase p across the reservoir boundary is determined from the knowledge of the upstream point between reservoir boundary b and boundary block bB. If b is upstream to bB (i.e., when $\Delta\Phi_w > 0$), the flow is from the aquifer to the reservoir boundary block, and Eq. (11.95) applies for water and $q_{osc_{b,bB}}^{n+1} = q_{fgsc_{b,bB}}^{n+1} = q_{gsc_{b,bB}}^{n+1} = 0$. If b is downstream to bB (i.e., when $\Delta\Phi_w < 0$), the flow is from the reservoir boundary block to the aquifer, and Eq. (11.94) applies for all phases. The water potential between the reservoir boundary and the reservoir boundary block is defined as $\Delta\Phi_w = (p_b - p_{bB}) - \gamma_w(Z_b - Z_{bB})$.

Example 11.9 Consider the single-well simulation problem presented in Example 11.7. Write the flow equations for boundary gridblock 3.

Solution

In this problem, the reservoir is subject to water influx. For gridblock 3, $n=3$, and $\psi_2 = \{2,4,7\}$. Gridblock 3 is a boundary block that falls on the reservoir lower boundary; therefore, $\xi_3 = \{b_L\}$, $\sum_{l\in\xi_3} q_{osc_{l,3}}^{n+1} = 0$, $\sum_{l\in\xi_3} q_{fgsc_{l,3}}^{n+1} = 0$, and

$\sum_{l\in\xi_3} q_{wsc_{l,3}}^{n+1} = q_{wsc_{b_L,3}}^{n+1}$, where $q_{wsc_{bL,\,3}}^{n+1}$ is estimated using Eq. (11.95) as

$$q_{wsc_{b_L,3}}^{n+1} = \left[\beta_c \frac{k_z A_z}{\mu_w B_w (\Delta z/2)}\right]_3^{n+1} \times (k_{rw})_{aq}^{n+1} \times \left[\left(p_{b_L} - p_{o_3}^{n+1}\right) - (\gamma_w)_3^n (Z_{b_L} - Z_3)\right]$$

or

$$q_{wsc_{b_L,3}}^{n+1} = \left[\beta_c \frac{k_z A_z}{\mu_w B_w (\Delta z/2)}\right]_3^{n+1} \left[\left(p_{WOC} - p_{o_3}^{n+1}\right) - (\gamma_w)_3^n \Delta z_3/2\right]$$

where $p_{b_L} = p_{WOC}$, $(k_{rw})_{aq}^{n+1} = 1$, and $(Z_{b_L} - Z_3) = \Delta z_3/2$.

Gridblock 3 has no wells; therefore, $q_{osc_3}^{n+1} = 0$, $q_{wsc_3}^{n+1} = 0$, and $q_{fgsc_3}^{n+1} = 0$. Observe also that $Z_2 = Z_3 = Z_4$.

The oil equation is obtained by substituting the given values into Eq. (11.54) and expanding the summation terms, yielding

$$\begin{aligned}
&T_{o_{2,3}}^{n+1}\left[\left(p_{o_2}^{n+1} - p_{o_3}^{n+1}\right) - \gamma_{o_{2,3}}^n \times 0\right] + T_{o_{4,3}}^{n+1}\left[\left(p_{o_4}^{n+1} - p_{o_3}^{n+1}\right) - \gamma_{o_{4.3}}^n \times 0\right] \\
&+ T_{o_{7,3}}^{n+1}\left[\left(p_{o_7}^{n+1} - p_{o_3}^{n+1}\right) - \gamma_{o_{7,3}}^n (Z_7 - Z_3)\right] + 0 + 0 \\
&= \frac{V_{b_3}}{\alpha_c \Delta t}\left\{\left[\frac{\phi(1 - S_w - S_g)}{B_o}\right]_3^{n+1} - \left[\frac{\phi(1 - S_w - S_g)}{B_o}\right]_3^n\right\}
\end{aligned} \tag{11.96}$$

The gas equation is obtained by substituting the given values into Eq. (11.55) and expanding the summation terms, yielding

$$
\begin{aligned}
&T_{g_{2,3}}^{n+1}\left[\left(p_{o_2}^{n+1}-p_{o_3}^{n+1}\right)+\left(P_{cgo_2}^{n+1}-P_{cgo_3}^{n+1}\right)-\gamma_{g_{2,3}}^{n}\times 0\right]\\
&+(T_oR_s)_{2,3}^{n+1}\left[\left(p_{o_2}^{n+1}-p_{o_3}^{n+1}\right)-\gamma_{o_{2,3}}^{n}\times 0\right]\\
&+T_{g_{4,3}}^{n+1}\left[\left(p_{o_4}^{n+1}-p_{o_3}^{n+1}\right)+\left(P_{cgo_4}^{n+1}-P_{cgo_3}^{n+1}\right)-\gamma_{g_{4,3}}^{n}\times 0\right]\\
&+(T_oR_s)_{4,3}^{n+1}\left[\left(p_{o_4}^{n+1}-p_{o_3}^{n+1}\right)-\gamma_{o_{4,3}}^{n}\times 0\right]\\
&+T_{g_{7,3}}^{n+1}\left[\left(p_{o_7}^{n+1}-p_{o_3}^{n+1}\right)+\left(P_{cgo_7}^{n+1}-P_{cgo_3}^{n+1}\right)-\gamma_{g_{7,3}}^{n}(Z_7-Z_3)\right]\\
&+(T_oR_s)_{7,3}^{n+1}\left[\left(p_{o_7}^{n+1}-p_{o_3}^{n+1}\right)-\gamma_{o_{7,3}}^{n}(Z_7-Z_3)\right]+0+\left[0+R_{s_7}^{n+1}\times 0\right]\\
&=\frac{V_{b_3}}{\alpha_c\Delta t}\left\{\left(\frac{\phi S_g}{B_g}\right)_3^{n+1}-\left(\frac{\phi S_g}{B_g}\right)_3^{n}+\left[\frac{\phi R_s(1-S_w-S_g)}{B_o}\right]_3^{n+1}-\left[\frac{\phi R_s(1-S_w-S_g)}{B_o}\right]_3^{n}\right\}
\end{aligned}
\tag{11.97}
$$

The water equation is obtained by substituting the given values into Eq. (11.56) and expanding the summation terms, yielding

$$
\begin{aligned}
&T_{w_{2,3}}^{n+1}\left[\left(p_{o_2}^{n+1}-p_{o_3}^{n+1}\right)-\left(P_{cow_2}^{n+1}-P_{cow_3}^{n+1}\right)-\gamma_{w_{2,3}}^{n}\times 0\right]\\
&+T_{w_{4,3}}^{n+1}\left[\left(p_{o_4}^{n+1}-p_{o_3}^{n+1}\right)-\left(P_{cow_4}^{n+1}-P_{cow_3}^{n+1}\right)-\gamma_{w_{4,3}}^{n}\times 0\right]\\
&+T_{w_{7,3}}^{n+1}\left[\left(p_{o_7}^{n+1}-p_{o_3}^{n+1}\right)-\left(P_{cow_7}^{n+1}-P_{cow_3}^{n+1}\right)-\gamma_{w_{7,3}}^{n}(Z_7-Z_3)\right]\\
&+\left[\beta_c\frac{k_zA_z}{\mu_wB_w(\Delta z/2)}\right]_3^{n+1}\left[\left(p_{WOC}-p_{o_3}^{n+1}\right)-(\gamma_w)_3^n\Delta z_3/2\right]+0\\
&=\frac{V_{b_3}}{\alpha_c\Delta t}\left[\left(\frac{\phi S_w}{B_w}\right)_3^{n+1}-\left(\frac{\phi S_w}{B_w}\right)_3^{n}\right]
\end{aligned}
\tag{11.98}
$$

11.4.4 Treatment of nonlinearities

The time linearization methods of the phase transmissibility terms in multiphase flow are similar to those presented in Section 8.4.1.2 for single-phase flow (explicit method, simple iteration method, and fully implicit method). There are other time linearization methods such as the linearized-implicit method (MacDonald and Coats, 1970) and the semiimplicit method of Nolen and Berry (1972); however, these methods deal with nonlinearities due to fluid saturation only. The time linearization methods of well production rates in multiphase flow are similar to those presented in Section 8.4.2 for single-phase flow (explicit transmissibility method, simple iteration on transmissibility method, and fully implicit method). It should be mentioned that the time linearization of well rate terms (production and injection) and fictitious well rates in multiphase flow are the same as those used for the treatment of flow terms between a block and its neighboring blocks (see Section 8.4.3).

The space linearization methods of phase transmissibility are different from those for single-phase flow. For phase transmissibility defined by Eq. (11.35),

$$T_{p_{l,n}} = G_{l,n}\left(\frac{1}{\mu_p B_p}\right)_{l,n} k_{rp_{l,n}} \tag{11.35}$$

the various space-weighting methods presented for single-phase flow (Section 8.4.1.1) work for the pressure-dependent terms, $\left(\frac{1}{\mu_p B_p}\right)_{l,n}$ and $\left(\frac{R_s}{\mu_o B_o}\right)_{l,n}$, but only the upstream-weighting method works for the saturation-dependent terms, $k_{rp_{l,n}}$. In fact, the function average-value method and the variable average-value method presented in Section 8.4.1.1 give erroneous results when applied to relative permeabilities. The most commonly used method for space linearization of pressure- and saturation-dependent terms is the upstream-weighting method.

11.4.5 Solution methods

In this section, we present the implicit pressure-explicit saturation (IMPES) and simultaneous solution (SS) methods as they apply to the two-phase oil/water flow model in multidimensional reservoirs. The flow equations (reduced set of equations) for block n in a multidimensional reservoir are presented in Section 11.3.1 as Eqs. (11.41) and (11.42).

The oil equation is

$$\sum_{l\in\psi_n} T^{n+1}_{o_{l,n}}\left[\left(p^{n+1}_{o_l} - p^{n+1}_{o_n}\right) - \gamma^n_{o_{l,n}}(Z_l - Z_n)\right] + \sum_{l\in\xi_n} q^{n+1}_{osc_{l,n}} + q^{n+1}_{osc_n} = \frac{V_{b_n}}{\alpha_c \Delta t}\left\{\left[\frac{\phi(1-S_w)}{B_o}\right]^{n+1}_n - \left[\frac{\phi(1-S_w)}{B_o}\right]^n_n\right\} \tag{11.41}$$

The water equation is

$$\sum_{l\in\psi_n} T^{n+1}_{w_{l,n}}\left[\left(p^{n+1}_{o_l} - p^{n+1}_{o_n}\right) - \left(P^{n+1}_{cow_l} - P^{n+1}_{cow_n}\right) - \gamma^n_{w_{l,n}}(Z_l - Z_n)\right] + \sum_{l\in\xi_n} q^{n+1}_{wsc_{l,n}} + q^{n+1}_{wsc_n} = \frac{V_{b_n}}{\alpha_c \Delta t}\left[\left(\frac{\phi S_w}{B_w}\right)^{n+1}_n - \left(\frac{\phi S_w}{B_w}\right)^n_n\right] \tag{11.42}$$

The $p_o - S_w$ formulation is used here; hence, the primary unknowns are p_o and S_w, and the secondary unknowns are p_w and S_o where $p_w = p_o - P_{cow}(S_w)$ and $S_o = 1 - S_w$. The expansions of the RHS of Eqs. (11.41) and (11.42) are

$$\frac{V_{b_n}}{\alpha_c \Delta t}\left\{\left[\frac{\phi(1-S_w)}{B_o}\right]_n^{n+1} - \left[\frac{\phi(1-S_w)}{B_o}\right]_n^{n}\right\}$$
$$= \frac{V_{b_n}}{\alpha_c \Delta t}\left\{-\left(\frac{\phi}{B_o}\right)_n^{n+1}\left(S_{w_n}^{n+1} - S_{w_n}^{n}\right) + \left(1 - S_{w_n}^{n}\right)\left[\phi_n^{n+1}\left(\frac{1}{B_{o_n}}\right)' + \frac{1}{B_{o_n}^{n}}\phi_n'\right]\left(p_{o_n}^{n+1} - p_{o_n}^{n}\right)\right\} \tag{11.99}$$

and

$$\frac{V_{b_n}}{\alpha_c \Delta t}\left[\left(\frac{\phi S_w}{B_w}\right)_n^{n+1} - \left(\frac{\phi S_w}{B_w}\right)_n^{n}\right]$$
$$= \frac{V_{b_n}}{\alpha_c \Delta t}\left\{\left(\frac{\phi}{B_w}\right)_n^{n+1}\left(S_{w_n}^{n+1} - S_{w_n}^{n}\right) + S_{w_n}^{n}\left[\phi_n^{n+1}\left(\frac{1}{B_{w_n}}\right)' + \frac{1}{B_{w_n}^{n}}\phi_n'\right]\left(p_{o_n}^{n+1} - p_{o_n}^{n}\right)\right\} \tag{11.62}$$

Eqs. (11.99) and (11.62) can be rewritten as

$$\frac{V_{b_n}}{\alpha_c \Delta t}\left\{\left[\frac{\phi(1-S_w)}{B_o}\right]_n^{n+1} - \left[\frac{\phi(1-S_w)}{B_o}\right]_n^{n}\right\}$$
$$= C_{op_n}\left(p_{o_n}^{n+1} - p_{o_n}^{n}\right) + C_{ow_n}\left(S_{w_n}^{n+1} - S_{w_n}^{n}\right) \tag{11.100}$$

and

$$\frac{V_{b_n}}{\alpha_c \Delta t}\left[\left(\frac{\phi S_w}{B_w}\right)_n^{n+1} - \left(\frac{\phi S_w}{B_w}\right)_n^{n}\right] = C_{wp_n}\left(p_{o_n}^{n+1} - p_{o_n}^{n}\right) + C_{ww_n}\left(S_{w_n}^{n+1} - S_{w_n}^{n}\right) \tag{11.101}$$

where

$$C_{op_n} = \frac{V_{b_n}}{\alpha_c \Delta t}\left\{\left(1 - S_{w_n}^{n}\right)\left[\phi_n^{n+1}\left(\frac{1}{B_{o_n}}\right)' + \frac{1}{B_{o_n}^{n}}\phi_n'\right]\right\} \tag{11.102a}$$

$$C_{ow_n} = \frac{V_{b_n}}{\alpha_c \Delta t}\left[-\left(\frac{\phi}{B_o}\right)_n^{n+1}\right] \tag{11.102b}$$

$$C_{wp_n} = \frac{V_{b_n}}{\alpha_c \Delta t}\left\{S_{w_n}^{n}\left[\phi_n^{n+1}\left(\frac{1}{B_{w_n}}\right)' + \frac{1}{B_{w_n}^{n}}\phi_n'\right]\right\} \tag{11.102c}$$

and

$$C_{ww_n} = \frac{V_{b_n}}{\alpha_c \Delta t} \left(\frac{\phi}{B_w} \right)_n^{n+1} \tag{11.102d}$$

A form of the reduced set of flow equations for the oil/water model that is suitable for applying a solution method is obtained by substituting Eqs. (11.100) and (11.101) into Eqs. (11.41) and (11.42).

The oil equation becomes

$$\sum_{l \in \psi_n} T_{o_{l,n}}^{n+1} \left[\left(p_{o_l}^{n+1} - p_{o_n}^{n+1} \right) - \gamma_{o_{l,n}}^n (Z_l - Z_n) \right] + \sum_{l \in \xi_n} q_{osc_{l,n}}^{n+1} + q_{osc_n}^{n+1} = C_{op_n} \left(p_{o_n}^{n+1} - p_{o_n}^n \right) + C_{ow_n} \left(S_{w_n}^{n+1} - S_{w_n}^n \right) \tag{11.103}$$

and the water equation becomes

$$\sum_{l \in \psi_n} T_{w_{l,n}}^{n+1} \left[\left(p_{o_l}^{n+1} - p_{o_n}^{n+1} \right) - \left(P_{cow_l}^{n+1} - P_{cow_n}^{n+1} \right) - \gamma_{w_{l,n}}^n (Z_l - Z_n) \right] + \sum_{l \in \xi_n} q_{wsc_{l,n}}^{n+1} + q_{wsc_n}^{n+1} = C_{wp_n} \left(p_{o_n}^{n+1} - p_{o_n}^n \right) + C_{ww_n} \left(S_{w_n}^{n+1} - S_{w_n}^n \right) \tag{11.104}$$

The coefficients C_{op_n}, C_{ow_n}, C_{wp_n}, and C_{ww_n} are defined in Eq. (11.102), and the derivatives $\left(\frac{1}{B_{o_n}} \right)'$, $\left(\frac{1}{B_{w_n}} \right)'$, and ϕ_n' are chord slopes that are defined as

$$\left(\frac{1}{B_{o_n}} \right)' = \left(\frac{1}{\overset{(v)}{B_{o_n}^{n+1}}} - \frac{1}{B_{o_n}^n} \right) \Big/ \left(\overset{(v)}{p_{o_n}^{n+1}} - p_{o_n}^n \right) \tag{11.105a}$$

$$\left(\frac{1}{B_{w_n}} \right)' = \left(\frac{1}{\overset{(v)}{B_{w_n}^{n+1}}} - \frac{1}{B_{w_n}^n} \right) \Big/ \left(\overset{(v)}{p_{o_n}^{n+1}} - p_{o_n}^n \right) \tag{11.105b}$$

and

$$\phi_n' = \left(\overset{(v)}{\phi_n^{n+1}} - \phi_n^n \right) \Big/ \left(\overset{(v)}{p_{o_n}^{n+1}} - p_{o_n}^n \right) \tag{11.105c}$$

The pressure dependence of the oil and water FVFs in the oil/water flow model is described by Eq. (7.6) and that of porosity is described by Eq. (7.11). Substitution of Eqs. (7.6) and (7.11) into Eq. (11.105) yields

$$\left(\frac{1}{B_{o_n}} \right)' = \frac{c_o}{B_o^\circ} \tag{11.106a}$$

$$\left(\frac{1}{B_{w_n}} \right)' = \frac{c_w}{B_w^\circ} \tag{11.106b}$$

and

$$\phi_n' = \phi_n^\circ c_\phi \tag{11.106c}$$

11.4.5.1 IMPES method

The IMPES method, as the name implies, obtains an implicit pressure solution followed by an explicit solution for saturation. In the first step, the transmissibilities, capillary pressures, and coefficients of pressure difference in the well production rates and fictitious well rates, in addition to the fluid gravities, are treated explicitly. The resulting water and oil equations using Eqs. (11.103) and (11.104) are combined to obtain the pressure equation for block n through the elimination of the saturation term $(S_{w_n}^{n+1} - S_{w_n}^{n})$ that appears on the RHS of equations. This is achieved by multiplying the oil equation (Eq. 11.103) by $B_{o_n}^{n+1}$, multiplying the water equation (Eq. 11.104) by $B_{w_n}^{n+1}$, and adding the two resulting equations. Then, the pressure equation is written for all blocks $n=1, 2, 3\ldots N$, and the resulting set of pressure equations is solved for block pressures at time level $n+1$ ($p_{o_n}^{n+1}$ for $n=1, 2, 3\ldots N$). The second step involves solving the water equation for block n (Eq. 11.104) explicitly for water saturation at time level $n+1$ ($S_{w_n}^{n+1}$). Capillary pressures are then updated ($P_{cow_n}^{n+1} = P_{cow}(S_{w_n}^{n+1})$ for $n=1, 2, 3\ldots N$) and used as $P_{cow_n}^{n}$ in the following time step.

For a volumetric reservoir (no-flow boundaries) with explicit well production rates, the pressure equation for block $n=1, 2, 3\ldots N$ is

$$\begin{aligned}
&\sum_{l\in\psi_n}\left(B_{o_n}^{n+1}T_{o_{l,n}}^{n} + B_{w_n}^{n+1}T_{w_{l,n}}^{n}\right)p_{o_l}^{n+1} - \left\{\left[\sum_{l\in\psi_n}\left(B_{o_n}^{n+1}T_{o_{l,n}}^{n} + B_{w_n}^{n+1}T_{w_{l,n}}^{n}\right)\right]\right.\\
&\left.+\left(B_{o_n}^{n+1}C_{op_n} + B_{w_n}^{n+1}C_{wp_n}\right)\right\}p_{o_n}^{n+1} = \sum_{l\in\psi_n}\left[\left(B_{o_n}^{n+1}T_{o_{l,n}}^{n}\gamma_{o_{l,n}}^{n} + B_{w_n}^{n+1}T_{w_{l,n}}^{n}\gamma_{w_{l,n}}^{n}\right)(Z_l - Z_n)\right]\\
&+\sum_{l\in\psi_n}B_{w_n}^{n+1}T_{w_{l,n}}^{n}\left(P_{cow_l}^{n} - P_{cow_n}^{n}\right) - \left(B_{o_n}^{n+1}C_{op_n} + B_{w_n}^{n+1}C_{wp_n}\right)p_{o_n}^{n}\\
&-\left(B_{o_n}^{n+1}q_{osc_n}^{n} + B_{w_n}^{n+1}q_{wsc_n}^{n}\right)
\end{aligned} \tag{11.107}$$

Solving Eq. (11.107) for oil-phase pressure distribution may, in general, require iterating on $B_{o_n}^{n+1}$, $B_{w_n}^{n+1}$, C_{op_n}, and C_{wp_n} to preserve material balance. For 1-D flow problems, Eq. (11.107) represents a tridiagonal matrix equation. In this case, the coefficients of the unknowns $p_{o_{n-1}}^{n+1}, p_{o_n}^{n+1}$, and $p_{o_{n+1}}^{n+1}$ in the equation for block n correspond to w_n, c_n, and e_n, respectively, and the RHS of the equation corresponds to d_n in Thomas' algorithm presented in Section 9.2.1.

The water saturation for individual blocks in a volumetric reservoir is obtained from Eq. (11.104) with explicit transmissibilities and capillary pressures as

$$\begin{aligned}
S_{w_n}^{n+1} = S_{w_n}^{n} + \frac{1}{C_{ww_n}}&\left\{\sum_{l\in\psi_n}T_{w_{l,n}}^{n}\left[\left(p_{o_l}^{n+1} - p_{o_n}^{n+1}\right) - \left(P_{cow_l}^{n} - P_{cow_n}^{n}\right) - \gamma_{w_{l,n}}^{n}(Z_l - Z_n)\right]\right.\\
&\left.+q_{wsc_n}^{n} - C_{wp_n}\left(p_{o_n}^{n+1} - p_{o_n}^{n}\right)\right\}
\end{aligned} \tag{11.108}$$

The water saturation for block n is solved for explicitly using Eq. (11.108) independent of the equations for other blocks. This new estimate of water saturation is used to update the capillary pressure for block n, $P^{n+1}_{cow_n} = P_{cow}(S^{n+1}_{w_n})$, and this updated value will be used as $P^n_{cow_n}$ in the calculations for the following time step.

11.4.5.2 SS method

The SS method, as the name implies, solves the water and oil equations simultaneously for the unknowns of the formulation. Although this method is well suited to fully implicit formulation, we demonstrate its application for a volumetric reservoir (no-flow boundaries, $\sum\limits_{l \in \xi_n} q^{n+1}_{osc_{l,n}} = 0$, and $\sum\limits_{l \in \xi_n} q^{n+1}_{wsc_{l,n}} = 0$) using explicit transmissibilities ($T^n_{o_{l,n}}$ and $T^n_{w_{l,n}}$), explicit well rates ($q^n_{osc_n}$ and $q^n_{wsc_n}$), and implicit capillary pressures. The capillary pressure terms ($P^{n+1}_{cow_l} - P^{n+1}_{cow_n}$) in the water equation (Eq. 11.104) are expressed in terms of water saturation. In addition, the fluid gravities are treated explicitly.

Therefore, for block n, the oil equation becomes

$$\begin{aligned} &\sum_{l \in \psi_n} T^n_{o_{l,n}} \left[\left(p^{n+1}_{o_l} - p^{n+1}_{o_n} \right) - \gamma^n_{o_{l,n}} (Z_l - Z_n) \right] + 0 + q^n_{osc_n} \\ &= C_{op_n} \left(p^{n+1}_{o_n} - p^n_{o_n} \right) + C_{ow_n} \left(S^{n+1}_{w_n} - S^n_{w_n} \right) \end{aligned} \tag{11.109}$$

and the water equation becomes

$$\begin{aligned} &\sum_{l \in \psi_n} T^n_{w_{l,n}} \left[\left(p^{n+1}_{o_l} - p^{n+1}_{o_n} \right) - \left[P^n_{cow_l} + P'^n_{cow_l} \left(S^{n+1}_{w_l} - S^n_{w_l} \right) - P^n_{cow_n} - P'^n_{cow_n} \left(S^{n+1}_{w_n} - S^n_{w_n} \right) \right] \right. \\ &\left. - \gamma^n_{w_{l,n}} (Z_l - Z_n) \right] + 0 + q^n_{wsc_n} = C_{wp_n} \left(p^{n+1}_{o_n} - p^n_{o_n} \right) + C_{ww_n} \left(S^{n+1}_{w_n} - S^n_{w_n} \right) \end{aligned} \tag{11.110}$$

The terms in Eqs. (11.109) and (11.110) are rearranged as follows:

$$\begin{aligned} &\sum_{l \in \psi_n} \left[T^n_{o_{l,n}} p^{n+1}_{o_l} + (0) S^{n+1}_{w_l} \right] - \left\{ \left[\left(\sum_{l \in \psi_n} T^n_{o_{l,n}} \right) + C_{op_n} \right] p^{n+1}_{o_n} + C_{ow_n} S^{n+1}_{w_n} \right\} \\ &= \sum_{l \in \psi_n} \left[T^n_{o_{l,n}} \gamma^n_{o_{l,n}} (Z_l - Z_n) \right] - q^n_{osc_n} - C_{op_n} p^n_{o_n} - C_{ow_n} S^n_{w_n} \end{aligned} \tag{11.111}$$

for the oil equation, and

$$\begin{aligned} &\sum_{l \in \psi_n} \left[T^n_{w_{l,n}} p^{n+1}_{o_l} - T^n_{w_{l,n}} P'^n_{cow_l} S^{n+1}_{w_l} \right] - \left\{ \left[\left(\sum_{l \in \psi_n} T^n_{w_{l,n}} \right) + C_{wp_n} \right] p^{n+1}_{o_n} \right. \\ &\left. + \left[\left(\sum_{l \in \psi_n} T^n_{w_{l,n}} P'^n_{cow_n} \right) + C_{ww_n} \right] S^{n+1}_{w_n} \right\} \\ &= \sum_{l \in \psi_n} T^n_{w_{l,n}} \left[\left(P^n_{cow_l} - P'^n_{cow_l} S^n_{w_l} \right) - \left(P^n_{cow_n} - P'^n_{cow_n} S^n_{w_n} \right) + \gamma^n_{w_{l,n}} (Z_l - Z_n) \right] \\ &- q^n_{wsc_n} - C_{wp_n} p^n_{o_n} - C_{ww_n} S^n_{w_n} \end{aligned} \tag{11.112}$$

for the water equation.

Eqs. (11.111) and (11.112) are written for all blocks ($n=1, 2, 3\ldots N$), and the $2N$ equations are solved simultaneously for the $2N$ unknowns. For 1-D flow problems, there are $2n_x$ equations that form a bitridiagonal matrix equation:

$$[\mathbf{A}]\vec{X}=\vec{b} \tag{11.113a}$$

or

$$\begin{bmatrix} [\mathbf{c}_1] & [\mathbf{e}_1] & & & & & \\ [\mathbf{w}_2] & [\mathbf{c}_2] & [\mathbf{e}_2] & & & & \\ & \cdots & \cdots & \cdots & & & \\ & & [\mathbf{w}_i] & [\mathbf{c}_i] & [\mathbf{e}_i] & & \\ & & & \cdots & \cdots & \cdots & \\ & & & & [\mathbf{w}_{n_x-1}] & [\mathbf{c}_{n_x-1}] & [\mathbf{e}_{n_x-1}] \\ & & & & & [\mathbf{w}_{n_x}] & [\mathbf{c}_{n_x}] \end{bmatrix} \begin{bmatrix} \vec{X}_1 \\ \vec{X}_2 \\ \cdots \\ \vec{X}_i \\ \cdots \\ \vec{X}_{n_x-1} \\ \vec{X}_{n_x} \end{bmatrix} = \begin{bmatrix} \vec{b}_1 \\ \vec{b}_2 \\ \cdots \\ \vec{b}_i \\ \cdots \\ \vec{b}_{n_x-1} \\ \vec{b}_{n_x} \end{bmatrix} \tag{11.113b}$$

where

$$[\mathbf{w}_i]=\begin{bmatrix} T^n_{ox_{i-1/2}} & 0 \\ T^n_{wx_{i-1/2}} & -T^n_{wx_{i-1/2}}P'^n_{cow_{i-1}} \end{bmatrix} \tag{11.114}$$

$$[\mathbf{e}_i]=\begin{bmatrix} T^n_{ox_{i+1/2}} & 0 \\ T^n_{wx_{i+1/2}} & -T^n_{wx_{i+1/2}}P'^n_{cow_{i+1}} \end{bmatrix} \tag{11.115}$$

$$[\mathbf{c}_i]=\begin{bmatrix} -\left(T^n_{ox_{i-1/2}}+T^n_{ox_{i+1/2}}+C_{op_i}\right) & -C_{ow_i} \\ -\left(T^n_{wx_{i-1/2}}+T^n_{wx_{i+1/2}}+C_{wp_i}\right) & \left[\left(T^n_{wx_{i-1/2}}+T^n_{wx_{i+1/2}}\right)P'^n_{cow_i}-C_{ww_i}\right] \end{bmatrix} \tag{11.116}$$

$$\vec{X}_i=\begin{bmatrix} p^{n+1}_{o_i} \\ S^{n+1}_{w_i} \end{bmatrix} \tag{11.117}$$

and

$$\vec{b}_i=\begin{bmatrix} T^n_{ox_{i-1/2}}\gamma^n_{o_{i-1/2}}(Z_{i-1}-Z_i)+T^n_{ox_{i+1/2}}\gamma^n_{o_{i+1/2}}(Z_{i+1}-Z_i)-q^n_{osc_i}-C_{op_i}p^n_{o_i}-C_{ow_i}S^n_{w_i} \\ \Big\{T^n_{wx_{i-1/2}}\gamma^n_{w_{i-1/2}}(Z_{i-1}-Z_i)+T^n_{wx_{i+1/2}}\gamma^n_{w_{i+1/2}}(Z_{i+1}-Z_i),-q^n_{wsc_i},-C_{wp_i}p^n_{o_i},-C_{ww_i}S^n_{w_i} \\ +T^n_{wx_{i-1/2}}\left[\left(P^n_{cow_{i-1}}-P^n_{cow_i}\right)-\left(P'^n_{cow_i}S^n_{w_{i-1}}-P'^n_{cow_i}S^n_{w_i}\right)\right] \\ +T^n_{wx_{i+1/2}}\left[\left(P^n_{cow_{i+1}}-P^n_{cow_i}\right)-\left(P'^n_{cow_{i+1}}S^n_{w_{i+1}}-P'^n_{cow_i}S^n_{w_i}\right)\right]\Big\} \end{bmatrix} \tag{11.118}$$

for $i=1, 2, 3\ldots n_x$.

The solution of the bitridiagonal matrix equation for 1-D flow problems is obtained using the same steps as in Thomas' algorithm, presented in Section 9.2.1, with scalar mathematical operations being replaced with matrix mathematical operations. Therefore, Thomas' algorithm for solving bitridiagonal matrix equation becomes.

Forward solution

Set

$$[\mathbf{u}_1] = [\mathbf{c}_1]^{-1}[\mathbf{e}_1] \tag{11.119}$$

and

$$\vec{g}_1 = [\mathbf{c}_1]^{-1}\vec{d}_1 \tag{11.120}$$

For $i=2, 3 \ldots n_x - 1$,

$$[\mathbf{u}_i] = [[\mathbf{c}_i] - [\mathbf{w}_i][\mathbf{u}_{i-1}]]^{-1}[\mathbf{e}_i] \tag{11.121}$$

and for $i=2, 3 \ldots n_x$,

$$\vec{g}_i = [[\mathbf{c}_i] - [\mathbf{w}_i][\mathbf{u}_{i-1}]]^{-1}\left(\vec{d}_i - [\mathbf{w}_\mathbf{i}]\vec{g}_{i-1}\right) \tag{11.122}$$

Backward solution

Set

$$\vec{X}_{n_x} = \vec{g}_{n_x} \tag{11.123}$$

For $i=n_x-1, n_x-2, \ldots, 3, 2, 1$,

$$\vec{X}_i = \vec{g}_i - [\mathbf{u}_i]\vec{X}_{i+1} \tag{11.124}$$

For a black-oil model, the resulting set of equations is a tritridiagonal matrix. The algorithm presented in Eqs. (11.119) through (11.124) can be used to obtain the solution, but note that the submatrices are 3×3 and the subvectors have dimensions of three.

11.5 Material balance checks

The incremental and cumulative material balance checks in multiphase flow are carried out for each component in the system. For oil and water ($p=o, w$), each component is contained within its phase; therefore,

$$I_{MB_p} = \frac{\displaystyle\sum_{n=1}^{N} \frac{V_{b_n}}{\alpha_c \Delta t}\left[\left(\frac{\phi S_p}{B_p}\right)_n^{n+1} - \left(\frac{\phi S_p}{B_p}\right)_n^{n}\right]}{\displaystyle\sum_{n=1}^{N}\left(q_{psc_n}^{n+1} + \sum_{l \in \xi_n} q_{psc_{l,n}}^{n+1}\right)} \tag{11.125a}$$

and

$$C_{MB_p} = \frac{\sum_{n=1}^{N} \frac{V_{b_n}}{\alpha_c} \left[\left(\frac{\phi S_p}{B_p} \right)_n^{n+1} - \left(\frac{\phi S_p}{B_p} \right)_n^{0} \right]}{\sum_{m=1}^{n+1} \Delta t_m \sum_{n=1}^{N} \left(q_{psc_n}^m + \sum_{l \in \xi_n} q_{psc_{l,n}}^m \right)} \tag{11.126a}$$

For the gas component, both free-gas and solution-gas components must be taken into consideration; therefore,

$$I_{MB_g} = \frac{\sum_{n=1}^{N} \frac{V_{b_n}}{\alpha_c \Delta t} \left\{ \left[\left(\frac{\phi S_g}{B_g} \right)_n^{n+1} - \left(\frac{\phi S_g}{B_g} \right)_n^{n} \right] + \left[\left(\frac{\phi R_s S_o}{B_o} \right)_n^{n+1} - \left(\frac{\phi R_s S_o}{B_o} \right)_n^{n} \right] \right\}}{\sum_{n=1}^{N} \left\{ \left[q_{fgsc_n}^{n+1} + \sum_{l \in \xi_n} q_{fgsc_{l,n}}^{n+1} \right] + \left[R_{s_n}^{n+1} q_{osc_n}^{n+1} + \sum_{l \in \xi_n} R_{s_{l,n}}^{n+1} q_{osc_{l,n}}^{n+1} \right] \right\}} \tag{11.125b}$$

and

$$C_{MB_g} = \frac{\sum_{n=1}^{N} \frac{V_{b_n}}{\alpha_c} \left\{ \left[\left(\frac{\phi S_g}{B_g} \right)_n^{n+1} - \left(\frac{\phi S_g}{B_g} \right)_n^{0} \right] + \left[\left(\frac{\phi R_s S_o}{B_o} \right)_n^{n+1} - \left(\frac{\phi R_s S_o}{B_o} \right)_n^{0} \right] \right\}}{\sum_{m=1}^{n+1} \Delta t_m \sum_{n=1}^{N} \left\{ \left[q_{fgsc_n}^{m} + \sum_{l \in \xi_n} q_{fgsc_{l,n}}^{m} \right] + \left[R_{s_n}^{m} q_{osc_n}^{m} + \sum_{l \in \xi_n} R_{s_{l,n}}^{m} q_{osc_{l,n}}^{m} \right] \right\}} \tag{11.126b}$$

11.6 Advancing solution in time

Pressure and phase saturation distributions in multiphase flow problems change with time. This means that the flow problem has an unsteady-state solution. At time $t_0 = 0$, all reservoir unknowns must be specified. Initially, fluids in the reservoir are in hydrodynamic equilibrium. Therefore, it is sufficient to specify the pressure at water-oil contact (WOC) and at oil-gas contact (OGC), and the initial pressure and saturations of all three phases can be estimated from hydrostatic pressure considerations, oil-water and gas-oil capillary pressure relationships, and phase saturations constraint equation. Details can be found elsewhere (Ertekin et al., 2001). The procedure entails finding phase pressures and saturations at discrete times (t_1, t_2, t_3, t_4, etc.) by marching the solution in time using time steps (Δt_1, Δt_2, Δt_3, Δt_4, etc.). The pressure and saturations solution is advanced from initial conditions at $t_0 = 0$ (time level n) to $t_1 = t_0 + \Delta t_1$ (time level $n+1$). The solution then is advanced in time from t_1 (time level n) to $t_2 = t_1 + \Delta t_2$ (time level $n+1$), from t_2 to $t_3 = t_2 + \Delta t_3$, and from t_3 to $t_4 = t_3 + \Delta t_4$, and the process is repeated as many times as necessary until the desired simulation time is reached. To obtain the pressure and saturations solution at time level $n+1$, we assign the pressures and saturations just obtained as pressures and saturations at time level n, write the flow equation for each component in every block (node) in the discretized reservoir, and solve the resulting set of linear equations for the set of unknowns. The calculation procedure within each time step for a black-oil model follows:

1. Calculate the interblock phase transmissibilities and coefficients C_{op}, C_{ow}, C_{og}, C_{wp}, C_{ww}, C_{wg}, C_{gp}, C_{gw}, and C_{gg}, and define the pressure and saturations at the old time level and at the old iteration of the current time level for all reservoir blocks. Note that the phase transmissibilities are calculated at the upstream blocks and are not necessarily constant.
2. Estimate the phase production rates (or write the phase production rate equations) at time level $n+1$ for each wellblock in the reservoir, as described in Section 11.4.2.
3. Estimate the phase flow rates (or write the phase flow rate equations) at time level $n+1$ for each fictitious well in the reservoir, that is, estimate the phase flow rates resulting from boundary conditions, as described in Section 11.4.3.
4. For every gridblock (or gridpoint) in the reservoir, define the set of existing reservoir neighboring blocks (ψ_n) and the set of reservoir boundaries that are block boundaries (ξ_n), expand the summation terms in the flow equations, and substitute for phase production rates from wellblocks obtained in **(2)** and phase flow rates from fictitious wells obtained in **(3)**.
5. Linearize the terms in the flow equations, as outlined in Section 11.4.4.
6. Factorize, order, and place the unknowns (at time level $n+1$) on the LHS, and place known quantities on the RHS of each flow equation.
7. Solve the resulting set of equations for the set of pressure and saturation unknowns (at time level $n+1$) using a linear equation solver.
8. Check for convergence of the solution. Proceed to **(9)** if convergence is achieved. Otherwise, update the interblock phase transmissibilities and the coefficients mentioned in **(1)**, define the pressure and saturations at the latest iteration at the current time level for all reservoir blocks, and start all over from **(2)**.
9. Estimate the wellblock production rates and fictitious well rates at time level $n+1$ if necessary by substituting for the pressures and saturations obtained in **(7)** into the phase flow rate equations obtained in **(2)** and **(3)**.
10. Perform incremental and cumulative material balance checks for all components (o, w, g) using the equations presented in Section 11.5.

11.7 Summary

In petroleum reservoirs, oil, water, and gas may coexist and flow simultaneously. In multiphase reservoirs, the phase saturations add up to one, capillary pressures between phases exist, and phase relative permeability and phase potential gradient among other things affect flow properties. Although volumetric and viscosity properties of water and gas phases are not different from those in single-phase flow, oil-phase properties are affected by both solution GOR and whether the pressure is below or above the oil bubble-point pressure. Simulation of multiphase flow involves writing the flow equation for each component in the system and solving all equations for the unknowns in the system. In black-oil simulation, the components are the oil, water, and gas all at standard conditions, and the flow model consists of one equation for each of the three components, the saturation constraint, and the oil/water and gas/oil capillary pressures. The model formulation dictates how the model equations are combined to produce a reduced set of equations. It also implies the choice of primary unknowns and secondary unknowns for the reservoir. The black-oil model formulation discussed in this chapter is the $p_o - S_w - S_g$ formulation, that is, the formulation that uses p_o, S_w, and S_g as the primary unknowns for the reservoir and p_w, p_g, and S_o as the secondary unknowns. The two-phase oil/water, oil/gas, and gas/water flow models can be considered subsets of the black-oil model presented in this chapter. To solve the model equations, the accumulation terms have to be expanded in a conservative way and expressed in terms of the changes of the primary unknowns over the same time step, the well production rate terms for each phase defined, and the fictitious well rate terms reflecting the boundary conditions need to be defined. In addition, all nonlinear terms have to be linearized. This process produces linearized flow equations, and the IMPES or SS solution methods can be used to obtain the linearized flow equations. The resulting set of linearized equations for all blocks can then be solved using any linear equation solver to obtain the solution for one time step. An extension to Thomas' algorithm can be used to solve simultaneously the equations of multiphase, 1-D flow problems.

11.8 Exercises

11.1 Consider the 1-D reservoir shown in Fig. 11.9. The reservoir has no-flow boundaries, gridblock 1 hosts a water injection well, and gridblock 4 hosts a production well. The reservoir contains oil and water only.

a. Name the four equations that constitute the flow model for this reservoir.

b. Name the four unknowns for a gridblock in his reservoir.

c. Write the general flow equations for gridblock n in this reservoir.

d. Write the saturation constraint equation and capillary pressure relationship in this reservoir.

e. If you use the $p_o - S_w$ formulation, name the primary unknowns and secondary unknowns for a gridblock in this reservoir.

f. Write the flow equations for gridblock n using the $p_o - S_w$ formulation.
g. Write the flow equations for gridblocks 1, 2, 3, and 4 using the $p_o - S_w$ formulation.

11.2 Complete the following problems that are related to Exercise 11–1.
a. If you use the $p_o - S_o$ formulation, name the primary unknowns and secondary unknowns for a gridblock in this reservoir.
b. Derive the flow equations for gridblock n using the $p_o - S_o$ formulation.
c. Write the flow equations for gridblocks 1, 2, 3, and 4 using the $p_o - S_o$ formulation.

11.3 Consider the 1-D reservoir shown in Fig. 11.9. The reservoir has no-flow boundaries, gridblock 1 hosts a gas injection well, and gridblock 4 hosts a production well. The reservoir contains oil and gas only.
a. Name the four equations that constitute the flow model for this reservoir.
b. Name the four unknowns for a gridblock in this reservoir.
c. Write the general flow equations for gridblock n in this reservoir.
d. Write the saturation constraint equation and capillary pressure relationship in this reservoir.
e. If you use the $p_o - S_g$ formulation, name the primary unknowns and secondary unknowns for a gridblock in this reservoir.
f. Write the flow equations for gridblock n using the $p_o - S_g$ formulation.
g. Write the flow equations for gridblocks 1, 2, 3, and 4 using the $p_o - S_g$ formulation.

11.4 Complete the following problems that are related to Exercise 11–3.
a. If you use the $p_o - S_o$ formulation, name the primary unknowns and secondary unknowns for a gridblock in this reservoir.
b. Derive the flow equations for gridblock n using the $p_o - S_o$ formulation.
c. Write the flow equations for gridblocks 1, 2, 3, and 4 using the $p_o - S_o$ formulation.

11.5 Consider the 1-D reservoir shown in Fig. 11.10. The reservoir has no-flow boundaries and gridblock 3 hosts a production well. The reservoir contains gas and water only.
a. Name the four equations that constitute the flow model for this reservoir.
b. Name the four unknowns for a gridblock in this reservoir.
c. Write the general flow equations for gridblock n in this reservoir.

d. Write the saturation constraint equation and capillary pressure relationship in this reservoir.
e. If you use the $p_g - S_g$ formulation, name the primary unknowns and secondary unknowns for a gridblock in this reservoir.
f. Write the flow equations for gridblock n using the $p_g - S_g$ formulation.
g. Write the flow equations for gridblocks 1, 2, 3, and 4 using the $p_g - S_g$ formulation.

11.6 Complete the following problems that are related to Exercise 11–5.
a. If you use the $p_g - S_w$ formulation, name the primary unknowns and secondary unknowns for a gridblock in this reservoir.
b. Derive the flow equations for gridblock n using the $p_g - S_w$ formulation.
c. Write the flow equations for gridblocks 1, 2, 3, and 4 using the $p_g - S_w$ formulation.

11.7 Consider the 1-D reservoir shown in Fig. 11.9. The reservoir has no-flow boundaries. Gridblock 1 hosts a water injection well, and gridblock 4 hosts a production well. The reservoir contains oil, gas, and water.
a. Name the six equations that constitute the flow model for this reservoir.
b. Name the six unknowns for a gridblock in this reservoir.
c. Write the general flow equations for gridblock n in this reservoir.
d. Write the saturation constraint equation and capillary pressure relationships in this reservoir.
e. If you use the $p_o - S_w - S_g$ formulation, name the primary unknowns and secondary unknowns for a gridblock in this reservoir.
f. Write the flow equations for gridblock n using the $p_o - S_w - S_g$ formulation.
g. Write the flow equations for gridblocks 1, 2, 3, and 4 using the $p_o - S_w - S_g$ formulation.

11.8 Complete the following problems that are related to Exercise 11.7.
a. If you use the $p_o - S_w - S_o$ formulation, name the primary unknowns and secondary unknowns for a gridblock in this reservoir.
b. Derive the flow equations for gridblock n using the $p_o - S_w - S_o$ formulation.
c. Write the flow equations for gridblocks 1, 2, 3, and 4 using the $p_o - S_w - S_o$ formulation.

11.9 Complete the following problems that are related to Exercise 11.7.
a. If you use the $p_o - S_o - S_g$ formulation, name the primary unknowns and secondary unknowns for a gridblock in this reservoir.

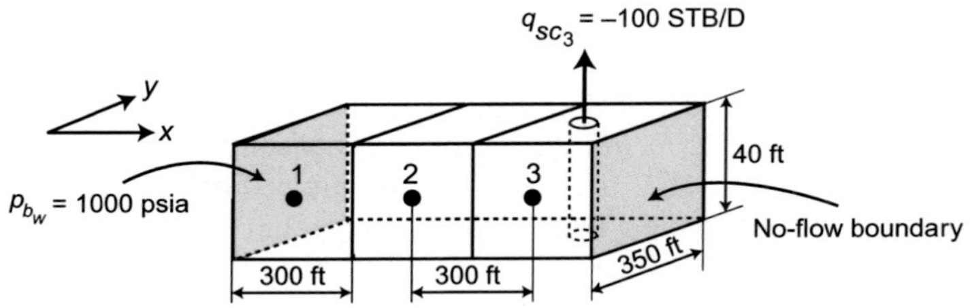

FIG. 11.15 Discretized 1-D reservoir for Exercises 11.11 and 11.12.

TABLE 11.5 Oil/water relative permeability data for Exercise 11.11.

S_w	k_{rw}	k_{row}	P_{cow} (psi)
0.130	0.000	1.0000	40
0.191	0.0051	0.9400	15
0.250	0.0102	0.8300	8.6
0.294	0.0168	0.7241	6.0
0.357	0.0275	0.6206	4.0
0.414	0.0424	0.5040	3.0
0.490	0.0665	0.3170	2.3
0.557	0.0910	0.2209	2.0
0.630	0.1148	0.1455	1.5
0.673	0.1259	0.0956	1.0
0.719	0.1381	0.0576	0.8
0.789	0.1636	0.0000	0.15

b. Derive the flow equations for gridblock n using the $p_o - S_o - S_g$ formulation.

c. Write the flow equations for gridblocks 1, 2, 3, and 4 using the $p_o - S_o - S_g$ formulation.

11.10 Derive the IMPES equations for the 1-D oil/gas flow model by executing the following steps:

a. Date the transmissibilities, capillary pressures, phase gravities, relative permeabilities, and phase properties in production rates at old time level t^n in Eqs. (11.50) and (11.51).

b. Expand the accumulation terms (the RHS of Eqs. 11.50 and 11.51) in terms of the $(p_{o_i}^{n+1} - p_{o_i}^n)$ and $(S_{g_i}^{n+1} - S_{g_i}^n)$.

c. Substitute the results of the second step into the equations of the first step.

d. Add the resulting oil equation from the third step multiplied by $(B_{o_i}^{n+1} - R_{s_i}^{n+1} B_{g_i}^{n+1})$ and the resulting gas equation from the third step multiplied by $B_{g_i}^{n+1}$ to obtain the pressure equation.

e. Solve for $S_{g_i}^{n+1}$ using the resulting oil equation from the third step for each block.

11.11 A 1-D horizontal, two-phase oil/water reservoir is described by three equal gridblocks as shown in Fig. 11.15. The reservoir rock is incompressible and has homogeneous and isotropic properties, $k = 270$ md and $\phi = 0.27$. Initially, the reservoir pressure is 1000 psia, and water saturation is irreducible, $S_{w_i} = 0.13$. gridblock dimensions are $\Delta x = 300$ ft, $\Delta y = 350$ ft, and $h = 40$ ft.

Reservoir fluids are incompressible with $B_o = B_o^\circ = 1$ RB/STB, $\mu_o = 3.0$ cP, $B_w = B_w^\circ = 1$ RB/STB, and $\mu_w = 1.0$ cP. Table 11.5 gives the oil/water relative permeability and capillary pressure data. The reservoir right boundary is sealed off to flow, and the reservoir left boundary is kept at constant pressure of 1000 psia because of a strong water aquifer. A 7-in vertical well at the center of gridblock 3 produces liquid at a rate of 100 STB/D. Using the IMPES solution method, find the pressure and saturation distributions in the reservoir at 100 and 300 days. Take single time steps to advance the solution from one time to another.

11.12 Consider the reservoir data presented in Exercise 11.11. Using the SS method, find the pressure and saturation distributions in the reservoir at 100 and 300 days. Take single time steps to advance the solution from one time to another.

Glossary

A

Accumulation terms The right-hand side term of the flow equation. Every flow equation consists of a flow term, plus a well term (source/sink) that equals the right side, accumulation term. For water, for instance, see Fig. G.1.

Water:

$$\Delta T_w \Delta \Psi_w + q_{w,i,j,k} = \frac{V_{i,j,k}}{\Delta t} \Delta_t \left(\Phi \frac{S_w}{B_w}\right)$$

Flow term + Well term = Accumulation term

Fig. G.1 Accumulation term

Alternating-direction implicit procedure (ADIP) Originally suggested in petroleum engineering applications in 1966 (Coats and Tarhune), this method solves the governing equation alternating between implicit and explicit modes.

Anisotropic permeability Because natural material is never homogenous or uniform, permeability varies significantly between the vertical and horizontal planes within a given formation. This variation in permeability in different planes or directions is known as anisotropic permeability.

Aphenomenal When a continuous logical train is not followed or first premise is false or illogical. Such conclusion or process is inherently spurious, meaning has no meaning or significance.

Areal discretization Any model has to be divided along space in order to find solutions that apply to that elemental volume. In Cartesian coordinate, this corresponds to assigning Δx and Δy to a particular grid. This process is called areal discretization.

B

Backward difference The accumulation term, in the finite difference flow equation, is backward difference in time if the remaining terms in the flow equation are dated at new time (t^{n+1}).

Black-oil model When the simulation process considers oil, gas, and water as discrete phases, disallowing any component exchange between phases.

Block-centered grid When grid properties are assigned to the center of a particular block (see Fig. G.2).

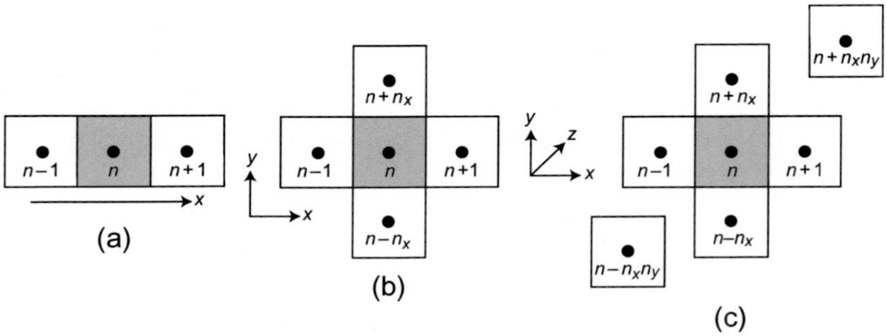

Fig. G.2 Block-centered grid. (a) $\psi_n = \{(n-1), (n+1)\}$, (b) $\psi_n = \{(n-n_x), (n-1), (n+1), (n+n_x)\}$, and (c) $\psi_n = \{(n-n_xn_y), (n-n_x), (n-1), (n+1), (n+n_x), (n+n_xn_y)\}$.

Block identification Numbering of grid blocks in order to assign them corresponding properties. There can be various ways of numbering the blocks, some yielding advantages over others.

Boundary conditions Because nature is continuous, but a reservoir model is not, every model has to have a specific values assigned to the boundary blocks, which may or may not correspond to original values of the block. These properties are assigned by the user of the model according to their expectation or knowledge of the prototype.

Block successive over relaxation (BSOR) It is an iterative technique for solving a set of linear algebraic equation, for which an entire block of properties are assumed/estimated simultaneously. See also, SOR.

C

Capillary pressure Capillary pressure (Pc) is the pressure difference across the interface between two immiscible fluids. The difference occurs because of the discontinuity between two fluids. The magnitude of capillary pressure depends on surface tension, interfacial tension, pore size, size distribution, and fluid properties.

Cartesian grid Discretization of the model in the Cartesian coordinate system.

Central difference The accumulation term, in the finite difference flow equation, is central difference in time if the remaining terms in the flow equation are dated at time ($t^{n+1/2}$) halfway between old time (t^n) and new time (t^{n+1}).

Compressible fluid In nature, everything is compressible, only variable being the amount of compression caused by certain pressure. In petroleum engineering, often incompressibility is assigned to fluid (such as water) by assigning a constant density (independent of pressure). Others, for which a small but constant compressibility factors are assigned, these fluids being called slightly compressible. Conventionally, slightly compressible fluid has a small but constant compressibility (c) that usually ranges from 10^{-5} to 10^{-6} psi^{-1}. Gas-free oil, water, and oil above bubble-point pressure are examples of slightly compressible fluids.

Conservation of mass Mass or energy cannot be created or destroyed, only transformed from one phase to another. In modeling, it means total mass entering a system or a block must equal total amount exiting plus (or minus) the amount retained (or extracted) in the block.

Constitutive equation The governing equation, along with boundary and initial conditions that are necessary to have the number of equations the same as the number of unknowns. For most reservoir simulators, the governing equation is the Darcey's law.

Crank-Nicolson formulation This formation is central difference in time and central difference in space.

D

Darcy's law It connects pressure drop across a porous body with flow rate with permeability of the porous medium as a proportionality constant. This is the most commonly used flow equation in petroleum reservoir engineering. The original equation was developed in the context of water flow through sand filters, but the law has been extended to include multiphase flow through multidimensional space.

E

Elementary volume The volume pertinent to the unit blocks or grids in a reservoir simulation model. It is synonymous with control volume.

Engineering approach This approach eliminates the partial differential representation of governing equations and use algebraic form of the governing equations directly.

This approach thus simplifies the reservoir modeling process without compromising accuracy or speed of computation.

Equation of state It is the functional form that connects fluid density with pressure and temperature. There are many equations of states, all of which are empirical but only a few are practical. The complexity arises from the fact that each reservoir has unique set of reservoir fluid and the compositions are such that different coefficients for equation of state should be used.

Explicit formulation When the governing equations are cast as an explicit function, for which pressures of each block can be calculated directly. This is the slowest and the most unstable solution scheme and is useless in the context of reservoir simulation.

F

Fictitious well This is a special technique for representing boundary conditions in the engineering approach. It involves replacing the boundary condition with a no-flow boundary plus a fictitious well having flow rate, which reflects fluid transfer between the gridpoint that is exterior to the reservoir and the reservoir boundary itself or the boundary gridpoint.

Flowing bottom-hole pressure (FBHP) The pressure measured in a well at or near the depth of the producing formation during production.

Formation volume factor (FVF) This is the ratio of the volume of a fluid under reservoir conditions to the volume at standard conditions. The ratio depends of the type of fluid and reservoir conditions (both pressure and temperature). For instance, for most oils, the FVF values are greater than 1.0. It means for water, this value is closer to 1.0, and for gas, it is only a fraction of 1.0, meaning gas would occupy much greater space under standard pressure and temperature conditions. It is the case because natural gas is highly compressible.

Forward difference The accumulation term, in the finite difference flow equation, is forward difference in time if the remaining terms in the flow equation are dated at old time (t^n).

G

Gas cap In oil reservoirs, where reservoir pressure is below the bubble point, natural gas escapes to be trapped by the caprock. The collection of this gas within the caprock is called gas cap.

Gas/oil contact (GOC) The bounding between a top gas layer and underlying oil layer within a petroleum formation. Such boundary exists because oil and gas are not miscible.

H

Heterogeneous Although Darcy's law and all other governing equations of mass and energy transport assume homogeneity, in reality, nature is inherently heterogeneous. In reservoir simulation, heterogeneity is recognized in the severe changes in permeability in space. Also, anisotropy can render a porous medium heterogeneous.

History matching The process involving the adjustment of reservoir rock/fluid parameters in order to match real production data and pressure of the reservoir. Because even best of models only have limited data available and the rest of the data have to be assumed/interpolated, the process of history matching is commonly used. However, history matching doesn't assure accuracy in predicting future performance. It is because different sets of properties would yield the same result, meaning the real properties remain elusive despite good history match.

I

Implicit formulation In this formulation, the algebraic equations are expressed in terms of pressure and saturation values, both sides of equations being at a future time. This system is inherently nonlinear, and linearization must be applied prior to solving the set of governing equations and constitutive relationships. Implicit formulation is unconditionally stable.

Implicit pressure and explicit saturation (IMPES) method In this formulation, the pressure terms are implicit, whereas saturation terms are explicit. It is easier to solve in this method. However, except for a narrow range of parameters and/or very small time steps, this method is unstable.

Incompressible fluid When the fluid density can be assumed to be constant or independent of pressure and temperature of confinement. For all practical purposes, only water and oil under certain conditions can be assumed to be incompressible.

Inflow performance relationship (IPR) It is the fluid flow as a function of flowing bottomhole pressure. The shape of the curve is determined by the quality of the reservoir. This curve is also used to determine at what stage pressure maintenance and other operations should take place to improve the production capability of the reservoir. Typically, the intersection between tubing performance versus production rate curve and the IPR marks the optimal operating conditions (Fig. G.3).

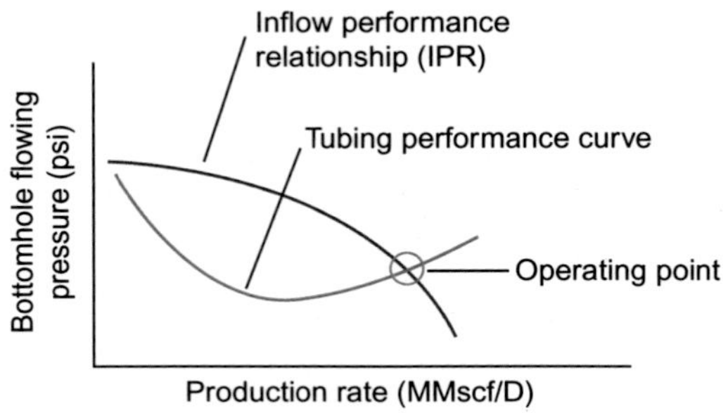

Fig. G.3 IPR and optimum operating conditions.

Initial conditions All values pressure and saturation prevailing at each grid block at the beginning of the reservoir simulation. Initial conditions are necessary for startup of a simulation procedure.

L

Line successive overrelaxation (LSOR) method See also BSOR. In this iterative technique, equations of each line are approximated and integrated.

Linearization Because all algebraic equations are nonlinear in every method other than explicit (which is practically irrelevant for its lack of stability in solving reservoir simulation equations), the algebraic equations have to be rendered linear prior to any attempt to solve them. This process is called linearization. Linearization is also necessary because of the boundary conditions and the presence of wells in a reservoir.

M

Mass accumulation term See accumulation term.

Mass balance Mass balance is a scientific way to verify if a given solution is not spurious. The process involves adding up total mass accumulated in each block and see there is an overall mass balance holds. If not, the iteration process has to restart for the given time step.

Mathematical approach This is the conventional method that uses partial differential equations, then discretizes using Taylor series approximations and finally derives the algebraic equations, to be solved with a numerical solver.

Mobility An expression containing permeability over viscosity. Typically, it represents ease of flow through a porous medium for a particular fluid.

Multiblock wells When a well penetrates more than one block in the reservoir simulator.

Multiphase flow Whenever more than one mobile phase exists. The most common scenario is the follow of three phases, namely, oil, water, and gas. Water is innate to petroleum reservoirs, whereas gas and oil are constantly separating from each other depending on the operating pressure.

N

Newton's iteration A linearization technique, in which the slope is taken in order to approximate the solution to a nonlinear equation.

No-flow boundaries This the assumption of a boundary through which no flow occurs. This is equivalent to perfect seal. Although it is absurd in nature, it is a good approximation for certain types of reservoirs.

P

Permeability It is the capacity of the rock to transmit fluid through it. It is assumed to be a constant and a strict property of the rock. Its dimension is L^2.

Point successive overrelaxation (PSOR) method It is the SOR method for which iterations are performed for each point. See SOR.

Pore volume Bulk volume multiplied by porosity. It represents void space of porous media.

Porosity Fraction of void volume over bulk volume of a porous medium.

R

Representative elemental volume (REV) This is the minimum sample volume for which the sample properties become insensitive to the sample size.

Reservoir characterization Detailed assignment of relevant rock and fluid properties and reservoir conditions for each blocked considered in a reservoir model. Conventionally, this is performed after numerous reservoir simulation runs in order to fine tune reservoir data to match the modeling data with the real history of the reservoir.

Residual oil saturation the saturation of oil that cannot be removed with any more waterflooding. This saturation is dictated by the oil/water interfacial tension and reservoir properties.

S

Sandface pressure This pertains to physical interface between the formation and the wellbore. This is the location where there is a discontinuity between porous medium flow and open flow in the production tubing, where Darcy's law ceases to apply. Pressure at this point is called sandface pressure.

Stability In a stable process, the errors subside and diminish with more number of interactions. Stability is measured by the outcome where a unique solution emerges.

Steady state When all parameters become insensitive to time.

Successive over relaxation (SOR) method It is an iterative technique for solving a set of linear algebraic equation. It starts off with an assumed value and then multiplies the new value with a factor to accelerate the convergence.

T

Transmissibility This is the product of formation rock and fluid properties. It expresses flow rate between two points per one psi pressure drop. It combines rock (k), fluid properties (β, μ), and blocks dimensions (Δx, Δy, Δz or Δr, $\Delta\theta$, Δz). For multiphase flow, it uses effective permeability to each phase and viscosity and formation volume factor of that corresponding phase.

U

Unsteady state When the flow parameters continue to change with time. A natural system is inherently dynamic, hence, in unsteady state.

W

Water/oil contact (WOC) This is the borderline between predominantly oil phase and water in the aquifer. Similar to gas-oil contact, WOC emerges because oil and water not miscible. In a porous medium, the WOC is not uniform and depends on the rock and fluid characteristics.

Appendix A

User's manual for single-phase simulator

A.1 Introduction

This manual provides information on data file preparation and a description of the variables used in preparing a data file, and it gives instructions for running the reservoir simulator on a PC. The simulator models the flow of single-phase fluid in reservoirs. Model description (flow equations and boundary conditions) for incompressible, slightly compressible, and compressible fluids; well operating conditions; and methods of solving the algebraic equations of the model is described in detail in the previous chapters of this book and by Ertekin et al. (2001). The simulator (written in FORTRAN) was developed to provide solutions to single-phase flow problems in undergraduate courses in reservoir simulation. The simulator can be used to model irregular rectangular reservoirs and single well in *r-z* radial-cylindrical coordinates using either a block-centered grid or a point-distributed grid. The purpose of presenting this simulator as part of this book is to provide the user with intermediate and final results so the user's solution for any given problem can be checked and any errors can be identified and corrected. Educators may use the simulator to make up new problems and obtain their solutions.

A.2 Data file preparation

The data required for the present simulator are classified into groups based on how the data within each group are related. A group of data could be as simple as defining a few related variables or as complicated as defining the variables for the well recursive data. These groups of data are classified, according to their format of input procedure, into five categories (A, B, C, D, and E). Categories A and B include 17 and 6 groups of data, respectively, whereas categories C, D, and E include one group of data each. The data of each category are entered using a specific format procedure; for example, category A uses format procedure A and category B uses format procedure B. Each group of data carries an identification name consisting of the word "DATA" followed by a number and an alphabet character; the number identifies the group and the alphabet character identifies the category and the format procedure. For example, DATA 04B identifies a group of data that belongs to category B that uses format procedure B and whose variables are defined under DATA 04B in Section A.3. Data file preparation, including format procedures and description of variables, follows the work of Abou-Kassem et al. (1996) for black-oil simulation. The folder available at www.emertec.ca contains four examples of data files prepared for the problems presented in Example 7.1 (ex7-1.txt), Example 7.7 (ex7-7.txt), Example 7.12 (ex7-12.txt), and Example 5.5 (ex5-5.txt).

Each format procedure is introduced by a title line (line 1), which includes the identification name and the group of data to be entered, followed by a parameter sequence line (line 2), which lists the order of parameters to be entered by the user. Only format procedure D has an additional parameter sequence line (line 3). The user, in each subsequent data line, enters the values of the parameters ordered and preferably aligned with the parameters shown in the parameter sequence line for easy recognition. Both format procedures B and E require a single-line data entry, whereas format procedures A, C, and D require multiple-line data entry and terminate with a line of zero entries for all parameters. The various groups of data and any specific instructions for each format procedure are presented in the following sections.

A.2.1 Format procedure A

This format procedure is suitable for entering data that describe the distribution of a grid block property over the whole reservoir. Such data include block size and permeability in the x-, y-, and z-directions; depth; porosity; modifiers for porosity, depth, bulk volume, and transmissibilities in the x-, y-, and z-directions; boundary conditions; and block identifiers that label a grid block as being active or inactive.

Each line of data (e.g., line 3) represents a property assignment for an arbitrary reservoir region having the shape of a prism with I1, I2; J1, J2; and K1, K2 being its lower and upper limits in the x-, y-, and z-directions. The data entered by each subsequent line (e.g., line 4) are superimposed on top of the data entered by all earlier lines; that is, the final distribution of a property is the result of the superposition of the entire arbitrary reservoir regions specified by all lines of data. This option is activated by setting the option identifier at the beginning of the parameter sequence line (line 2) to 1. *DATA 25A has no option identifier, but it implicitly assumes a value of 1*. This is a powerful method for entering data if a block property is distributed into well-defined (not necessarily regular) reservoir regions. For a homogeneous property distribution, only one line of data is needed (with I1 = J1 = K1 = 1, I2 = n_x, J2 = n_y, and K2 = n_z). If; however, a block property is so heterogeneous that it varies from block to block and regional property distribution is minimal, this method loses its effectiveness. In such cases, the option identifier at the beginning of the parameter sequence line (line 2) is set to 0, and the data for all blocks are entered sequentially in a way similar to natural ordering of blocks along rows (i.e., i is incremented first, j is incremented second, and k is incremented last). In this case, both active and inactive blocks are assigned property values, and the terminating line of zero entries is omitted.

A.2.2 Format procedure B

This format procedure is suitable for entering data involving a combination of integer and/or real variables. Groups of data of this type include options for the method of solution, block ordering scheme, and units of input and output; control integers for printing options and number of grid blocks in the x-, y-, and z-directions; fluid density; fluid and porosity compressibilities and reference pressure for porosity; and simulation time. Note that the values of the parameters are entered in line 3. They are ordered and aligned with the parameters shown in the parameter sequence line (line 2) for easy recognition.

A.2.3 Format procedure C

This format procedure is suitable for entering a PVT property table for a natural gas. The parameter sequence line (line 2) lists pressure as the independent variable followed gas FVF and gas viscosity as the dependent variables. *It is important to note that the range of pressure in the PVT table must cover the range of pressure changes expected to take place in the reservoir and that the pressure entries in the table must have equal pressure intervals.* Each line of data represents one entry in the table of data that corresponds to a specified value of the independent variable. The data in the table are entered in order of increasing value of the independent variable (pressure). Note that entries in each line of data (e.g., line 3) are ordered and aligned with the parameters specified on the parameter sequence line (line 2) for easy recognition.

A.2.4 Format procedure D

This format procedure is suitable for entering well recursive data. As mentioned earlier, this format procedure has two parameter sequence lines. The parameters in the first parameter sequence line (line 2) include a time specification that signals a new user's request (SIMNEW), an override time step to be used (DELT), the number of wells changing operating conditions (NOW), the minimum flowing bottom-hole pressure for producing wells (PWFMIN), and the maximum bottom-hole pressure for injection wells (PWFMAX). This line of data can be repeated, but each subsequent line must have a time specification larger than the last time specification. The parameters in the second parameter sequence line (line 3) include data for individual wells such as the well identification number (IDW), wellblock coordinates (IW, JW, and KW), the well type and well operating condition (IWOPC), the wellblock geometric factor (GWI), a specified value of condition (SPVALUE), and the well radius (RADW). There must exist NOW lines describing NOW individual wells immediately following the line where NOW specification appears if NOW > 0. Using this format procedure, any number of wells can be introduced, shut-in, reopened, recompleted, etc., at any number of key times.

A.2.5 Format procedure E

This format procedure is used to enter one line of information, such as the name of the user and the title of the computer run, consisting of up to 80 alphanumeric characters.

A.3 Description of variables used in preparing a data file

There are 26 data groups in the data file. The descriptions of the variables within each data group are given under the data group itself. Follows is a list of all 26 data groups, starting with DATA 01E and ending with DATA 26D:

DATA 01E	**Title of Simulation Run**
TITLE	Name of user and title of simulation run (one line having up to 80 alphanumeric characters)

Note

For identification purposes, the name of user and title of simulation run appear immediately after acknowledgement in all four output files.

DATA 02B	**Simulation Time Data**
IPRDAT	Option for printing and debugging input data file =0, do neither print nor debug input data file =1, print input data file and activate messages to debug data file
TMTOTAL	Maximum simulation time, D [d]
TMSTOP	Time to stop this simulation run, D [d]

DATA 03B	**Units**
MUNITS	Option for units of input data and output =1, customary units =2, SPE preferred metric units =3, laboratory units

DATA 04B	Control Integers for Printing Desired Output (1, print; 0, do not print)
BORD	Block order
MLR	Left and right half width
BASIC	Basic intermediate results used in simulation
QBC	Results and intermediate results related to boundary conditions
EQS	Block equations and details of solution method
PITER	Block pressure every outer iteration
ITRSOL	Detailed results related to method of solving linear equations and block pressure every inner iteration for iterative methods

Note

The results of simulation appear in four separate files. Description of reservoir and results related to PITER, QBC, reservoir production rates, and material balance checks appear in MY-OUT1.LIS. Those related to BORD, MLR, BASIC, QBC, EQS, PITER, and ITRSOL appear in MY-OUT2.LIS. Tabulation of reservoir pressure as a function of time appears in MY-OUT3.LIS. Tabulation of reservoir and well performances appear in MY-OUT4.LIS.

DATA 05B	Reservoir Discretization and Method of Solving Equations
IGRDSYS	Type of grid system used in reservoir discretization. =1, block-centered grid =2, point-distributed (or node) grid
NX	Number of gridblocks (or gridpoints) in the *x*-direction (or the *r*-direction if NY=0)
NY	Number of gridblocks (or gridpoints) in the *y*-direction. For single-well simulation, set NY=0
NZ	Number of gridblocks (or gridpoints) in the *z*-direction
RW	Well radius for single-well simulation, ft [m]
RE	External radius of reservoir for single-well simulation, ft [m]
NONLNR	Linearization of nonlinear terms. The options that apply to the mathematical approach (MA) or engineering approach (EA) are as indicated in the succeeding text. =1, explicit treatment of transmissibility and production term (MA) (×) =2, simple iteration on transmissibility and production term (MA) (×) =3, explicit treatment of transmissibility and coefficient of pressure drop in production term (EA) =4, simple iteration on transmissibility and coefficient of pressure drop in production term (EA) =5, Newton's iteration (MA and EA)
LEQSM	Method of solving linear equations =1, Thomas' algorithm for 1-D flow problems =2, Tang's algorithm for 1-D flow problems where blocks form a ring (×) =3, Jacobi iterative method for 1-D, 2-D, and 3-D flow problems =4, Gauss-Seidel iterative method for 1-D, 2-D, and 3-D flow problems =5, PSOR iterative method for 1-D, 2-D, and 3-D flow problems

Continued

DATA 05B	Reservoir Discretization and Method of Solving Equations
	=6, LSOR iterative method for 2-D and 3-D flow problems =7, BSOR iterative method for 3-D flow problems (×) =8, g-band using natural ordering for 1-D, 2-D, and 3-D flow problems
TOLERSP	User's specified value for maximum absolute relative deviation between two successive outer iterations. *A value of 0.001 is recommended*
DXTOLSP	User's specified value for maximum absolute pressure deviation between two successive inner iterations. *A value of 0.0001 psi is recommended*

Notes

1. TOLERSP and DXTOLSP are convergence tolerances. Compressible flow problems use TOLERSP. Iterative linear equation solvers use DXTOLSP. *The correct solution to a simulation problem is obtained using the recommended tolerances.*
2. If stricter convergence tolerances are specified, the program uses the recommended tolerances to save on iterations. Stricter tolerances do not improve the pressure solution but rather increase iterations. Relaxed tolerances, however, influence the solution and may result in unacceptable material balance errors.
3. Options marked with (×) are not active in this version

DATA 06A to DATA 21A	Reservoir Description and Initial Pressure Distribution
I1, I2	Lower and upper limits in the *x*-direction of a parallelepiped region or the *r*-direction of a reservoir region in single-well simulation
J1, J2	Lower and upper limits in the *y*-direction of a parallelepiped region; for single-well simulation, set J1 = J2 = 1
K1, K2	Lower and upper limits in the *z*-direction of a parallelepiped region or a reservoir region in single-well simulation
IACTIVE	Block indicator for active and inactive blocks =0, inactive gridblock or gridpoint =−1, inactive gridblock or gridpoint to identify constant pressure block =1, active gridblock or gridpoint
DX	Block size in the *x*-direction for block-centered grid (or gridpoint spacing in the *x*-direction for point-distributed grid), ft [m]
DY	Block size in the *y*-direction for block-centered grid (or gridpoint spacing in the *y*-direction for point-distributed grid), ft [m]
DZ	Block size in the *z*-direction for block-centered grid (or gridpoint spacing in the *z*-direction for point-distributed grid), ft [m]
KX	Block permeability in the *x*- or *r*-direction, md [μm^2]
KY	Block permeability in the *y*-direction if NY > 0, md [μm^2]
KZ	Block permeability in the *z*-direction, md [μm^2]

Continued

DATA 06A to DATA 21A	Reservoir Description and Initial Pressure Distribution
DEPTH	Elevation of top of gridblock for block-centered grid (or elevation of gridpoint for point-distributed grid) below selected datum, ft [m]
PHI	Block porosity, fraction
P	Block pressure, psia [kPa]
RATIO	Property modifier, dimensionless =0.0, property is not modified >0.0, property is increased by that ratio <0.0, property is decreased by that ratio

Notes

1. A number of gridblocks (or gridpoints) that are part of the reservoir are deactivated on purpose to simulate a specified gridblock (or gridpoint) pressure.
2. DX, DY, and DZ are supplied for all gridblocks (or gridpoints) whether active or inactive.
3. Ratio is the desired fractional change of a property value entered by the user or internally calculated by the simulator. Modifiers can be applied to the block porosity, block elevation, block bulk volume, and transmissibilities in the *x*-, *y*-, and *z*-directions.
4. For a point-distributed grid, define the gridpoint spacing in a given direction (DX in *i*-direction, DY in *j*-direction, and DZ in *k*-direction) by setting the upper limit of a parallelepiped region in that direction only equal to the coordinate of the upper limit gridpoint in the same direction minus one

DATA 22B	Rock Data and Fluid Density
CPHI	Porosity compressibility, psi^{-1} [kPa^{-1}]
PREF	Reference pressure at which porosities are reported, psia [kPa]
RHOSC	Fluid density at reference pressure and reservoir temperature, lbm/ft^3 [kg/m^3]
DATA 23B	**Type of Fluid in the Reservoir**
LCOMP	Type of fluid indicator =1, incompressible fluid =2, slightly compressible fluid =3, compressible fluid (natural gas)
IQUAD	Interpolation within gas property table =1, linear interpolation =2, quadratic interpolation
DATA 24B	**Fluid Properties for LCOMP = 1 (Incompressible Fluid)**
FVF	Formation volume factor at reservoir temperature, RB/STB
MU	Fluid viscosity, cP [mPa.s]

Continued

DATA 24B	Fluid Properties for LCOMP = 2 (Slightly Compressible Fluid)
FVF0	Formation volume factor at reference pressure and reservoir temperature, RB/STB
MU0	Fluid viscosity at reference pressure and reservoir temperature, cP [mPa.s]
CO	Fluid compressibility, psi^{-1} [kPa^{-1}]
CMU	Rate of relative change of viscosity with respect to pressure, psi^{-1} [kPa^{-1}]
PREF	Reference pressure at which FVF0 and MU0 are reported, psia [kPa]
MBCONST	Handling of liquid FVF and liquid viscosity in transmissibility terms =1, constant values independent of pressure =2, values that depend on pressure

DATA 24C	Fluid Properties for LCOMP = 3 (Natural Gas)
PRES	Pressure, psia [kPa]
FVF	Gas formation volume factor, RB/scf [m^3/std m^3]
MU	Gas viscosity, cP [mPa.s]

Note

Gas FVF and viscosity are supplied in a table form. The pressure is entered in increasing order using equal intervals.

DATA 25A	Boundary Conditions
I1, I2	Lower and upper limits in the *x*-direction of a parallelepiped region or the *r*-direction of a reservoir region in single-well simulation
J1, J2	Lower and upper limits in the *y*-direction of a parallelepiped region; for single-well simulation, set J1 = J2 = 1
K1, K2	Lower and upper limits in the *z*-direction of a parallelepiped region or a reservoir region in single-well simulation
IFACE	Block boundary subject to boundary condition =1, block boundary in the negative direction of *z*-axis =2, block boundary in the negative direction of *y*-axis =3, block boundary in the negative direction of *x*-axis or *r*-direction =5, block boundary in the positive direction of *x*-axis or *r*-direction =6, block boundary in the positive direction of *y*-axis =7, block boundary in the positive direction of *z*-axis
ITYPBC	Type of boundary condition =1, specified pressure gradient at reservoir boundary, psi/ft [kPa/m] =2, specified flow rate across reservoir boundary, STB/D or scf/D [std m^3/d] =3, no-flow boundary =4, specified pressure at reservoir boundary, psia [kPa] =5, specified pressure of the block on the other side of reservoir boundary, psia [kPa]
SPVALUE	Specified value of boundary condition
ZELBC	Elevation of center of boundary surface for block-centered grid (or elevation of boundary node for point-distributed grid) below selected datum, ft [m]

Continued

DATA 25A	Boundary Conditions
RATIO	Property modifier for area open to flow or geometric factor between reservoir boundary and boundary gridblock (or gridpoint), dimensionless =0.0, property is not modified >0.0, property is increased by that ratio <0.0, property is decreased by that ratio

Notes

1. All reservoir boundaries are assigned a no-flow boundary condition as a default. Therefore, there is no need to specify no-flow boundaries.
2. For ITYPBC=5, ZELBC is the elevation of the point (node) that represents the block whose pressure is specified.
3. DATA 25A has no option identifier at the beginning of the parameter sequence line (line 2).
4. For single-well simulation using point-distributed grid, a specified FBHP must be simulated as a specified pressure boundary condition

DATA 26D	Well Recursive Data
NOW	Number of wells that will change operational conditions =0, no change in well operations >0, number of wells that change operational conditions
SIMNEW	Time specification signaling user's new request, D [d]; well data entered here will be active starting from previous time specification until this time specification and beyond
DELT	Time step to be used, D [d]
PWFMIN	Minimum BHP allowed for production well, psia [kPa]
PWFMAX	Maximum BHP allowed for injection well, psia [kPa]
IDW	Well identification number; each well must have a unique IDW =1, 2, 3, 4…
IW, JW, KW	(i, j, k) location of wellblock
IWOPC	Well operating condition IWOPC for production well =−1, specified pressure gradient at well radius, psi/ft [kPa/m] =−2, specified production rate, STB/D or scf/D [std m^3/d] =−3, shut-in well =−4, specified bottom-hole pressure, psia [kPa] IWOPC for injection well =1, specified pressure gradient at well radius, psi/ft [kPa/m] =2, specified injection rate, STB/D or scf/D [std m^3/d] =3, shut-in well =4, specified bottom-hole pressure, psia [kPa]
GWI	Wellblock i geometric factor, RB-cP/D-psi [m^3.mPa.s/(d.kPa)]
SPVALUE	Specified value of the operating condition
RADW	Well radius, ft [m]

Notes

1. The NOW line can be repeated for different times, but each subsequent line must have a time specification larger than the previous time specification.
2. The NOW line can be used to specify new values for DELT, PWFMIN, or PWFMAX at desired times during simulation.
3. The specified value of PWFMIN and PWFMAX must be within the range of the pressure specified in the PVT table. For realistic simulation of slightly compressible and compressible fluids, these two parameters need to be specified. However, setting **PWFMIN** $\leq -\mathbf{10^6}$ and **PWFMAX** $\geq \mathbf{10^6}$ deactivates the function of these two parameters.
4. This data group terminates with a line of zero entries.
5. Each IWD line enters specifications for one well. This line must be repeated NOW times if NOW > 0.
6. Both IWOPC and the specified rate are positive for injection well, and both are negative for production well.
7. For single-well simulation using point-distributed grid, a specified FBHP must be simulated as a specified pressure boundary condition.
8. RADW is specified here to handle options IWOPC = 1 or −1

A.4 Instructions to run simulator

The user of the simulator is provided with a copy of a reference data file (e.g., REF-DATA.TXT) similar to the one presented in Section A.6. The user first copies this file into a personal data file (e.g., MY-DATA.TXT) and then follows the instructions in Section A.2 and observes the variable definitions given in Section A.3 to modify the personal data file such that it describes the constructed model of the reservoir under study. The simulator can be run by clicking on the compiled version (SinglePhaseSim.exe). The computer responds with the following statement requesting file names (with file type) of one input and four output files:

```
ENTER NAMES OF INPUT AND OUTPUT FILES
'DATA.TXT' 'OUT1.LIS' 'OUT2.LIS' 'OUT3.LIS' 'OUT4.LIS'
```

The user responds using the names of five files, each enclosed within single quotes separated by a blank space or a comma as follows and then hits the "Return" key.

```
'MY-DATA.TXT','MY-OUT1.LIS','MY-OUT2.LIS','MY-OUT3.LIS','MY-
OUT4.LIS'
```

The computer program continues execution until completion.

Each of the four output files contains specific information. 'MY-OUT1.LIS' contains debugging information of the input data file if requested and a summary of the input data, block pressure, production and injection data including rates and cumulatives, rates of fluid across reservoir boundaries, and material balance checks for all time steps. 'MY-OUT2.LIS' reports intermediate results, equations for all blocks, and details specific to the linear equation solver every iteration in every time step. 'MY-OUT3.LIS' contains concise reporting in tabular form of block pressures at various times. 'MY-OUT4.LIS' contains concise reporting in tabular form of reservoir performance as well as individual well performances.

A.5 Limitations imposed on the compiled version

The compiled version of SinglePhaseSim is provided here for demonstration and student training purposes. The critical variables were therefore restricted to the dimensions given next.

1. Number of gridblocks (or gridpoints) in x- or r-direction ≤ 20
2. Number of gridblocks (or gridpoints) in y-direction ≤ 20
3. Number of gridblocks (or gridpoints) in z-direction ≤ 10
4. Number of entries in PVT table ≤ 30
5. Number of wells = 1 well/block
6. Unrestricted number of times, wells change operational conditions
7. Maximum number of time steps = 1000 (precautionary measure)

A.6 Example of a prepared data file

The following data file was prepared as a benchmark test problem:

```
'*DATA 01E* Title of Simulation Run'
'TITLE'
J.H. Abou-Kassem. Input data file for Example 7.1 in Chap. 7.
'*DATA 02B* Simulation Time Data'
'IPRDAT TMTOTAL TMSTOP'
   1     360     10
'*DATA 03B* Units'
'MUNITS'
  1
'*DATA 04B* Control Integers for Printing Desired Output'
'BORD  MLR  BASIC  QBC  EQS  PITER  ITRSOL'
 1     1     1     1    1     1      1
'*DATA  05B*  Reservoir  Discretization  and  Method  of  Solving
Equations'
'IGRDSYS  NX NY NZ   RW    RE    NONLNR  LEQSM  TOLERSP  DXTOLSP'
   1       4  1  1 0.25 526.604   4       8       0.0       0.0
'*DATA  06A*  RESERVOIR  REGION  WITH  ACTIVE  OR  INACTIVE  BLOCK
IACTIVE'
```

```
1,'I1 I2    J1 J2    K1 K2    IACTIVE'
   1  4      1  1     1  1       1
   0  0      0  0     0  0       0
'*DATA  07A*  RESERVOIR  REGION  HAVING  BLOCK  SIZE  DX  IN  THE
X-DIRECTION'
1,'I1 I2    J1 J2    K1 K2    DX (FT)'
   1  4      1  1     1  1      300
   0  0      0  0     0  0        0.0
'*DATA  08A*  RESERVOIR  REGION  HAVING  BLOCK  SIZE  DY  IN  THE
Y-DIRECTION'
0,'I1 I2    J1 J2    K1 K2    DY (FT)'
350 350 350 350
'*DATA  09A*  RESERVOIR  REGION  HAVING  BLOCK  SIZE  DZ  IN  THE
Z-DIRECTION'
0,'I1 I2    J1 J2    K1 K2    DZ (FT)'
4*40
'*DATA  10A*  RESERVOIR  REGION  HAVING  PERMEABILITY  KX  IN  THE
X-DIRECTION'
1,'I1 I2    J1 J2    K1 K2    KX (MD)'
   1  4      1  1     1  1      270
   0  0      0  0     0  0        0.0
'*DATA  11A*  RESERVOIR  REGION  HAVING  PERMEABILITY  KY  IN  THE
Y-DIRECTION'
1,'I1 I2    J1 J2    K1 K2    KY (MD)'
   1  4      1  1     1  1       0
   0  0      0  0     0  0       0.0
'*DATA  12A*  RESERVOIR  REGION  HAVING  PERMEABILITY  KZ  IN  THE
Z-DIRECTION'
1,'I1 I2    J1 J2    K1 K2    KZ (MD)'
   1  4      1  1     1  1       0
   0  0      0  0     0  0       0.0
'*DATA 13A* RESERVOIR REGION HAVING ELEVATION Z'
1,'I1 I2    J1 J2    K1 K2    DEPTH (FT)'
   1  4      1  1     1  1       0.0
   0  0      0  0     0  0       0.0
'*DATA 14A* RESERVOIR REGION HAVING POROSITY PHI'
1,'I1 I2    J1 J2    K1 K2    PHI (FRACTION)'
   1  4      1  1     1  1       0.27
   0  0      0  0     0  0       0.0
'*DATA 15A* RESERVOIR REGION HAVING INITIAL PRESSURE P'
1,'I1 I2    J1 J2    K1 K2    P (PSIA)'
   1  4      1  1     1  1       0
   0  0      0  0     0  0       0.0
```

```
'*DATA 16A* RESERVOIR REGION WITH BLOCK POROSITY MODIFICATION
RATIO'
1,'I1 I2    J1 J2    K1 K2    RATIO'
   0  0     0  0     0  0      0.0
'*DATA 17A* RESERVOIR REGION WITH BLOCK ELEVATION MODIFICATION
RATIO'
1,'I1 I2    J1 J2    K1 K2    RATIO'
   1  4     1  1     1  1      0.0
   0  0     0  0     0  0      0.0
'*DATA 18A* RESERVOIR REGION WITH BLOCK VOLUME MODIFICATION
RATIO'
1,'I1 I2    J1 J2    K1 K2    RATIO'
   1  4     1  1     1  1      0.0
   0  0     0  0     0  0      0.0
'*DATA 19A* RESERVOIR REGION WITH X-TRANSMISSIBILITY MODIFICATION
RATIO'
1,'I1 I2    J1 J2    K1 K2    RATIO'
   1  4     1  1     1  1      0.0
   0  0     0  0     0  0      0.0
'*DATA 20A* RESERVOIR REGION WITH Y-TRANSMISSIBILITY MODIFICATION
RATIO'
1,'I1 I2    J1 J2    K1 K2    RATIO'
   1  4     1  1     1  1      0.0
   0  0     0  0     0  0      0.0
'*DATA 21A* RESERVOIR REGION WITH Z-TRANSMISSIBILITY MODIFICATION
RATIO'
1,'I1 I2    J1 J2    K1 K2    RATIO'
   1  4     1  1     1  1      0.0
   0  0     0  0     0  0      0.0
'*DATA 22B* Rock and Fluid Density'
'CPHI    PREF    RHOSC'
 0.0      14.7    50.0
'*DATA 23B* Type of Fluid in the Reservoir'
'LCOMP IQUAD'
  1     1
'*DATA 24B* FOR LCOMP= 1 AND 2 OR *DATA 24C* FOR LCOMP= 3 ENTER
FLUID PROP'
'LCOMP=1:FVF,MU;LCOMP=2:FVFO,MUO,CO,CMU,PREF,MBCONST;LCOMP=3:
PRES,FVF,MU TABLE'
1.0 0.5
'*DATA 25A* Boundary Conditions'
'I1 I2  J1 J2  K1 K2  IFACE  ITYPEBC  SPVALUE  ZELBC  RATIO'
  1  1   1  1   1  1    3       4       4000     20     0.0
  4  4   1  1   1  1    5       3          0     20     0.0
  0  0   0  0   0  0    0       0          0.      0.0  0.0
```

```
'*DATA 26D* Well Recursive Data'
'NOW    SIMNEW    DELT    PWFMIN    PWFMAX'
'IDW   IW   JW    KW   IWOPC      GWI          SPVALUE    RADW'
  1        10.0          10.0 -10000000.0  100000000.0
  1     4    1     1    -2           11.0845       -600    0.25
  0         0.0           0.0         0.0            0.0
```

References

Abdelazim, R., Rahman, S.S., 2016. Estimation of permeability of naturally fractured reservoirs by pressure transient analysis: an innovative reservoir characterization and flow simulation. J. Pet. Sci. Eng. 145, 404–422.

Abdel Azim, R., Doonechaly, N.G., Rahman, S.S., Tyson, S., Regenauer-Lieb, K., 2014. 3D Poro-thermo-elastic numerical model for analyzing pressure transient response to improve the characterization of naturally fractured geothermal reservoirs. Geotherm. Res. Coun. 38, 907–915.

Abou-Kassem, J.H., 1981. Investigation of Grid Orientation in Two-Dimensional, Compositional, Three-Phase Steam Model (Ph.D. dissertation). University of Calgary, Calgary, AB.

Abou-Kassem, J.H., 1996. Practical considerations in developing numerical simulators for thermal recovery. J. Pet. Sci. Eng. 15, 281–290.

Abou-Kassem, J.H., 2006. The engineering approach versus the mathematical approach in developing reservoir simulators. J. Nat. Sci. Sustain. Tech. 1 (1), 35–67.

Abou-Kassem, J.H., Aziz, K., 1985. Analytical well models for reservoir simulation. Soc. Pet. Eng. J. 25 (4), 573–579.

Abou-Kassem, J.H., Ertekin, T., 1992. An efficient algorithm for removal of inactive blocks in reservoir simulation. J. Can. Pet. Tech. 31 (2), 25–31.

Abou-Kassem, J.H., Farouq Ali, S.M., 1987. A unified approach to the solution of reservoir simulation equations. SPE # 17072. In: Paper presented at the 1987 SPE Eastern Regional Meeting, Pittsburgh, Pennsylvania, 21–23 October.

Abou-Kassem, J.H., Osman, M.E., 2008. An engineering approach to the treatment of constant pressure boundary condition in block-centered grid in reservoir simulation. J. Pet. Sci. Tech. 26, 1187–1204.

Abou-Kassem, J.H., Ertekin, T., Lutchmansingh, P.M., 1991. Three-dimensional modeling of one-eighth of confined five- and nine-spot patterns. J. Pet. Sci. Eng. 5, 137–149.

Abou-Kassem, J.H., Osman, M.E., Zaid, A.M., 1996. Architecture of a multipurpose simulator. J. Pet. Sci. Eng. 16, 221–235.

Abou-Kassem, J.H., Farouq Ali, S.M., Islam, M.R., 2006. Petroleum Reservoir Simulations: A Basic Approach. Gulf Publishing Company, Houston, TX, USA, 2006, 480 pp. ISBN: 0-9765113-6-3.

Abou-Kassem, J.H., Osman, M.E., Mustafiz, S., Islam, M.R., 2007. New simple equations for inter-block geometric factors and bulk volumes in single-well simulation. J. Pet. Sci. Tech. 25, 615–630.

Aguilera, R., 1976. Analysis of naturally fractured reservoirs from conventional well logs. J. Pet. Tech. 28 (7), 764–772. Document ID: SPE-5342-PA. https://doi.org/10.2118/5342-PA.

Aguilera, R., Aguilera, M.S., 2003. Improved models for petrophysical analysis of dual porosity reservoirs. Petrophysics 44 (1), 21–35.

Aguilera, R.F., Aguilera, R., 2004. A triple porosity model for petrophysical analysis of naturally fractured reservoirs. Petrophysics 45 (2), 157–166.

Appleyard, J.R., Cheshire, I.M., 1983. Nested factorization. SPE # 12264. In: Paper Presented at the 1983 SPE Reservoir Simulation Symposium, San Francisco, 15–18 November.

Aziz, K., 1993. Reservoir simulation grids: opportunities and problems, SPE # 25233. In: Paper Presented at the 1993 SPE Symposium on Reservoir Simulation, New Orleans, LA, 28 February–3 March.

Aziz, K., Settari, A., 1979. Petroleum Reservoir Simulation. Applied Science Publishers, London.

Babu, D.K., Odeh, A.S., 1989. Productivity of a horizontal well. SPERE 4 (4), 417–420.

Barnum, R.S., Brinkman, F.P., Richardson, T.W., Spillette, A.G., 1995. Gas Condensate Reservoir Behaviour: Productivity and Recovery Reduction Due to Condensation, SPE-30767.

Bartley, J.T., Ruth, D.W., 2002. A look at break-through and end-point data from repeated waterflood experiments in glass bead-packs and sand-packs, SCA 2002-17. In: International Symposium of Monterey, California, September 22–25, 2002, 12 pp.

Bear, J., 1972. Dynamics of Flow in Porous Media. American Elsevier Publishing Co., New York.

Bear, J., 1988. Dynamics of Fluids in Porous Media. Dover Publications, New York.

Behie, A., Vinsome, P.K.W., 1982. Block iterative methods for fully implicit reservoir simulation. SPEJ 22 (5), 658–668.

Bentsen, R.G., 1985. A new approach to instability theory in porous media. Soc. Petrol. Eng. J. 25, 765.

Bethel, F.T., Calhoun, J.C., 1953. Capillary desaturation in unconsolidated beads. Pet. Trans. AIME 198, 197–202.

Bourbiaux, B., Granet, S., Landereau, P., Noetinger, B., Sarda, S., Sabathier, J.C., 1999. Scaling up matrix-fracture transfers in dual-porosity models: theory and application. In: Proceedings of Paper SPE 56557 Presented at the SPE Annual Technical Conference and Exhibition, 3–6 October, Houston.

Bourdet, D., 2002. Well Test Analysis: The Use of Advanced Interpretation Models. Handbook of Petroleum Exploration & Production, vol. 3. (HPEP). Elsevier.

Breitenbach, E.A., Thurnau, D.H., van Poollen, H.K., 1969. The fluid flow simulation equations. SPEJ 9 (2), 155–169.

Brons, F., Marting, V., 1961. The effect of restricted fluid entry on well productivity. J. Pet. Tech. 13 (02), 172–174.

Burland, J.B., 1990. On the compressibility and shear strength of natural clays. Géotechnique 40, 329–378.

Chatzis, I., Morrow, N.R., Lim, H.T., 1983. Magnitude and detailed structure of residual oil saturation. Soc. Pet. Eng. J. 23 (2), 311–326.

Choi, E., Cheema, T., Islam, M., 1997. A new dual-porosity/dual-permeability model with non-Darcian flow through fractures. J. Pet. Sci. Eng. 17 (3), 331–344.

Chopra, S., Castagna, J.P., Portniaguine, O., 2006. Seismic resolution and thin-bed reflectivity inversion. CSEG Recorder 31, 19–25.

Chopra, S., Sharma, R.K., Keay, J., Marfurt, K.J., 2012. Shale gas reservoir characterization workflows. In: SEG Las Vegas 2012 Annual Meeting. https://blog.tgs.com/hubfs/Technical%20Library/shale-gas-reservoir-characterization-workflows.pdf.

Chuoke, R.L., van Meurs, P., van der Poel, C., 1959. The instability of slow, immiscible, viscous liquid-liquid displacements in permeable media. Trans. AIME 216, 188–194.

Coats, K.H., 1978. A highly implicit steamflood model. SPEJ 18 (5), 369–383.

Coats, K.H., George, W.D., Marcum, B.E., 1974. Three-dimensional simulation of steamflooding. SPEJ 14 (6), 573–592.

Coats, K.H., Ramesh, A.B., Winestock, A.G., 1977. Numerical modeling of thermal reservoir behavior. In: Paper Presented at Canada-Venezuela Oil Sands Symposium, 27 May 27–4 June, Edmonton, Alberta.

Cortis, A., Birkholzer, J., 2008. Continuous time random walk analysis of solute transport in fractured porous media. Water Resour. Res. 44, W06414. https://doi.org/10.1029/2007WR006596.

Crovelli, R.A., Schmoker, J.W., 2001. Probabilistic Method for Estimating Future Growth of Oil and Gas Reserves. U.S. Geological Survey Bulletin 2172-C.

Dranchuk, P.M., Islam, M.R., Bentsen, R.G., 1986. A mathematical representation of the Carr, Kobayashi and burrows natural gas viscosity. J. Can. Pet. Tech. 25 (1), 51–56.

Duguid, J.O., Lee, P.C.Y., 1977. Flow in fractured porous media. Water Resour. Res. 26, 351–356.

Dyman, T.S., Schmoker, J.W., 2003. Well production data and gas reservoir heterogeneity–reserve growth applications. In: Dyman, T.S., Schmoker, J.W., Verma, M.K. (Eds.), Geologic, Engineering, and Assessment Studies of Reserve Growth. U.S. Geological Survey, Bulletin 2172–E.

Dyman, T.S., Schmoker, J.W., Quinn, J.C., 1996. Reservoir Heterogeneity as Measured by Production Characteristics of Wells—Preliminary Observations. U.S. Geological Survey Open-File Report 96–059, 14 p.

Ehrenberg, S.N., Nadeau, P.H., Steen, Ø., 2009. Petroleum reservoir porosity versus depth: influence of geological age. AAPG Bull. 93, 1263–1279.

Eisenack, K., Ludeke, M.K.B., Petschel-Held, G., Scheffran, J., Kropp, J.P., 2007. Qualitative modeling techniques to assess patterns of global change. In: Kropp, J.P., Scheffran, J. (Eds.), Advanced Methods for Decision Making and Risk Management in Sustainability Science. Nova Science Publishers, New York, pp. 83–127.

Ertekin, T., Abou-Kassem, J.H., King, G.R., 2001. Basic Applied Reservoir Simulation. SPE Textbook Series, vol. 7. SPE, Richardson, TX.

Farouq Ali, S.M., 1986. Elements of Reservoir Modeling and Selected Papers. Course Notes, Petroleum Engineering, Mineral Engineering Department, University of Alberta.

Fatti, J., Smith, G., Vail, P., Strauss, P., Levitt, P., 1994. Detection of gas in sandstone reservoirs using AVO analysis: a 3D seismic case history using the Geostack technique. Geophysics 59, 1362–1376. https://doi.org/10.1190/1.1443695.

Fishman, N.S., Turner, C.E., Peterson, F., Dyman, T.S., Cook, T., 2008. Geologic controls on the growth of petroleum reserves. In: U.S. Geological Survey Bulletin 2172–I, 53 p.

Gale, J.F.W., 2002. Specifying lengths of horizontal wells in fractured reservoirs. In: Society of Petroleum Engineers Reservoir Evaluation and Engineering, SPE Paper 78600, pp. 266–272.

Gilman, J., 1986. An efficient finite-difference method for simulating phase segregation in the matrix blocks in double-porosity reservoirs. SPE Reserv. Eng. 1 (04), 403–413.

Gong, B., Karimi-Fard, M., Durlofsky, L.J., 2008. Upscaling discrete fracture characterizations to dual-porosity, dual-permeability models for efficient simulation of flow with strong gravitational effects. Soc. Pet. Eng. J. 13 (1), 58.

Gupta, A.D., 1990. Accurate resolution of physical dispersion in multidimensional numerical modeling of miscible and chemical displacement. SPERE 5 (4), 581–588.

Gupta, A., Avila, R., Penuela, G., 2001. An integrated approach to the determination of permeability tensors for naturally fractured reservoirs. J. Can. Pet. Tech. 40 (12), 43–48.

Hamilton, D.S., Holtz, M.H., Ryles, P., et al., 1998. Approaches to identifying reservoir heterogeneity and reserve growth opportunities in a continental-scale bed-load fluvial system: Hutton Sandstone, Jackson Field, Australia. AAPG Bull. 82 (12), 2192–2219.

Hariri, M.M., Lisenbee, A.L., Paterson, C.J., 1995. Fracture control on the Tertiary Epithermal-Mesothermal Gold-Silver Deposits, Northern Black Hills, South Dakota. Explor. Min. Geol. 4 (3), 205–214.

Hoffman, J.D., 1992. Numerical Methods for Engineers and Scientists. McGraw-Hill, New York.

Islam, M.R., 1992. Evolution in oscillatory and chaotic flows in mixed convection in porous media in non-Darcy regime. Chaos, Solitons Fractals 2 (1), 51–71.

Islam, M.R., 2014. Unconventional Gas Reservoirs. Elsevier, Netherlands, 624 pp.

Islam, M.R., Mustafiz, S., Mousavizadegan, S.H., Abou-Kassem, J.H., 2010. Advanced Reservoir Simulation. John Wiley & Sons, Inc./Scrivener Publishing LLC, Hoboken, NJ/ Salem, MA.

Islam, M.R., Hossain, M.E., Islam, A.O., 2018. Hydrocarbons in Basement Formations. Scrivener-Wiley, 624 pp.

Jadhunandan, P.P., Morrow, N.R., 1995. Effect of wettability on waterflood recovery for crude-oil/ brine/rock systems. SPE Res. Eng. 10, 40–46.

Keast, P., Mitchell, A.R., 1966. On the instability of the Crank-Nicolson formula under derivative boundary conditions. Comput. J. 9 (1), 110–114.

Lake, L.W., 1989. Enhanced Oil Recovery. Prentice Hall, Englewood Cliffs, NJ, USA.

Lake, L.W., Srinivasan, S., 2004. Statistical scale-up of reservoir properties: concepts and applications. J. Pet. Sci. Eng. 44 (1–2), 27–39.

Landereau, P., Noetinger, B., Quintard, M., 2001. Quasi-steady two-equation models for diffusive transport in fractured porous media: large-scale properties for densely fractured systems. Adv. Water Resour. 24 (8), 863–876.

Lee, A.L., Gonzalez, M.H., Eakin, B.E., 1966. The viscosity of natural gases. Trans. AIME 237, 997–1000.

Leverett, M.C., Lewis, W.B., 1941. Steady flow of gas/oil/water mixtures through unconsolidated sands. Trans. AIME 142, 107–116.

Liu, R., Wang, D., Zhang, X., et al., 2013. Comparison study on the performances of finite volume method and finite difference method. J. Appl. Math. 2013, 10 pp.

Lough, M.F., Lee, S.H., Kamath, J., 1998. A new method to calculate the effective permeability of grid blocks used in the simulation of naturally fractured reservoirs. In: Proceedings of Paper Presented at the SPE Annual Technical Conference.

Lucia, 2007. Carbonate Reservoir Characterization: An Integrated Approach. Springer, 332 pp.

Lutchmansingh, P.M., 1987. Development and Application of a Highly Implicit, Multidimensional Polymer Injection Simulator (Ph.D. dissertation). The Pennsylvania State University.

MacDonald, R.C., Coats, K.H., 1970. Methods for numerical simulation of water and gas coning. Trans. AIME 249, 425–436.

McCabe, P.J., 1998. Energy Sources—cornucopia or empty barrel? Am. Assoc. Pet. Geol. Bull. 82, 2110–2134.

McDonald, A.E., Trimble, R.H., 1977. Efficient use of mass storage during elimination for sparse sets of simulation equations. SPEJ 17 (4), 300–316.

Mousavizadegan, H., Mustafiz, S., Islam, M.R., 2007. Multiple solutions in natural phenomena. J. Nat. Sci. Sustain. Tech. 1 (2), 141–158.

Mustafiz, S., Islam, M.R., 2008. State-of-the-art of petroleum reservoir simulation. Pet. Sci. Tech. 26 (10–11), 1303–1329.

Mustafiz, S., Belhaj, H., Ma, F., Satish, M., Islam, M.R., 2005a. Modeling horizontal well oil production using modified Brinkman's model. In: ASME International Mechanical Engineering Congress and Exposition (IMECE), Orlando, Florida, USA, November.

Mustafiz, S., Biazar, J., Islam, M.R., 2005b. The Adomian solution of modified Brinkman model to describe porous media flow. In: Third International Conference on Energy Research & Development (ICERD-3), Kuwait City, November, 2005.

Mustafiz, S., Mousavizadegan, H., Islam, M.R., 2008a. Adomian decomposition of Buckley Leverett equation with capillary terms. Pet. Sci. Tech. 26 (15), 1796–1810.

Mustafiz, S., Mousavizadegan, S.H., Islam, M.R., 2008b. The effects of linearization on solutions of reservoir engineering problems. Pet. Sci. Tech. 26 (10–11), 1224–1246.

Mustafiz, S., Zaman, M.S., Islam, M.R., 2008. The effects of linearization in multi-phase flow simulation in petroleum reservoirs. J. Nat. Sci. Sustain. Tech. 2 (3), 379–398.

Noetinger, B., Estebenet, T., 2000. Up-scaling of double porosity fractured media using continuous-time random walks methods. Transp. Porous Media 39, 315–337.

Nolen, J.S., Berry, D.W., 1972. Tests of the stability and time-step sensitivity of semi-implicit reservoir simulation techniques. Trans. AIME 253, 253–266.

Odeh, A.S., 1982. An overview of mathematical modeling of the behavior of hydrocarbon reservoirs. SIAM Rev. 24 (3), 263.

Odeh, A.S., Babu, D., 1990. Transient flow behavior of horizontal wells pressure drawdown and buildup analysis. SPE Form. Eval. 5 (01), 7–15.

Okiongbo, K.S., 2011. Effective stress-porosity relationship above and within the oil window in the North Sea Basin. Res. J. Appl. Sci. Eng. Technol. 3 (1), 32–38.

Park, Y., Sung, W., Kim, S., 2002. Development of a FEM reservoir model equipped with an effective permeability tensor and its application to naturally fractured reservoirs. Energy Sources 24 (6), 531–542.

Passey, Q.R., Creaney, S., Kulla, J.B., Moretti, F.J., Stroud, J.D., 1990. A practical model for organic richness from porosity and resistivity logs. AAPG Bull. 74, 1777–1794.

Peaceman, D.W., 1983. Interpretation of wellblock pressures in numerical reservoir simulation with nonsquare gridblocks and anisotropic permeability. SPEJ 23 (3), 531–534.

Peaceman, D.W., 1987. Interpretation of Wellblock Presures in numerical reservoir simulation: Part 3 off-center and multiple Wells within a Wellblock. SPE-16976. In: Paper Presented at the 1987 SPE Annual Technical Conference and Exhibition, Dallas, USA, September 27–30.

Peaceman, D.W., Rachford Jr., H.H., 1955. The numerical solution of parabolic and elliptic equations. J. SIAM 3 (1), 28–41.

Pedrosa Jr., O.A., Aziz, K., 1986. Use of hybrid grid in reservoir simulation. SPERE 1 (6), 611–621.

Peters, E.J., Flock, D.L., 1981. The onset of instability during two phase immiscible displacement in porous media. Soc. Petrol. Eng. J. 21, 249.

Price, H.S., Coats, K.H., 1974. Direct methods in reservoir simulation. SPEJ 14 (3), 295–308.

Pride, S.R., Berryman, J.G., 2003. Linear dynamics of double-porosity dual-permeability materials. I. Governing equations and acoustic attenuation. Phys. Rev. E. 68 (3), 036603.

Pruess, K., 1985. A practical method for modeling fluid and heat flow in fractured porous media. Soc. Pet. Eng. J. 25 (01), 14–26.

Puryear, C.I., Castagna, J.P., 2008. Layer-thickness determination and stratigraphic interpretation using spectral inversion: theory and application. Geophysics 73 (2), R37–R48. https://doi.org/10.1190/1.2838274.

Rickman, R., Mullen, M., Petre, E., Grieser, B., Kundert, D., 2008. A practical use of shale petrophysics for stimulation design optimization: all shale plays are not clones of the Barnett Shale. In: Annual Technical Conference and Exhibition, Society of Petroleum Engineers, SPE 11528.

Rogers, J.D., Grigg, R.B., 2000. A Literature Analysis of the WAG Injectivity Abnormalities in the CO_2 Process. Society of Petroleum Engineers. https://doi.org/10.2118/59329-MS.

Root, D.H., Attanasi, E.D., 1993. A primer in field-growth estimation. In: Howell, D.G. (Ed.), The Future of Energy Gases: U.S. Geological Survey Professional Paper no. 1570, pp. 547–554.

Rose, W., 2000. Myths about later-day extensions of Darcy's law. J. Pet. Sci. Eng. 26 (1–4), 187–198.

Saad, N., 1989. Field Scale Simulation of Chemical Flooding (Ph.D. dissertation). University of Texas, Austin, TX.

Saghir, M.Z., Islam, M.R., 1999. Viscous fingering during miscible liquid-liquid displacement in porous media. Int. J. Fluid Mech. Res. 26 (4), 215–226.

Sarkar, S., Toksoz, M.N., Burns, D.R., 2004. Fluid Flow Modeling in Fractures. Massachusetts Institute of Technology, Earth Resources Laboratory, USA.

Scher, H., Lax, M., 1973. Stochastic transport in a disordered solid. I. Theory. Phys. Rev. B 7, 4491–4502.

Schmoker, J.W., 1996. A resource evaluation of the Bakken Formation (Upper Devonian and Lower Mississippian) continuous oil accumulation, Williston Basin, North Dakota and Montana. Mt. Geol. 33, 1–10.

Settari, A., Aziz, K., 1975. Treatment of nonlinear terms in the numerical solution of partial differential equations for multiphase flow in porous media. Int. J. Multiphase Flow 1, 817–844.

Shanley, K.W., Cluff, R.M., Robinson, J.W., 2004. Factors controlling prolific gas production from low-permeability sandstone reservoirs: implications for resource assessment, prospect development, and risk analysis. AAPG Bull. 88 (8), 1083–1121.

Sheffield, M., 1969. Three phase flow including gravitational, viscous, and capillary forces. SPEJ 9 (3), 255–269.

Skempton, A.W., 1970. The consolidation of clays by gravitational compaction. Q. J. Geol. Soc. 125 (1–4), 373–411.

Spillette, A.G., Hillestad, J.G., Stone, H.L., 1973. A high-stability sequential solution approach to reservoir simulation. SPE # 4542. In: Paper presented at the 48th Annual Fall Meeting, Las Vegas, Nevada, 30 September–3 October.

Stell, J.R., Brown, C.A., 1992. Comparison of production from horizontal and vertical wells in the Austin Chalk, Niobrara, and Bakken plays. In: Schmoker, J.W., Coalson, E.B., Brown, C.A. (Eds.), Geological Studies Relevant to Horizontal Drilling—Examples from Western North America. Rocky Mountain Association of Geologists, pp. 67–87.

Sudipata, S., Toksoz, M.N., Burns, D.R., 2004. Fluid Flow Modeling in Fractures. Massachusetts Institute of Technology, Earth Resources Laboratory, Report no. 2004–05.

Tang, I.C., 1969. A simple algorithm for solving linear equations of a certain type. Z. Angew. Math. Mech. 8 (49), 508.

Teimoori, A., Chen, Z., Rahman, S.S., Tran, T., 2005. Effective permeability calculation using boundary element method in naturally fractured reservoirs. Pet. Sci. Technol. 23 (5–6), 693–709.

Thomas, G.W., Thurnau, D.H., 1983. Reservoir simulation using an adaptive implicit method. SPEJ 23 (5), 759–768.

Vinsome, P.K.W., 1976. Orthomin, an iterative method for solving sparse banded sets of simultaneous linear equations. SPE # 5729. In: Paper Presented at the 1976 SPE Symposium on Numerical Simulation of Reservoir Performance, Los Angeles, 19–20 February.

Warren, J., Root, P.J., 1963. The behaviour of naturally fractured reservoirs. Soc. Pet. Eng. J. 3 (3), 245–255.

Woo, P.T., Roberts, S.J., Gustavson, S.G., 1973. Application of sparse matrix techniques in reservoir simulation. SPE # 4544. In: Paper Presented at the 48th Annual Fall Meeting, Las Vegas. 30 September–3 October.

Yang, Y., Aplin, A.C., 2004. Definition and practical application of mudstone porosity-effective stress relationships. Petrol. Geosci. 10, 153–162.

Author Index

Note: Page numbers followed by *f* indicate figures and *t* indicate tables.

Subject Index

Note: Page numbers followed by *f* indicate figures and *t* indicate tables.

V

W

Printed in the United States
By Bookmasters